Proceedings of the Seventh Conference on

# MAGNETISM AND
# MAGNETIC MATERIALS

Edited by

## J. A. Osborn

Originally published as a Supplement to the
Journal of Applied Physics, Vol. 33, 1962

Springer Science+Business Media, LLC
1962

# CONFERENCE ON MAGNETISM AND MAGNETIC MATERIALS

Papers presented at the Conference on Magnetism and Magnetic Materials, Phoenix, Arizona, November 13–16, 1961

Sponsored by the American Institute of Electrical Engineers and the American Institute of Physics

Cooperating Societies:

Institute of Radio Engineers

Metallurgical Society of the A. I. M. E.

Office of Naval Research

## March, 1962

ISBN 978-1-4899-6193-8   ISBN 978-1-4899-6391-8 (eBook)
DOI 10.1007/978-1-4899-6391-8

Copyright © 1962 Springer Science+Business Media New York
Originally published by the American Institute of Physics, New York in 1962
Softcover reprint of the hardcover 1st edition 1962

**American Institute of Electrical Engineers Headquarters**

**33 West 39th Street, New York 18, New York**

# CONTENTS

* Abstract only.

(*continued on page iv*)

CONTENTS (*continued*)

## Spin Configurations and Anisotropy

## Thin Films-2

## Magnetization

\* Abstract only.

(*continued on page v*)

CONTENTS (*continued*)

## Oxides-1

## Soft Magnetic Materials

## Rare Earths

## Devices and Phenomena

———
* Abstract only.

(*continued on page vi*)

CONTENTS (*continued*)

## *Antiferromagnetism and Resonance*

## *Permanent Magnets and Micromagnetics*

## *Alloys and Compounds*

(*continued on page vii*)

CONTENTS (*continued*)

## Oxides-2 and Crystals

# CONFERENCE ORGANIZATION

## General Conference Chairman

L. R. Bickford, Jr.

## Steering Committee

L. R. Bickford, Jr., *Chairman*

| | | |
|---|---|---|
| R. S. Gardner | J. A. Krumhansl | J. S. Smart |
| C. L. Hogan | J. A. Osborn | H. C. Wolfe |
| | T. O. Paine | |

## Technical Program Committee

I. S. Jacobs
F. E. Luborsky    *Chairmen*

| | | |
|---|---|---|
| P. A. Albert | J. M. Hastings | H. Rubenstein |
| J. F. Dillon | F. Keffer | W. L. Shevel |
| S. Foner | C. J. Kriessman | M. T. Weiss |

## Local Committee

P. B. Myers, *Chairman*

| | | |
|---|---|---|
| D. B. Cachelin | F. R. Gleason | J. B. Picone |
| J. F. Cubbage | C. S. Graniere | J. D. Raile |
| Mrs. D. L. Fresh | L. J. Ittenbach | N. P. Rowley |
| D. L. Fresh | T. H. Kemp | A. J. Weslowski |
| | J. M. Lommel | |

## Exhibits

John Leslie Whitlock Associates

## Publications Committee

J. A. Osborn, *Chairman*

| | | |
|---|---|---|
| J. H. Crawford, Jr. | I. S. Jacobs | J. S. Smart |
| | J. A. Krumhansl | |

## Magnetic Materials Literature Digest

| | |
|---|---|
| J. C. Slonczewski, *Editor* | J. E. Goldman |
| W. Palmer, *Assistant Editor* | C. J. Kriessman |
| W. H. Meiklejohn, *Committee Chairman* | J. A. Krumhansl |

## AIEE Subcommittee on Magnetics

L. R. Bickford, Jr., *Chairman*

| | | |
|---|---|---|
| A. C. Beiler | E. A. Gaugler | M. F. Littman |
| E. Both | J. E. Goldman | L. R. Maxwell |
| R. M. Bozorth | D. M. Grimes | W. H. Meiklejohn |
| F. G. Brockman | C. L. Hogan | W. Morrill |
| H. Brooks | I. S. Jacobs | J. A. Osborn |
| H. B. Callen | C. J. Kriessman | T. O. Paine |
| R. A. Chegwidden | V. E. Legg | G. T. Rado |

## Sponsoring Society Representatives

American Institute of Electrical Engineers
R. S. Gardner
American Institute of Physics
H. C. Wolfe

## Cooperating Society Representatives

The Institute of Radio Engineers
T. N. Anderson
The Metallurgical Society of A. I. M. E.
P. A. Albert
Office of Naval Research
M. A. Garstens

# Journal

# of

# Applied Physics

**Supplement to Volume 33, Number 3**          **March, 1962**

## "Let Thy Words Be Few"

*Ecclesiastes 5:2*

THOSE venturesome magnetikers who made the trek west found Phoenix with the perfect weather of the winter southwest: clear skies and invigorating air. For this the local Chamber of Commerce can thank their formidable ally C. L. Hogan who persuasively guided the Conference party to a very civilized frontier indeed. And now, as every year, the many who worked to make the Conference interesting and lively have asked themselves, "Was it a good meeting; was all the work worthwhile?" This is something that those attending the Conference must decide, but in this case many are both witness and jury as a large portion of those attending the meeting also participated in some part of the program machinery. A reasonable estimate would be that more than 150 people were required to ready the Conference, including nearly 100 referees. As this meeting is run by volunteers (some in the army sense), it is a many-body system whose state is briefly observable but once a year. Although this sometimes seems like organizational anarchy, it is in fact just this wide participation each year of new groups with fresh points of view that keep the Conference lively.

As the Conference has evolved its committees have experimented with ways in which it might be of greater service to those in the field of magnetism. This year the 1960 Magnetic Materials Literature Digest[1] prepared with scholarly devotion by J. Slonczewski, W. Meiklejohn, and their associates, was printed with Conference support and distributed to registrants.

In previous years we have attempted to communicate some of the excitement of science to high school students in the Conference city through speakers and a special high school day arranged in conjunction with the meet-ings. This year the Conference through its Phoenix colleagues (and some fine high school science teachers) sponsored student exhibits at the meeting which demonstrated magnetic phenomena.

The variety of topics covered in the sessions was so large as to make a review here inappropriate. One subject was explored rather thoroughly by a number of authors: the nonlinear compressibility of contributed papers. While many, no doubt, experienced the sensation of being fitted to a procrustean bed, authors have made it easier to meet the restrictive time and proceedings format each year. Hopefully because it is realized that these requirements are necessitated by the desire for an early appearance of the proceedings, the large size of the resulting volume, and economics. The issuance of the Proceedings as a supplement of the Journal of Applied Physics and the essential help of its Publications Staff would have been impossible had the Conference not only been self-supporting but, additionally, been able to subsidize the Proceedings. The means which have been sought for Conference financial support then have been essential to its success: the exhibits, the advertising, and the bound version of the Proceedings by Plenum Press this year.

Thoughtful contributions and willing hands were many. The impersonal listing of Committee members elsewhere gives little idea of the long hours and late nights needed for these preparations. In particular, the Program Committee, under I. Jacobs and F. Luborsky, and the Local Committee (and the hardworking wives) headed by P. Myers. The editorial, refereeing gymkhana this year was successfully completed due to the generous participation of the many scientists and engineers who acted as referees and the willing labors of my colleagues on the Publication Committee.

J. A. OSBORN
*Editor*

---

[1] Copies at $2.00 each may be obtained from the American Institute of Physics, 335 East 45th Street, New York 17, New York.

JOURNAL OF APPLIED PHYSICS　　SUPPLEMENT TO VOL. 33, NO. 3　　MARCH, 1962

# General Session

## L. R. Bickford, *Chairman*

## The State of the Art of Magnetic Memories

Q. W. Simkins

*Development Laboratories, Components Division, International Business Machines Corporation, Poughkeepsie, New York*

A broad range of magnetic memories finds extensive use in today's data processing equipment. The most significant factors in evaluating a memory for a given application are reliability, cost, size, speed, and function. Ferrite core memories with capacities of $10^3$ to $10^6$ bits and with cycle times as short as 0.5 $\mu$sec are in use. Recent developments in this area, including partial switching, 2 cores/bit, and new core fabrication techniques, will be discussed.

There have been extensive development efforts recently in many forms of magnetic metal film devices. Many geometries, modes of operation, fabrication techniques, substrate materials, and film compositions are being used. A comparison of some of the resulting devices will be made, and the effects of the factors listed above will be discussed. The application of metal film memory devices will be considered and a comparison drawn with ferrite devices.

Techniques for achieving special purpose functions such as nondestructive readout, read-only and associative memory using magnetic elements, will also be described and compared.

THE pre-eminent position of magnetic memories in today's computer technology hardly need be pointed out. While the field of magnetic surface recording on tapes, disks, drums, cards, etc. is of much interest, my discussion will be limited to the field of electronically addressable magnetic memories. Memories of this sort cover a wide range of size, speed, and application. Registers and buffers storing as few as 1000 bits have successfully exploited magnetic techniques, while today's most powerful data processors have one or more random access memories in the million-bit category.

Figure 1 shows the size and speed of a number of high performance memories. Complete read-write cycle times, including address decoding, are used, except as indicated. This chart is far from complete and lists only representative high-speed memories. For many applications, other parameters such as cost, reliability, size or power dissipation may be more important.

Ferrite core memories covering a wide range of speeds and sizes are available. A number of thin film memories are also represented in Fig. 1.

Most of the larger ferrite core memories employ coincident-current read and write,[1,2] and are often called bit-organized or three-dimensional memories. Many of the smaller and faster memories, including nearly all those using metal film devices, have coincident-current write but non-coincident read and are called word-organized or two-dimensional memories.

In the remainder of this paper, I shall discuss the ferrite and metal film technologies, including recent device and memory system developments. Finally extensions of the art will be considered, with emphasis on the external circuits and the transmission lines required to link these circuits to the memory elements.

Since about 1955, ferrite cores have been the standard random access memory element. Many factors have contributed to the increased performance of ferrite core memories in terms of speed, size, reliability, and reduction in cost. Among the most significant have been the reduction in size of the cores and improvements in the manufacturing, automatic testing, and assembly techniques. Most of the newer memories employ 30/50 or 19/30 (inside diameter/outside diameter in mils) size cores and recent work indicates the feasibility of pressing[3] or extruding and cutting[4] 13/21 cores. Smaller cores permit several things: a reduction in the drive requirements; closer packing, hence shorter drive and sense lines for a given memory size; lower drive line impedances and back voltages, simplifying driver design; and improved surface-to-volume ratios, giving better thermal characteristics. Some of these characteristics are illustrated in Table I. The effect of core size on the array characteristics is shown in Table II.

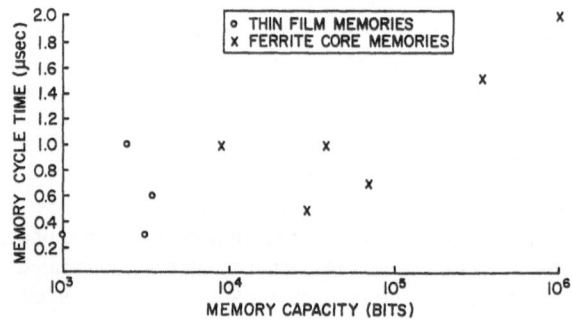

FIG. 1. Characteristics of existing high-performance memories.

[1] J. W. Forrester, J. Appl. Phys. **22**, 44–48 (1951).
[2] J. A. Rajchman, Proc. Inst. Radio Engrs. **41**, 1407–1421 (1953).
[3] M. H. Cook and E. C. Schuenzel, "A diminutive ferrite memory core," Electronics Division Fall Meeting, Am. Ceramic Soc., October 27, 1961.
[4] W. L. Shevel, Jr., J. M. Brownlow, O. A. Gutwin, and K. R. Grebe, "New ferrite core arrays for large capacity storage," Symposium on Large Capacity Memory Techniques for Computing Systems, Washington, D. C., May 23, 1961.

Ferrite apertured plates in which many bits can be stored in a single pressed piece have also been developed.[5] Although the successful operation of memories employing such plates has been reported, they are not extensively used.

An interesting core plane assembly technique which permits partial machine wiring of memory planes is shown in Fig. 2. With this device all the $X$ wires in a plane are fed through hollow needles and then simultaneously threaded through the cores. $Y$ wires are similarly machine threaded before hand wiring of the $Z$ and sense lines.

While three-dimensional memories are, in general, more economical,[6] particularly in the larger sizes, two-dimensional core memories do offer a number of advantages for high-speed operation. First, since very large read currents may be used, fast reading and high output signals may be obtained. Second, partial switching techniques[8] may also be employed. Several schemes have been developed for efficient operation in the partial switching mode. For example, in the 2 core/bit arrangement[9,10] bipolar output signals are obtained by the differential connection of the sense line. This arrangement has the further advantage that the drive line loading is independent of the stored information. The speed advantage of partial switching is shown in the plot of Fig. 3. This plot, which is especially useful in evaluating devices for 2-D applications, is obtained by the repeated switching of a core from its saturation state to a partially switched state by applying pulse fields of a given width and of increasing amplitude. Between each point obtained on the curve, the core is reset with a strong read pulse. Partial switching also reduces the core heating problem and permits the use of lower coercivity and less square core material. There are a number of problems, however, associated with partial switching of ferrite cores, including drive tolerances, in-

TABLE II. Memory transmission line characteristics.

| Memory elements | Density (bits/inch) | | $Z_0$ | $1/t_0$ (bits/nsec) |
| | $B$ | $W$ | | |
|---|---|---|---|---|
| Ferrite cores: | | | | |
| 50/80 | 10 | 10 | 180 | 40 |
| 30/50 | 16 | 16 | 150 | 60 |
| 19/30 | 28 | 28 | 140 | 80 |
| 13/21 | 45 | 45 | | |
| Flat thin film[a] | 50 | 5 | $\approx 20$ | $\approx 300$ |

[a] These data are taken or inferred from reference 18.

creased line impedance, and a time dependent threshold effect. The latter is to be discussed in a paper by Shahan and Gutwin[11] at this conference.

The drive and sense lines through the cores of a very large high-density array behave like transmission lines. These lines, which may be as long as 70 ft, have a characteristic impedance of about 150 ohms and may have delays of up to 200 nsec. Considerable care must be taken to circumvent the effect of line delays and to avoid the coupling of energy from one drive line to another and from drive to sense line. Measures taken include the systematic placement of wires through cores, transposition between planes, the inversion of alternate core planes to reduce inductive coupling loops, the use of rectangular sense and inhibit segments, the accurate termination of all long lines, staggering of $X$ and $Y$ drivers, and staggering of the inhibit drivers and sense strobes corresponding to the $X$ and $Y$ line delays.[12]

Another development of major importance in large ferrite core memories is the load-sharing switch.[13,14] By means of a core decoding switch that is transformer

TABLE I. Ferrite core characteristics.

| Core size (mils) | $H_c$ (oersteds) | Half select current (ma) | $T_{sw}$ ($\mu$ sec) | $S/V$ (in.$^{-1}$) |
|---|---|---|---|---|
| 50/80 | 1.5 | 400 | 1.5 | 220 |
| 30/50 | 1.7 | 250 | 1.0 | 360 |
| 30/50 | 3.6 | 600 | 0.4 | 360 |
| 19/30 | 3.6 | 400 | 0.4 | 640 |
| 13/21 | 3.6 | 240 | 0.4 | 1000 |

FIG. 2. Machine wiring of core plane.

[5] J. A. Rajchman, Proc. Inst. Radio Engrs. **45**, 325–334 (1957).

[6] An interesting technique for reducing the number of diodes required in a 2-D memory by utilizing the minority carrier storage effect is described by Melmed and Shevlin.[7]

[7] A. Melmed and R. Shevlin, IRE Trans. on Electronic Computers, **EC-8**, No. 4, 474–478 (1959).

[8] R. E. McMahon, "Impulse switching of ferrites," Solid State Circuits Conference, February, 1959, Digest of Technical Papers, pp. 16–17.

[9] C. J. Quartly, Elec. Eng. **31**, 756–758 (1959).

[10] W. H. Rhodes, L. A. Russell, F. E. Sakalay, and R. M. Whalen, IBM J. Research Develop. **4**, No. 2, 189–196 (1960).

[11] V. T. Shahan, and O. A. Gutwin, J. Appl. Phys. **33**, 1049 (1962), this issue.

[12] C. A. Allen, E. D. Councill, and G. D. Bruce, Electronics **34**, 68–71 (May 12, 1961).

[13] G. Constantine, Jr., IBM J. Research Develop. **2**, 204–211 (1958).

[14] N. G. Vogl, "A new load-sharing matrix switch," 1961 International Solid State Circuits Conference, February, 1961, Digest of Technical Papers, pp. 104–105.

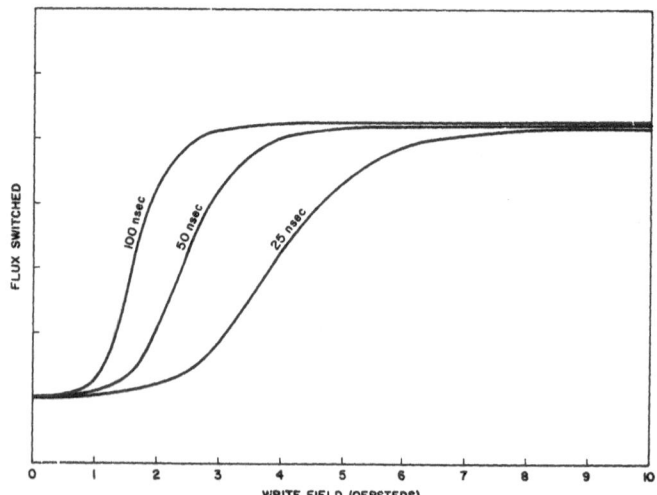

FIG. 3. Partial switching characteristics of a ferrite core.

coupled to the drive lines, the required drive power may be shared among a number of transistors. Since the addressing of a memory is arbitrary and programming restrictions are generally unacceptable, the memory must be capable of repeatedly interrogating a single address. By means of the load-sharing switch, the maximum power requirement on a single driver is greatly reduced, although the average drive power is unchanged. This makes possible the use of high-speed, medium-power transistors as memory core drivers.

While in limited use to date, memories employing thin magnetic metal film devices,[15-19] have received much attention. Thin film memories are potentially attractive due to their very fast switching, excellent thermal properties and bulk fabrication. Most memories of this type involve the deposition of a thin film

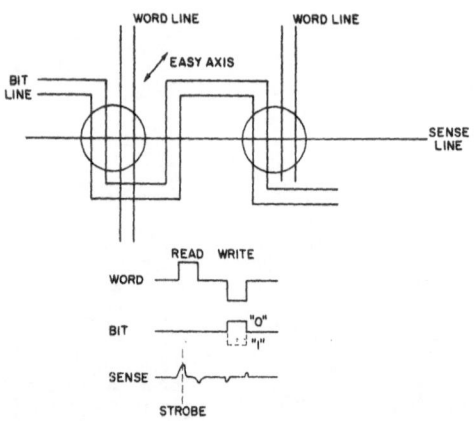

FIG. 4. Thin film memory operation—parallel mode.

[15] A. V. Pohm and S. M. Rubens, Proc. of the Eastern Joint Computer Conference, December, 1956, p. 120.

[16] J. I. Raffel, J. Appl. Phys. **30**, 608–618 (1959).

[17] E. E. Bittman, 1959 Solid State Circuits Conference, Digest of Technical Papers, pp. 22–23.

[18] J. I. Raffel, T. S. Crowther, A. H. Anderson, and T. O. Herndon, Proc. Inst. Radio Engrs. **49**, No. 1, 155–164 (1961).

[19] E. M. Bradley, J. Brit. Inst. Radio Engrs. **20**, No. 10, 765–784 (October, 1960); Electronics **33**, 78–81 (September, 1960).

(1000–2000 A) of nickel-iron (usually the non-magnetostrictive 81% Ni−19% Fe composition) on a flat substrate of glass, mica, or metal. The films may be deposited by plating, by vacuum evaporation, or by sputtering in the presence of an orienting field to achieve uniaxial anisotropy. Individual spots of either circular or rectangular shape with dimensions varying from 10 to 100 mils may be either etched or obtained by evaporation through a mask. Bradley has reported work[19] in which the individual bits in a large sheet were defined only by the crossings of the drive lines. This arrangement has the added advantage of simplifying the registration problem.

Thin film memories have been operated in both parallel[17] and perpendicular[18] drive modes. In the parallel mode, one arrangement of which is illustrated in Fig. 4, the word and bit fields in a 2-D array are applied parallel but off the easy axis to assure fast rotational switching. If sufficiently good film properties could be obtained, a memory in this mode could be operated in the bit-organized or 3-D fashion, but to date this has not been feasible. Dietrich and Proebster have shown[20] that irreversible flux changes or "creeping" occur for fields much less than the wall motion coercive force $H_c$ for fields applied at 45°, while this does not occur for fields applied in the easy direction. This phenomenon makes the realization of a parallel mode thin film memory very difficult.

The perpendicular drive mode thin film memory is illustrated in Fig. 5. Here the word field is applied in the hard direction, rotating the magnetization through 90° either in the clockwise or counterclockwise direction depending on the initial state of magnetization which corresponds to the stored information. This rotation induces either a positive or a negative signal in the sense line which runs in the easy direction (perpen-

[20] W. E. Proebster and W. Dietrich, International Solid State Circuits Conference, February, 1961, Digest of Technical Papers, pp. 66–67.

dicular to the word line). A small bit field in the easy direction applied at the end of the word pulse determines the new rest state of the magnetization and therefore controls the writing process. A variation on the perpendicular drive mode involves a slight rotation of the easy axis, so that there is a preferred direction after reading and a unipolar bit pulse may be used.

Individual films of the type described above have been switched in as little as 1 nsec.[21] In principle, large numbers of bits can be fabricated simultaneously (the current state of the art seems to be about 1000[18,19]). Recent progress in obtaining greater film uniformity is significant in this regard; however, uniformity, particularly edge effects and dispersion and variation in the easy axis presently limit plane sizes to about 10 square inches or less. Another major problem in memories of this type is the relatively low signal level obtained. A simple calculation shows that with a 40-mil spot, 1000 A thick, the maximum signal attainable is about $10^{-4}$ v-μsec even with perfect coupling of the drive and sense lines. A typical high-density film array might have a 1-mv signal level with 20-nsec switching. This implies that film devices of this geometry must be switched very rapidly, and therefore fast rising drive fields must be used. The difficulty of generating such fields, together with the noise problem in the sense line, has limited the field of application of thin film memories to high-speed, low-capacity memories. Since these devices do not have a closed flux path, they are also relatively sensitive to fields from all sources. This must be taken into account in their design, and usually dictates the minimum spacing of the bits in a plane.

The sandwich or dual film geometry, in which a second film with an easy axis parallel to the first is placed over the first with the stripline wiring between, reduces the demagnetization problem somewhat but introduces additional fabrication and uniformity problems.[18]

A number of film memory elements employing other geometries have also been described. While most of the twistor[22] memories use a thin Permalloy tape wrapped on a copper wire, they can be fabricated by plating on a wire[23] and are considered here as film type memory devices. Both the twistor, which has a helical easy axis, and the magnetic rod[24] memory element, which has a longitudinal easy axis, have an open flux path. This is shown in Fig. 6. In the twistor, the word and bit drive fields are orthogonal (both at 45° to the easy axis). The rod, which has a high (14 oe) coercivity to reduce the self-demagnetization, makes use of parallel drive fields in the easy direction supplied by currents flowing in 10-turn solenoids.

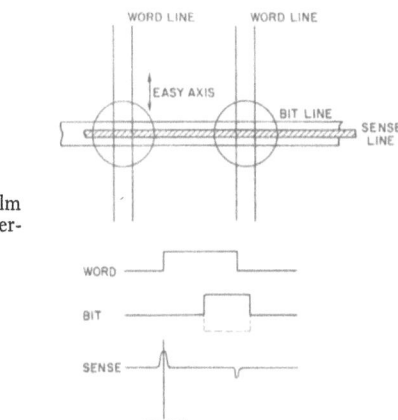

FIG. 5. Thin film memory operation—perpendicular mode.

Cylindrical films[25–28] deposited either on metallic or glass substrates with a circumferential easy axis have also been reported (see Fig. 6). Memory elements of this type may be operated in either the parallel or perpendicular mode. An advantage of this geometry is the absence of a demagnetizing field in the static remanent state, permitting the use of thicker films with resultant higher signal levels. High-speed switching has also been reported with these devices, and recent work in our laboratory indicates that switching constants as low as 0.01 oe-μsec can be achieved in this geometry.

The switching properties of magnetic memory elements play a major role in determining memory performance; however, in many applications, particularly in the submicrosecond speed range, the associated logic, driver, and sense circuits and the interconnecting transmission lines are controlling. Many film memory ele-

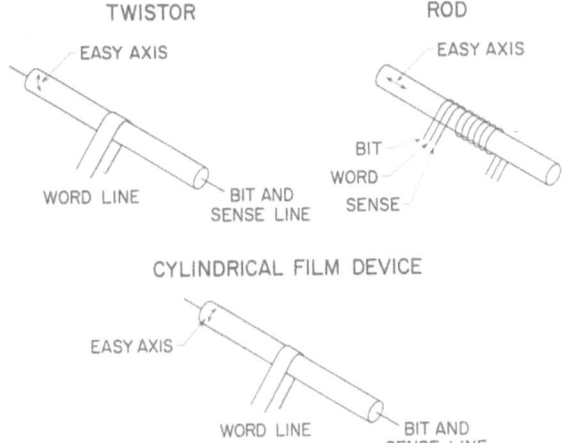

FIG. 6. Cylindrical geometry film memory devices.

[21] W. E. Proebster, W. Dietrich, and P. Wolf, IBM J. Research Develop. 4, No. 2, 189–196 (1960).
[22] A. H. Bobeck, Bell System Tech. J. 1319–1340 (1957).
[23] S. J. Schwartz and J. S. Sallo, IRE Trans. on Electronic Computers, EC 8, No. 4, 465–469 (1959).
[24] D. A. Meier, Proc. of the Electronic Components Conference, May, 1960, pp. 122–128.

[25] T. R. Long, J. Appl. Phys. 31, 123S–124S (1960).
[26] T. R. Long, "A discussion of electrodeposited cylindrical film memory elements," Electrochemical Society Meeting, October, 1961.
[27] G. R. Hoffman, J. A. Turner, and T. Kilburn, J. Brit. Inst. Radio Engrs. 20, No. 1, 31–36 (1960).
[28] G. Rostky, Electronic Design 9, 26–27 (August 1961).

ments switch in the rise time of the drive field; thus, not only are very wide band pass drive and sense circuits required, but in addition the array must be so designed that tolerable loss, distortion, and cross coupling will occur in the transmission paths. If the signal delay is comparable to the rise time, it is generally necessary to terminate the lines in their characteristic impedances to prevent excessive ringing and distortion; this increases the required drive power. Strip transmission lines, in which the width-to-separation ratio is high, are attractive because of their low impedance, well-defined field pattern, and relative insensitivity to inductive coupling. These considerations have led to the use of very thin glass[18] or metal[19] substrates for flat film memories. In some cases, loss and distortion in these lines may limit the size of memories or at least of sense segments. The importance of high density memory element packaging is apparent from the foregoing. Comparative data for a number of memory elements are shown in Table II.

Much of the present research and development effort in the memory area is aimed at extending memory capacity or speed and at special function memories. Memory capacity limitations are primarily economic, thus any technique for reducing the cost of memory elements or reducing the drive requirements is beneficial in this area.

As described above, many of the film devices switch very rapidly; consequently, memory speed is largely determined by the band width and delay of the drivers, sense amplifiers, transmission lines, and logic stages in the regeneration loop. In two-dimensional read-write memories, cycle time is often limited by the sense circuits' recovery time from writing. This effect may be limited by reducing the coupling from the bit to sense line,[19] by termination of the lines, or by write noise cancellation in a bridge circuit.[18,29]

Another means of achieving high speed in a memory system is the use of nondestructive read out (NDRO). Since there are usually several times as many read and regenerate operations as there are write operations, a fast NDRO read, "slow" write memory has many potential applications. The potential speed advantage of an NDRO memory is considerable; not only does it save the switching time associated with the read operation, but also it saves the time required for the regenerative sense loop to function and avoids the troublesome sense circuit write recovery problem. A

number of NDRO memory techniques have been shown to work, including cross field interrogation of ferrite cores,[30] permeability sensing of cores,[31,32] ac sensing of cores,[33] multipath ferrite devices,[34] bias restoration of cores,[35] hard film restoration of soft films in film pairs,[36,37] and less than 90° rotation of thin film memory elements. Most of these techniques, particularly those relying on the reversible switching of flux, are troubled by low output signals and small margins.

Other types of special function memory include read-only and associative memories. Read-only memories are a special class of NDRO memories in which the information must be altered mechanically. Magnetic read-only memories fall into two classes; those which involve linear coupling elements such as Kilburn's peg memory,[38] and those which involve nonlinear coupling elements such as the permanent magnet twistor memory.[39] The advantage of the nonlinear coupling element is a reduction in noise sensitivity. An associative memory[35] is not normally addressed in the conventional manner but rather a word is called out based on a match of portions of the stored words. A means of achieving this type of memory with magnetic elements based on the bias restored core NDRO cell is described by Kiseda et al.[35]

In summary, compact, highly reliable ferrite core memories are used widely today in data processing systems. Present work in the field, including a substantial effort in the development of a variety of thin film devices, promises higher speed and perhaps more versatile memories for future use.

[29] W. B. Gaunt and D. C. Weller, "A 12-kilobit, 5-microsecond twistor variable store," 1961 International Solid-State Circuits Conference, February, 1961, Digest of Technical Papers, pp. 106–107.

[30] R. M. Tillman, IRE Trans. on Electronic Computers **EC-9**, 323–328 (1960).

[31] G. H. Perry and S. J. Widdows, "Low coercive-force ferrite ring cores for a fast non-destructively read store," International Solid-State Circuits Conference, February, 1960, Digest of Technical Papers, pp. 58–59.

[32] W. L. Shevel, Jr. and O. A. Gutwin, "Partial switching, nondestructive-readout storage systems," International Solid-State Circuits Conference, February, 1960, Digest of Technical Papers, pp. 62–63.

[33] R. E. McMahon, "ac and impulse switching techniques for fixed, random access and analog memory use," International Solid-State Circuits Conference, February, 1961, Digest of Technical Papers, pp. 68–69.

[34] J. A. Rajchman and A. W. Lo, Proc. Inst. Radio Engrs. 44, 321–322 (1956).

[35] J. R. Kiseda, H. E. Petersen, W. C. Seelbach, and M. Teig, IBM J. Research Develop. **5**, No. 2, 106–121 (1961).

[36] L. J. Oakland and T. D. Rossing, J. Appl. Phys. **30**, 54S–55S (1959).

[37] R. J. Petschauer and R. D. Turnquist, "A nondestructive readout film memory," Proceedings of the Western Joint Computer Conference, May, 1961, pp. 411–425.

[38] I. L. Auerbach, Proc. Inst. Radio Engrs. **49**, No. 1, 330–331 (1961).

[39] W. A. Barrett, F. B. Humphrey, J. A. Ruff, and H. L. Stadler, IRE Trans. on Electronic Computers **EC-10**, No. 3, 451–461 (1961).

# High Magnetic Field Research

Benjamin Lax

*National Magnet Laboratory,\* Massachusetts Institute of Technology, Cambridge 39, Massachusetts and*
*Lincoln Laboratory,† Massachusetts Institute of Technology, Lexington 73, Massachusetts*

The use of high magnetic fields as a research tool for a wide variety of physical phenomena is clearly recognized today and the subject of an International Symposium at Cambridge this fall. Another important step in promoting research with the help of large magnetic fields has been the sponsorship of the M. I. T. National Magnet Laboratory by the Air Force. This paper will review the highlights of the conference which included papers on research in plasma physics, low temperatures, solid state and the latest developments for generating high magnetic fields. The plans and objectives of the National Magnet Laboratory and description of the physical facilities will be presented. In addition, a brief review will be given of a number of experiments already performed in the existing Magnet Laboratory at M. I. T.

## INTRODUCTION

SCIENCE, like many other creative efforts of mankind, is characterized by fashions and trends. In the proper course of evolution certain ideas and tools achieve that state of maturity which inevitably makes them ripe for fruitful exploitation. The early period of this century was marked by the intensive productivity of spectroscopy and related atomic theory. During the late 1920's and 30's research was strongly influenced by quantum theory and nuclear physics stimulated by accelerators, which even today are important tools in high energy physics. After World War II solid-state physics emerged as a major field of science due to the impact of microwave and low temperature techniques and the advances in materials science. Today we stand on the threshold of a new era which may become known as the decade of high magnetic field research. Within the last year two events may be regarded as milestones which herald the coming of this era. The first International Conference on High Magnetic Fields, which was attended by more than 700 conferees at the Massachusetts Institute of Technology, indicated a growing enthusiasm and interest in this subject. A more concrete evidence of progress and enhanced future activity was the beginning of the construction of the National Magnet Laboratory.

The conference which was sponsored by the Solid State Sciences Division of the Air Force Office of Scientific Research had two major themes. The first of these was centered specifically around the art of generating high magnetic fields and involved research, design, and development of solenoids of different types. The second focused attention upon the physical research which utilizes these solenoids and which covered a broad spectrum including solid-state physics, low temperatures, plasma physics, fusion research, and biomagnetics. It is not possible in a short review such as this to give a proper coverage of the conference, consequently only the highlights of some of the areas will be discussed.

## HIGH MAGNETIC FIELDS

Magnetic fields represent an old tool for research. However, only a few attempts beyond the 10 000 gauss range have been made until relatively recently. In general there are now four major approaches to the generation of high magnetic fields, each probably complementary to the other; namely, pulsed magnets, water-cooled coils, cryogenic magnets and superconducting coils.

### Pulsed Fields

Pulsed systems are relatively inexpensive and easy to construct from condenser banks, a switch and some type of wound or rigid coil design. These can generate *useful* fields from 100 000 gauss to perhaps 500 000 gauss depending on the particular problem to be pursued. Usually the measurement techniques are fairly intricate with such transient systems. Nevertheless, with ingenuity pulsed magnets have progressed beyond the stage of an exploratory instrument. One of the early applications of pulsed fields was made by Kapitza[1] who has pioneered in the measurement of the magnetoresistance of a number of materials. This type of study has been revived by the Harvard group[2] and others and the work on metals has been reported by Olsen[3] at this conference. The beautiful experiments of Schoenberg and coworkers on the de Haas van Alphen[4] effects in copper and other metals was discussed by Priestley.[3] More modern experiments on cyclotron resonance[5] and millimeter wave spin resonance[6] have been carried out at Lincoln Laboratory and were reported by Foner.[3] Other examples of pulsed field experiments are those in magnetism, the susceptibility measurements of paramagnetic materials by Stevenson,[3] and the novel spin-flop experiments on antiferromagnets by Jacobs.[3] Pulsed fields are

\* Supported by the Air Force through the Air Force Office of Scientific Research.

† Operated with support from the U. S. Army, Navy, and Air Force.

[1] P. Kapitza, Proc. Roy. Soc. (London) **A123**, 292 (1929).

[2] H. P. Furth and R. W. Waniek, Phys. Rev. **104**, 343 (1956).

[3] Papers presented at the 1961 International High Magnetic Field Conference will appear in a copy of the Proceedings to be published by the Technology Press (1962).

[4] D. Shoenberg, *The Fermi Surface* (John Wiley & Sons, Inc., New York, 1960), p. 74.

[5] B. Lax, J. G. Mavroides, H. J. Zeiger, and R. W. Keyes, Phys. Rev. **122**, 31 (1961).

[6] S. Foner, Phys. Rev. **107**, 683 (1957); J. phys. radium **20**, 336 (1959).

also becoming very important for determining the transition fields of hard superconductors in the hundred thousand gauss range. Bubble chambers and other devices for analysis of high energy particles are now employing pulsed magnets. The production of shock waves in plasmas by large pulsed fields for producing high temperatures was described by Kolb[3] of the Naval Research Laboratory. High energy storage in inductors was considered by Carruthers[3] of the Atomic Energy Research Establishment, England. Another interesting phenomenon that was described by Furth[3] during the proceedings was the plastic deformation of soft metals such as copper and aluminum by the large magnetic forces at the center of a pulsed coil. Finally when pulsed magnets are discussed the technique of implosion developed by Fowler[3] must be included. With this technique he is able to generate momentarily fields in excess of 10 megagauss.

## Water-Cooled Magnets

The second method for generating high magnetic fields involves the use of water-cooled coils. Such systems at approximately 2-megawatt level, capable of generating fields of the order of 100 000 gauss now exist in a number of laboratories throughout the world. In this country, in addition to the present magnet laboratory at M. I. T., there is one at the Naval Research Laboratory, the University of California, N. A. S. A. Lewis Research Center, and the one at Bell Telephone Laboratories. Others are being planned for the near future. In England there now are three; Oxford, Cambridge, and at the Radar Research Establishment in Great Malvern. There is a two-megawatt installation at Leiden, Holland. There is also one in Wrocklaw, Poland, and two in Japan, one at Tokyo University and one at Tohoku. Larger installations, of course, are being built at M. I. T., which I shall discuss shortly, and one of comparable magnitude is being planned at the Lebedev Institute in Moscow. The possibility also exists that in England a third such large facility will be built.

The areas of research that have been first exploited by the use of water-cooled magnets are spectroscopy, low temperature, and magnetism. More recently, solid-state physicists and plasma physicists have also become good customers of high magnetic fields. A number of the speakers, who discussed various fields of research that are being carried on with magnetic fields, made an excellent case for the steady-state water-cooled magnets. One of these was Kurti[3] from Oxford who presented cogent arguments why larger magnetic fields of the order of 200 000 gauss or more would be invaluable in adiabatic nuclear polarization and adiabatic nuclear cooling. Similarly, Professor Bloembergen[3] of Harvard, who talked about spin resonance, clearly demonstrated that larger magnetic fields would permit greater sensitivity in nuclear resonance, higher resolution in paramagnetic resonance, and in particular, would permit the

observation of resonance in paramagnetic and antiferromagnetic systems where the natural frequencies due to crystalline fields or internal effective fields occur in the far infrared region of the spectrum. This latter possibility has already been partially exploited and demonstrated by the pulsed technique in a number of such resonance experiments discussed by Foner.[3] John Galt[3] of Bell Telephone Laboratories reviewed cyclotron resonance experiments in solids and showed that larger magnetic fields would be extremely helpful in exploring the absorption curves of metals at microwave frequencies whereby the anomalous skin effect could be minimized by the magnetic field. Cyclotron resonance of ions at high magnetic fields were experimentally illustrated by Buchsbaum[3] of Bell Telephone Laboratories where the high fields permitted studies of the linewidth and resolution of multi-ion systems which otherwise could not be observed. In the solid-state area we also showed that high continuous fields was extremely important for the study of magneto-optical studies of semiconductors, metals and magnetic materials. Water-cooled magnets have also been applied advantageously to measurement of the magnetization of rare earths, which was reviewed by W. Henry. Oscillatory effects in solids in high fields was considered by Kahn of the National Bureau of Standards. Another area where steady magnetic fields are important for the study of physical properties, is in plasma physics. Professor Allis[3] of M. I. T. demonstrated from theoretical considerations how large magnetic fields permit the study of plasma waves over a wide range of the parameters involved and in which a variety of phenomena, including cyclotron resonance, Alfvèn waves, whistler modes, and magnetohydrodynamic waves could be profitably investigated. The problem of ambi-polar diffusion in plasmas, which is not only of fundamental interest, but is also inevitably encountered in any plasma studies was discussed by Lehnert[3] of the Royal Institute of Technology in Stockholm. In addition, the role of high magnetic fields in producing magnetohydrodynamic phenomena such as "wakes" was considered by Sears[3] of Cornell University.

The practical problems of generating magnetic fields with water-cooled systems was one of the important areas considered in this conference. Recent advances on water-cooled systems was discussed by Professor Bitter[3] of M. I. T. who described the basic problems involved in the design of coils from consideration of heat transfer, mechanical strength of materials and power availability. He predicted that water-cooled systems could ultimately reach the one-half million gauss mark with powers of the order of 50 megawatts on a continuous basis or with a low-duty cycle. In addition to the paper presented by Professor Bitter on the ultimate limits to be achieved with water-cooled coils, a number of other people discussed different systems and different power supplies utilized in energizing the coils. Parkinson[3] of the Radar Research Establishment discussed a battery system which for a short period of the order of 15 min or so was capable of achieving fields as large as 128 000 gauss. Another interesting system which was presented was the one at N. A. S. A. Lewis Research Center, which utilized a homopolar generator[3] at low voltage and high currents and a relatively simple water-cooled copper magnet capable of achieving 100 000 gauss quite readily. Perhaps one of the most attractive systems discussed during the conference was that presented by Adkins[3] of Cambridge who made a good case for a silicon rectifier system capable of supplying two megawatts with good stability and low ripple if operated during the night. The attractive part was that the cost was much less than that of some of the other systems, which balances some of the disadvantages. In addition to these systems, various other large magnet installations involved with the fusion work were presented, such as the Stellarator at Princeton by Mills,[3] the large mirror machine at the Lawrence Radiation Laboratory, California, by Coensgen.[3] Other large water-cooled magnets for fusion will be presented in fair detail in the Proceedings of the Conference.

## Cryogenic and Superconducting Magnets

The next major class of magnets that received a great deal of attention during the conference were the cryogenic systems, and this may be divided into two major categories, superconducting and nonsuperconducting. In the latter category several efforts were reported among which the most advanced was that of Laquer[3] at Los Alamos on hydrogen-cooled copper magnets which have attained fields of the order of 60 000 gauss or more. Aluminum magnets were also discussed by Taylor[3] of Lawrence Radiation Laboratory and the group at the National Bureau of Standards in Boulder, Colorado. One of the most ambitious undertakings in this regard was the neon-cooled system[3] to be used with the homopolar generator by the group at N. A. S. A., Cleveland. An extremely attractive system used by the Philips group at Eindhoven involved a liquid nitrogen precooled magnet[3] which was then pulsed by 100-kw generator to provide long pulses of the order of seconds.

The real excitement of the conference came from the superconducting magnets which will be discussed in detail by Dr. Kunzler[3] in a succeeding paper. Without stealing his thunder, let me just merely state that already four different groups at Bell Telephone Laboratories, Westinghouse, Lincoln Laboratory, and Atomics International have superconducting magnets of either niobium tin or niobium zirconium in excess of 50 000 gauss. An interesting paper on the field analysis of superconducting coils was given by Gauster[3] of the Oak Ridge National Laboratory. In some respects the superconducting magnet appears to be the poor-man's instrument from the initial capital cost and a useful one provided he has available helium and cryogenic techniques in his laboratory. There is no question that superconducting magnets will play an exceedingly important role

in high field magnet research in the future. One can speculate on its particular role in high field magnet research in the future. One can speculate on its particular role at the moment, however, like many new developments, it will in all likelihood complement the other systems and in individual situations will offer advantages, and in some cases such as large volume magnets, it will be the only practical solution.

## NATIONAL MAGNET LABORATORY

The National Magnet Laboratory is sponsored by the Air Force Office of Scientific Research and operated by M. I. T. In spirit and operation it will be similar to the Brookhaven National Laboratory, but on a smaller scale. The research activities will primarily be in the solid state and low temperature areas and will include an active program of research and development of magnets. The construction of the Laboratory has just begun and all the electrical equipment, motor generators, bus bars and other hardware are on order. The system will utilize large water-cooled magnets which will be capable of achieving 250 000 gauss in a one-inch working internal diameter. In order to achieve this field with copper coils, a total of 8 megawatts of continuous power will be supplied by four 2-megawatt dc generators operating from two synchronous motors, each with a fly wheel. The fly wheels will provide a pulse capability at four times the normal power or 32 megawatts for several seconds. In addition, we will be able to sweep the magnetic field and operate any of four of the machines individually or in suitable parallel-series combinations. The current fluctuations will be maintained at less than 0.015%. The ripple voltage at the fundamental frequency of 6 cps will be less than 0.02%.

The floor plan for the operating portion of the Magnet Laboratory is shown in Fig. 1, including the motor generator, sets, etc., which will be located in a new building with special foundations to minimize vibrations. There will be eight special magnet rooms, each with a coil for solid state and other experiments which require

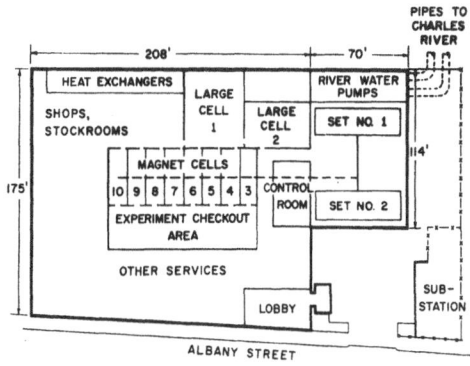

FIG. 1. Floor plan of the National Magnet Laboratory now under construction, showing the location of two motor generator sets with ten magnet rooms to house water-cooled solenoids for high field research.

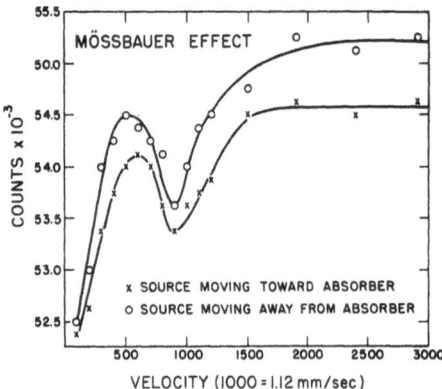

FIG. 2. Mössbauer effect in stainless steel in a longitudinal field of 64 kgauss showing a single Zeeman line. (After N. Blum.)

large space. The building will allow for expansion in the event that it cannot accommodate all the participants anticipated in the near future. The bus bars and the water lines will run in the basement underneath the magnet rooms and the power will be automatically switched from one coil to another from a central control room. In addition to the water-cooled magnets, there will be pulsed magnets and ultimately a combination of large superconducting magnets with the water-cooled magnets which will permit a wide variety of possibilities. The present aim of the Laboratory is to achieve fields of the order of 250 000 gauss at the 8-megawatt level but eventually to exceed these and aim for the 300 000 to 400 000 gauss range with the fly wheel pulse system, and with the combination of superconducting and water-cooled magnets.

### Present Laboratory and Research Program

We have remodeled the present magnet laboratory which has a power supply of 1.7 megawatts and a variety of magnets capable of providing fields up to 126 000 gauss. The most convenient magnet has been the one with the two-inch bore, capable of achieving 90 000 gauss and has been used for a large number of experiments. The list of experiments that have been performed by the staff of the National Magnet Laboratory and graduate and post-doctoral students includes a variety of magneto-optical effects in semiconductors, paramagnetic and antiferromagnetic materials, and magnetic properties of antiferromagnetics, which were discussed by Dr. Foner[3] at this meeting, and measurements of critical fields in hard superconductors. Among these I also want to mention the Mössbauer effect at high fields which is of particular pertinence to this conference. Cooperative experiments with members of Lincoln Laboratory, faculty and staff of other laboratories at M. I. T. and those in the local area are listed as follows: magnetotunneling in semiconducting diodes, magneto-reflection experiments in metals, cyclotron resonance in diamond, magnetic susceptibility measurements, and behavior of conducting fluids in magnetic fields.

About two years ago when Pound[7] first reported his Mössbauer Zeeman effect in $Fe^{57}$, it occurred to me that our high magnetic fields might possibly make an important contribution in this area. At that time, the plan was to use the external magnetic field to measure the $g$ factor of both the excited and ground state of the nucleus and the internal field, with one set of measurements at high fields. In the meantime, however, this was accomplished by Hanna and his group[8] at Argonne with low field polarization experiments and combined with nuclear resonance. Nevertheless, Norman Blum at the National Magnet Laboratory proceeded with the experiment and indeed the results which he obtained together with some more recent data in cooperation with Professor Grodzins of M. I. T. are beginning to give some interesting information about internal fields in these materials. Figure 2 shows the simple Zeeman spectrum obtained in $Fe^{57}$, in this particular case in stainless steel, where it is necessary to use a large external field and indeed we obtained a single peak consistent with the theory. The object is to study the behavior of this peak as a function of magnetic field as shown in Fig. 3. From this plot of the Zeeman shift it turns out that with the knowledge of the $g$ factors, it is possible to measure the actual field seen by the nucleus which is equal to the external field minus a hyperfine field. From this one can then measure the hyperfine constant. This experiment has now been carried out in stainless steel and also iron. The promising future for this type of experiment is that it can be done in many materials which have no internal fields and will permit the measurement of the spin factors for the ground and excited states of the nucleus as well as the hyperfine constants in these materials. In general it appears to be a very powerful technique for studying the internal field problem.

Figure 4 shows the results of the four-millimeter cyclotron resonance in diamond carried out by Rauch of Lincoln Laboratory. Some months ago he reported the

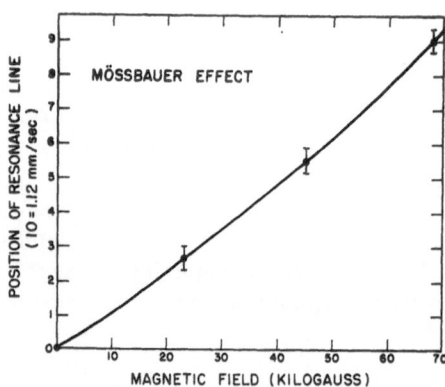

FIG. 3. Energy shift of Zeeman line with applied magnetic field in stainless steel. (After N. Blum.)

[7] R. V. Pound and G. A. Rebka, Jr., Phys. Rev. Letters 3, 439 (1959); 3, 554 (1959).
[8] S. S. Hanna, J. Heberle, C. Littlejohn, G. J. Perlow, R. S. Preston, and D. H. Vincent, Phys. Rev. Letters 4, 177 (1960).

first two peaks[9] which he was unable to identify. By using monochromatic excitation he showed that the 1.07 $m_0$ peak was that due to the spin orbit split-off valence band. Hence from the 0.7 $m_0$ light hole it was possible to estimate the heavy hole to be 2.2 $m_0$. At 70 kMc this required fields of the order of 60 kgauss. Hence the experiments were repeated with higher fields to obtain all three masses as shown.

### CONCLUSIONS

In order to carry out the experiments to date, we have had to work on a two-shift basis and are already starting on a three-shift operation in order to accommodate additional experiments that are being planned for the existing magnet laboratory. It is expected that in about one and a half years the new laboratory will be in oper-

[9] C. Rauch, Phys. Rev. Letters **7**, 83 (1961).

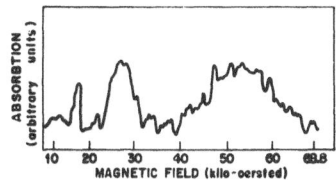

FIG. 4. Cyclotron resonance in diamond at 70 kMc in high magnetic field. Peaks correspond to 0.7 $m_0$ light hole, 1.07 $m_0$ split-off hole, 2.2 $m_0$ heavy hole. (After C. Rauch.)

ation and will permit many more researchers from other organizations as well as graduate students and visiting scientists to participate and perform experiments with high fields. Since the sponsorship comes from the Solid State Sciences Division of the Air Force Office of Scientific Research, it is expected that the bulk of the program will be in solid state. However, the new facilities will be made available to research workers of other fields as well, with the object of making this a truly National Laboratory and with an international flavor as well.

JOURNAL OF APPLIED PHYSICS     SUPPLEMENT TO VOL. 33, NO. 3     MARCH, 1962

# Modification of Spin Screw Structure due to Anisotropy Energy and Applied Magnetic Field

TAKEO NAGAMIYA

*Department of Physics, Osaka University, Osaka, Japan*

Recent development in the theory of screw structure of spins and its modification due to anisotropy energy and applied magnetic field is reviewed, with reference to neutron diffraction work on rare-earth metals and MnAu₂.

### 1. INTRODUCTION

THE screw structure of spins, first predicted by Yoshimori[1] and later by Villain[2] and Kaplan,[3] has found its application in a number of examples. Rare-earth metals with more than half-filled $4f$ shells, ranging from Tb to Tm (Tb, Dy, Ho, Er, Tm), present intricate but interesting examples. For these metals magnetic measurements have been done by the people at Ames[4] and neutron diffraction measurements by the people at Oak Ridge.[5] They all show ferromagnetism at low temperatures, not always a simple ferromagnetism in actuality, and apparently antiferromagnetism at intermediate temperatures. This apparent antiferro-

magnetism was actually found to be a screw in Dy and Ho (possibly also in Tb) and a sinusoidally varying spin arrangement in Er and Tm. It is very likely that the crystalline field anisotropy energy of the $4f$ shell of these metal atoms has modified the simple screw arrangement of spins which these metals would have had in virtue of a certain characteristic exchange interaction if there were no anisotropy energy. A study of the crystalline field anisotropy energy and its effect on the spin arrangement has recently been done by Miwa and Yosida,[6] Elliott,[7] and Kaplan,[8] independently, and they succeeded in interpreting the magnetic structures of these metals.

A magnetic field acting on the crystal should also affect the magnetic structure. A magnetic field is, so to say, the source of the simplest anisotropy energy which can be varied at will. The effect of an applied field on the screw structure of spins has been studied at Saclay

[1] A. Yoshimori, J. Phys. Soc. Japan **14**, 807 (1959); T. Nagamiya, J. phys. radium **20**, 70 (1959)—Proc. International Conference on Magnetism, Grenoble, 1958.

[2] J. Villian, J. Phys. Chem. Solids **11**, 303 (1959).

[3] T. A. Kaplan, Phys. Rev. **116**, 888 (1959).

[4] D. R. Behrendt, S. Legvold, and F. H. Spedding, Phys. Rev. **109**, 1544 (1958); B. D. Rhodes, S. Legvold, and F. H. Spedding, Phys. Rev. **109**, 1547 (1958); J. F. Elliott, S. Legvold, and F. H. Spedding, Phys. Rev. **100**, 1595 (1955); R. W. Green, S. Legvold, and F. H. Spedding, Phys. Rev. **122**, 122 (1961).

[5] M. K. Wilkinson, W. C. Koehler, E. O. Wollan, and J. W. Cable, J. Appl. Phys. **32**, 48S (1961); W. C. Koehler, J. Appl. Phys. **32**, 20S (1961); J. W. Cable, E. O. Wollan, W. C. Koehler, and M. K. Wilkinson, J. Appl. Phys. **32**, 49S (1961).

[6] H. Miwa and K. Yosida, Proc. International Conference on Magnetism and Crystallography, Kyoto, 1961 (to be published).

[7] R. J. Elliott, Proc. International Conference on Magnetism and Crystallography, Kyoto, 1961 (to be published); Phys. Rev. **124**, 346 (1961).

[8] T. A. Kaplan, Proc. International Conference on Magnetism and Crystallography, Kyoto, 1961 (to be published); Phys. Rev. **124**, 329 (1961).

by Herpin and Mériel[9] with MnAu$_2$ by neutron diffraction measurements. Rare-earth metals are also being investigated at Oak Ridge by Koehler.[10] The corresponding theory, applicable to simplest cases only at the present moment, has been developed by Herpin and Mériel,[9] Enz,[11] and the present writer.[12]

It is the purpose of this memorandum to briefly review the information at present available and to outline the theories mentioned above.

## 2. OUTLINE OF THE THEORY OF SCREW STRUCTURE

Before going over to the main part of the present article, it might be helpful to briefly outline the theory of screw structure. Consider a set of layers of atoms whose spins are coupled ferromagnetically within each layer with an exchange constant $J_0$, between adjacent layers with $J_1$, between next-nearest-neighboring layers with $J_2$, and so on. Let the spins in the same layer be parallel and their direction be specified by an angle $\theta_n$, measured in one plane from a certain specified direction, where $n$ is the number of the layer under consideration. Then the interaction energy can be written

$$E = -S^2 \sum_n [J_0 + 2J_1 \cos(\theta_{n+1} - \theta_n) + 2J_2 \cos(\theta_{n+2} - \theta_n) + \cdots], \quad (1)$$

where, for the sake of simplicity, the number of pairs interacting in the same way are supposed to be included in the exchange constants. If we put $\theta_n = nq + \text{const}$, we have

$$E = -NS^2 J(q),$$
$$\text{where } J(q) = J_0 + 2J_1 \cos q + 2J_2 \cos 2q + \cdots, \quad (2)$$

$N$ being the number of layers. Thus, the minimum of the energy corresponds to the maximum of $J(q)$, and if $J(q)$ is the largest at $q = 0$ or $\pi$ the system will show ferromagnetism or antiferromagnetism, respectively. If, however, $J(q)$ is the largest at $q_0$ different from 0 and $\pi$, the system will have a screw structure, in which the spin vectors rotate uniformly with an angle $q_0$ as one goes from layer to layer. Yoshimori has proved rigorously that for crystals consisting of equivalent magnetic atoms—equivalent in the sense that the environment of every atom, as regards the magnetically interacting neighbors, is the same apart from translation—the screw structure represents the stable solution of the problem of minimum energy if $J(q)$ is the largest at $q_0$. The example he took was MnO$_2$, for which he explained beautifully the neutron diffraction lines observed by

Erickson,[13] with $q_0 = 5\pi/7$. His proof also applies to hexagonal close-packed structures if the screw axis is parallel to the hexagonal axis, which is the case for rare-earth metals with more than half-filled $4f$ shell. (If the screw axis is perpendicular, or in general oblique, to the hexagonal axis, the uniform rotation has to be modified in such a way that, corresponding to two atoms in each unit cell, the spin vectors of one sublattice is rotated by a certain angle with respect to those of the other sublattice.)

The discovery of screw structure was made through the following consideration. Several years ago, Nakamura and Nagai[14] at Kyushu University were investigating the spin wave spectrum of MnF$_2$, which has the same crystal structure as MnO$_2$. The spin wave frequency starts at a finite value at zero wave number, corresponding to the antiferromagnetic resonance frequency, but it goes down with increasing wave number $q_z$ along the $z$ axis if the exchange constant between atoms neighboring along the $c$ axis is greater than the exchange constant between a corner atom and a neighboring body-center atom. If the ratio of these exchange constants exceeds 1.2, the frequency becomes negative in a certain interval of $q_z$. This suggested that for such a value of the ratio the simple antiferromagnetic structure known for MnF$_2$ must be unstable and that instead a certain static spin wave must be realized. This idea of the present writer lead to Yoshimori's discovery of screw structure. It is noted in passing that a recent experiment by Owen[15] of the ESR in MnF$_2$ diluted with ZnF$_2$ revealed that the ratio mentioned above is very small and is negative.

The neutron diffraction pattern to be obtained from a screw structure is as follows. As known by the theory of magnetic scattering of neutrons, only the component of the spin vector parallel to the plane of reflection is effective for magnetic lines. In this plane, the component along one coordinate axis and that along the other coordinate axis give rise independently to the line intensity if the neutrons are unpolarized. Each component oscillates sinusoidally in a screw structure so that the plus and minus phase differences between consecutive planes due to this sine wave, namely $+q_0$ and $-q_0$, come into the scattered wave. Therefore the Bragg condition is modified as

$$\mathbf{k}' - \mathbf{k} \pm \mathbf{q}_0 = \mathbf{K}, \quad (3)$$

where $\mathbf{k}'$ and $\mathbf{k}$ are, as usual, the wave vectors of the scattered and incident neutrons, respectively, and $\mathbf{K}$ is a reciprocal lattice vector. The direction of the vector $q_0$ represents the direction of the screw axis. The ordinary Bragg reflection lines are therefore each split

[9] A. Herpin and P. Mériel, C. R. Acad. Sci. 250, 1450 (1960); preprint from Service de Physique du Solide et de Résonance Magnétique, Centre d'Etudes Nucléaires de Saclay.
[10] W. C. Koehler, Proc. International Conference on Magnetism and Crystallography, Kyoto, 1961 (to be published).
[11] U. Enz, Physica 26, 698 (1960); J. Appl. Phys. 32, 22S (1961).
[12] T. Nagamiya, K. Nagata, and Y. Kitano, Proc. International Conference on Magnetism and Crystallography, Kyoto, 1961 (to be published); Progr. Theor. Phys., Kyoto (to be published).

[13] R. A. Erickson, unpublished work.
[14] T. Nakamura and O. Nagai, unpublished work; O. Nagai and A. Yoshimori, Progr. Theoret. Phys. (Kyoto) 25, 595 (1961).
[15] J. Owen, Proc. International Conference on Magnetism and Crystallography, Kyoto 1961 (to be published); M. R. Brown, B. A. Coles, J. Owen, and R. W. H. Stevenson, Phys. Rev. Letters 7, 246 (1961).

into two, corresponding to $\pm q_0$, and by measuring the magnitude of the splitting one can determine the value of $q_0$. It is interesting to note that the reflection lines with $K = 0$ can also be observed.

As can be seen from the above argument, the period of the screw has nothing to do with the lattice spacing. This is in fact a new feature characteristic of the screw structure. Furthermore, as long as only exchange interactions are concerned, the plane in which the spins rotate is indeterminate. If this plane is perpendicular to the direction along which the rotation propagates, i.e., perpendicular to the direction of the screw axis, then the screw is called "proper." On the other hand, if this plane contains the screw axis, then the screw may rather be called "cycloidal." Now, when one takes into account the anisotropy energy or an applied field, the situation is changed in the following way. In the case where the anisotropy energy stabilizes the plane perpendicular to the screw axis, the screw is proper; when it stabilizes a plane which contains the screw axis, the screw is a cycloid. When there is a large uniaxial anisotropy energy which stabilizes the screw axis, as in the case of Er and Tm, the cycloid may happen to be squeezed to become a collinear array of spin vectors. If the anisotropy energy is such that it stabilizes a circular cone whose axis coincide with the screw axis, the spin vectors may rotate on a cone, giving rise to a ferromagnetic component along the screw axis, as in the low temperature phase of Ho and that of Er. For this anisotropy energy, it is also possible that the spin vectors rotate on a plane which is oblique or parallel to the spin axis, as in the intermediate temperature phase of Er, since then the spin vectors can point alternately near the upper cone and the lower cone and at the same time can lower the exchange energy by rotating on a plane. In the case of a proper screw, the anisotropy energy within the plane will modify the uniform rotation of the spin vectors. If this anisotropy energy has a six-fold symmetry and is large enough, the spin structure will be one of the following four: (1) ferromagnetic, when the turn angle of the spin vectors in the original screw is less than 30°, as realized in the low temperature phase of Dy, (2) a screw of pitch six, when the turn angle is between 30° and 90°, (3) a screw of pitch three, when the turn angle is between 90° and 150°, (4) antiferromagnetic, when the turn angle exceeds 150°. For moderately large anisotropy energy in the plane, the spin arrangement will be complex; for the case where the system is going to be ferromagnetic, the spin vectors will oscillate sinusoidally in the neighborhood of one of the easy axis, as one may guess from the result to be described later in the study of the effect of strong applied field.

The effect of an applied field is different depending on whether the anisotropy energy is small or large. In the case where a small anisotropy energy stabilizes the plane perpendicular to the screw axis and a magnetic field is applied in the same plane, the initial magnetization corresponds to a small tilt of the spin vectors toward the field direction. Then, by increasing the field, one will find that the proper screw changes into a cycloid whose plane is perpendicular to the field, since the susceptibility of the cycloid is larger than that of the proper screw and, therefore, by converting the system from proper screw to cycloid above a certain critical field, one gains more by the decrease in the energy of exchange and interaction with the magnetic field of the system than one loses by the increase of the anisotropy energy. This change is similar to that known in antiferromagnetic crystals. In this case, a further increase of the field will make the spin vectors point closer to the field direction and finally make all of them parallel to this direction. In the case where the anisotropy energy confines the spin vectors firmly in the plane, the effect of the applied field is somewhat peculiar, as will be described later. The effect of the applied field for various cases of anisotropy energy and for finite temperature has not yet been studied fully.

The origin of the exchange interaction that gives rise to a screw structure is not yet fully understood. In compounds, such as $MnO_2$ and $FeCl_3$, the ratio of the superexchange constants may happen to conform to the condition of stabilizing a screw structure. In $MnAu_2$ and in rare-earth metals, indirect exchange interaction through $s$-$d$ and $s$-$f$ interaction, respectively, might be responsible for their screw structures; this interaction is, as known well, extends over a number of atomic distances and changes sign alternately as a function of distance. The general trend of the observed Néel temperature in the second half of the rare-earth series is in agreement with that expected from this interaction.[6,7,16]

## 3. ANISOTROPY ENERGY AND MAGNETIC STRUCTURE IN RARE-EARTH METALS

Elliott[7] and Miwa and Yosida[6] have studied the crystalline field anisotropy energy for the second half of the rare-earth series. The high-temperature susceptibilities of the metals give the Curie constants appropriate to an assembly of tripositive rare-earth ions. They have a well-defined total angular momentum $J$ composed of the total spin angular momentum $S$ and the total orbital momentum $L$, $L$ and $S$ taking the maximum possible values in accordance with the Hund rules and pointing parallel to each other. These ions are subject to a crystalline electric field of six-fold symmetry in the hexagonal crystal of the metals, and the crystalline field potential as a function of the orientation of $J$ can be expanded in spherical harmonics, $Y_2^0$, $Y_4^0$, $Y_6^0$, and $Y_6^6$. The magnitude and sign of the coefficients of these spherical harmonics depend on the form of the charge cloud of the ion and, in particular, the coefficient of $Y_2^0$ depends on the quadrupole moment of the charge cloud, as well as on the deviation of the crystalline

[16] P. G. de Gennes, Compt. rendus 247, 1836 (1958).

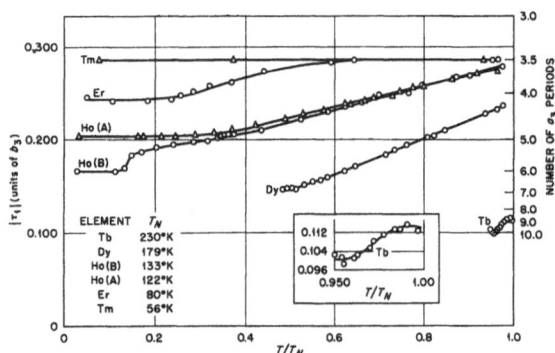

FIG. 1. Frequency of oscillation $\tau_1$ along the $c$ axis of the spin arrangement in Er and Tm, and period of the screw arrangement along $c$ axis ($a_3$ axis) in Tb, Dy, and Ho (by courtesy of Dr. W. C. Koehler).

TABLE I.[a]

|  | Tb | Dy | Ho | Er | Tm |
|---|---|---|---|---|---|
| $J$ | 6 | 15/2 | 8 | 15/2 | 6 |
| $S$ | 3 | 5/2 | 2 | 3/2 | 1 |
| $L$ | 3 | 5 | 6 | 6 | 5 |
| $Y_2^0$ | + | + | + | − | − |
| $Y_4^0$ | − | + | + | + | − |
| $Y_6^0$ | − | + | − | + | − |
| $Y_6^6$ | + | − | + | − | + |

[a] The + and − signs indicate the signs of the coefficients of the spherical harmonics in the first column; for $Y_6^6$ the + sign means that the anisotropy energy is maximum in the direction of the $a$ axis and the − sign minimum.

structure from a perfect hcp lattice. For the metals under consideration the axial ratio $c/a$ is about 4% smaller than the ratio for a perfect hcp lattice, 1.633, and the quadrupole moment of the ion is negative for Tb, Dy, and Ho (the charge cloud is flat, perpendicular to the direction of $\mathbf{J}$), while for Er and Tm it is positive (long along the direction of $\mathbf{J}$). This means that in the first three metals the vector $\mathbf{J}$ is stabilized by being perpendicular to the $c$ axis, while in the second two it is stabilized by being parallel to $c$.

Potentials of the fourth and sixth degrees are all important, but the corresponding anisotropy energies should decrease with increasing temperatures more rapidly than the anisotropy energy corresponding to the second-order potential, so that at high temperatures the anisotropy energy should be uniaxial in the main and one would expect a proper screw structure for Tb, Dy, and Ho and a sinusoidally varying linear array of moment vectors for Er and Tm. The theory for the latter will be given in the next section. At low temperatures more complex structures are expected depending on the sign and magnitude of the higher order potentials. Table I shows the signs of these potentials obtained by Elliott and Miwa and Yosida, together with the $J$, $S$, and $L$ values of the ions which determine these signs; it is assumed that the crystalline electric field is the same for all the metals and it arises from point charges put on nearest neighbor ions.

For Tm it is seen from Table I that all the coefficients of the axially symmetric spherical harmonics are negative, which means that the $c$ axis is stabilized by all these harmonics as the direction of easy magnetization. In fact, neutron diffraction experiments show that the moment vectors line up along the $c$ axis with a sinusoidally varying magnitude below the Néel temperature, 53°K, down to 4.2°K. The period of this sine wave was found to be $3.5c$ over the whole temperature range.

For Dy the signs are all positive for the axially symmetric harmonics, and one expects the moment vectors to lie in the $c$ plane. In actuality, a screw structure

persists from the Néel temperature of 179°K down to 87°K, the turn angle varying from 43.2° to 26.5°, but below 87°K ferromagnetism appears, presumably because the anisotropy energy of $Y_6^6$ increases.

Tb also has a screw structure[10] between the Néel temperature of 230° and 218°K with a turn angle varying from 20° to 18°. Below 218°K it becomes ferromagnetic with moment vectors pointing parallel and lying in the hexagonal plane, as one would expect from the small turn angle and the + sign for $Y_2^0$.

For Ho and Er the sign of the coefficient of $Y_6^0$ is different from those of the coefficients of $Y_2^0$ and $Y_4^0$. At high temperatures, where the second-order anisotropy energy predominates, the structure in Ho should be similar to that in Dy and the structure in Er to that in Tm. At low temperatures, where the anisotropy energy of the sixth degree comes into play, the minimum of the total anisotropy energy would be at an intermediate angle between 0° and 90°, and therefore the moment vectors would rotate on a cone or on a plane oblique or parallel to the $c$ axis. The neutron diffraction experiments and magnetic measurements show that in Ho a screw structure persists from the Néel temperature of 133°K (in another specimen 122°K[10]) down to 20°K, below which a small ferromagnetic component appears along the $c$ axis, while the perpendicular component still rotate; in Er in the range between the Néel temperature of 80° and 53°K the moment vectors are parallel to the $c$ axis and oscillate sinusoidally, in the range between 53° and 20°K both the components of the moment vectors parallel and perpendicular to the $c$ axis oscillate, and below 20°K they rotate on a narrow cone whose axis coincides with the $c$ axis. (See Fig. 1.)

## 4. THEORY OF THE MODIFICATION OF SCREW STRUCTURE BY ANISOTROPY ENERGY

In the preceding two sections a qualitative discussion was given on the modification of screw structure by anisotropy energy of various types. A more exact treatment of this problem will be outlined in this section, along the line followed by Miwa and Yosida, as well as by Elliott and Kaplan.

First, we shall be concerned with the magnetic structure near the Néel point, basing our argument on the approximation of Weiss molecular field. To avoid confusion between the total angular momentum $J$ and exchange constants $J_i$ or their Fourier transform $J(q)$, the former will be denoted by $S$ and will be called spin. Let us first consider the case of Er or Tm and assume that the spins have ordered component along the $c$ or $z$ axis. The thermal average of the $z$ component of the spins on the $n$th layer will be denoted by $\sigma_n$. Then the Weiss field acting on $S_n$ of the $n$th layer is

$$H_{ex,n} = 2J_0\sigma_n + 2J_1(\sigma_{n-1} + \sigma_{n+1}) + 2J_2(\sigma_{n-2} + \sigma_{n+2}) + \cdots, \quad (4)$$

corresponding to Eq. (1), provided that it is understood that $S$ represents itself the magnetic moment, i.e., the factor $g\mu_B$ is put equal to unity. The equation of self-consistency to determine $\sigma_n$ is then the following:

$$\sigma_n = \frac{\mathrm{Tr}\, S_{nz} \exp[H_{ex,n} S_{nz}/kT]}{\mathrm{Tr}\, \exp[H_{ex,n} S_{nz}/kT]}. \quad (5)$$

Near the Néel temperature, where $H_{ex,n}$ is small, we can expand the exponential function into power series. If we retain the terms linear in $H_{ex,n}$, we have form (5)

$$\sigma_n = \frac{S(S+1)}{3kT}[2J_0\sigma_n + 2J_1(\sigma_{n-1} + \sigma_{n+1}) + 2J_2(\sigma_{n-2} + \sigma_{n+2}) + \cdots]. \quad (6)$$

To solve this equation, we put $\sigma_n = \exp[inq]$ and have

$$1 = [S(S+1)/3kT]2J(q). \quad (7)$$

Here $2J(q)$ represents the magnitude of the molecular field, so that in order to have the maximum stability $J(q)$ must take its maximum value, i.e., $q$ must be equal to $q_0$. Then, the Néel temperature is obtained from (7) to be

$$T_N = S(S+1)2J(q_0)/3k. \quad (8)$$

The above argument shows that $\sigma_n$ varies sinusoidally along the $z$ axis below $T_N$, though its amplitude is still undetermined. To determine the amplitude, terms proportional to $(H_{ex,n}/kT)^3$ must be taken into account, and such a calculation has actually been done.[6,7,8] One can naturally expect the variation of the amplitude to be proportional to $(T_N - T)^{\frac{1}{2}}$. At the same time, however, the harmonic of the third degree comes into oscillation along the $z$ axis. In general, harmonics of odd degrees should come in as the temperature is lowered and the sine wave should become squared.

In a more general treatment, one has to assume the three components of the Weiss field. Furthermore, a uniaxial anisotropy energy of the form

$$D[S_z^2 - \tfrac{1}{3}S(S+1)] \quad (9)$$

may be assumed for the rare-earth metals near the Néel temperature. Then, Eq. (5) is replaced by

$$\sigma_n = \frac{\mathrm{Tr}\, S_n \exp[(\mathbf{H}_{ex,n} \cdot \mathbf{S}_n + D[S_{nz}^2 - \tfrac{1}{3}S(S+1)])/kT]}{\mathrm{Tr}\, \exp[(\mathbf{H}_{ex,n} \cdot \mathbf{S}_n + D[S_{nz}^2 - \tfrac{1}{3}S(S+1)])/kT]}. \quad (10)$$

Expanding as before in powers of the argument of the exponential function and calculating the terms which are proportional to $H_{ex,n}/kT$ and terms proportional to $H_{ex,n}D/(kT)^2$, one obtains linear equations like (6) separately for the three components, except that now one has in place of the factor $S(S+1)/3kT$ the following factors:

$$\frac{S(S+1)}{3kT}\left[1 - \frac{D}{kT}\frac{4S(S+1)-3}{15}\right]$$

for the $z$ component,

$$\frac{S(S+1)}{3kT}\left[1 + \frac{D}{kT}\frac{4S(S+1)-3}{30}\right]$$

for the $x$ and $y$ components. (12)

The Néel temperature for the $z$ component and that for the $x$ and $y$ components are obtained by putting the above quantities equal to $[2J(q_0)]^{-1}$, and if $D$ is negative, as in the case of Er and Tm, the Néel temperature thus obtained for the $z$ component is higher than that for the $x$ and $y$ components, while the reverse is true if $D$ is positive. Thus, in Er and Tm the $z$ component will start to oscillate when the temperature is decreased, but the perpendicular component will begin to oscillate only at a lower temperature which, however, is not the temperature obtained in the above way, since at such a temperature higher order terms with respect to $H_{ex,nz}$ would already be nonnegligible. Miwa and Yosida calculated this second transition temperature by using the trial function of the form

$$\sigma_{nx} = 0, \quad \sigma_{ny} = B\sin nq, \quad \sigma_{nz} = C\cos nq \quad (13)$$

and minimizing the free energy based on the Weiss approximation (taking terms up to the fourth order in the amplitudes) with respect to $B$, $C$, and $q$. They found that the temperature where $\sigma_{ny}$ starts to oscillate is lower than the temperature given from (12). They further added in $\sigma_{nz}$ a term proportional to $\cos 3nq$ and calculated its amplitude. Kaplan[8] also dealt with the second transition, obtaining series solution of the self-consistency equations.

In the other three rare-earth metals, where $D$ is positive, the Néel temperature calculated from the $x$ and $y$ components is higher, so that these components first begin to oscillate, and since evidently the free energy is the lowest when the magnetic moment rotates on a circle, one has here a screw structure. In this case, there is no reason why the $z$ component must begin to oscillate at lower temperatures, so that the screw persists down to the temperature where higher order anisotropy energies begin to play a role.

Now, at low temperatures where higher order anisotropy energies are important, one can make the following approximate treatment. There the spins have their full values since the anisotropy energies, though important, are still small compared with the exchange interactions.

On the other hand, the anisotropy constants are rapidly varying functions of temperature. Thus, one can confine oneself at absolute zero and consider the varying anisotropy constants as parameters. The actual calculation is complex, depending on the relative importance of different spherical harmonics, but it is not a difficult matter to see what would occur. As already mentioned qualitatively in the preceding section, there are a number of cases. When the anisotropy energy has a deep minimum on a cone whose axis is the $c$ axis, the layer magnetization vectors rotate on a cone which is somewhat broader than the cone of the anisotropy energy since exchange interactions among these vectors have a trend of making them rotate on a plane. In Ho and Er this situation is realized at lowest temperatures and the vertex angle of the cone, as measured by neutron diffraction, is $77°$ for Ho and $30°$ for Er. When this minimum becomes less pronounced, the exchange interactions overcome the anisotropy energy, and the magnetization vectors begin to rotate on a nearly planar surface which is inclined to the $c$ axis. The transition between these two arrangements is of the first kind. If the cone gets narrower by the increasing relative importance of the anisotropy energy of negative $Y_2^0$ and $Y_4^0$, the plane may contain the $c$ axis. This joins continuously to the elliptic oscillation given by (13). It is hard to distinguish by neutron diffraction study whether the plane contains the $c$ axis or not. In any case, the transition observed in Er at $20°K$ can be interpreted as being the transition between a conical structure and a planar or cycloidal structure.

The screw-ferro transition in Dy was also referred to in the preceding section. If the anisotropy energy within the hexagonal plane of six-fold symmetry gets large at low temperatures, transition from screw to ferromagnetism will occur, since the turn angle decreases below $30°$ with lowering temperature. According to the theory to be described in the next section, however, a direct transition to parallel alignment of the magnetization vectors will not take place, but instead one expects a transition first to a structure of sinusoidally oscillating moments about the easy direction and then to a perfect parallel alignment. Enz[11] remarks, however, that a large magnetostriction accompanying the ferromagnetic state—presumably due to an alignment of the quadrupole moments of the Dy ions—may accelerate the transition to ferromagnetism.

## 5. VARIATION OF THE PERIOD OF OSCILLATION WITH TEMPERATURE

Neutron diffraction measurements at Oak Ridge show pronounced variations of the period of oscillation with temperature when magnetization vectors rotate in the $c$ plane or when there are rotating components of these vectors in the $c$ plane (Fig. 1). In Tm and Er in the range where there is no $c$ component the period remains constant.

Yosida and Miwa[17] considered these variations to

[17] K. Yosida and H. Miwa, J. Appl. Phys. 32, 8S (1961).

arise from the anharmonic variation of $J(q)$ in the neighborhood of $q=q_0$ and calculated them by the method of spin waves. Elliott[7] ascribed the variations to the quadrupole-quadrupole coupling between ions and could satisfactorily explain the observation in Dy, together with the observed change of the screw-ferro transition temperature as a function of applied magnetic field. Near the Néel temperature the ions appear to be spherical when averaged over their thermal motion so that the quadrupole-quadrupole interaction is unimportant, but as the temperature is lowered it gradually becomes important and makes the magnetization vectors of neighboring layers tend to align parallel. Thus the turn angle decreases. A similar calculation was made by Miwa and Yosida, but they pointed out a difficulty for the case of Er and Tm. In these metals the moments align parallel to the $c$ axis and vary sinusoidally in magnitude, which means that at the nodal points of the sine wave the ions look like spheres but at the loops they appear as ellipsoids elongated along the $c$ axis. Therefore the quadrupole-quadrupole coupling acts in them as in Dy, but actually there is no observable change in the period in Er and Tm. It may be said at the present moment that we have no convincing explanation of the observed variation of the period.

## 6. MODIFICATION OF SCREW STRUCTURE BY EXTERNAL MAGNETIC FIELD

An external magnetic field applied to the crystal can be regarded as an origin of the simplest anisotropy energy, namely the unidirectional anisotropy energy. In this section, a brief account of the theory of the effect of applied field will be given. As mentioned in the beginning of this paper, Herpin and Mériel studied this effect in $MnAu_2$ by neutron diffraction experiment and they developed a theory. Enz also treated the same problem for the special case of small turning angle, referring to the Herpin-Mériel theory. The present writer independently developed a similar theory, though stimulated by Herpin-Mériel's experiment, and made the most extensive consideration of the problem.

The starting point is the energy expression given by (1). To this, the field term

$$-\mu H \sum_n \cos\theta_n \qquad (14)$$

is now to be added, assuming that the field $H$ is applied in the plane in which the moment vectors rotate. The consideration is confined to absolute zero. The anisotropy energy which confines the moment vectors in the plane perpendicular to the screw axis is at first assumed to be so large that it confines these vectors in this plane for any field strength, but later this assumption will be removed.

The case where the applied field is weak was already treated by Yoshimori.[1] He gave the formula for the initial susceptibility, which, with the use of the general expression of the exchange interaction constant $J(q)$ defined by (2) and the energy expression of the form (1) and (14), can be written as

$$\chi_0 = \frac{\mu^2}{2S^2[2J(q_0) - J(2q_0) - J(0)]}.$$ (15)

Correspondingly, the $\theta_n$ value varies with $n$ as

$$\theta_n = nq_0 + \alpha - 2\chi_0\mu^{-1}H\sin(nq_0 + \alpha), \quad \alpha = \text{arbitrary}.$$ (16)

These results are valid when the variation of $\theta_n$ is considered to the first order of $H$, or when the energy as an implicit function of $H$ through $\theta_n$'s is considered to the second order of $H$. Even in this approximation, one can determine the variation of $q$ with $H$ by at first minimizing the energy with respect to $\theta_n$'s and then with respect to $q$. Then, one has:

$$q - q_0 = \frac{-2J'(2q_0)\mu^2H^2}{4S^4J''(q_0)[2J(q_0) - J(2q_0) - J(0)]^2}.$$ (17)

The writer has calculated the energy term proportional to $H^4$. It was found that this term becomes infinite for $q = \pi$ and $q = \pi/2$ and vanishes for $q = 2\pi/3$. The fact that the expansion of the energy in powers of $H$ diverges $q = \pi$ indicates that for this case the antiferromagnetic structure having the spin axis perpendicular to the applied field is more stable than the structure in which $q$ starts from $\pi$ and varies with $H$ according to (17). This is expected since the susceptibility of the antiferromagnetic structure is twice as large as that of the latter. In Fig. 2 is shown how the stability region of the antiferromagnetic structure extends in the $q$, $H$ plane; a detailed account of this figure will be given later. A similar situation is encountered with $q = \pi/2$, where two interpenetrating sets of antiferromagnetic arrangement of spins is stabilized, as shown schematically in the same figure. In this case, the stability boundary depends sensitively on the functional form of $J(q)$; in the figure it is assumed that $J(q)$ is expressed only in terms of $J_0$, $J_1$, and $J_2$. For $q = q_0 = 2\pi/3$, the right-hand side of (17) vanishes, since in this case $2q_0$ is essentially the same as $q_0$ and both $2q_0$ and $q_0$ correspond to the maximum of $J(q)$ so that $J'(2q_0)$ in the denominator vanishes. It is in fact easy to show that the triangular structure persists up to the highest field $H_0$ for $q_0 = 2\pi/3$, where the moment vectors become parallel to the field direction. That there is an extended region of stability of the triangular structure in Fig. 2 can provisionally be understood by the fact that the $q$ value, which starts at $H = 0$ from $q_0$ near $2\pi/3$, converges to $2\pi/3$ as $H$ is increased. It is expected that no other singularities than these three can actually appear.

At sufficiently high fields, moment vectors will all align parallel to the applied field. When the field is decreased, they will begin to deviate from the field direction. It can be shown that there is a critical field $H_0$ above which they are parallel to the field direction but below which they oscillate sinusoidally about the field direction with number $q_0$. This can be seen most easily by expanding the energy in powers of $\theta_n$'s which may be assumed to be small. The writer expanded the

energy in powers of $\sin(\theta_n/2)$ and further put

$$\sin(\theta_n/2) = \sum_q \xi_q \exp[inq]. \quad (\xi_{-q} = \xi_q^*).$$ (18)

Then the energy can be written

$$\frac{E}{N} = -S^2J(0) - \mu H + 4\sum_q [S^2J(0) - S^2J(q) + \tfrac{1}{2}\mu H]|\xi_q|^2 + O(\xi_q^4).$$ (19)

If the coefficient of $|\xi_q|^2$ is positive for all $q$, then $\xi_q$ must vanish for all $q$ in order that the energy be minimized. This coefficient is minimum for $|q| = q_0$, so that, when $H$ decreases below $H_0$ defined by

$$\tfrac{1}{2}\mu H_0 = S^2J(q_0) - S^2J(0),$$ (20)

there will appear a range of $q$ in the neighborhood of $|q| = q_0$ in which the coefficient is negative. Herpin and Mériel, and also Enz, assumed that there is in this case a sinusoidally oscillating solution. The present writer found the same result but with a mathematical proof. For this proof and for the calculation of the amplitude, the fourth-order terms have to be calculated. We shall, however, not go into the mathematical details. The result we have is that $|\xi_{q_0}|$ is the only nonvanishing amplitude and this is given by

$$|\xi_{q_0}| = \frac{\mu(H_0 - H)^{\frac{1}{2}}}{2S[3J(q_0) - J(2q_0) - 2J(0)]^{\frac{1}{2}}}.$$ (21)

Correspondingly, the oscillation of the angle is expressed as

$$\sin(\theta_n/2) = 2|\xi_{q_0}|\sin(nq_0 + \text{const}),$$ (22)

and the magnetization is given by

$$m = \mu - \frac{\mu^2(H_0 - H)}{S^2[3J(q_0) - J(2q_0) - 2J(0)]}.$$ (23)

It is easy to determine the variation of $q$ with $H$ if we assume a sine wave with wave number $q$ and minimize the energy with respect to $q$. This variation is proportional to $H_0 - H$.

It is noted that the value of $H_0$ is small when $q_0$ is small. $J(q)$ is maximum at $q_0$ and should be minimum at $q = 0$ since $J(q)$ is an even function of $q$, so that one sees from (20) that $H_0$ is proportional to $q_0^4$ when $q_0$ is small. In the approximation of taking $J_0$, $J_1$, and $J_2$ only, one has

$$\tfrac{1}{2}\mu H_0 = -8S^2J_2(1 - \cos q_0)^2, \quad (J_2 < 0)$$ (24)

with $\cos q_0 = -J_1/4J_2$. In MnAu$_2$, $q_0 = 2\pi/7$ and $H_0$ is 10 koe at room temperature. In Ho, it has a lower value.

As we have seen in the above, at weak fields the moment vectors rotate in the plane with a little tilt toward the field direction and at strong fields below $H_0$ they oscillate sinusoidally about the field direction. It is not yet possible to predict precisely what would occur at intermediate fields, but it is most likely that a sudden

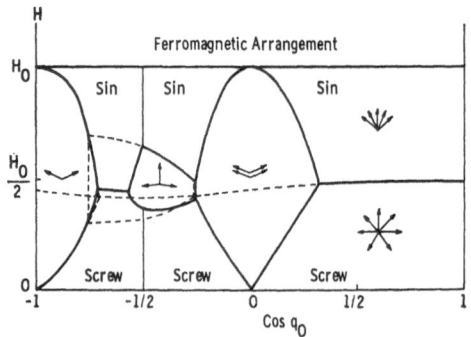

FIG. 2. Domains of stability for various spin structures represented schematically, calculated for $J(q) = 2J_1 \cos q + 2J_2 \cos 2q$. The domain for triangular spins enclosed with solid curves was obtained with the energy expression for low field calculated up to terms of $H^4$ and that enclosed with broken curves by that calculated up to $H^2$. High anisotropy energy confining the spin vectors in the basal plane is assumed.

jump between these two structures occurs at a certain intermediate field, with a corresponding jump in the magnetization. This field, or the transition field $H_t$, can be calculated by comparing the energy expressions obtained for weak and strong fields. If they are assumed to be $-S^2 J(q_0) - \frac{1}{2}\chi_0 H^2$ and $-S^2 J(0) - \mu H - \frac{1}{2}\chi_1 (H_0 - H)^2$, respectively, where $\chi_0$ is given by (15) and $\chi_1$ is the coefficient of $-(H_0 - H)$ of the last term in (23), then we have

$$H_t/H_0 = [(1+\beta)(2+\beta)]^{\frac{1}{2}} - (1+\beta), \quad (25)$$

where

$$\beta = 2S^2 [J(q_0) - J(2q_0)]/\mu H_0. \quad (26)$$

For small $q_0$, namely for large $\beta$, $H_t/H_0$ tends to $\frac{1}{2}$. In Fig. 2, the slightly curved horizontal line near $H = H_0/2$ represents $H_t$, calculated with $J_1$ and $J_2$ only. This curve separates the regions in which the screw structure and the sine wave structure are respectively stable. However, there are regions in which the doubly antiferromagnetic, triangular, and singly antiferromagnetic structures are respectively stable, as shown schematically in the same figure; these regions have been obtained by comparing the corresponding energies with those of the screw and sine structures. The vertical line at $\cos q_0 = -\frac{1}{2}$ indicate that along it the triangular structure persists up to $H_0$.

Now, when the anisotropy energy is no longer assumed to be large enough, the situation will be changed in the following way. For simplicity, let us assume the anisotropy energy of the form (9) with positive $K$. Consider the structure in which the moment vectors are in the plane which contains the screw axis and is perpendicular to the applied field, namely the cycloidal structure. In this structure the average anisotropy energy is greater by $\frac{1}{2}K$ than in the screw structure, so that the total energy in the presence of an external field is expressed as

$$\frac{E}{N} = -S^2 J(q_0) + \frac{1}{2}K - \frac{1}{2}\chi_1 H^2, \quad (27)$$

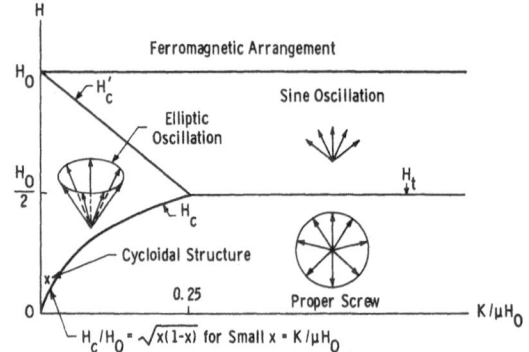

FIG. 3. Change of spin structure as a function of applied magnetic field and anisotropy constant $K$; the anisotropy energy is assumed to be of the form of $K \cos^2$ (spin vector, $z$) with positive $K$. $\beta = [J(q_0) - J(2q_0)]/[J(q_0) - J(0)]$ is assumed to be large, i.e., $q_0$ to be small.

where $\chi_\perp$ can be calculated to be[12]

$$\chi_\perp = \mu^2 / [2S^2 J(q_0) - 2S^2 J(0) - K]. \quad (28)$$

Comparing (27) with the energy of the screw structure, namely $-S^2 J(q_0) - \frac{1}{2}\chi_0 H^2$, and considering that $\chi_\perp$ is greater than $\chi_0$, one sees that above the critical field, defined by

$$H_c = [K/(\chi_\perp - \chi_0)]^{\frac{1}{2}}, \quad (29)$$

the cycloidal structure is more stable.

Since, however, the sinusoidal wave structure must be stable just below $H_0$, there must be another transition at a field higher than $H_c$. To calculate this field, $H_c'$, one may assume that there is an elliptic oscillation with $q_0$ about the field direction, with both parallel and perpendicular components to the easy plane of magnetization. For the parallel component, one finds the energy (19), with $q = q_0$, but for the perpendicular component the coefficient of the square of the amplitude in (19) is replaced by

$$S^2 J(q_0) - S^2 J(0) + K + \frac{1}{2}\mu H. \quad (30)$$

Therefore, putting this quantity equal to zero, we have

$$H_c' = H_c - 2K/\mu. \quad (31)$$

These critical fields are plotted in Fig. 3 for the limiting case of vanishing $q_0$ or infinite $\beta$, together with the schematical representation of the spin structures in the three regions; here the plane of the figure is supposed to be the easy plane and the field is supposed to be applied along the ordinate axis.

*Note added in proof.* In Section 2, following Eq. (2), it is stated that Yoshimori gave the rigorous proof that for crystals consisting of equivalent magnetic atoms, the screw structure represents the stable solution of the problem of minimum energy. Although Yoshimori considered only a single special structure of rutile type with restricted interactions and he did not explicitly claim to have given the rigorous proof, still the above statement in its essence is correct. D. H. Lyons and T. A. Kaplan, Phys. Rev. **120**, 1580 (1960), consciously gave this proof and they generalized the problem.

JOURNAL OF APPLIED PHYSICS    SUPPLEMENT TO VOL. 33, NO. 3    MARCH, 1962

# Some Magnetic First-Order Transitions

D. S. RODBELL AND C. P. BEAN

*General Electric Research Laboratory, Schenectady, New York*

Some transitions between ordered and disordered magnetic states that take place via transitions of first order are tabulated. Among these transitions are some of the order-disorder type. The theoretical basis for such a transition to occur is indicated using a simple physical model that assumes: (1) a strain dependent exchange interaction and (2) a compressible lattice. The necessary condition for first-order behavior is determined, and the theory is compared to the behavior of a real material, MnAs. It is found to be in substantial agreement with the behavior of that compound. The system represented by the simple model employed here possesses ferro-, antiferro-, and paramagnetic states. The equilibrium boundaries between these states are determined in the pressure-temperature plane as an illustration of the possible transitions that may be observed with this model.

## INTRODUCTION

FIRST-ORDER transition is used here in its usual thermodynamic sense, i.e., the system under consideration possesses an equilibrium free energy that is a function of temperature and pressure and the first derivative of that free energy function with either temperature or pressure exhibits a discontinuity at a critical value of temperature or pressure. Thus the entropy and/or the volume of the system has a discontinuity. This is not the situation for a second-order transition in which case it is the second derivatives that suffer discontinuities, i.e., the thermal expansion, specific heat, and compressibility.

There is rapidly accumulating experimental evidence that some magnetic systems exhibit magnetic transitions that are thermodynamically of first order. The systems that have been observed include both metallic and nonmetallic types and range from elemental metals (some rare-earths) to multiple constituent compounds. The accompanying Table I lists some of the materials so far studied with some pertinent references. The transitions that are involved here include the ferromagnetic (F), antiferromagnetic (AF), ferrimagnetic ($F_i$), spiral (S), and paramagnetic (P) states. The lattice symmetry changes that occur at the transitions are not usually large, but some of the transitions involve sizable volume changes. While detailed study is needed for each case, we believe that these transitions are driven by magnetic interaction free energies rather than nonmagnetic energies that normally cause structural transitions. (Examples of primarily structural transitions are the hexagonal to fcc transition in cobalt metal and the martensitic fcc to bcc transition in some nickel-iron alloys.) The evidence for the listed transitions being of first order is not necessarily complete for all the systems of Table I. Each listing however satisfies at least one of the following conditions: (a) there is a latent heat, (b) there is a hysteresis, (c) there is a discontinuous change of magnetization with temperature or pressure.

The characterization that may be imposed on these transitions falls into two convenient groupings[1] which

are:

(1) Order-order and (2) Order-disorder.

## ORDER-ORDER

The term "exchange inversion" was initially coined[2] to describe the AF to $F_i$ transition in Cr-doped $Mn_2Sb$ and has been explored by Kittel[3] in a theoretical effort to explain this phenomenon.[4] In this treatment an exchange interaction is presumed to change sign, crossing zero at a critical value of a lattice parameter. Thermal expansion will, at a critical temperature, bring the lattice parameter to this value and the transition will then occur. This is certainly a reasonable way to view the experimental facts of $(Mn_{1-x}Cr_x)_2Sb$.

When a more detailed view is made of the individual exchange interactions in a system that possesses a group of competitive exchange interactions, it is found that spiral spin configurations are to be expected for certain values of the exchange constants and their existence is also dependent upon the geometry of the various interactions. Yoshimori[5] and Villain[6] have discovered these properties in the case of rutile-type structures while Kaplan *et al.*[7] have found these types of arrangements in the spinel geometry. Different geometries of course provide different exchange interactions to consider, but in addition, provide different anisotropies. The main effect of anisotropy is to stabilize certain directions for the spiral—when it exists—over other directions. Both Kaplan[8] and Elliott[9] have considered this question with regard to the rare-earth

[1] We do not include the purely metamagnetic type in these groupings because that transition is only first order with an imposed magnetic field and is usually of second order with temperature the variable. It is of peripheral interest to note that this is also the case for the superconducting transition.

[2] T. J. Swoboda, W. H. Cloud, T. A. Bither, M. S. Sadler, and H. S. Jarrett, Phys. Rev. Letters **4**, 509 (1960).

[3] C. Kittel, Phys. Rev. **120**, 335 (1960).

[4] Some earlier considerations of transitions between ordered spin states have been given by: J. S. Smart, Phys. Rev. **90**, 55 (1953); L. Néel, Compt. rend. **242**, 1824 (1956); G. W. Pratt, Jr., Phys. Rev. **108**, 1233 (1957); and S. H. Liu, D. R. Behrendt, S. Legvold, and R. H. Good, Phys. Rev. **116**, 1464 (1959).

[5] A. Yoshimori, J. Phys. Soc. Japan **14**, 807 (1959).

[6] J. Villain, J. Chem. Phys. Solids **11**, 303 (1959).

[7] T. A. Kaplan, K. Dwight, D. Lyons, and N. Menyuk, J. Appl. Phys. **32**, 13S (1961).

[8] T. A. Kaplan, Phys. Rev. **124**, 329 (1961).

[9] R. J. Elliott, Phys. Rev. **124**, 346 (1961).

TABLE I. Some magnetic systems that exhibit transitions of first order.

| Material | Critical temperatures °K | Some references |
|---|---|---|
| $MnBr_2$ | AF $\xrightarrow{2.16}$ P | E. O. Wollan, W. C. Koehler, and M. K. Wilkinson, Phys. Rev. **110**, 638 (1958). |
| $MnSn_2$ | AF $\xrightarrow{73}$ AF | J. S. Kouvel and C. C. Hartelius, Phys. Rev. **123**, 124 (1961). |
| $TiCl_3$ | AF $\xrightarrow{210}$ P | S. Ogawa, J. Phys. Soc. Japan **15**, 1901 (1960). |
| $Mn_3Ge_2$ | AF $\xrightarrow{113}$ F | N. G. Fakidov and Yu. N. Tsiovkin, Fiz. Metl. i Metalloved. Akad. Nauk. S.S.S.R. **7**, 685 (1959), p. 47 in translation. |
| $Cr_3As_2$ | $F_i$ $\xrightarrow{213}$ P | M. Yuzuri, J. Phys. Soc. Japan **15**, 2007 (1960). |
| FeRh | AF $\xrightarrow{350}$ F | M. Fallot and R. Hocart, Rev. sci. **77**, 498 (1939); F. deBergevin and L. Muldawer, Comptes rend. **252**, 1347 (1961). |
| $KMnF_3$ | AF $\xrightarrow{88.3}$ P | A. J. Heeger, O. Beckman, and A. M. Portis, Phys. Rev. **123**, 1652 (1961). |
| MnAs | F $\xrightarrow{318}$ P | F. Heusler, Z. angew Chem. **1**, 260 (1904); L. F. Bates Phil. Mag. **8**, 714 (1929), A. Smits, H. Gerding, and F. VerMast, Z. physik. Chem. 357 (1931). |
| $(Mn_{1-x}Cr_x)_2Sb$ | AF $\xrightarrow{T(x)}$ $F_i$ | T. J. Swoboda, W. H. Cloud, T. A. Bither, M. S. Sadler, and H. S. Jarrett, Phys. Rev. Letters **4**, 509 (1960); W. H. Cloud, H. S. Jarrett, A.E. Austin, and E. Adelson Phys. Rev. **120**, 1969 (1960); H. S. Jarrett, P. E. Bierstedt, F. J. Darnell, and M. Sparks, J. Appl. Phys. **32**, 57S (1961). |
| $x \leqslant 0.045$ | AF $\rightarrow$ S $\rightarrow$ $F_i$ | P. Bierstedt, F. Darnell, W. Cloud, R. Flippin and H. Jarrett, Phys. Rev. Letters **8**, 15 (1962). |
| $Mn_{0.9}Li_{0.1}Se$ | AF $\xrightarrow{71}$ F | S. J. Pickart, R. Nathans, and G. Shirane, Phys. Rev. **121**, 707 (1961); R. R. Heikes, T. R. McGuire, and R. J. Happel, Phys. Rev. **121**, 703 (1961). |
| Dy | F $\xrightarrow{85}$ S | D. R. Behrendt, S. Legvold, and F. H. Spedding, Phys. Rev. **109**, 1544 (1958); M. K. Wilkinson, W. C. Koehler, E. O. Wollan, and J. Cable, J. Appl. Phys. **32**, 485 (1961). For a review of other rare-earth metals and intermetallic compounds see Wilkinson, Child, Koehler, Cable, and Wollan papers 311 & 312 in Proceedings of the International Conference on Magnetism and Crystallography (Kyoto, 1961). |
| NiO[a] | | |

AF    antiferromagnetic
F      ferromagnetic
$F_i$      ferrimagnetic
S      spiral
P      paramagnetism

[a] In the antiferromagnetic compound NiO there is a lack of critical scattering of neutrons at the transition temperature [W. L. Roth (private communication) and P. G. deGennes, (private communication concerning unpublished work of T. Riste at Saclay)]. This fact coupled with the distorted Brillouin function observed by W. L. Roth (unpublished work) and consistant with the observations quoted at this conference by Professor M. Tinkham suggest strongly that NiO may be in the category of materials that we are concerned with in this paper.

metals' spin ordering, while Yosida and Miwa[10] have calculated the transition between the ferromagnetic and spiral states in the case appropriate to dysprosium and find the possibility of a transition of first order to exist provided that a sufficient anisotropy is also present. Yosida and Miwa derive the first-order phase transition without the assumption of strain dependent exchange.

## ORDER-DISORDER

Let us consider the question of transitions of first order that occur between states of ordered and disordered spin. We consider a single average exchange

interaction as that which considerably simplifies the problems encountered with multiple exchange interactions and allows us to focus our attention on the primary features of the order-disorder transition.

To view the physics of this treatment, let us consider the usual method of treating magnetic disorder in which there is assumed to be a ferromagnetic interaction between the magnetic moments localized on each atom site and further, usually implicit, this interaction is not a function of lattice spacing. At low temperatures there exists substantially complete magnetization or perfect long range order. As the temperature is raised, the thermal randomization tends to destroy this order and the magnetization falls. The course of the magnetization is described more or less by application of the molecular

---

[10] K. Yosida and H. Miwa, J. Appl. Phys. **32**, 8S (1961).

field concept to the Brillouin function; and at the Curie temperature $T_c$ the spontaneous magnetization becomes zero without discontinuity. This transformation is of second order; and the cooperative nature of the system is reflected by a discontinuity in the specific heat at $T_c$; there is not however the latent heat or discontinuous density change that characterize a first-order transition.

Now let us consider what results if we assume that the exchange energy (or Curie temperature) is a strong function of interatomic spacing. We show such a dependence in Fig. 1(a). At absolute zero, the system's free energy may be lowered by a distortion of the lattice in the direction of increasing the Curie temperature. This distortion will introduce to the free energy a term in strain energy which will increase the free energy, and thus, a compromise between distortion and exchange may be found that minimizes the free energy. This volume is indicated in Fig. 1(a) as $v'$. If we compare the course of the magnetization with temperature of such a system that is free to distort itself (free) with that of a system whose volume is fixed (clamped), we find the behavior sketched in Fig. 1(c), which omits the usual thermal expansion in order to focus upon magnetic effects. At low temperatures the free system

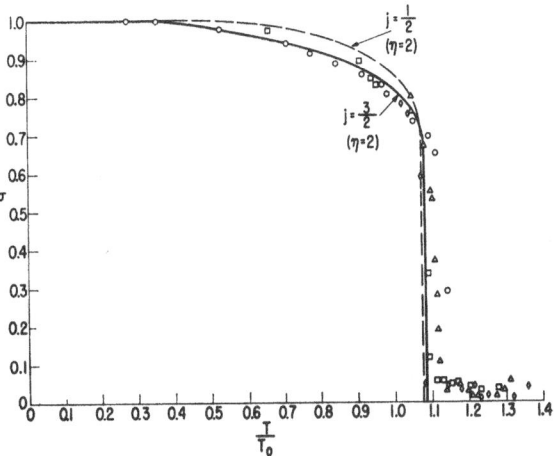

FIG. 2. A comparison of the observed data for MnAs with the theory for $j = \frac{3}{2}$; $\eta = 2$ is estimated from the volume change at the transition. The $j = \frac{1}{2}$ prediction is indicated in dash. $T_0 = 285°K$.

will lose less of its magnetization than the clamped system because its effective Curie temperature (close to $T'$) is greater than that of the clamped system. If this change in effective Curie temperature is large enough, then as the temperature is raised it is possible for the free system to have net magnetization at temperatures greater than $T_0$, the Curie temperature of the clamped system. Since the driving force to lattice expansion is caused by the ordered magnetic spins, the loss in magnetic moment or order at higher temperatures causes a diminution of the lattice strain as indicated in Fig. 1(b). As the temperature is further increased, the magnetization can find no way to decrease smoothly to zero since it is stable at these temperatures only because the magnetization exists. At a critical temperature, $T_{crit}$, the magnetization and lattice distortion go to zero discontinuously.

The properties of a system in which the exchange energy dependence is given by

$$T_c = T_0(1 + \beta(v - v_0)/v_0) \qquad (1)$$

are calculated. $T_c$ is the Curie temperature appropriate to a lattice volume $v$ while $v_0$ is the equilibrium volume in the absence of magnetic interactions. The course of the magnetization with temperature of such a system depends upon the steepness $\beta$ of the exchange interaction dependence on interatomic distance, the compressibility $K$, and upon $T_0$. The behavior may be the usual second-order transition to paramagnetism, but it can in fact become a first-order transition with the properties usually associated therewith, e.g., latent heat and discontinuous density change. In the absence of an externally applied pressure, the transition will be of first order if[11]

$$\eta \equiv 40NkKT_0\beta^2[j(j+1)]^2/[(2j+1)^4-1] > 1. \qquad (2)$$

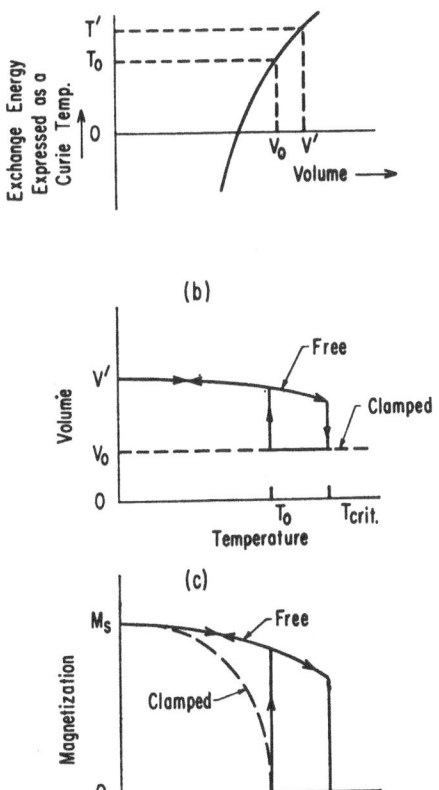

FIG. 1. A schematic representation of the volume dependent exchange interaction (a) and some effects that result in (b) the sample volume and (c) the magnetization.

---

[11] This condition results when consideration of the detailed behavior of the free energy is made; a complete account is given elsewhere [C. P. Bean and D. S. Rodbell, Phys. Rev. (to be published)].

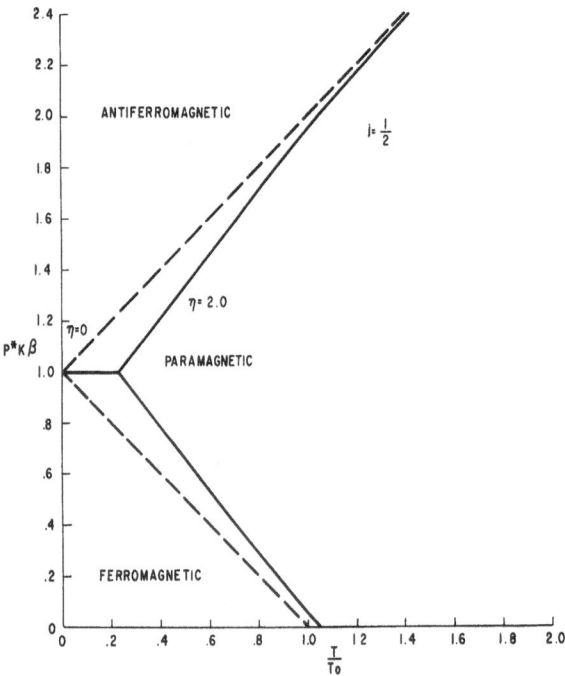

FIG. 3. The equilibrium phase diagram for a $j=\frac{1}{2}$ system. Shown in dash is the $\eta=0$ case and in solid the case of $\eta=2$.

In order to illustrate the connection of this theory with a real system, we compare it to the compound MnAs in Fig. 2. There are no adjustable parameters. The data are from several different samples and some independent measurements. $T_0$ is taken from Serres[12] as the intercept of $1/x$ vs $T$, while the slope of that curve indicates $j=\frac{3}{2}$ to be the most likely spin state. $\eta$ is determined to be 2 from the observed[13] 1.8% volume change at the transition. The latent heat is largely attributable to the change in entropy of the spin system. The detailed theory and experimental evidence are contained in the paper by Bean and Rodbell[11] mentioned above from which we shall quote results in the next section.

## MAGNETIC PHASE DIAGRAMS

The equilibrium magnetic states of the model employed here are examined below to see how they depend upon temperature and pressure and to determine the ranges of stability of these states. Contained within the fundamental assumptions of our model [i.e., Eq. (1)] are the possibilities of several magnetic states and transitions between them. The parameter $\eta$ that we have already found to be of importance remains so in the calculation of the phase diagrams; however, as an introduction it is instructive to consider the rather transparent situation of $\eta=0$. Let us take $T_0$ positive, corresponding to ferromagnetism, if $\beta[(v-v_0)/v_0]=-1$,

then the Curie temperature $T_c$ becomes zero corresponding to paramagnetism, while for $\beta(v-v_0)/v_0<-1$ the state becomes antiferromagnetic. Thus for the equilibrium states on a pressure vs temperature plot (neglecting thermal expansion) we have a stable ferromagnetic state forming a triangular area for $T<T_0$ and $P<(K\beta)^{-1}$ with a straight line boundary connecting the $T=T_0$ and $P=(K\beta)^{-1}$ points. The antiferromagnetic phase is the reflection of this area in the line of constant $P=(K\beta)^{-1}$. The area between these phases is the paramagnetic phase. Figure 3 plots this phase diagram for $\eta=0$ in dashed lines. Since $\beta$ may be positive or negative and $T_0$ may be positive or negative, we must consider the four possible combinations of these two parameters. In addition we shall include the effects of thermal expansion and should not be surprised to find that this parameter enters as a negative pressure.

The result of a previous calculation[11] for the case of a system of $j=\frac{1}{2}$ particles is the implicit dependence of the magnetization on temperature. Considered as a two sublattice configuration, so as to allow antiferromagnetism to occur, there is found that

$$\frac{T}{T_0\cos\phi}=\frac{\sigma}{\tanh^{-1}\sigma}\left[1-\beta(PK-\alpha_l T)+\frac{\eta\sigma^2}{3}\cos\phi\right], \quad (3)$$

where $\alpha_l$ is the coefficient of thermal expansion for the lattice and $\phi$ is the angle between the sublattice magnetizations. The determination of the equilibrium phase diagrams proceeds from (3) above by dividing both sides of that expression by $[1-\beta(PK-\alpha_l T)]$ and obtaining thereby

$$\frac{T}{T_0^*}=\frac{\sigma}{\tanh^{-1}\sigma}\left[1+\frac{\eta^*\sigma^2}{3}\right],$$

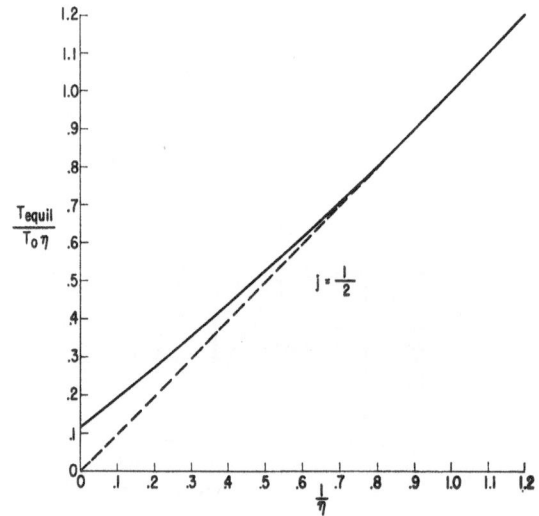

FIG. 4. The equilibrium temperature, normalized by $T_0\eta$, as a function of $\eta^{-1}$ for a $j=\frac{1}{2}$ system.

[12] A. Serres, J. phys. radium **8**, 146 (1947).
[13] B. T. M. Willis and H. P. Rooksby, Proc. Phys. Soc. (London) **B67**, 290 (1954).

where

$$T_0{}^* \equiv T_0 \cos\phi(1 - P^*K\beta), \qquad (4)$$
$$\eta^* \equiv \eta \cos\phi/(1 - P^*K\beta),$$

and

$$P^* \equiv P - \alpha_l T/K.$$

Evaluating the free energy for $j = \frac{1}{2}$ and various $\eta$'s, $T_{\text{equil}}$ as a function of $\eta$ is determined (recalling that $T_{\text{equil}}$ is the temperature at which the magnetic and nonmagnetic states have the same free energy). In Fig. 4 the results of these evaluations are given as a plot of $T_{\text{equil}}\eta/T_0$ vs $1/\eta$. As no restrictions have been put upon $\eta$ and since $T_0\eta = T_0{}^*\eta^*$, we may proceed with Eq. (3) above, and with Fig. 4 evaluate for any $\eta$ the equilibrium states of the spin $\frac{1}{2}$ system that is of interest here. The arbitrary spin case proceeds in an exactly analogous way. Figure 3 shows the phase diagram as a plot of $P^*K\beta$ vs $T/T_0$ for $\eta = 0$ and 2.0. In these plots we have taken $T_0$ and $\beta$ positive, while in Fig. 5 we have indicated the modifications resulting from $(T_0, \beta)$ taking other sign combinations. The transition across a phase boundary will be of first order if $\eta^* > 1$ is satisfied, this being the general criterion of which Eq. (2) is a special case.

The effect of thermal expansion may be seen by viewing the phase diagrams as plotted. For increasing $T/T_0$ in the absence of an applied pressure one leaves the origin with a slope of $-\alpha_l\beta T_0$. This can result in phase changes occurring if thermal expansion carries the system across phase boundaries. These phase changes may be from an ordered magnetic state to a disordered one, or they can be changes between ordered states[14] followed by another phase change to the disordered state, exactly what happens will depend upon $\eta$ and $\alpha_l$. In Fig. 6 is illustrated the possible change from antiferromagnetism to ferromagnetism to paramag-

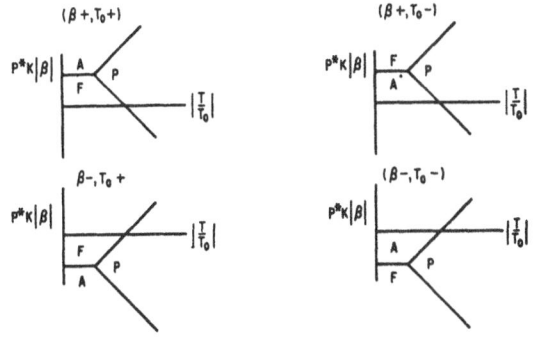

FIG. 5. The modification of the equilibrium phase diagram that results for the different possible sign combinations for $T_0$ and $\beta$.

[14] This is the transition between ordered states calculated by C. Kittel (reference 3).

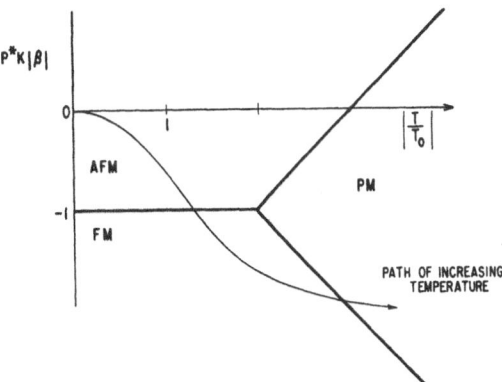

FIG. 6. The course, with increasing temperature, of transit through the phase diagram with $\beta$, $T_0$ both negative. The thermal expansion coefficient determines the trajectory of the course followed. The transitions shown are from antiferromagnetism to ferromagnetism and from ferromagnetism to paramagnetism.

netism with increasing temperature. The minimum value of $\eta$ that can give rise to such behavior is $\eta_{\min} = (0.111\alpha_l\beta T_0)^{-1}$ in the approximation that $\alpha_l$ and $\beta$ are constants. The other possible triple state change with our model and increasing temperature is from ferromagnetism to antiferromagnetism to paramagnetism. We note also that at fixed temperature a large variety of transitions may be induced with applied hydrostatic pressure. This model is undoubtedly a great oversimplification. We do not expect any real material to be quantitatively described by these considerations, and we put forth these ideas only to stimulate experiment and its qualitative interpretation.

## CONCLUSION

The existence of several first-order transitions that are intimately connected with magnetism has provided the impetus to examine the role of magnetic energies in these transitions. There are two distinct cases. One is concerned with transitions between ordered spin states; the other is between ordered and disordered states of the spin system. Examination of the latter case shows that the spin system entropy plays a dominant role. We also find that both of the cases considered may occur in one simple model in which a strain-dependent exchange interaction exists.

## ACKNOWLEDGMENTS

This paper is a result of the interest in many aspects of magnetic phase transitions by our colleagues at the General Electric Research Laboratory. We wish to thank R. W. DeBlois, I. S. Jacobs, J. Kasper, J. Kouvel, P. E. Lawrence, and W. L. Roth for helpful discussions and continuing stimulation.

JOURNAL OF APPLIED PHYSICS     SUPPLEMENT TO VOL. 33, NO. 3     MARCH, 1962

# Superconducting Materials and High Magnetic Fields

J. E. KUNZLER

*Bell Telephone Laboratories, Incorporated, Murray Hill, New Jersey*

The recent discovery that $Nb_3Sn$ remains superconducting in magnetic fields exceeding 88 kgauss while carrying current densities in excess of 100 000 amp/cm$^2$ has stimulated widespread activity directed toward the construction of superconducting magnets and the understanding of "hard" superconductivity. Some of the relevant events that have transpired since the discovery of superconductivity by Onnes in 1911 are reviewed. The critical current versus magnetic field ($I_c$ vs $H$) characteristics of several materials are discussed and the justification of using such measurements on small samples as a basis for magnet design is presented. Nb-Zr alloys appear useful for magnets capable of fields as large as 80–100 kgauss while $Nb_3Sn$ appears to be useful for fields of 200 kgauss. Superconducting magnets of $Nb_3Sn$ and of Nb-Zr have been constructed and tested by several laboratories. Fields as large as about 70 kgauss have been generated with $Nb_3Sn$ magnets and 60 kgauss with Nb-Zr magnets. Fields exceeding 100 kgauss have been attained by augmenting the field produced by a superconducting $Nb_3Sn$ magnet with the field from a conventional Bitter solenoid. Superconducting magnets capable of fields of at least 100 kgauss are almost a certainty and it is quite likely that magnets will eventually be constructed to produce fields of 200 kgauss. Fields of several hundred kgauss are an intriguing possibility.

## I. INTRODUCTION

THE discovery announced less than a year ago,[1] that $Nb_3Sn$ remains superconducting in magnetic fields exceeding 88 kgauss while carrying large current densities has stimulated widespread activity directed towards the construction of superconducting magnets and the understanding of "hard" superconductivity. Since then, several other superconductors, notably Nb–Zr,[2-5] $V_3Ga$,[6] and $V_3Si$,[6] have been found to have promising properties useful for superconducting magnets. Magnets have been constructed of several superconducting materials and have been found to operate in a manner consistent with the properties observed using samples of the construction material.[7-9] The object of this paper is to summarize very briefly and correlate some of the more important points that have been previously reported and to discuss recent progress pertaining to both superconducting magnets and an understanding of the properties of the materials.

Although the current interest in high field superconductors is primarily a consequence of the recent finding that they have applications for magnets, the idea of a superconducting magnet is almost as old as superconductivity itself. Soon after superconductivity was discovered by Onnes in 1911,[10] interest developed with respect to the possibility of using superconductors for the windings of magnets. However, this dream was shattered when it was discovered[11] that a fairly modest field (less than a few hundred gauss) destroyed superconductivity in the superconductors than known. Hopes for superconducting magnets were rekindled by de Haas and Voogd[12] in 1930 when they found that Pb-Bi retained some superconductivity in fields of at least 20 kgauss. However, this hope also died when Keesom[13] found that the current carrying capacity of Pb–Bi decreased very rapidly with increasing field in the region of a few hundred gauss. It appears that these measurements were not extended to higher fields. This is unfortunate, since it is now known[5,14] that Pb–Bi can sustain current densities in the 10–15 kgauss range that are large enough to be useful for applications to magnets. The above historical background and its consequences have been discussed more extensively elsewhere[5]. A discussion of the distinction between "hard" and "soft" superconductors is also included there.

The more recent chain of events that resulted in superconducting magnets becoming a reality began in 1955 when Yntema[15] reported the operation at about 7 kgauss of an electromagnet having superconducting windings. In 1960, Autler[16] reported the operation of a superconducting solenoid at 4.3 kgauss and was able to later increase the maximum field to about 10 kgauss. During this latter period the properties of Mo–Re were investigated and a superconducting solenoid was constructed and found to operate in a manner consistent

[1] J. E. Kunzler, E. Buehler, F. S. L. Hsu, and J. H. Wernick Phys. Rev. Letters **6**, 89 (1961).

[2] J. E. Kunzler, Bull. Am. Phys. Soc. **6**, 298 (1961).

[3] T. G. Berlincourt, R. R. Hake, and D. H. Leslie, Phys. Rev. Letters **6**, 671 (1961).

[4] J. E. Kunzler, B. T. Matthias, E. Buehler, and F. S. L. Hsu (to be published).

[5] J. E. Kunzler, Revs. Modern Phys. **33**, 1 (1961).

[6] J. H. Wernick, F. J. Morin, F. S. L. Hsu, J. P. Maita, D. Dorsi, and J. E. Kunzler, 1961 International Conference on High Magnetic Fields, Massachusetts Institute of Technology, November 1–4, 1961 (to be published).

[7] J. E. Kunzler, E. Buehler, F. S. L. Hsu, B. T. Matthias, and C. Wahl, J. Appl. Phys. **32**, 325 (1961).

[8] J. E. Kunzler, 1961 International Conference on High Magnetic Fields, Massachusetts Institute of Technology, November 1–4, 1961.

[9] J. K. Hulm, M. J. Fraser, H. Riemersma, A. J. Venturino, and R. A. Wien, 1961 International Conference on High Magnetic Fields, Massachusetts Institute of Technology, November 1–4, 1961.

[10] H. K. Onnes, Commun. Phys. Lab. Univ. Leiden **120b** (1911); **122b** (1911).

[11] H. K. Onnes, Commun. Phys. Lab. Univ. Leiden **133d** (1913); **139f** (1914).

[12] W. J. de Haas and J. Voogd, Commun. Phys. Lab. Univ. Leiden, **208b** (1930); **214b** (1931).

[13] W. H. Keesom, Commun. Phys. Lab. Univ. Leiden **234f** (1935); Physica **2**, 35 (1935).

[14] F. S. L. Hsu, P. R. White, E. Buehler, and J. E. Kunzler (to be published).

[15] G. B. Yntema, Phys. Rev. **98**, 1197 (1955).

[16] S. H. Autler, Rev. Sci. Instr. **31**, 369 (1960); Bull. Am. Phys. Soc. **6**, 64 (1960).

with the observations on short samples of wire.[7] When superconductivity in Nb$_3$Sn was found to persist at high current densities in magnetic fields exceeding 88 kgauss and a means of fabricating Nb$_3$Sn into "wire" was developed,[1] it was apparent that high field superconducting magnets would become a reality. The practicality of high field superconducting magnets was even more certain following the initial disclosure[2] that ductile Nb–Zr has the properties needed for magnets capable of fields up to 80–100 kgauss. The Mo–Re studies, which had demonstrated the close relationship between the properties of short lengths of wire and those required for magnet construction, were essential in arriving at this assurance.

The flurry of activity which developed during the past eight months has uncovered several promising new materials and has resulted in considerable progress in superconducting magnet construction. In addition, significant progress has been made toward the understanding of the "hard" superconductors. However, new phenomena have been observed that appear to require considerable revision in our thinking concerning the nature of "hard" superconductivity.

## II. EXPERIMENTAL CONSIDERATIONS

The property of superconductors of interest for magnets that is most frequently studied is the maximum current carrying capacity (while in the superconducting state) as a function of applied magnetic field. These observations are usually made at temperatures that are accessible with the use of liquid helium (1.5–4.2°K). This particular type of measurement has the merit of providing nearly the same environment to the test sample that it would have if it were part of the central region of the inner layer of windings of a superconducting solenoid. Figure 1(a) shows, in a schematic fashion, the sample arrangement used for determining the variation of critical current vs applied magnetic field ($I_c$ vs $H$). These curves generally have shapes such as is illustrated in Fig. 1(b). The samples are pieces of wire or small rods having a rectangular cross section. Measurements are made by either keeping the magnetic field strength constant and gradually increasing the current until a voltage is detected or, alternatively, by keeping the current fixed and increasing the magnetic field strength. The existence of a detectable potential drop along the sample indicates the disappearance of a continuous superconducting path. The regions under the curves represent the conditions for which superconductivity exists while those above (or to the right of) the curves of Fig. 1(b) represent the lack of superconductivity.

In general, optimum current characteristics of "hard" superconductors are obtained by subjecting them to a maximum degree of mechanical deformation. The three curves in Fig. 1(b) illustrate typical behavior for three different physical states of the wire. The heavy curve

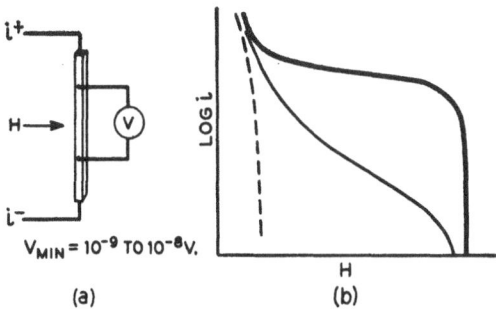

Fig. 1. Schematic representation of the experimental arrangement (a) and an idealized representation of typical results (b). In (a) the cryostat is omitted and the magnetic field is shown normal to the current direction. The presence of a voltage is detected with a sensitive dc amplifier having a sensitivity of $10^{-9}$ to $10^{-8}$ v depending on the experimental arrangement. The curves of (b) represent the boundaries between the absence of a detectable voltage (superconducting) and the presence of the smallest observable voltage (nonsuperconducting). The heavy curve represents typical behavior of a "hard" superconductor saturated with strain by extreme mechanical deformation (hard working), the light curve represents the behavior of a partially worked alloy and the dashed curve illustrates the effect of annealing.

illustrates the behavior observed for samples that are severely worked or highly strained. The relatively flat plateau and the "knee" are characteristic features that have been consistently observed for such materials. The lighter solid curve represents the type of behavior observed on partially deformed samples. The dashed curve illustrates the behavior observed following the annealing of many alloys, such as Mo–Re. However, it has recently been observed[17,18] that annealing Nb–Zr alloys can produce an increase in the critical current. This phenomena is believed to be a result of phase dissociation or precipitation hardening and the increase in $I_c$ can be interpreted as being due to increased strain.

Our present understanding of "hard" superconductors suggests that all three of the curves of Fig. 1(b) will coincide at sufficiently low values of the critical current density, $I_c$.

For considerations involving the selection of possible materials for superconducting magnet construction, it is desirable to know $I_c$ vs $H$ characteristics at several temperatures. However, it is not possible to obtain this information for many materials since steady state magnetic fields are only available up to fields of about 100 kgauss. Pulse fields can be extended to higher fields but not to sufficiently high fields and their use has been subject to difficulties. An alternative procedure[5,19] has been used to determine another useful parameter, the critical field as a function of temperature ($H_c$ vs $T$).

[17] R. G. Treuting, J. H. Wernick, and F. S. L. Hsu, 1961 International Conference on High Magnetic Fields, Massachusetts Institute of Technology, November 1–4, 1961.

[18] G. D. Kneip, Jr., J. O. Betterton, Jr. D. S. Easton, and J. O. Scarbrough, 1961 International Conference on High Magnetic Fields, Massachusetts Intitute of Technology, November 1–4, 1961.

[19] J. E. Kunzler, F. S. L. Hsu, and E. Buehler, Bull. Am. Phys. Soc. 6, 123 (1961) (post-deadline paper F16).

Observations are made of the critical field (the maximum magnetic field which is observed in the limit of zero current) as a function of temperature, in the range of accessible fields, and the curves are extrapolated to lower temperatures and higher fields. These measurements are made in a manner similar to those for determining the $I_c$ vs $H$ characteristics except that the temperature is varied and the current is kept fixed. Although additional useful information can be obtained by determining $H_c$ vs $T$ curves for several values of current density, the use of a small current density has the experimental advantage of avoiding serious thermal conduction and heat dissipation problems in the awkward temperature range between liquid helium and liquid hydrogen temperatures. Also, the curve determined for the limit of zero current density appears to be a characteristic property of the particular "hard" superconductor and is not sensitive to changes in the physical state of the material (providing the chemical composition remains constant on a microscopic scale).

### III. $I_c$ VS $H$ CHARACTERISTICS

Critical current versus applied magnetic field strength ($I_c$ vs $H$) characteristics have been determined for a number of "hard" superconductors that are of interest for superconducting magnets. Among these are Mo-Re,[7] $Nb_3Sn$,[1,5,19-22] Nb-Zr,[2-5,23,24] and Pb-Bi.[5,14] Superconductivity of Mo-Re was reported by Hulm,[25] of $Nb_3Sn$ by Matthias and co-workers,[26] of Nb-Zn by Matthias,[27] while superconductivity of Pb-Bi dates back at least to de Haas and Voogd.[12] Curves that illustrate typical $I_c$ vs $H$ characteristics for several "hard" superconductors are shown in Fig. 2. The curves are very sensitive to the physical state of the superconductor and the critical current is generally found to increase with increasing mechanical deformation (e.g., compare the "as cast" 3Nb:Zr curve and the one for 3Nb:Zr "hard worked wire"). Similarly the critical current density for $Nb_3Sn$ "wire" at 4.2°K is about 50 times that observed for bulk $Nb_3Sn$.[1] The enhanced behavior of the "wire" is believed to be associated with increased numbers of dislocations that are produced because of the presence of considerable chemical inhomogeneity and physical strain in the "wire".

A significant property of "hard" superconductors and an important consideration in magnet design is the

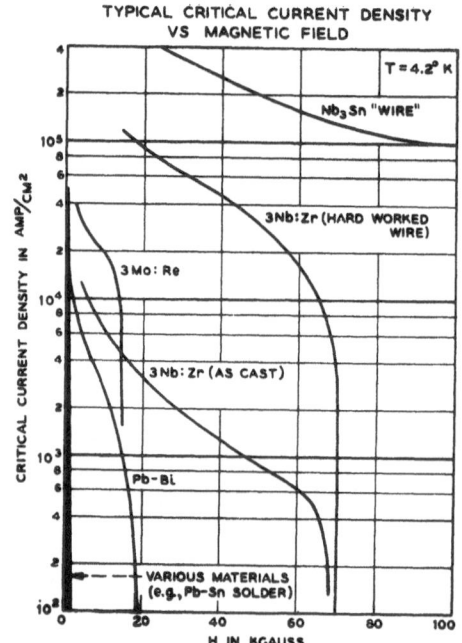

FIG. 2. Typical $I_c$ vs $H$ curves for several "hard" superconductors. The "knee" in the curves is a feature that is characteristic of "hard" superconductors in a highly strained or mechanically deformed state. The critical current density is very sensitive to the physical state (amount of strain and mechanical deformation) of the material. The critical field (magnetic field above the "knee" at the limit of zero current density) appears to be a constant which is characteristic of the material.

variation of the maximum current carrying capacity of a superconductor (at constant magnetic field strength) with the size of the conductor. In the case of soft superconductors, the maximum current carrying capacity varies as the diameter of a wire since the current is being carried by a thin surface layer having an approximate thickness equal to the penetration depth $\lambda$. Although the variation of maximum current carrying capacity with sample size has not been studied extensively under conditions that eliminate other variables (such as the degree of mechanical deformation), the data available at present indicates that the critical current (for samples in the same state of deformation and strain) is proportional to the cross-sectional area of the sample, i.e., the critical current density is nearly independent of the size of the wire. This is the behavior that would be expected if the so called superconducting "filaments" are uniformly distributed throughout the superconductor. The idea of superconducting "filaments" has received widespread usage to explain the existence of superconducting paths in "hard" superconductors in fields sufficiently high that magnetic susceptibility measurements indicate the presence of only a small amount of superconducting material.

It has usually been observed that increased mechanical working increases the critical current density (at constant field), and that annealing produces the opposite effect. However, Kniep, Betterton, Easton, and Scar-

[20] V. D. Arp, R. H. Kropschot, J. H. Wilson, W. F. Love, and R. Phelan, Phys. Rev. Letters 6, 452 (1961).

[21] J. O. Betterton, Jr., R. W. Boom, G. D. Kneip, R. E. Worsham, and C. E. Roos, Phys. Rev. Letters 6, 532 (1961).

[22] H. R. Hart, Jr., I. S. Jacobs, C. L. Kolbe, and P. E. Lawrence, Bull. Am. Phys. Soc. 6 (1961) (post-deadline paper); 1961 International Conference on High Magnetic Fields, Massachusetts Institute of Technology, November 1-4, 1961.

[23] R. W. Boom (private communication).

[24] P. R. Aron and H. C. Hitchcock (to be published).

[25] J. K. Hulm, Phys. Rev. 98, 1539 (1955).

[26] B. T. Matthias, T. H. Geballe, S. Geller, and E. Corenzwit, Phys. Rev. 95, 1435 (1954).

[27] B. T. Matthias, Phys. Rev. 92, 874 (1953).

brough[18] and Treuting, Wernick, and Hsu[17] have observed substantial increases (in some cases over a factor of 10) in critical current density of heat treated Nb-Zr wires that had previously been hard worked. This behavior is believed to be associated with phase separation during the heat treatment as was mentioned above and will require considerable additional study before a complete understanding will be possible. However, these observations have practical implications. For example, the critical current density of hard worked Nb-Zr alloys appears to be a maximum at a composition between 25 and 35 atomic % Zr. Considerable difficulty has been encountered in the fabrication of hard worked wire of these compositions while a composition of 50% Zr appears to be considerably easier to handle. It has been possible[17] to enhance the critical current density of Nb-50% Zr by a combination of heat treatment and mechanical working so that it is comparable to the values for the 25–30% Zr alloys. The 50% Zr alloy has the further advantage of having an appreciable higher critical field.[2,4,5]

## IV. SUPERCONDUCTING MAGNETS

At least four different materials have been produced thus far in sufficient quantities to be tested in magnets capable of producing magnetic fields greater than about 10 kgauss. These are Nb, Mo-Re, $Nb_3Sn$, and Nb-Zr. Several laboratories are in the process of constructing and studying $Nb_3Sn$ and Nb-Zr magnets.[9,28–30] A few results have been reported and will be discussed. Results on test solenoids using these materials have also been obtained at Bell Telephone Laboratories in collaboration with E. Buehler, C. F. Hempstead, J. H. Wernick, Y. B. Kim, F. S. L. Hsu, and K. M. Olsen. They are summarized below and will be reported in detail elsewhere.[31]

$Nb_3Sn$ appears to be useful for magnetic fields as large as about 200 kgauss and Nb-Zr for fields up to 80–100 kgauss.

$Nb_3Sn$ "wire" having an outside diameter of 0.030 in. has been produced in continuous lengths as long as about 10 000 ft.[32] Several small solenoids (test coils) have been constructed using this wire and they were found to operate in a manner consistent with the measurements made on short test samples. A few of these coils have been tested in external fields in order to determine their performance at high fields. Figure 3 shows the results

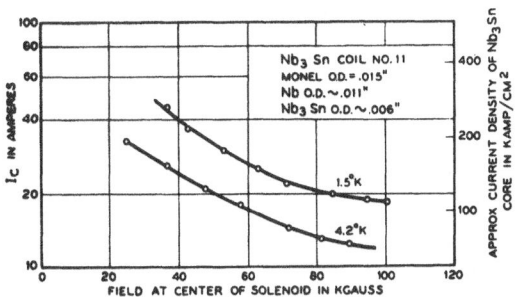

FIG. 3 Critical current vs magnetic field ($I_c$ vs $H$) characteristics of superconducting solenoid (test coil) wound with $Nb_3Sn$ "wire" The characteristics plotted are presumably those of the middle of the inner layer. The magnet field value for each point is the sum in an external field and that produced by the test solenoid. At 1.5°K and with no applied field the superconducting solenoid produced 37 kgauss whereas it produced 16 kgauss when in an external field of 85 kgauss. The characteristics shown confirm the predictions based on short test samples.

obtained with a coil wound with monel jacketed "wire" (0.015 in. o.d.) which was fired for about 15 hr at 1000°C after being wound. The core of the wire contained excess Nb (24 wt % Nb 76% Sn) which seems to be desirable for processing. The coil has an o.d. of about $1\frac{1}{2}$ in., is about 2-in. long, has a $\frac{1}{4}$-in. core, and contains about 4000 turns of $Nb_3Sn$ "wire". The monel jacket was used for the insulation between turns and quartz fibrecloth was used for the insulation between layers. The field corresponding to the point on the 1.5°K curve at 37 kgauss (Fig. 3) was generated by the coil alone and the fields for the points at higher fields are the sum of the coil field plus an external field generated by a conventional copper Bitter solenoid. The field at the center of the solenoid (coil) was measured with a magnetoresistance probe[33] (e.g., the field at 101 kgauss consisted of 85 kgauss of external field and 16 kgauss from the $Nb_3Sn$ coil).

The $I_c$ vs $H$ characteristics of the coil are remarkably similar to those obtained on the shortest lengths of the wire and the current density is within a factor of two of that of the best samples we have reported.[1] Furthermore, the fractional increase in critical current density between 4.2 and 1.5°K is nearly the same as that for the early samples.[1] A similar $Nb_3Sn$ coil of the same size but wound with 0.020-in. o.d. "wire" gave almost identical curves as those on Fig. 3 except that the current values were proportionately higher (but the current densities were the same). The maximum fields generated by this latter coil were slightly higher due to the more favorable packing factor.

A larger $Nb_3Sn$ solenoid ($3\frac{1}{2}$ in.-o.d.$\times$4-in. long) has recently been wound and tested.[31] It was wound from the same original length of wire as the two smaller coils described above. It would be expected to generate a maximum field of about 75 kgauss if it performed with the same characteristics as the smaller coils. The solenoid has produced a field of about 70 kgauss without

[28] D. J. Rose, 1961 International Conference on High Magnetic Fields, Massachusetts Institute of Technology, November 1–4, 1961.

[29] J. O. Betterton, Jr., and D. S. Easton, 1961 International Conference on High Magnetic Fields, Massachusetts Institute of Technology, November 1–4, 1961.

[30] R. R. Hake, T. G. Berlincourt, and D. H. Leslie, 1961 International Conference on High Magnetic Fields, Massachusetts Institute of Technology, November 1–4, 1961.

[31] J. E. Kunzler, E. Buehler, C. F. Hempstead, J. H. Wernick, F. S. L. Hsu, Y. B. Kim, and K. M. Olsen, (to be published).

[32] K. M. Olsen, E. O. Fuchs, and R. F. Jack, J. Metals **6**, 724 (1961).

[33] F, S, L, Hsu and J. E. Kunzler (to be published).

operating at maximum capacity which represents completely satisfactory performance.

In summary, we have found that the Nb₃Sn "wire" solenoids have performed as consistently with earlier predictions[1,2,5] (based on short wire samples or small coils) as could be reasonably expected. Furthermore, we have found no evidence (at least in the case of Nb₃Sn) to support Le Blanc's[34] objections to the correlation of the $I_c$ vs $H$ characteristics of short samples with those expected for magnets. Preliminary reports have also been made by other laboratories[28,29] concerning Nb₃Sn test magnets. They have reported fields as high as 28 kgauss.

At the present, fields of about 60 kgauss have been achieved using superconducting Nb-Zr magnets.[9,30] The behavior of Nb-Zr solenoids has been observed to be considerably more complicated than that of Nb₃Sn solenoids and there are significant differences among the observations from different laboratories.[9,30,31,35] Insufficient information is available to attempt to correlate these results but some conclusions are possible.

We have tested superconducting solenoids wound with 2Nb:Zr and with 3Nb:Zr wire in a manner similar to that described for the Nb₃Sn solenoids (i.e., the total field at the center of the solenoid was the sum of the externally applied field and the field generated by the test solenoid). Typical coils wound with 10-mil wire purchased from Wah Chang Corporation, are about 2½-in. long, have an o.d. of 2½-in. and a ¼-in. core and contain about 10 000 turns of wire which is insulated with Formvar. The field was measured in the core of the solenoids using a magnetoresistance probe.[33] These solenoids (both 2Nb:Zr and 3Nb:Zr) were found to have critical currents in the absence of an external field of about 20 amp and to generate maximum fields of about 40 kgauss. It was observed that when the total field at the center of the solenoid is high enough and the critical current is less than about 10 amp, the $I_c$ vs $H$ characteristics of the solenoids are essentially identical to those of short wire samples. However, for higher values of current the curves diverge and the $I_c$ vs $H$ curve for the coil becomes nearly flat. Furthermore, at high currents the difference in critical current between 4.2° and 1.5°K is only about 10–20%, whereas for short samples it is almost a factor of two. Considerable work will be required before this strange behavior is understood, but it appears that it is caused in part by "weak regions" in the wire. Such a region can return to the normal state before the coil as a whole does, and at high current densities its properties become further deteriorated by the heat generated. By repeatedly driving the coil normal, we have burned-out (i.e., melted a short section of wire) the windings of such solenoids. The region of failure was usually near a bend or kink and not in the center of the inner layer where the magnetic field

is the largest as might have been expected. Furthermore, the performance of the solenoid demonstrated anomalous behavior prior to the failure which had not been observed in earlier operation.

Although difficulties have been encountered when Nb-Zr solenoids are operated under conditions near the maximum field that is possible, the critical current density is sufficiently high that it is practical to avoid these difficulties by operating at somewhat reduced currents. We have found no departures from the solenoid relationship between current and magnetic field generated for Nb-Zr or Nb₃Sn solenoids.

We have not made extensive studies in fields of a few kgauss, but we have not observed the "training" effects reported by Le Blanc[36] for Mo-Re solenoids.

Test magnets have also been wound with 0.020-in. diam 3Nb:Zr wire. This wire is believed to have the same history as the 0.010-in. wire except for the lesser degree of mechanical working. However, the critical current density is almost a factor of two less than for the 0.010-in. wire. It appears that the additional mechanical working, that is required to produce the 0.010-in. wire, has a pronounced effect. This observation suggests that there may be appreciable gain by cold working the wire to still smaller diameter as was found to be the case for Mo-Re.[5,7]

## V. SUPERCONDUCTORS AND MAGNETIC SHIELDING

In addition to being useful for magnets, high-field superconductors are also valuable for shielding and shaping magnetic fields. Two applications are briefly discussed below.

One of the troublesome problems associated with superconducting magnet construction is the connections between several lengths of wire. If the connections are made in the absence of a magnetic field, or in a sufficiently low field, a variety of materials (including Pb-Sn solder) can be used for the connections. Often this is not convenient, but there is an alternative method that then can be used which involves the shielding of the connections from the magnetic field.[37] This can be done by using a small "thick-walled" cylinder of a suitable superconducting material or with a small solenoid. The controlling characteristics of a solenoid for shielding are similar to those for generating a field, i.e., a superconducting solenoid can produce a field difference between the outside and its center by passing a current through the windings, or alternatively it will react to the production of a field difference (if the leads to the windings are shorted) by generating a current through the windings. The same $I_c$ vs $H$ characteristics apply to either case providing proper consideration is given to the magnetic field distribution. Solenoids of wire are

[34] M. A. R. Le Blanc, 1961 Wescon Meeting, San Francisco, California, August 14–26, 1961.

[35] H. C. Hitchcock and P. R. Aron (private communication).

[36] M. A. R. Le Blanc and W. A. Little, 1961 Conference on Fundamental Research in Superconductivity, IBM, Yorktown Heights, New York, June 15–17, 1961.

[37] J. E. Kunzler (to be published).

usually preferred to thick-walled cylinders for magnetic shielding for the same reasons that they are preferred for magnetic field generation (higher critical current density).

Figure 4 illustrates how a small superconducting solenoid might be used for shielding the connections in the vicinity of a large magnet. Since, in general, the field at the location of the shielding solenoid is but a fraction of that of the large magnet, a relatively small coil can be used for shielding. Two points must be considered in the use of such shielding solenoids: (1) The resistance of the windings including connections must be zero or so near zero that appreciable flux will not leak through the solenoid during the period of its use. (2) If the field in the center of the shielding solenoid is to be maintained at its initial values, either a solid cylinder (one turn coil) or solenoid with each layer of windings shorted separately must be used (in a shorted series wound solenoid, the field at the center decreases as the external field increases). If necessary, either of these difficulties can be eliminated by using a series wound solenoid connected to an external current source.

Cioffi has used superconductors for shielding the air gap of an electromagnet.[38] He was able to reduce the leakage flux by more than a factor of ten and was able to obtain a high degree of field homogeneity. The method he used is illustrated by Fig. 5. The gap and the windings of the electromagnet were shielded by wrapping them with several layers of Pb-Bi sheet. The continuous layers were insulated to avoid the formation of a shorted turn. The windings were also superconducting and the field was not generated until the system was cooled to liquid helium temperatures. The torturous path, through the spiral layers of the superconductor shield reduced the leakage flux to a negligible amount. A magnet of this type containing less than 10 lb of magnetic

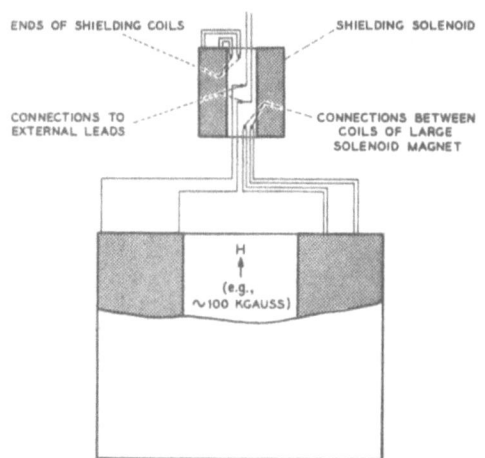

FIG. 4. Schematic illustration of the use of a small superconducting solenoid for shielding superconducting connections in the vicinity of a magnet providing a high field.

[38] P. P. Cioffi, J. Appl. Phys. (to be published).

FIG. 5. Schematic of the arrangement used by Cioffi[38] for avoiding leakage flux and improving the homogeneity of an electromagnet by the use of superconductor shielding.

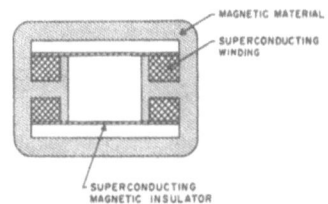

material in liquid helium can do the equivalent job of a 750-lb Alnico V maser magnet[38] operating at room temperature.

## VI. OTHER PROPERTIES OF HIGH-FIELD SUPERCONDUCTORS

The information thus far available indicates that the maximum field at which "hard" superconductors remain superconducting (in the limit of zero current) is a characteristic property of the material. As long as the chemical composition (on a microscopic basis) and crystal structure remain unchanged, this critical field value appears to remain constant regardless of the means of preparation or the degree of mechanical deformation. However, the critical current density is very sensitive to the mechanical history of the samples. It has been pointed out[5] that this behavior is consistent with suggestions[39,40] that the "filaments" in "hard" superconductors are associated with dislocations. The dislocation explanation is consistent with the formation of increased numbers of "filaments" with mechanical deformation and the requirement of a constant effective diameter for the filaments (required for a constant value of the critical field). The dislocation concept has the further advantage of providing continuous paths throughout the material in the form of a network. Hauser and Buehler[41] have obtained additional evidence for the dislocation model deformation studies on single crystals of superconductors.

It is generally assumed that the amount of material that is superconducting at high fields in "hard" superconductors is quite small because the magnetic susceptibility indicates that the magnetic field penetrates most of the sample.[42] If this is true, the current densities in the "filaments" are much larger than the several hundred thousand amp/cm² average current densities values that have been observed. However, the magnetic susceptibility is not necessarily a suitable measure of the amount of material that is contained in filaments having diameters that are appreciably less than the penetration depth. It is conceivable that the combination of the numbers of "filaments" (dislocations) and their "effective" diameters is great enough that a substantial part of a sample can be superconducting (in the sense of

[39] R. Shaw and D. E. Mapother, Phys. Rev. 118, 1474 (1960).
[40] J. J. Hauser and E. Helfand (to be published).
[41] J. J. Hauser and E. Buehler (to be published).
[42] R. M. Bozorth, H. J. Williams, and D. D. Davis, Phys. Rev. Letters 5, 148 (1960).

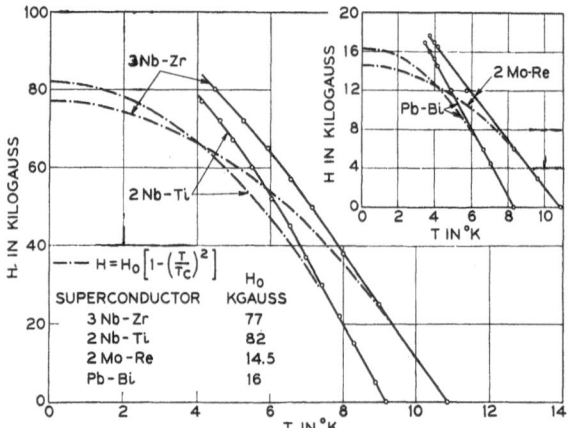

FIG. 6. The experimentally observed variation of the critical field with temperature for 3Nb:Zr, 2Nb:Ti, 2Mo:Re, and Pb-Bi. The data were observed at low current density ($\sim$1–10 amp/cm$^2$). In all cases the variation of the field with temperature is far from being parabolic. The data for 2Nb-Ti can be reasonably well fitted with a $(T/T_c)^{1.5}$ dependence, while those for the other materials are more nearly linear.

having zero resistivity) without this effect being reflected in the susceptibility measurements. This possibility is suggested by the recently reported[6] agreement between the change in the temperature of the heat capacity anomaly for V$_3$Ga (which has been observed in fields up to 19 kgauss) and the analogous electrical measurements. The entropy associated with the anomally in a magnetic field of 19 kgauss is essentially the same as that in zero fields.

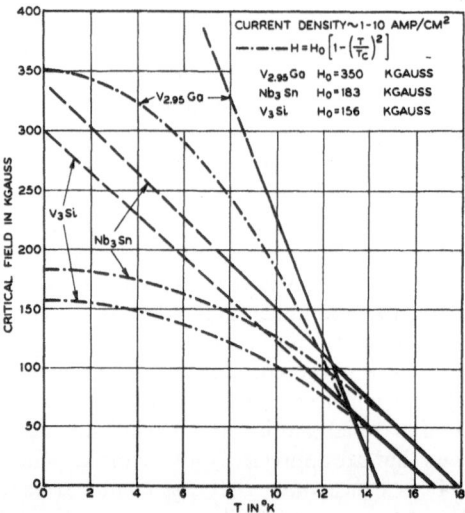

FIG. 7. The variation of $H_c$ with temperature for V$_{2.95}$Ga, Nb$_3$Sn, and V$_3$Si. The experimental observations are represented by the heavy straight lines. Although the critical temperature of V$_{2.95}$Ga is somewhat lower than that of Nb$_3$Sn, the high slope of the $H_c$ vs $T$ curve suggests that V$_3$Ga has a critical field at 0°K (and at liquid helium temperatures) about twice that of Nb$_3$Sn and in excess of 500 kgauss. Even a parabolic extrapolation of the V$_{2.95}$Ga data leads to a value higher than that obtained for a linear extrapolation of the Nb$_3$Sn data.

The thermodynamics of "hard" superconductivity appears to be very complicated. The variation of the critical field with temperature is far from parabolic.[5,19,43] Recent observations on several materials[6] show that the variation of the critical field with temperature is more nearly linear than parabolic over the temperature and magnetic field ranges thus far investigated. Some typical results are shown in Fig. 6 where the broken curves are parabolas. The 2Nb:Ti experimental curve is the nearest to parabolic in slope and it is reasonably well fitted by $(T/T_c)^{1.5}$ over the range of the observations. The other experimental curves are more nearly linear.

There are several materials whose critical fields at 4.2°K are greater than currently available steady-state magnetic fields. The $H_c$ vs $T$ characteristics for V$_{2.95}$Ga, Nb$_3$Sn, and V$_3$Si are shown in Fig. 7. Superconductivity of V$_3$Ga was reported by Matthias and co-workers[44] and of V$_3$Si by Hardy and Hulm.[45] The linear heavy lines represent the experimental observations. In the case of Nb$_3$Sn, a parabolic curve would give a critical field at 0°K of about 180 kgauss. From the variation of critical current with temperature[1] (also see Fig. 3), it is apparent that the curve cannot be as flat in the 1–4°K range as the parabolic relationship would predict. Furthermore, Hart and co-workers[22] have observed superconductivity at useful current densities in Nb$_3$Sn at fields as large as 190 kgauss at 4.2°K by using pulsed field techniques. It is clear that the critical field (in the limit of zero current) at 0°K is well over 200 kgauss. The data for V$_{2.95}$Ga suggest that the critical field of V$_3$Ga is above 500 kgauss. It is of interest that even a parabolic curve for V$_{2.95}$Ga gives a field at 0°K of about 350 kgauss which is higher than a linear extrapolation for Nb$_3$Sn.

## VII. CONCLUSION

Superconducting magnets capable of fields of at least 100 kgauss are almost a certainty and it is quite likely that magnets will eventually be constructed to produce fields of 200 kgauss. Fields of several hundred kgauss are an intriguing possibility. The earlier conclusion[2,5] that Nb-Zr alloys are likely to be useful for fields as large as 80–100 kgauss and Nb$_3$Sn for fields of at least 200 kgauss, subject to the satisfactory solution of engineering problems, still appears valid.

## ACKNOWLEDGMENTS

The technical context of this article contains contributions from many of my collaborators as is pointed out in the text. The writer is particularly indebted to M. Tanenbaum for stimulating discussions and several valuable comments.

[43] L. J. Donadieu, 1961 Conference on Fundamental Research in Superconductivity, IBM, Yorktown Heights, New York, June 15–17, 1961.
[44] B. T. Matthias, E. A. Wood, E. Corenzwit, and V. B. Bala, J. Phys. Chem. Solids 1, 188 (1956).
[45] G. F. Hardy and J. K. Hulm, Phys. Rev. 93, 1004 (1954).

# Magnetic Devices

## W. L. Shevel, Jr., *Chairman*

## Threshold Properties of Partially Switched Ferrite Cores

V. T. Shahan and O. A. Gutwin

*IBM Research Center, Yorktown Heights, New York*

A switching threshold relaxation effect has been observed which is detrimental to high-speed selection of partially switched (time-limited switching) ferrite memory cores. This paper discusses this threshold effect as determined by measurements on cores of 1- to 5-oe coercivity. Two aspects of switching threshold are observed: A static threshold $(H_f)$ which is effective at a time long after switching, and a relaxation effect which is a reduction in threshold immediately after switching. The reduction is considerable and may limit the amplitude of the digit-disturb field that a core may withstand to a small fraction of the coercive force. A first approximation to the switching threshold dependence on time may be given as: $H_t = H_f(1 - e^{-3t/\tau})$ for $t > o$. This representation is reasonably accurate for most ferrite cores, although a sum of exponentials is necessary for an improved description. The ratio of the static threshold to the coercivity $(H_f/H_c)$ is influenced by the duration and polarity (relative to the switching field) of the disturb field, the amplitude of the switching field, the flux state, the ambient temperature, and the ferrite composition. The static threshold of a core switched to the 50% flux state by a 10-nsec pulse may be only half the value of the threshold after switching to the 50% state with a 2-μsec pulse. The relaxation time also depends on the above mentioned parameters (except relative polarity) and, in addition, is influenced by the coercivity of the sample, lower coercivity samples having longer relaxation times. Measurements at ambient temperatures ranging from −195°C to near the Curie temperature (+250°C) show the product of coercivity and relaxation time to be nearly constant for a given core. The relaxation time is a maximum at the 40–50% flux state and approaches zero at the major loop remanent flux states.

**P**ARTIAL switching techniques allow higher speed operation of word-organized ferrite core memories[1]. When such random-access memories are employed having cycle times less than a microsecond, a serious attendant problem is the disturb sensitivity of the core that is exhibited as a reduction of switching threshold during a time interval after writing, which ranges from a few nanoseconds to several microseconds. This disturb sensitivity is described by an instantaneous switching threshold $H(t)$, which may be characterized by a final value $(H_f)$ and a relaxation time $(\tau)$ which is defined as the time interval after writing during which $H(t) < 0.95 \ H_f$. The dependence of $H_f$ and $\tau$ on pertinent parameters are reported below.

### EXPERIMENT

The threshold of interest here is defined as the amplitude of some arbitrarily long disturb which will cause an arbitrarily small irreversible flux change. In this experiment a single 100-μsec disturb pulse was used and the arbitrarily small flux change was chosen equal to 5% of a complete flux reversal, i.e., $|\Delta\Phi| = 0.1 \ |\Phi_R|$. The procedure used for measurement of the threshold was: (1) Establish the major loop remanent flux states $(\pm\Phi_R)$; (2) return to negative remanence $(-\Phi_R)$; (3) apply a pulse $I_w$ to establish the flux state $\Phi_w$; (4) return to negative remanence; (5) apply sequentially $I_w$ and a disturb pulse $I_2$ separated by time $T_d$, $I_2$ being of such an amplitude that the core is switched to the flux state $\Phi_w + 0.1\Phi_R$ (or $\Phi_w - 0.1\Phi_R$ if $I_2$ is negative with respect to $I_w$). The field at the mean radius caused by $I_2$ is by

definition the instantaneous threshold $H(t)$ for the flux state $\Phi_w$ at time $t = T_d$. Thresholds were measured in this manner to determine time and amplitude dependence on flux state, ambient temperature, disturb polarity, switching amplitude and duration, physical dimensions, and Curie temperature. The current pulse $I_w$ had rise and fall times of less than 5 nsec. The data presented are for a 30-mil i.d., 50-mil o.d. copper manganese ferrite of 2-oe coercivity. Cores having other dimensions were also checked to eliminate dimensional reasonance as a possible cause of the temporary reduction of threshold.

The instantaneous threshold for a given set of parameters may be reasonably approximated by a simple exponential, i. e., $H(t) = H_f(1 - e^{-3t/\tau})$, $t > 0$. Figure 1(a) gives $H(t)$ for two flux states, both obtained with 100-nsec write pulses. These curves show the general shape of the threshold curves obtained throughout this series of measurements.

In time-limited switching, a given flux state is established by any set of $I_w$, $T_w$ values from the switching curve for that particular flux state. It is found that there is no unique final threshold $(H_f)$ for a particular state. This is shown on the plot of $H_f$ versus flux state with switching time as a parameter [Fig. 1(b)]. The shorter the $T_w$ value used to reach a given flux state, the lower the value of $H_f$ for that state. At the 50% state, the value of $H_f$ decreases from 1.8 oe for a 1-μsec pulse to 1.1 oe for a 10-nsec pulse. It is interesting to note that this latter threshold is only 10% higher than the threshold after thermally demagnetizing the core.

In every case the instantaneous threshold was higher in the direction of $H_w$ than in the opposite direction, this difference being greater for times less than $\tau$. This

[1] R. E. McMahon, Solid State Circuits Conference, Digest of Technical Papers, pp. 16-17 (1959).

[2] M. J. Freiser (private communication).

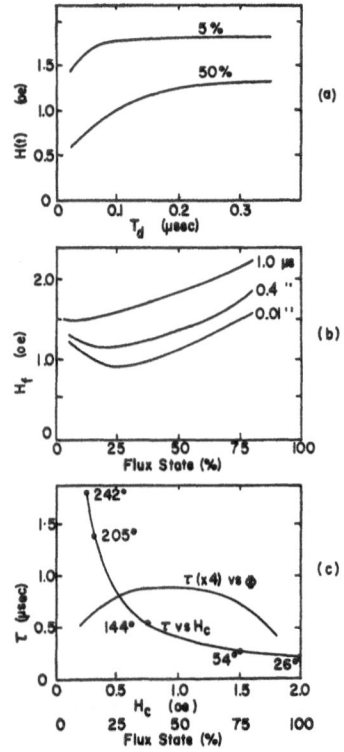

FIG. 1. Threshold properties of partially switched copper manganese ferrite core. (a) Variation of instantaneous threshold $H(t)$ with time $T_D$ after writing for 5% and 50% flux states; (b) threshold final value $H_f$ (in the direction of $I_w$) versus flux state for different values of writing time $T_w$; (c) relaxation time ($\tau$) versus coercivity for various core temperatures and relaxation time versus flux state with $T_w$ held constant.

pulse. The plot of $\tau$ versus flux state [Fig. 1(c)] for constant $H_w$ shows a maximum $\tau$ near the 40% state. This is the same point at which $H_f$ was found to be minimum for a given $H_w$. From this point $\tau$ decreases for both increasing and decreasing flux state. For any given flux state $\tau$ is maximum for the shortest $T_w$ and becomes small for $T_w$ of one microsecond or more. $\tau$ was observed to have no dependence on relative direction of the disturb field.

The above mentioned dependencies were measured at an ambient temperature of 26°C. Thresholds were also measured for the 50% flux state (100 nsec $T_w$) with ambient temperature ranging from −195°C to +242°C. The lower temperatures were obtained using constant temperature baths; the elevated temperature measurements were made both in oil baths and in air, with good agreement. $H_f$ showed much the same behavior as the coercivity ($H_c$) while $\tau$ increased markedly as ambient temperature increased. It is interesting to note that the product of $\tau$ and $H_c$ is nearly constant at 0.46 oe $\mu$sec independent of temperature. Other samples of the same material were prepared with firing schedules altered sufficiently to give a range of coercivity from 1 to 4 oe. Over this range the product of $\tau$ and $H_c$ did not change.

Measurements on magnesium-manganese, copper zinc manganese, and nickel ferrites, and measurements on YIG showed the same general type of behavior reported for the copper manganese ferrite, although the $\tau H_c$ product varied to a limited extent.

### ACKNOWLEDGMENTS

The authors wish to acknowledge W. L. Shevel, Jr., and J. M. Brownlow for their advice and suggestions, and N. J. Mazzeo for his assistance with the measurements.

asymmetry was also seen in plots of flux versus disturb amplitude with $T_d$ as a parameter, the asymmetry decreasing with time until the final value of threshold is reached. This can be shown to be consistent with a model of Bloch walls moving across energy barriers, the relaxation time being associated with the time required for a wall to relax back to the nearest energy minimum after termination of the writing pulse.[2]

The relaxation time also depends on flux state and on the amplitude ($H_w$) and duration ($T_w$) of the writing

JOURNAL OF APPLIED PHYSICS    SUPPLEMENT TO VOL. 33, NO. 3    MARCH, 1962

# Properties of Magnetic Films for Memory Systems

E. M. BRADLEY

*International Computers and Tabulators (Engineering) Limited, Stevenage, Hertfordshire, England*

The technique of using magnetic thin films in computer memory devices is reviewed, and conclusions drawn as to the properties of the film that are required for successful operation of a memory. A description is then given of a series of experiments which were designed to determine the best compromise in properties. Particular importance is attached to the easy axis dispersion, and it is claimed that the addition of a small quantity of cobalt to the nickel iron alloy reduces this despersion by about a factor 3.

## 1. INTRODUCTION

It is now well known that magnetic films can be made in which the anisotropy energy (expressed as a function of $\theta$, which is the angle between the easy axis and the magnetization) is given by $E = K \sin^2\theta$, where $K$ is a constant. It is found further that the magnetization vector will rotate coherently if fields are applied away from the easy direction. This process has been studied extensively[1-3] in the past few years and is found to occur with a very good approximation to the simple theory, but only in those films in which the preparation conditions are carefully controlled.

A consequence of the simple theory is that the hysteresis loop exhibited by the film will be rectangular when a magnetic film is applied in the easy direction. It is this property of the film which first attracted the attention of workers in the computer device field, following the work of Rajchman[4] and Forrester[5] on the use of rectangular hysteresis loop ferrite material for immediate access memory devices. However, it turns out that it is very difficult to exploit this property of magnetic films for matrix memories because it is difficult to control the coercivity and the rectangularity to the required accuracy, and the switching time can be very long.[6] The reason for this is that the switching process is not by coherent rotation when fields are applied near to the easy axis, but by wall motion which is dependent very critically on the detailed structure of the film.

Several ways have been proposed for exploiting the coherent rotation process for matrix selection.[7-11] In all these arrangements the storage element can switch extremely quickly, perhaps in the order of one nanosecond.[12] and there are large tolerances which allow either variation of driving currents or variation of film properties.

The organization of elements into a matrix memory has, in most cases, been of the type where word selection is external to the matrix, so that the rows of the matrix are word locations and the columns digit locations. The common feature of the proposed systems is that the field produced by a current flowing in a word selection wire is arranged to be nearly at right angles to the easy axis of the magnetic film, the field produced by a current in the digit line is almost parallel to easy axis, and the sense line is parallel to the digit line. It is particularly convenient for the construction of the matrix that these fields may be generated by two straight strip line systems running at right angles, and no bends are required in the lines.

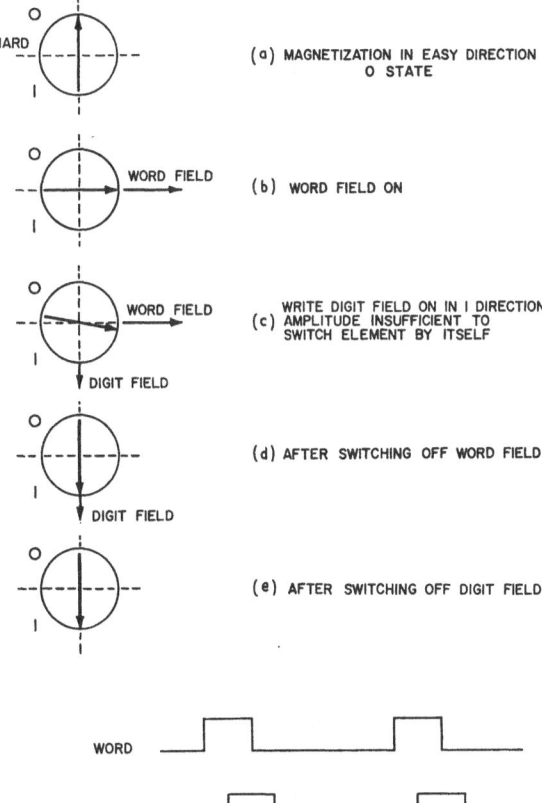

FIG. 1. Memory operation in the bidirectional digit pulse mode.

[1] E. M. Bradley and M. Prutton, J. Electronics and Control 6, 81 (1959).

[2] M. Prutton and E. M. Bradley, Proc. Phys. Soc. (London) 75, 557 (1960).

[3] D. O. Smith, J. Appl. Phys. 29, 264 (1958); 30, 264S (1959); 32, 70S (1961).

[4] J. A. Rajchman, Proc. Inst. Radio Engrs. 41, 1407 (1953).

[5] J. W. Forrester, J. Appl. Phys. 22, 44 (1951).

[6] E. M. Bradley and M. Prutton, J. Appl. Phys. 31, 285S (1960).

[7] J. I. Raffel, J. Appl. Phys. 30, 60S (1959).

[8] E. E. Bittman, IRE Trans. on Electronic Computers EC8, 92 (1959).

[9] E. M. Bradley, J. Brit. Inst. Radio Engrs. 20, 765 (1960).

[10] M. Williams, International Conference on Components and Materials used in Electronic Engineering, Proc. Inst. Elec. Engrs. (London) (to be published).

[11] British Patent Specification No. 880,383.

[12] W. Dietrich, W. E. Proebster, and P. Wolf, IBM J. Research Develop. 4, 189 (1960).

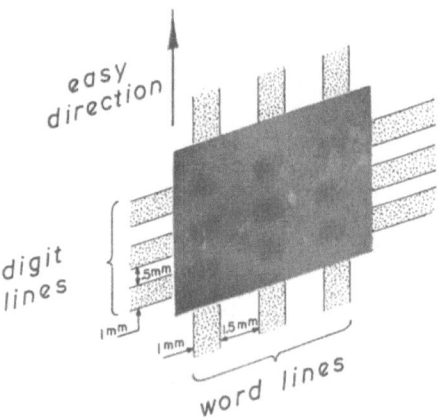

FIG. 2. Photograph of domain patterns
formed under strip line system.

The mode of operation in which the largest tolerances are obtained,[10] when using one magnetic field element for storage of each bit, is the bidirectional digit pulse mode[11] in which the fields are arranged as shown in Fig. 1, which also shows the timing of the pulse trains. It will be noted that, nominally, the pulses are applied exactly in the easy and hard directions, hence to avoid the zero torque condition which can cause the break-up of the film into a fine domain pattern; the word pulse must not be turned off before the digit pulse is turned on. This mode will work with the film easy axis displaced a few degrees either side of the field direction.

A similar arrangement has been proposed by Raffel[7] in which a permanent digit field is maintained on every element to provide the return torque which is necessary after the word field has been turned off. This permanent field can be provided by passing a dc current through each digit selection strip (as proposed by Raffel), or by placing the whole matrix in the magnetic field from a permanent magnet or coil system. A further way in which the return torque may be provided has been called the tilted mode,[9] where the word and digit fields are applied at right angles to one another but at a small angle to the hard and easy magnetization directions. This system requires better control over the easy axis skew, which is the angle between the average anisotropy axis of a storage element and the direction of the deposition field, probably as small at $\pm 2°$ about a nominal angle, conveniently 5°. The tolerances in these systems of operation have been examined in detail by M. Williams.[10]

The simple theory can therefore be used to predict possible modes of operation, and it suggests, in the modes proposed, that very large tolerances in digit current should be obtainable. In practical films, one reason that the theoretical tolerances cannot be obtained is that the discontinuous transitions predicted to occur at a critical applied field are not discontinuities, and some change of field is required to effect change from one state to the other. It is found that the reason

for this is that the film will break into a domain pattern, and that different areas of the film have different critical fields.[13] The consequence of this in the design of memory elements is that larger digit fields are required to effect saturation of the element than are predicted by the simple theory.

Some films behave very closely to the theory in that the change of field required to change from the saturated state on one side of the discontinuity to saturation on the other is only a few millioersted for a specimen of 2 cm diameter. Films made under different processing conditions, however, exhibit no net coherent rotation for the field applied at any angle. These films usually have similar hysteresis loops for any direction of the applied field. It must therefore be the objective in the manufacture of magnetic films to arrive at a process which produces consistently films having properties very near to those predicted by the simple coherent rotation process, together with low-angle skew over a large area and sufficiently low $H_k$ (anisotropy field) for small driving currents to be used.

## 2. PERFORMANCE OF MEMORY ELEMENTS

It has been found that, although some films prepared on bulk aluminum substrates can be very close to the simple model when they are several centimeters in size, the same film when etched into a small dot of millimeter size may be considerably changed in properties. (Memory elements must be of this small size to reduce driving ampere turns, drive line length, and hence delay times.) Generally the proportion of flux rotating is considerably reduced, and thus the flux output from an element working in one of the rotational storage modes is much lower than expected from the total flux content of the element. In addition it is found that the discontinuous transitions predicted by the model are no longer even approximate

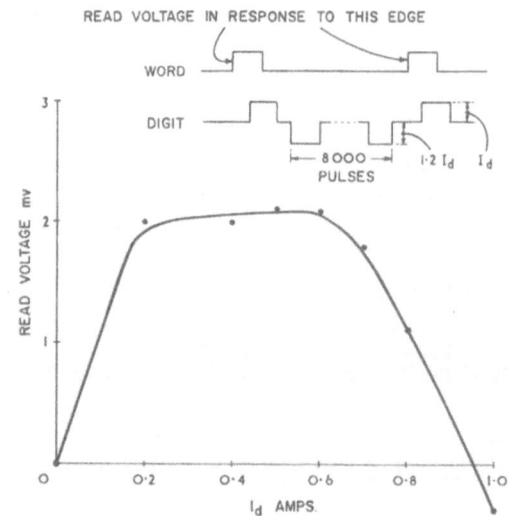

FIG. 3. Read output voltage as function of digit write current.

[13] M. Prutton, Brit. J. Appl. Phys. 11, 335 (1960).

discontinuities, since a large change of field is required to effect change from saturation in one state to saturation in the other.

This behavior is thought to depend greatly on the detailed structure of the edge of the film, for instance some films etched into a dot pattern were found to have increased $H_k$ by three times compared with the complete sheet before etching, while others showed no increase. The properties of the films are thought to be governed by the detailed nature of the edge domains which are in turn controlled by the shape of the edge.

The other technique commonly used[7] for the formation of a dot pattern is to mask during deposition of the film. The author has no first-hand detailed experience of this technique, but microscopic examination of the edges of dots formed in this way show that they are much more diffuse. This perhaps alters entirely the character of the edge domain formation, since memory devices using masked dots have been made successfully.

A different approach to the problem has been adopted in which a complete sheet of film is used, and the individual memory elements are defined by the applied magnetic field pattern instead of the edge of the magnetic material. This can easily be achieved by passing the selection current through a strip line mounted very close to the film. In this arrangement the horizontal component of magnetic field, which is the only component that affects the magnetization, generated by a current in the line falls very rapidly away from the edge of the line. Calculations show that it is possible to define the field sufficiently close to the region under the strip line that no interference between fields from adjacent strip lines will be found for small strip spacings. Figure 2 shows a photograph of an area of magnetic film used to store a nine bits. The zig-zag nature of the wall at right angles to the easy direction is clearly seen.

Experiments made on storage elements selected in this way show that the flux rotating is much higher, and the transitions are much sharper, than obtained with

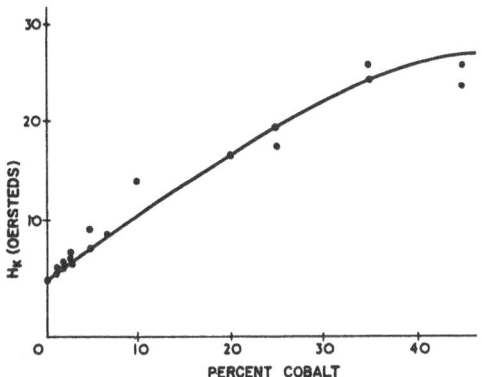

FIG. 5. $H_k$ as function of cobalt content with nickel iron ratio adjusted for zero magnetostriction.

etched dots of the same size as the strip lines. However, the field required to rotate the magnetization under the line into the hard direction is some three times (for strip lines of 1 mm width and films of 1000 A thickness) that expected from a value of $H_k$ measured for the whole plate. Further study has been made of this effect, which is probably due to magnetostatic interaction between the film under the strip and the unaffected part outside. This interpretation suggests that the magnitude of $H_k$ plays a secondary part in determining the drive fields. Another paper[14] presented to this conference describes measurements of the distribution of magnetization angle under a strip line.

The behavior of a typical film element used for storage is shown in Fig. 3. This is a curve of read output voltage as a function of digit write current using the pulse train shown. The curve was obtained for a film 800 A thick, having $H_k = 5.5$ oe and $H_c = 5.0$ oe, with the digit field applied along the easy axis and the word field at right angles to this. The word and digit strips used were 1 mm wide, and word current was 1.5 A, which is well above the threshold value. This type of curve is perhaps the most useful devised so far for the analysis of the behavior of a film element as a storage device.

For zero digit current the word pulse is applied by itself, and since this is along the hard direction, the film is split into a fine domain pattern after the first pulse, and thus zero output results at the edge of the following pulse. As the digit current is increased the output increases since a larger proportion of the flux rotates under the influence of the easy direction field. When the digit current is increased still further the output level saturates and remains constant because the magnetization of the element falls into the easy direction in a saturated state. As the value of digit current rises even further the disturb digit pulses can switch the magnetization back into the opposite state without the coincident word pulse. It has been found that this process can be very slow and may not be fully complete at any given level of

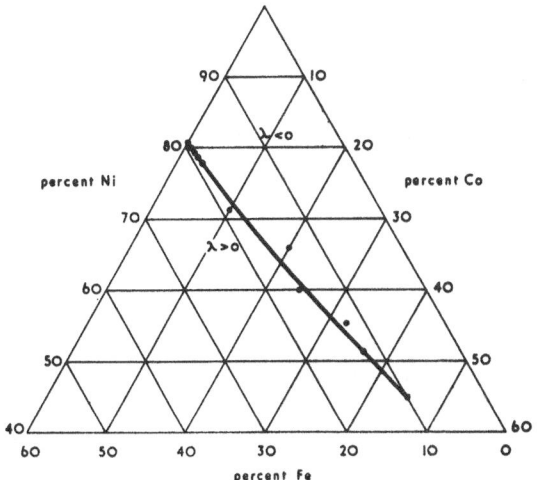

FIG. 4. NiFeCo composition having zero magnetostriction.

[14] K. D. Leaver and M. Prutton, J. Appl. Phys. 33, 1095 (1962), this issue.

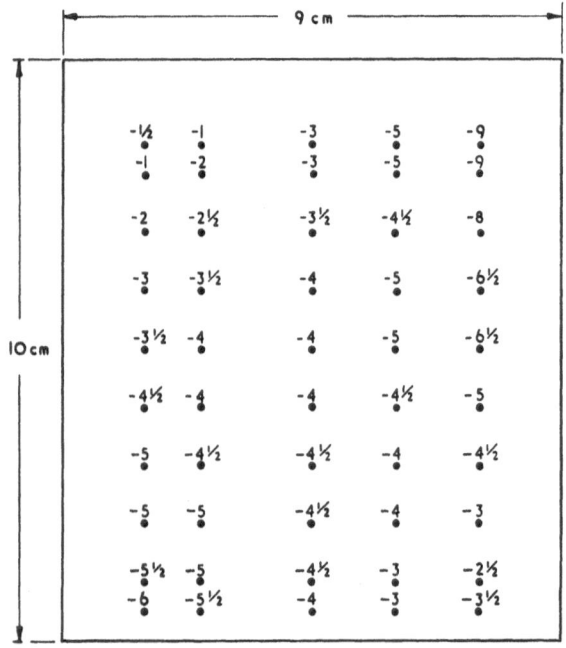

ingot alloy — 80/17/3 Ni Fe Co
film        — 810 A thick
            $H_c$ = 3·9 oe
            $H_k$ = 6·0 oe

(a)

81/19 Ni Fe
990 A thick
$H_c$ = 2·3 oe
$H_k$ = 3·0 oe

(b)

FIG. 6. Easy axis skew plot for films prepared from 80/17/3 NiFeCo and 81/19 NiFe. Figures are angle between easy axis and edge of plate. Aligning field is 3° to edge of plate.

current until several thousand digit pulses have been passed along the strip, hence the standard test devised uses 8000 pulses. Since this is a test to determine tolerances in a storage device the digit disturb pulse amplitude is set 1.2 times the digit write amplitude.

The disturb digit current level ,which is that current necessary to reduce the read voltage to 90% of the maximum value, is found to be proportional to the easy direction coercivity, and the disturb field is approximately 1.5 times the coercivity. The digit current for zero output is dependent mainly on the angle between the easy axis and the drive lines, and a satisfactory skew in this angle is about ±10°. The slope of the output voltage digit current curve is dependent on the film characteristics and is found to be sharpest for those films which exhibit hysteresis loop properties nearest to the simple model.

Summarizing, the film properties required for a storage device are:

(1) The hysteresis loops exhibited by the film should be a close approximation to the simple theory.

(2) The easy axis skew over the area of the film should be smaller than ±10°.

(3) The film should have high coercivity and low $H_k$; in practice with films having $H_k = 5-6$ oe the $p$ ratio ($H_c/H_k$) should be greater than 0.8.

## 3. ALUMINIUM SUBSTRATE

The planar film memories so far described which use the rotational mode of selection have either used glass or aluminum as a substrate for the film. It would appear from our experiments that there is a number of disadvantages in the use of glass which may be overcome by using a metallic substrate without introducing further disadvantages peculiar to the aluminum substrate. It is desirable to reduce the inductance of strips wound around the substrate to a minimum. If glass is used, considerations of mechanical strength dictate a lower limit of about 0.004 in. for the thickness. This means that the spacing of any strip line to a ground plane or a return line must be at least 0.004 in. However, with an aluminum substrate this can be very small indeed, limited only by the practical lower thickness of plastic materials, which may be as low as 0.0005 in. The self-inductance of a single word drive line wrapped around two aluminium plates of size such that each word contains 50 bits is 30 nh without film and 50 nh with a suitable magnetic film. The inductance of this line, wrapped around a glass plate 0.004-in. thick, is 250 nh. Thus, the device with film on aluminium requires about $\frac{1}{5}$ of the driving power of the glass-based film memory.

The second advantage is that the sense line may have lower mutual inductance with the digit drive line, hence causing lower noise. It has been pointed out by Raffel[15]

[15] J. I. Raffel, T. S. Crowther, A. H. Anderson, and T. O. Herndon, Proc. Inst. Radio Engrs. 49, 155 (1961).

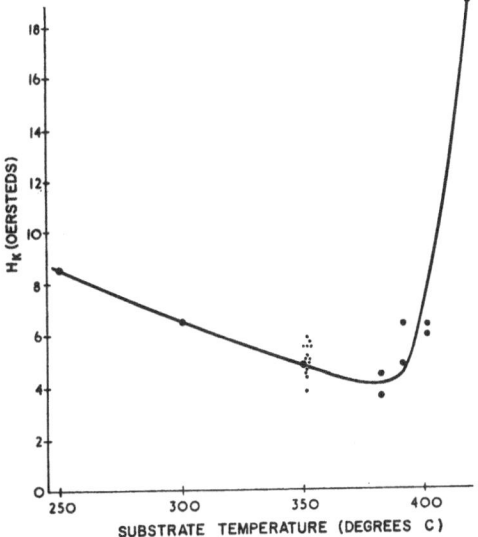

FIG. 7. $H_k$ as function of substrate temperature during deposition for films prepared from 80/17/3 NiFeCo alloy.

that if the impedance of the sense line is made too small, by very close spacing, the attenuation of the line may become too high, due to the resistive loss of the line. B. Jinman[16] has observed that for a line 9 ft long of 1-mm wide, 0.002-in copper strip, spaced 0.002 in. from the aluminium plate by "Melinex" or "Mylar," an original pulse rise time of 0.25 nsec has decayed to 7 nsec and suffered a delay of 13 nsec. It is concluded from these results that the very small spacings possible with an aluminium substrate can be used without increasing the attenuation of the line to a high value, and thus the noise generated in the sense line during the write process can be smaller with the metal substrate.

No damping effect has been noted on the rotational reversal due to the presence, so close to the film, of the metallic conducting substrate. In these experiments we have used pulses with approximately 10 nsec rise time.

### 4. PROPERTIES OF MAGNETIC FILMS

The experiments leading to the development of a large area aluminium plate coated with magnetic film of suitable properties have been divided into two parts. First, an investigation was made of hysteresis loop properties of film samples approximately 2 cm square prepared on glass microscope slides from various alloys and under various conditions. This has enabled us to decide the preparation conditions required to produce films having loops near to the simple model. These processing conditions were then used to coat aluminium plates of size 4 in. $\times$ 3 in. and were found to give films of approximately similar hysteresis loop properties, provided the plate was first highly polished.

The angle of the easy axis to the edge of the plate, the skew angle, was measured by a pulse technique[9] for

these films, and the processing modified to give the smallest variation of angle of easy axis over the plate.

### Glass Microscope Slide Specimens

Early experiments showed[2] that films prepared from 81/19 NiFe alloy with the slide held at a critical temperature obeyed, with a fair approximation, the simple rotational model, provided that the thickness of the film was between 1000 and 2000 A.

It was found in later experiments that if a small quantity of cobalt was added to the nickel iron, say 3%, then the irreversible transitions predicted by the simple model became much sharper, and that the temperature of the substrate during deposition was not as critical. It was also found that in thicker films (up to 6000 A) the proportion of flux rotating when fields were applied near the hard axis was very much higher.[17] However, it was found that the anisotropy field was somewhat increased by the addition of cobalt. Following these results the analysis was extended of the properties of NiFeCo films up to 40% Co maintaining the nickel iron ratio so that the films had no magnetostriction. This it was found could be achieved by maintaining the ingot iron content at about 20% for low Co content and dropping to about 10% Fe for 50% Co. Figure 4 shows the zero magnetostriction composition line. Figure 5 shows that $H_k$ varies approximately linearly with cobalt content up 35%. It is to be noted that similar behavior has been reported by Wolf[18] for electroplated zero magnetostrictive films of NiFeCo.

It has been thought important that films for storage devices should be zero magnetostrictive for two reasons: First, that mechanical bending of a substrate carrying a magnetostrictive film causes an extra stress which results in rotation of the easy axis. Such bending might be

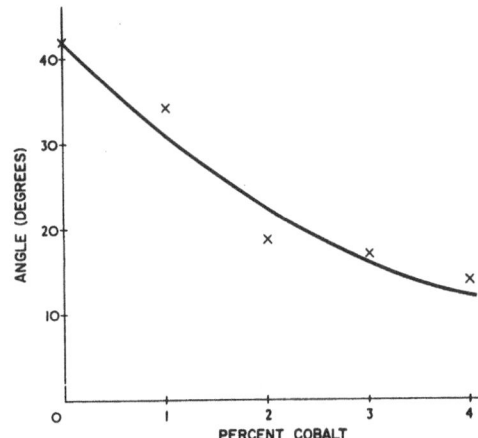

FIG. 8. Mean total angle skew plotted as function of ingot cobalt content.

16 B. Jinman (private communication).

17 French Patent Specification No. 1,274,802.
18 I. W. Wolf, paper in the Symposium on the Electric and Magnetic Properties of Thin Layers, Leuven, Belgium, September, 1961 (to be published).

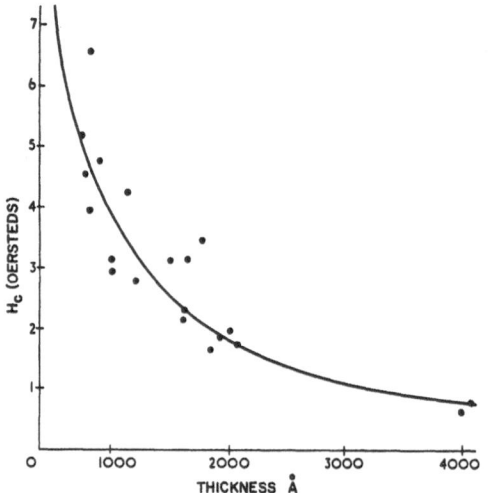

FIG. 9. Easy direction coercivity as function of film thickness.

caused in assembly of a magnetic film storage device. Second, and more important, it has been found from experiments with 2 cm squares of magnetostrictive alloys etched from films on microscope slides, that the easy axis angle is a function of the magnitude of the magnetostriction. The explanation for this behavior might be that during the condensation of the film an anisotropic stress is built into the film which causes the observed rotation of the easy axis.

### Properties of Films on Aluminium Substrates 4 in. × 3 in.

Further experiments with specimens 2 cm square deposited onto aluminium showed that very similar behavior was obtained provided that the aluminium surface was well polished. A series of experiments was then designed to examine the suitability for a storage device of films deposited onto 4-in. × 3-in. aluminium plates, and of particular relevance in this study was the skew of easy axis over the plate. This was measured by a pulse technique using of a pair of orthogonally mounted strip lines, which could be rotated to provide pulsed magnetic fields at any angle to the edge of the plate, and could be placed at any position over the plate. The measurement of the angle between the easy axis and the edge of the plate has been described in detail in a previous paper.[9]

Figure 6 shows the easy axis skew plot which has been obtained for a film 810 A thick of 80% Ni 17% Fe 3% Co (weight %) alloy referred to in a previous publication[9] as "Gyralloy" contained in the starting ingot. Figure 6 also shows, in contrast, a typical plot obtained for a film of 990 A thickness deposited from an alloy of 81% Ni 19% Fe ingot composition. Both films are zero magnetostrictive and were deposited on a substrate maintained at 350°C with all other processing conditions maintained the same.

We have varied many processing parameters in the

investigations on the effects on the film properties of this processing, and the effects of the most important are reported below.

#### Substrate Temperature

Variation of $H_k$ with substrate temperature is very similar to that observed with the glass microscope slide specimens, and no appreciable dependence of easy axis angle skew on temperature has been found for the temperature range 250° to 380°C, i.e., the range near the minimum $H_k$. The variation of $H_k$ is plotted in Fig. 7.

#### Cobalt Content

The cobalt content of the ingot has been varied from 0% Co to 4% Co, while maintaining the nickel iron ratio to achieve zero magnetostrictive films. Figure 8 shows a plot of the mean total angle skew which is the average of the angle skew at the extreme corners of the plates. As has been shown, $H_k$ increases with increased Co content so that it is desirable to keep the cobalt content as low as possible in order to keep the driving currents low. Hence a compromise has to be sought between high $H_k$ and low easy axis angle skew. The best compromise is thought to be 3% Co.

#### Thickness

The easy direction coercivity is found to depend markedly on thickness. Figure 9 shows a plot of the variation of $H_c$ against thickness for films prepared from 80/17/3 NiFeCo at 350°C. $H_k$ is found to be independent of thickness but angle skew is slightly dependent. For instance, the skew is on average slightly larger for 800 A thickness than it is for 1200 A. $H_{ch}$, coercivity measured in the hard direction (which is some measure of the angle dispersion within a small area since low $\alpha_q$, the angle between the hard direction and that applied field direction at which $\frac{3}{4}$ of the available flux will rotate coherently in one sense,[19] is never found with high $H_{ch}$), is also larger for 800 A thick films than for 1200 A. For thinner films still, $H_{ch}$ increases, and the net rotating flux reduces to zero for films of about 400 A thickness.

It is thus necessary to determine a compromise for thickness so that the film will have good flux rotational properties and sufficiently high coercivity. To some extent the coercivity is interchangeable with the angle dispersion because the digit current used for every element has to set those elements with the least favorable easy axis orientation. However, in the range 500–1500 A, the coercivity reduces much more quickly than the angle skew increases so that the choice of the coercivity is the overriding consideration. It has been found empirically that the digit disturb threshold field for 1-mm wide strips using 8000, 0.5 $\mu$sec pulses, is 1.5 times

[19] T. Crowther, MIT Lincoln Laboratory Group Report 51-2 (February, 1959).

the coercivity measured for the whole plate. With these considerations in mind, the best thickness range for this type of storage plate, when using 1-mm strip lines to select the area of film, was found to be 700–900 A.

### Vapor Beam Angle of Incidence

The pattern of easy axis skew in most film specimens tends to be barrel-shaped, suggesting that if the angle of incidence of the vapor beam were reduced, the skew would be reduced. However, superimposed on the barrel pattern, which is generally of the order ±5° at the extremities of the 4-in.×3-in. plate (for an angle of incidence of 10° with 80/17/3 NiFeCo alloy), is a variation which has no distinct pattern, but which is about the same size. This means that most plates will have a total angle skew of less than 10°, and some will have almost zero. It ought to be possible to reduce the mean total skew by reducing the angle of incidence still further.

## 5. CONCLUSIONS

The addition of 3% cobalt to nickel iron, while still maintaining zero magnetostriction in the resulting film, enables the deposition onto a large area (4 in.×3 in.) aluminium plate of a film having characteristics suitable for immediate access storage devices. Many processing factors influence the performance of the film, and a number of compromises have to be sought.

## ACKNOWLEDGMENTS

The author would like to acknowledge that the points discussed in this paper have been the subject of many discussions with J. B. James, B. J. Steptoe, and P. Astwood of I. C. T. and M. Williams and K. Bingham of G. E. C. Hirst Research Centre. The film deposition experiments described were made by P. Astwood, Miss J. M. Farrer, and Mrs. J. Phillips of I. C. T.

The author would also like to thank the Directors of I. C. T. for permission to publish this paper.

---

JOURNAL OF APPLIED PHYSICS    SUPPLEMENT TO VOL. 33, NO. 3    MARCH, 1962

# Magneto-Optically Measured High-Speed Switching of Sandwich Thin Film Elements

J. C. SUITS AND E. W. PUGH
*IBM Research Center, Yorktown Heights, New York*

A technique employing the Kerr magneto-optic effect has been used for measuring switching times as short as 10 nsec in thin ferromagnetic films, both individually and when coupled together in a sandwich structure. The switching speed of individual 1-mm-diam Permalloy bits has been found to be dependent on film thickness, the thicker bits switching more slowly. It was further observed that the time to switch 90% of a film is much more sensitive to film thickness than the time to switch 50% of the same film. When two similar films, magnetized in a head-to-tail fashion, are placed on either side of a drive line strip, the time to switch 50% of the films is essentially the same as for individual bits. However, the time required to switch 90% of the material is considerably shorter for sandwich elements than for the corresponding single bits. These results suggest that the slower switching of thick bits may be attributed to an incoherent rotational mechanism associated with the inhomogeneous demagnetizing field.

TO achieve high packing density and low drive currents, individual thin film memory elements are being designed with smaller and smaller areas. This in turn requires that they be made correspondingly thinner in order to maintain low self-demagnetizing fields. However, a reduction in linear dimensions by a factor of $n$ results in an $n^2$ reduction in integrated output signal. Sandwich elements, with two films magnetized in a head-to-tail fashion and placed on either side of a drive line, have been proposed as a means for utilizing thicker films without having excessive internal demagnetizing fields.[1] In this paper a magneto-optical method for making high-speed switching measurements on such elements is described together with switching results on 1-mm-diam films *before* and *after* being incorporated into the proposed sandwich configuration.

## MEASUREMENT OF SWITCHING SPEED

The bits were driven by pulsing an evaporated aluminum drive strip. The pulse repetition rate was maintained at 60 pps, with pulses alternated in polarity. The negative pulses were always 1 μsec. in length and 12 oe in height, and sufficient to completely saturate the film. The positive pulses were varied in height $H$ from 0 to 12 oe and in width $T$ from 1 μsec to 10 nsec. The rise time of all pulses was about 3 nsec.

The state of magnetization of the bits was sensed by means of the Kerr magneto-optic effect. Polarized light was directed through the glass substrate and reflected off the film. It was then directed through a nearly

---

[1] See, for example, the survey article by A. V. Pohm and E. N. Mitchell, IRE Trans. on Electronic Computers **EC-9**, 308 (1960).

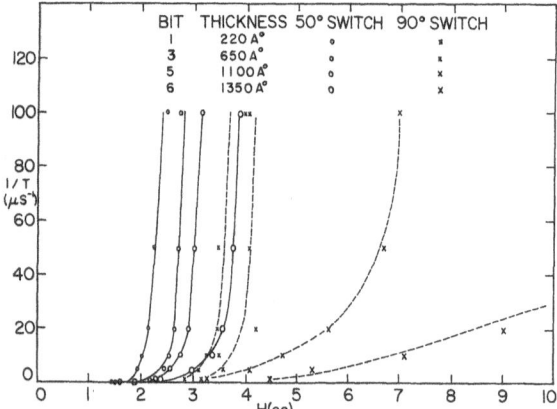

FIG. 1. Switching times for single bits 1, 3, 5, and 6 of series A 14.

crossed analyzer into an EMI 9502B photomultiplier tube. The change in photomultiplier tube output is proportional to the flux change $\Delta\phi$ of the film, and not to $d\phi/dt$ as in the usual inductive pickup sensing scheme. When driven by the above described set of pulses, the photomultiplier output is a 60-cps square wave, the amplitude of which is proportional to the flux change caused by the particular shape of positive pulse chosen. This output square wave is sent through a noise reducing, narrow bandpass filter tuned to 60 cps and then to an oscilloscope for display.

## SANDWICH FABRICATION

The film sandwiches were made by first evaporating 1-mm-diam circular bits of 80-20 Permalloy on a microscope slide maintained at 300°C in a vacuum of $10^{-4}$ mm Hg and in a magnetic field of 100 oe directed in the plane of the bits. The deposition rate (about 1000 A/min) and film thickness were controlled by a quartz crystal oscillator monitor. A shutter was arranged to slide over the bit mask so that bits of various thicknesses could be obtained during a single evaporation.

After a row of seven bits of various thicknesses was evaporated, a 1000 A layer of silicon oxide was evaporated onto all the bits. Following this a 3-mm-wide, 6000-A-thick aluminum strip was evaporated onto the

silicon oxide. This was followed by another layer of silicon oxide and finally a second row of Permalloy bits was evaporated over the first row in order to complete the film sandwich.

Switching measurements were made after evaporation of the drive strip and after evaporation of the second row of bits. A current through the evaporated aluminum strip produces a field in opposite directions on opposite sides of the strip, thus magnetizing the sandwich bits always in a head-to-tail fashion. The easy axis of all the bits was at 30° to the magnetic field direction produced by a current through the aluminum strip.

## EXPERIMENTAL RESULTS

Films twice as thick as reported here will remain saturated in the remanent state when in the sandwich structure. However, only data for films thin enough to remain in a saturated remanent state before sandwiching are presented here in order to provide a direct measure of the effect of sandwiching on switching speeds. Some high-speed switching curves are shown in Fig. 1 for various thicknesses of single bits. It is seen that the thinnest bits switch most rapidly, and that the field required to switch 50% of the film is much less than that required to switch 90%. It is further seen that the difference between the 50% and 90% switching characteristics becomes much larger as the films become thicker. The slope of the 50% switching curves (approximately $3.6\times10^8$oe$^{-1}$ sec$^{-1}$) is the same order of magnitude as is expected for pure rotation. A value calculated by Olson and Pohm[2] for pure rotation using resonance data is $2.2\times10^8$oe$^{-1}$ sec$^{-1}$.

It is postulated that 50% or more of the film switches by relatively pure rotation, while over 10% of the film, near the edges, switches by a slower incoherent process, abetted by the inhomogeneities of the demagnetizing field. This accounts for the strong thickness dependence of the 90% switching curves and the relative insensitivity to film thickness of the 50% switching curves.

Switching curves for bits 1 and 6 of Fig. 1 are shown in Fig. 2 both before and after evaporation of the second film of the sandwich element. It is seen that the 50% curves are little changed by the sandwiching process, while the 90% switching curves are substantially improved. This result further substantiates the postulated model for the thickness dependence of switching speeds, in individual films.

## ACKNOWLEDGMENTS

The authors are indebted to W. C. Kateley for help in film preparation, G. E. Keefe for assistance in some of the measurements, and especially to N. C. Ford, Jr., who was instrumental in the early development and utilization of magneto-optic techniques in this laboratory.

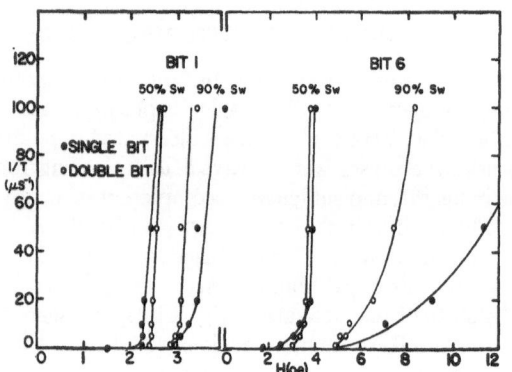

FIG. 2. Switching times for single and double bits of series A 14.

[2] C. D. Olsen and A. V. Pohm, J. Appl. Phys. **29**, 274 (1958).

JOURNAL OF APPLIED PHYSICS     SUPPLEMENT TO VOL. 33, NO. 3     MARCH, 1962

# Magneto-Optic Readout for Computer Memories*

R. L. CONGER AND J. L. TOMLINSON

*U. S. Naval Ordnance Laboratory, Corona, California*

Experimental measurements have shown that nondestructive magneto-optic readout from magnetic film computer elements by means of the longitudinal Kerr-effect can operate at rates of at least one million bits per second. If the steady light source of the magneto-optic apparatus is sufficiently bright, a signal-to-noise ratio of four or better can be obtained from the photomultiplier readout tube. Since the reversal time of thin ferromagnetic films is less than a microsecond, writing speed can be as fast as readout. Information readout from a magnetic film can be obtained by scanning with a high-intensity light spot. Spurious variations in surface reflectivity of the films are sufficiently small so that parts of the film magnetized in opposite directions can be clearly determined. Experiments with an electronic strobing magneto-optic method make it possible to observe the details of the magnetization reversal process in very small parts of a film with a time resolution of the order of a nanosecond. Since this strobing process involves sampling and integration, the information bandwidth is too small to be used for digital computer readout. However, this method is useful in studying fine details of the magnetization reversal process. Measurements show that the reversal is repeatable in the different parts of the film.

## INTRODUCTION

THE possibility was investigated of using a thin ferromagnetic film as the basis for a nondestructive readout computer memory based on optical interrogation methods. This type of readout would make use of the magneto-optical Kerr effect, by which an incident beam of plane-polarized light reflected from a magnetized surface becomes rotated.[1] The degree of rotation, which is a function of the direction of magnetization, is detected by an optical polarization analyzer and a photomultiplier.

Kerr-effect readout has several potential advantages. (1) It is completely nondestructive. (2) Readout from a very small area of the film can be obtained. Since the theoretical limit of magneto-optical resolution is a few microns,[2] the minimum size of the area that can be read magneto-optically is about $10^{-8} cm^2$. (3) Writing densities of $10^6$ bits/$cm^2$ on high coercive force MnBi films appear feasible.[3] (4) Writing currents and other sources of spurious magnetic fields have no effect on the magneto-optical readout. This is in contrast to thin film memories using pickup coils for readout where the signal obtained from the flux change of the film is often small compared to voltages induced by reading and writing currents. (5) The bandwidth of this readout system is large.

## EXPERIMENTAL

To determine if photomultiplier noise would limit the speed of magneto-optic readout, an experiment was performed in which the magnetization of a film of 80% Ni 20% Fe was reversed rapidly by a magnetic field pulse applied to the film. The output of the photomultiplier tube that sensed the magnetization change was amplified in a wide-band amplifier and observed on an oscilloscope. The process of magnetization change in one direction appeared as a step when viewed versus time on the oscilloscope. It is of interest that the signal was propor-

tional to the magnetization, not to the rate of magnetization change.

In order to find out whether a single observation would provide sufficient information to distinguish the change in light intensity, the oscilloscope was operated with the single sweep, which was initiated from the oscilloscope control panel. Figure 1 shows a reversal of magnetization as viewed on the oscilloscope, with a time base of 1 $\mu$sec per division. The illustration indicates that a thin film memory with optical readout should work at a rate of at least 1 Mc. It can be seen that the signal-to-noise ratio, which is approximately proportional to the light intensity, is of the order of 4 to 1. In this experiment the light source was a 120-v, 12-amp, carbon arc. The signal-to-noise ratio is also proportional to the area of the film seen by the photomultiplier, which in this case was 5 by 10 mm.

A present limitation of the system is imposed by the large size of the film element used. The phototube noise could be reduced by operation of the tube at a lower temperature, or it could be made smaller in comparison with the signal by increasing the light intensity. For some applications, the bandwidth could be reduced in order to reduce the noise.

To test the feasibility of scanning a film containing information in the form of a domain pattern, the magneto-optic apparatus was modified by the addition of a scanning disk driven by an air turbine at a rate of about 2000 cps. A small slit in the edge of the disk produced a narrow beam of light which scanned across the film. A

[1] C. Fowler, Phys. Rev. 100, 746 (1955).
[2] D. Treves, J. Appl. Phys. 32, 358 (1961).
[3] H. J. Williams, R. C. Sherwood, F. G. Foster, and E. M. Kelley, J. Appl. Phys. 28, 1181 (1957).

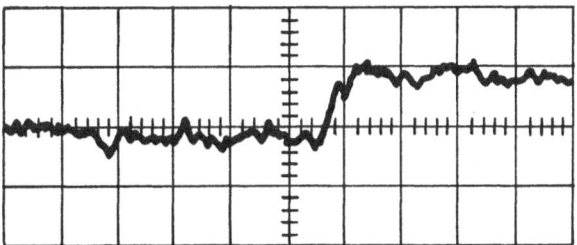

FIG. 1. Magneto-optic readout signal.

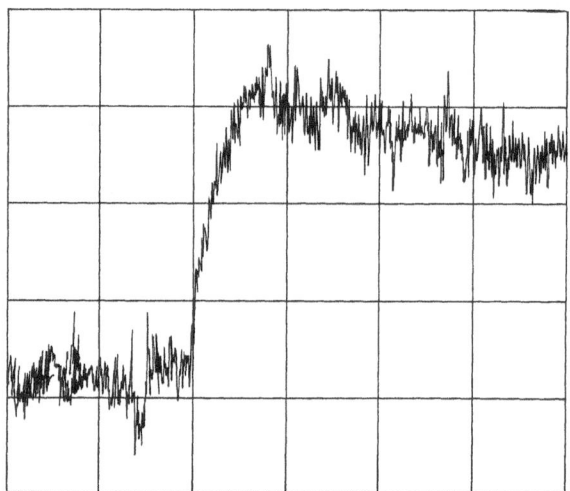

FIG. 2. Magneto-optic strobe output.

diffuser was placed in front of the photomultiplier to prevent the signal obtained from being a function of the direction of the light striking the tube. For these tests, the sun was used as a light source, and a photocell-actuated motor moved a mirror to keep a beam of sunlight shining on the apparatus. The output was again displayed on a cathode-ray oscilloscope, and the trace clearly showed a map of the domain pattern of the film. Differences in the reflectivity of the film were smaller in magnitude than the changes produced by the change of magnetization. The maximum scanning rate of this experiment was limited to about $10^3$ cm/sec by the upper speed of the air turbine and scanning disk used.

Other methods of scanning are possible. For example, the flying spot scanner could be used, or the magnetic film could be on a moving tape or drum, and the light beam could be stationary. Readout would produce no wear.

To produce domain patterns for experiments, a small ferrite rod was used to write on a magnetic film. In films of 80% Ni 20% Fe evaporated in a magnetic field, there is a tendency for long domains to form in the easy vertical direction; thus vertical lines are easily written, but horizontal lines tend to be a series of short vertical lines. To overcome this difficulty, a horizontal grid pattern of aluminum can be evaporated onto the glass substrate before the Fe-Ni alloy is evaporated. The aluminum causes horizontal lines of high-coercive force material to form in the film; these effectively break up the vertical domains and make it possible to write in either the horizontal or vertical direction. It was found that some films of Fe, Co, and Ni performed better than the 80% Ni 20% Fe films. Under the best conditions, it was possible to write clearly legible words on the evaporated films.

Also possible are other types of writing such as Curie-point writing, in which it is proposed that an electron beam be used to change the magnetization by raising the temperature at electron impact above the Curie point.[4] Or the writing could be done by current-carrying drive windings of the type used in the usual thin film memory.

An electronic strobing magneto-optic method has been used as a research tool for studying small regions of films magneto-optically at very short time intervals; this apparatus has been described previously.[5] Hundreds of samples of the output of a photomultiplier are sampled by an electronic gate with a pulse width of $10^{-9}$ sec, and these samples are averaged to remove noise effects. The time resolution is about $10^{-9}$ sec, even though the information bandwidth is much less. With this apparatus, the steps in the reversal process of a film can be determined with a high degree of resolution. It is found that the reversal process is repeatable. This apparatus, like the other magneto-optical equipment, shows the magnetization rather than the rate of change of magnetization; thus there is no doubt about the amount of magnetization reversed.

Figure 2 shows a typical trace of the X–Y recorder output of the apparatus. This trace shows magnetization of a square which is 1/64th of the area of the 1-cm² magnetic film. The time scale is 1 μsec per division. It is frequently observed, as in Fig. 2, that the first part of the reversal is very fast, but the fast reversal accounts for less than half the total magnetic flux, and the remaining reversal takes several times as long.

From these tests, it appears that magneto-optic readout for computer memory is feasible. Information rates of greater than 1 Mc could be obtained, and reading and writing density of $10^6$ bits per cm² seems quite feasible at the lower information rates. The readout is completely nondestructive and is not affected by drive and other currents.

[4] L. Mayer, J. Appl. Phys. **29**, 1003 (1958).
[5] P. C. Archibald, R. L. Conger, R. W. Sharp, and J. L. Tomlinson, Rev. Sci. Instr. **31**, 653 (1960).

JOURNAL OF APPLIED PHYSICS     SUPPLEMENT TO VOL. 33, NO. 3     MARCH, 1962

# Switching Properties of Multilayer Thin Film Structures

A. Kolk, L. Douglas, and G. Schrader

*The National Cash Register Company, Electronics Division, Hawthorne, California*

Structures formed by depositing different ferromagnetic thin metal films in direct contact with each other were found to display a square hysteresis loop without the step discontinuity usually associated with gross heterogeneous magnetic composites. The high-speed switching characteristics of a bilayer structure composed of an electroplated 97 Fe-3 Ni layer in contact with a uniaxially oriented 79 Ni-21 Fe electroplate were intensively investigated. The switching characteristics of the bilayer were compatible with a nucleation and growth mechanism.

THE magnetic switching properties of structures formed by depositing in direct contact thin films of two different ferromagnetic materials were studied. Both electrodeposition and vacuum deposition were employed for forming the layers. The investigation included structures composed of various combinations of iron, nickel, cobalt, or the alloys of these materials, and qualitatively similar behavior of the hysteresis loop was noted in all cases. However, the data in this paper will be restricted to a configuration consisting of an electrodeposited film of uniaxially oriented permalloy in contact with an electrodeposited film of a magnetically isotropic 97Fe-3Ni alloy on a 10-mil Be-Cu wire substrate.

The Permalloy alloy was deposited from a plating bath described by Wolf[1] in the presence of a 160-oe, axially-oriented field. The 97Fe-3Ni alloy was deposited from a modified Fischer-Langbein bath.[2]

The hysteresis characteristics of the deposits were measured at 170 cycles in a modified version of the instrument described by Crittenden *et al.*[2,3]

The films composed of layers a few thousand angstroms, or less, in thickness displayed a square hysteresis loop in the easy direction with no evidence of the step observed in gross, heterogeneous, magnetic structures. The saturation flux was the sum of the flux of each of the individual layers.

The coercive force $H_c$ showed complicated behavior when the ratio of thickness of the layers was varied. The $H_c$ of the composite was sometimes less than the $H_c$ of either layer measured in the absence of the other layer. However, if the $H_c$ of the material used in each layer was measured at the same thickness as the composite thickness, then the $H_c$ of the composite fell between those of the two materials comprising it.

The switching response to a step-function magnetic drive field was studied. The apparatus and the experimental method are described elsewhere.[4] Since the switching signal included the eddy current damping in the substrate and the ringing in the test circuit, the switching pulse was obtained by subtracting the signal obtained with a bare Be-Cu wire from the signal obtained with a plated wire.

A typical set of switching pulses is shown in Figs. 1 and 2 for the bilayer and for each individual material measured separately at the same thickness as in the composite. For ease of comparison, the value of the applied field was chosen to be twice the $H_c$ of the respective material. It can be seen that the switching pulses for the bilayer are intermediate in shape as compared with the 97Fe-3Ni and the 79Ni-21Fe alloy.

A transverse field was applied by driving current down the wire substrate. The effect of the transverse field on the switching transient is shown in Figs. 1 and 2.

The coercive force and switching coefficients of the three samples are given in Table I together with corresponding data for 79Ni-21Fe and 97Fe-3Ni, plated to the same total thickness as the bilayer plate. The values of the switching coefficient and the coercive force depend on the exact plating conditions and substrate treatment. Both were maintained constant for comparison of the deposits presented in the table.

The switching time was measured between the points where the voltage was 10% of the peak value. The

[1] I. W. Wolf and V. P. McConnell, Tech. Proc. Am. Electroplaters' Soc., 43rd Ann. Con., Wash., **1956**, p. 215.

[2] A. J. Kolk, "Research on magnetic rod storage and switching device," The National Cash Register Company, Final Research Report, July, 1960.

[3] E. C. Crittenden, Jr., A. A. Hudinac, and R. I. Strough, Rev. Sci. Instr. **22**, 872 (1951).

[4] D. A. Meier and A. J. Kolk, Office of Naval Research Symposium on Large Capacity Memory Techniques for Computing Systems, May, 1961 (to be published).

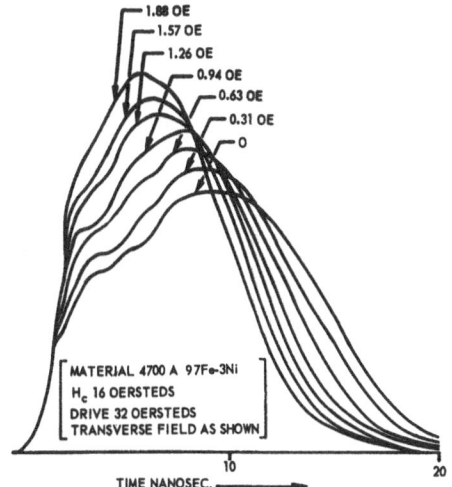

FIG. 1. Effect of a transverse field on the switching pulse shape for a 97Fe-3Ni electroplate.

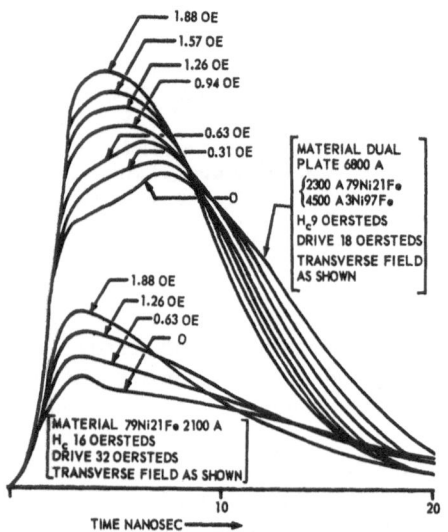

FIG. 2. Effect of a transverse field on the switching pulse shape for a 79Ni-21Fe electroplate and for a layered structure.

TABLE I. Switching properties of alloy and bilayer films.

| Material | Thickness | $H_c$ oe | $S_w$ oe-$\mu$sec |
|---|---|---|---|
| 79Ni-21Fe | 6800 A | 4 | 0.21 |
| 79Ni-21Fe | 2100 A | 16 | 0.43 |
| 97Fe-3Ni | 6800 A | 14.5 | 0.35 |
| 97Fe-3Ni | 4700 A | 16 | 0.21 |
| Bilayer | | | |
| 97Fe-3Ni | 4700 A } | 9 | 0.21 |
| 79Ni-21Fe | 2100 A } | | |

switching coefficient was obtained assuming the usual linear relation

$$S_w = \tau(H - H_0).$$

In the case of the bilayer, the use of the percentage of peak-voltage method is subject to question because of a change in shape of the curve as the drive field is increased. The peak-voltage method gives values of the switching coefficient about 100% larger than those obtained by the volt-sec area.

At values of drive field near the $H_c$ of the bilayer, the shape of the switching curves is similar to that of the permalloy. As the drive field is increased, the shape begins to resemble that of the 97Fe-3Ni alloy. At the lower drives an initial spike is evident (which is masked, when the drive field is increased), by the response time of the test apparatus, and by the finite rise time of the applied field.

The switching curves of 97Fe-3Ni films were measured separately and were similar to those of ferrite cores. The switching of 97Fe-3Ni plates has also been investigated by the interrupted pulse technique, and data qualitatively identical to that of McKay and Smith[5]

were obtained. The quantitative development of Haynes[6] for the case of random nucleation and three-dimensional wall motion has been applied to two-dimensional growth and was found to fit the experimental data.

Since both separate films involved in the bilayer show switching behavior which can be explained with a model based on nucleation and growth by wall motion,[7] it is reasonable to assume that the change in coercive force of the bilayer structures occurs by modifying the critical nucleation and growth parameters. The nuclei, once formed, apparently include both layers and grow by two-dimensional wall motion.

In summary, it has been found that the switching properties of magnetic films can be controlled by forming layered structures. The threshold switching fields can be reduced without affecting the rectangularity of the hysteresis loop. The phenomenon is of a general nature and is applicable to either electrodeposited or vacuum deposited films. It is anticipated that layered structures will permit design choice of magnetic properties for computer application of thin films.

### ACKNOWLEDGMENTS

The authors would like to acknowledge the assistance of M. Flavin in switching measurements and the assistance of M. Orlovic in the preparation of evaporated film structures. The helpful discussions of H. Wieder, Naval Ordnance Laboratory, Corona, California, and D. Meier are gratefully acknowledged.

[5] R. W. McKay and K. C. Smith, J. Appl. Phys. **31**, 133S (1960).

[6] M. K. Haynes, J. Appl. Phys. **29**, 472 (1958).

[7] N. Menyuk and J. B. Goodenough, J. Appl. Phys. **26**, 8 (1955).

JOURNAL OF APPLIED PHYSICS    SUPPLEMENT TO VOL. 33, NO. 3    MARCH, 1962

# Magnetostatic Interactions between Thin Magnetic Films*

Harrison W. Fuller and Donald L. Sullivan

*Computer Products Division, Laboratory for Electronics, Inc., Boston, Massachusetts*

This paper discusses some magnetostatic interactions between two thin films that are separated by a small distance between parallel film planes. The external field of a domain wall in one film is calculated using a simple wall model, and the effect of this field on a second adjacent film is determined. The conditions are investigated under which the wall field of the first film together with an external field can cause nucleation of a domain in the second film. The interaction energy between a movable wall in one film and a stationary wall or a linear imperfection in a second adjacent film is calculated using simple models. The equilibrium position of the movable wall as a function of an externally applied field is then determined. Means for electrically modifying the degree of interaction are investigated. Experimental results are presented illustrating the aforementioned interactions. The nucleation of long narrow domains in one film is achieved by the use of a domain wall, a scratch imperfection, or a film edge in a second adjacent film. The influence of a stationary wall in one film on the controlled wall motion of a wall in a second film is demonstrated. A comparison of theory and experiment is made. Mention is made of how the subject interactions might be utilized in new information storage devices.

MAGNETOSTATIC interactions between two or more thin magnetic films in a layered structure have been suggested[1-3] as means for usefully modifying the switching characteristics of thin film memory elements. The present paper considers locallized magnetostatic interactions between two films in a layered structure, such as the effect of the localized fringing field of a domain wall in one film on the neighboring magnetization distribution of the adjacent film. It appears that use could be made of such locallized interactions for making new thin magnetic film memory devices.[4]

An edge view of a layered film structure is shown in Fig. 1 where the lower and upper films have saturation magnetizations $M_s$ and $M_s'$, and thicknesses $\tau$ and $\tau'$, respectively, and where the films are separated by a distance $s$. The easy axes of the films are taken perpendicular to the end view. The films are assumed to be sufficiently thin so that the walls are of a simple Néel type. A linear dipole is used as a wall model consisting of two lines of opposite charge, $\pm\rho=\pm M_s\tau$, at a wall-width distance $2a$.

The magnetostatic interaction energy per unit wall length of two straight parallel 180° walls separated by a distance $x$ as shown in Fig. 1 is

$$E_i = M_s M_s' \tau\tau' \ln\frac{[s^2+(x+a-b)^2][s^2+(x-a+b)^2]}{[s^2+(x+a+b)^2][s^2+(x-a-b)^2]}, \quad (1)$$

where $s \gg a,b$ favors the linear dipole approximation,

and where the effect on the shape of one wall from the field of the other is neglected. If the lower wall in Fig. 1 is assumed to be stationary, and if a field $H$ is applied to the upper film, then the magnetostatic energy per unit wall length of the upper film associated with a translation of the upper wall is

$$E_m = -2\tau' M_s' H x. \quad (2)$$

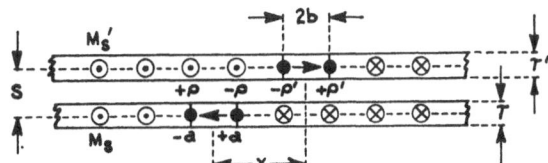

FIG. 1. Geometry for calculation of interaction between layered film structure.

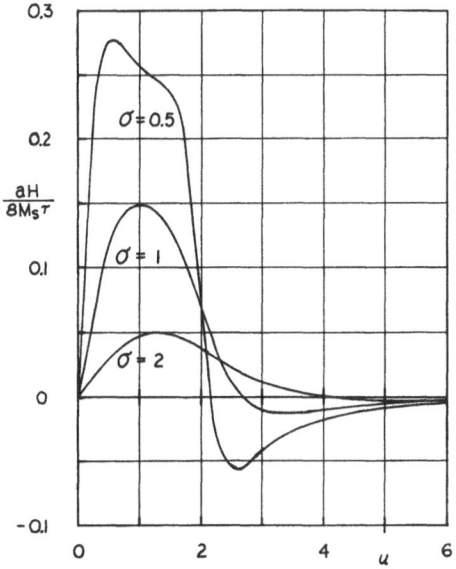

FIG. 2. Plots of equilibrium position of movable wall interacting with stationary wall.

* This work was supported in part by contracts with the Electronics Research Directorate of the Air Force Cambridge Research Laboratory, and the Information Systems Branch of Office of Naval Research.

[1] L. J. Oakland and T. D. Rossing, J. Appl. Phys. **30**, 54S (1959).
[2] J. I. Raffel, T. S. Crowther, A. H. Anderson, and T. O. Herndon, Proc. Inst. Radio Engrs. **49**, 155 (1961).
[3] K. D. Broadbent, Proc. Inst. Radio Engrs. **48**, 1728 (1960).
[4] H. W. Fuller and H. Rubinstein, "Methods of utilizing thin magnetic film properties for large capacity files," Symposium on Large Capacity Memory Techniques for Computing Systems, sponsored by Information Systems Branch, ONR, May, 1961 (to be published).

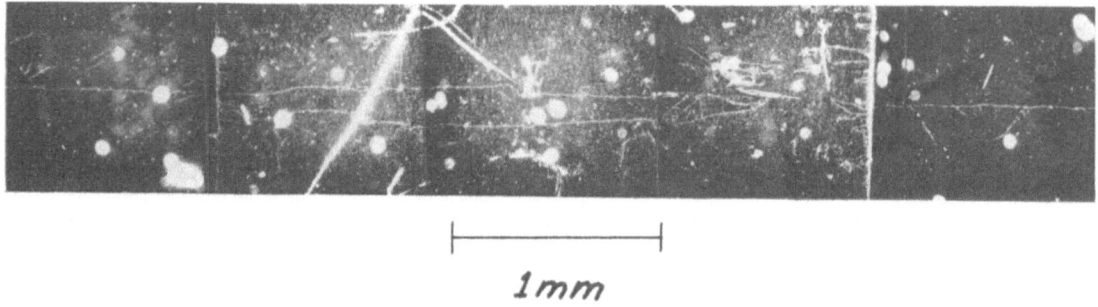

*1mm*

Fig. 3. Bitter pattern showing narrow domain nucleated in low–$H_c$ film by field of Néel wall in high-$H_c$ film plus applied field.

The equilibrium position for the movable upper wall is found from $\partial(E_i+E_m)/\partial x=0$, giving an equilibrium force function

$$-\frac{aH}{8M_s\tau}=\frac{u[u^2-(3\sigma^2+4)]}{[\sigma^2+(u+2)^2][\sigma^2+(u-2)^2][\sigma^2+u^2]}, \quad (3)$$

with $u=x/a$, $\sigma=s/a$, where wall coercive force is neglected and where $a=b$ is assumed for simplicity. The force function, Eq. (3), is plotted, for positive values of $u$, in Fig. 2 for $\sigma=0.5$, 1, and 2. The function is symmetrical about the origin. It is seen from Fig. 2 that as a wall is moved from left to right under the influence of an increasing $H$ field, the wall moves gradually from $u=-\infty$ to the position $u=-5$ ($\sigma=2$), at which time the wall jumps abruptly to $u\sim0$. The wall then moves slowly to $u=1.5$ as $aH/8M_s\tau$ approaches 0.05, whereupon the wall escapes the interaction field and travels indefinitely to the right. Taking sample values $M_s=10^3$, $\tau=3\times10^{-6}$, $a=b=5\times10^{-6}$, and $s=10^{-5}$, the escape field, from Eq. (3), is $H\sim20$ oe.

Wall-wall interaction of the above kind was measured experimentally. The multi-layer film, vacuum deposited on a 1-$\times$1-in. glass substrate, consisted of a $\frac{1}{8}$-$\times$1-in. 81/19 NiFe film of 400 A thickness, and a 1-$\times$1-in. 50/50 NiCo film of 470 A thickness, the two separated by a 900 A film of MgF$_2$. The lower film had $H_c=1.3$ oe and the upper film had $H_c=15$ oe. A single wall was established in the high coercive force film as the "stationary" wall, and a wall was moved in a controlled manner along the 1-in. length of the low coercive force film with the use of a tapered sheet conductor.[5] The moving wall position was observed magneto-optically. An escape field of 0.3 oe (beyond the wall coercive force of 0.83 oe) was measured. The walls used in this interaction experiment were found by Bitter-pattern observations to be

cross-tie walls. Interaction calculations were not made for the case of cross-tie walls, but it is reasonable to expect that the interaction would be much smaller than for Néel walls.

The strength of the Néel wall interaction calculated above can be electrically modified by applying a field along the hard axis of the film containing the "stationary" wall. The magnetization of the film rotates toward the direction of the applied field and increases or decreases the interaction depending on whether the magnetization direction at the center of the "stationary" wall is antiparallel or parallel to the hard-axis field.[6]

Another kind of interaction exists between a wall in one film and the (initially saturated) magnetization of the adjacent film. The fringing field of the wall furnishes a locallized transverse field in the adjacent film in the neighborhood of the wall. The presence of the wall field causes the domain-nucleating field to be locally lowered in the adjacent film, and a long narrow domain can be formed with an externally applied field. Figure 3 shows a Bitter-pattern view of a long narrow domain nucleated in an uppermost one-eighth inch wide low-coercive-force film by an external field with the aid of the field from a stationary Néel wall in the underlying high-coercive-force film. The multi-layer structure is similar in composition and construction to that previously described. The edges of the $\frac{1}{8}$-in. wide strip of low-$H_c$ material show up as bright vertical lines at the left and right of Fig. 3. The "stationary" Néel wall in the high-$H_c$ film is visible as a bright horizontal line on either side of the $\frac{1}{8}$-in. wide strip. The nucleated domain runs horizontally across the $\frac{1}{8}$-in. wide strip. Similar domain nucleation is obtained from film edges (with magnetization perpendicular to the edge), linear scratch imperfections, and jagged irregular domain walls. In these cases the interaction is of longer range, and the two films can be on separate substrates.

[5] H. Rubinstein, T. McCormack, and H. Fuller, 1961 International Solid-State Circuits Conference Proceedings, February, 1961.

[6] H. Rubinstein, H. W. Fuller, and M. E. Hale, J. Appl. Phys. **31**, 437 (1960).

JOURNAL OF APPLIED PHYSICS     SUPPLEMENT TO VOL. 33, NO. 3     MARCH, 1962

# Permalloy-Sheet Transfluxor-Array Memory

G. R. Briggs and J. W. Tuska

*RCA Laboratories, Radio Corporation of America, Princeton, New Jersey*

Annealed molybdenum Permalloy sheets of $\frac{1}{8}$–1 mil thickness have been photoetched to form transfluxor-element memory arrays utilizing inhibited-flux storage. Operating characteristics are summarized for single elements switched in 21 nsec-3.2 $\mu$sec. Some findings are: *one* to *zero* information flux ratios of 5:1 to 16:1, *one* to *disturb noise* flux ratios greater than 30:1, and $S_w$ values of 1.7 to 0.14 oe $\mu$sec. In addition, operation is possible over a temperature range of $-70°$ to $+150°$C without drive-current compensation. Also, for a $\frac{1}{8}$-mil thick element switched in times less than 0.1 $\mu$sec, $S_w$ values less than 0.2 oe $\mu$sec are obtained, indicating that fast switching may occur partially by uniform flux rotation. Array photoetching is described. Arrays containing 128 and 1152 elements have been fabricated. The latter permit a storage density of 250 000 bits/cu in. to be attained. The arrays receive a stress-relief treatment which results in a signal uniformity of $\pm3\%$. Photofabrication of windings has been accomplished also. This process generates continuous windings linking the transfluxor apertures.

THE development of the inhibited-flux mode of memory storage[1,2] in the ferrite transfluxor,[3] has made it possible to eliminate the need for a coercive-force threshold in the magnetic material. As a result, there remains a need only for a degree of retentivity sufficient to distinguish between the two storage states. In this paper are presented data which show that the inhibited-flux mode can be applied to small 3-apertured transfluxor elements photoetched in rolled, annealed, 4-79 Mo Permalloy sheets. For these elements it has been found possible to obtain useful memory storage properties over a wide range of memory cycle times. As will be described, the use of magnetic metal makes possible the construction of memory arrays, complete with winding conductors, using photofabrication techniques exclusively.

Permalloy-sheet transfluxors have been evaluated in the word-organized and XY-coincidence modes of memory selection. The word-organized system evaluated is similar to the design considered by Baldwin and Rogers. The XY-coincidence system evaluated [Fig. 1(a)] is a modification of the dc-biased design considered by Lawrence. Separate read driving is used, utilizing windings linking the leftmost apertures. This location insures minimum coupling between the read driving and sense windings and minimizes half-read noise. For Permalloy elements, it has been found desirable to elongate the sensed apertures to obtain improved discrimination against sensed-leg flux change caused by disturbs. An aperture length of 1.5–2 times the aperture width has been found satisfactory.

Operating data obtained for single (1-bit) elements photoetched in Permalloy of various thicknesses are summarized in Table I. A useful *one* to *zero* information signal ratio is observed, even for switching in 21 nsec. Repeated inhibit and half-select disturbs have been found not to reduce this ratio. Disturb signals are

sufficiently small that digit-plane capacities greater than 1000 bits are practical. A value of $S_w$, for $\frac{1}{8}$-mil Permalloy, less than the minimum of about 0.2 oe $\mu$sec expected for nonuniform flux rotation switching[4] is obtained. This indicates that switching may occur partially by uniform flux rotation.[5]

Temperature-range data have been obtained for a 1-mil thick element operated in the word-organized mode. Using fixed drives, a temperature run from $-67°$ to $+200°$C changed the switching time from 0.7 to 0.6 $\mu$sec, reduced the *one* flux 20%, and produced a minimum in *zero* flux at 100°C. Most of the flux decrease occurred between 150–200°C.

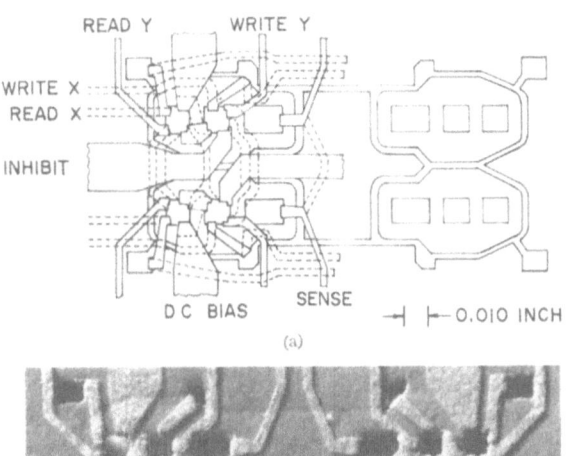

FIG. 1. (a) Diagram of a 4-transfluxor group of a Permalloy sheet transfluxor memory array. (b) Photograph of a section of an array with photofabricated windings.

[1] J. A. Baldwin, Jr., and J. L. Rogers, J. Appl. Phys. **30**, 58S (1959).

[2] W. W. Lawrence, Jr., *Proceedings of the Eastern Joint Computer Conference* (New York, December, 1956), p. 101.

[3] J. A. Rajchman and A. W. Lo, Proc. Inst. Radio Engrs. **44**, 321 (1956).

[4] E. M. Gyorgy, J. Appl. Phys. **31**, 110S (1960).

[5] W. Lee Shevel, Jr., J. Appl. Phys. **30**, 47S (1959).

TABLE I. Operation of Permalloy sheet transfluxor elements having smaller apertures of 0.010×0.010 in., sensed apertures of 0.010 ×0.020 in., and leg widths of 0.006 in. Data for which a dc bias is listed are for the XY coincident mode. I, R, and W denote inhibit, read, and write, respectively.

| Permalloy thickness (mil) | Drives (amp)[b] | | | Bias (amp) | Switching flux (mv μsec) | Switching time[c] (μsec) | | Flux ratios | | | $S_w$[d] (oe μsec) |
|---|---|---|---|---|---|---|---|---|---|---|---|
| | $I_I$ | $I_R$ | $I_W$ | | | Read | Write | 1/0 | 1/$I$ | 1/½$R$ | |
| 1 | 0.14 | 0.33 | 0.30 | 0.17 | 4.4 | 3.2 | 2.8 | 12 | 111 | 64 | 1.7 |
| 0.5 | 0.7 | 0.96 | 0.8 | ... | 2.4 | 0.10 | 0.19 | 11 | ... | ... | 0.52 |
| 0.5 | 0.67 | 1.2 | 1.0 | 0.55 | 2.2 | 0.4 | 0.4 | 16 | 180 | 81 | 0.9 |
| 0.5[a] | 0.12 | 0.28 | 0.28 | 0.15 | 2.2 | 2.5 | 2.1 | 9 | 60 | 30 | 1.0 |
| 0.25 | 0.9 | 0.80 | 1.1 | ... | 1.2 | 0.05 | 0.09 | 8 | ... | ... | 0.24 |
| 0.125 | 0.7 | 1.50 | 0.8 | ... | 0.45 | 0.021 | 0.09 | 4.5 | ... | ... | 0.19 |
| 0.125 | 0.7 | 0.92 | 0.8 | ... | 0.45 | 0.027 | 0.09 | 4.7 | ... | ... | 0.14 |

[a] Sensed aperture 0.010×0.015 in.
[b] Single-turn windings. Pulse rise time less than 10% of switching time, except minimum of 7 nsec for switching time less than 0.1 μsec. Currents are full-select values.
[c] For reading, time for 90% switching-flux change. For writing, duration of writing-current pulse.
[d] Computed for reading operation. Reading path mean length assumed.

Permalloy-transfluxor arrays have been fabricated by photoetching. In this process, annealed magnetic metal[6] is first coated on both sides with photoresist plastic.[7] The coated sheet is then exposed to uv illumination through registered photopatterns. In this step, aperture patterns are exposed on both sides of the sheet and transfluxor-delineating channel patterns are exposed on one side. The photoresist is next developed and the sheet is electroetched. The channels etch from one side of the sheet leaving the photoresist on the opposite side as a bridging medium to hold the elements in place. The apertures etch faster than the channels. Completion of the channel surrounding an element causes the electroetching circuit to that element to be broken. This process acts as an automatic etching control and produces high element dimensional accuracy; a precision of ±4 μ in element leg width has been obtained consistently. Experimental arrays fabricated include 128-element sheets for XY-mode use, having the element dimensions shown in Fig. 1, and 1152-element sheets for word-organized use. The elements of the latter are one half the size of the former.

Storage parameters are sensitive to mechanical stress of the elements. Coatings in general produce compressive surface stress, and this has been observed to cause an increase in $S_w$ and reductions in the one to zero and noise ratios. A process for relieving the moderate surface stress produced by the photoresist has been found. In this process, arrays, after the electroetching step, are coated on both sides with silicone rubber to a thickness of 0.4–0.7 mil. The arrays are then baked at a temperature of 325°C for several minutes. This treatment partially decomposes the photoresist and produces permanent relief of the surface stress. The silicone rubber is unaltered by the baking and assumes the role of holding the elements in position. The rubber itself produces negligible stress. Sheets stress relieved as described exhibit a one-signal uniformity of ±3%, for all elements of typical arrays fabricated of 0.5-mil Permalloy.

In addition to the photoetching of arrays, photofabrication of windings linking the apertures of the transfluxor elements has been accomplished [Fig. 1(b)]. In this process, a thin copper layer is first vacuum evaporated onto the silicone rubber and then increased in thickness to 0.5–1.0 mil by electroplating. At this stage, the sheet is encased in copper. The sheet is next coated with photoresist and exposed to uv illumination through registered winding photopatterns placed against both surfaces. The winding patterns overlap the apertures, shadowing the photoresist-coated copper through-plating. Finally, the photoresist is developed and the copper is $FeCl_3$ etched to form the windings.

Arrays with windings photofabricated as described are less than 0.004-in. thick. Tests in the 1-μsec switching time region have shown that spacing the arrays 0.001 in. apart does not produce serious interaction. For the 1152-element sheets mentioned, this spacing would yield a bit density of 250 000/cu in.

### ACKNOWLEDGMENTS

The authors are indebted to R. C. Ballard, C. W. Henderson, and B. Denton for aiding this research. They wish also to record the unstinting contributions made to the project by the late J. Truska.

[6] Magnetics, Inc., Hymu 80 Lamination No. 65307.
[7] Eastman Kodak Company, KPR.

JOURNAL OF APPLIED PHYSICS     SUPPLEMENT TO VOL. 33, NO. 3     MARCH, 1962

# Demonstration of Magnetic Domain-Wall Storage and Logic

J. M. BALLANTYNE

*Lincoln Laboratory,\* Massachusetts Institute of Technology, Lexington 73, Massachusetts*

The operations of read-in, shifting, conditional erase, and fan-out in a system of domain-wall storage and logic proposed by Smith have been experimentally demonstrated. The several operations were performed by controlling the current flowing through a configuration of wires placed closely under a 5 to 40 mil wide, 50 to 300 A thick Permalloy strip, and the sequence of events was observed on the top side of the film using the Bitter technique. The basic wiring geometry was a narrow-spaced grid of series-connected parallel wires that made an angle of 50° to 75° with the strip of film, the smaller angles being necessary for the thinner films. Conditional erase was most successful in films below 100 A thick, as in this thickness range walls of the same sense form double walls which require relatively large fields for erasure.

## INTRODUCTION

AN experimental investigation of the system of domain-wall storage and logic proposed by Smith[1] is reported. In this system binary information is stored directly in the domain walls of a narrow Permalloy strip by controlling the sense of rotation of the magnetization vector within the wall. Information is carried from one point to another by shifting the walls along the strip of film, and logical operations may be performed by using the operation of conditional erase. Information in a single strip may be fanned-out into two strips at a junction, and likewise walls in separate strips may be juxtaposed into a single strip. The various logical and storage operations were performed by controlling the current flowing through an appropriate configuration of wires placed closely under a thin strip of evaporated Permalloy film, and the quasi-static sequence of events was observed on the top side of the film using the Bitter technique. The films used were evaporated in strips 5 to 40-mils wide from a melt of 80% Ni and 20% Fe. They range from 50 to 300 A thick and were deposited in an orienting field perpendicular to the length of the strips on 0.0035-in. thick glass cover slips.

## RESULTS AND DISCUSSION

### Read-In

Néel walls of known sense were read into the films with sets of wires like those shown in Fig. 1, the smaller wiring angles being necessary for the thinner films. The convention of Smith[1] is used for denoting the sense of the walls, that is, a *P* wall is one in which the magnetization vector rotates in a clockwise direction as one proceeds through the wall from left to right, and an *N*

wall is one in which the magnetization rotates in the counterclockwise direction. As shown in Fig. 1, the wiring geometry alone determines the sequence of the read-in walls. Much better results were obtained with the triple-pass read-in configuration as shown than with a single-pass wire, since in the triple-pass arrangement the walls are most favorably located on the lines of zero applied field between the wires, while no such constraint is placed on the position of the read-in walls by the single-pass arrangement.

Walls can be, and were, read in singly (at the end of a strip), in pairs or as a series. Figure 2(a) shows walls read into a 300-A thick film using the triple-pass geometry and a current of about 1 amp. It is interesting to note that even though the wires made an angle of about

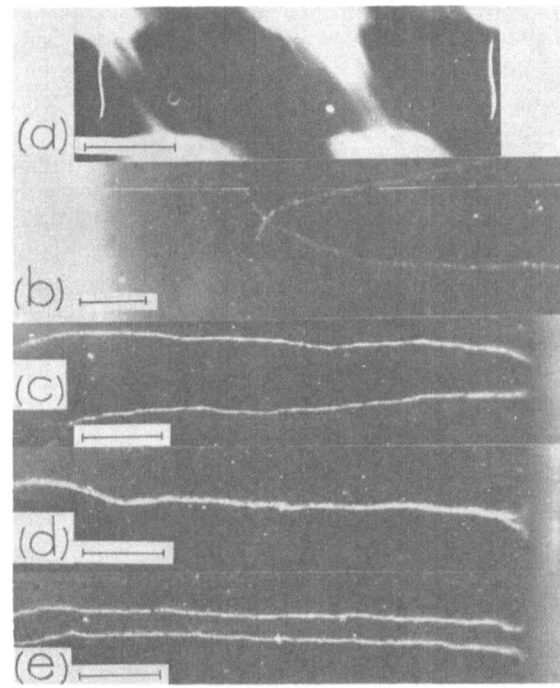

FIG. 2. Observed domain-wall patterns. (a) Two vertical walls read into a horizontal strip. The walls are near the ends of the figure, the broad sloping bands in the center being read-in wires. (b) Walls of like sense "unwinding" in a 200-A thick film. (c), (d), (e) Walls of like sense before, during, and after contact in a 50-A thick film. The scale markers represent $300 \mu$ in (a) and $30 \mu$ in (b), (c), (d), and (e).

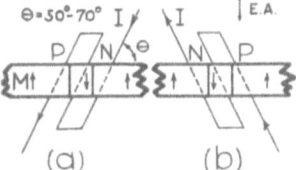

FIG. 1. Wiring configurations for read-in.

\* Operated with support from the U. S. Army, Navy, and Air Force.

[1] D. O. Smith, IRE Transactions on Electronic Computers (published, December 1961).

60° with the film, the resulting walls were practically perpendicular to the strip. This remains true until the angle between the wires and the film becomes about 50° or less, at which point the walls take on some of the slope of the wires.

### Shifting

Series of walls were shifted in either direction using the slanted double winding proposed by Smith.[1] There appears to be little difference in magnitude between the field required to read in a domain and the field required to shift a series of walls along the strip, thus it is necessary that no "spaces" be left in the train of information, as an extraneous wall would be read into any such space during the shifting operation. In order to observe all phases of domain-wall motion the current in the shifting wires was increased very slowly, and under these conditions the walls appeared to drift uniformly during shifting, as though being pushed through a viscous medium. This observation of the smoothness of wall motion, plus evidence from the conditional erase operation, confirmed the belief that the sense of a wall could be preserved on shifting.

### Conditional Erase

Conditional erase is the name given the operation of bringing two adjacent walls in contact, resulting in either a double wall, or erasure of both walls, depending respectively on whether the walls are of the same or of opposite sense. It was found that the success of the operation of conditional erase depends strongly on the uniformity of the walls and the thickness of the films. Walls must be straight and free of cross ties, breaks, or perturbation trails,[2] since if any of the preceding are present when two walls are touched, the walls will break at the imperfection and erase each other regardless of sense. In films thicker than 100 A practically no differ-

ence was found between the field required to erase a pair of walls of the same sense and the field required to erase a pair of walls of opposite sense. This was due to the fact that in films of this thickness walls of the same sense will "unwind" each other from the ends, as illustrated in Fig. 2(b).

In films from 50 to 100 A thick, however, the situation was quite different. In this thickness range the field required to erase a pair of walls of the same sense was 19 to 40 oe, 9 to 20 times as large as the 2-oe field necessary to shift walls and erase pairs of walls of the opposite sense. This margin is quite satisfactory, and arises because walls of the same sense do not "unwind" from the ends in these thinner films. Figures 2(c), (d), and (e) show the nonerasure of a pair of walls of the same sense on being touched. It was found that the closest spacing at which walls of opposite sense were stable without erasure was about 0.0005 in.

### Fan-Out and Juxtaposition

The same wire configuration used for shifting was employed without difficulty to fan out information in a $Y$ strip. The operation of juxtaposing data from the double legs of a $Y$ strip into the single leg is a great deal more difficult than the fan-out operation, since in juxtaposition each leg must be controlled separately. Although separate control of walls in one or the other input leg of the $Y$ strip has been achieved, and the existence of a remanent wall in the parallel input leg verified, the complete juxtaposition operation described by Smith[1] demands a more complicated wiring and switching arrangement than we have so far constructed successfully on the small scale required.

### ACKNOWLEDGMENT

The author wishes to express appreciation and thanks to Dr. D. O. Smith for suggesting the problem and for many enlightening discussions as the work progressed.

[2] D. O. Smith and K. J. Harte, J. Appl. Phys. (to be published).

JOURNAL OF APPLIED PHYSICS    SUPPLEMENT TO VOL. 33, NO. 3    MARCH, 1962

# Flux Reversal in Ferrite Cores under the Effect of a Transverse Field*

KAM LI

*RCA Laboratories, Princeton, New Jersey*

It is known that flux reversal in ferrites involves three switching mechanisms, namely, domain wall motion, nonuniform rotation, and uniform rotation. The experimental results presented in this paper demonstrate that a steady transverse field in conjunction with set and reset pulses is capable of separating these mechanisms into three regions of flux dispersion. Each region is designated by flux components $\phi_1$, $\phi_2$, and $\phi_3$ with corresponding characteristic transverse fields $H_1$, $H_2$, and $H_3$. The output waveform of the longitudinal sense signal during flux reversal is characterized by its first peak, second peak, and the long tail. The 50% value of the second peak occurs at the same characteristic transverse field $H_2$. The transverse field has no appreciable effect on their distribution. The experimental results agree with theoretical predictions in which less flux is switched by domain wall motion at higher drive currents. Furthermore, a proper transverse field has the effect of speeding up the switching time of a ferrite core without sacrificing the amplitude of the output signal due to the fast reversal mechanisms.

## INTRODUCTION

SQUARE loop ferrites have been widely used for the manipulation of digital information.[1] The characteristics and mechanisms of flux reversal in polycrystalline ferrites have been investigated intensively. Domain wall motion following nucleation and growth of the 180° Bloch walls, together with nonuniform and uniform rotation have been proposed by Menyuk and Goodenough,[2] Gyorgy[3,4] and Shevel[5] as the mechanisms of flux reversal. These mechanisms compete with one another, with domain wall motion being predominant at slow speed and the other two at intermediate and high speeds. It is the intent of this paper to show experimentally that the mechanisms for flux reversal may be separated by the application of transverse fields. In addition the switching time is reduced by the applied transverse field.

## EXPERIMENTAL RESULTS

The characteristic of the flux reversal under various transverse fields has been investigated. In these experiments, mercury relay pulses with a rise time of 0.4 nsec are used for drive and reset pulses. A sampling oscilloscope with an effective response time of 0.7 nsec is used for sensing. Several ferrite cores with the composition of 30% MnO, 30% MgO, and 40% $Fe_2O_3$ are used for the tests. A transverse field is provided by a pair of permanent magnets. The intensity of the transverse field may be varied from zero to several thousand oersteds by changing the separation of the poles.

The output waveforms are shown in Fig. 1 at constant pulse amplitude of drive and reset with the intensity of the transverse field as a parameter. The first peak, second peak, and the long tail of the output waveform have significant meaning in the later discussions. Integration of the output waveform gives the flux $\phi$. The

flux switched, expressed as a percentage of the flux at zero transverse field, is shown vs the transverse field in Fig. 2. Three regions of flux dispersion are apparent. Each region is designated by flux components $\phi_1$, $\phi_2$, and $\phi_3$ with corresponding characteristic transverse fields $H_1$, $H_2$, and $H_3$. The amplitude of the driving pulse influences the reversal mechanisms. The threshold field for the tested ferrite core is 1.6 oe. A drive field in the range from 9 to 12.3 oe is used for studying the reversal behavior under various transverse fields. The flux

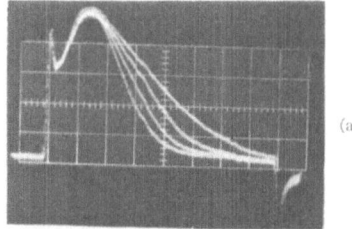

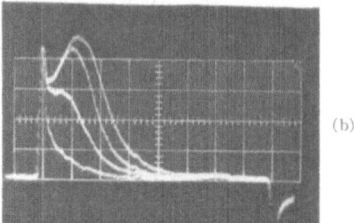

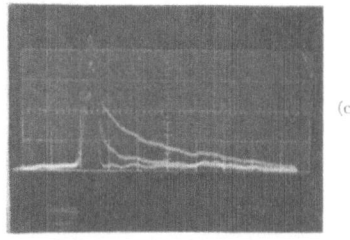

FIG. 1. Output waveforms at various transverse fields. (a) $H_t=0$, 300, 390, 450 oe; (b) $H_t=540$, 675, 850, 1000 oe; (c) $H_t=1000$, 1300, 2000 oe. Vertical scale 3 v/division. Horizontal scale 5 nsec/division for (a) and (b), and 2 nsec/division for (c). Core size o.d.=0.080 in., i.d.=0.040 in., $d=0.025$ in., drive pulse=9 oe, reset pulse=12.3 oe.

* This work was supported by the Bureau of Ships, U. S. Navy.

[1] J. A. Rajchman, RCA Rev. **20**, 92 (1959).
[2] N. Menyuk and J. B. Goodenough, J. Appl. Phys. **26**, 8 (1955).
[3] E. M. Gyorgy, J. Appl. Phys. **28**, 1011 (1957).
[4] E. M. Gyorgy and F. B. Hagedorn, J. Appl. Phys. **30**, 1368 (1959).
[5] W. Lee Shevel, Jr., J. Appl. Phys. **30**, 47S (1959).

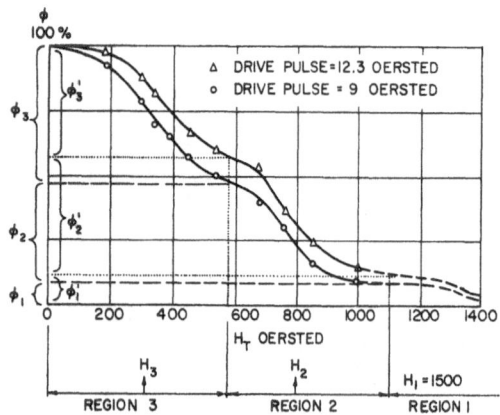

FIG. 2. Flux of output vs transverse field with two different amplitudes of drive as parameter. Three regions of dispersion characterized by $\phi_1$, $\phi_2$, $\phi_3$, $H_1$, $H_2$, and $H_3$. Reversible flux accounts for 6% of the total flux reversal.

component $\phi_3$ decreases to $\phi_3'$ as the drive amplitude increases as shown in Fig. 2. The characteristic transverse fields $H_1$, $H_2$, and $H_3$ at different drive amplitudes remains unchanged within experimental accuracy.

## DISCUSSION

The experimental results show three regions of flux dispersion at various transverse field intensities. Within the framework of the present theories, the interpretation of these three components $\phi_3$, $\phi_2$, and $\phi_1$ (the measured values include the reversible component) are due to the mechanisms of domain wall motion, nonuniform rotation and uniform rotation, respectively. In region 3, the second peak is slightly affected, while the first peak remains unchanged. A significant amount of the flux component $\phi_3$ in Fig. 2 is eliminated in this region by the transverse field. It is the slowest component of flux reversal and corresponds to the long tail of the output waveform at zero transverse field [Fig. 1(a)]. Further support of this interpretation is from the

redistribution of the flux components $\phi_1$, $\phi_2$, and $\phi_3$ at different amplitudes of drive (Fig. 2). The component $\phi_3$ drops from 55% to 40% while the drive magnetomotive force increases from 9 to 12.3 oe. This behavior fits the theories which predict less flux reversal by domain wall at higher drive.

In region 2, the second peak decreases while the first peak remains constant as the transverse field is increased. This medium speed component corresponding to nonuniform rotation coincides with the second peak of the output waveform with the characteristic field intensity $H_2$. The amount of the flux $\phi_2$ increases as the drive increases. The fastest mode of flux reversal corresponding to uniform rotation takes place in region 1. The first peak of the output waveform characterizes this component and the reversible flux. For a drive field of five to seven times the threshold field this mode accounts for about only a few percent of the total flux reversal. Its speed is comparable to the response time of the sensing equipment.

If the transverse field affects appreciably the distribution of flux reversal mechanism in favor of nonuniform rotation over domain wall motion, the second peak should be increased at low transverse fields. Experimentally, the second peak remains practically constant over regions 3. Therefore, the transverse field acts to eliminate the slow reversal mechanism component first, but does not change the distribution among the mechanisms appreciably. For the tested core, a transverse field of 500 oe reduces the switching time by a factor of 2 with 50% of the total flux reversal while the second peak decreases by only 10%.

## ACKNOWLEDGMENTS

The author wishes to express his gratitude to J. A. Rajchman and R. A. Shahbender for their encouragement, to R. C. Ricci for discussion, and to A. M. Monsen for laboratory assistance.

JOURNAL OF APPLIED PHYSICS    SUPPLEMENT TO VOL. 33, NO. 3    MARCH, 1962

# Approximate Solution of the Equations of Magnetization Reversal by Coherent Rotation*

R. F. ELFANT

*IBM Research Center, Yorktown Heights, New York*

AND

F. J. FRIEDLAENDER

*Purdue University, Lafayette, Indiana*

A complete solution of the modified Landau-Lifshitz equation of magnetization reversal by coherent rotation is presented for $\alpha^2 \gg H_a/4\pi M_s$, where $\alpha$ is the phenomenological damping constant, $H_a$ is the applied field, and $M_s$ is the saturation magnetization. This solution extends previous work in that it is valid for $t \geq 0$, whereas, other solutions have not been valid over this entire range. In addition, from this solution the time dependence of the demagnetizing field is found.

THE rotational theory of flux reversal is based on the Gilbert modification of the Landau-Lifshitz equation[1]

$$\frac{d\mathbf{M}}{dt} = -|\gamma|\mathbf{M}\times\mathbf{H} + \frac{\alpha}{M_s}\left(\mathbf{M}\times\frac{d\mathbf{M}}{dt}\right), \quad (1)$$

where $|\gamma|$ is the magnitude of the gyromagnetic ratio, $\alpha$ is the phenomenological damping constant, and $M_s$ is the saturation magnetization. If the toroidal shape of the ferrite core is approximated by an infinite cylinder where the $z$ axis is chosen to correspond to the axis of the cylinder, the magnetization vector $\mathbf{M}$ can be written as

$$\mathbf{M} = M_z\mathbf{e}_z + M_\theta\mathbf{e}_\theta + M_r\mathbf{e}_r, \quad (2)$$

where $\mathbf{e}_z$, $\mathbf{e}_\theta$, and $\mathbf{e}_r$ are the unit vectors in cylindrical coordinates.

If the crystalline anisotropy is neglected, then $\mathbf{H}$ is given by the vector sum of the applied, demagnetizing, and exchange fields,[2] i.e.,

$$\mathbf{H} = \mathbf{H}_a + \mathbf{H}_d + \mathbf{H}_{ex} = H_a\mathbf{e}_z - 4\pi M_r\mathbf{e}_r - S(M_r\mathbf{e}_r + M_\theta\mathbf{e}_\theta),$$

where $S = 2A/M_s^2 r^2$ and $A$ is the exchange constant. Making the following substitutions

$$\mathbf{m} = \frac{\mathbf{M}}{M_s}, \quad \mathbf{h}_{ex} = \frac{\mathbf{H}_{ex}}{M_s}, \quad \mathbf{h}_d = \frac{\mathbf{H}_D}{M_s}, \quad \mathbf{h}_a = \frac{\mathbf{H}_a}{M_s}$$

and $\tau = M_s|\gamma|t/(1+\alpha^2)$ into Eq. (1), one finds

$$\frac{dm_r}{d\tau} = m_\theta(Sm_z + h_a)$$
$$\qquad - \alpha m_r[4\pi(1 - m_r^2) + m_z(h_a + Sm_z)] \quad (3a)$$

$$\frac{dm_\theta}{d\tau} = -m_r[h_a + m_z(S + 4\pi)]$$
$$\qquad - \alpha m_\theta[m_z(h_a + Sm_z) - 4\pi m_r^2] \quad (3b)$$

* This investigation was supported in part by the National Science Foundation.
[1] T. L. Gilbert, Phys. Rev. **100**, 1243 (1955).
[2] E. M. Gyorgy and F. B. Hagedorn, J. Appl. Phys. **30**, 1368 (1959).

$$\frac{dm_z}{d\tau} = 4\pi m_\theta m_r$$
$$\qquad + \alpha[(1 - m_z^2)(h_a + Sm_z) + 4\pi m_z m_r^2]. \quad (3c)$$

Equations (3a), (3b), and (3c) are the same equations appearing in Gyorgy's and Hagedorn's paper[2] (correcting typographical errors). As Gyorgy and Hagedorn point out, it is reasonable to neglect all terms containing $S$. Hence making this simplifying assumption and eliminating $m_\theta$ from Eqs. (3a) and (3c) by employing the condition that $m_\theta = (1 - m_z^2 - m_r^2)^{\frac{1}{2}}$, one obtains

$$\frac{dm_r}{d\tau} = (1 - m_r^2 - m_z^2)^{\frac{1}{2}}(h_a)$$
$$\qquad - \alpha m_r[4\pi(1 - m_r^2) + h_a m_z] \quad (4a)$$

$$\frac{dm_z}{d\tau} = (1 - m_r^2 - m_z^2)^{\frac{1}{2}}(4\pi m_r)$$
$$\qquad + \alpha[(1 - m_z^2)h_a + 4\pi m_z m_r^2]. \quad (4b)$$

Since the demagnetizing energy is proportional to $m_r^2$, one would expect that $m_r^2 \ll 1$. Assuming that $\alpha^2 \gg h_a/4\pi$, neglecting all terms containing $m_r^2$, and making the transformations;

$$m_r = F(\tau)\,\text{sech}[\alpha h_a\tau + G(\tau)],$$
$$m_z = \tanh[\alpha h_a\tau + G(\tau)]$$

into Eqs. (4a), (4b), and using the initial conditions that $m_z = -|m_i|$ and $m_r = 0$ at $\tau = 0$, one obtains:

$$m_r = \frac{h_a}{4\pi\alpha}(1 - e^{-4\pi\alpha\tau})\,\text{sech}\left[h_a\tau\left(\alpha + \frac{1}{\alpha}\right) - \tanh^{-1}|m_i|\right.$$
$$\qquad\qquad \left. - \frac{h_a}{4\pi\alpha^2}(1 - e^{-4\pi\alpha\tau})\right] \quad (5)$$

$$m_z = \tanh\left[h_a\tau\left(\alpha + \frac{1}{\alpha}\right) - \tanh^{-1}|m_i|\right.$$
$$\qquad\qquad \left. - \frac{h_a}{4\pi\alpha^2}(1 - e^{-4\pi\alpha\tau})\right]. \quad (6)$$

These two equations completely describe the time de-

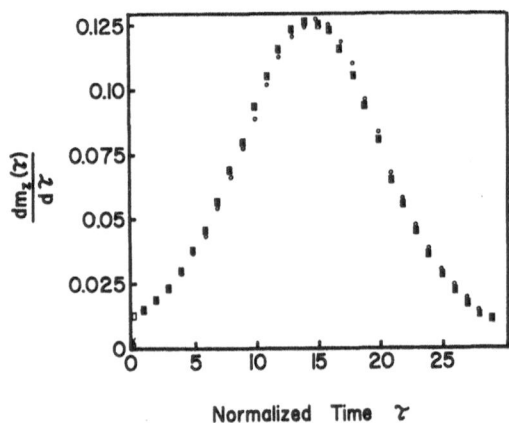

FIG. 1. $dm_z/d\tau$ as a function of $\tau$ from computer solution of Eqs. (4a) and (4b) (circles), from Eq. (6) (crosses) and from Eq. (8) (squares).

pendence of the vector components of the magnetization during the reversal process.

For $t$ such that $4\pi\alpha\tau > 3$ [i.e., $t > t_0 = 3(1+\alpha^2)/4\pi M_s\alpha|\gamma|$] Eqs. (5) and (6) reduce to

$$m_r = \frac{h_a}{4\pi\alpha} \operatorname{sech}\left[h_a\tau\left(\alpha+\frac{1}{\alpha}\right) - \tanh^{-1}|m_i| - \frac{h_a}{4\pi\alpha^2}\right] \quad (7)$$

$$m_z = \tanh\left[h_a\tau\left(\alpha+\frac{1}{\alpha}\right) - \tanh^{-1}|m_i| - \frac{h_a}{4\pi\alpha^2}\right]. \quad (8)$$

Equation (8) above is essentially that given by Conger and Essig[3] as their Eq. (15). Olson and Pohm[4] have indicated the general time dependence of the demagne-

---

[3] R. L. Conger and F. C. Essig, Phys. Rev. **104**, 915 (1956).
[4] C. D. Olson and A. V. Pohm, J. Appl. Phys. **29**, 274 (1958).

tizing field, employing the condition that $m_z$ is constant for $t < t_0$, a condition not required here. In principle, one can also obtain Eq. (7) above from the work of Olson and Pohm.[4] Smith[5] obtains results similar to those of Olson and Pohm.[4]

For the material considered (MgMn ferrite), it was found that $\alpha = 0.48$.[6] Figure 1 gives a comparison of the approximate solutions of $dm_z/d\tau$ to the digital computer solution of the basic equations [(4a), (4b)] for $H_a = 10$ oe, $M_s = 200$ gauss, and $\alpha = 0.48$.

From Eq. (8) it is readily found that

$$H_aT_s = \frac{2\alpha}{|\gamma|}\tanh^{-1}|m_i|, \quad (9)$$

where $T_s$ is the time to switch $m_z$ from $-|m_i|$ to $|m_i|$. From Eq. (8) the maximum output voltage is found to be

$$\left(\frac{d\Phi}{dt}\right)_{\max} = \frac{4\pi A_T}{\alpha}|\gamma|M_sH_a \text{ abvolts}, \quad (10)$$

where $A_T$ is the total cross-sectional area of the core. Equations (9) and (10) have been previously obtained[3,4] and are presented here only for completeness.

An approximate solution of the equations of magnetization reversal by coherent rotation has been found for $\alpha^2 \gg H_a/4\pi M_s$. This solution [Eqs. (5), (6)] is valid for $t \geq 0$ where as other investigators have obtained solutions for $t \geq 3(1+\alpha^2)/4\pi M_s\alpha|\gamma|$. The time dependence of the demagnetizing field [Eq. (5)] has been found.

---

[5] D. O. Smith, J. Appl. Phys. **29**, 264 (1958).
[6] R. F. Elfant and F. J. Friedlaender, International Conference on Magnetism and Crystallography, Kyoto, Japan, September, 1961.

# All-Magnetic Logic Elements Using Strained Permalloy Wire

H. R. IRONS

*U. S. Naval Ordnance Laboratory, White Oak, Silver Spring, Maryland*

An element for performing digital logic which employs Permalloy wire, solenoids, and resistors and utilizes a two-phase current pulse source has been studied experimentally. The re-entrant hysteresis loop exhibited by a torsionally strained Permalloy wire provides the mechanism for power gain. A balanced input circuit assures unidirectional information flow and provides means for obtaining logic functions. Each element consists of two 0.6-in. bifilar-wound solenoids for bias and reset pulse currents, one or several 0.025-in. solenoids to which logic inputs are applied, an output resistor and a Permalloy wire core. The application of a bias current pulse to the logic element will cause the magnetization to change from the *zero* to the *one* state provided a signal is present simultaneously at the input to nucleate the magnetization reversal. The reset current pulse switches the magnetization back again and produces an output across the bias coil which is used to control one or several following elements. A buffer zone, the magnetization of which is controlled by the reset pulse, separates adjacent logic elements to minimize element interaction. Measurements on elements employing a 1.5-mil, 5–79 Permalloy core show that a switching time of 15 $\mu$sec and a power gain of 10 can be achieved. Preliminary tests indicate satisfactory operation over the temperature range of $-55°$ to $+185°$C. These elements have been used to construct flip-flops, binary counters and a serial adder which employs majority logic. The circuits operated successfully when the pulse currents from the power supply were varied by $\pm20\%$. Because an open flux structure is used, it is believed that similar techniques will have application in circuits employing deposited magnetic films.

## I. INTRODUCTION

SEVERAL systems for performing logic with all-magnetic devices have been described in recent years. Except for the shift registers employing thin films[2] or the twistor[3], these devices have employed closed magnetic circuits of ferrite material. The logic element to be described here employs a Permalloy core in an open magnetic circuit and should be adaptable to fabrication methods using thin films. The Permalloy core material permits operation over a wider temperature range than is possible with ferrite cores.

## II. BASIC LOGIC ELEMENT

The logic element evaluated makes use of a phenomenon that has been described in detail by Sixtus and Tonks[4], namely that mechanically strained nickel-iron wires exhibit a re-entrant hysteresis loop. The magnetization reversal in such wires can be nucleated by a strong field applied to a small section of the wire and then propagated through a large region of the wire by a bias field having a value insufficient to cause the initial nucleation.

The logic element is shown schematically in Fig. 1. Assume the magnetization has been directed to the left by a reset current in the winding $N_R$. The application of the currents $I_B$ and $I_{\mathrm{IN}}$ will cause the nucleation of a reverse region of magnetization under the coil $N_S$ which will then be extended to the ends of the coil $N_B$ by the field $H_B$. If the input signal $I_{\mathrm{IN}}$ had not been present, the field produced by $I_B$ would not have been sufficient to nucleate the magnetization reversal and the magnetization would have remained directed to the left. Since the magnetization reversal is controlled by the small amount of energy needed to assist the field $H_B$ in nucleating the reverse magnetization, a power gain can be realized in the element. The coil $N_B$ is made sufficiently long to reduce demagnetization effects and provide reliable storage of the input information after the current $I_B$ is removed. An output signal for driving other logic elements is obtained from coil $N_B$ when the reset current is applied to the coil $N_R$. Since the magnitude of the reset current is not limited by threshold considerations, large output signals can be obtained. Close coupling to the Permalloy wire is obtained by winding $N_B$ and $N_R$ in bifilar fashion. This gives an output *one* to *zero* ratio of 40 on a flux basis. Unidirectional information flow is assured by using a balancing coil $N_s'$ connected in series opposition to coil $N_S$ so that no signal appears at the input terminals during the reseting operation. If the balancing coil is placed 0.025 in. or more from the coil $N_S$, it will not affect the ability of the input signal to nucleate reverse magnetization during the input phase. The winding $N_R$ is made longer than $N_B$ and provides a saturated buffer region which minimizes interactions between adjacent logic elements.

## III. LOGIC OPERATIONS

The logic elements are serially connected in two groups, I and II, and are driven by a two-phase source so that in phase 1 group I receives a bias pulse and logic input signals and group II receives a reset pulse to produce logic output signals. In phase 2, group II receives a bias pulse and the logic output signals from group I, etc. Each output winding $N_B$ is connected to the proper input windings by a resistor which transmits enough

[1] U. F. Gianola, J. Appl. Phys. **32**, 27S (1961).
[2] K. D. Broadbent and F. J. McClung, *Proceedings of the Solid State Circuits Conference*, Philadelphia, February, 1960.
[3] A. H. Bobeck and R. F. Fischer, J. Appl. Phys. **30**, 43S (1959).
[4] K. J. Sixtus and L. Tonks, Phys. Rev. **42**, 419 (1932).

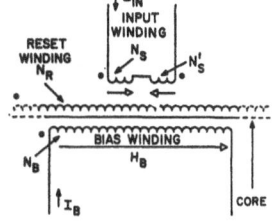

FIG. 1. Schematic diagram of Permalloy-wire logic element.

signal current to cause reliable nucleation. This resistor must not load the output winding to the extent that the current $I_B$ cannot reverse the magnetization in a reasonable time.

Logic functions are readily obtained by using multiple input windings either superposed or adjacent. A simple *exclusive or* circuit is formed by two superposed input windings of the balancing type so arranged that when both inputs are present their magnetic fields cancel, but when either input is present alone, switching occurs. The uniformity of the components used is sufficiently good that such cancellation is reliable.

Majority logic elements for which threshold-type operations are necessary have also proven successful. The threshold can be provided by the field value required to nucleate the switching process or by a bias current in an additional solenoid.

## IV. EXPERIMENTAL RESULTS

Elements have been fabricated using a core of 1.5-mil, 5-79 Permalloy wire which has been annealed at 350°C and then placed under a torsional strain of $\pi$ rad/in. The single layer solenoids $N_R$ and $N_B$ are bifilar wound with No. 43 wire and have an inside diameter of 3 mils. The coil $N_B$ and the buffer coil used between adjacent elements are each 0.6 in. in length. The input coil $N_S$ is a single layer solenoid 0.025 in. long. The resistance of the coupling loop is about 2 ohms.

The maximum permissible value of the field $H_B$ (the field at which switching occurs with no input signal applied) is approximately 6 oe. This field switches the core in about 7 $\mu$sec under normal loading conditions. When the field is reduced to 3 oe, the only change in performance is an increase in the switching time to 20 $\mu$sec. The switching due to the reset current occurs in 1.5 $\mu$sec and produces an output voltage of 150 mv. Approximately $10^{-7}$ w-sec of energy is required to switch a single element. Measurements of input and output characteristics indicate that one logic element can control as many as ten other elements and tolerate a 2 to 1 change in the bias and reset currents. In majority logic operations where the amplitude of the output signal must be closely controlled, a branching of three seems possible.

Preliminary temperature tests indicate that the magnetic properties of the core are sufficiently stable to permit operation over the temperature range of $-55°$ to $+185°$C. Possible changes in coupling loop resistance can reduce the fan-out possible under wide variations in temperature. A unit which has been tested for two weeks at 150°C has not shown any measurable change in its characteristics.

A 3-mil core was used in additional experiments. The results were similar to those obtained with the 1.5-mil core except for a two-fold increase in the switching times. Logic elements employing the 3-mil-diam core have been used to construct flip-flops, binary counters, and a full serial adder. The adder required only three logic elements (one majority logic element and two *exclusive or* elements) and two delaying elements. These circuits operated successfully when the pulse currents from the power supply were varied by $\pm 20\%$. A fan-out of 3 and a fan-in of 8 *or* circuits have been demonstrated experimentally.

A limiting factor in the performance of these devices is the nonuniform magnetic properties of the Permalloy wire. Local imperfections apparently act as nucleation points for magnetization reversal. When such imperfections are present, $H_B$ must be reduced to assure that nucleation is started by the input signal and not by the imperfections.

## V. CONCLUSIONS

The advantages of the logic element described are: large power gain, good logic capability, operation over wide temperature ranges, and the simplicity of the two phase power source required. The obvious disadvantages are the slow speed, the difficulties involved in making the small coils, and the susceptibility of the open magnetic circuit to ambient fields.

Perhaps the most important advantage of these circuits is that their essential features, namely, the open flux structure, a gain mechanism which does not depend on turns ratio, and a simple method of obtaining unilateral information flow, should permit the application of thin film techniques in their fabrication. Such a process should make them competitive in cost and speed with other all-magnetic devices.

### ACKNOWLEDGMENTS

The author would like to acknowledge the help of R. Alben, L. Salba, and E. Rowe in the construction and test of the experimental devices. The magnetic wire was prepared by H. Helms, Jr.

JOURNAL OF APPLIED PHYSICS    SUPPLEMENT TO VOL. 33, NO. 3    MARCH, 1962

# A Permanent Magnet Twistor Memory Element of Improved Characteristics

E. J. Alexander, J. C. McAlexander and L. H. Young

*Bell Telephone Laboratories, Inc., Whippany, New Jersey*

AND

R. J. Salhany

*University of Rhode Island, Kingston, Rhode Island*

A card changeable permanent magnet twistor memory element of improved characteristics has been designed. The desired characteristics were obtained by:

(1) Halving the solenoid width to enhance the drive field.

(2) Replacing the copper return wire in the twistor pair with a reverse wrapped twistor wire.

(3) Installing a Permalloy sheet between the word solenoid and the solenoid board to reduce interaction fields within the memory.

(4) Annealing the twistor wire to lower the coercive force and to increase the squareness ratio of the moly-Permalloy tape.

The chief result of these changes is that the net switching field at the bit is substantially increased and therefore the outputs of the memory are less dependent on the properties of the magnetic materials.

A 210 000 bit card changeable permanent magnet twistor memory utilizing the new element design has been constructed and evaluated.

In comparison to previously reported permanent magnet twistor memories, the new design is characterized by twice the speed, 2.5 times the bit density, 3 times the output voltage level, 2.5 times the signal available for "1–0" discrimination, and 4 times the strobe time tolerance.

## INTRODUCTION

A CARD changeable permanent magnet twistor memory element of improved characteristics has been designed. The structure of the basic twistor bit has been modified to obtain higher and more uniform outputs at a faster switching speed. The design changes include:

(1) Reducing the solenoid width by a factor of 2.

(2) Replacing the copper return wire in the twistor pair with a reverse wrapped twistor wire.

(3) Installing a Permalloy sheet between the word solenoid and the solenoid board.

(4) Annealing the twistor wire to lower the coercive force and increase the squareness ratio of the Permalloy tape.

The chief result of these changes is that for the same solenoid read current the net switching field at the bit is 2.5 times larger than in earlier designs. As a consequence, the outputs are less dependent on the characteristics of the magnetic materials employed in the memory.

## DESIGN FEATURES

Several memories incorporating these changes have been built. The organization of the new memory and the method of accessing are the same as for previously reported PMT memories.[1-3] The salient features of the new plane are shown in Fig. 1.

The word solenoid width has been cut in half. A given drive current applied to a solenoid of half the original width will produce an approximately doubled drive field at the twistor bit.

The copper return wire of the conventional twistor pair was replaced by a reverse-wrapped twistor wire.

For a given solenoid width, twice as much material is switched and the output is approximately doubled. A balanced transmission line is also obtained as both wires of the twistor pair are identically loaded. The resulting decrease in noise level together with the increased output improved signal discrimination.

A thin Permalloy sheet was sandwiched between the word solenoid and the solenoid board. The chief effect of this sheet is to provide a low reluctance path for the external fields of the magnetized bars on the program card which represent stored zeros. Hence the field is localized to the vicinity of the magnet and does not interact as much with adjacent bits. Reducing this interaction effect increases the signal-to-noise ratio significantly and makes memory operation less sensitive to errors in positioning the magnet with respect to the bit. In addition, the Permalloy sheet enhances the bit switching field within the solenoid.

The Permalloy sheet also acts as a distributed shield. Hence the effects of large external fields such as the earth's magnetic field and stray fields due to power apparatus are appreciably reduced.

The twistor wire was annealed to lower the coercive force and increase the squareness ratio of the moly-

[1] D. H. Looney, Proc. Western Joint Computer Conf. **1959**, 36.

[2] W. A. Barrett, F. B. Humphrey, J. A. Ruff, and H. L. Stadler, IRE Trans. on Electronics Computers **EC-10**, 451 (1961).

[3] U. F. Gianola, Proc. Large Capacity Memory Symposium, 1961.

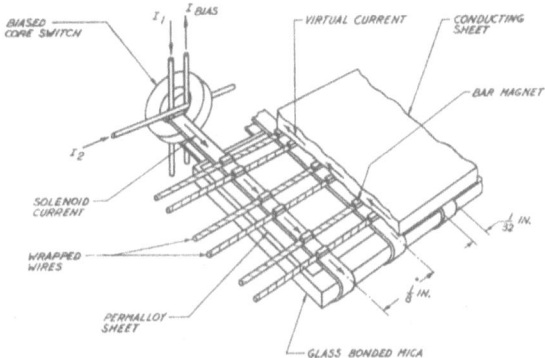

FIG. 1. Salient features of the new plane.

Permalloy tape. One effect of the annealing is to improve the aging characteristics of the magnetic material.[4] As a result of the reduced coercive force and increased squareness, the output of a switched wire ("one") is increased and the output of the shuttled wire ("zero") is decreased for a given drive field.

To ensure the uniformity of one's and zero's, twistor wire is annealed to a constant coercive force. As the wire is annealed, the coercive force is monitored and the annealing temperature is modified as required to obtain the desired coercive force.

Since the effective resistance and inductance of the word solenoid were increased by these modifications in the plane it was also necessary to increase the minimum flux requirement on the word access core by approximately 30%.

## MEMORY CONSTRUCTION

Modular memories of 1024, 2048, 4096, and 8192 words, each of 26 bits per word, have been designed. The basic building block is the word plane. Each plane consists of 64 word solenoids attached to a glass-bonded mica substrate. A continuous twistor cable containing 26 pairs of wire is bonded to the planes and the appropriate number of planes is stacked up to build a memory of the desired size. The active bit density is 352 bits per cu in. Isolation diodes and damping resistors for the horizontal and vertical drive windings are mounted on the memory module.

The 1024 word, 26 bit per word, memory is $6\frac{1}{4}$-in. high, $10\frac{3}{4}$-in. wide, 8-in. deep, and weighs 16 lb.

## TEST RESULTS

Modules of 1024 and 8192 words have been built and evaluated. For interrogation the memories require drive

--- 

[4] E. M. Tolman, J. Appl. Phys. **32**, 360S (1961).

currents of $1.25 \pm 7\%$ amp applied in coincidence to two-turn horizontal and vertical drive windings. The core bias current is $2.4 \pm 3\%$ amp applied to a one-turn winding. The drive pulses have a rise time of $0.25 \pm 20\%$ $\mu$sec and a pulse width of $1.25 \pm 5\%$ $\mu$sec. A random access read-cycle time of 2.5 $\mu$sec is achieved. The output is read during a 0.2-$\mu$sec-wide strobe interval.

The memories were tested by simulating extreme operating conditions by varying the net drive current $\pm 17\%$ with nominal pulse width and rise time. Every bit in each memory was examined as a one and a zero under the influence of positive and negative interaction fields. The external field of the magnetized bar representing a stored zero on the magnet card can either aid or oppose switching of the twistor bit depending on its position relative to the bit. Card patterns are readily devised which will subject a stored one to maximum positive or negative interaction fields due to stored zeros. A positive interaction field aids switching and opposes reset of the one. Hence the one pulse peaks earlier and its amplitude may be rapidly falling off during the strobe interval. The effect of a negative interaction field is the opposite of a positive field.

With worst case interactions fields and $\pm 17\%$ variation of drive current, the 1024 word, 26 bit per word memory has a minimum peak one of 6.5 mv and a maximum peak zero of 1.5 mv when the twistor pair is terminated for maximum power transfer. A signal-to-noise ratio of 2.5:1 is maintained for a strobe interval of 0.2 $\mu$sec with the extreme operating conditions described.

Under nominal drive conditions, but with worst case interactions, a signal-to-noise ratio of better than 6:1 for a 0.2-$\mu$sec strobe interval is obtained.

JOURNAL OF APPLIED PHYSICS    SUPPLEMENT TO VOL. 33, NO. 3    MARCH, 1962

# Internal Fields

P. J. WOJTOWICZ, *Chairman*

## Anisotropy of the Hyperfine Interaction in Magnetite

E. L. BOYD AND J. C. SLONCZEWSKI
*IBM Research Center, Yorktown Heights, New York*

This paper describes effects of anisotropy of the hyperfine interaction on nuclear magnetic resonance in a ferromagnetic or ferrimagnetic material. It is shown how the line shape in a polycrystalline sample depends on whether the excitation is through domains or domain walls. The absorption spectrum of $Fe^{57}$ in powdered magnetite ($Fe_3O_4$) is observed, and the results show that, in this case, the excitation is predominantly through domains. They also establish a value for the hyperfine anisotropy on octahedral sites.

THIS paper reports evidence of anisotropy of the hyperfine interaction in powdered ferrimagnetic magnetite ($Fe_3O_4$). Since the beginning of NMR experiments in ferromagnetic powders, effects of this anisotropy on the line shape have been predicted,[1] but until now unambiguous experimental confirmation has been lacking.[2] We describe here the line shapes expected under various conditions of nuclear excitation. By comparing these shapes with those observed, we establish that the Fe nuclei on octahedral sites are excited by domain rotation and determine a numerical value of the anisotropy.

The relevant feature of the cubic spinel structure of magnetite is that an iron nucleus on an octahedral site is exposed to a local crystal field of trigonal symmetry. On phenomenological grounds one expects the nuclear resonance frequency $\nu_r$ for such a site to obey the relation

$$\nu_r = \nu_\perp + (\nu_{||} - \nu_\perp)\cos^2\theta, \tag{1}$$

where $\theta$ is the angle between the spontaneous magnetization $\mathbf{M}$ and the local trigonal axis, and where $\nu_\perp$ and $\nu_{||}$ are constants.

A powder sample will have a distribution density $\rho(\nu_r)$ of values of $\nu_r$ because the values of $\theta$ are distributed. The absorption $A$ may be written in the form

$$A(\nu) = \int_{\nu_{||}}^{\nu_\perp} I(\nu_r - \nu)F[H_1(\nu_r)]\rho(\nu_r)d\nu_r. \tag{2}$$

Here $\nu$ is the frequency of applied power, and $I(\nu_r - \nu)$ is the absorption intensity characteristic of the resonant frequency $\nu_r$. The function $F[H_1(\nu_r)]$ takes into account the dependence of absorption on the intensity of the transverse component $H_1$ of the rf field. The largest part of $H_1$ is due to the alternating component of the hyperfine field excited by forced motion of $\mathbf{M}$.[1] In general $H_1$ depends on $\theta$ and through Eq. (1) on $\nu_r$. If

$I(\nu_r - \nu)$ is the imaginary part of the nuclear susceptibility and its width is small compared to $|\nu_{||} - \nu_\perp|$, then $A$ is proportional to the absorption weight $F[H_1(\nu)]\rho(\nu)$.

The shape of the line represented by $A(\nu)$ is sensitive to the mode of nuclear excitation. The following limiting cases may be considered:

(1) Discrete orientation. This occurs if the excitation is by domain rotation and the equilibrium orientation of $\mathbf{M}$ is parallel to an axis of minimum crystalline anisotropy energy. The absorption for [100] and [111] axes consists of sharp lines at the positions indicated by arrows in Fig. 1.

(2) Random orientation. This occurs if the crystalline anisotropy is overwhelmed by shape anisotropy or some other random influence, and the excitation is through domain rotation. In this case $F$ is constant, and $F\rho$ varies in proportion to $\rho$ which is proportional to $(\nu_\perp - \nu)^{-\frac{1}{2}}$ for $\nu_{||} < \nu < \nu_\perp$. This case is shown by curve a in Fig. 1. This line shape is well known in connection with the effect of quadrupole interactions on nonferromagnetic nuclear resonance.[3]

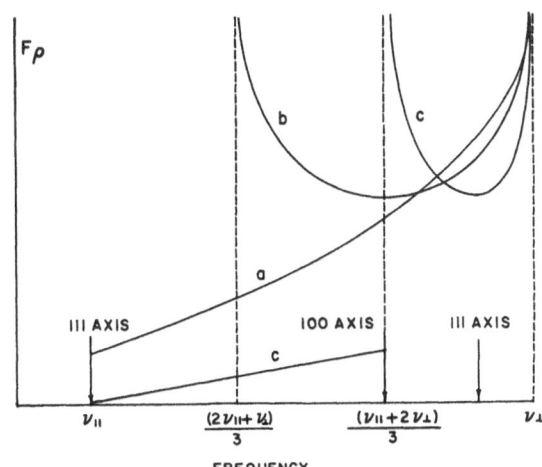

FIG. 1. Frequency dependence of the nuclear absorption weight $F\rho$ for several conditions of nuclear excitation.

[1] A. M. Portis and A. C. Gossard, J. Appl. Phys. **31**, 205S (1960); and A. C. Gossard, thesis, University of California, Berkeley, 1960 (unpublished).

[2] See, however, positive results for YIG obtained by F. Boutron and C. Robert, Compt. rend. **253**, 433 (1961).

[3] A. Abragam, *The Principles of Nuclear Magnetism* (Oxford University Press, Oxford, 1961).

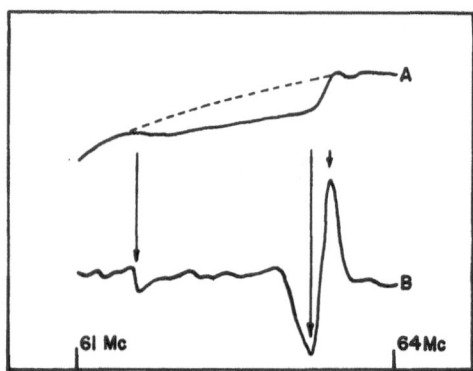

FIG. 2. Data traces of the nuclear resonance from octahedral sites in Fe$_3$O$_4$. Curve A direct absorption vs frequency; curve B the derivative of the absorption vs frequency.

(3) Domain wall excitation. In this case **M** is confined to the plane of a domain wall, and one can easily show that $\rho(\nu)$ has infinite peaks at $\nu = \nu_\perp$ and $\nu_\alpha$. Here $\nu_\alpha$ is a constant given by

$$\nu_\alpha = \nu_\perp + (\nu_{||} - \nu_\perp) \sin^2\psi, \qquad (3)$$

where $\psi$ is the angle between the trigonal axis and the normal to the domain wall. For nonvanishing $F$ the product $F\rho$ also has such singularities. Qualitative line shapes for (100) planes ($\sin^2\psi = \frac{2}{3}$) and (110) planes ($\sin^2\psi = \frac{1}{3}$, 1) are shown by curves b and c, respectively, in Fig. 1. In general, a domain wall produces at least two peaks while a random distribution produces only one.

In the experiment the absorption was observed in an isotopically enriched powder by continuous wave techniques[4] at temperatures ($T$) of 100° to 300°K. One sharp line is attributed to the hyperfine interaction on tetrahedral sites where anisotropy is not expected from symmetry.

A broad second line was detected by a decrease in marginal oscillator level shown in curve A of Fig. 2. The dotted curve indicates a nominal base line. Curve B of Fig. 2 shows the first derivative of absorption. The important persistent features of this trace are the three peaks indicated by arrows. Allowing for broadening due to isotropic effects, this line closely resembles curve a of Fig. 1 for a random distribution. This line is attributed to octahedral sites, and it is concluded that in this case the nuclei are excited predominantly by domain rotation and that the observed absorption is proportional to the imaginary part of the nuclear susceptibility. In many runs no reproducible evidence of any other form of excitation was observed even though the ferromagnetic anisotropy changes sign in the temperature range studied.

The resonance frequency pattern is linear in $T^{\frac{3}{2}}$. The extrapolated internal fields at 0°K are $H$ (tet.) = 508 koe, $H_\perp$(oct.) = 498 koe, and $H_{||}$ (oct.) = 477 koe. These values are consistent with Mössbauer counting rates.[5]

The internal fields at octahedral sites are considered to be mean values characteristic of Fe$^{2+}$ and Fe$^{3+}$ ions since these ions appear in equal numbers on octahedral sites. Electric quadrupole interactions are not involved because the nuclear spin is $\frac{1}{2}$. A dipolar contribution of 19 koe to $H_\perp$ (oct.) − $H_{||}$ (oct). was calculated from published lattice sums[6] for the spinel lattice. The difference between this value and the measured total of 12 koe is probably attributable to specific crystal-field effects.

[4] The authors wish to thank R. J. Blume for the circuit diagrams for the oscillating limiter prior to publication.

[5] I. Solomon, Compt. rend. **251**, 2675 (1960); R. Bauminger, S. G. Cohen, A. Marinov, S. Ofer, and E. Segal, Phys. Rev. **122**, 1447 (1961). K. Ono, Y. Ishikawa, and A. Ito, International Conference on Magnetism and Crystallography, Kyoto, September 1961. The authors wish to thank Dr. S. Ruby of Westinghouse and Dr. G. K. Wertheim of Bell Labs. for information concerning their further Mössbauer studies of Fe$_3$O$_4$.

[6] R. A. Johnson and D. W. Healy, Jr., J. Chem. Phys. **26**, 1031 (1957).

JOURNAL OF APPLIED PHYSICS    SUPPLEMENT TO VOL. 33, NO. 3    MARCH, 1962

# de Haas van Alphen Effect in Zinc Manganese Alloys*

F. T. Hedgcock and W. B. Muir†

*The Franklin Institute Laboratories, Philadelphia, Pennsylvania*

Observations of the long period de Haas van Alphen effect in pure zinc and in a zinc alloy containing 0.01 at.% manganese have been made to determine whether the Fermi surface of the alloy changes in the temperature region where the zinc manganese alloy exhibits an electrical resistance minimum. On the basis of a nearly free electron interpretation it has been shown that: (i) Within the experimental error the Fermi surface of pure zinc and the 0.01 at.% zinc manganese alloy are identical. (ii) The number of electrons added or subtracted from the conduction band on alloying cannot be greater than 1 electron per impurity atom. (iii) If the resistance minimum is due to a change in the density of states in the conduction band, this change is less than 0.0015%. (iv) If the ionic state of the manganese ions changes in the temperature range of the resistance minimum, then less than half of the ions are involved. In order to obtain a consistent interpretation of the variation of the amplitude of the de Haas van Alphen oscillations in the alloy it was necessary to assume that the collision damping (Dingle) factor varies with magnetic field. A simple extension of the Schmitt scattering model [R. W. Schmitt, Phys. Rev. 103, 83 (1956)] would predict this behavior both for the relaxation time derived from magnetoresistance and the de Haas van Alphen effect.

* To be submitted for publication in full to The Physical Review.

† Submitted as partial fulfillment of the requirement for the Ph.D. degree at the University of Ottawa, Ottawa, Canada.

---

JOURNAL OF APPLIED PHYSICS    SUPPLEMENT TO VOL. 33, NO. 3    MARCH, 1962

# Nuclear Resonance in Ferromagnetic Alloys

Toshimoto Kushida and A. H. Silver

*Scientific Laboratory, Ford Motor Company, Dearborn, Michigan*

AND

Yoshitaka Koi and Akira Tsujimura

*Faculty of Engineering, Tokushima University, Tokushima, Japan*

The internal fields $H_i$ in ferromagnetic alloys were measured at both the solvent and the solute nuclei at magnetic and nonmagnetic atoms using NMR techniques. $Co^{59}$ resonances in Co-rich alloys (Co-Fe, -Ni, -Cu, -Cr, -Mn, -Al) have fine structures which depend on the kind and the concentration of impurity metals. Similar structures are found for the $Fe^{57}$ resonances in Fe-rich alloys. These structures are tentatively interpreted as caused by the anisotropy of $H_i$ at the nearest neighbors. The $Co^{59}$ resonance is also observed in Fe-rich Co-Fe alloys, and it is found that $H_i$ seen at the Co site is lower by about 50 koe than that at the Fe site. The $Co^{59}$ line width is about 400 kc for (Fe+1% Co) alloy. Both the temperature and the pressure dependences of the $Co^{59}$ frequency were measured. $H_i$ at $Cu^{63}$ and $Cu^{65}$ was measured in Co-rich Co-Cu and in Fe-rich Fe-Cu ferromagnetic alloys. The magnitude of these internal fields at Cu and their observed pressure dependence are, at least, not inconsistent with the contention that the observed $H_i$ is mainly produced by the $4s$ conduction-electron polarization, although $H_i$ at the Cu nucleus, 217.7 koe for Fe-Cu alloy and 157.5 koe for Co-Cu alloy, is considerably higher than the usual theoretical prediction.

## INTRODUCTION

SINCE Portis and Gossard[1] discovered a strong nuclear resonance signal of $Co^{59}$ in ferromagnetic fcc cobalt metal, nuclear magnetic resonance (NMR) technique has been used to investigate the internal field $H_i$ which is the magnetic field seen at the nucleus in many ferromagnetic materials.

High accuracy of this method enables us to measure the pressure dependence of $H_i$[2,3] as well as to measure its temperature dependence very precisely. The inhomogeneity of $H_i$ inside the samples caused by either alloy-ing[4,5] or by mechanical defects[5,6] has been observed as an increase in line width or as additional lines. Thereby more detailed information about the distribution of $H_i$ can be obtained than from a single averaged value of $H_i$ which is obtained from a low-temperature-specific-heat measurement,[7] although if the inhomogeneity exceeds a certain amount, the lines are smeared out and unable to be observed.

The present article will deal with $H_i$ measured at the

¹ A. M. Portis and A. C. Gossard, J. Appl. Phys. 31, 205S (1960).
² Y. Koi, A. Tsujimura, and T. Kushida, J. Phys. Soc. Japan 15, 2100 (1960).
³ G. B. Benedek and J. Armstrong, J. Appl. Phys. 32, 106S (1961).

⁴ Y. Koi, A. Tsujimura, T. Hihara, and T. Kushida, J. Phys. Soc. Japan 16, 574 (1961); R. Street, D. S. Rodbell, and W. L. Roth, Phys. Rev. 121, 84 (1961).
⁵ R. C. LaForce, S. F. Ravitz, and G. F. Day, Phys. Rev. Letters 6, 226 (1961).
⁶ W. A. Hardy, J. Appl. Phys. 32, 122S (1961).
⁷ V. Arp, D. Edmonds, and R. Petersen, Phys. Rev. Letters 3, 212 (1959).

nucleus of both magnetic and non-magnetic impurity metal atoms in ferromagnets, and the distribution of $H_i$ among solvent ferromagnetic atoms.

## EXPERIMENTAL METHODS

NMR in the alloys has been measured using a marginal-oscillator spectrometer in zero external field. The spectrometer is frequency modulated and is operated either regeneratively or superregeneratively.

The samples were powders of dilute alloys prepared from high purity metals in an induction furnace. The pressure dependence of the resonance frequency has been measured in a steel bomb in conjunction with a modified Bridgman press[8] in the range 1 to 8000 kg/cm².

## EXPERIMENTAL RESULTS AND DISCUSSION

### $H_i$ at Magnetic Solute Atoms in Ferromagnets

The NMR line of $Co^{59}$ in iron-rich Fe-Co alloy was observed as a function of temperature[9] and pressure. The resonance frequency in (Fe+1% Co) was measured from liquid $N_2$ temperature to 650°K. The resonance frequency at 0°K, $\nu(0)$, was extrapolated from these values using a $T^{\frac{3}{2}}$ law, as 289.2 Mc, corresponding to $H_i(0)$ equal to 289.7 koe. This value is about 30 koe smaller than the value deduced from specific-heat measurements,[7] which is much less accurate than NMR measurements.

It is noted that although $H_i(0)$ seen at Co in Fe is about 50 koe lower than that at Fe, 339 koe, it is appreciably higher than that at Co nuclei in any of fcc and hcp Co metals.

$H_i(T)/H_i(0)$ at the Co nucleus decreases with increasing temperature more rapidly than that at the Fe nucleus. Recently a similar discrepancy between the temperature dependences of $Ni^{61}$ and $Co^{59}$ resonance frequencies in Ni-rich Ni-Co alloy was noted by Bennett and Streever,[10] although the Co frequency drops off more *slowly* with temperature than the Ni resonance frequency.

It is interesting to see whether this discrepancy between the temperature dependences of the Fe and the Co resonance frequencies in Fe-Co alloy is caused simply by an *implicit* effect of temperature via thermal expansion or by an *explicit* temperature effect, namely, to see whether or not this discrepancy still exists at constant volume.

Measurement of the pressure dependence of the resonance frequency makes this point clear. Since the pressure dependence of the $Fe^{57}$ frequency in pure Fe has been measured by Benedek and Armstrong[3] and this pressure dependence is not expected to change

TABLE I. The pressure dependence of the internal field $H_i$ at $Co^{59}$ nucleus in iron-rich Fe-Co alloy, $Cu^{63}$ nucleus in iron-rich Fe-Cu alloy, and $Fe^{57}$ in pure iron.

| Nucleus | $d\ln H_i/dP$ (kg/cm²)⁻¹ |
|---|---|
| $Co^{59}$ in Fe | $+1.6\times10^{-7}$ |
| $Cu^{63}$ in Fe | $-3.0\times10^{-7}$ |
| $Fe^{57}$ in Fe | $-1.6\times10^{-7}$[a] |

[a] See reference 3.

appreciably in our alloy (Fe+1% Co), only the Co resonance frequency in this alloy was measured as a function of pressure. The Co resonance frequency increases linearly with increasing pressure. On the other hand, the Fe resonance frequency decreases with pressure. The results are shown in Table I. The pressure dependence of $Cu^{63}$ NMR frequency in Fe, which will be mentioned in the next section, is also shown.

Using the pressure dependence of $H_i$ in conjunction with the values of compressibility and thermal expansion coefficient, the temperature dependence of the internal field $H_i(T)$ at constant pressure can be converted into that at constant volume.[11] The results are shown in Fig. 1 in a reduced scale, namely as $H_i(T)/H_i(0)$ versus $T/T_C$, where $T_C$ is the Curie temperature of each ferromagnet.

There is a definite discrepancy between $H_i(T)/H_i(0)$ at Co and Fe even at constant volume. This difference in explicit temperature dependence may suggest that the numbers of localized $d$ electrons at Co atom may be slightly increased at higher temperature, though a unique interpretation is very difficult as will be shown later.

If the similar discrepancy between $H_i(T)$ at Ni and Co in the Ni-rich Ni-Co alloy found by Bennett and Streever[10] is also predominantly due to an explicit temperature effect, we might speculate that the explicit temperature effect tends to push the $d$ electrons at Fe atoms into adjacent Co atoms and to expel the Co $d$ electrons into surrounding Ni atoms.

The linewidth of the Co resonance was measured as a function of Co concentration from 0.5 to 5%. The width for 0.5% sample is 370 kc and increases with the increasing concentration to 900 kc for the 5% alloy. The center frequency does not change with concentration within the experimental error.

According to Marshall[12] the internal field at the nucleus in iron group ferromagnets arises mainly from the contact interaction of the nucleus with the 4s electrons and with the inner core electrons. The contribution from the 4s electrons consists of two parts. The first part is due to the 4s conduction electrons polarized by the spins of the 3d electrons, and the second part is due to some mixing of the 4s wave function into the 3d band.

[8] T. Fuke, J. Phys. Soc. Japan **16**, 266 (1961).

[9] Y. Koi, A. Tsujimura, T. Hihara, and T. Kushida, J. Phys. Soc. Japan **16**, 1040 (1961).

[10] L. H. Bennett and R. L. Streever, Jr., J. Appl. Phys. **33**, 1093 (1962), this issue. One of the authors, T. K., is grateful to Dr. Bennett for sending the manuscript prior to the publication.

[11] T. Kushida, G. B. Benedek, and N. Bloembergen, Phys. Rev. **104**, 1364 (1956).

[12] W. Marshall, Phys. Rev. **110**, 1280 (1958).

The magnitude of $H_i$ due to this mechanism is expressed by

$$H_i = -(8\pi/3)\mu|\psi(0)|^2 n\tilde{p},$$
$$\tilde{p} = p + 2\bar{S}a^2/n, \tag{1}$$

where $|\psi(0)|^2$ is the average probability density of $4s$ electrons at nucleus, $n$ is the number of conduction electrons per atom, $p$ is their polarization, $\bar{S}$ is the mean spin per atom, and $a^2$ is an average of the amount of $4s$ wave function mixed into the $3d$ wave function. Both parts produce a positive field, i.e., a field parallel to the direction of the magnetization, since the electron $g$ value is negative. On the other hand, the inner core $s$ electrons polarized by the spin of the $3d$ electrons produce a negative field. The actual field has a negative sign.

In the case of dilute ferromagnetic alloys, the main internal field at a solute atom nucleus may come from the following sources: (1) $4s$ electron polarization at the solute atom. The $4s$ electrons are polarized by the $3d$ electrons of the solvent atoms as well as those of the solute atom itself. The internal field from this origin is expressed by Eq. (1). (2) Transfer of $3d$ electrons between the solvent and the solute atoms on alloying. The change in the number of $3d$ electrons affects both the $4s$ electron polarization and the core polarization at the solute atom. (3) Other contributions from the $3d$ electrons of the solvent atoms surrounding the solute atom. For instance, the core $s$ electrons at the solute atom could be polarized by the $3d$ spins of the surrounding atoms directly[13] and/or indirectly through the $3d$ shell of the solute atom.

TABLE II. The internal field $H_i$ at the Cu nucleus in iron and in cubic cobalt. The data for pure iron and cobalt are included for comparison.

| Nucleus | Temperature °C | Resonance frequency Mc | Internal field koe |
|---|---|---|---|
| Cu$^{63}$ in Fe | | 240.0 | |
| | 0 | | 212.7 |
| Cu$^{65}$ in Fe | | 257.1 | |
| Cu$^{63}$ in Co | | 177.9 | |
| | 9 | | 157.5 |
| Cu$^{65}$ in Co | | 190.7 | |
| Fe$^{57}$ in Fe | 0 | 46.65 | 330.5[a] |
| Co$^{59}$ in Co | 0 | 213.2 | 213.5[b] |

[a] See reference 3.
[b] See reference 1.

These interactions make a unique interpretation of the origins of the internal field at the ferromagnetic impurities in ferromagnets very difficult.[14]

### $H_i$ at a Nonmagnetic Solute Atom in Ferromagnets

The measurement of the internal field at nonmagnetic solute atoms, however, could eliminate some of the causes mentioned above and would help to elucidate the various contributions to $H_i$ in ferromagnets. For instance, a proper choice of nonmagnetic atoms as a field probe could minimize the sources (2) and (3). One could hope to measure the magnitude of conduction electron polarization as in the case of Knight shift experiments in nonmagnetic metals.

$H_i$ at relatively heavy nonmagnetic metals in ferromagnets have been measured using different methods.[13,15,16] The internal field caused by (3), however, seems to make an estimate of conduction electron polarization difficult.[13] For instance, the relative magnitudes of $H_i$ at Sn$^{119}$ measured by means of Mössbauer effect in Fe, Co, and Ni give inconsistent values with those expected from conduction electron polarization picture.[13] As a part of a systematic study of $H_i$ at nonmagnetic metal atoms in ferromagnets, $H_i$ at the Cu nucleus in iron and cubic cobalt has been measured using NMR techniques. Results of the measurements are shown in Table II together with data for pure iron[3] and cubic cobalt[1] for comparison. Unfortunately the sign of the field has not yet been determined. The pressure dependence of $H_i$ at Cu in iron has also been measured, and the results are given in Table I. The resonance frequency of Cu$^{63}$ decreases linearly with increasing pressure.

It is noted that the internal fields at the Cu atoms

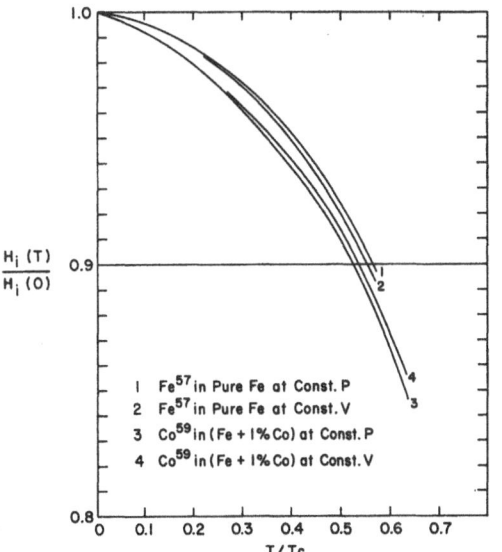

FIG. 1. Temperature dependence of the internal field $H_i(T)$ at Fe$^{57}$ and Co$^{59}$ in iron both at constant pressure and at constant volume in a reduced scale.

[13] A. J. F. Boyle, D. St. P. Bunbury, and C. Edwards, Phys. Rev. Letters **5**, 553 (1960).

[14] G. K. Wertheim, J. Appl. Phys. **32**, 110S (1961).
[15] C. T. Wei, C. H. Cheng, and P. A. Beck, Phys. Rev. **122**, 1129 (1961).
[16] B. N. Samoilov, V. V. Sklyarevskii, and E. P. Stepanov, Soviet Phys.—JETP **11**, 261 (1960).

in iron and in cobalt are comparable in magnitude ($\sim 70\%$) with the fields in pure iron and in cobalt; the ratio of the fields at the Cu nucleus in iron and in cobalt is nearly equal to the ratio of the saturation magnetizations of iron and cobalt; and the pressure dependence of the field at the Cu nucleus in iron is more negative than that at the Fe in iron.

These results suggest that $H_i$ at Cu in these ferromagnets is mainly produced by the conduction electron polarization in ferromagnets. Namely, the mechanism (1) mentioned in the previous section seems to be the dominant one, where the second part of the effective polarization $\tilde{p}$ in Eq. (1), may not be important. Since $|\psi(0)|^2$ is practically unchanged[13] for Fe and Co, the value of $H_i$ arising from the conduction electron polarization is roughly proportional to the saturation magnetization. Furthermore, $H_i$ at Cu in Co agrees roughly with our preliminary measurement using Al$^{27}$ as a field probe in fcc Co.[17] This is in agreement with the model $\tilde{p} \simeq p$. The disturbances caused by the mechanisms (2) and (3) are expected to be much smaller in this case. One of the possible objections to this model is that the magnitude of $H_i$ itself is appreciably larger than the usual theoretical estimates[12]; but it could be within the uncertainty in the theoretical estimations,[18,19] if we assume that these measured values have the same sign.

The second mechanism (2)[20] is, at least, not likely to be the *dominant* source of the measured $H_i$. If (2) were the main effect, we may expect $d \ln H_i/dP$ for $H_i$ at Cu to be positive or at least equal to that at Fe. The observed pressure dependence is, however, in disagreement with this expectation. Furthermore, there seems to be no positive evidence for the Cu impurity in Fe possessing a localized magnetic moment.

The third mechanism (3) cannot be the main source of the field. If it were the dominant term, the same mechanism would play an important role in pure ferromagnets also. This makes it difficult to understand the experimental results that $H_i$ at Cu is roughly 70% of that at solvent nuclei, since there is a large additional (subtractive) core-polarization field as expected from the localized magnetic moment at ferromagnetic atoms. In addition, if this assumption is true, the presence of a Cu atom in Co metal will greatly affect the field at Co atoms surrounding the Cu atom. The observed change,[4] however, is less than 1%.

Although the model that the conduction electron polarization is mainly responsible for the observed $H_i$ is not inconsistent with the present experimental results, a more systematic investigation using different nonmagnetic atoms as the field probes and a determination

of the sign of $H_i$ at these atoms are highly desirable in order to clarify this problem.

## $H_i$ at Solvent Atoms in Ferromagnets

Internal fields at solvent ferromagnetic atoms which have been disturbed by the presence of impurity atoms can also be observed using NMR techniques. Co resonance lines in fcc Co-rich alloys have been observed in Co-Fe, Co-Ni, Co-Cu, Co-Al, Co-Mn, and Co-Cr alloys as a function of impurity concentration.[4] Essentially the same results for Co-Fe and Co-Ni alloys, with somewhat better resolution probably because of higher purity of the constituent metals, have been reported by La Force *et al.*[5] Similar but less systematic observations have been made for the Fe resonance in Fe-rich alloys.[9,21]

The alloy NMR lines generally have structures, which are less pronounced in highly-doped specimens. Resolved satellites are observed for cubic Co-Fe and Co-Ni, whereas for the other Co-rich alloys investigated the lower-frequency tails (some of which have fine structures) spread more strongly than the higher-frequency tails upon alloying. Since the quadrupole broadening of Co lines in the alloys can be estimated to be of the order of 1 Mc,[4] the observed structures of the lines essentially describe the distribution of the internal field in the alloys. The shift of the center of gravity of the line is compatible with the average $H_i$ in Co-Fe and Co-Ni alloys observed by means of a low-temperature specific-heat measurement,[7] although the intensity of the far wings in the NMR lines is difficult to measure because of the uncertainty in the base line.

Fe$^{57}$ NMR lines in (Fe+0.9% Cr) and (Fe+0.5% Ni) alloys have longer tails at the higher frequency wings, indicating poorly resolved satellites. Since the Fe$^{57}$ nucleus has no nuclear quadrupole moment, the line shape change upon alloying has a magnetic origin.

Some of the satellite lines observed in very dilute Co alloys have been identified as produced by the presence of stacking faults[5,6] and of the hcp phase.[2,5,6] These lines are also observed in pure Co samples, and their relative intensities depend on the metallurgical history of the samples.

The other satellites in the alloys are produced by the spacial distribution of internal field around the impurity. However, the assignment of each satellite to a particular neighboring site around the impurity is ambiguous at present.

It is known that the disturbance of the electronic structure in metals caused by an impurity is confined to the immediate neighborhood of the impurity.[22–24] The spacial distribution of $H_i$ around the impurity is divided into two parts: (1) a radial distribution, i.e., $H_i$

[17] The preliminary value of Al$^{27}$ NMR frequency in fcc Co (Co+1% Al) is 189.1 Mc, which corresponds to 170.4 koe.

[18] R. E. Watson and A. J. Freeman, J. Appl. Phys. **32**, 118S (1961).

[19] D. A. Goodings and V. Heine, Phys. Rev. Letters **5**, 370 (1960).

[20] W. M. Lomer and W. Marshall, Phil. Mag. **3**, 185 (1958).

[21] J. I. Budnick, L. J. Bruner, E. L. Boyd, and R. J. Blume, Bull. Am. Phys. Soc. **5**, 491 (1960).

[22] J. Friedel, Nuovo cimento, Suppl. **7**, 287 (1958).

[23] N. Bloembergen and T. J. Rowland, Acta Met. **1**, 731 (1953).

[24] K. Yosida, Phys. Rev. **106**, 893 (1957).

varies sharply with the distance, $r_j$, between the impurity and the pertinent nucleus $j$; and (2) an angular distribution, i.e., $H_i$ at the neighboring atoms with the same distance $r_j$ from the impurity might not be equal and could depend on the angle $\theta$ between the direction of the magnetization and the radius vector of the position of the pertinent nucleus. In the case of the NMR lines in the nonmagnetic or the paramagnetic alloys, the hyperfine field in these metals, Knight shift, has been successfully interpreted in terms of (1) and (2).[25,26]

The density of conduction electrons $\rho(\mathbf{r})$ about the impurity is modulated as [25,26]

$$\frac{\Delta\rho(\mathbf{r})}{\rho_0} = \sum_l (2l+1)\{[n_l(k_m r_j)^2 - j_l(k_m r_j)^2]\sin^2\delta_l - j_l(k_m r_j)n_l(k_m r_j)\sin 2\delta_l\}.$$

Here $j_l$ and $n_l$ are Bessel and Neumann functions, $\delta_l$ is a scattering phase shift, and $k_m$ is the wave number at the Fermi level. The oscillatory nature of Eq. (2) predicts the presence of satellites in the NMR lines in these metals.[25] Usually the linewidth of each component smears out the expected structure.

When the impurity atom has a magnetic moment, the polarization of the conduction electron is also modulated though a spin dependent interaction between the magnetic electrons at the impurity and $s$-band electrons. The internal field at the neighboring atoms will be doubly modulated in this case.[24,26]

In the case of ferromagnetic alloys an impurity will affect the internal field at its neighbors in a more complicated manner. The transfer of $3d$ electrons between the impurity and the neighboring atoms will strongly affect the value of $H_i$ at the neighbors as well as at the impurity. The $3d$ electrons at the impurity would directly polarize the inner core of the neighboring atoms. $H_i$ at the nearest neighbors may have a strong angular

dependence as well. A classical dipole field from the impurity and anisotropic indirect-coupling[26] through the conduction electron may produce this angular dependence in $H_i$. A crude estimate of this effect at the nearest neighbors gives about the same order of magnitude as the observed satellite separation.[27]

Since the total intensity of the satellite lines in Co+1% Fe or Co+1% Ni alloys is of the order of 10% of the intensity of the entire spectral line and the relative intensity among the satellites are essentially independent of the impurity concentration, it is suggested that the satellite lines are mainly caused by the anisotropy of $H_i$ at the nearest neighbors, the number of the nearest neighbors being 12 in a fcc structure. The radial distribution of $H_i$ and its anisotropy at the further neighbors may produce the broadening of the component lines, although some of the fine structure could come from the radial part.

## CONCLUSION

The internal field $H_i$ in ferromagnetic dilute alloys was measured at both the solvent and the solute metal nuclei using NMR techniques. $H_i$ at Co in Fe-rich Fe-Co alloy lies about midway between $H_i$ at Fe-in-Fe and Co-in-Co. $H_i$ at Cu was measured in Co-rich Co-Cu and in Fe-rich Fe-Cu alloys. The magnitude of these internal fields at Cu and their pressure dependence are, at least, not inconsistent with the contention that the observed $H_i$ is mainly produced by the $4s$ conduction-electron polarization. The structures usually observed in the solute-atom NMR lines are tentatively interpreted as caused by the anisotropy of $H_i$ at the nearest neighbors.

## ACKNOWLEDGMENTS

The authors would like to express their appreciation to Professors A. M. Portis, T. Nagamiya, and K. Yosida for valuable discussions.

[25] A. Blandin and E. Daniel, J. Phys. Chem. Solids 10, 126 (1959).

[26] D. L. Weinberg and N. Bloembergen, J. Phys. Chem. Solids 15, 240 (1960).

[27] The difference between $H_i$ at Co in fcc Co and in hcp Co has been explained in terms of the dipole field.[28]

[28] Y. Koi, A. Tsujimura, T. Hihara, and T. Kushida, Report of International Conference of Magnetism in Japan (1961), to be published.

JOURNAL OF APPLIED PHYSICS    SUPPLEMENT TO VOL. 33, NO. 3    MARCH, 1962

# Crystalline Electric Fields in Spinel-Type Crystals

V. J. Folen

*U. S. Naval Research Laboratory, Washington 25, D. C.*

Electron paramagnetic resonance (EPR) measurements are reported for the $Fe^{3+}$ ion in long range "ordered" and "disordered" single crystals of the spinel-type compound lithium aluminate ($Li_{0.5}Al_{2.5}O_4$) which is isomorphous with lithium ferrite ($Li_{0.5}Fe_{2.5}O_4$). In the "disordered" aluminate most of the EPR linewidths were found to be very broad, whereas in the "ordered" aluminate the linewidths were comparatively narrow. The broadness of the linewidths in the "disordered" aluminate prevented quantitative determination of the crystalline electric fields in this state. From comparison of the "ordered" and "disordered" EPR linewidths, it was deduced that large fluctuations in the crystalline electric fields may exist in other "disordered" spinels also. In the "ordered" aluminate, the EPR spectrum was measured for the tetrahedral sites (which have trigonal symmetry), and the values $|D| = 0.104$ cm$^{-1}$, $|a-F| = 0.0166$ cm$^{-1}$, and $g_{\parallel} = 2.006 \pm 0.002$ were obtained. The magnetocrystalline anisotropy of $Li_{0.5}Fe_{2.5}O_4$ was interpreted on the basis of the crystalline electric field parameters measured in the aluminate.

## INTRODUCTION

THE "one-ion" model is known to be applicable to the theory of the magnetocrystalline anisotropy[1-4] and other properties of ferrimagnetic compounds. Since the magnetic properties that are predictable by this model depend only on the kind of ion under consideration and on the crystal structure of the ferrimagnetic compound, one can interpret these properties with the use of the crystalline electric field parameters obtained from electron paramagnetic resonance (EPR) measurements on doped diamagnetic single crystals which are isomorphous with the ferrimagnetic compound. In particular, EPR measurements on the $Fe^{3+}$ ion (which is the major cation constituent in the ferrimagnetic compounds) are required for the interpretation of many of the magnetic properties in these compounds. Such measurements have already been performed on single crystals of yttrium gallate[5] which is isomorphous with yttrium iron garnet. Previous EPR measurements[6] of the crystalline electric fields acting on $Fe^{3+}$ in spinels involved the determination of the cubic crystalline electric field parameter ($a$) by an interpretation of the linewidths in polycrystalline $MgAl_2O_4$ and $ZnAl_2O_4$. Attempts[6] to determine the crystalline field parameters in single crystals of these spinels were unsuccessful because of the broadness of the $Fe^{3+}$ EPR lines in the single crystals of $MgAl_2O_4$ and because of the difficulty in obtaining single crystals of $ZnAl_2O_4$.

In the present paper, measurements of the EPR spectra are reported for $Fe^{3+}$ in both long range "ordered" and "disordered" single crystals of lithium aluminate ($Li_{0.5}Al_{2.5}O_4$) which is isomorphous with lithium ferrite ($Li_{0.5}Fe_{2.5}O_4$). The two states of long

range "order" were produced by using the heat treatment procedures given by Braun.[7]

The space groups[7] for the "disordered" and "ordered" aluminate are $O_h(7)$-$Fd3m$ and $O(6)$-$P4_33$ or $P4_13$, respectively. Thus the trivalent cation site symmetries in the "disordered" structure are $\bar{3}m$ and $\bar{4}3m$ for octahedral ($B$) and tetrahedral ($A$) sites, respectively, whereas in the "ordered" structure the trivalent cation site symmetries are 2 and 3 for $B$ and $A$ sites, respectively.

## EXPERIMENTAL

The $Li_{0.5}Al_{2.5}O_4$ single crystals in which 0.05 mole percent $Al^{3+}$ was replaced by $Fe^{3+}$ were synthesized using a flux[8] technique. A combined PbO and $PbF_2$ flux and a lowering rate of 1°C per hour from 1300°C were utilized in the synthesis. The degree of long range order that was produced by various heat treatments was determined by measurement of the intensities of the superstructure lines in the x-ray patterns obtained with single crystals which had been powdered.

The measurements of the EPR spectra were made at 24 kMc using a Pound stabilized klystron and a cylindrical $TE_{011}$ transmission cavity. The cavity output was detected with a silicon diode, amplified in a narrow band 1000-cps amplifier, phase detected and finally displayed as the first derivative of an absorption curve on an X-Y recorder. The magnetic field was sinusoidally modulated at 1000 cps and was measured using proton and lithium nuclear resonance.

## THEORY

The spin Hamiltonian for the $Fe^{3+}$ ion in a combined cubic and axial crystalline field is given by[9]

$$\mathcal{H} = g\beta \mathbf{H} \cdot \mathbf{S} + D[S_z^2 - \tfrac{1}{3}S(S+1)]$$
$$+ \tfrac{1}{6}a[S_\xi^4 + S_\eta^4 + S_\zeta^4 - \tfrac{1}{5}S(S+1)(3S^2+3S-1)]$$
$$+ (1/180)F[35S_z^4 - 30S(S+1)S_z^2 + 25S_z^2$$
$$- 6S(S+1) + 3S^2(S+1)^2], \quad (1)$$

[1] G. T. Rado and V. J. Folen, Bull. Am. Phys. Soc. **1**, 132 (1956); G. T. Rado, V. J. Folen, and W. H. Emerson, Proc. Inst. Elec. Engr. (London) **104B**, Suppl. No. 5, 198 (1957); V. J. Folen and G. T. Rado, J. Appl. Phys. **29**, 438 (1958).

[2] K. Yosida and M. Tachiki, Progr. Theoret. Phys. (Kyoto) **17**, 331 (1957).

[3] W. P. Wolf, Phys. Rev. **108**, 1152 (1957).

[4] J. C. Slonczewski, Phys. Rev. **110**, 1341 (1958).

[5] S. Geschwind, Phys. Rev. **121**, 363 (1961).

[6] Y. Sugiura, J. Phys. Soc. Japan **15**, 1217 (1960).

[7] P. B. Braun, Nature **170**, 1123 (1952).

[8] J. W. Nielson and E. F. Dearborn, J. Phys. Chem. Solids **5**, 202 (1958).

[9] B. Bleaney and R. S. Trenam, Proc. Roy. Soc. (London) **A223**, 1 (1954).

where $D$ and $F$ are the quadratic and quartic axial field parameters, respectively, $a$ is the cubic crystalline electric field splitting parameter, and the $z$ axis is in the direction of the axial field.

For $H$ at an angle $\theta$ with respect to the $z$ axis, the magnetic field (to second order in $D$) corresponding to the $\frac{1}{2} \rightarrow -\frac{1}{2}$ transition is obtained from

$$H = H_0 + 8D^2 \sin^2\theta \cos^2\theta \left[ \frac{1}{H+q} + \frac{1}{H-q} \right]$$

$$-\frac{1}{2} D^2 \sin^4\theta \left[ \frac{9}{2H-q} + \frac{9}{2H+q} - \frac{5}{2H-3q} - \frac{5}{2H+3q} \right]. \quad (2)$$

Here $H_0 = h\nu/g\beta$, $q = D(3\cos^2\theta - 1)$, and $D/g\beta$ is written as $D$.

The magnetic fields of EPR transitions for $H$ along trigonal axis ($\theta = 0$) are obtained from[5]

$$\pm\frac{3}{2} \rightarrow \pm\frac{5}{2}: H_{1,5} = H_0 \mp 4D \pm \frac{4}{3}(a-F) - \frac{20}{27} \frac{a^2}{H_{1,5} \pm 2D}$$

$$\pm\frac{1}{2} \rightarrow \pm\frac{3}{2}: H_{2,4} = H_0 \mp 2D \mp \frac{5}{3}(a-F) + \frac{20}{27} \frac{a^2}{H_{2,4} \mp 2D} \quad (3)$$

$$-\frac{1}{2} \rightarrow +\frac{1}{2}: H_3 = H_0 - \frac{20a^2}{27} \left( \frac{1}{H_3 + 2D} + \frac{1}{H_3 - 2D} \right),$$

where all the crystalline field parameters are now divided by $g\beta$.

## RESULTS

The EPR spectra of the $Fe^{3+}$ ion in the "disordered" aluminate were measured with $H$ along the [111], [110], and [100] directions. Except for one $\frac{1}{2} \rightarrow -\frac{1}{2}$ transition in the [111] spectrum which had a linewidth of 40 gauss (determined by the separation between points of maximum slope in an absorption curve), the EPR linewidths were very broad in the [111], [110], and [100] spectra. This prevented quantitative determination of the crystalline electric field parameters in the "disordered" crystals. The spectrum for $H\|[111]$ consisted of a broad $\frac{1}{2} \rightarrow -\frac{1}{2}$ transition (corresponding to $Fe^{3+}$ ions in trigonal fields along the [$\bar{1}11$], [$1\bar{1}1$], and [$11\bar{1}$] directions for which $\theta = 70.5°$) on which was superimposed the 40-gauss linewidth $\frac{1}{2} \rightarrow -\frac{1}{2}$ transition whose trigonal axis coincided with the [111] direction. In addition, $\Delta M = \pm 2$ transitions were observed in this spectrum. From the appearance of the spectra for $H$ in the [111], [110], and [100] directions, as well as the angular variation of the linewidths and positions of the $\frac{1}{2} \rightarrow -\frac{1}{2}$ transitions, it was concluded that the spectra have trigonal symmetry. Since the $B$ sites possess this symmetry, it appears that a substantial number of $Fe^{3+}$ ions are located on $B$ sites in the "disordered" aluminate. Using Eq. (2) and measurements of the angular variation of the $\frac{1}{2} \rightarrow -\frac{1}{2}$ transitions in the

vicinity of the [111] direction, $|D|$ was found to be approximately 0.13 cm$^{-1}$. When the linewidths in the "disordered" single crystals were compared with the relatively narrower linewidths observed in the "ordered" single crystals, it was deduced that there are large fluctuations in the crystalline electric fields in "disordered" spinels. Such fluctuations may be responsible for the broadness of the EPR spectrum of $Fe^{3+}$ in $MgAl_2O_4$[6] which is a "disordered" spinel. In addition, the magnitude of these fluctuations is relevant to the Callen-Pittelli theory[10] of ferromagnetic resonance linewidth which is based on spin wave scattering by spatial fluctuations of the crystalline fields associated with the $D$ term in "disordered" magnetic materials.

In the "ordered" aluminate, the observed EPR spectrum showed that there are four types of sites per unit cell having trigonal symmetry along the [111], [$\bar{1}11$], [$1\bar{1}1$], and [$11\bar{1}$] directions, respectively. This symmetry is in accord with the symmetry of the $A$ sites (3). Thus, a substantial number of $Fe^{3+}$ ions are located on the $A$ sites in the "ordered" aluminate. The spectrum corresponding to the octahedral sites (which have twofold distortion axes along the $\langle 110 \rangle$ directions) was not observed.

Using Eq. (3), the results of the measurements for $H\|[111]$ yielded the following values: $|D| = 0.104$ cm$^{-1}$, $|a-F| = 0.0166$ cm$^{-1}$, and $g_{\|} = 2.006 \pm 0.002$. On the basis of the "one-ion" model of magnetocrystalline anisotropy in ferrimagnetics, one can calculate the $D$ contribution to the first-order cubic anisotropy constant in the isomorphous compound $Li_{0.5}Fe_{2.5}O_4$. This is obtained from[3]

$$k_1 = \gamma(D^2/kT)t(y). \quad (4)$$

Here $k_1$ is the first-order anisotropy constant per ion, $t$ is a certain function[3] of $y$, and $y = \exp(-g\beta H_m/kT)$, where $H_m$ is the molecular field. $\gamma$ is 4/9 when the axial fields are along the four $\langle 111 \rangle$ directions. Using the molecular fields that have been evaluated[11] from saturation magnetization measurements on $Li_{0.5}Fe_{2.5}O_4$ and the $D$ value in $Li_{0.5}Al_{2.5}O_4$, one obtains $k_1 = 1.3 \times 10^{-4}$ cm$^{-1}$ for $T = 126°K$. At this temperature the measured[12] value of $k_1$ for the tetrahedral sites in lithium ferrite is about 65 times larger than the value calculated from the $D$ measured in the aluminate, in agreement with the conclusion stated in reference 12 that (on the basis of the observed temperature dependence of the anisotropy) the $D$ contribution is small. Although the $D$ obtained from EPR measurements on the aluminate may be different from that in the isomorphous ferrite, differences sufficiently large to account for the factor of 65 are not expected.

Measurements of the magnetic fields occurring in Eq. (3) yielded 0.01 cm$^{-1}$ as a preliminary value of the magnitude of $a$ in the aluminate. Using this value of $a$

[10] H. B. Callen and E. Pittelli, Phys. Rev. **119**, 1523 (1960).
[11] G. T. Rado and V. J. Folen, J. Appl. Phys. **31**, 62 (1960).
[12] V. J. Folen, J. Appl. Phys. **31**, 166S (1960).

and $|a-F| = 0.0166$ cm⁻¹, it is seen that the $F$ term is comparable with the $a$ term in the aluminate. This is in accord with the conclusions[12] obtained from the effects of the "order-disorder" transition on the magnetocrystalline anisotropy in lithium ferrite. Thus, it is reasonable to attribute a part of the anisotropy of the tetrahedral sublattice in "ordered" $Li_{0.5}Fe_{2.5}O_4$ to the $F$ term in the spin Hamiltonian.

The determination of the signs of the crystalline electric field parameters in the aluminate and of more precise (using computer diagonalizations) value of the $a$ parameter is underway. In addition, single crystals of $Li_{0.5}Ga_{2.5}O_4$ have been synthesized, and it was found that this material is isomorphous with $Li_{0.5}Fe_{2.5}O_4$ and $Li_{0.5}Al_{2.5}O_4$. EPR studies of the crystalline electric fields in the gallate will be reported elsewhere together with a fuller account of the present work.

## ACKNOWLEDGMENTS

The author wishes to express his appreciation to Dr. G. T. Rado for many useful discussions and to Mr. R. A. Becker for assistance in the single crystal synthesis and x-ray analysis.

JOURNAL OF APPLIED PHYSICS    SUPPLEMENT TO VOL. 33, NO. 3    MARCH, 1962

# Hyperfine Fields, Spin Orbit Coupling, and Nuclear Magnetic Moments of Rare-Earth Ions

R. E. Watson*

*Avco, RAD, Wilmington, Massachusetts*

AND

A. J. Freeman

*Materials Research Laboratory, Ordnance Materials Research Office, Watertown, Massachusetts*

An investigation is presented of the related problems of hyperfine fields, spin orbit coupling, and the determination of nuclear magnetic moments, all of which are strongly dependent on the $4f$ charge distributions in the rare earths and play an important role in Mössbauer, NMR, paramagnetic resonance, and optical absorption measurements. The discussion is based on a set of conventional approximate nonrelativistic Hartree-Fock wave functions which were determined for nine ions of the series. Results are presented, and the various approximations and assumed relationships between these quantities (listed in the title) are discussed in the light of comparisons between theoretical predictions and experiment.

## INTRODUCTION

THE magnetic properties of the rare earths are of course determined by the behavior of the $4f$ electrons. Of the many properties which have been investigated experimentally, we shall concern ourselves in this paper with hyperfine fields, spin orbit coupling, and the determination of nuclear magnetic moments, all of which are strongly dependent on the $4f$ distributions and play an important role in Mössbauer, NMR, paramagnetic resonance, and optical absorption measurements. (In another paper[1] at this Conference we are reporting on the problem of neutron magnetic scattering from the rare earths.) Our discussions are based on a set of conventional approximate nonrelativistic Hartree-Fock (H-F) wave functions which were determined for nine ions of the series.

Since hyperfine measurements are generally used to determine nuclear magnetic moments (n.m.m.), accurate theoretical estimates of the hyperfine fields are needed for accurate values of the n.m.m. Reciprocally, uncertainties in our knowledge of n.m.m.'s are responsible for the current uncertainties in determinations of the effective magnetic field at the nucleus of rare-earth ions by Mössbauer and NMR techniques. Direct measurements of the n.m.m. would obviously eliminate these difficulties, but to date this has been possible only for the special case of Eu. In general, most of the n.m.m. data are based on knowledge of $4f$ orbital behavior, and specifically on the $4f$ expectation value of $r^{-3}$ (i.e., $\langle r^{-3} \rangle$) which represents the (dominant) orbital contribution to the hyperfine field. Obviously, theoretical computations of this quantity represent serious potential sources of error in the resulting moments, particularly in view of the lack of accurate wave functions for these ions. For this reason, spin-orbit coupling parameters, $\zeta_{4f}$'s, have been utilized to estimate values of $\langle r^{-3} \rangle$—a procedure which assumes, however, a relationship between the two quantities. We have been concerned with exploring this relationship, particularly in regard to the behavior of $\langle r^{-3} \rangle$, and the ways in which the $\langle r^{-3} \rangle$ *parameter* may differ from an $\langle r^{-3} \rangle$ *integral*

* Part of the work of this author was done as a National Science Foundation Post-Doctoral Fellow at the author's present address: Theoretical Physics Division, AERE, Harwell, Berkshire, England.

[1] M. Blume, A. J. Freeman, and R. E. Watson, J. Appl. Phys. 33, 1242 (1962), this issue.

computed for a $4f$ shell. Furthermore, hydrogenic[2,3] or modified hydrogenic[4] behavior was previously assumed for these quantities, and, while this has yielded "fairly good" results, it is of theoretical interest to investigate the predictions based on H-F functions.

## SPIN ORBIT COUPLING

$\zeta_{4f}$'s have been calculated for those nonspherical rare-earth ions for which we have obtained conventional H-F results using the well-known formula:

$$\zeta_{4f} = e^2 \hbar^2 (2m^2 c^2)^{-1} \int_0^\infty P_{4f}^2(r) [r^{-1} dV/dr] dr, \quad (1)$$

where $P_{4f}(r)$ is the $4f$ radial orbital, and $V(r)$ is the potential seen by the $4f$ electron. We used $V(r)$'s calculated only with coulomb effects specifically included, i.e., exchange contributions to an "effective" $V(r)$ were not included. It has been a common procedure to somehow relate $\zeta$'s to $\langle r^{-3} \rangle$; in this way measured values of $\zeta$ have been used to determine $\langle r^{-3} \rangle$ and in turn n.m.m.'s. This shows the relationship between the various quantities considered in this paper and accounts in part for our interest in $\zeta$. Some of the resultant $\zeta_{4f}$'s are listed in Table I, along with experimental values and the $\zeta_{4f}$'s obtained[5] by Ridley for her $Pr^{3+}$ and $Tm^{3+}$ Hartree functions. The experimental values are those tabulated by Jørgensen, Wybourne, and McClure with compromises where necessary and with uncertainties of at least 50 cm$^{-1}$. Inspection of the table shows the Hartree $\zeta_{4f}$'s to be in excellent numerical agreement with experiment, whereas the H-F values are appreciably larger. More accurate relativistic or nonrelativistic H-F solutions should not substantially affect this observation. The good agreement of the Hartree $\zeta_{4f}$'s is, we believe, due to a cancellation of errors. The Hartree $P_{4f}(r)$ is radially expanded relative to the Hartree-Fock function and hence has a smaller $\zeta_{4f}$. One expects the "true" many-electron eigenfunction to have a charge density which is, if anything, more contracted than the H-F density; thus the source of the smaller magnitude of the Hartree $\zeta_{4f}$'s is apparently not the key to the discrepancies for the H-F values. Similar discrepancies have been observed elsewhere.[6]

As already noted, the $V(r)$'s used when computing the H-F $\zeta_{4f}$'s did not include exchange effects. The inclusion of exchange, as simply a modification of $V(r)$ as a one-electron potential, would further increase the computed values and result in poorer agreement with experi-

TABLE I. Theoretical and experimental 4 f spin-orbit coupling parameter ($\zeta_{4f}$'s) for some rare-earth ions. All energies are in cm$^{-1}$.

| | $Ce^{3+}$ $(4f^1)$ | $Pr^{3+}$ $(4f^2)$ | $Sm^{3+}$ $(4f^5)$ | $Dy^{3+}$ $(4f^9)$ | $Er^{3+}$ $(4f^{11})$ | $Tm^{3+}$ $(4f^{12})$ |
|---|---|---|---|---|---|---|
| Hartree-Fock | 830 | 980 | 1480 | 2315 | 2830 | ... |
| Hartree | | 785 | | | | 2740 |
| Experimental | 644 | 800 | 1200 | 1850 | 2400 | 2750 |

ment. Naive estimates of those correlation effects which could be written as perturbing the $P(r)$ or $V(r)$ of Eq. (1) suggest that these also would increase the $\zeta_{4f}$'s. There are, in addition, other correlation effects which cannot be straightforwardly inserted into Eq. (1), such as the possibility that the correlation energy is different for different $J$ states. It is not obvious what the behavior of such differences would be, if they exist, but it should be noted that in the case of multiplet spectra the discrepancies between computed and observed Slater $F^k$ parameters appear to be largely due to this. To the extent that such correlation effects do play a role in the LS coupling spectra, the observed $\zeta_{4f}$ cannot be thought to take the form of Eq. (1).

Other possible causes of the disagreement between computed and observed $\zeta_{4f}$'s can only briefly be mentioned here. First there is the effect arising from any partial breakdown of Russell-Saunders coupling, and secondly relativistic effects (e.g., "spin-other orbit" coupling) could contribute to an "effective" $\zeta_{4f}$. Zeeman effect calculations[7] for rare-earth atoms which have taken such matters into account in the process of obtaining good agreement between observed and computed $g_J$'s suggest that such effects do not make large contributions to the observed $\zeta_{4f}$'s. Judd has investigated deviations from the Landé interval rule due to spin-spin coupling and the partial breakdown of Russell-Saunders coupling due to the mixing of excited multiplet states (of the same configuration) into the ground multiplet state. (Such mixing can be classed as a "correlation" effect.) In his investigation $\zeta_{4f}$'s were parameters to be adjusted so as to match experiment and their resultant values deviated but slightly from $\zeta_{4f}$ values obtained by assuming pure Russell-Saunders coupling. Further investigation of these matters seems desirable. Another possible contributor to the $\zeta_{4f}$ discrepancies would be any variation in $P_{4f}(r)$ with $J$. Any such variation could contribute to the energy difference between $J$ states and thus affect an observed $\zeta_{4f}$. This matter will be taken up in a forthcoming publication which reports calculations for $Ce^{3+}(4f^1)$ in its two different $J$ states [$J = \frac{5}{2}$ (ground) and $J = \frac{7}{2}$] in which LS coupling was explicitly included in the H-F formalism. We believe that Eq. (1) does not define the parameter $\zeta_{4f}$ actually obtained from experiment; this matter too will be taken up in a forthcoming publication

[2] R. J. Elliott and K. W. H. Stevens, Proc. Roy. Soc. (London) 219A, 387 (1953).

[3] B. Bleaney, Proc. Phys. Soc. (London) 68A, 937 (1955).

[4] B. R. Judd and I. Lindgren, Phys. Rev. 122, 1802 (1961), see also references therein; and I. Lindgren (to be published).

[5] E. C. Ridley, Proc. Cambridge Phil. Soc. 56, 41 (1960).

[6] K. Kambe and J. H. Van Vleck, Phys. Rev. 96, 66 (1954), have observed that a $\zeta_{2p}$ computed for H-F oxygen functions lies 10% higher than experiment, and M. J. D. Powell (unpublished) has observed 30% discrepancies for iron series H-F functions, quite similar to what is seen here.

[7] R. A. Satten, J. Chem. Phys. 21, 637 (1953), and references therein.

TABLE II. Hartree-Fock, Hartree, Thomas-Fermi, and parametrized $\langle r^{-3} \rangle$ integrals for rare-earth ions (in units of $a_0^{-3}$).

| | $Ce^{3+}$ | $Pr^{3+}$ | $Nd^{3+}$ | $Sm^{3+}$ | $Eu^{2+}$ | $Gd^{3+}$ | $Dy^{3+}$ | $Er^{3+}$ | $Tm^{3+}$ | $Yb^{3+}$ |
|---|---|---|---|---|---|---|---|---|---|---|
| Hartree-Fock $\langle r^{-3} \rangle =$ | 4.72 | 5.39 | 6.03 | 7.36 | 7.53 | 8.84 | 10.34 | 12.01 | $\cdots$ | 13.83 |
| Hartree $\langle r^{-3} \rangle =$ | | 4.33 | | | | | | | 11.42 | |
| Thomas-Fermi $\langle r^{-3} \rangle =$ | | | | | 6.95 | | | | | |
| Parametrized $\langle r^{-3} \rangle$'s | | | | | | | | | | |
| Elliott and Stevens | $4\pm1$ | $5\pm1.2$ | $6\pm1.5$ | $7.5\pm2$ | | $9.2\pm2.3$ | $10.5\pm2.6$ | $12\pm3$ | $13\pm3.3$ | $13.5\pm3.5$ |
| Bleaney | 4.8 | 5.5 | 6.2 | 7.5 | | 9.1 | 10.9 | 12.7 | 13.6 | 14.4 |
| Judd and Lindgren | 3.64 | 4.24 | 4.83 | 6.02 | | 7.27 | 8.60 | 10.10 | 10.95 | 11.89 |

as we cannot discuss this in the limited space available here.

Finally, it should be noted that, with the exception of the $Ce^{3+}$ data, we are dealing with values obtained for the trivalent ions in salts or solutions. It is generally accepted that parameters such as the $\zeta_{4f}$'s or $F^k$'s are but slightly affected by the ion's environment. The fact that the discrepancy for $Ce^{3+}$ is as great as for the other ions supports this. Despite the small magnitude of the effect, the fact that we are generally not dealing with free ion data should be borne in mind.

## HYPERFINE FIELDS AND NUCLEAR MAGNETIC MOMENTS

Let us now consider the calculated $\langle r^{-3} \rangle$. It is seen from Table II that the H-F $\langle r^{-3} \rangle$'s are similar in behavior to those of Bleaney[3] or of Elliott and Stevens[2]; discrepancies between the HF and Bleaney values are under 5%. Ridley's Hartree $\langle r^{-3} \rangle$'s and Judd and Lindgrens' values are approximately 20% smaller. Their close agreement is not surprising because Judd and

Table III. The magnitudes of the nuclear magnetic moments ($\mu$'s) of a few rare-earth isotopes as estimated by Bleaney and Lindgren, from optical (other than $4f$ shell) hyperfine parameters and by using the current Hartree-Fock $\langle r^{-3} \rangle$'s.

| | $Pr^{141}$ | $Nd^{143}$ | $Sm^{147}$ | $Ho^{165}$ | $Yb^{173}$ |
|---|---|---|---|---|---|
| $\mu$ as estimated by Bleaney | 3.9 n.m. | 1.0 n.m. | 0.83 n.m. | 3.3 n.m. | 0.68 n.m. |
| $\mu$ as estimated by Lindgren | 5.0 | 1.2 | 1.03 | 4.1 | 0.72 |
| Optical $\mu$'s | 4.0 | 1.1 | 0.76 | 3.7 | 0.67 |
| $\mu$ obtained using Hartree-Fock $\langle r^{-3} \rangle$'s | 3.9 | 1.0 | 0.85 | 3.5[a] | 0.63 |

[a] Interpolated value.

Lindgren relied in part on matching Ridley's functions and in part on the observed $\zeta_{4f}$'s.

The implications of using the various $\langle r^{-3} \rangle$'s in estimates of the $\mu$'s can be seen by inspecting Table III, where $\mu$'s are listed for some isotopes for which we also have estimates from Bleaney, and Lindgren, *and* from optical hyperfine parameters involving other than $4f$ electrons, i.e., unfilled $6s$, $6p$, and/or $5d$ shells. The $\mu$ values reflect what we have already seen in Table II, i.e., both the Bleaney and the Hartree-Fock $\mu$'s are quite similar, while those of Lindgren are consistently larger since the observed hyperfine parameter $a_{4f}$ is of the form $a_{4f} \sim |\mu| \langle r^{-3} \rangle / I$, where $I$ is the nuclear spin. The "optical" $\mu$'s generally lie between the other values and are closest to those of Bleaney. If we assume, as is usual, that only the $4f$ shell contributes to the hyperfine field and that an average $4f$ orbital $\langle r^{-3} \rangle$ suffices to describe the field, then the Hartree-Fock based $\mu$'s are, *in principle*, the best estimates in Table III. We do not believe that such assumptions are valid and that one must talk of and use "*effective*" $\langle r^{-3} \rangle$'s which are not identical with the $\langle r^{-3} \rangle$ *integrals*. While the "optical" values are also subject to theoretical uncertainty, those obtained with $a_{6s}$'s (i.e., all but $Nd^{143}$) and the Fermi-Segre-Goudsmit formula have at least been obtained by different means. These values suggest smaller deviations than we would otherwise expect between "effective" $\langle r^{-3} \rangle$'s and the integrals. We believe that rather substantial uncertainties are associated with the $\mu$ values given in any row of Table III.

A fuller treatment of this subject which presents further data, more discussion, and a more complete referencing to pertinent work[8] will be submitted for publication.

[8] A. J. Freeman and R. E. Watson (to be published).

JOURNAL OF APPLIED PHYSICS    SUPPLEMENT TO VOL. 33, NO. 3    MARCH, 1962

# Magnetic Hyperfine Interaction and Electronic Relaxation in Sm³⁺ in EuIG*

M. E. Caspari, S. Frankel, and G. T. Wood

*Department of Physics, University of Pennsylvania, Philadelphia, Pennsylvania*

The attenuation and rotation of the integral correlation of the 1415–122 kev $\gamma$-$\gamma$ cascade following the $K$ capture of $Eu^{152}$ was measured in neutron irradiated polycrystalline samples EuIG from $-25°C$ to above the Néel temperature with and without magnetizing field perpendicular to the counter plane. The existence of time-dependent hyperfine interactions of short relaxation times was taken into account in the evaluation of the average component of the effective hyperfine field $\langle H_{int}{}^z \rangle$ along the direction of the magnetizing field. The sign, magnitude, and temperature dependence of $\langle H_{int}{}^z \rangle$ was in excellent agreement with that calculated from molecular field theory under the assumption that the electronic configuration of $Sm^{152}$ following the $K$ capture of $Eu^{152}$ was that of the $Sm^{3+}$ ion. The contribution to the magnetic hyperfine field arising from exchange between the $4f$ and inner core $s$ electrons was found to be small compared with that from the orbital and spin moments of the $4f$ shell. The absolute magnitude of the hyperfine field was $4.7 \times 10^6$ oe. Under the assumption that the time-dependent interactions are largely magnetic, the electronic relaxation time was found to be $3.2 \times 10^{-12}$ sec at 400°C and varied inversely with the absolute temperature.

THE method of investigating hyperfine interactions in magnetic atoms or ions in magnetic materials by $\gamma$-$\gamma$ angular correlation techniques has been discussed previously.[1] In this paper we will give the results of measurements of the 1415- to 122-kev $\gamma$-$\gamma$ angular correlation in $Sm^{152}$ following the $K$ capture of $Eu^{152}$ in neutron irradiated polycrystalline samples of EuIG enriched in $Eu^{151}$ over the temperature range from $-25°C$ to above the Néel temperature.

A consistent interpretation of our data was only possible under the assumption that the electronic relaxation time of the rare-earth ions $\tau_s$ is smaller than the mean lifetime $\tau_n$ of the intermediate excited nuclear state ($2+$ state in $Sm^{152}$) over the entire temperature range studied in this paper. All of the attenuation of the integral correlation above the Néel temperature and most of the attenuation below the Néel temperature is then due to transitions among the magnetic sublevels of the intermediate nuclear state induced by the time-dependent magnetic and/or quadrupole hyperfine interactions.[2]

Below the Néel temperature with a small magnetizing field applied perpendicular to the plane of the counters ($z$ direction), there exists a rotation of the angular correlation pattern due to the precession of the nuclear magnetic moment of the intermediate excited state in the average component of the magnetic hyperfine field along the $z$ direction. This average component of the magnetic hyperfine field $\langle H_{int}{}^z \rangle$ stems from the unequal statistical distribution among the electronic magnetic substates which are split by the exchange field $H_{exch}$ acting on the spins of the rare-earth ions. According to the application of the Néel theory of ferrimagnetism to rare-earth iron garnets, $H_{exch}$ arises mainly from the antiferromagnetic exchange interaction between the $Fe^{3+}$ and rare-earth ion sublattices, is proportional to the $Fe^{3+}$ sublattice magnetizations, and can, therefore, be oriented by an applied magnetizing field.[3]

By an extension of a theory of Abragam and Pound,[2] it is possible to show that,[4] within approximations satisfied by our experimental conditions, the integral correlation function $W(\theta)$ (i.e., the coincidence rate for an angle $\theta$ between the counters integrated over $\tau_n$) below the Néel temperature with an applied field perpendicular to the counter plane becomes

$$W(\theta) = 1 + A_2 G_2 P_2 [\cos(\theta - \Delta\theta)], \qquad (1)$$

where the rotation

$$\Delta\theta = \tfrac{1}{2} \tan^{-1}\beta, \qquad (2)$$

with $\beta = 2\omega\tau_N/(1+\lambda\tau_N)$ and $\omega = \mu_N{}^0 g_N \langle H_z{}^{int} \rangle/\hbar$. Here, $A_2$ is the coefficient in the Legendre polynomial expansion of the unperturbed $W(\theta)$, $\mu_N{}^0$ is the nuclear Bohr magneton, $g_N$ and $\tau_N$ are the "$g$" factor and lifetime of the intermediate nuclear state, respectively, and the parameter $\lambda$ represents the influence of the time-dependent perturbations on the correlation. The attenuation factor $G_2$ is $(1+\lambda\tau_N)^{-1}(1+\beta^2)^{-\frac{1}{2}}$. Above the Néel temperature $\beta = 0$, $G_2 = (1+\lambda\tau_N)^{-1}$, and $\Delta\theta = 0$.

From the measured values of the rotation and attenuation of the integral angular correlation, the

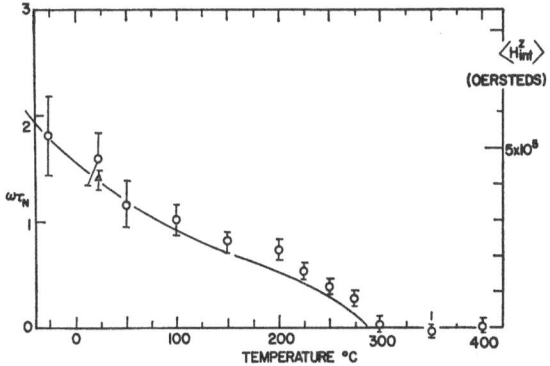

FIG. 1. The parameter $\omega\tau_N$ and the average component of the magnetic hyperfine field acting at the $Sm^{152}$ nuclei in the direction perpendicular to the counter plane, $\langle H_{int}{}^z \rangle$ as a function of temperature. The solid curve represents a theoretical relation of $\langle H_{int}{}^z \rangle$ as a function of temperature as explained in the text.

* Supported in part by the AROD, the U. S. Air Force, and the N.S.F.

[1] M. E. Caspari, S. Frankel, and M. A. Gilleo, J. Appl. Phys. **31**, 320S (1960).
[2] A. Abragam and R. V. Pound, Phys. Rev. **92**, 943 (1953).
[3] R. Pauthenet, Ann. Phys. **3**, 424 (1958).
[4] A more detailed account of this development will be published in the immediate future.

values of the parameters $\omega\tau_N$ and $\lambda\tau_N$ were obtained as a function of temperature. The value of $\langle H_{int}{}^z\rangle$ as a function of temperature was obtained using the independently measured values of $g_N = 0.351$ [5] and $\tau_N = 2\times10^{-9}$ sec.[6] The parameter $\omega\tau_N$ and $\langle H_{int}{}^z\rangle$ are plotted as a function of temperature in Fig. (1).

Under the assumption that the effect of the hyperfine interaction on the 1415- to 122-kev angular correlation in $Sm^{152}$ following the $K$ capture of $Eu^{152}$ is determined by the $Sm^{3+}$ electronic configuration, we have calculated $\langle H_{int}{}^z\rangle$ from molecular field theory, using standard perturbation theory and neglecting crystalline field and saturation effects.[7] In this calculation we have used the magnetic hyperfine interaction Hamiltonian given by Abragam and Pryce[8] and have neglected the "indirect" contribution to the Hamiltonian arising from the exchange interaction between the $4f$ and inner core $s$-electrons. The value of $\langle r^{-3}\rangle_{av}$ ($r$ = radius of $4f$ electrons) for $Sm^{3+}$ was taken from Elliot and Stevens[9] and that of $H_{exch}$ as a function of temperature in EuIG from Wolf and Van Vleck.[10] The theoretical value of $\langle H_{int}{}^z\rangle$ is indicated by the curve in Fig. 1 and is seen to fit the experimental points well. The calculated sign of $\langle H_{int}{}^z\rangle$ also agrees with the observed sense of the rotation of the angular correlation.[11]

In addition, the absolute magnitude of the magnetic hyperfine field $\langle|H_{int}|\rangle$ was calculated and was found to be $4.7\times10^6$ oe almost independent of temperature.

While the excellent agreement between theory and experiment may be partly fortuitous due to the uncertainties in the value of $\langle r^{-3}\rangle_{av}$ and the neglect of the crystalline field in the calculation, it nevertheless lends experimental support to the following conclusions:

(1) The electronic configuration following the $K$ capture decay of $Eu^{152}$ reaches that of the $Sm^{3+}$ ground state configuration within a time less than $10^{-9}$ sec.

(2) Calculations based on the molecular field theory appear to predict adequately the value of the magnetic hyperfine field acting on the $Sm^{3+}$ nucleus in EuIG. In contrast, the experimentally determined sublattice magnetization of the $Sm^{3+}$ ion sublattice in SmIG appears to be zero,[3] a fact which does not seem to be compatible with conclusions reached from molecular field calculations.[12]

(3) In contrast to the transition elements, the "indirect" contribution to the magnetic hyperfine interaction arising from the exchange interaction between the $4f$ electrons and inner core $s$ electrons in $Sm^{3+}$ is small compared with the "direct" contribution.

(4) The presence of short electronic relaxation times ($<10^{-9}$ sec) in the rare-earth ions in rare-earth iron garnets, first postulated in order to explain the results of ferrimagnetic resonance experiments,[13] has received experimental confirmation.

Neglecting quadrupole interactions, the value of $\tau_s$ can be calculated from $\lambda$, following the development of Abragam and Pound.[2] At 400°C we find $\tau_s = 3.2\times10^{-12}$ sec in good agreement with the value for $\tau_s$ postulated by de Gennes, Kittel, and Portis[14] to explain the linewidth of ferrimagnetic resonance experiments. Within experimental accuracy $\tau_s$ was found to vary inversely with the absolute temperature over the temperature range investigated ($-25°$ to 400°C).

### ACKNOWLEDGMENTS

The authors are indebted to the Bell Telephone Laboratories for supplying the samples used in this experiment. We would also like to thank Dr. M. A. Gilleo for his efforts in the sample preparation and for his valuable advice and help during the initial stages of this investigation.

[5] R. Stiening and M. Deutsch, Phys. Rev. 121, 1484 (1961).

[6] A. W. Sunyar, Phys. Rev. 98, 653 (1955).

[7] J. Kanamori and K. Sugimoto, J. Phys. Soc. Japan 13, 754 (1958).

[8] A. Abragam and M. H. L. Pryce, Proc. Roy. Soc. (London) A205, 135 (1951).

[9] R. J. Elliot and K. W. H. Stevens, Proc. Roy. Soc. (London) A219, 387 (1953).

[10] W. P. Wolf and J. H. Van Vleck, Phys. Rev. 118, 1490 (1960).

[11] M. E. Caspari, S. Frankel, D. Ray, and G. T. Wood, Phys. Rev. Letters 6, 345 (1961).

[12] J. A. White and J. H. Van Vleck, Phys. Rev. Letters 6, 412 (1961).

[13] C. Kittel, Phys. Rev. 115, 1587 (1959).

[14] P. G. de Gennes, C. Kittel, and A. M. Portis, Phys. Rev. 116, 323 (1959).

JOURNAL OF APPLIED PHYSICS    SUPPLEMENT TO VOL. 33, NO. 3    MARCH, 1962

# The Atomic Moments and Hyperfine Fields in Fe₂Ti and Fe₂Zr*

C. W. KOCHER AND P. J. BROWN

*Brookhaven National Laboratory, Upton, New York*

The magnetic properties of $Fe_2Ti$ and $Fe_2Zr$, intermetallic compounds with Laves type structures, were studied using the techniques of neutron diffraction and nuclear resonance fluorescence (Mössbauer effect). Both compounds are ferromagnetic, with magnetic moments of 0.35 and 2.56 Bohr magnetons, respectively, at room temperature, and 0.92 and 3.12 Bohr magnetons, respectively, at liquid helium temperatures. A magnetic form factor for the iron atoms in $Fe_2Zr$ was determined. The nuclear resonance fluorescence experiments gave measures of the magnetic fields at the iron nuclei. At room temperature, the field for $Fe_2Zr$ was 190±10 kgauss, while that for $Fe_2Ti$ was very low, less than 5 kgauss. The relation between the atomic magnetic moments and the magnetic fields at the nuclei is discussed.

COMPOUNDS $Fe_2Ti$ and $Fe_2Zr$ are members of the group of intermetallic compounds possessing structures of the type first described by Laves and his co-workers,[1] and commonly known as Laves phases. Laves phases are found at the composition $AB_2$ in very many alloy systems where the radius ratio of $A$ to $B$ is near 1.2 to 1. This is thought to occur because geometrically the Laves structures provide an efficient means of filling space with spheres whose radii are in the ratio $(\frac{3}{2})^{\frac{1}{2}}:1$ when the small spheres are twice as numerous as the large ones.

There are three distinct Laves structures all of which, ideally, have the same space filling properties; all three are built up from double layers of hexagonal arrays of the large $A$ atoms. These layers may be stacked in the same way as the single layers in the close-packed hexagonal structure so that alternate layers superpose—this is the arrangement adopted by $Fe_2Ti$; alternatively, the double layers may be stacked as in the cubic close-packed structure so that superposition occurs every third layer—this arrangement is that of $Fe_2Zr$. The third Laves phase may be regarded as being transitional between the other two. In all the Laves phases the smaller $B$ atoms are located at the vertices of tetrahedra which form a framework surrounding the $A$ atoms. The arrangement is such that each $A$ atom has four $A$ neighbors and twelve $B$ neighbors, whereas each $B$ atom has six $A$ and six $B$ neighbors.

It is thought that the occurrence of Laves phase is determined largely by size considerations and that the bonding is substantially the same as in the metals; the stoichiometry of the phases and their hard brittle character arise from geometrical factors, and do not indicate electrochemical interaction between the component metals.

A large number of Laves phases are found among the alloys between transition elements. Here the size requirements are often met when $A$ is one of the early transition metals in a series and $B$ is one of the later metals in either the same or a different series. The literature contains little or no information on the magnetic properties of such alloys and the present neutron diffraction and Mössbauer studies were undertaken not only to remedy this situation but also to investigate the dependency of the internal magnetic fields of the iron atoms on their magnetic moment.

Samples of $Fe_2Zr$ and $Fe_2Ti$ were prepared from materials of better than 99.9% purity by melting the metals together in the correct proportions under vacuum in zirconia crucibles. The furnace cooled ingots were examined metallographically and by x-ray powder diffraction. The lattice parameters obtained from the powder photographs were consistent with those found by Elliot[2]; in neither case were detectable amounts of a second phase present. The amount of zirconium contamination of the $Fe_2Ti$ was found spectrographically to be less than 0.3%. Saturation magnetization measurements show that both compounds are ferromagnetic.[3] The magnetic moments in Bohr magnetons per molecule derived from these measurements are:

|  | Room temperature | Liquid He |
|---|---|---|
| $Fe_2Ti$ | 0.35 | 0.92 |
| $Fe_2Zr$ | 2.56 | 3.12 |

The coherent magnetic scattering of neutrons from the two phases has been measured using the polarized beam method.[4] This technique enables the magnetic scattering amplitude to be determined with fair precision even when the magnetic moment is as low as that in $Fe_2Ti$. The measurements were made on polycrystalline samples cut from the ingots; the polarization ratios were obtained for the reflections from large grains which were in favorable orientations. In $Fe_2Ti$ the diffraction measurements led to an assignment of 0.10±0.03 $\mu_B$ for the iron atoms in the $2(a)$ sites and 0.20±0.02 $\mu_B$ for the iron atoms in the $6(h)$ sites of space group $P6_3/mmc$. These values lead to a moment of 0.35 $\mu_B$ for the magnetic moment per molecule, in good agreement with the magnetization measurements. In assigning these moments the iron metallic form factor was used, since the limited number of measurements and the uncertainty of the structural parameters did not permit the derivation of a form factor.

* Work performed under the auspices of the U. S. Atomic Energy Commission.

[1] F. Laves and H. Witte, Metallwirtschaft 14, 645 (1935).
[2] R. P. Elliot, OSR Technical Note OSR-TN-247 (1954).
[3] T. R. McGuire (private communication).
[4] R. Nathans, C. G. Shull, G. Shirane, and A. Andresen, J. Phys. Chem. Solids 10, 138 (1959).

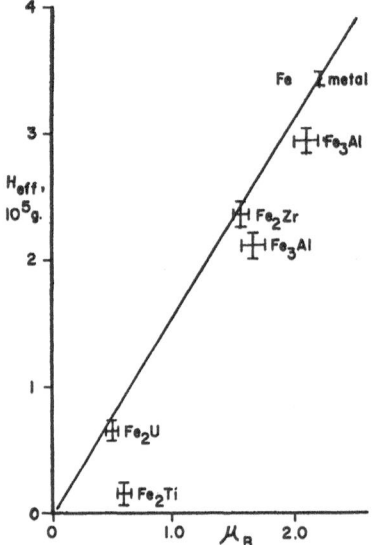

FIG. 1. The relation between $H_{eff}$, the magnetic field seen by the nucleus, and $\mu$, the atomic magnetic moment, both given for zero degrees absolute temperature, for various materials.

In the simpler cubic structure of $Fe_2Zr$ all the iron atoms are equivalent and there are no structural parameters, it was therefore possible to use the diffraction data on this phase to derive a partial magnetic form factor for the iron atoms. This form factor lies above that for iron metal and thus indicates a more compact distribution of magnetic electrons in $Fe_2Zr$ than in $\alpha$-iron. In both materials the assignment of any moment to Ti or Zr atoms can be ruled out.

The Mössbauer patterns of the $Fe_2Ti$ at room temperature showed two nonsymmetric absorption lines. As an unsplit source $Co^{57}$ in copper was used. The average half-width of the lines, source plus absorber, was found to be $0.350\pm0.010$ mm/sec with the center shifted by $-0.186\pm0.006$ mm/sec compared to stainless steel. While the principal origin of the $Fe_2Ti$ Mössbauer pattern lies in electric quadrupole effects, the apparent lack of symmetry can be explained by the presence of a small additional Zeeman splitting of the iron nuclei levels. The estimate of the Zeeman field based on the difference in peak shapes was $5\pm3$ kg. Data taken at nitrogen temperatures are consistent with this analysis and estimate of the room temperature Zeeman field; at this temperature the complication of two different iron moments plus a quadrupole splitting comparable with the hyperfine effects does not permit a unique set of parameters to be derived from the data. The room temperature values for the Zeeman field were extrapolated to 0°K using the data for temperature dependence of the saturation magnetization given above. In Fig. 1, the value plotted for the magnetic moment represents the weighted average of the two iron sites; their relative magnitudes are determined from the neutron data.

For $Fe_2Zr$ a more typical absorption pattern is found. A result of the regular environment of the iron atoms is the absence of significant quadrupole effects. The

Zeeman field at room temperature gives a value of $190\pm10$ kg for the internal field at the iron sites. Again the 0°K data shown in Fig. 1 represents an extrapolation using the saturation magnetization data.

Freeman and Watson[5] have proposed that the nuclear magnetic fields are quite sensitive to changes in the mean radius of the $3d$ electrons, and also the extent to which there are orbital momentum contributions to the total magnetic moment. All other contributions to the Zeeman fields should be directly proportional to the intrinsic moment of the iron atoms. In Fig. 1, a plot is shown of the field at the iron nucleus as a function of the moment on the iron atom. In addition to the $Fe_2Ti$ and $Fe_2Zr$ data, we have included the results of Mössbauer work on $Fe_2U$ taken by Komura et al.[6] and the work of Johnson et al.[7] on ordered $Fe_3Al$, where there are two distinct iron moments.[8] The line drawn represents the behavior if there is a purely linear relation between the hyperfine field and the intrinsic moment at the iron sites. It is seen that over a substantial range of magnetic moment values the linear relationship holds up fairly well. At very low moment values the points fall below the curve. In the case of $Fe_2Ti$ the field is quite low in contrast to the result for $Fe_2U$. Both of these alloys are Laves phases; they differ, however, in $Fe_2Ti$ being of the hexagonal type discussed above, while $Fe_2U$ is of the cubic type along with $Fe_2Zr$. The nonmetals shown, as expected, depart considerably from the straight line dependence characterizing the metals.

The surprising character of the data in Fig. 1 is not the departure of the individual points from the straight line but the extent to which the simple proportionality to the net moment is valid. For the $Fe_3Al$ points we know that the asphericity of the unpaired $3d$ electrons is different for the two iron sites. The neutron form factor data for $Fe_2Zr$ mentioned above demonstrates that the density of the unpaired electron is somewhat different from $\alpha$-iron. No similar data is as yet available on $Fe_2U$ and $Fe_2Ti$. It appears, however, that the magnitude of the magnetic fields at the nucleus is essentially established only by the net unpaired spin. Other factors, such as the spatial distribution of the spin density, orbital contributions, and the effects of the $4s$ electrons would appear to play secondary roles in influencing the hyperfine field. Thus it can be expected that the problem of utilizing the experimental results to give significant information on the configurations of the outer electrons must await further clarification of the theoretical situation.

[5] A. J. Freeman and R. E. Watson, Phys. Rev. 123, 2027 (1961).
[6] S. Komura, N. Kunitomi, P. Tseng, N. Shikazono, H. Takekoshi, J. Phys. Soc. Japan 16, 1479 (1961).
[7] C. E. Johnson et al., Second International Mössbauer Conference, Saclay, France, September, 1961.
[8] R. Nathans, T. Pigott, and C. G. Shull, J. Phys. Chem. Solids 6, 38 (1958).

JOURNAL OF APPLIED PHYSICS     SUPPLEMENT TO VOL. 33, NO. 3     MARCH, 1962

# Internal Magnetic Fields in Nickel-Rich Nickel-Cobalt Alloys

LAWRENCE H. BENNETT AND RALPH L. STREEVER, JR.
*National Bureau of Standards, Washington 25, D. C.*

The nuclear magnetic resonances of both the Ni$^{61}$ nuclei and the Co$^{59}$ nuclei have been observed in a series of ferromagnetic nickel alloy powders containing up to 2.0 atomic percent cobalt. The room temperature value of the magnetic field at the site of the Co$^{59}$ nucleus in any of these alloys is found to be about 111 koe, much smaller than the 211-koe field at a Co$^{59}$ site in pure cobalt. This is also considerably smaller in magnitude than the magnetic field at the nickel site in either the alloy or in pure nickel, which, assuming a Ni$^{61}$ nuclear moment of 0.30 nuclear magnetons, is about 170 koe. The linewidths of both the nickel and cobalt resonances increase with concentration with no detectable shift in frequency. The temperature dependence of the Co$^{59}$ resonance frequency in a nickel alloy containing 0.59 atomic percent cobalt has been measured, and found to be different fron that of Ni$^{61}$ in pure nickel and in the alloy.

A S part of a continuing study of nuclear magnetic resonance in nickel and nickel alloys, dilute alloys of cobalt in nickel were prepared. An object was to determine the hyperfine field at the cobalt nucleus and study the nuclear resonances of both the Co$^{59}$ and the Ni$^{61}$ as a function of cobalt concentration. An estimate of the internal magnetic field at the Co$^{59}$ nucleus obtained from a crude linear extrapolation of the low temperature specific heat data[1] is 80 koe. Measurements of the beta decay from cobalt in nickel[2] did not provide an accurate value of the internal field but did determine that the sign of the cobalt hyperfine field in nickel is negative. One can use the semiempirical rule[3] that the hyperfine field at the nucleus will be linear in total spin of the atom, along with the assumption that the local spin of the cobalt in the alloy will be less than its local spin in pure cobalt metal. In that case the field at the Co$^{59}$ nucleus in pure cobalt (211 koe) places an upper limit of approximately 213 Mc on the resonance frequency of Co$^{59}$ in the alloy. For these reasons a search was made for the cobalt resonance from a frequency of 213 Mc down.

## EXPERIMENTAL METHODS AND RESULTS

Powder samples were prepared by an atomization process.[4] Particles 10 $\mu$ and less were annealed for one hour at 1100°C under hydrogen.

The nickel resonances in a pure and in a 0.59 percent Co in Ni sample were observed on the oscilloscope using a conventional self-quenched super-regenerative oscillator-detector originally designed for the observation of nuclear quadrupole resonance absorption by Dean.[5] A similar oscillator, but modified to operate at higher frequencies, using a 6AF4 triode in place of a triode connected 6AK5 pentode, was used to search for the cobalt resonance. The cobalt resonance was observed on the oscilloscope in a 0.016, 0.59, and 2.0 percent Co in Ni sample at room temperature at a frequency of 112 Mc, corresponding to an internal field of 111 koe.

Because it is difficult to obtain accurate frequency measurements with the super-regenerative oscillator, a push-pull marginal oscillator[6] was used to record the nuclear resonances of both the cobalt and the nickel in these alloys. This marginal oscillator operates over a wide range of rf levels from 20 to 200 Mc. It is very stable and has been found very satisfactory for this work. A double frequency detection scheme is employed which presents, in the limit of low modulation, the second derivative of the resonance line shape. With this circuit we have observed the nickel and the cobalt resonances in the same samples as above. The intensity of the cobalt signal was a maximum in the half-percent sample.

The line shape will be discussed in more detail elsewhere. For the purpose of presenting the data here, we have defined the width as the frequency difference between the main positive and negative peaks of the second derivative and the frequency of the resonance as the crossover point between them. The room temperature widths of both the Ni$^{61}$ and the Co$^{59}$ lines increase as function of cobalt concentration as shown in Fig. 1. The linewidth of the Co$^{59}$ resonance in the 0.59 percent alloy

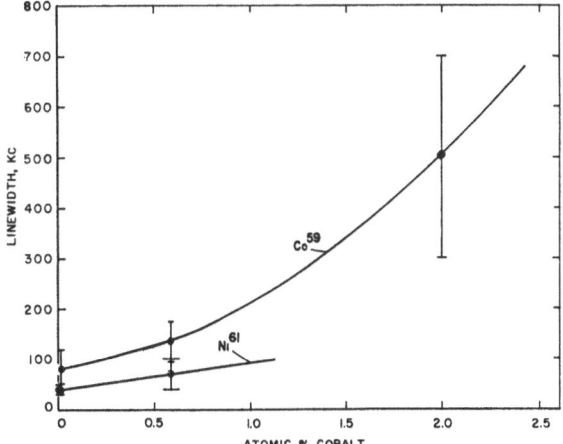

FIG. 1. Linewidths of Co$^{59}$ and Ni$^{61}$ resonances in nickel cobalt alloys at room temperature.

[1] V. Arp, D. Edmonds, and R. Petersen, Phys. Rev. Letters **3**, 212 (1959).
[2] D. D. Hoppes and R. W. Hayward (private communication).
[3] R. E. Watson and A. J. Freeman, Phys. Rev. **123**, 2027 (1961).
[4] We wish to thank the Federal-Mogul Corporation for the preparation of these alloys.
[5] C. Dean, Ph.D. thesis, Harvard University (1952).
[6] Similar to that described in G. Benedik and Y. Kushida, Phys. Rev. **118**, 46 (1960), and R. G. Shulman, Phys. Rev. **121**, 125 (1961).

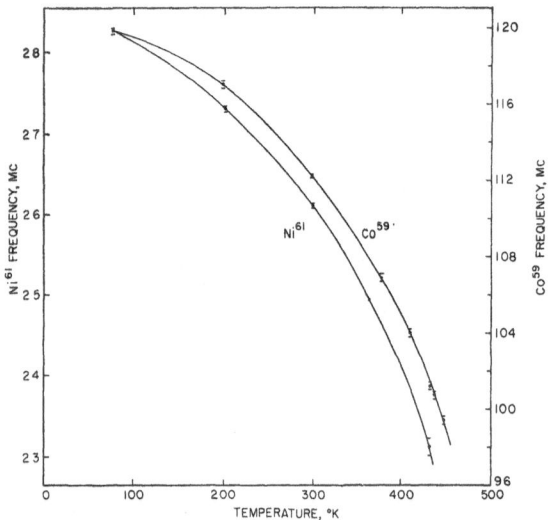

FIG. 2. Temperature dependence of Ni[61] resonance frequency in nickel, and Co[59] resonance frequency in half-percent alloy.

increased from 140 kc at room temperature to 300 kc at 77°K. Shifts, if any, in the resonance frequency as a function of cobalt concentration were too small to be detected. The precision in detecting shifts in the Ni[61] resonance is about 20 kc; in the Co[59] about 60 kc.

A comparison of the temperature dependence of the resonance frequency of Co[59] in the 0.59 percent alloy with the frequency of the Ni[61] resonance[7] in pure nickel is shown in Fig. 2. No attempt was made to normalize to the appropriate Curie temperatures because the correction would be small compared with the observed difference and because the Ni[61] resonance in the half-percent alloy was not shifted in frequency from the pure nickel at room or liquid-nitrogen temperatures.

[7] R. L. Streever, Jr., and L. H. Bennett, Bull. Am. Phys. Soc. **5**, 491 (1960), and to be published. The temperature dependence of the Ni[61] resonance has also been reported by L. J. Bruner, J. I. Budnick, R. J. Blume, and E. L. Boyd, Bull. Am. Phys. Soc. **5**, 491 (1960); and T. Hihara, T. Kushida, Y. Koi, and A. Tsujimura, J. Phys. Soc. Japan (to be published).

## DISCUSSION

The internal field at the cobalt nucleus found here is much smaller than the field at the nickel nucleus, provided, of course, that the nuclear moment of nickel is assumed to be 0.30 nm.[8]

That there is no detectable shift in the nickel resonance frequency as a function of concentration can be explained on a localized moment model by assuming that any nickel atom having cobalt as a neighbor does not contribute to the main line, but forms a "satellite" line (which, however, we have not yet observed). The cobalt case is different in that, for low concentrations, nearly all of the cobalt atoms have only nickel as close neighbors. Thus the fact that there is no shift in the cobalt frequency reinforces the localized moment viewpoint.

The temperature dependence of the cobalt frequency in the half-percent sample is not identical to the temperature dependence of the nickel, in either pure nickel or in the alloy. The cobalt frequency does not fall as fast as the nickel frequency with increasing temperature. The difference in the explicit temperature dependence between the two sites may be due to a different temperature dependence of the excitation of electrons higher in the *d* band. In addition, the effects of thermal expansion are probably different at the two sites.

*Note added in proof.* It has come to our attention since the submission of this paper that the spectra of small amounts of cobalt in nickel was studied by S. F. Ravitz, R. C. LaForce, and G. F. Day, International Conference on Magnetism and Crystallography, Kyoto, Japan, September, 1961.

### ACKNOWLEDGMENTS

We are indebted to I. L. Cooter, Dr. R. E. Howard, and Dr. Y. Kushida for helpful comments. We thank D. Brown and G. Uriano for their assistance in making measurements and V. M. Johnson for invaluable electronics contributions.

[8] J. W. Orton, P. Auzins, and J. E. Wertz, Phys. Rev. **119**, 1691 (1961).

# Thin Films–1

## C. J. Kriessman, *Chairman*

## Inhomogeneous Coherent Magnetization Rotation in Thin Magnetic Films

K. D. Leaver and M. Prutton

*International Computers and Tabulators (Engineering) Limited, Stevenage, Hertfordshire, England*

Reversal processes in a continuous, thin, uniaxial magnetic film have been studied. The applied field was produced by a narrow strip conductor laid over the film and parallel to its easy axis. A hysteresis loop technique is described for measuring the magnetization distribution under the strip. Gradual changes in the orientation of the magnetization over distances comparable to the strip width are reported. Internal demagnetizing fields are postulated as being significant in causing this distribution.

### INTRODUCTION

DURING an investigation into the reversal processes in thin ferromagnetic films under the influence of inhomogeneous applied fields, a situation was found in which an expected 90-deg domain wall was not in fact observed. The inhomogeneous field was applied in the hard direction of a continuous, thin, uniaxial magnetic film by means of a long narrow strip conductor laid over its surface. On observing the magnetization distribution under this strip by means of the Kerr magneto-optic effect, it was found that, when the field was applied, no well-defined 90-deg domain walls were observed near the edges of the strip. The variation across the strip in the orientation of the magnetization appeared to be gradual. This observation suggested the experiment now to be described, in which loop plotter techniques are employed to discover the nature of the magnetization distribution.

### APPARATUS

A strip of copper 0.5 cm wide was laid over the surface of the film and along its easy direction. The strip was supplied with alternating current at 400 cps, creating an inhomogeneous magnetic field in the plane of the film. Between this strip and the film were placed a number of fine wires, also parallel to the easy axis, and at different distances from the center of the strip. These were to act as pickup lines, and, together with a return line placed in zero field, formed a series of pickup loops which could be connected to the vertical amplifier of an M-H loop plotter. A variable transformer was used to cancel out the direct flux pickup from the strip line; the loop plotter was otherwise similar to that described by Crittenden *et al.*[1]

The ideal would have been to present to the film a field which was constant over the width of the strip and a step function at either edge. In practice the field dropped to 90% of its peak value at a distance of 0.230 cm from the center of the strip. The effect of this will be discussed later.

The observations described below were made on film specimen No. 587, evaporated onto a polished aluminium substrate measuring 11.5 by 8 cm, at a temperature of 350°C. The film material was Ni-Fe-Co (melt composition 80/17/3) deposited to a thickness of 3600 A at a rate of 1600 A/min. The highest pressure reached during evaporation was $3 \times 10^{-5}$ mm Hg. A steady magnetic field of 470 oe was applied along the shorter edge of the substrate. The values of anisotropy field $H_k$ and easy direction coercivity $H_c$, measured in a homogeneous field, were 5.4 oe and 0.7 oe, respectively.

### EXPERIMENTAL RESULTS

Each pickup line was connected in turn to the input to the loop plotter, and in each case a straight line hysteresis loop, similar to that observed in a homogeneous field, was obtained. The magnitude of $H_k$, derived from the hysteresis loop, was seen to increase with the distance $y$ of the pickup line from the center of the strip.

The character of the reversal process can be deduced from the following four points.

(1) Magneto-optical observations of the reversal process under such a strip in similar films on glass substrates have indicated that the magnetization under the strip reverses by rotation. This is confirmed by observations of transverse flux changes in films on both types of substrate.

(2) Straight line hysteresis loops have been observed for all positions of the pickup loop.

(3) The film has a uniaxial anisotropy.

(4) The effective field is everywhere and at all times in the hard direction.

Hence it may be inferred that coherent rotation of spins occurred along a line parallel to the length of the strip conductor. It follows that the signal displayed along the $Y$ axis of the loop plotter was proportional to the sine of the angle $\theta$ between the magnetization and the easy axis. Since the pickup line was very close to the film, it effectively detected changes in the mag-

[1] E. C. Crittenden, A. A. Hudimac, and R. I. Strough, Rev. Sci. Instr. **22**, 872 (1952).

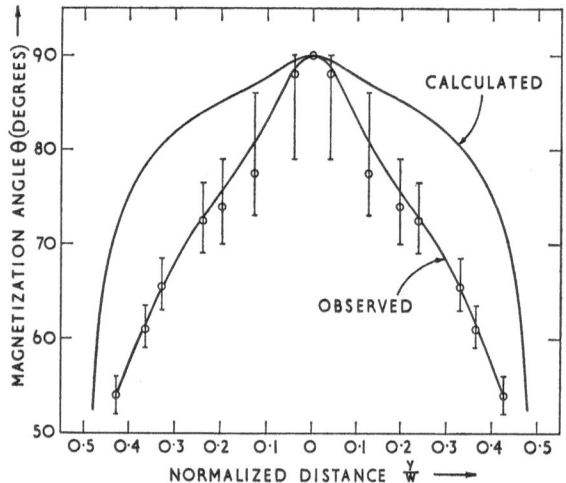

Fig. 1. Angle between magnetization and easy axis as a function of the distance $y$ from the center of a strip of width $w$. *Upper curve is calculated assuming that $\theta_y$ depends solely upon the anisotropy and the local value of the applied field.*

netization of only that part of the film immediately beneath it. Hence the value of $\theta$ could be derived for the magnetization at various distances from the strip center.

If $\theta_y$ is the value of $\theta$ under the pickup line at position $y$, and $\theta_0$ is the value of $\theta_y$ at $y=0$, then, because each hysteresis loop is a straight line, the value of $\sin\theta_y/\sin\theta_0$ must be independent of $\theta_0$ and hence of the applied field. This leads to the conclusion that, for all values of $\theta_0$,

$$H_{k0}/H_{ky} = \sin\theta_y/\sin\theta_0 \quad (H < H_{k0}).$$

In this equation, $H_{ky}$ is the effective anisotropy field at position $y$, and $H_{k0}=H_{ky}$ at $y=0$.

The graph of Fig. 1 shows values of $\theta_y$, derived from the value of $H_{ky}$, plotted against $y/w$ ($w=$strip width) for an applied field which just makes $\theta_0$ equal to 90 deg.

With it is compared the value of $\theta_y$ calculated on the assumption that the orientation of the spins depends only upon the anisotropy and the local value of the applied field. This curve also is computed for an applied field which just makes $\theta_0$ equal to 90 deg.

### DISCUSSION

That these two curves do not agree may be anticipated from the following argument. For the sake of simplicity, consider the case in which the field is uniform immediately under the strip and zero elsewhere. If the magnetization were to rotate homogeneously, free poles would appear at the domain walls separating the rotating region from the rest of the film. The magnitude of the demagnetizing field due to these free poles would tend to infinity near the domain walls, and since it opposes the applied field, this is inconsistent with the assumption of homogeneous spin rotation.

We may conclude that, even in a region in which the applied field is substantially homogeneous and which is finite in extent, the orientation of the magnetization can be inhomogeneous over much longer distances than has hitherto been postulated[2] or observed.[3]

All films so far observed, covering a thickness range of 500 to 6000 A, show this effect, and the plausibility argument outlined above makes it reasonable to expect similar behavior in any film which displays coherent spin rotation in homogeneous fields.

It is found that $H_{k0}$, the magnitude of the field required to make $\theta_0$ equal to 90 deg, is not equal to the value of $H_k$ measured in a homogeneous field. The presence of demagnetizing fields might lead us to expect this, so that the occurrence of a form of inhomogeneous rotation is of interest in an investigation into the cause of this change in effective anisotropy field.

[2] D. O. Smith, J. Appl. Phys. **32**, 70S (1961).
[3] E. Fuchs, Naturwissenschaften **48**, 450 (1961).

JOURNAL OF APPLIED PHYSICS    SUPPLEMENT TO VOL. 33, NO. 3    MARCH, 1962

# Effect of Substrate Cleanness on Permalloy Thin Films*

JOHN S. LEMKE

*Remington Rand Univac, Division of Sperry Rand Corporation, Saint Paul, Minnesota*

The use of mechanical and chemical cleaning techniques in the production of a clean, random, and smooth microslide glass substrate surface for the vacuum deposition of uniaxial thin Permalloy film elements has been studied. The principal cleaning techniques intensively investigated were: (1) a mild detergent wash followed by a vapor degreasing cycle in isopropyl alcohol; (2) a chalk paste scrub followed by an ultrasonically agitated distilled water rinsing cycle and a forced hot air drying process; and (3) hydrofluoric acid etching followed by a distilled water rinsing cycle and a forced air-drying process. The surface condition of the cleaned microslide glass was assessed with electron micrographs of preshadowed carbon replicas (magnification 88 000✕). The micrographs show that the chalk-cleaned glass substrate has the smoothest surface. The variously cleaned substrates were then used for the vacuum deposition of 1700-A 4-mm-diam Permalloy film elements (melt composition 83% nickel, 17% iron). This work was carried out in a $10^{-6}$ mm Hg vacuum system. Twenty-five evaporations were made with a total of 54 elements per evaporation. The Permalloy film elements deposited on the differently cleaned glass substrates were then examined by means of a 1000-cycle hysteresis loop apparatus and the following measurements were made: (1) coercive force $H_C$; (2) saturation flux $\phi_s$; (3) orientation of the anisotropy axis $\theta$; (4) magnetoelastic strain coefficient $\eta$; and (5) anisotropy field $H_K$. Dispersion measurements of the easy axis were made on a 1000-cycle crossed field hysteresis loop apparatus. Measured in this manner, the chalk-cleaned glass substrates yielded consistently the lower average value of $H_C$; i.e., 1.4 oe. The average value of $H_K$ for the three different cleaning techniques was 2.5 oe. Within the total range of values of coercive force and the anisotropy field, the chalk-cleaned and the acid-etched glass substrates yielded the same values; i.e., $H_C \pm 0.2$ oe, $H_K \pm 0.3$ oe. Measurements of angular dispersion varied from 8° to less than 2° with an average value of less than 3°. The chalk-cleaned substrates yielded the lowest values of angular dispersion.

## INTRODUCTION

A VARIETY of mechanical and chemical cleaning techniques has been used for the preparation of an absolutely clean, random, and smooth glass substrate surface for the vacuum deposition of thin magnetic films. T. Putner[1] has reported that a detergent washing combined with vapor degreasing produces glass surface cleanliness nearly equal to that of discharge cleaning. K. Behrndt and F. S. Maddocks[2] report that chemical cleaning alone may not be sufficient to produce the homogeneity needed for the reproducibility of magnetic parameters of thin magnetic films. They found an SiO coating after chemical cleaning decreased the scattering of coercive force $H_C$. S. Nielson[3] has reported that vacuum melted glass yields a smooth surface and better reproducibility of magnetic parameters for thin magnetic films. J. C. Lloyd and R. S. Smith[4] have reported a correlation between substrate roughness and coercive force $H_C$ and anisotropy field $H_K$ for electrodeposited films.

The cleaning media studied in this investigation are the following: a solution of hydrofluoric acid and water or nitric acid; a sulfuric acid and chromic acid solution[5]; a solution of sodium hydroxide, water, and alcohol[6]; sodium carbonate[7]; a detergent wash and vapor degreasing[1] and a chalk scrub. Each of the foregoing cleaning techniques was investigated to determine the relative cleanness and smoothness it is capable of producing on a glass substrate and the resultant effect on the magnetic parameters of the uniaxial Permalloy thin films.

## EXPERIMENTAL

The test substrates were 1✕3-in. soda lime microslide glass. The surface condition of the cleaned microslide glass substrates was assessed by their wetting properties[1] and electron microscopy. The preshadowed carbon replicas (magnification 88 000✕) indicate that the chalk-cleaned and the sulfuric-chromic acid etched glass substrates yield the smoothest surfaces. The chalk cleaning consisted of rubbing a chalk paste onto the glass substrate surface, then rinsing the substrate with ultrasonically agitated distilled water, and finally drying with forced hot air. From the resulting micrographs, wetting properties, and preliminary vacuum studies, three different cleaning procedures were chosen for intensive investigation of the magnetic parameters of vacuum-deposited Permolloy thin films. The procedures used were: (1) an acid etching (2% hydrofluoric acid at room temperature); (2) a simple washing (mild detergent and vapor degreasing); (3) a mechanical cleaning (chalk scrub). Micrographs of these three glass surfaces are shown in Fig. 1.

After a series of slides was cleaned by each of the three procedures outlined above, an array of eighteen 4-mm and one 8-mm-diam Permalloy film element was then vacuum deposited on each slide. The evaporation was made in a glass chamber in which oil diffusion pumps yield a $10^{-6}$ mm Hg ultimate pressure as measured by an ionization gauge tube. The evaporation

* This work supported by the Department of Defense. (See ASTIA reports on Project Lightning. Second Phase, Fourth Quarterly Progress Report through Third Phase, Fourth Quarterly Progress Report.

[1] T. Putner, Brit. J. Appl. Phys. **10**, 332 (1959).
[2] K. Behrndt and F. S. Maddocks, J. Appl. Phys. **30**, 276S (1959).
[3] S. Nielson, "Clean substrates for Permalloy films," 1960 Vacuum Symposium Transactions, p. 293 (1960).
[4] J. C. Lloyd and R. S. Smith, J. Appl. Phys. **30**, 274S (1959).
[5] Charles D. Hodgman, Editor, *Handbook of Chemistry and Physics* (Chemical Rubber Publishing Company, Cleveland, Ohio, 1950), Vol. 32, p. 2710.
[6] Reference 5, p. 2708.
[7] Raymond E. Kirk, Editor, *Encyclopedia of Chemical Technology* (Interscience Publishers, Inc., New York, 1960), Vol. 10, p. 1242.

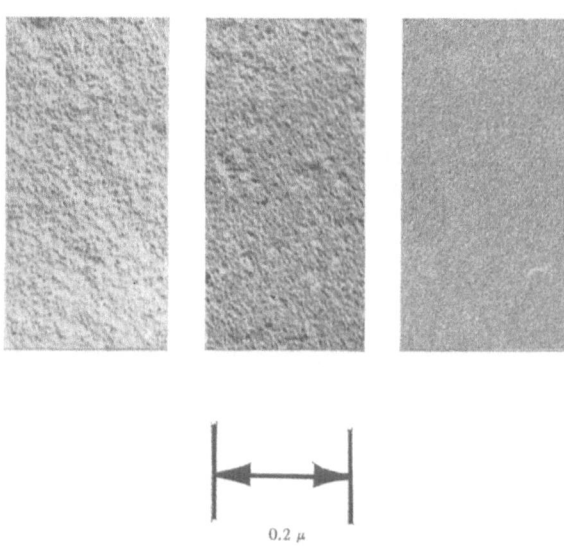

FIG. 1. Electron micrographs of glass surfaces (magnification 88 000×). Left—acid-cleaned glass. Center—detergent-cleaned glass. Right—chalk-cleaned glass.

TABLE I. The effect of substrate cleanness on Permalloy thin films.

| | Acid ‡ | Acid † | Detergent ‡ | Detergent † | Chalk ‡ | Chalk † |
|---|---|---|---|---|---|---|
| **Coercive force** | | | | | | |
| Average value (oe) | 1.7 | 1.7 | 1.7 | 1.9 | 1.5 | 1.5 |
| Maximum value (oe) | 1.9 | 2.9 | 2.0 | 3.1 | 1.7 | 2.0 |
| Minimum value (oe) | 1.4 | 1.2 | 1.4 | 1.1 | 1.4 | 1.1 |
| Standard deviation | 0.12 | 0.50 | 0.14 | 1.03 | 0.09 | 0.40 |
| **Anisotropy field** | | | | | | |
| Average value (oe) | 2.6 | 2.6 | 2.7 | 2.7 | 2.6 | 2.6 |
| Maximum value (oe) | 2.7 | 3.6 | 2.8 | 4.1 | 2.7 | 3.8 |
| Minimum value (oe) | 2.4 | 2.2 | 2.5 | 2.2 | 2.5 | 2.2 |
| Standard deviation | 0.09 | 0.37 | 0.11 | 0.58 | 0.08 | 0.35 |
| **Angular dispersion** | | | | | | |
| Average value (deg) | | 3.9° | | 4.4 | | 3.2 |
| Maximum value (deg) | | 7.3° | | 8.4 | | 4.3 |
| Minimum value (deg) | | 2.2° | | 2.1 | | 1.8 |
| Standard deviation | | 1.1 | | 1.4 | | 0.7 |

‡—Single evaporation.
†—Series of 25 evaporations.

source was 83% nickel, 17% iron wire wrapped around a radial tungsten wire. The test substrates were baked in vacuum for two hours by three radiative infrared bulbs at a recorded temperature of 375°C. An iron-Constantan thermocouple junction cemented with Sauereisen cement to a microslide glass was used as a temperature sensor. Deposition was initiated by passing a current through the tungsten filament and was continued for approximately 30 sec in a 100-oe magnetic field. Film thickness was approximately 1700 A as measured by a crystal monitor.[8]

Next, 24 consecutive evaporations were made, and the resulting films were measured in a 1000-cycle hysteresis loop tracer.[9] A crossed field 1000-cycle hysteresis loop tracer[10] was used to measure dispersion of the "easy" axis for the one 8-mm-diam Permalloy film.

## RESULTS

Table I shows the average value, the maximum value, the minimum value, the standard deviation, and the percentage of standard deviation for the coercive force $H_C$ and the anisotropy field $H_K$ for a single evaporation and for a series of 25 evaporations. All of the eighteen 4-mm-diam Permalloy films were tested on each test microslide glass substrate for the first evaporation, whereas for the next 24 evaporations, only six of the eighteen 4-mm-diam Permalloy films were tested in the

1000-cycle hysteresis loop tracer. These six were chosen in a sequence so that after every third evaporation all of the Permalloy films had been tested for a given substrate area. With this procedure any variation in film thickness or temperature is accounted for in the mathematic analysis. The cleaning techniques did not appear to alter $H_K$ values significantly. The chalk-cleaned microslide substrate produced the lowest measured average value, the least deviation, and the lowest percent of standard deviation for the coercive force $H_C$. The acid-etched substrate has a measured deviation of $H_C$ equal to the chalk-cleaned substrate. (Hysteresis loop tracer reliability ±5% for $H_C$.) These values are typical for more than one hundred evaporations in which a variety of vacuum systems and vacuum procedures were used.

Table I also shows the measured average value, standard deviation, and percentage of standard deviation of $\theta_{90}$, the angular dispersion of the "easy" axis. The chalk-cleaned microslide glass substrates yielded the lowest average (3.2°) value and lowest percent of standard deviation (21%) of the three cleaning techniques, i.e., acid 3.9°, 27% detergent 4.4° 30%. However, the values of $\theta_{90}$, the angular dispersion of the "easy" axis, are not of a magnitude as great as the values of coercive force.

## CONCLUSIONS

Thin Permalloy magnetic films have been vacuum deposited onto glass substrates which have been cleaned in a variety of ways. Chalk-cleaned glass substrates have a very smooth and random surface as viewed by an electron microscope. In measured values of coercive force, the chalk-cleaned glass substrates yield the lowest average value, the least amount of scattering, and even more important better reproducibility from evaporation

[8] S. J. Lins and H. S. Kukuk, "Resonance frequency shift thin-film thickness monitor," 1960 Vacuum Symposium Transactions, 333 (1961).

[9] E. M. Bradley and M. Prutton, J. Electronics and Control 6, 81 (1959).

[10] T. Crowther, "Techniques for measuring the angular dispersion of the easy axis of magnetic films," Lincoln Laboratory Group Report No. 51–2.

run to evaporation run. These results are consistent with the reported work of J. C. Lloyd and R. S. Smith and their correlation of substrate roughness and coercive force. Although the differences of $\theta_{90}$ for the different cleaning techniques are within the limitations of the crossed field hysteresis loop checker ($\pm 1°$), nevertheless the results are significant because the known variables of $\theta_{90}$ (composition, substrate deposition temperature, and evaporation rate) were not varied in the experimentation.

### ACKNOWLEDGMENTS

The author wishes to express his appreciation to E. Liepa for his measurements, W. J. Simon for his work with the electron microscope, and the valuable technical assistance of R. W. Olmen.

---

JOURNAL OF APPLIED PHYSICS          SUPPLEMENT TO VOL. 33, NO. 3          MARCH, 1962

# Support and Extension of the Rotational Model of Thin Film Magnetization

J. E. SCHWENKER AND T. R. LONG
*Bell Telephone Laboratories, Inc., Murray Hill, New Jersey*

This paper describes the magnetic behavior of some thin permalloy films in terms of the rotational model of Stoner and Wohlfarth. Experimental evidence is shown for the existence of a single domain whose orientation is dependent upon the applied field as postulated in the model. Measured orientations agree closely tations agree closely with the predicted ones. Phenomenological explanations for certain observed deviations from the model are given. A modification of the model can be used to account for the deviations and leads to a qualitative understanding of the low-speed region of switching behavior. The films used were flat, evaporated, 81-19 NiFe films about 1000 A thick and films of similar composition about 10 000 A thick, electroplated on a 5-mil-diam wire.

**M**ANY workers have used the Stoner–Wohlfarth model for rotational behavior of magnetization as a standard for comparison of experimental magnetic material in the form of films of the order of 1000 A thick. The purpose of this paper is to give evidence of behavior following this equilibrium model in planar evaporated films about 1000 A thick and cylindrical films about 10 000 A thick, to discuss some of the techniques for comparison with the model, and to suggest a way of augmenting the model to explain dynamic behavior in the low and intermediate speed range. For the materials studied, typical parameter values are $H_c = 1$ oe, $H_k = 2$–5 oe, $I_s = 10^4$ emu.

The Stoner–Wohlfarth model[1] expresses the potential energy of a planar film as

$$E = k_0 + k_1 \sin^2\theta - \mathbf{H} \cdot \mathbf{I}_s,$$

which assumes that the material consists of a single domain at saturation magnetization. This expression can be solved for the equilibrium orientation of magnetization $\theta$ in the presence of an applied field $\mathbf{H}$. By a series of such solutions, one can predict the shape of a conventional hysteresis loop obtained in repsonse to an alternating field applied in some arbitrary direction. This equilibrium model seems to apply in the dynamic case for frequencies less than the order of kc. In particular, the predicted loop obtained in the so-called hard direction consists of a central linear region and two regions of complete saturation. The hysteresis loop that one observes with existing materials shows a slightly open central region instead of the closed one predicted by the model. This is presumed to be primarily due to the formation of many small domains with subsequent losses.

If we simultaneously apply a small bias field in the easy direction (a few percent of the anisotropy field), we can avoid formation of these domain walls and, as a result, can observe a hysteresis loop which corresponds to the rotational model. Experiments show that the bias field required to cause completely rotational behavior is dependent upon the particular material being used. A slight modification of the Stoner–Wohlfarth model enables us to account for this observed variation. Assume that we have an ensemble of independent regions in the film whose individual easy axes are distributed over the range $0 \pm \alpha$ with respect to the mean easy direction. In the absence of a bias field, when the alternating field has the value $H_k$, the magnetization in this ensemble will be oriented in the range $\pi/2 \pm \beta$. When the applied field is now reduced, half of the distribution will rotate counterclockwise, and the other half will rotate clockwise, producing the multi-domain situation. Suppose we now repeat the experiment with a bias field sufficient that when the alternating field has a value $H_k$, the resulting orientations for the ensemble fall approximately in the range $\pi/2$ to $\pi/2 - 2\beta$. Now when the alternating field is reduced, all of the magnetization rotates in a clockwise sense, and we consequently observe a closed hysteresis loop.

Several techniques have been developed for measuring

---

[1] E. C. Stoner and E. P. Wohlfarth, Phil. Trans. Roy. Soc. London **240A**, 599 (1948).

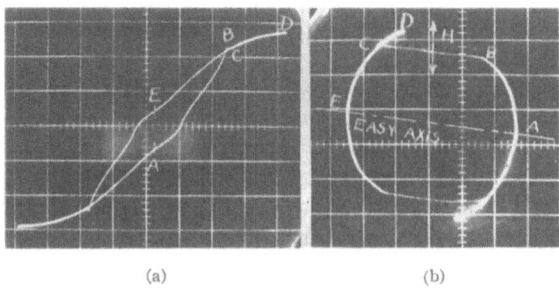

FIG. 1. (a) $M$–$H$ loop for a cylindrical film with the field applied at 83° with respect to the easy axis. (b) Polar plot of the magnetization for the same applied field of Fig. 1(a).

this presumed distribution in orientation. Typical measurements of both types of material show that nearly all of the easy axes fall within ±1° of the nominal easy axis. One can assume a coexisting distribution in the value of the anisotropy field $H_k$. In flat films it is possible to make a measurement of the distribution of easy axis directions which is independent of a coexisting distribution in $H_k$. Unfortunately, this is more difficult in the cylindrical films of interest to us. However, it can be shown that a small distribution in $H_k$ has only a small effect on the apparent distribution of easy axis orientations. ($\Delta\theta/\theta \approx -\Delta H_k/H_k$) We would, of course, like to be able to measure the existing distribution in $H_k$. One can do this if one is prepared to make a rather strong assumption regarding the independence of these two distributions. We have not pursued this further because the uncertainty in the value of $H_k$ is a strong function of the distribution in easy axis orientations. (An uncertainty of about 1° in easy axis orientation can lead to an apparent dispersion as large as 8% in the value of $H_k$.)

It is particularly interesting to observe hysteresis loops obtained by following trajectories of applied field which do not lie along the easy or hard directions, and in fact some of these loops enable a more critical evaluation of the material in terms of the model. Figure 1(a) shows a hysteresis loop (magnitude of applied field versus parallel component of magnetization) obtained from a cylindrical film by applying a field at an angle of about 83° from the nominal easy axis. The small cusps which occur as the magnetization approaches saturation are predicted by the Stoner–Wohlfarth model. The model predicts the appearance of cusps for orientations in the range of 76° to 90° with a broad maximum around 86°. The bulge in the center of the loop has yet to be satisfactorily explained. The appearance of these cusps was reported by Bradley and Prutton in 1959[2]. However, they found these cusps only in comparatively thin films, and they did not observe the rapid decrease in magnetization at the trailing edge of the cusp. This can be

[2] E. M. Bradley and M. Prutton, J. Electronics Control 6, 81 (1959).

explained if we assume that the distribution of orientation in their sample was of the order of 5° or more.

The cusps can be explained in the following manner: As the applied field increases, the magnetization rotates reversibly along the lower branch of the loop toward the direction of the applied field until the applied field reaches the value corresponding to the peak of the cusp. (This applied field lies on the astroid $h_l^{\frac{2}{3}}+h_t^{\frac{2}{3}}=1$.) At this point a second domain is formed whose orientation corresponds to the second-equilibrium solution at that point. Then as the field slightly increases, the new domain grows at the expense of the previous domain.

This process will occur over a finite, predictable range of applied field, depending on the dispersion of easy axis directions. As the model predicts, this results in a decrease in the component of magnetization in the direction of the applied field because of the relative orientation of the two domains with respect to the applied field[2]. As the applied field continues to increase, the magnetization in the new domain rotates toward the applied field direction.

This process is dramatically demonstrated by Fig. 1(b). This is a polar plot of the magnitude and direction of the resultant magnetization in the presence of the applied field of Fig. 1(a). Here we plot dynamically the components of magnetization parallel and perpendicular to the applied field direction. The applied field direction is vertical on the plot, and the easy axis is not quite horizontal as indicated by the dashed line. In this picture the magnetization rotates counterclockwise in general. Assume, for example, that when the applied field is zero, the magnetization lies along the easy axis at point A. As the applied field is slowly increased, the magnetization rotates counterclockwise coherently toward point B. At point B the second domain begins to form. The orientation of the second domain is indicated approximately by point C. It is obvious from the picture that the process occurring between points B and C cannot be a single domain process, since the resultant magnitude is seen to be less than the saturation value. Having reached point C, the magnetization of the new domain rotates clockwise to D toward the direction of the applied field, which is vertical in this picture. When the applied field is subsequently reduced, the magnetization rotates counterclockwise coherently, reaching point E when the applied field is zero. Point-by-point measurements agree with the calculated orientations within $1\frac{1}{2}°$, except in the region close to the easy axis where deviations run to 4°.

The domain wall process just described is characteristic of what occurs when the applied field crosses the astroid from the inside to the outside. The speed of this process is experimentally related to the angle between the equilibrium orientations.

JOURNAL OF APPLIED PHYSICS    SUPPLEMENT TO VOL. 33, NO. 3    MARCH, 1962

# Induced Magnetic Anisotropy of Evaporated Films Formed in a Magnetic Field

Minoru Takahashi

*The Research Institute for Iron, Steel, and Other Metals, Tōhoku University, Sendai, Japan*

The origin of the induced magnetic anisotropy in evaporated films of pure iron and nickel-iron alloys formed at normal incidence in a magnetic field was investigated. The temperature dependence of the anisotropy $K_u$, its relations to film thickness, substrate temperature, and substrate material, and the relaxation phenomena of the anisotropy during isothermal annealings with and without a magnetic field were systematically studied. The temperature dependence of $K_u$ is neither explained by the crystal anisotropy nor by shape anisotropy. The observed relations between $K_u$ and film thickness etc. suggest that the stress induced in the interior of the film, rather than the stress of limited parts of the film, is responsible for the anisotropy if the anisotropy due to the atomic pair orientation is not considered. The interior stress may be caused by the structure defects formed during evaporation. The relaxation phenomena of $K_u$ in the alternate isothermal heat treatments reveals that the anisotropy has originated partly from a directional alignment of such imperfections as vacancies, dislocations, and impurities. It is also suggested that magnetostriction may play an important role in this directional alignment.

## I. INTRODUCTION

FERROMAGNETIC thin films have been extensively investigated as switching elements in future computers. There are many problems, however, from practical as well as physical view points. One of these problems is the origin of the induced magnetic anisotropy in the evaporated films. This is an essential and fundamental problem, and it may be thought that the quickest way to estimate the value of the films as magnetic materials is to clarify the origin of this anisotropy.

As is well known, the anisotropy in films formed in a magnetic field occurs with oblique incidence[1-3] as well as with normal incidence of the vapor deposition beams.[4-6] It has been made clear by Smith,[7] and also by others, that a greater part of the anisotropy caused by oblique incidence is due to the microstructure of the films. On the other hand, the origin of the induced magnetic anisotropy in the case of normal incidence is not yet clear. Few systematic studies have sought to determine the influence of the conditions of film preparation upon the magnetic anisotropy. Therefore, a systematic and quantitative investigation of this influence of evaporation conditions upon the uniaxial anisotropy and on the change of the anisotropy with various heat treatments was undertaken in the present work, in order to clarify the origin of the anisotropy in iron and nickel-iron films.

## II. EXPERIMENTAL PROCEDURE

The films of iron and nickel-iron were prepared by evaporation onto substrates of a special shape, usually made of quartz. The distance between tungsten filament and substrate was 5 cm and the degree of vacuum $10^{-5}$ mm Hg. A magnetic field of 250 oe was always applied during evaporation.

The induced uniaxial anisotropy was measured by using a highly sensitive magnetometer. Examples of the torque curves are shown in Fig. 1(a).[8] As seen from this figure, the curves change gradually from twofold to fourfold symmetry with increasing measuring field $H_m$. The rotational hysteresis loss, $W_r = \frac{1}{2}\int_0^{2\pi}(\mathbf{L}-\mathbf{L}')d\theta$, has its maximum in the neighborhood of the anisotropy field [Fig. 1(b)]. In the expression for $W_r$, $\theta$ is an angle between the direction of easy magnetization and that of the intrinsic magnetization, and $\mathbf{L}$ and $\mathbf{L}'$ are torques measured in clockwise and counterclockwise directions, respectively. A maximum field attainable by a Helmholtz coil used in the present study is 773 oe, and so the coefficient of $\sin 2\theta$ obtained by the Fourier analysis of the $\frac{1}{2}(\mathbf{L}+\mathbf{L}')$ curves was not fully saturated in some cases [cf., Fig. 1(b)]. The anisotropy constant $K_u$, therefore, was determined by extrapolation of the coefficient of $\sin 2\theta$ as $H_m \rightarrow \infty$.

It was assumed at the beginning of this study that some of the following three factors would be possible as the origin of the induced anisotropy: (i) Traces of fibrous structures or anisotropic aggregations of crystallites and the anisotropic shape of the crystallites,[7] (ii) stress localized at the film surface or at the film bottom contacting the substrate, (iii) stress induced by lattice imperfections.[9,10] In order to investigate these three factors, (1) the temperature dependence of $K_u$, (2) the relation of $K_u$ to film thickness, substrate temperature,

[1] T. G. Knorr and R. W. Hoffman, Phys. Rev. **113**, 1039 (1959).

[2] D. O. Smith, J. Appl. Phys. **30**, 264S (1959).

[3] V. Kamberský, Z. Málek, Z. Frait, and M. Ondris, Czechoslov. J. Phys. (September 1960).

[4] M. Takahashi, D. Watanabe, T. Sasagawa, and S. Ogawa, J. Phys. Soc. Japan **14**, 1459 (1959); M. Takahashi, D. Watanabe, T. Kōno, and S. Ogawa, J. Phys. Soc. Japan **15**, 1351 (1960).

[5] C. D. Graham, Jr. and J. M. Lommel, G. E. Research Laboratory, Schenectady (private communication).

[6] W. Schüppel, O. Stemme, W. Andrä, and Z. Málek, Inst. Magnetishe Werkstoffe der Deutschen Akademie der Wissenschaften, Berlin, Jena (private communication).

[7] D. O. Smith, M. S. Cohen, and G. P. Weiss, J. Appl. Phys. **31**, 1755 (1960).

[8] The direction of the applied field during deposition coincides with $\theta = 0$ in all torque curves in the present paper.

[9] E. W. Pugh, E. L. Boyd, and J. F. Freedman, IBM J. Research Develop. **4**, 163 (1960).

[10] M. Prutton and E. M. Bradley, Proc. Phys. Soc. (London) **75**, 557 (1960).

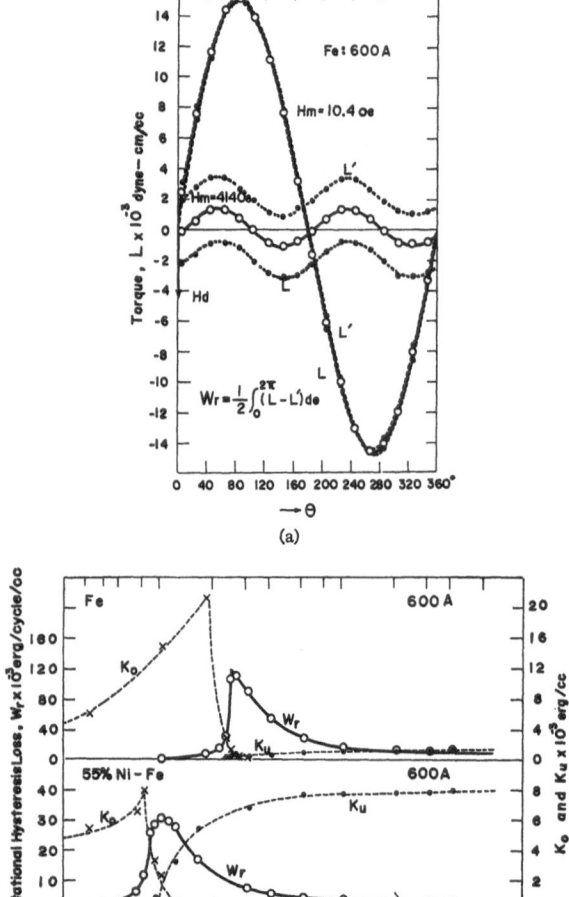

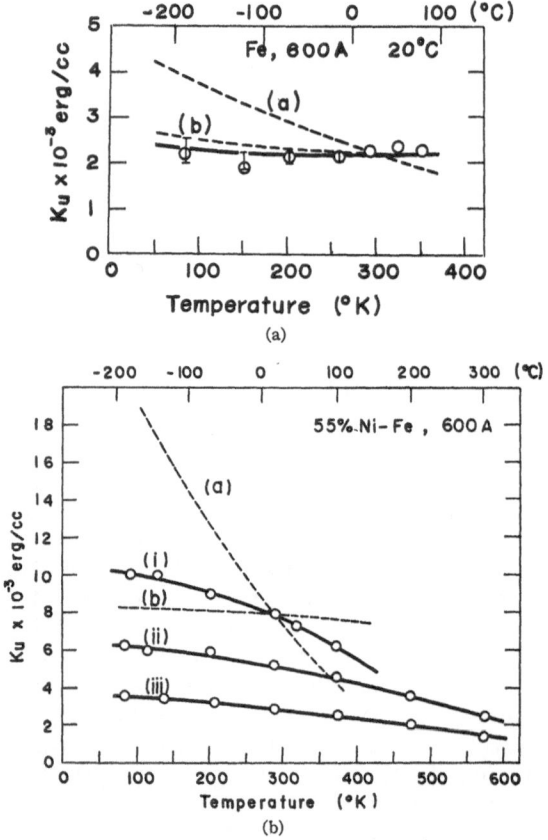

FIG. 1 (a) Examples of the torque curve for an Fe film formed at 20°C. The solid curves are means of **L** and **L'** curves. (b) The relations between the rotational hysteresis loss and the measuring field for Fe and 55% Ni-Fe films formed at 20°C. $K_0$ and $K_u$ are the coefficients of $\sin\theta$ and $\sin2\theta$, respectively.

and substrate material, and (3) the change of $K_u$ with isothermal magnetic annealing were experimentally studied.

## III. EXPERIMENTAL RESULTS AND DISCUSSION

### 1. Temperature Dependence of $K_u$

#### a. Fibrous Structure

If the fibrous structure exists in the film, and the fibrous axis which was determined to be [111] in the case of iron and nickel-iron films[11] is tilted by $\gamma$ from the film normal toward the vapor source, the magnetic anisotropy in the plane of the film can be expressed[2] in the following equation, provided that the magneto-elastic effect of stress is neglected: $E_u = -(\tau K_1/2) \times [\sin^2\gamma \cos^2\theta - (7/6) \sin^4\gamma \cos^4\theta]$, where $\theta$ is the angle between the intrinsic magnetization and the projection

of the fibrous axis on the film plane and $\tau$ is the degree of texture which is described by $0 < \tau < 1$. In the present study, no appreciable fibrous structure was detected by electron diffraction. If a trace of the fibrous structure not detected by electron diffraction exists, say $\tau \simeq 0.1$, $E_u$ is roughly evaluated to be $1 \times 10^3$ erg/cc for iron, provided that the value of the crystalline anisotropy constant $K_1$ is $4.2 \times 10^5$ erg/cc for iron and $\gamma$ is 10°. This anisotropy value can be readily measured by the torquemeter used in the present study. The temperature dependence of $K_u$ due to the fibrous structure should correspond to that for the crystal anisotropy. According to the equations obtained by Carr.[12] $K_u$ must be proportional to $(M_{ST})^{10}$ for iron films and to $(1-1.74T/T_c) \cdot (M_{ST})^{10}$ for iron-nickel films. $M_{ST}$ is the saturation magnetization at the temperature $T$ and $T_c$ is the Curie temperature.

FIG. 2. (a) The temperature dependence of $K_u$ for Fe films: (a) curve, $K_{uT} = K_{uT_R}(M_{ST}/M_{ST_R})^{10}$; (b) curve, $K_{uT} = K_{uT_R}(M_{ST}/M_{ST_R})^2$.
(b) The temperature dependence of $K_u$ for 55% Ni-Fe films: (a) curve, $K_{uT} = K_{uT_R}[(1-1.74T/T_c)/(1-1.74T_R/T_c)](M_{ST}/M_{ST_R})^{10}$. (b) curve, $K_{uT} = K_{uT_R}(M_{ST}/M_{ST_R})^2$. (i) Substrate temperature was held at 20°C during evaporation. (ii) The substrate was held at 300°C during the evaporation. The film was annealed at 350°C for 5 hr in a magnetic field immediately after the evaporation. (iii) The substrate was held at 300°C during the evaporation. The film was annealed at 450°C for 5 hr in a magnetic field immediately after the evaporation.

[11] D. M. Evans and H. Wilman, Acta Cryst. **5**, 731 (1952).

[12] W. J. Carr, Jr., J. Appl. Phys. **29**, 436 (1958).

### b. Anisotropic Shape of Crystallites or Anisotropic Aggregation of Crystallites

If the origin of the anisotropy is ascribed to the anisotropic shape or anisotropic aggregation of crystallites such as the chain structure of crystallites proposed by Smith,[7] then the magnetic anisotropy should be expressed by the following equation: $E_u = -\frac{1}{2} \Delta N M_{ST}^2 \times \cos^2\theta$, where $\Delta N$ is the difference between the demagnetizing factors along the long and short axes of crystallites. $K_u$ now depends on the temperature through $M_{ST}^2$.

### c. Atomic Pair Orientation

If atomic pair orientation is assumed to be the anisotropy origin, it is expressed by the following expression[13-15]: $E_u = -A \cdot [(M_{ST}^2/M_0^2)(M_{ST'}^2/M_0^2)/kT'] \times (\cos^2\theta)$, where $A$ is a function of both the coefficients of the dipole-dipole interactions for atom pairs and the concentration of atoms, $M_{ST'}$ the saturation magnetization at the annealing temperature $T'$ at which the field is applied, and $k$ the Boltzmann factor. $K_u$ depends the temperature through $M_{ST}^2$ in the same way as in the case of (b).

### d. Measurements and Interpretation

The temperature dependence of $K_u$ obtained from the torque curves is shown in Figs. 2 (a) and (b). The theoretical curves normalized at the observed value for 20°C are also shown. The temperature change of the saturation magnetization was assumed to be equal to that of the bulk metal, and the equation obtained by Bloch[16] was used.

As is clearly seen from these figures, there is a distinct discrepancy between the observed results and the prediction due to the crystalline anisotropy in all cases. It is surprising that the value of $K_u$ is almost constant for the iron film. Recently, the same results have been reported by Graham[17] for a nickel film. On the other

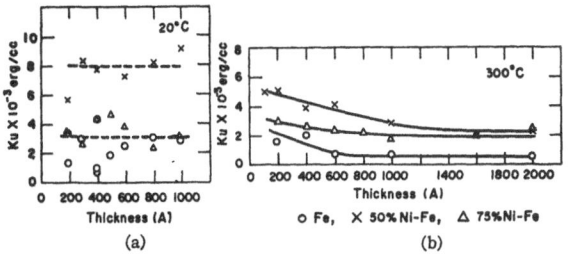

FIG. 3. (a) The thickness dependence of $K_u$ for Fe and Ni-Fe films formed at 20°C in a magnetic field. (b) The thickness dependence of $K_u$ for films formed at 300°C.

[13] L. Néel, Compt. rend. 237, 1468, 1613 (1953); J. Phys. radium 15, 225 (1954).
[14] S. Taniguchi, Sci. Repts. Research Insts., Tohoku Univ. A7, 269 (1955).
[15] S. Chikazumi, J. Phys. Soc. Japan 11, 551 (1956).
[16] F. Bloch, Z. Physik 61, 206 (1930).
[17] C. D. Graham, Jr., and J. M. Lommel, reported in the Kyoto Conference, September 1961.

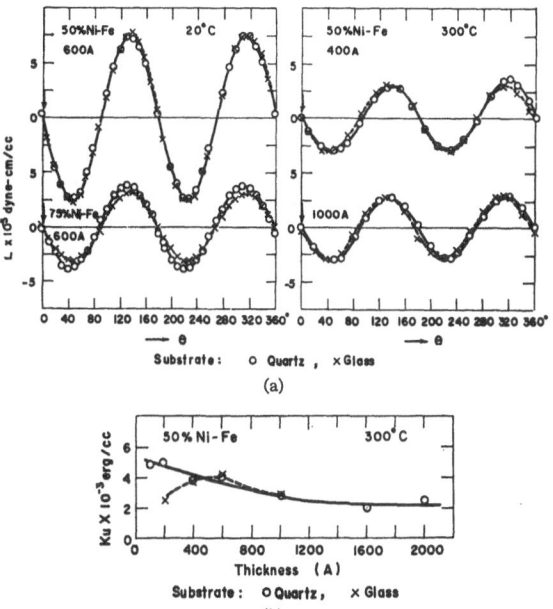

FIG. 4. (a) The torque curves for 50% and 75% Ni-Fe films evaporated on quartz and glass substrates, respectively. (b) The relation of $K_u$ to the film thickness.

hand, $K_u$ depends on the temperature roughly through $(M_{ST})^{17}$ for nickel-iron films in curve (i) and roughly through $(M_{ST})^{14}$ in curves (ii) and (iii). The atomic pair orientation theory may be partially responsible for the origin of $K_u$ in these films, because, by annealing at 350° or 450°C for several hours, the temperature dependence of $K_u$ approaches what is expected by this theory. But a greater part of $K_u$ seems to have another origin.

## 2. The Relation of $K_u$ to Film Thickness, Substrate Material and Substrate Temperature

As is well known stress occurs[18] in evaporated films, and it has been observed[19-21] that the distribution of this stress is localized in a surface layer of the film. According to Kinoshita, the stress averaged in the whole film, $\sigma$, is related to the stress $\sigma_0$ localized in a surface layer by the following equation: $\sigma = \sigma_0 d_0/D$, where $d_0$ and $D$ are the thickness of this surface layer and the film thickness, respectively. If the stress localized in the film surface causes the magnetic anisotropy, this is expressed by the following formula: $E_u = -\frac{3}{2}\lambda(\sigma_0 d_0)/D$, where $\lambda$ is the isotropic magnetostriction. $K_u$, therefore, should decrease with an increasing thickness, since $K_u$ is a volume effect, while the localized stress is a surface effect.

[18] R. W. Hoffman, R. D. Daniels, and E. C. Crittenden, Jr., Proc. Phys. Soc. (London) LXVII, 6-B, 497 (1954).
[19] H. P. Murbach and H. Wilman, Proc. Phys. Soc. (London) LXVI, 11-13, 905 (1953).
[20] J. D. Finegan and R. W. Hoffman, J. Appl. Phys. 30, 597 (1959).
[21] K. Kinosita and H. Kondo, J. Phys. Soc. Japan 15, 1339 (1960).

## a. The Relation between $K_u$ and Film Thickness

In Figs. 3(a) and (b), the thickness dependence of $K_u$ is given for iron and nickel-iron films formed at 20° and at 300°C. As is clearly seen in this figure, the observed values vary considerably from experiment to experiment in the case of 20°C, while the anisotropy slightly decreases with thickness increase in the case of 300°C.

If the $K_u$ is related to such an internal stress induced in the limited parts of the film, its magnitude should be dependent upon the film thickness in the way expressed in the last equation, and this tendency should become stronger with decreasing substrate temperature. The experimental results, however, contradict this prediction.

## b. The Relation of $K_u$ to Substrate Material

Next, in order to determine whether the stress arises from the difference in the thermal expansion coefficients of the film and the substrate or not, the relationship, between the anisotropy constant and substrate material was studied. Figure 4(a) shows examples of the torque curves for 50% and 75% nickel-iron films evaporated on quartz and glass substrates. As is well known, the coefficients of thermal expansion are very different for quartz and glass, that is, it is almost zero for quartz, while it is almost the same for glass as in metals. As is seen from this figure, however, the anisotropy scarcely changes with different substrate materials at different substrate temperatures and different film compositions. Different compositions correspond to different magnitudes of magnetostriction. The situation can clearly be seen in Fig. 4(b). It may be concluded from these figures that there is no relation between $K_u$ and the substrate material within the limits of experimental error except for extremely thin films.

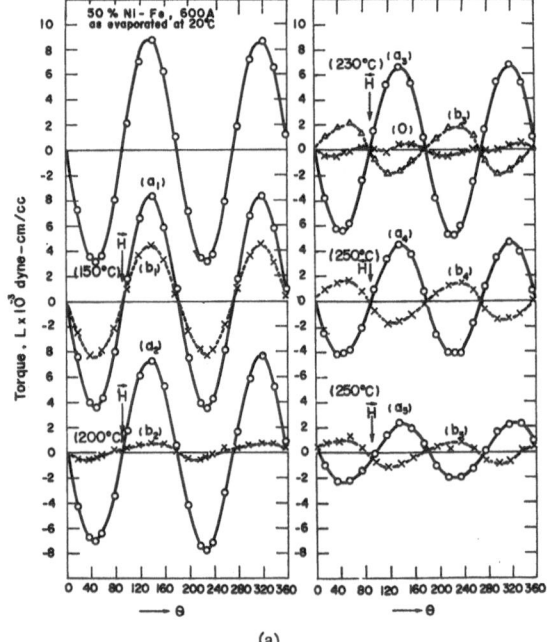

(a)

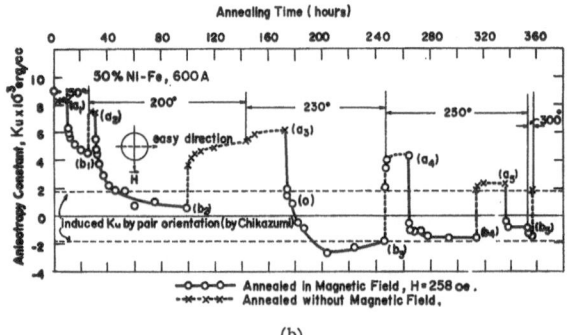

(b)

FIG. 6. (a) The change of the torque curve with the isothermal annealings with and without a magnetic field for 50% Ni-Fe film formed at 20°C in a magnetic field of 250 oe. (b) The anisotropy constant as a function of the isothermal annealings.

## c. The Dependence of $K_u$ upon the Substrate Temperature

Examples of the torque curves for the films formed at various substrate temperatures and the relations of $K_u$ to substrate temperature are shown in Figs. 5(a) and (b). The torque curve of an iron film formed at 20°C shows that the direction of the field during evaporation does not coincide with the one of easy magnetization but considerably deviates from this. The direction of the field is likely to approach the direction of easy magnetization when the substrate temperature increases. The amplitudes of the torque curves for 50% nickel-iron films remarkably decrease with increasing substrate temperature, just as is reported by many researchers, and the ultimate value of $K_u$ becomes $1.4\times10^3$ ergs/cc, that is, the value obtained by Chikazumi and Oomura[22] from a bulk alloy of 50%

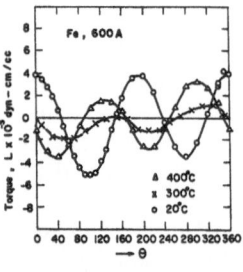 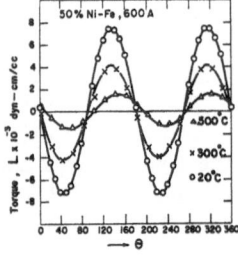

(a)

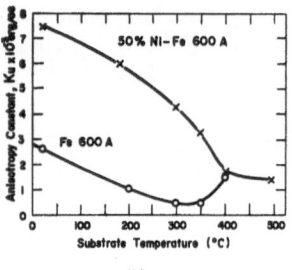 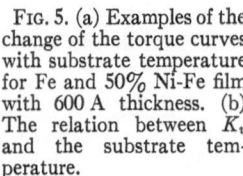

(b)

FIG. 5. (a) Examples of the change of the torque curves with substrate temperature for Fe and 50% Ni-Fe film with 600 A thickness. (b) The relation between $K_u$ and the substrate temperature.

[22] S. Chikazumi and T. Oomura, J. Phys. Soc. Japan **10**, 842 (1955).

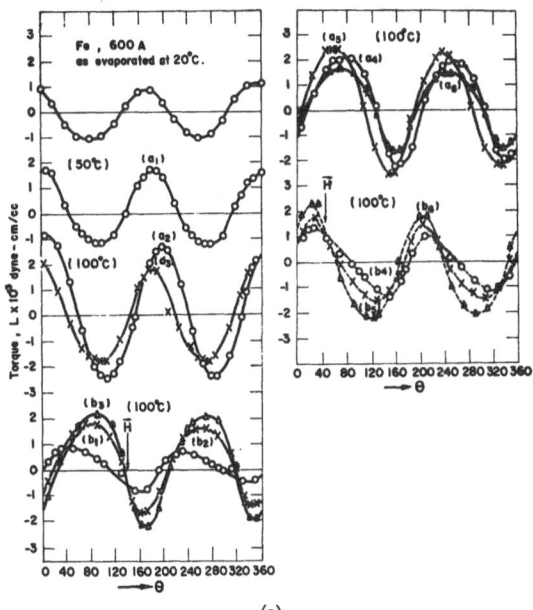

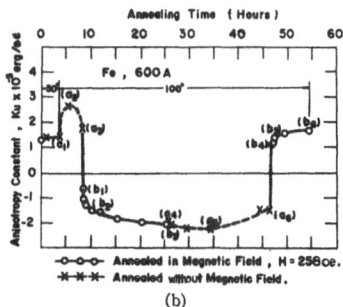

FIG. 7. (a) The change of the torque curve with the isothermal annealings with and without a magnetic field for Fe film formed at 20°C in a magnetic field of 500 oe. (b) The anisotropy constant as a function of the isothermal annealings.

nickel-iron. This tendency contradicts neither the stress model of the limited parts nor the interior stress model. The latter is a structure defects model.

### d. Discussion

Putting the above facts together, the structural defects formed during evaporation, that is, such imperfections as vacancies, dislocations, impurity atoms, etc., may be the strongest origin of the induced anisotropy, except that due to the atomic pair orientation, if these imperfections are directionally aligned. In order to study much more about the nature of the imperfection alignment, annealing experiments were performed.

### 3. The Changes of $K_u$ with Isothermal Heat Treatment

In order to verify the assumption of the directional alignment of imperfections, it is necessary to examine whether $K_u$ is changed with isothermal magnetic annealing at relatively low temperatures or not. Similar experiments were done by Mitchell[23] and Segmüller[24] for 82% nickel-iron films, which were annealed only in

[23] E. N. Mitchell, J. Appl. Phys. **29**, 286 (1958).
[24] A. Segmüller, J. Appl. Phys. **32**, 89S (1961).

the intermediate temperature range between 75° and 200°C, and also by Graham[25] for nickel films. But the effect of the magnetic annealing has never been observed above 200°C in these experiments.

The specimens used in the present experiment were 50% nickel-iron and iron films, 600 A thick, formed at 20°C in a magnetic field of 250 oe. The field was applied along the direction of difficult magnetization during the magnetic annealing. The torques were always measured at room temperature.

### a. 50% Nickel-Iron Film

Torque curves of an as-evaporated film and those after heat treatments are shown in Fig. 6(a). The change of $K_u$ obtained from these curves is shown in Fig. 6(b) as a function of the annealing time at various temperatures. The curves marked $(a_i)$ and $(b_i)$ in Fig. 6(a) correspond to $(a_i)$ and $(b_i)$ in Fig. 6(b). The abrupt change of the amplitude of the torque curve at the beginning of the magnetic annealing increases gradually with the annealing temperature, while the shape of the torque curve is deformed with the lapse of time and, finally, the easy direction is almost exchanged with the difficult direction [c.f., $(a_3)$, $(b_3)$; $(a_4)$, $(b_4)$]. The reversed torque curves do not have a complete fourfold symmetry. Furthermore, it is very surprising that the reversed torque curve is again deformed and the original form of the curve is recovered with a smaller amplitude than that of the initial curve, and, hence, the initial easy direction is again recovered by the annealing without a magnetic field, as seen in the changes from $(b_3)$ to $(a_4)$ and from $(b_4)$ to $(a_5)$. The amplitude of variation of $K_u$ is gradually decreased with repeated cycles of the annealing at a definite temperature or with the increasing annealing temperature, as seen from such points as $(a_1)$, $(a_2)$, $(a_3)$ $\cdots$, and $(b_3)$, $(b_4)$, $(b_5)$ $\cdots$.

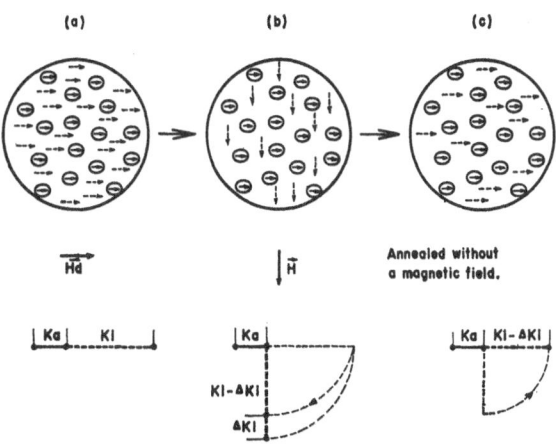

FIG. 8. Structure models for explaining the relaxation and recovering phenomena of the anisotropy for Ni-Fe films formed in a magnetic field.

[25] C. D. Graham, Jr., and J. M. Lommel, J. Appl. Phys. **32**, 83S (1961).

### b. Iron Film

A similar experiment was done for an iron film and the results are shown in Figs. 7(a) and (b). As seen from the former figure, the amplitude of the torque curve increases first by the annealing without a magnetic field at 100°C and then decreases [c.f., $(a_2)$, $(a_3)$]. A similar behavior is slightly shown in nickel-iron films also. The iron film from its torque curves behaves in almost the same way as the nickel-iron films; but the torque of the former was changed more easily than that of the latter with the magnetic annealing at a low temperature. It should be noted that such a recovering phenomenon demonstrated by the nickel-iron films could not be observed in the iron film and that only the amplitude of the iron torque curve slightly decreased while its shape was changed little by the prolonged annealing without a magnetic field even after the reversal [c.f., $(a_4)$, $(a_6)$]. The original direction of easy magnetization, however, is very easily recovered, when the magnetic field is applied along the initial easy direction, as seen in the change from $(a_6)$ to $(b_4)$.

### c. Proposed Model

The above relaxation curves and, particularly, the fact that the amplitude of the $K_u$ variation decreases with the annealing time enables us to propose the following model. Figure 8 shows structure models of an evaporated film which has the uniaxial anisotropy. A schematic explanation for the experimental results obtained will be assisted by these models.

Let it be assumed that the uniaxial anisotropy in nickel-iron films is composed of two different factors, as shown in $(a)$, that is, $K_a$ which is due to the magnetic orientation of pairs of atoms occupying small round regions in this figure and $K_i$ which is caused by the ordering of those imperfections such as vacancies, dislocations, impurities, and others which occupy the remaining parts of the film. The direction of easy magnetization in the small directional regions of atom pairs, possibly, does not change, and so the magnitude of $K_a$ in these regions is also unchanged, by the isothermal magnetic annealing below 300°C, as shown in (b). The reason for this is as follows. The activation energy required for the atomic pair orientation is usually thought to be 2 ev, while this energy as obtained from the relaxation curves is only about 1 ev. The latter value is almost the same as that for vacancy displacement. The direction of easy magnetization in the regions of the ordering of imperfections, on the other hand, will easily change from its initial easy direction to a perpendicular direction with a prolonged magnetic anneal at relatively low temperatures, while the magnitude of $K_i$ will decrease by $\Delta K_i$, in (b), due to diminished imperfections with annealing, whether the magnetic field is applied or not. That is, the directional ordering of imperfections will follow the direction of the applied field gradually, and the anisotropy will be $K_i - \Delta K_i$. In films which contain only a few lattice defects as a result of annealing at a sufficiently high temperature, the relaxation phenomenon at such low temperatures would not take place.

The recovery of the direction of easy magnetization is shown in the last figure (c). The reason for this recovery is not yet fully known, but it is clear that this phenomenon is related to the existence of the atomic pair orientation and, possibly, is caused by some interactions between $K_a$ and $K_i$, that is, magnetostatical or magnetoelastical interactions, because this phenomenon was not confirmed in pure iron films. It is also suggested from the experimental relation of $K_u$ to film composition that the origin of $K_i$ may be closely related to magnetostriction.[26] The relation of $K_u$ to composition will be described elsewhere.

### IV. CONCLUSION

The origin of the induced anisotropy in films formed at normal incidence is very complicated and it is not fully understood at present. It is suggested, however, that the origin other than that due to atomic pair orientation may be ascribed to a directional ordering of such imperfections as impurities, vacancies, and dislocations formed during evaporation and that this ordering may be closely related to magnetostriction. A detailed physical explanation should be made after more conclusive experiments.

[26] W. Andrä, Z. Málek, W. Schüppl, and O. Stemme, J. Appl. Phys. **31**, 442 (1960).

JOURNAL OF APPLIED PHYSICS     SUPPLEMENT TO VOL. 33, NO. 3     MARCH, 1962

# Angular Dispersion and its Relationship with Other Magnetic Parameters in Permalloy Films*

ROBERT W. OLMEN AND SIDNEY M. RUBENS

*Remington Rand Univac, Division of Sperry Rand Corporation, St. Paul, Minnesota*

An experimental investigation has been made of the dispersion of the easy axis in Permalloy films deposited in an orienting field. The dispersion was studied as a function of the following parameters: reversible rotation limit $H_{tr}$; easy direction coercive force $H_c$; anisotropy field $H_k$; and hard direction hysteresis loop squareness ratio $(M_r/M_s)_d$, and coercive force $(H_{cd})$. Observations were made with a Kerr magneto-optic apparatus and with a 1000-cycle hysteresigraph. Angular dispersion of the easy axis, measured by the hysteresigraphic method, was found to increase with $H_c/H_k$. Reversible rotation limit was not found to decrease with increasing angular dispersion as predicted by theory, but rather to decrease more rapidly initially with increasing dispersion and then to increase again at larger dispersion. The initial effect is believed to be due to the influence of local variation of $H_k$ on the reversible rotation measurement. There is some indication that hard direction coercive force increases with increasing dispersion, but there seems to be no simple correlation between dispersion and hard direction hysteretic properties. Reversible rotation limit was found to be roughly proportional to coercive force in most films, except those which exhibit an unusual type of wall motion switching behavior.

## INTRODUCTION

ANGULAR dispersion of the easy axis and its relationship to other magnetic parameters in anisotropic Permalloy films has been studied by several investigators. Smith[1] has reported that angular dispersion varies directly with $H_k$ in iron-rich Permalloy films. Alexander,[2] Spain and Rubinstein,[3] and West[4] report that squareness ratio of the hard direction hysteresis loop $(M_r/M_s)_d$ is a function of easy axis dispersion. West also reported that the coercive force $H_{cd}$ measured in the hard direction is a function of an average dispersion angle $\beta$. Crowther[5] has discussed the application of the static theory to magnetic films with angular dispersion. White[6] has extended this theory to show the relationship to be expected between reversible rotation limit and $\alpha_{90}$.[7] West has used the same model and the Stoner-Wohlfarth theory to derive the relationships between certain difficult direction hysteretic parameters and angular dispersion. The adequacy of these theories is examined by measuring a number of different magnetic parameters with a hysteresigraph and a Kerr magneto-optic apparatus for a group of Permalloy films with a wide range of $\alpha_{90}$. This is an extension of investigations of the relationship between $\alpha_{90}$ and some difficult direction hysteretic parameters reported by Rubens and Olmen.[8]

## EXPERIMENTAL EQUIPMENT AND PROCEDURE

Vacuum-deposited films with a composition of 80–81% nickel and the balance iron were made by evaporating an iron-nickel alloy from a tungsten ring source in a vacuum chamber maintained at about $1 \times 10^{-5}$ torr. The vapor was deposited through a mask onto glass microslides maintained at 300°C. An orienting magnetic field of 80 oe was maintained in the plane of the substrate during deposition and cooling. A crystal thickness monitor was used to determine film thickness and rate of deposition. The film specimens were 8 mm in diameter and approximately 1500 A thick.

The Kerr magneto-optic technique[9] was used to measure the reversible rotation limit for 57 Permalloy film samples with values of $\alpha_{90}$ ranging from 1.0° to 17°. The film samples were positioned in the Kerr apparatus with their easy axes in coincidence with the optical axis. The alignment was made by successively applying and removing a transverse field (applied in the hard direction) $H_t \gg H_k$ to a film specimen initially saturated in the "easy direction" and rotating the sample until it was observed to be demagnetized for either sense of $H_t$. The adjustment was found to be critical in many cases. After the sample was properly aligned, it was saturated in one sense along the easy axis. A transverse field, whose magnitude was increased incrementally, was applied and removed until the first reversed domain was observed. Even if no breakup was observed, the sample was again saturated along the easy axis. The average value of the transverse field at which reversed domains were first observed (for each sense) in a saturated film was recorded as $H_{tr}$, the reversible limit. The domain patterns observed are very reproducible for any one sense of the transverse field, but, in general, different for the opposite sense. The easy direction coercive force $H_c$ was also measured with the Kerr apparatus, and general observations of the wall-switching behavior were made.

* A portion of the work reported herein was sponsored by the U. S. Department of Defense.

[1] D. O. Smith, J. Appl. Phys. **32**, 70S (1961).
[2] R. G. Alexander, J. Appl. Phys. **30**, 266S (1959).
[3] R. J. Spain and H. Rubinstein, J. Appl. Phys. **32**, 288S (1961).
[4] F. G. West, J. Appl. Phys. **32**, 290S (1961).
[5] T. S. Crowther, Lincoln Laboratory Group Report No. 51-2 (February, 1959).
[6] R. A. White (private communication).
[7] In the hysteresigraphic method of measuring angular dispersion, the pickup axis is always at right angles to the drive field. $\alpha$ is the angle between the drive-field direction and the difficult direction of the sample, and $\alpha_{90}$ is that value of $\alpha$ for which 90% of maximum remanence is detected.
[8] S. M. Rubens and R. W. Olmen, Symposium on the Electric and Magnetic Properties of Thin Metallic Layers, Louvain, Belgium, 1961.

[9] C. A. Fowler, Jr., and E. M. Fryer, Phys. Rev. **100**, 746 (1955).

TABLE I. Average values of magnetic parameters for a selected group of 8 mm Permalloy films.
(All values are in oe, unless otherwise specified.)

| Number of samples | Dispersion angle, $\alpha_{90}$ (degrees) | Reversible limit $H_{tr}$ | $H_c/H_k$ | $\sin\alpha_{90}$ | $(M_r/M_s)_d$ | $\alpha_{90}H_k$ | Diff. direction coercive field $H_{cd}$ | Reversible limit $\theta_r = \sin^{-1} H_{tr}/H_k$ |
|---|---|---|---|---|---|---|---|---|
| 7 | 1.2 | 2.3 | $\dfrac{1.5}{3.1}=0.5$ | 0.021 | 0.68 | 0.065 | 0.65 | 56° |
| 9 | 2.1 | 2.0 | $\dfrac{1.5}{3.1}=0.5$ | 0.036 | 0.43 | 0.111 | 0.67 | 48° |
| 10 | 3.0 | 1.5 | $\dfrac{1.6}{2.8}=0.6$ | 0.052 | 0.55 | 0.145 | 0.64 | 44° |
| 8 | 4.1 | 2.0 | $\dfrac{1.8}{2.6}=0.7$ | 0.072 | 0.72 | 0.187 | 0.62 | 52° |
| 7 | 5.1 | 1.8 | $\dfrac{2.1}{2.1}=1.0$ | 0.089 | 0.92 | 0.187 | 0.69 | 53° |
| 4 | 5.9 | 2.5 | $\dfrac{2.6}{2.6}=1.0$ | 0.104 | 0.73 | 0.296 | 0.80 | 77° |

Measurements of dispersion angle $\alpha_{90}$ were made on a 1000-cycle crossed-field hysteresis loop checker in a manner similar to method No. 1 of Crowther. Measurements of $H_c$, $H_k$, $(M_r/M_s)_d$, and $H_{cd}$ were made on a 1000-cycle hysteresis loop apparatus in the conventional manner. $(M_r/M_s)_d$ and $H_{cd}$ were measured at a transverse drive field of approximately $2H_k$, since these parameters varied with drive below that field. Table I summarizes the average values obtained for the film samples with smaller values of $\alpha_{90}$. The results for samples with large values of $\alpha_{90}$ are omitted, although the results obtained do not differ materially from those listed.

## DISCUSSION OF RESULTS

The relationships between dispersion angle $\alpha_{90}$ and the magnetic parameters shown in Table I can be summarized as follows:

$H_c/H_k$ increases from 0.5 to 1.0 as $\alpha_{90}$ increases from 1.2° to 5.9°. This trend continues to even larger values of $\alpha_{90}$. Since this correlation is not associated with a systematic change in magnetoelastic strain coefficient $\eta$, it is independent of average composition. Smith has shown for inverted films $(H_c/H_k>1)$ "locked-in" regions of negative anisotropy become increasingly common, in agreement with the observation that $\alpha_{90}$ is larger for such films. The use of a ring source to produce the films for this study introduces complicated angle of incidence deposition effects, which could cause inversion, although all of the films were made under similar conditions.

West reported that $H_{cd}=\beta H_k$ and $(M_r/M_s)_d=\sin\beta$ for small $\beta$ within the limit of his measurements. For the 1500-A thick Permalloy samples studied by us, there is no simple correlation of this type between the hard direction hysteretic parameters and $\alpha_{90}$ measured by the Crowther method. The magnitudes of $H_{cd}$ and $(M_r/M_s)_d$ are much larger than predicted by relationships of the type proposed. The largest average value of $H_{cd}$ is observed for the group with the largest average value of $\alpha_{90}$ but the differences in $H_{cd}$ for groups with widely different $\alpha_{90}$ is generally insignificant. The average hard

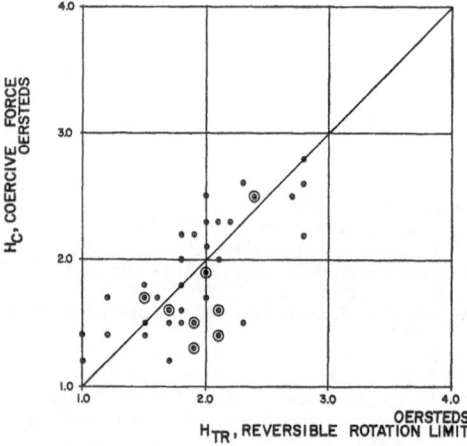

FIG. 1. Reversible rotation limit $H_{tr}$ measured with a Kerr magneto-optic apparatus, as a function of easy axis coercive field $H_c$.

direction hysteresis loop squareness ratio $(M_r/M_s)_d$ in Table I is nearly the same for the groups with the largest and smallest average values of $\alpha_{90}$. Between these extremes, $(M_r/M_s)_d$ does tend to increase directly with $\sin\alpha_{90}$.

$H_{tr}$ (or reversible rotation angle $\theta_r = \sin^{-1} H_{tr}/H_k$) decreases with increasing $\alpha_{90}$, and then increases again for larger values of $\alpha_{90}$. White predicted on the basis of the simple static theory and the Crowther model, that $\theta_r$ should decrease rapidly with increasing $\alpha_{90}$, and then decrease at a slow rate with further increases in $\alpha_{90}$. $\theta_r$ decreases more rapidly than predicted initially. This is believed to be due to local variation of $H_k$ since the measurement of $H_{tr}$ is affected by both amplitude dispersion and easy axis dispersion, whereas the measurement of $\alpha_{90}$ involves only easy axis dispersion. The cause of the large reversible limits associated with the samples with large $\alpha_{90}$ is not known. $H_{tr}$ was found to be better correlated with $H_c$ than with the other magnetic properties. This relationship is illustrated in Fig. 1 for 45 film samples. The line $H_c = H_{tr}$ is drawn through the points. The double circles represent points for two different samples. Some films were studied for which $H_c$ deviated from $H_{tr}$ by 50 to 100%. However, these are all films which exhibit a threshold wall motion switching, i.e., at some threshold value of an applied longitudinal field they are remagnetized completely, or nearly so, and it is very difficult to introduce domain walls in such films. They also have large values of $H_{tr}$ ($\sim 3$ oe).

We have compared the magnetic properties of a selected group of vacuum-deposited Permalloy films with the properties predicted on the basis of simple static models including the effects of easy-axis dispersion. The lack of agreement between the predicted and observed behavior indicates that the methods of measurement and/or the theoretical models require further refinement.

### ACKNOWLEDGMENTS

The authors are indebted to E. Maas and Z. Grietens for much of the data in this study. They also wish to acknowledge the assistance and advice of R. A. White.

---

JOURNAL OF APPLIED PHYSICS     SUPPLEMENT TO VOL. 33, NO. 3     MARCH, 1962

# Anisotropy Sources for Electrodeposited Permalloy Films

M. LAURIENTE AND J. BAGROWSKI

*Westinghouse Electric Corporation, Air Arm Division, Baltimore 3, Maryland*

Experimental evidence has been found to indicate that electrodeposited iron-nickel films are sensitive to the geometric anisotropy of the chain-like growth of crystallites that compose the conducting film of the substrate. Chromium-gold films were vacuum evaporated on glass slides at normal and oblique incidence. Fired gold films were also made to be used as examples of minimum geometric anisotropy. Using Wolf's solution as an electrolyte, Permalloy films were electrodeposited, some with an some without a magnetic field. Uniaxial anisotropy was found for all the specimens, including the controls, with the exception of those specimens plated without a magnetic field on both, normal incidence substrates, and fired gold substrates. Angles as small as 2° were found sufficient to induce a marked anisotropy. From the knowledge that evaporation at oblique incidence creates a geometric anisotropy by a process of self-shadowing of the crystallites, it is believed that this geometric anisotropy is replicated in the magnetic film as a shape anisotropy which in turn induces a magnetic anisotropy. In support of this theory, directional polishing of the substrates was also found to induce an anisotropy, although in these cases usually with higher $H_K$ values. The epitaxy model can be helpful for applying data from evaporating techniques to plating techniques. Magnetic films in this study averaged 3000 A in thickness. The $B-H$ tests were made at 400 cps. Conclusions are that, in this experiment, the anisotropy induced by the magnetic field can be separate from the one induced by the substrate, and sensitivity to geometric anisotropy is pronounced.

### INTRODUCTION

MUCH has been written on the sensitivity of vacuum evaporated magnetic films to the angle of incidence.[1-3] This work was undertaken to determine if a like effect could be exerted by the geometric anisotropy present in the conductor used as a substrate for electrodeposited magnetic films.

### EXPERIMENT

Glass slides were used as substrates to receive conducting films of either vacuum evaporated chromium-gold films or fired gold films. The chromium is present only to improve the adhesion of gold to glass. All of the evaporated slides were made on a plane parallel to the base plate of the evaporator and at a distance of 16 in. from the evaporating source which was a resistance heated molybdenum boat. Because of this distance it was possible to cut out 12-mm square samples to an accuracy of a degree in obliquity. Specimens with deviations from the normal of less than 2° were considered to be a normal incidence. The fired gold films were made by dipping the slides in a solution,[4] heating slowly to 600°C, and then furnace cooling. This surface was included in the experiment because the amorphous-like film characteristic of vitrified medium

[1] D. O. Smith, J. Appl. Phys. **32**, 70S (1961).
[2] T. G. Knorr and R. W. Hoffman, Phys. Rev. **113**, 1039 (1959).
[3] M. S. Blois, Jr., J. Appl. Phys. **26**, 975 (1955).

[4] This solution is marketed by Hanovia Chemical & Manufacturing Company East Newark, New Jersey, as liquid bright gold.

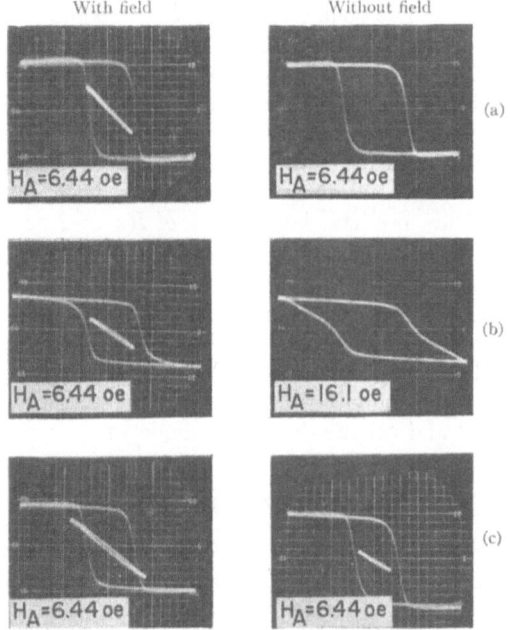

With field                    Without field

FIG. 1. 400-cps loops for films on substrates: (a) fired gold; (b)
(b) normal incidence; (c) oblique incidence. $H_A$ refers to applied
field.

was considered to be ideal as an example of minimum
geometric anisotropy.

Wolf's solution[5] was used as the electrolyte for the
electrodeposition process. All films were plated at room
temperature. Graphite was used as the anode. Serious
depletion of the electrolyte was avoided by limiting the
number of specimens for each bath. Simple iron titra-
tions were also taken before and after each series of
specimens to insure control. Permanent magnets were
used to provide a field strength of 200 oe in the plating
cell. Nominal composition of the films was 83% Fe-17%
Ni and thickness averaged 3000 A.

### EXPERIMENTAL RESULTS

The loop characteristics of the films measured at 400
cps are shown in Fig. 1. The loop in the hard direction
has been superimposed on the easy direction where
evidence of uniaxial anisotropy was found. Note that
uniaxial anisotropy was absent for films made without a
field for both the normal incidence and the fired gold;
also, the exceptional loop width shown for Fig. 1(b).
Slight disregistry could be found for the normal inci-

[5] I. W. Wolf and V. P. McConnell, 43rd Proceedings American
Electroplaters Society, 215 (1956).

dence specimen if the easy and hard directions were
superimposed, indicative of a slight anisotropy. Such
was not the case for the fired gold sample, pointing out
the sensitivity of the magnetic film to the geometric
anisotropy and the isotropy of the amorphous gold film.
At 2° obliquity, a pronounced uniaxial anisotropy was
observed as further confirmation of this hypothesis. The
maximum angle of incidence examined was 10°. Over
this range a wide variation of loop shapes were found,
several of rhombic shape. In general, the forms were
inconsistent and $H_K$, the anisotropy field, likewise
varied. The particular sample shown in Fig. 1(c) made
at 5° obliquity was selected as one of the better examples
of uniaxial anisotropy in this category. An explanation
is offered that the randomness found is a reflection of
the source of the anisotropy, referring of course to the
statistical distribution of the crystallites of the substrate.

In sharp contrast, the films made in the field displayed the characteristic rectangular loop shown in
Fig. 1(b) and $H_K$ was much more consistent.

Directional polishing of substrates was also found to
induce anisotropy, although in these cases usually with
higher $H_K$ value. In view of the obvious results from the
oblique incidence, this bit of information would appear
to be redundant.

In summary, a case has been shown where anisotropy
has been induced by the magnetic field alone, i.e. the
fired gold example, Fig. 1(a); another example has been
shown for oblique incidence alone, i.e. evaporated
specimen, Fig. 1(b). These results show that the contributions of the respective sources, magnetic induction
and geometry, can be separate.

### CONCLUSION

From the knowledge that evaporation at oblique
incidence creates a geometric anisotropy by a process of
self-shadowing of the crystallites,[1] it is postulated that
this geometric anisotropy is replicated in the magnetic
film as a shape anisotropy which in turn induces a magnetic anisotropy. With this model the theories for
evaporated films could then be extended to include
electrodeposited films leading to the conclusion that
electrodeposited films are very sensitive to the geometric
anisotropy in the substrate.

### ACKNOWLEDGMENTS

The authors wish to express their gratitude to G. E.
Lynn, J. Hurt, and J. Winter for their assistance on
instrumentation envolved.

JOURNAL OF APPLIED PHYSICS    SUPPLEMENT TO VOL. 33, NO. 3    MARCH, 1962

# Constriction of Hard Direction Hysteresis Loops in Thin Permalloy Films

S. MIDDELHOEK

*International Business Machines Corporation Research Laboratory, Zurich, Switzerland*

According to single domain theory, thin Permalloy films with uniaxial anisotropy should have straight hard direction M-H loops without hysteresis. Practically, however, the M-H loops of almost all films are open, when the ac field is applied exactly in the hard direction and the earth's magnetic field is carefully compensated. Moreover, films thicker than about 800 A often show hard direction hysteresis loops which are constricted at the origin.

As is well known, magnetization reversal in the hard direction is associated with the splitting up of the film into a great number of long domains parallel to the easy axis. Whether the walls separating these domains are of the Bloch or Néel type depends upon the strength of the applied field. It will be shown that the constricted loops can theoretically be constructed when, in addition to the anisotropy energy and the energy of the magnetization in the applied field, the energy of these walls is considered. Bitter observations of the reversal process confirm the theory, showing that the constriction is actually a result of Neél-Bloch-Néel wall transitions.

THIN Permalloy films evaporated in a magnetic field exhibit uniaxial anisotropy. Under the influence of an external magnetic field the magnetization of the film can be turned out of the easy direction, and the resulting magnetization direction $\theta$ is found by minimization of the total energy with respect to $\theta$. Assuming that the film behaves as a single domain one obtains theoretically a straight hysteresis loop in the hard direction. Experimentally, however, in the hard direction a closed straight hysteresis loop is seldom obtained. The hysteresis loops often show a certain squareness or constriction.

Earlier Olson and Pohm[1] and Smith[2] pointed out the connection between the open hysteresis loops in the hard direction and the splitting up of the film into a large number of domains, which was later on confirmed by many authors.[3-7]

For the explanation of the constricted loop we assume that the film is divided by $N$ walls/cm and that the magnetization in the domains is homogeneous. The total energy density of such a film under the influence of a field $H_y$ in the hard direction, neglecting the wall width, can be written as:

$$E = K \sin^2\theta - H_y M_s \sin\theta + N\gamma_w(\theta),$$

in which the first term is the anisotropy energy, the second term is the energy of the magnetization in the applied field, and the third term is the wall energy.[6] Equilibrium exists when:

$$\frac{\partial E}{\partial \theta} = 2K \sin\theta \cos\theta - H_y M_s \cos\theta + N\frac{\partial \gamma_w(\theta)}{\partial \theta} = 0.$$

To calculate the shape of the hysteresis loop the wall energy as a function of the angle $\theta$ must be known.

The energy of a Bloch wall in a thin film consists mainly of the stray field energy due to the free poles at the intersection of the wall with the film surface. The stray field energy is generally proportional to the square of the magnetization component which causes these free poles. When under the influence of a magnetic field, the magnetization of the surrounding domains makes an angle $\theta$ with the easy axis, this component is $M_s \cos\theta$. For the wall energy we then assume

$$\gamma_B(\theta) = \gamma_B(180) \cos^2\theta.$$

The stray field energy of a Néel wall is proportional to the square of the component which causes the volume charges. This component is equal to $M_s[1 \mp \sin\theta]$ and therefore we assume

$$\gamma_N(\theta) = \gamma_N(180)[1 \mp \sin\theta]^2,$$

where the plus and the minus signs indicate the two possible directions of the magnetization in the wall. A more precise calculation shows that the above approximations for the wall energy are permitted.

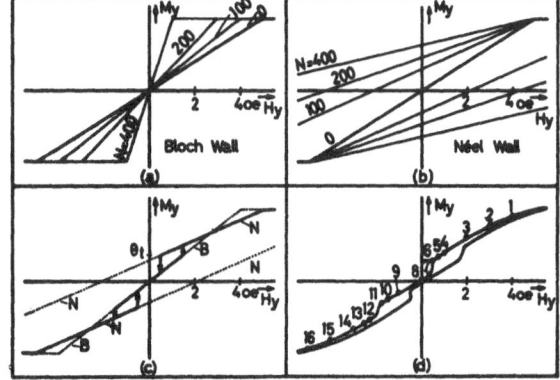

FIG. 1. Hard direction hysteresis loops (a) for different Bloch wall densities and (b) for different Néel wall densities ($D=1800$ A, $H_k=5$ oe). (c) Theoretically constructed hard direction hysteresis loop ($D=1800$ A, $H_k=5$ oe, $N=100$ walls/cm) and (d) the experimentally observed loop ($D=1800$ A, $H_k=5$ oe, $N=400$ walls/cm before the N-B-N transition and $N=200$ walls/cm after N-B-N transition).

[1] C. D. Olson and A. V. Pohm, J. Appl. Phys. 29, 274 (1958).
[2] D. O. Smith, J. Appl. Phys. 29, 264 (1958).
[3] E. Fuchs, Z. angew. Phys. 13, 157 (1961).
[4] E. Feldtkeller, Z. angew. Phys. 13, 161 (1961).
[5] S. Middelhoek, Helv. Phys. Acta 33, 519 (1960).
[6] S. Middelhoek, Z. angew. Phys. 13, 151 (1961).
[7] R. J. Spain and H. Rubinstein, J. Appl. Phys. 32, 288S (1961).

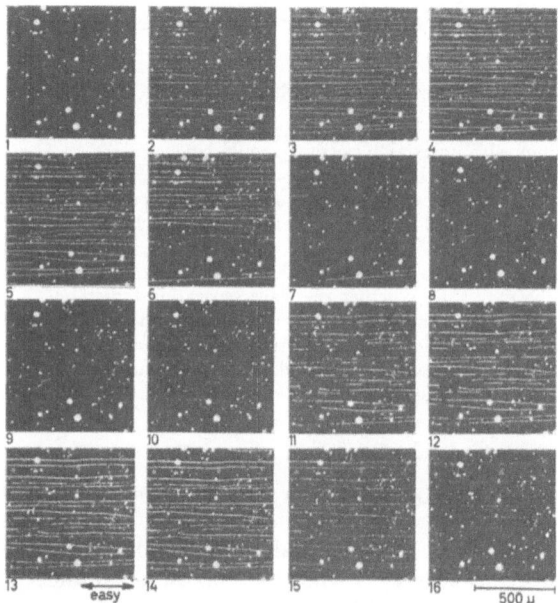

FIG. 2. Bitter patterns as a function of the field in the hard direction. The numbers under the pictures correspond to the numbers on Fig. 1(d).

If we substitute these expressions in the equation for equilibrium we obtain for the Bloch wall:

$$\frac{\partial E}{\partial \theta} = 2K \sin\theta \cos\theta - H_y M_s \cos\theta$$
$$- 2N\gamma_B(180) \sin\theta \cos\theta = 0.$$

The magnetization component in the field direction $M_y = M_s \sin\varphi$, so that the hysteresis loop becomes:

$$M_y = \frac{M_s^2}{2K - 2N\gamma_B(180)} H_y,$$

which is a straight line [Fig. 1(a)] with saturation field:

$$H_k' = \frac{2K - 2N\gamma_B(180)}{M_s}.$$

For the Néel wall we obtain:

$$\frac{\partial E}{\partial \theta} = 2K \sin\theta \cos\theta - H_y M_s \cos\theta$$
$$\mp 2N\gamma_N(180)[1\mp\sin\theta][\cos\theta] = 0$$

and the hysteresis loops assume the form:

$$M_y = \frac{H_y M_s^2 \pm 2N\gamma_N(180)M_s}{2K + 2N\gamma_N(180)}$$

which are plotted for different $N$ in Fig. 1(b).

The reversal process in the hard direction for a film which splits up into a hundred elongated domains per cm is described with the help of Fig. 1(c). For relatively large fields in the hard direction the magnetization makes a large angle $\theta$ with the easy direction. The angle through which the magnetization rotates is small and then the Néel wall is energetically the most favorable one. Decreasing the field increases this angle. At a certain angle $\theta_t$ the total energy of the film is the same whether the walls are of the Néel or of the Bloch type, and transition from Néel to Bloch type will occur. The angle $\theta_t$ is given by:

$$\gamma_B(180) \cos^2\theta_t = \gamma_N(180)[1-\sin\theta_t]^2.$$

When we substitute $\gamma_B(180)=4.0$ ergs/cm² and $\gamma_N(180) =8.5$ ergs/cm² ($D=1800$ A) we obtain: $\sin\theta_t=0.36$.

For decreasing fields the Bloch wall hysteresis curve is followed until at negative fields a Bloch-Néel wall transition again occurs at the same $\theta_t$. The constriction of the hard direction loop is thus a result of the Néel-Bloch-Néel wall transitions.

In Fig. 2 the Bitter patterns of a film ($D=1800$ A, $H_c=0.7$ oe, $H_k=5$ oe, 74–26 Ni-Fe) are shown which belong to the experimental loop of Fig. 1(d). For large fields, the film is just split up, and the Néel walls are not clearly visible. Visibility, however, continuously increases for decreasing fields until at a still positive field the visibility again decreases although discontinuously.

From domain wall observations as a function of the film thickness[8] it is known that the visibility of a Bloch wall is much less than that of a Néel wall, thus justifying the assumption that the above discontinuous decrease of the visibility represents a Néel-Bloch wall transition. Further decreasing the field to negative fields, a discontinuous increase in visibility again occurs, predicted by theory, which now represents the Bloch-Néel wall transition.

### ACKNOWLEDGMENT

The author wishes to thank O. Voegeli for making the Bitter observations.

[8] S. Methfessel, S. Middelhoek, and H. Thomas, IBM J. Research Develop. 4, 96 (1960).

# Bitter Patterns on Single-Crystal Thin Films of Iron and Nickel

H. SATO, R. S. TOTH, AND R. W. ASTRUE

*Scientific Laboratory, Ford Motor Company, Dearborn, Michigan*

Basic domain structures of epitaxially grown single crystals of Fe and Ni were investigated by the Bitter pattern technique. Films of these materials were obtained by evaporation onto freshly cleaved MgO single crystals heated to an appropriate temperature in a vacuum of $10^{-5}$mm Hg or less. The nature of the resulting films was examined using reflection electron diffraction. The Bitter patterns were obtained with the metal films on the substrate. From the mode of epitaxy, the structure of the domains and the direction of the moments were determined with respect to the orientation of the substrate. The cleavage surface of the substrate is not perfectly flat, but contains rather large steps as well as tear lines, and these defects are clearly observable in the powder patterns. The relations between such imperfections and the domain patterns are discussed. In addition, the patterns show signs of the existence of matching strain between the metal film and the substrate. The effect of the strain becomes more important as the film thickness decreases, as shown by small intricate domains that are formed. The importance of the 90° wall over the 180° wall in the case of thin films with cubic symmetry is emphasized.

## INTRODUCTION

THE single-crystal thin film technique has many advantages, chiefly because single crystals of some materials which are very difficult to obtain in bulk form can be easily and quickly grown as thin films. In order to utilize this technique advantageously for the study of magnetic materials, it is necessary to know the characteristics of the epitaxially grown single crystal. There seem to be several differences between bulk crystals and thin film crystals. For example, it has been reported that the anisotropy constants of some epitaxially grown single crystals, as long as the film adheres to the substrate, have higher values than those obtained in bulk material.[1] In order to investigate the origin of these deviations from bulk behavior, basic domain configurations of epitaxially grown single crystals of iron and nickel were investigated by the Bitter pattern technique. Although the domain structures of polycrystalline films have been extensively investigated,[2] very little domain work has been reported on single-crystal thin films.[3] In the present research the Bitter pattern was observed on the film when it adhered to the substrate. This was done for two reasons: first, the characteristics of the film while on the substrate are those which are of interest; and second, the stripping process would introduce further imperfections into the film.

## IRON

Iron was evaporated onto freshly cleaved MgO crystals which had been heated to about 450°C in a vacuum of $1 \times 10^{-5}$mm Hg. MgO was chosen for the reason that it is not attacked by the magnetic colloidal suspension. Since there is no way of stripping the deposited film from the MgO substrate, the film was examined by reflection-electron diffraction to determine whether it was indeed

a single crystal. The mode of epitaxy was found to be (001) MgO ∥ (001) Fe, and [110] MgO ∥ [100] Fe, in agreement with previous observations. Single crystals of iron can also be grown on cleaved NaCl surfaces.

Due to the mode of epitaxy of iron on MgO, the crystal orientation of iron is determined with respect to that of the MgO substrate, and consequently it is possible to determine the orientation of the Bitter figures with respect to the crystallographic directions of the iron film. Figures 1(a) and 1(b) represent an example of the Bitter pattern obtained on an iron film, about 1000 A thick together with its interpretation. Since the cleaved surface is not perfect, tear lines and steps appear on the surface, and these are revealed by the accumulations of magnetic powder. The parallel dark lines in the figure indicate steps (of the order of 1000 A in height), and the thin lines indicate tear lines on the substrate. The appearance of the surface can be substantially improved by cleaving with a shorter stroke, but polishing and etching have the disadvantage of destroying the atomic flatness of the surface.

Domain walls are indicated by lines which make an angle of about 45° with respect to the big steps. Since such steps serve as sites for flux leakage, spike domains occur in order to compensate for this. The general trend of the steps is to make an angle of 45° with respect to the [100] direction of the MgO crystal. In other words, they are in the [100] direction of the Fe crystal. Superficially, therefore, the pattern resembles that observed on a bulk single crystal, whose face is tilted slightly from the (001) plane. Closer examination, however, reveals a situation which is different and much more complicated. In the case observed, the magnetic field was approximately in the [$\bar{1}$10] direction and had been reversed several times. A small remanent field was still on in the [$\bar{1}$10] direction. The main direction of the domains is [010] at the center. The lines in the [110] direction and [$\bar{1}$10] direction indicate 90° domain walls, and those in the [100] direction indicate 180° walls. At the steps the domain is continuous, with the magnetic powder accumulating on the edge of the step because of the flux leakage. However, one can notice a change in the thickness

[1] S. Chikazumi, J. Appl. Phys. **32**, 81S (1961).

[2] For example: H. J. Williams and R. C. Sherwood, J. Appl. Phys. **28**, 548 (1957); E. E. Huber, Jr., D. O. Smith, and J. B. Goodenough, J. Appl. Phys. **29**, 294 (1958); S. Methfessel, S. Middelhoek, and H. Thomas, IBM J. Research Develop. 4, 96 (1960).

[3] B. Elschner and D. Unangst, Z. Naturforsch. **11a**, 98 (1956).

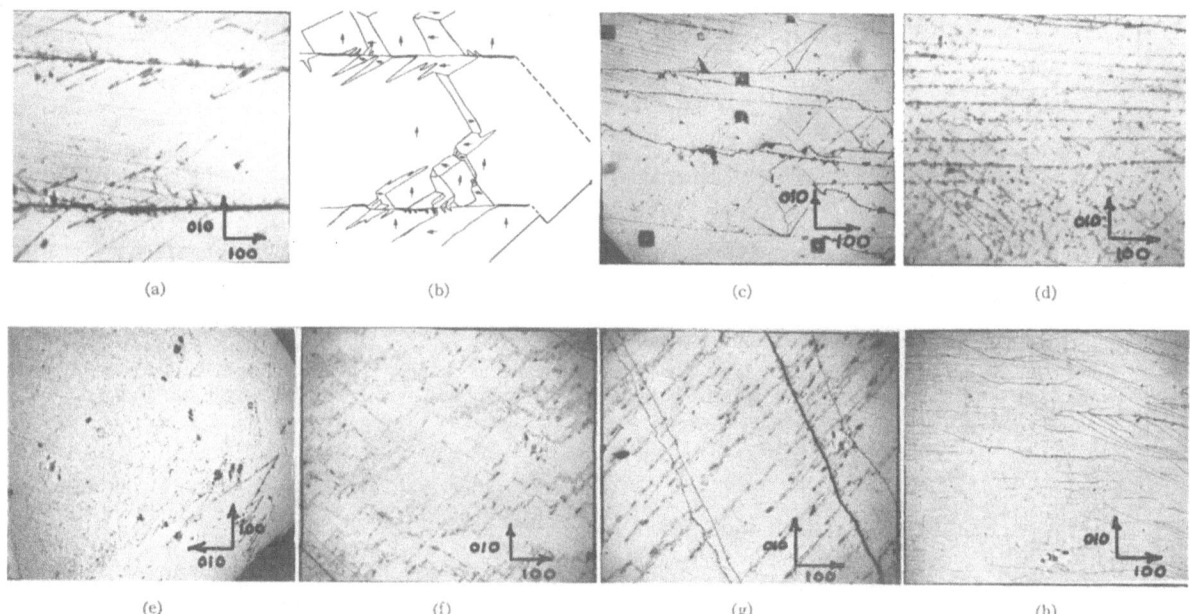

FIG. 1. Bitter patterns of epitaxially grown single-crystal thin films; arrows indicate two principal directions, [100], [010] of the metal crystals. (a) iron, 1000 A (138×); (b) interpretation of (a); (c) iron, 330 A, virgin state (138×); (d) iron, 120 A, virgin state (540×); (e) iron, 270 A, virgin state (69×); (f) iron, 270 A, virgin state (258×); (g) iron, 270 A, virgin state (258×); (h) nickel, 150 A, virgin state (240×). The long axis of the domain is in the [010] direction. (Very thin.) Dark lines approximately in the [100] direction are tear lines of the substrate.

of this accumulation, depending on whether the magnetic moment of the domain is parallel or perpendicular to the step. From this change in thickness one can also infer the direction of the spins.

Further, two important features are clearly noticeable. One is the predominance of the 90° wall, and the other is that these 90° walls are not always exactly parallel. As the film becomes thinner, the spins tend to rotate inside the plane of the film at the wall. This causes the change from the Bloch type wall to the Néel type in the case of uniaxial films.[4] However, in the case where the film has cubic symmetry, a 90° wall is formed before the spins rotate 180° in the plane of the film. Therefore 180° walls become unfavored in very thin films. The deviation from parallelism is also brought about by the thinness of the film, causing the decrease in wall energy per unit length.[2]

The tendency of avoiding formation of 180° walls becomes more conspicuous as the films become thinner, as illustrated in Figs. 1(c) and 1(d). Thus, the formation of domains predominantly with 90° walls is one of the principal ways in which single-crystal thin films differ from bulk crystals.

Moreover, as the films become thinner, the domains become smaller and more complicated. In some cases, the patterns observed become very similar to maze patterns [Figs. 1(e) and 1(f)]. This is probably due to strain, presumably caused by the difference in expansion coefficients of the film and the substrate, and by the

presence of irregularities on the surface of the substrate. Where maze-like patterns are formed. [Fig. 1(f)], the boundaries of the tiny domains are all in one of the [110] directions, indicating that only 90° walls exist.

When the films become thinner, the role of the demagnetizing field to form closure domains decreases. Depending upon the condition of the film, therefore, unidirectional domains such as those shown in Fig. 1(g) can be observed. It is remarkable that these elongated domains are all in the [110] direction. The spin direction is not that of the long axis of the domain, but instead makes an angle of 45° with it. This shows that the domain walls formed are 90° walls. Both Figs. 1(f) and 1(g) are portions of Fig. 1(e). No cross-tie walls were observed by changing the thickness of the films. This also is due to the ease of forming 90° walls in the presence of cubic symmetry as explained above.

### NICKEL

Single-crystal nickel thin films can easily be grown epitaxially on both MgO and NaCl. The mode of epitaxy is (001) || (001), [100] || [100]. In this case, the domain patterns are not easy to see, and domains have not yet been seen on thick films of nickel. From the analogous case of iron films it is inconceivable that nickel films have only a single domain when in the virgin state. In the first place, the spins do not lie inside the plane of the film since the easy direction is [111]. Also the anisotropy constant is small, and this makes the wall wider. In the second place, the magnitude of the magnetic moment is small compared to that of iron. By making the film

[4] L. Néel, Compt. rend **241**, 533 (1955).

thinner, however, one causes the spins to tend to lie in the plane of the film, and it becomes easier to reveal the magnetic domains. The magnetic domains can indeed be observed under such conditions in virgin films (that is, before any magnetic field has been applied), as shown in Fig. 1(h).

The observed pattern is very similar to that obtained in thin iron films [Fig. 1(g)]. However, the direction of the long axes of the domains is [010]. Since from the above reasoning the spins are supposed to lie in the [110] direction, the long axes of the domains make an angle of 45° with the magnetic moments, and the boundaries are 90° walls. This situation is exactly equivalent to the case of the thin iron film.

## CONCLUSION

Ferromagnetic domain patterns have been observed on single-crystal thin films of iron and nickel, and their orientations relative to the principal crystallographic directions of the film have been determined. The main difference between the single-crystal thin film and the bulk crystal is the tendency of the thin crystal to form domains having 90° walls, in preference to the formation of domains having 180° walls. The 90° walls are also preferred to 180° Néel walls, and cross-tie walls have not been found in single-crystal thin films. The essential difference between domains of single-crystal thin films and those of uniaxial polycrystalline films is thus apparent.

---

JOURNAL OF APPLIED PHYSICS          SUPPLEMENT TO VOL. 33, NO. 3          MARCH, 1962

# Electron Microscope Study of the Roughness of Permalloy Films Using Surface Replication*

ALFRED BALTZ
*The Franklin Institute Laboratories, Philadelphia, Pennsylvania*

Permalloy films evaporated at normal and oblique incidence to a substrate were examined by replication of their surfaces. Electron micrographs revealed chains of particles perpendicular to the incident beam. The degree of alignment and the lengths of the chains depended upon the angle of incidence and the film thickness. A statistical analysis of the micrographs was made. The magnetic data showed a discrepancy between the magnetic thickness and the optical thickness for oblique incidence films.

THE presence or absence of any particular orientation in the surface roughness and its correlation to the magnetic properties of evaporated 17 wt.% Fe-83 wt.% Ni films was investigated. The results which are in substantial agreement with those reported by Smith et al.[1] were obtained by an entirely different technique. The films, evaporated at normal and oblique incidence with respect to a substrate were replicated by various techniques in an effort to rule out any artifacts which might have been introduced by the replicating material. Replication of the surface, rather than a study by transmission electron microscopy, was chosen since by this method films could be examined regardless of thickness.

Electron micrographs taken of carbon replicas[2] of Permalloy films evaporated at normal incidence to the substrate exhibited generally a randomly oriented granular structure [Fig. 1(a)]. This random structure was observed in films regardless of their thickness. However, when the micrographs were examined very carefully, an alignment of surface structure was seen occasionally

[Fig. 1(b)]. When an alignment was observed in films evaporated at normal incidence, the chains of particles did not run in a particular direction but rather, clusters of chains were present running in one direction in one

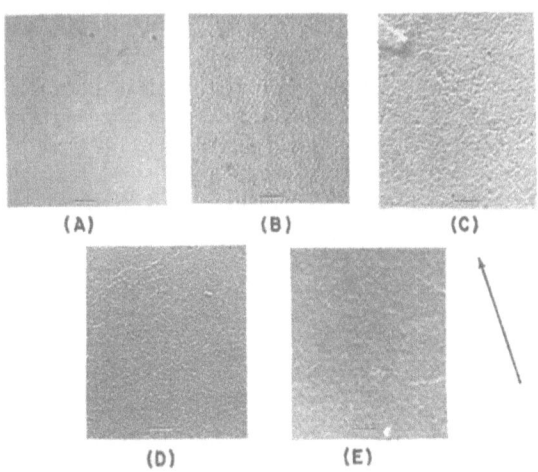

FIG. 1. Electron micrographs of replicas of the surface of Permalloy films deposited on carbon covered micro-cover-slides at room temperature: (a) normal incidence, 800 A thick, (b) normal incidence (alignment observed visually) 800 A thick, (c) 55°, 100 A thick, (d) 60°, 250 A thick, (e) 40°, 850 A thick. Direction of arrow=direction of vapor stream. Shadowing material tungsten oxide. Shadowing angle 30°. Magnification indicator=0.1 μ.

---

* This research was supported in whole or in part by the United States Air Force under contract, monitored by The Aeronautical Systems Division, Air Force Systems Command.
[1] D. O. Smith, M. S. Cohen, and G. P. Weiss, J. Appl. Phys. **31**, 1755 (1960).
[2] D. E. Bradley, Brit. J. Appl. Phys. **5**, 65 (1954).

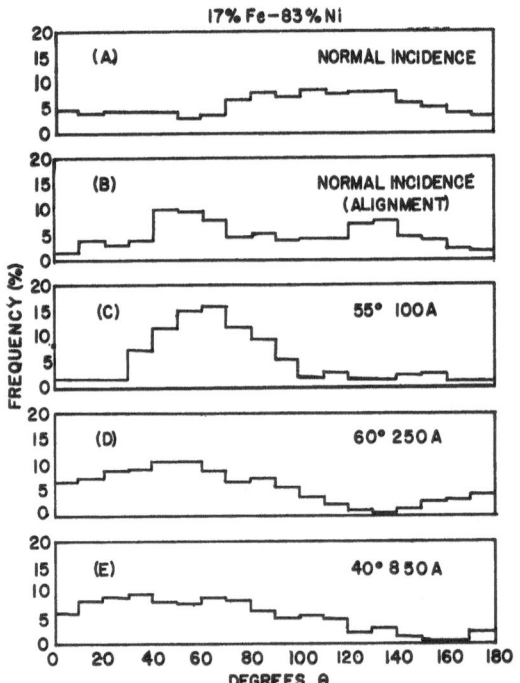

FIG. 2. Frequency of particle chains vs angle of chain direction ($\theta$). Direction of metal shadow $\theta = 0°$.

area and in another direction in another area. This type of alignment will be referred to as regional alignment.

For films evaporated at oblique incidence to the substrate the degree of alignment was dependent upon the angle of incidence and the thickness of the films, the thickness being more influential than the angle of incidence. One could observe a better alignment of particle chains in thin films evaporated at a smaller angle of incidence than in thick films evaporated at a larger angle. Directional alignment of 100-A films evaporated at 55° was obvious [Fig. 1(c)]. The chains of particles having an average diameter of approximately 200–250A, ran perpendicular to the incident beam as has been observed by Smith[1] and König.[3] As can be seen from the micrograph, the replica was shadowed in the direction of the vapor stream of the Permalloy, making the shadow run approximately perpendicular to the particle chains. It was observed that the particle size decreases with increasing film thickness. Films 250 A thick evaporated at 60° exhibited an average particle size of approxi-

[3] H. König and G. Helwig, Optik **6**, 111 (1950).

mately 100–150 A. Chains of aligned particles, much shorter than the ones observed in the 100-A films, could be seen [Fig. 1(d)].

The tendency of Permalloy films to become smoother with thickness was evident on films evaporated to 850 A thickness at 40° incidence. Here an alignment of chains of particles which were very short was observed [(Fig. 1(e)]. As for the normal incidence films, in which an alignment of topography was present, these films also exhibited a regional alignment.

In an attempt to secure quantitative data a statistical count of chain directions was made. Any chain with more than two particles adjacent to each other was counted for this purpose. The chain direction was plotted as degrees $\theta$ vs frequency of direction in % (Fig. 2). The direction of the shadow casting was taken as 0°. From these graphs it can be shown that the thin films (100 A) have a very high uniaxial peak [Fig. 2(c)]. The 250 A film [Fig. 2(d)] also exhibited a uniaxial tendency, so that one can conclude that an alignment of surface topography is rather evident for films up to 250 A thick. For thicker films and smaller angles of incidence the graphs show a more and more randomly oriented pattern as predicted by visual inspection of the micrographs Fig. 2(a), (b), (e). It is also evident that there is a directional trend in the few films evaporated at normal incidence, as had been predicted [Fig. 2(b)].

The magnetic data generally agreed with the results obtained by Smith[1], Cohen,[4] and Kambersky[5] including a 90° change in the easy direction for angles > 70°. The most important magnetic result was an obvious discrepancy for oblique incidence films between the magnetic thickness, i.e., thickness determined from magnetic moment measurements assuming a value of $I_s$, and the optical thickness as determined by a multiple beam interferometer. This is in contrast to normal incidence films for which good agreement was found.

## ACKNOWLEDGMENTS

The author acknowledges the contributions made by J. A. Johnson during the earlier stages of this work and wishes to thank W. D. Doyle and Dr. F. R. L. Schoening for valuable discussion and cooperation.

[4] M. S. Cohen, J. Appl. Phys. **32**, 87S (1961).
[5] V. Kambersky, Z. Málek, Z. Frait, and M. Ondris, Czech. J. Appl. Phys. **B11**, 171 (1961).

JOURNAL OF APPLIED PHYSICS     SUPPLEMENT TO VOL. 33, NO. 3     MARCH, 1962

# A Theoretical Model for Partial Rotation

H. Thomas

*International Business Machines Corporation, Research Laboratory, Zurich, Switzerland*

A simple model which consists of a set of parallel strips with different anisotropy constants, interacting by the magnetic stray field, is investigated. It displays an incoherent reversal process similar to the partial rotation process observed in thin Permalloy films, and shows under certain conditions a unidirectional rotational hysteresis.

IN an investigation of the magnetization reversal in thin Permalloy films at an angle to the easy direction, a new reversal process was observed which is different from either wall motion or coherent rotation and which has been termed partial rotation.[1] At a field value near the coherent rotation threshold, a transition takes place from the initial single domain state to a partially reversed state, consisting of a bandlike pattern of long, narrow domains with their long axes approximately perpendicular to the magnetization direction in the single domain state immediately before the transition.[2] This intermediate state remains stable for further increasing field, until finally the magnetization reversal is completed by a wall motion process. The suggested magnetization distribution at the various steps of the magnetization reversal is in good agreement with electron microscopical observations by Feldtkeller.[3]

According to the qualitative interpretation given,[1] this reversal process was assumed to be a consequence of the magnetization ripple,[4,5] which occurs in the single domain state before the transition because of the anisotropy variations in the film. When this magnetization ripple becomes unstable at a field value near the coherent rotation threshold, it was expected that the stray fields arising in this rippling pattern cause the observed incoherent reversal process. In this paper, we present a simple model which displays an incoherent process similar to the one observed.

## PARTIAL ROTATION

The response of the magnetization to the anisotropy variations in the film is restricted to a longitudinal ripple around the average magnetization by the exchange and stray field coupling.[4,5] Lateral variations are practically completely suppressed by the stray fields arising, and the exchange forces cause a cutoff below a certain wavelength for the remaining longitudinal variations. It can be shown that this cutoff wavelength increases strongly when the applied field approaches the average coherent rotation threshold $H_r$. Therefore, when the magnetization ripple becomes unstable at a field value near $H_r$, the subsequent irreversible process

will be governed by a certain long wavelength component of the anisotropy variations with the wave normal parallel to the average magnetization direction immediately before the transition. If this magnetization direction is approximated by that of a homogeneous film for $H = H_r$, the angle between the wave normal and the easy direction $\delta$ is related to the angle between the field and the easy direction $\gamma$ by[2]

$$tg^3\delta = tg\gamma. \qquad (1)$$

Since we are only interested in the general features of this irreversible process, we simplify the model as much as possible. The most important rippling wave is replaced by a system of parallel bands. We assume for alternating bands slightly different anisotropy constants $K_1$ and $K_2$ and, in general, different widths $b_1$ and $b_2$, but for convenience the same easy direction. The thickness is taken equal to the film thickness $D$.

We neglect exchange forces between the bands and assume uniform magnetization in the film plane in each band. If the angle between the magnetization direction and the band normal is $\psi_1$ and $\psi_2$, respectively, the interfaces between the bands carry the constant charge density

$$\sigma = \pm M(\cos\psi_2 - \cos\psi_1). \qquad (2)$$

While the anisotropy energy and the external field energy are simply the sum of the contributions of the single bands, the stray field energy will be proportional to $\sigma^2$ and establishes therefore a coupling between adjacent bands. By introducing a coupling factor $g(b_1, b_2, D)$, the average energy density can be written

$$\langle E \rangle = b_1/(b_1 + b_2)[K_1 \sin^2(\psi_1 + \delta) + MH \cos(\gamma + \delta + \psi_1)]$$
$$+ b_2/(b_1 + b_2)[K_2 \sin^2(\psi_2 + \delta) + MH \cos(\gamma + \delta + \psi_2)]$$
$$+ g(b_1, b_2, D) \cdot M^2 (\cos\psi_2 - \cos\psi_1)^2. \quad (3)$$

The coupling factor $g$ can be evaluated by writing the stray field energy as the sum of the self-energies and the mutual energies of uniformly charged strips of width $D$. Using the expansions given by Rhodes and Rowlands,[6] one finds in the limit $b_1, b_2 \gg D$

$$g(b_1, b_2, D)$$
$$= 2\frac{D}{b_1 + b_2}\left\{2\ln\frac{b_1 + b_2}{D} + \ln\left[\frac{1}{\pi}\sin\frac{\pi b_1}{b_1 + b_2}\right] + \frac{3}{2}\right\}. \quad (4)$$

[1] S. Methfessel, S. Middelhoek, and H. Thomas, J. Appl. Phys. **32**, 1959 (1961).

[2] S. Middelhoek, Symposium on the Electric and Magnetic Properties of Thin Films (IUPAP), Louvain, Belgium, September 4–7, 1961.

[3] E. Feldtkeller, Elektron. Rechenanlagen **3**, 167 (1961).

[4] H. W. Fuller and M. E. Hale, J. Appl. Phys. **31**, 238 (1960).

[5] E. Fuchs, Z. angew. Physik **13**, 157 (1961).

[6] P. Rhodes and G. Rowlands, Proc. Leeds Phil. Soc. **6**, 191 (1954).

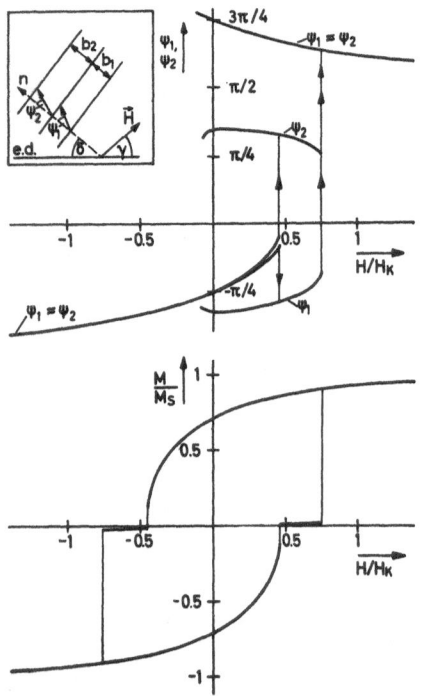

FIG. 1. Stable magnetization directions as functions of the applied field and corresponding hysteresis loop for the parameter values given in the text.

For $D \sim 500 A$, $b_1 \sim b_2 \sim 25 \mu$, $g$ is of the order $10^{-2}$ to $10^{-1}$, so that for Permalloy films with $M \sim 800$ gauss, $K = \frac{1}{2}(K_1+K_2) \sim 10^3$ ergs, the quantity

$$q = 4g(M^2/K), \qquad (5)$$

which is a measure for the ratio between the stray field and anisotropy energies, is still of the order 25 to 250, i.e., large compared to 1.

The equilibrium directions of the magnetization are obtained by minimizing the energy (3) with respect to $\psi_1$ and $\psi_2$. Figure 1 shows as an example the stable equilibrium directions and the corresponding hysteresis loop obtained by numerical solution for the case $b_1 = b_2$, $\gamma = \delta = \pi/4$, $(K_1 - K_2)/(K_1 + K_2) = 0.1$, $q = 10$. In addition to the "single domain" states $\psi_1 \approx \psi_2$, there exists in fact a stable, partially reversed state. Starting at $H = 0$ with the single domain state $\psi_1 = \psi_2 = -\pi/4$, the magnetization directions remain almost equal until they become unstable near the coherent rotation threshold $H_r = 0.5 \langle H_K \rangle$, where $\langle H_K \rangle = (K_1 + K_2)/M$. Then a transition may occur to the partially reversed state, the magnetization rotating in opposite directions in adjacent bands. This state remains stable up to a second critical field $H = 0.75 \langle H_K \rangle$, where finally the transition to the reversed single domain state takes place. For $q = 100$, this second critical field is $H = 1.1 \langle H_K \rangle$.

There is a stable partially reversed state even in the case $K_1 = K_2$. For $b_1 = b_2$, $\gamma = \delta = \pi/4$ one finds in addition to the single domain state

$$\psi_1 = \psi_2 = \psi; \quad 2H \cos\psi = H_K \cos 2\psi \qquad (6)$$

a solution, which has for $q \gg 1$ the parametric representation

$$\psi_1 = \frac{\beta}{2} + \frac{1}{q}\left(1 + \frac{\cos\beta}{2}\right); \quad \psi_2 = -\frac{\beta}{2} + \frac{1}{q}\left(1 + \frac{\cos\beta}{2}\right)$$

$$H = H_K \cos\frac{\beta}{2}\left(1 - \frac{\cos\beta}{2}\right) \qquad (7)$$

and is stable for $|H| < H_K/\sqrt{2}$. This result shows that the incoherent reversal mode occurs in this model for arbitrarily small differences between the anisotropy constants $K_1$ and $K_2$.

## UNIDIRECTIONAL ROTATIONAL HYSTERESIS

It is interesting to note that this model also shows the effect of unidirectional rotational hysteresis, such as was observed in certain Permalloy films by Doyle, et al.[7] In these experiments, a rotational loss is observed for small fields at field directions antiparallel to the initial magnetization direction. Thus, for a reversing field antiparallel to the easy direction, there must be two stable states symmetrical to the single domain state with a nonvanishing magnetization component perpendicular to the easy direction.

We therefore consider the case $\gamma = \delta = 0$, $K_1 > K_2$ in the limit $q \gg 1$. For fields $H < H_{K2}$, the initial single domain state $\psi_1 = \psi_2 = 0$ remains stable, but for fields $H_{K2} < H < (b_1 H_{K1} + b_2 H_{K2})/(b_1 + b_2) = \langle H_K \rangle$ we find in fact two stable states

$$\psi_1 = 0, \quad \psi_2 = \pm\left[4 \frac{(b_1 + b_2)}{b_1 q} \frac{H - H_{K2}}{H_{K1} + H_{K2}}\right]^{\frac{1}{2}} \qquad (8)$$

and it will depend on the history, which one is occupied. If the difference between $H_{K1}$ and $H_{K2}$ is large, so that $H_{K2} < \frac{1}{2}\langle H_K \rangle$, then by rotating the field $\mathbf{H}$ through 360°, for $H_{K2} < H \lesssim \frac{1}{2}\langle H_K \rangle$ the only loss occurring will be the unidirectional loss caused by the transition from one of the two states (8) to the other.

## ACKNOWLEDGMENTS

The author thanks W. F. Brown and S. Shtrikman for a valuable discussion, and J. H. Hohl for help with the programming and numerical work on the IBM 1620 computer.

[7] W. D. Doyle, J. E. Rudisill, and S. Shtrikman, J. Appl. Phys. 33, 1162 (1962), this issue.

JOURNAL OF APPLIED PHYSICS    SUPPLEMENT TO VOL. 33, NO. 3    MARCH, 1962

# Factors Influencing Coercive Force Values in Sputtered Permalloy Films

A. J. Noreika and M. H. Francombe

*Philco Research Division, Blue Bell, Pennsylvania*

Depending upon the experimental conditions used, the mean composition of a sputtered Permalloy film can change progressively with thickness or can be maintained almost constant. In the first case, the extent to which a homogeneous composition is achieved depends upon the ease with which diffusion and mixing can proceed during film growth. It appears that chemical inhomogeneity, resulting from incomplete mixing, leads to the appearance of an anomalous maximum in the plot of wall coercive force $H_c$ versus thickness within the range 600–1000 A. This maximum is similar to that observed in evaporated films and has previously been identified with the transition from Néel to Bloch-type domain walls. The anisotropy coercive force $H_k$ is found to be relatively insensitive to average film composition and homogeneity, but varies systematically with sputtering power and substrate temperature. A similar variation occurs for $H_c$ in those films which are chemically homogeneous, i.e., in which the mean composition remains substantially constant during growth. The equilibrium temperature attained at the substrate after a few seconds and the resulting values of $H_k$ and $H_c$ can thus be effectively controlled by regulating the electrode potential $V$ and current $I$.

## INTRODUCTION

RECENT studies[1,2] of the properties of Permalloy (81% Ni, 19% Fe) films prepared by cathodic sputtering have shown that this technique is capable of yielding films possessing a strongly developed uniaxial anisotropy, with values of wall and anisotropy coercive forces, $H_c$ and $H_k$, similar to those obtained in evaporated or electrodeposited films. Films sputtered in argon are free from the oblique-incidence effects observed with evaporated films[3] and, for a given set of experimental conditions, the values of $H_c$ and $H_k$ can be accurately reproduced between successive sputtering runs. Systematic adjustment of the magnetic properties, however, requires a knowledge of the dependence of these coercive forces on factors such as film thickness, compositional homogeneity and annealing temperature.

A preliminary investigation of the influence of these factors[2] has shown that the variation of $H_c$ with film thickness is sensitive to changes in film composition and to small amounts of oxygen contamination, while $H_k$ variations can be closely correlated with changes in sputtering power and substrate surface temperature. In the present paper further data are included which indicate that the homogeneity of the films, and hence the variations in $H_c$, can be controlled by regulating the temperature. Also, a more quantitative relationship between the magnetic properties, sputtering power, and substrate temperature is given which provides a simple means of predetermining the coercive force values.

## RESULTS AND DISCUSSION

The present studies were made with a 4-in. square vacuum-melted cathode of composition 81.04% nickel, 18.96% iron, using the sputtering apparatus described previously[1,2]. Substrate temperatures during test sputtering runs recorded by means of fine-wire Chromel-Alumel thermocouples pressed to the upper surface of the substrate. The wall and anisotropy coercive forces were determined from 1000 cps $B$–$H$ measurements, while the film composition, relative to the nonmagnetostrictive composition (assumed here to be 81% nickel, 19% iron), was obtained by observing the change in $H_k$ produced by the application of a known strain along the easy axis of the film.

The results for two different sets of sputtering conditions are shown in Fig. 1. The films made in series A

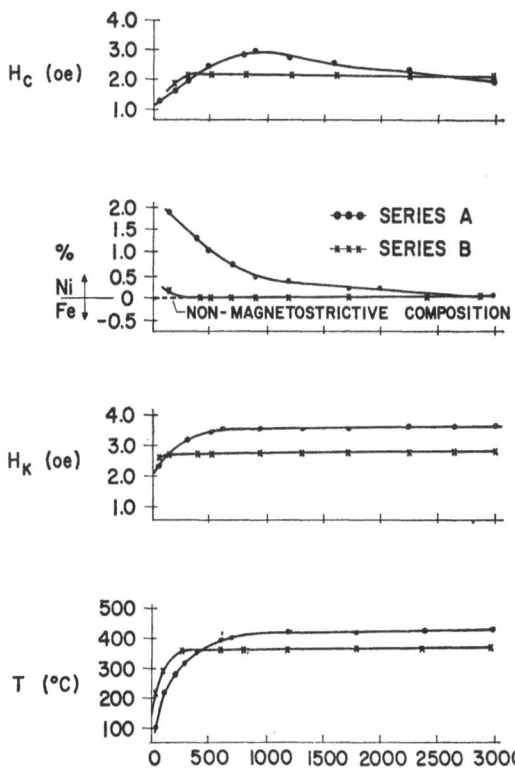

Fig. 1. Variation of wall and anisotropy coercive forces, film composition and substrate temperature with thickness of sputtered films. Series A, sputtered at 3500 v, 120 ma, from cold cathode; series B, sputtered at 4000 v, 90 ma, from hot cathode.

[1] M. H. Francombe and A. J. Noreika, J. Appl. Phys. **32**, 97S (1961).

[2] M. H. Francombe and A. J. Noreika, Symposium on Electrical and Magnetic Properties of Thin Metallic Layers, Louvain, Belgium September 4–7, 1961.

[3] D. O. Smith, J. Appl. Phys. **30**, 264S (1959).

were deposited after the freshly cleaned cathode had been allowed to cool, whereas those in series B were sputtered while the cathode surface was still hot. In the latter case, radiation from the cathode surface raises the initial substrate temperature to about 120°C, and the equilibrium maximum temperature is therefore established at a lower film thickness than for series A. When series B was repeated using the same sputtering voltage and current but starting with a cold cathode, the constant maximum value of $H_k$ was unchanged from that shown in Fig. 1, but the dependence of the coercive forces and film composition on thickness was found to be closely similar in form to that shown for series A.

The variation of film composition with thickness has been explained[2] by assuming that, after the first Ni-rich layer is formed, a constant composition alloy approximating the nonmagnetostrictive composition is deposited and diffuses into the Ni-rich under layer. The manner in which this Ni-rich layer forms, and in particular the observation that it is almost completely suppressed when the equilibrium substrate temperature is more rapidly established, is not yet completely understood. However, electron microscope studies now in progress suggest that at film temperatures below 350°C, and thus at small film thicknesses, the alloy is preferentially depleted in iron due to the formation of nickel ferrite, $NiFe_2O_4$. At higher temperatures, Ni and Fe appear to oxidize at the same rate to form a rock-salt type monoxide solid solution with the same metal atom ratio as the original alloy. The results of this investigation, and its bearing on film composition and homogeneity, will be described in a separate paper.

The anomalous maximum observed in $H_c$ at a thickness of about 800 A is similar to that reported for evaporated films, and has been correlated with the transition from Néel- to Bloch-type domain walls.[4] Comparison of the dependence of $H_c$ and composition on thickness shown in Fig. 1 leads us to suggest that, at least in sputtered films, this maximum can be attributed to chemical inhomogeneity arising from incomplete diffusion in the composite film.[5] In series B, where practically no compositional variation with thickness occurs, and where the homogeneity is therefore greatly improved, no maximum in $H_c$ is observed. This interpretation is consistent with the recent studies of Smith and Harte[6] who demonstrated that inhomogeneity, with associated local variations in anisotropy, leads to increased values of $H_c$ relative to $H_k$.

We may infer from the data shown in Fig. 1 that for films in which the alloy composition remains almost constant with increasing thickness, such as those in series B, the values of $H_k$ and $H_c$ appear to depend primarily upon sputtering power and probably upon substrate temperature. Since the equilibrium substrate temperature can be accurately set by adjusting the sputtering voltage and current, it should be a relatively simple matter to make films with chosen values of $H_c$ and $H_k$. To achieve controlled values of $H_c$ and $H_k$ an experimental relationship between coercive forces and sputtering power has been obtained by varying this power in the range 300 to 420 w. With the experimental conditions used it was found that the maximum substrate temperature in °C corresponded almost exactly to the power in watts. The observed changes in coercive forces within this power range were 1.9 to 3.0 oe for $H_c$ and 2.2 to 3.9 oe for $H_k$; both parameters increased approximately linearly with power and thus with temperature.

In adjusting the equilibrium substrate temperature by regulating the power, it should be noted that the sputtering conditions and deposition rate are also being altered. Therefore, it might not be permissible to relate the observed changes in $H_c$ and $H_k$ only to variations in temperature. It was found possible, however, to extrapolate the temperature relationship down to 200°C by using values of $H_c$ and $H_k$ measured at small film thicknesses where the temperature is rising rapidly. The sputtering power and deposition rate in this range do not vary with film thickness and are the same as for thicker films for which the temperature has attained an equilibrium value. Conclusive proof of the dependence of the coercive forces upon temperature alone over the entire range considered here must await the results of further experiments in which the substrate surface temperature is controlled by means independent of sputtering power.

If, as these results imply, $H_k$ is a simple function of substrate temperature, its increase with temperature is unexpected in that it conflicts with the behavior anticipated from the Fe-pair ordering model proposed by Néel[7] and Taniguchi.[8] However, experimental results similar to those given above have been reported for evaporated films by Prutton and Bradley[9] who found a marked increase of $H_k$ with substrate temperature in the range above 350°C. It is possible that this departure from theory may be attributed to differences in the time required to completely attain a pair-ordered condition at different temperatures.

[4] S. Methfessel, S. Middelhoek, and H. Thomas, IBM J. Research Develop. 4, 96 (1960).

[5] We have previously demonstrated that when crystal growth and diffusion in the films are inhibited by oxide contamination the $H_c$ maximum is emphasised, leading in some cases to inverted characteristics, with $H_c > H_k$.

[6] D. O. Smith and K. J. Harte, Lincoln Lab. Tech. Rept. No. 53G-0061 (1961).

[7] L. Néel, J. phys. radium 15, 225 (1954); Compt. Rend. 238, 305 (1954); J. Appl. Phys. 30, 355 (1959).

[8] S. Taniguchi, Sci. Repts. Research Inst. Tohoku Univ. A7, 269 (1955).

[9] M. Prutton and E. M. Bradley, Proc. Phys. Soc. (London) 75, 557 (1960).

JOURNAL OF APPLIED PHYSICS    SUPPLEMENT TO VOL. 33, NO. 3    MARCH, 1962

# Spin Configurations and Anisotropy

## I. S. Jacobs, *Chairman*

## The Nature of One-Ion Models of the Ferrimagnetic Anisotropy

Peter J. Wojtowicz
*RCA Laboratories, Princeton, New Jersey*

Considerable progress has been made toward the understanding of the origin and characteristics of the ferrimagnetic anisotropy through the introduction of the so-called "one-ion models." In this paper the nature of the one-ion model is examined from the point of view of formal statistical mechanics. The purpose is to establish the relationship between the one-ion models and the rigorous theory, to derive a formulation for correction terms, and to determine the conditions under which one-ion theories will be applicable. Under the assumption that the Hamiltonian contains isotropic pair-exchange interactions plus single-ion anisotropy operators alone, the formal theory demonstrates that the anisotropic free energy *cannot*, in general, be decomposed into a sum of single-ion contributions. It is shown, however, that a one-ion theory based on the correct single-spin density operator will be adequate in certain restricted circumstances. The correction terms to this one-ion theory are found to consist of fluctuations in the expectation values of multiplets of ions and thus represent *n*-ion contributions. A criterion for the applicability of the one-ion theory has been constructed in terms of the magnitudes of the anisotropy energy and the Curie temperature.

WITHIN recent years considerable progress has been made toward the understanding of the origin and characteristics of the ferrimagnetic anisotropy through the introduction and development of the so-called "one-ion models." Comprising the results of several authors,[1-3] the one-ion model proposes that the observed anisotropy is simply a sum over all crystal sites of single-ion anisotropies, the origin of which lies in the interaction of the individual magnetic ions with the crystalline electric field. The central assumption which permits the independence of the ions and the consequent additivity of single-ion anisotropies is that the discreet near-neighbor exchange interactions can be replaced by a Weiss molecular field. In this investigation the one-ion model is examined from the point of view of formal statistical mechanics. The purpose is to establish the relationship between one-ion models and the rigorous theory, to derive a formulation for correction terms, and to determine the conditions under which one-ion theories will be applicable. The present article is only a brief outline of the theory; a more complete mathematical treatment will be published elsewhere.

The derivation of a one-ion theory of the ferrimagnetic anisotropy begins with the choice of a Hamiltonian operator of the general form

$$\mathcal{3C} = \mathcal{3C}_0 + \epsilon \sum_i A_i. \tag{1}$$

It is not necessary to specify the structure of the isotropic portion $\mathcal{3C}_0$ except to require the presence of at least a set of isotropic pair-exchange interactions to provide a mechanism for the ferrimagnetism. The

anisotropic part of $\mathcal{3C}$ is a sum of anisotropic single-ion operators which represent the interaction between individual magnetic ions and the crystal field. The $A_i$ are chosen to be of order unity so that the magnitude of $\epsilon$ is a measure of the strength of the interaction and determines the relative size of the anisotropy energy. For simplicity in this presentation, it is assumed that the $A_i$ and $\mathcal{3C}_0$ commute. With the Hamiltonian expressed in the usual direct-product matrix representation the partition function is just the trace of the density operator: $Z = \text{Tr}[\exp(-\beta\mathcal{3C})]$, where $\beta = 1/kT$; $k$ is the Boltzmann constant, and $T$ is the temperature. Regarding the anisotropy as a form of perturbation on the basic isotropic ferrimagnetism, it is convenient to re-express the partition function as

$$Z = Z_0 \langle \Pi_i \exp(-\beta\epsilon A_i) \rangle, \tag{2}$$

where $Z_0 = \text{Tr}[\exp(-\beta\mathcal{3C}_0)]$ is the partition function of the purely isotropic ferrimagnet, while

$$\langle X \rangle = Z_0^{-1}\text{Tr}[X \exp(-\beta\mathcal{3C}_0)] \tag{3}$$

is the thermal expectation value of the operator $X$ evaluated in the ensemble whose Hamiltonian is the isotropic $\mathcal{3C}_0$. The free energy, equal to $-kT \ln Z$ may be divided into a sum of two components, $F^I = -kT \ln Z_0$ and

$$F^A = -kT \ln\langle \Pi_i \exp(-\beta\epsilon A_i) \rangle. \tag{4}$$

$F^I$ is simply the free energy of the unperturbed ferrimagnet and is completely isotropic. The component $F^A$ will have both isotropic and anisotropic contributions; it will be called the anisotropic free energy because its origin lies in the anisotropic interaction of the magnetic ions with the crystal field.

The development leading to Eq. (4) demonstrates that the anisotropic free energy *cannot*, in general, be

[1] K. Yosida and M. Tachiki, Progr. Theoret. Phys. (Kyoto) **17**, 331 (1957).
[2] W. P. Wolf, Phys. Rev. **108**, 1152 (1957).
[3] J. C. Slonczewski, Phys. Rev. **110**, 1341 (1958); J. Appl. Phys. **32**, 253S (1961).

decomposed into a sum of single-ion contributions. Such a decomposition would require that the expectation value of the product of the operators $\exp(-\beta\epsilon A_i)$ be equivalent to the product of the expectation values. But this is not the case here since the different ions are not independent; their behavior is highly correlated through the agency of the exchange interactions in $\mathfrak{IC}_0$. For the purposes of examining a one-ion theory of the anisotropy, however, $F^A$ can be further separated into the terms $\sum_i F_i^A$ and $\Delta F^A$:

$$F_i^A = -kT \ln\langle\exp(-\beta\epsilon A_i)\rangle, \qquad (5)$$

$$\Delta F^A = -kT \ln\frac{\langle\Pi_i \exp(-\beta\epsilon A_i)\rangle}{\Pi_i\langle\exp(-\beta\epsilon A_i)\rangle}, \qquad (6)$$

where $F_i^A$ is the single-ion contribution of ion $i$ to $F^A$, and where $\Delta F^A$ is a defect function which measures the error introduced by attempting to separate $F^A$ into one-ion terms alone. Equation (5) may now be used as a proper basis for the development of a one-ion theory of the anisotropy. If the correlations are not too important, $\Delta F^A$ will be small, and the $F_i^A$ should provide an adequate approximation; if $\Delta F^A$ is shown to be large, correction terms based on Eq. (6) will have to be included.

Expressed completely in one-ion language, Eq. (5) becomes

$$F_i^A = -kT \ln \mathrm{Tr}_i[\rho_i \exp(-\beta\epsilon A_i)], \qquad (7)$$

where only the operators for ion $i$ appear explicitly, and where $\mathrm{Tr}_i$ denotes the trace within the submatrix of ion $i$. The quantity $\rho_i$ is the single-spin density operator for ion $i$ given by $Z_0^{-1} \mathrm{Tr}_{N-i}[\exp(-\beta\mathfrak{IC}_0)]$, where $\mathrm{Tr}_{N-i}$ denotes the trace within the submatrix of the $N-1$ ions that are not $i$. The operator $\rho_i$ plays the role of a distribution function describing the thermal behavior of ion $i$ under the influence of the interactions in $\mathfrak{IC}_0$. As such it completely determines all of the one-ion properties in the $\mathfrak{IC}_0$ ensemble and reflects the influence of many-body correlations on the single-ion behavior. Substitution of the molecular-field approximation for $\rho_i$ into Eq. (7) gives the one-ion models of Wolf,[2] Slonczewski,[3] and (in the limit of small $\epsilon$) of Yosida and Tachiki.[1] Since the molecular-field approxi-

mation does not account for the effects of exchange-induced correlations, the one-ion models can only be expected to provide qualitative results even when the correct one-ion theory Eq. (7) is adequate.[4]

A discussion of the properties of the defect function at finite temperatures ($1/3T_c$ and above) is facilitated by expanding $\Delta F^A$ in a power series in the parameter $-\beta\epsilon$:

$$-\beta\Delta F^A = \sum_{r=1} \frac{(-\beta\epsilon)^r}{r!}\Omega_r. \qquad (8)$$

The coefficients $\Omega_r$ can be computed from Eq. (6) through the use of the well-known moment-invariant transformation. The first $\Omega_1$ is found to be identically zero, showing that Eq. (7) contains all of the one-ion anisotropy effects of the problem. The second, third, and higher order coefficients appear in the form of correlation functions or fluctuation functions for the thermal expectation values of pairs, triplets, and higher multiplets of anisotropy operators evaluated in the $\mathfrak{IC}_0$ ensemble. These terms therefore represent $n$-ion contributions to $F^A$. The explicit form of $\Omega_2$ is

$$\Omega_2 = \sum_{i\neq j} [\langle A_i A_j\rangle - \langle A_i\rangle\langle A_j\rangle], \qquad (9)$$

while higher $\Omega_r$ are progressively more complex and cannot be reproduced here.

Detailed considerations of the temperature dependence of the $\Omega_r$ and of the convergence properties of the series, Eq. (8) enable the construction of a criterion for the relative importance of $\Delta F^A$. If $\epsilon/k$ is much smaller than $T_c$, then the $F_i^A$ alone will account for the anisotropy. If $\epsilon/k$ is of the same order of magnitude as $T_c$, then $\Delta F^A$ will be important, and one-ion theories will not be adequate. Representative values of $\epsilon/k$ for ions in octahedral spinel sites are 0.2°, 5°, 60°, and 350°K for $Fe^{3+}$, $Cr^{3+}$, $Fe^{2+}$ and $Co^{2+}$, respectively. For the special case of large anisotropy ions highly diluted in an otherwise isotropic ferrimagnet, however, Eq. (9) shows that $\Delta F^A$ will always be small independent of the magnitude of $\epsilon/kT_c$, and the one-ion theory Eq. (7) will suffice.

---

[4] See also L. R. Walker, J. Appl. Phys. 32, 264S (1961).

JOURNAL OF APPLIED PHYSICS    SUPPLEMENT TO VOL. 33, NO. 3    MARCH, 1962

# Magnetic Structure Work at the Nuclear Center of Grenoble

E. F. Bertaut, A. Delapalme, F. Forrat, and G. Roult

*Centre d'Etudes Nucléaires, Grenoble, France*

AND

F. de Bergevin and R. Pauthenet

*Laboratoire d'Electrostatique et de Physique du Métal, Institut Fourier, Grenoble, France*

Three structures are briefly reported: FeRh is antiferromagnetic at low temperatures and ferromagnetic above 60°C. Fe has in both region a high moment of about $3\mu_B$; Rh has a definite moment in the ferromagnetic region of about $0.6\mu_B$. $Cr_3Se_4$, a nonstoichiometric compound with ordered holes, has a magnetic structure closely related to that of $MnBr_2$. $UCoO_4$ isomorphous of $UMgO_4$ (space group Imma) is antiferromagnetic with Co-O-O-Co interactions and a Néel temperature of 12°K.

A BRIEF survey of work done on 3 magnetically interesting compounds FeRh, $Cr_3Se_4$, and $UCoO_4$ is given. A fuller account will be published in separate papers.

*FeRh*: The alloy is prepared in powder form by reduction under hydrogen of the ilmenite type compound $Fe_2O_3$, $Rh_2O_3$ which is easily prepared (this preparation is easier to carry out than melting Fe and Rh together). Atomic proportions were 0.53 at. % Rh, 0.47 at. % Fe. The compound is of the ordered CsCl type: Fe in (000); Rh in $(\frac{1}{2}\frac{1}{2}\frac{1}{2})$.[1] Fallot[2] has shown that there is a transition from a nonmagnetic low-temperature state to a ferromagnetic state at a critical temperature which varies from 150° to 400°K when the Rh concentration increases from 50 to 55%.

The low-temperature state is antiferromagnetic with a magnetic unit cell doubled in all directions. The appearance of only (odd, odd, odd) magnetic reflections indicates a G-type[3] structure, which means that each iron atom is surrounded by 6 iron spins of opposite directions. It is not clear if there is one af lattice of Fe or 2 uncoupled af lattices of iron and rhodium because the only quantity which may be derived from intensity measurements is

$$S_{Fe}^2 + S_{Rh}^2 = 2.72.$$

If Rh had no moment, we should have $S_{Fe} = 1.65$, and the iron moment would be 3.3 $u_B$.

In the ferromagnetic region above 60°C in our sample, the magnetic unit cell becomes identical to the chemical one and at least theoretically the quantities $S_{Fe} - S_{Rh}$ (from super lattice lines) and $S_{Fe} + S_{Rh}$ may be independently estimated from intensity measurements. We preferred to use neutron diffraction for the determination of

$$S_{Fe} - S_{Rh} = 1.20 \pm 0.15$$

and the direct determination from the magnetic saturation of $S_{Fe} + S_{Rh} = 1.83 \pm 0.01$.

One has in the ferromagnetic region $S_{Fe} = 1.52 \pm 0.07$; $S_{Rh} = 0.31 \pm 0.07$.

The clear result is that Rh has a small moment in the ferromagnetic region and that the moment of Fe in $Fe_{0.47}Rh_{0.53}$ is much higher ($\sim 3\mu_B$) than in $\alpha-Fe$ ($2.2\mu_B$). Let us recall that Rh in the periodic table is just below Co and that Shull[4] has found in the 50–50% alloy CoFe moments of 2.0 and 2.9 $\mu_B$. Diffraction alone does not allow to conclude which moment goes with which atom. From our result it is quite plausible that here too Fe is in the high moment state.

*$Cr_3Se_4$*: This compound was prepared by reaction of powdered Cr and Se. It derives from the NiAs structure by a monoclinic deformation. Parameter values are $a = 6.32$ A; $b = 3.62$ A; $c = 11.77$ A; $\beta = 91°28'$. $Cr_3Se_4$ is nonstoichiometric with an ordering of holes on the chromium lattice.[5] The Curie constant is 6.52 compared to 6.75, which is the value expected for $2Cr^{3+} + 1Cr^{2+}$.

The Néel temperature is 80°K and the paramagnetic temperature is unusually small ($\Theta_p = -3°K$ compared to $\sim -300°K$ in CrSe).[6]

The study by neutron diffraction was intended to get information on the distribution of three- and bivalent Cr atoms. The magnetic unit cell requires doubling of

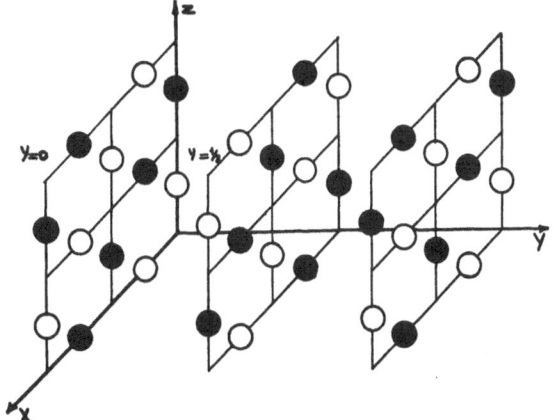

Fig. 1. Magnetic structure of $UCoO_4$. Full and empty circles denote opposite spin orientations of the cobalt atoms.

[1] F. de Bergevin and L. Muldawer, Compt. rend. **252**, 1347 (1961).

[2] M. Fallot, Ann. phys. **10**, 291 (1938).

[3] E. O. Wollan and W. C. Koehler, Phys. Rev. **100**, 545 (1955).

[4] C. G. Shull (private communication).

[5] M. Chevreton and E. F. Bertaut, Compt. rend. **253**, 145 (1961).

[6] I. Tsubokawa, J. Phys. Soc. Japan **11**, 662 (1956).

the $a$ and $c$ axis. Although there is a qualitative agreement with a spin configuration closely related to MnBr$_2$,[7] the determination of spin moments and orientations has not yet been completed.

$UCoO_4$: This compound is prepared by reaction of UO$_3$ and CoO. The orthorhombic, but pseudotetragonal compound, is isomorphous with UMgO$_4$[8] and belongs to space group Imma. Parameters are $a=c=6.497$ and $b=6.952$ A. Co is in parameterless positions $(4b)$ $00\frac{1}{2}$; $0\frac{1}{2}\frac{1}{2}$; U is in $(4e)\pm(0\frac{1}{4}z)$ with $z=0.025$; 8 0$_I$ are in $(8i)$ $\pm(x\frac{1}{4}z;\bar{x}\frac{1}{4}z)$ with $x=0.296$ and $z=0.033$; 8 0$_{II}$ are in $(8h)$ $\pm(0yz;0\frac{1}{2}-y,z)$ with $y=-0.016$ and $z=0.201$.

---

[7] W. C. Koehler, M. K. Wilkinson, J. W. Cable, and E. O. Wollan, J. phys. radium **20**, 180 (1959).

[8] W. H. Zachariasen, Acta Cryst. **7**, 788 (1954).

The magnetic structure observed at liquid helium temperature is shown in Fig. 1. It requires doubling of $a$ and $c$. The spin orientation is either along $a$ or $c$. The moment value of Co, determined by susceptibility measurements in the paramagnetic region is near to $S=4/2$. Neutron diffraction intensities in the af region agree well with $S=2.1$, so that here the usually observed orbital contribution in Co$^{2+}$ is conserved at low temperatures. The af ligands between Co atoms are all of the kind Co-O-O-Co. This explains the low Néel temperature of 12°K.

The authors are indebted to Mm. Daniel Bloch and Yves Barnier for magnetic measurements on the powder sample of FeRh.

---

JOURNAL OF APPLIED PHYSICS     SUPPLEMENT TO VOL. 33, NO. 3     MARCH, 1962

# Neutron Diffraction Study of Magnetic Ordering in Thulium

W. C. KOEHLER, J. W. CABLE, E. O. WOLLAN, AND M. K. WILKINSON
*Oak Ridge National Laboratory, Oak Ridge, Tennessee*

Neutron diffraction measurements have been made on a single-crystal specimen of metallic thulium at temperatures ranging from 293° to 4.2°K. In the temperature range between 38° and 56°K the magnetic structure of thulium is similar to that observed for the high temperature form of erbium; namely, a simple oscillating $z$-component type antiferromagnetic structure. Below 38°K additional satellite reflections are observed in the diffraction patterns from which it is inferred that the sinusoidal modulation is modified below this temperature. At 4.2°K the normal lattice reflections show an increase in intensity in zero field which is consistent with a mean magnetic moment per thulium atom of approximately one Bohr magneton directed parallel to the $c$ axis. The low temperature magnetic structure which is proposed for thulium is one in which the moments are parallel to the $c$ axis of the crystal but change their orientations according to the sequence 4, 3, 4, 3 $\cdots$. The fundamental period of the modulation remains constant over the entire range of temperature at a value corresponding to 3.5 $a_3$ periods.

MAGNETIC measurements on polycrystalline thulium have revealed the existence of a magnetic ordering transition to an antiferromagnetic state at a temperature variously reported as 51°K[1] or as 60°K.[2] At low temperatures, Rhodes *et al.*[1] noted behavior of the measured magnetization which was indicative of ferromagnetism. More recently Davis and Bozorth,[2] from hysteresis measurements, reported a Curie point of 22°K and a net magnetic moment per atom of 0.5 Bohr magnetons. In very high magnetic fields Henry[3] found a magnetic moment per atom of 3.4 Bohr magnetons, e.g., approximately half the free ion value for trivalent thulium. The high temperature transition has also been detected in specific heat measurements.[4] A broad $\lambda$-type anomaly peaked near 55°K was observed, but there was no evidence for a secondary

anomaly near 22°K. In powder neutron diffraction experiments on metallic thulium[5] a transition to a complex antiferromagnetic structure was observed at 53°K. The elucidation of this structure from powder data alone was not considered feasible. Recently a single crystal specimen of thulium has been made available to us, and neutron diffraction studies on this specimen have now been carried out at temperatures ranging from room temperature to 4.2°K. The description of these studies and of the results deduced therefrom is the subject of this report.

The neutron diffraction data are most conveniently described in terms of the scattering density in reciprocal space. Between approximately 56° and 38°K, the scattering density map shows a single pair of magnetic satellite reflections associated with each of the allowed normal lattice reflections with the important exception that magnetic satellites of all (OOH$_3$) reflections are

---

[1] B. L. Rhodes, S. Legvold, and F. H. Spedding, Phys. Rev. **109**, 1547 (1958).

[2] D. D. Davis and R. M. Bozorth, Phys. Rev. **118**, 1543 (1960).

[3] W. E. Henry, J. Appl. Phys. **31**, 323S (1960).

[4] L. D. Jennings, E. Hill, and F. H. Spedding, J. Chem. Phys. **34**, 2082 (1961).

[5] W. C. Koehler, E. O. Wollan, M. K. Wilkinson, and J. W. Cable, Rare Earth Research Developments Conference, Lake Arrowhead, California, October, 1960.

absent. As in the high temperature magnetic structure of erbium[6] one is led by these observations to a simple oscillating z-component type antiferromagnetic structure. The magnitude of the wave vector of the oscillation is found to be constant over the above mentioned temperature range at a value of $2/7\ b_3$ which corresponds to a repeat period of 3.5 $a_3$ periods.

Below 38°K the scattering density map is more complicated. In addition to the satellites observed at high temperatures there are found other satellites which correspond to the second and third harmonics of the primary satellites. As in the high temperature range, no magnetic reflections associated with $(OOH_3)$ reflections are observed. Further, the normal lattice reflections, again excepting $(OOH_3)$ reflections, show a small but definite increase in intensity which is suggestive of a small net magnetic moment.

The above mentioned observations are illustrated in Fig. 1 in which the temperature dependence of the intensities of representative reflections is shown. At the Néel temperature, near 56°K, only the primary satellite of the (110) reflection is observed. With decreasing temperature the intensity of this reflection increases smoothly and approaches saturation at 4.2°K. Within the limits of detection the second and third harmonics develop at the same but lower temperature, e.g., about 38°. In contrast to erbium[6] where a second harmonic was never observed, and to dysprosium[7] and holmium[5] where the second harmonic was always very weak relative to the first, the second harmonic in thulium is found to be relatively strong. In the figure is shown also the temperature variation of intensity of the (100) and (110) normal lattice reflections. A small magnetic contribution is superimposed upon the nuclear peak which fact makes a precise measurement of the Curie point difficult. As can be seen from the figure, however, the temperature at which these reflections first have a magnetic contribution appears to be somewhat greater than 22°K. The magnitudes of the magnetic intensities superposed on the normal lattice reflections are consistent with a ferromagnetic structure in which the average moment per atom is one Bohr magneton and in which this average moment is parallel to the c axis.

According to theory,[8,9] all orders of the anisotropy energy in thulium favor strongly a moment alignment along the c axis. In such a case the value of one Bohr magneton measured here is consistent with the value of 0.5 Bohr magneton reported by Davis and Bozorth,

[6] J. W. Cable, E. O. Wollan, W. C. Koehler, and M. K. Wilkinson, J. Appl. Phys. **32**, 49S (1961).
[7] M. K. Wilkinson, W. C. Koehler, E. O. Wollan, and J. W. Cable, J. Appl. Phys. **32**, 48S (1961).
[8] R. J. Elliott, International Conference on Magnetism and Crystallography, Kyoto, Japan, September, 1961.
[9] K. Yosida and H. Miwa, International Conference on Magnetism and Crystallography, Kyoto, Japan, September, 1961.

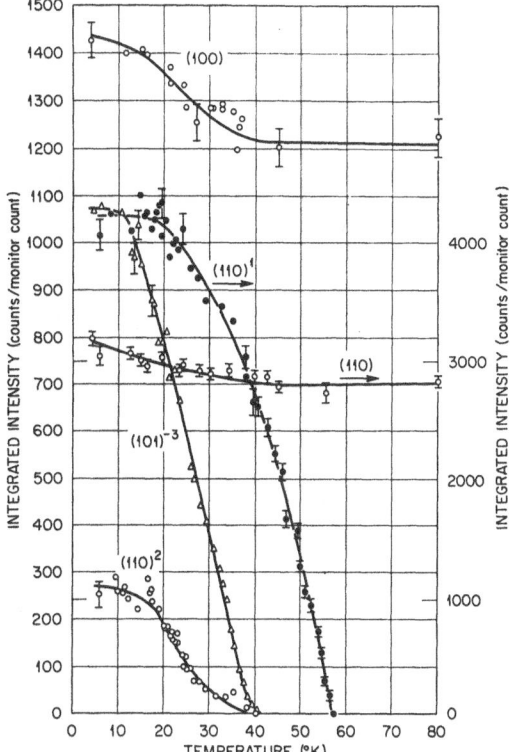

FIG. 1. Temperature dependence of reflections of thulium. For (110) and (110)¹ read scale at right.

for only half the moment will be measured with a polycrystalline sample.

From the low temperature diffraction data a magnetic structure has been derived for which the predicted intensities are in excellent agreement with those observed. In this structure each atom has a magnetic moment of 7.0 Bohr magnetons parallel to the c axis. That is to say the sinusoidal modulation of the high temperature structure is completely "squared up," and each atom has its maximum moment. The sense varies from layer to layer according to the sequence: + + + + − − − + + + + − − −, i.e., in a 4, 3, 4, 3, etc., arrangement.

As Davis and Bozorth have already noted, Henry's result for polycrystalline thulium implies that in very high fields it is difficult to turn the moments away from the c axis, but possible to flip moments originally antiparallel to the field.

We may note finally that the breadth of the λ anomaly in the specific heat data may be attributed to the existence of the transition at 38°K at which temperature the high-order satellites first appear.

It is a pleasure for the authors to express their thanks to Professor S. Legvold and Professor F. H. Spedding and to their students for making available to them their single-crystal specimen.

# Temperature Dependence of the Magnetocrystalline Anisotropy of Face-Centered Cubic Cobalt*†

D. S. RODBELL

*General Electric Research Laboratory, Schenectady, New York*

Single crystals of face-centered cubic cobalt have been examined between 4.2° and 850°K. The usual transformation to the hexagonal structure below 700°K is avoided by using two special forms of samples: (a) thin films evaporated onto MgO substrates, and (b) the precipitated cobalt-rich phase in a 2% Co-Cu single crystal. In both cases the fcc cobalt is stabilized by and has a close correspondence with the host lattice. Standard ferromagnetic (electron-spin) resonance techniques have been used to determine the magneto-crystalline anisotropy parameters $K_1/M$ and $K_2/M$ over the temperature range indicated, and, in addition, the spectroscopic splitting factor $g$ is found to be $2.06 \pm 0.03$ independent of temperature. The temperature dependence of the anisotropy constants is in accord with the relation $K(T)/K(o) = [M(T)/M(o)]^n$, where for $M(T)/M(o)$ we have taken the determination by V. Jaccarino [Bull. Am. Phys. Soc. 4, 461 (1959)] of the temperature dependence of the cobalt-nuclear magnetic resonance frequency which we assume to be proportional to the magnetization. The results that are obtained indicate that the power $n$ of the dependence noted is nearly 10 for $K_1$ and much higher for $K_2$. At 0°K samples of type (a) have $-K_1/M = 600$ oe.

* This work was supported in part by Wright Air Development Division, Air Research and Development Command, United States Air Force.

† The detailed paper for this abstract will be published in the Proceedings of the International Conference on Magnetism, held at Kyoto, Japan in 1961.

---

# Magnetoelectric Effects in Antiferromagnetics*

G. T. RADO AND V. J. FOLEN

*U. S. Naval Research Laboratory, Washington 25, D. C.*

Spin-ordered materials may exhibit a magnetic polarization which is proportional to an applied electric field and an electric polarization which is proportional to an applied magnetic field. In this paper a comprehensive discussion is given of the present knowledge of these magnetoelectric (ME) effects. The specific topics covered include the thermodynamic and magnetic symmetry considerations which are relevant to the ME effects and also to the piezomagnetic and piezomagnetoelectric effects. The major part of the paper is a review of the experimental and theoretical work on ME effects in $Cr_2O_3$ carried out by the present authors. This includes measurements of the anisotropy of the ME effects, theory of the temperature dependence and atomic mechanism of the ME effects, observation of the magnetically as well as of the electrically induced ME effect, experiments on magnetic annealing and other structure sensitive aspects of ME effects, and the role of antiferromagnetic domains in the interpretation of the results.

## I. INTRODUCTION

UNTIL quite recently, the subjects of magnetostatics and electrostatics were considered to be independent, and in the textbooks they are usually treated in separate chapters. The magnetic induction (**B**) was thought to be a function of the magnetic field (**H**) but not of the electric field (**E**), and the electric displacement (**D**) was thought to be a function of the electric field but not of the magnetic field. Today, however, the situation is somewhat different. Experiments performed during the last two years have revealed the existence of two new effects which are known as "magnetoelectric" (ME) effects. One is the appearance of a magnetic polarization $[\mathbf{M} = (\mathbf{B} - \mathbf{H})/4\pi]$ which is proportional to an applied electric field, and the other is the appearance of an electric polarization $[\mathbf{P} = (\mathbf{D} - \mathbf{B})/4\pi]$ which is proportional to an applied magnetic field. We shall refer to these effects as the electrically induced ME effect $[(ME)_E]$ and the magnetically induced ME effect $[(ME)_H]$, respectively.

Typical experimental arrangements used for the production and detection of these and other effects are shown schematically in Fig. 1. In this array, the magnetic and electric fields are considered to be the independent variables and the magnetic and electric polarizations the dependent variables. Both of the effects depicted along the diagonal are well known and will not be discussed further. The relatively new effects, of course, are those illustrated in the off-diagonal spaces. The $(ME)_E$ effect (shown in the upper right-hand space) was first observed by Astrov,[1] and the $(ME)_H$ effect (shown in the lower left-hand space) was first

* Presented at the Seventh Conference on Magnetism and Magnetic Materials, Phoenix, Arizona, November 13–16, 1961.

[1] D. N. Astrov, J. Exptl. Theoret. Phys. (U.S.S.R.) **38**, 984 (1960) [translation: Soviet Phys.—JETP **11**, 708 (1960)].

observed by Rado and Folen.[2] Single crystals of anti-ferromagnetic $Cr_2O_3$ were used in both experiments.

Even before these experiments were performed, Landau and Lifshitz[3] had pointed out that ME effects may, in principle, exist in spin-ordered materials. Their arguments are based on thermodynamic and symmetry considerations rather than on atomic mechanisms. Subsequently, Dzyaloshinskii[4] showed on the basis of similar but more detailed thermodynamic and symmetry arguments that ME effects may exist in the specific case of $Cr_2O_3$.

In Sec. II we discuss the thermodynamic and symmetry considerations which are relevant to the ME effects and also to the piezomagnetic[3,5,6] effect and the recently proposed piezomagnetoelectric[7] effect. We then review, in Sec. III, the experimental and theoretical work on ME effects carried out by the present authors.[2,8–10]

## II. THERMODYNAMICS AND SYMMETRY

### A. Thermodynamic Potential

A convenient starting point is the thermodynamic potential $\Phi$ of a sample of unit volume. We may define $\Phi$ by

$$\Phi = F + \sum_i X_i x_i - (1/4\pi)\mathbf{E}\cdot\mathbf{D} - (1/4\pi)\mathbf{H}\cdot\mathbf{B}, \quad (1)$$

where $F$ is the Helmholtz free energy. The quantities $X_i$ represent intensive variables (e.g., stresses) other than components of $\mathbf{E}$ and $\mathbf{H}$, and the quantities $x_i$ represent extensive variables (e.g., strains) other than components of $\mathbf{D}$ and $\mathbf{B}$. Next we recall the equations

$$F = U - TS, \quad (2)$$

$$dU = dQ + dW, \quad (3)$$

$$dQ = T\,dS, \quad (4)$$

$$dW = -\sum_i X_i dx_i + (1/4\pi)\mathbf{E}\cdot d\mathbf{D} + (1/4\pi)\mathbf{H}\cdot d\mathbf{B}, \quad (5)$$

where $U$, $T$, and $S$ denote, respectively, the internal

[2] G. T. Rado and V. J. Folen, Phys. Rev. Letters 7, 310 (1961).

[3] L. D. Landau and E. M. Lifshitz, *Electrodynamics of Continuous Media* (Addison-Wesley Publishing Company, Inc., Reading, Massachusetts, 1960), p. 119. (English translation of a 1958 Russian edition.)

[4] I. E. Dzyaloshinskii, J. Exptl. Theoret. Phys. (U.S.S.R.) 37, 881 (1959) [translation: Soviet Phys.—JETP 10, 628 (1960)].

[5] I. E. Dzyaloshinskii, J. Exptl. Theoret. Phys. (U.S.S.R.) 33, 807 (1957) [translation: Soviet Phys.—JETP 6, 621 (1958)].

[6] A. S. Borovik-Romanov, J. Exptl. Theoret. Phys. (U.S.S.R.) 36, 1954 (1959) and 38, 1088 (1960) [translations: Soviet Phys.—JETP 9, 1391 (1959) and 11, 786 (1960)]. These papers contain experimental results on the piezomagnetic effect and further references on relevant magnetic symmetry considerations.

[7] G. T. Rado (to be published).

[8] V. J. Folen, G. T. Rado, and E. W. Stalder, Phys. Rev. Letters 6, 607 (1961).

[9] G. T. Rado, Phys. Rev. Letters 6, 609 (1961).

[10] G. T. Rado and V. J. Folen, paper presented at the International Conference on Magnetism and Crystallography, Kyoto, Japan, September, 1961, to be published in J. Phys. Soc. Japan. This paper is a somewhat condensed version of Sec. III of the present paper.

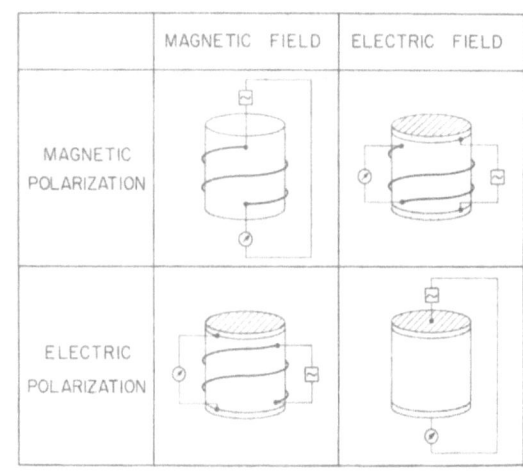

FIG. 1. Schematic explanation of typical experimental arrangements used for investigating the magnetic and electric polarizations produced by magnetic and electric fields. The arrangements show in the off-diagonal spaces involve magnetoelectric (ME) effects.

energy, temperature, and entropy of the sample under consideration, and $dQ$ represents an infinitesimal amount of heat supplied to the sample. Equation (2) is the definition of $F$, Eqs. (3) and (4) express, respectively, the first and second laws of thermodynamics, and Eq. (5) represents an infinitesimal amount of work performed on the sample. With the help of Eqs. (2) through (5), we obtain from Eq. (1):

$$d\Phi = -S\,dT + \sum_i x\,dX_i - (1/4\pi)\mathbf{D}\cdot d\mathbf{E} - (1/4\pi)\mathbf{B}\cdot d\mathbf{H}, \quad (6)$$

which shows that $d\Phi$ is zero if the temperature and all the other intensive variables are held constant. It can be shown, in fact, that if $\Phi$ is a minimum, then the sample is in a state of stable equilibrium. The condition that $d\Phi$ be an exact differential leads to the thermodynamic relations

$$x_i = \frac{\partial \Phi}{\partial X_i}\bigg|_{T,\mathbf{E},\mathbf{H}}, \quad (7)$$

$$\mathbf{D} = -4\pi\frac{\partial \Phi}{\partial \mathbf{E}}\bigg|_{T,X_i,\mathbf{H}}, \quad (8)$$

$$\mathbf{B} = -4\pi\frac{\partial \Phi}{\partial \mathbf{H}}\bigg|_{T,X_i,\mathbf{E}}, \quad (9)$$

which include some related results given by Landau and Lifshitz.[3]

The basic method[3] for deriving the thermodynamics of ME effects (and of some other phenomena) may now be expressed in the form of a three-part procedure: (1) Write down a sufficiently general formal expression for the $\Phi$ of the crystal under consideration. (2) Reduce the number of independent coefficients by requiring that $\Phi$ be invariant with respect to all the symmetry operations which leave the magnetic structure of the crystal invariant. (3) Use Eqs. (7) through (9) to

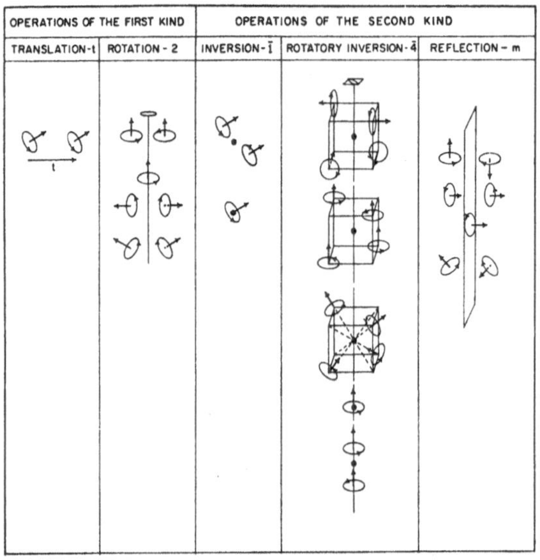

| OPERATIONS OF THE FIRST KIND | | OPERATIONS OF THE SECOND KIND | | |
|---|---|---|---|---|
| TRANSLATION - t | ROTATION - 2 | INVERSION - 1̄ | ROTATORY INVERSION - 4̄ | REFLECTION - m |

Fig. 2. Effect of various symmetry operations on axial vectors. Operations of the first kind affect an axial vector in the same way as if it were a polar vector, whereas operations of the second kind affect an axial vector differently than a polar vector. (After Donnay *et al.*, reference 11.)

obtain the constitutive equations and appropriate general expressions for the other extensive variables.

## B. Thermodynamics of the ME Effects in $Cr_2O_3$

Following Dzyaloshinskii,[4] we now apply the above-mentioned procedure to the case of $Cr_2O_3$. The magnetic structure of this particular corundum type crystal is invariant with respect to the symmetry operations $2C_3$, $3U_2$, and $IR$. Here $C_3$ is a 3-fold vertical axis (which is parallel to the $c$ axis), $U_2$ is a 2-fold horizontal axis, $I$ is an inversion (i.e., a change of sign of all the spatial coordinates), and $R$ is the time reversal ($t \rightarrow -t$) transformation. If we now restrict $\Phi$ to terms involving powers no higher than the second in the field components $E_k$ and $H_l$ (where both $k$ and $l$ denote $x$, $y$, or $z$), then the magnetic symmetry requires that $\Phi$ be of the form

$$\Phi = \Phi_0 - (1/8\pi)[\epsilon_\perp(E_x^2 + E_y^2) + \epsilon_{||}E_z^2 \\ + \mu_\perp(H_x^2 + H_y^2) + \mu_{||}H_z^2] + \Phi_{ME}, \quad (10)$$

$$\Phi_{ME} = -(1/4\pi)[\alpha_\perp(E_xH_x + E_yH_y) + \alpha_{||}E_zH_z], \quad (11)$$

where the $z$ axis is chosen to be along the principal axis ($c$ axis) of the $Cr_2O_3$ crystal, and the subscripts $||$ and $\perp$ indicate, respectively, parallel and perpendicular directions with respect to this axis. The term $\Phi_0$ in Eq. (10) denotes the value of $\Phi$ for $E = H = 0$, the term in brackets comprises the usual electric and magnetic contributions to $\Phi$, and $\Phi_{ME}$ represents the magnetoelectric term. Thus, the coefficients $\epsilon$ and $\mu$ are the usual dielectric constant and permeability, respectively, while $\alpha$ may be called the magnetoelectric parameter or

[11] G. Donnay, L. M. Corliss, J. D. H. Donnay, N. Elliott, and J. M. Hastings, Phys. Rev. **112**, 1917 (1958).

$4\pi$ times the "magnetoelectric susceptibility". [The numerical factors involving $\pi$ which appear in Eqs. (10) and (11) were inserted for convenience.] By applying the relations (8) and (9) to Eq. (10), we obtain the constitutive equations

$$B_{x,y} = \mu_\perp H_{x,y} + \alpha_\perp E_{x,y}, \quad (12a)$$

$$B_z = \mu_{||}H_z + \alpha_{||}E_z, \quad (12b)$$

$$D_{x,y} = \epsilon_\perp E_{x,y} + \alpha_\perp H_{x,y}, \quad (13a)$$

$$D_z = \epsilon_{||}E_z + \alpha_{||}H_z, \quad (13b)$$

where the subscript $x,y$ denotes the component of a vector in the $x,y$ plane. Since the magnetoelectric terms appearing in Eqs. (12) and (13) arise from $E_kH_l$-type terms in $\Phi$, as shown by Eqs. (10) and (11), we may deduce from the present example of $Cr_2O_3$ a necessary condition for the occurrence of ME effects in an arbitrary material: The magnetic structure of the material must be invariant with respect to one or more symmetry operations which permit the existence of $E_kH_l$-type terms in the thermodynamic potential. As we have indicated, the combined $IR$ transformation is one such operation: $I$ changes the sign of $E_k$ but not of $H_l$, and $R$ changes the sign of $H_l$ but not of $E_k$.

It should be mentioned, in this connection, that a paramagnetic material cannot exhibit ME effects.[3,4] The reason is that the magnetic structure of such a material is invariant with respect to $R$ alone (because the average properties of a paramagnet are unchanged if all spins are inverted), whereas an $E_kH_l$-type term obviously changes sign when subjected to the $R$ transformation by itself.

## C. Remarks on Magnetic Symmetry

Throughout our discussion of ME effects and related phenomena, we are interested in the symmetry of the

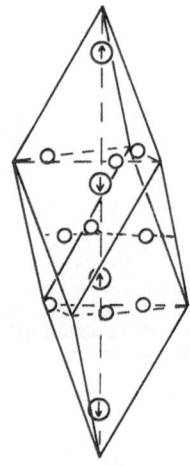

Fig. 3. Possible magnetic structure of $Cr_2O_3$. A different but equally possible structure is obtained by reversing all the spins.

○ – OXYGEN
◯ – CHROMIUM

distribution of the magnetic moment rather than just in the symmetry of the charge distribution. Thus we are concerned with the magnetic symmetry and not just with the crystallographic symmetry. Specifically, the effects under discussion involve macroscopic directional properties and hence the magnetic point group. Thus it is useful to recall that certain quantities, such as the magnetic moment and the magnetic field, possess two unusual symmetry properties: They transform like axial vectors (i.e., like antisymmetric tensors) rather than like ordinary (polar) vectors, and they reverse their sign under the $R$ transformation. The first of these properties is illustrated in Fig. 2. Although this figure appears to be complicated, the information contained in it may be condensed into the following two statements: (1) An axial vector should be represented geometrically by a circular loop possessing a direction of circulation and not by a straight arrow. (2) On the basis of this representation, the various transformations of an axial vector may be carried out on a point by point basis.

It is now easy to verify that the magnetic structure of $Cr_2O_3$, shown in Fig. 3, is indeed invariant under the combined $IR$ transformation. We note, in this connection, that either the midpoint between the two upper $Cr^{+++}$ ions or the midpoint between the two lower $Cr^{+++}$ ions may be regarded as the center of inversion.

## D. Some "Unusual" Contributions to the Thermodynamic Potential

We now summarize the general form of the piezoelectric ($\Phi_{PE}$), piezomagnetic ($\Phi_{PM}$), magnetoelectric ($\Phi_{ME}$), and piezomagnetoelectric ($\Phi_{PME}$) contributions to the thermodynamic potential $\Phi$.

$$\Phi_{PE} = \sum_{ijk} \gamma_{ijk}\sigma_{ij}E_k; \tag{14}$$

$$\Phi_{PM} = \sum_{ijk} \lambda_{ijk}\sigma_{ij}H_k; \tag{15}$$

$$\Phi_{ME} = \sum_{kl} \alpha_{kl}E_kH_l; \tag{16}$$

$$\Phi_{PME} = \sum_{ijkl} \pi_{ijkl}\sigma_{ij}E_kH_l. \tag{17}$$

In these equations, the stress tensor components are denoted by $\sigma_{ij}$, and each of the indices $i$, $j$, $k$, and $l$ is understood to be summed over the coordinates $x$, $y$, and $z$. The coefficients $\gamma_{ijk}$, $\lambda_{ijk}$, $\alpha_{kl}$, and $\pi_{ijkl}$ are subject to the restriction that they be compatible with the magnetic symmetry of the crystal under consideration. The piezoelectric effect is well known, the existence of the piezomagnetic and magnetoelectric effects was suggested by Landau and Lifshitz,[3] and the piezomagnetoelectric effect was recently proposed by Rado.[7] He showed on the basis of symmetry considerations that in $Cr_2O_3$ the PME effect may occur even though the PM effect, for example, cannot occur, and that in this

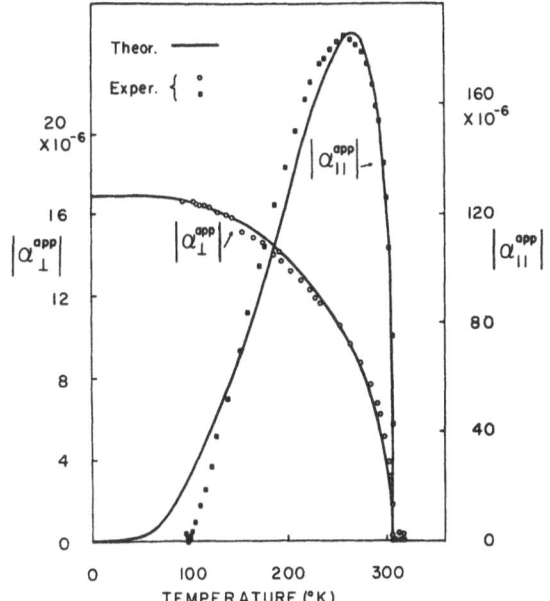

FIG. 4. Comparison of measured and calculated magnetoelectric susceptibilities of $Cr_2O_3$. The *experimental points* (after reference 8) represent $|\alpha_{||}{}^{app}|$ and $|\alpha_\perp{}^{app}|$ as measured by the $(ME)_E$ effect. The *theoretical curves* (after reference 9) represent $|\alpha_{||}|$ and $|\alpha_\perp|$ on the basis of Eqs. (21) and the following numerical values: $T_N = 306°K$, $\Theta = 0.38T_N$, $a_{||} = 38.7 \times 10^{-3}$, and $a_\perp = 1.32 \times 10^{-3}$. Here $T_N$ is the Néel temperature, and $\Theta$ is the "Curie-Weiss" constant defined by Eq. (4.16) of reference 13. The $a$'s as well as the $\alpha$'s are dimensionless in the Gaussian units used.

crystal the maximum number of independent coefficients $\pi_{ijkl}$ is reduced from 81 to 9. Attempts to observe the PME effect experimentally are underway.

## III. RECENT STUDIES OF MAGNETOELECTRIC EFFECTS

### A. Anisotropy of the $(ME)_E$ Effect

The first experimental observation of the $(ME)_E$ effect, reported by Astrov,[1] was made on a single crystal of $Cr_2O_3$ which had not been crystallographically oriented. It seemed of interest, therefore, to determine whether the $(ME)_E$ effect depends on crystallographic direction. In our early experiments[8] on $Cr_2O_3$ we used two x-ray oriented single-crystal disks whose cylindrical axes, denoted by $z'$, were along $z$ and $x$, respectively. An alternating (1000 cps) electric field was applied to the silvered plane surfaces of the disks, and the magnetic flux density was measured by amplifying the voltage induced in a pickup coil. No magnetic field was applied to the disks, and the quantities measured experimentally were $\alpha_{||}{}^{app} = B_z/E_z$ and $\alpha_\perp{}^{app} = B_x/E_x$. As indicated by Eqs. (12), these apparent $\alpha$'s differ from the true $\alpha$'s unless the demagnetizing field as well as the applied magnetic field is zero. We found that the approximation $\mu_{||} \approx \mu_\perp \approx 1$ (which is well fulfilled in antiferromagnetics) permits expressing the demagnetizing corrections by the relations[2]

$$\alpha_{||}{}^{app} \approx [1 - (N_{z'}/4\pi)]\alpha_{||}; \quad \alpha_\perp{}^{app} \approx [1 - (N_{z'}/4\pi)]\alpha_\perp, \tag{18}$$

where $N_{z'}$ is the axial demagnetizing factor of an oblate spheroid having the same axial ratio as the disk being measured. The numerical value of the bracket is about $\frac{1}{2}$ for the disks used in our experiments.

Our experimental results (Fig. 4) show that $|\alpha_{||}{}^{app}|$ and $|\alpha_{\perp}{}^{app}|$ differ strongly in temperature dependence.[8] Similar results were obtained independently by Astrov and reported in a recent paper.[12] It will be seen in Secs. IIIB and IIIC that this anisotropic temperature dependence constitutes an important "constraint" for the theory of the ME effect. Another aspect of the data shown in Fig. 4 is that the maximum value of $|\alpha_{||}{}^{app}|$ is considerably larger than that of $|\alpha_{\perp}{}^{app}|$. Although this result was found to be generally valid if both quantities refer to domain structures resulting from the same magnetic annealing procedure (see Sec. IIIE), it should be pointed out that the $|\alpha_{||}{}^{app}|$ and $|\alpha_{\perp}{}^{app}|$ shown in Fig. 4 were measured on different samples and that their domain structures were probably different. Thus the ratio $|\alpha_{||}{}^{app}|/|\alpha_{\perp}{}^{app}|$ indicated by these early data is not a fundamental characteristic of $Cr_2O_3$. Measurements of both $\alpha$'s on the same spherical sample subjected to various magnetic annealing treatments are now nearing completion. In regard to the signs of the $\alpha$'s (see also Sec. IIIE), we found that in the particular cases corresponding to Fig. 4 the sign of $\alpha_{||}$ is positive for $T > 97°K$ and negative for $T < 97°K$, whereas the sign of $\alpha_{\perp}$ is negative at all the temperatures investigated. These absolute determinations of sign are based arbitrarily (but consistently) on a right-handed coordinate system. Recent experiments indicate that the "cusp" observed at $97°K$ is only partly due to a slight error ($\approx 2°$) in the orientation of the sample used for the $\alpha_{||}$ measurements of Fig. 4.

## B. Theory of the Temperature Dependence of the $(ME)_E$ Effect

Following reference 9, we now describe a phenomenological theory of the temperature dependence of the $\alpha$'s. It is convenient to begin by introducing a fictitious magnetic field $h$, which we define by requiring that the magnetization $M$ induced by $h$ be identical with that induced by the applied electric field $E^a$. The introduction of $h$ allows us to predict the temperature dependence of the $\alpha$'s without having to perform statistical mechanical calculations. If the applied $H$ is zero, and if demagnetizing effects are neglected, we obtain from Eq. (12b)

$$\alpha_{||} = 4\pi M_z / E_z{}^a = 4\pi \chi_{||} h_z / E_z{}^a, \tag{19}$$

where $\chi_{||}$ denotes the parallel antiferromagnetic sus-

ceptibility. If we recall that experimentally $M_z$ is proportional to $E_z{}^a$, and that the $(ME)_E$ effect vanishes unless the material possesses an ordered spin arrangement, it seems reasonable to assume that $h_z$ has the form

$$h_z = a_{||} E_z{}^a \langle S_z \rangle_{av}, \tag{20}$$

where $a_{||}$ is an essentially temperature-independent constant of the material and $\langle S_z \rangle_{av}$ (the thermal average of the expectation value of $S_z$) is proportional to the zero-field sublattice magnetization $M_0 = 2\mu_B \langle S_z \rangle_{av} (N/2)$. Here $\mu_B$ is the Bohr magneton and $N/2$ is the number of magnetic ions (e.g., $Cr^{+++}$) per unit volume situated on each of the two sublattices. By combining Eqs. (19) and (20) and using similar arguments for $\alpha_{\perp}$, we obtain

$$\alpha_{||} = 4\pi a_{||} \chi_{||} \langle S_z \rangle_{av} \quad \text{and} \quad \alpha_{\perp} = 4\pi a_{\perp} \chi_{\perp} \langle S_z \rangle_{av}. \tag{21}$$

The temperature dependence of $\alpha_{||}$ and $\alpha_{\perp}$ is now predictable by calculating $\chi_{||} M_0$ and $\chi_{\perp} M_0$ by means of the Néel-Van Vleck molecular field theory[13] of a two-sublattice antiferromagnet. As shown in Fig. 4, the results are in satisfactory over-all agreement with experiment. In the case of $\alpha_{\perp}$ the agreement is actually very good so that the $\alpha_{\perp}$ vs $T$ curve may be regarded as a measure of the temperature dependence of the sublattice magnetization. Although the agreement between theory and experiment is less good in the case of $\alpha_{||}$, it is no worse than that found[14] in the case of $\chi_{||}$. Thus the existing discrepancies in the case of $\alpha_{||}$ may reflect the limitations of the two-sublattice model rather than those of Eqs. (21). Attempts to resolve this problem are underway.

## C. Atomic Mechanism of the ME Effect

The next problem is to find an atomic mechanism which yields essentially the same temperature dependence as the phenomenological Eqs. (21) and which may explain the order of magnitude of the observed $\alpha$'s. One possible mechanism[9] satisfying these requirements is the combined action of $E^a$, $E^c$ and the spin-orbit coupling, where $E^c$ denotes the electric field resulting from the linear term in the crystalline potential at a sublattice site. For the purpose of estimating the $\alpha$'s, we may equate the fictitious Zeeman shift $|W| = 2\mu_B S \cdot h$ to a suitable fourth order energy perturbation whose order of magnitude is given by

$$|W| \approx (eE^a \cdot r)(eE^c \cdot r)(\lambda L \cdot S)^2 / (\Delta_1 \Delta_2 \Delta_3). \tag{22}$$

Here $\lambda$ is the spin-orbit coupling parameter and the $\Delta$'s are appropriate splittings. The expression for $h$ obtained in this way is consistent with (although not rigorously equal to) Eq. (20) and leads to theoretically

[12] Some of the experimental results reported in references 8 and 2 [but not, for example, the $(ME)_H$ effect] were obtained independently by Astrov in a paper whose translation appeared very recently (after references 8 and 2) and thus could not be fully discussed in the present review. See D. N. Astrov, J. Exptl. Theoret Phys. (U.S.S.R.) **40**, 1035 (1961) [translation: Soviet Phys.—JETP **13**, 729 (1961)].

[13] See, for example, T. Nagamiya, K. Yosida, and R. Kubo, Advances in Phys. **4**, 1 (1955).

[14] T. R. McGuire, E. J. Scott, and F. H. Grannis, Phys. Rev. **102**, 1000 (1956). More recent measurements of the $\chi_{||}$ and $\chi_{\perp}$ of $Cr_2O_3$ by S. Foner (to be published) extend the data of McGuire et al. to lower temperatures. Foner's results indicate that $\chi_{||}$ is not zero at $T = 0°K$.

estimated $a$'s which agree within an order of magnitude with the $a$'s required to fit the experimental curves of $|\alpha_{11}{}^{app}|$ vs $T$ and $|\alpha_{\perp}{}^{app}|$ vs $T$ by means of Eqs. (21).[9] Calculations of the relevant matrix elements[15] occurring in $W$ have justified these estimates. We also note that by using the experimentally determined signs of the $\alpha$'s, the direction of $\langle S_z \rangle_{av}$ with respect to $E_z{}^c$ may be obtained. Other possible mechanisms (e.g., contributions arising from orbital moments) have also been investigated.

It may be mentioned that Eq. (22) represents a kind of quadratic Stark effect which is, however, linear with respect to $E^a$. The role of the spin-orbit coupling is to make the spins respond to the distortion produced by $E^a$ in the electron cloud of each magnetic ion. Although the fictitious field concept is not rigorously applicable to the model discussed above, the predicted temperature dependence of the $\alpha$'s is close to (and at sufficiently low temperatures identical with) that given by Eqs. (21). It has also been shown[7] that the mechanism described by Eq. (22) applies to the $(ME)_H$ effect as well as to the $(ME)_E$ effect.

In a modified mechanism recently proposed by Date, Kanamori, and Tachiki,[16] the spin-orbit coupling is replaced by an intrasublattice exchange interaction. If this interaction turns out to be as large as estimated by these authors, then their exchange mechanism (which is ineffective[16] in the case of $\alpha_{\perp}$) may yield a larger $\alpha_{11}$ than that obtained on the basis of Eq. (22). The temperature dependence of $\alpha_{11}$ predicted by the

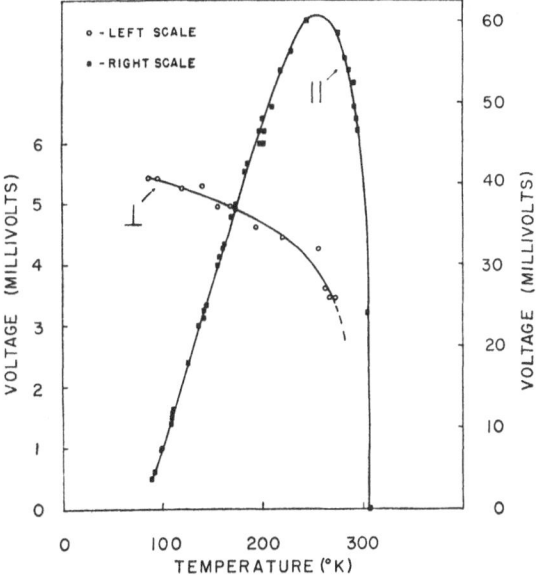

FIG. 5. Temperature dependence of the voltages $|V_{11}|$ (right-hand scale) and $|V_{\perp}|$ (left-hand scale) measured by means of the $(ME)_H$ effect in $Cr_2O_3$. The quantities $|V_{11}|$ and $|V_{\perp}|$ are proportional to $|\alpha_{11}|$ and $|\alpha_{\perp}|$, respectively. (After reference 10.)

[15] V. J. Folen and G. T. Rado (to be published).
[16] M. Date, J. Kanamori, and M. Tachiki, J. Phys. Soc. Japan 16, 2589 (1961).

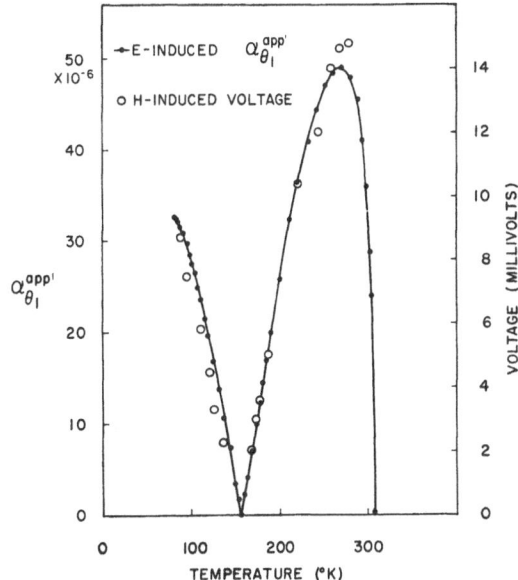

FIG. 6. Comparison of the temperature dependence of $|\alpha_{\theta_1}{}^{app}|$ and $|V_{\theta_1}|$. These quantities were measured by means of the $(ME)_E$ and $(ME)_H$ effects, respectively. Both $\alpha_{\theta_1}{}^{app}$ and $V_{\theta_1}$ change sign at about 154°K. The angle $\theta_1$ is about 30°. (After reference 2.)

exchange mechanism[16] is exactly the same as that exhibited by the $\alpha_{11}$ of Eq. (21).

*Note added in proof.* On the basis of their negative result for the electric field splitting of a paramagnetic resonance line in ruby, Date *et al.*[16] deduced that the spin-orbit mechanism described above is too small to account for the measured values of $\alpha_{11}$ in $Cr_2O_3$. However, J. O. Artman and J. C. Murphy, Bull. Am. Phys. Soc. 7, 14 (1962), and N. Bloembergen and co-workers (to be published) did observe such a splitting, and their result (which is equivalent to $a_{11} = 28.6 \times 10^{-3}$ as compared to the value $a_{11} = 38.7 \times 10^{-3}$ deduced[9] from our ME experiments) shows that the measured $\alpha_{11}$ in $Cr_2O_3$ can, in fact, be accounted for by the spin-orbit mechanism.

### D. Observation of the $(ME)_H$ Effect

Up to this point we have discussed the $(ME)_E$ effect only. However, the thermodynamic theory (see Sec. IIB) suggests that the $(ME)_H$ effect also exists, and it is clearly necessary to test this prediction. The experiments described below constitute the first observation of the $(ME)_H$ effect.[2] By switching on a magnetic field along the geometrical axis ($z'$) of each of the disks mentioned in Sec. IIIA, a voltage $V$ proportional to this field was induced between the silvered disk surfaces. Although $V$ decays with a relaxation time $\tau$ determined by the conductivity and dielectric constant of $Cr_2O_3$, the value of $\tau$ was found to be sufficiently large to permit us to measure $V$ with a vacuum tube electrometer. In the vicinity of room temperature the low voltage measurements could not be made because of drift in the apparatus. The actual voltage $V$ may not be equal to the open-circuit voltage $V^0$ (because the resistance of the disks may not be small compared to the input impedance of the electrometer), but it is clear that $V$ must be proportional to $V^0$ and hence, because of Eqs. (13), proportional to $\alpha$.

Our experimental results on the $(ME)_H$ effect (Fig. 5) show that the temperature dependence of $|V_{11}|$ and $|V_\perp|$ is essentially the same as that of the $|\alpha_{11}|$ and $|\alpha_\perp|$, respectively, which we deduced from our $(ME)_E$ effect (Fig. 4) experiments. The equality of the temperature dependences of the two ME effects is shown even more clearly in Fig. 6. The data plotted in this figure were taken on a disk for which the angle $\theta$ between $z'$ and $z$ has the value $\theta_1 \approx 30°$. It is seen that $|\alpha_{\theta_1}{}^{app}|$ and $|V_{\theta_1}|$, which were measured by means of the $(ME)_E$ and $(ME)_H$ effects, respectively, are proportional to each other throughout the temperature range under study. Furthermore, we observed that the signs of $\alpha_{\theta_1}{}^{app}$ and $V_{\theta_1}$ reverse at the same temperature. More generally, we found experimentally that the three ratios $|V_{11}|/|\alpha_{11}|$, $|V_\perp|/|\alpha_\perp|$ and $|V_{\theta_1}|/|\alpha_{\theta_1}|$, which we noted to be independent of temperature, are actually equal to each other provided the $V$ and $\alpha$ contained in a given ratio are measured on the same sample. [As before, the $V$'s and $\alpha$'s are determined by means of the $(ME)_H$ and $(ME)_E$ effects, respectively.] Thus we obtained a necessary verification of an essential aspect of ME effects and of the existence of $E_k H_l$-type terms in the thermodynamic potential of $Cr_2O_3$.

### E. Structure Sensitive Influences on ME Effects

Soon after our early experiments, we observed that some aspects of the ME effects are structure sensitive.[2] Using both the $(ME)_E$ and the $(ME)_H$ effects, we found that the signs and magnitudes of $\alpha_{11}$ and $\alpha_\perp$, but *not* their temperature dependences, may vary (a) from sample to sample, and (b) upon cooling a sample through the Néel temperature either with or without the presence of a magnetic field. In view of "(a)", it is understandable that the relation

$$\alpha_\theta = \alpha_{11} \cos^2\theta + \alpha_\perp \sin^2\theta, \qquad (23)$$

which follows from the form of the $\alpha$ tensor of $Cr_2O_3$, is not obeyed if the $\alpha$'s are measured on different samples. It is interesting to note that the magnetic annealing mentioned in "(b)" caused the $\alpha$'s of some weakly magnetoelectric samples to increase by factors as large as 300. When applied perpendicular to the $c$ axis, a given magnetic field (fields up to 13 000 oe were used) was found to produce a larger annealing effect than when it was applied parallel to this axis.[15] In addition, an almost complete "erasure" of $\alpha_{11}$ was accomplished by cooling a $\theta=0$ (i.e., $z' \| z$) sample through the Néel temperature in the presence of a 60-cps magnetic field. Similar but less extensive observations of structure

sensitive aspects of the $(ME)_E$ effect [but not of the $(ME)_H$ effect] were made independently by Astrov and reported in a recent paper.[12] He proposes the same domain structure as that suggested by the present authors[2] on the basis of different arguments.

All the above-mentioned observations provide strong evidence for the existence of antiferromagnetic "domains." We find, in fact, that by postulating a specific domain structure we can give a simple and natural explanation of the experimental facts. Our tentative suggestion[2] is that in $Cr_2O_3$ there are two kinds of domains and that they are characterized, respectively, by each spin of a domain pointing toward, or away from, the nearest oxygen plane which is perpendicular to the $c$ axis. A unit cell of the latter kind of domain is shown in Fig. 3. Both kinds of domains are consistent with the existing susceptibility[14] and neutron diffraction[17] data, and the difference between the two kinds of domains results from the asymmetry of the $Cr^{+++}$ sites with respect to the oxygen planes. This asymmetry produces odd terms in the crystalline electric potential, and in the possible mechanism[9] of the ME effects described in Sec. IIIC these terms play an essential role and determine the signs of the $a$'s and hence of the $\alpha$'s. We note, in this connection, that the mechanism described in Sec. IIIC leads to

$$a \approx E_z{}^c (er\lambda L)^2 / (2\mu_B \Delta_1 \Delta_2 \Delta_3) \qquad (24)$$

as an approximate measure of either of the (unequal) quantities $a_{11}$ and $a_\perp$. Thus the suggested domain structure, in conjunction with Eqs. (21), explains the fact that the observed $\alpha$'s can have either sign and that (due to partial cancellations resulting from the presence of two kinds of domains) they can have variations in their magnitudes. Since the two kinds of domains differ solely by a 180° reversal of all the spins, it is clear that the intrinsic $\alpha$'s associated with each domain should have the same temperature dependence. Thus we have a natural explanation of the striking fact that the temperature dependence of each observed $\alpha_\theta$ does not vary even though its sign and magnitude may vary. Finally, we note that the *large* changes in the magnitudes of the $\alpha$'s which can be produced by magnetic annealing from a temperature just above the Néel point are difficult to explain without postulating antiferromagnetic domains.

### ACKNOWLEDGMENTS

We wish to thank Mr. M. J. Marrone, Miss J. W. Radue, and Mr. E. W. Stalder for help in taking the data during various phases of this investigation, Mr. R. A. Becker for technical assistance in connection with sample preparation, and Dr. R. D. Myers and Dr. E. Prince for useful discussions.

[17] B. N. Brockhouse, J. Chem. Phys. **21**, 961 (1953).

JOURNAL OF APPLIED PHYSICS    SUPPLEMENT TO VOL. 33, NO. 3    MARCH, 1962

# Observation of Antisymmetric Exchange Interaction in Yttrium Orthoferrite

D. Treves* and S. Alexander*

*Bell Telephone Laboratories, Incorporated, Murray Hill, New Jersey*

Sensitive static torque measurements were carried out on single crystals of $YFeO_3$ in order to determine if weak ferromagnetism in this material is caused by single ion magnetocrystalline anisotropy or by anisotropic exchange interaction.

The following symmetry considerations show *a priori* which cubic and higher terms in the torque can be nonzero. The magnetic energy $F$ is expanded in a series in the applied field $H$

$$F = \sum \sigma_i(0)H_i + \tfrac{1}{2} \sum \chi_{ij}H_iH_j + \sum C_{ijk}H_iH_jH_k + \cdots.$$

Using the fact that $F$ must be invariant under all the symmetry operations of the magnetic point group appropriate to these crystals, one finds that the only cubic coefficients that may be nonzero are $C_{xxz}$, $C_{yyz}$, $C_{zzz}$. The expression for the torque $\mathbf{T}$ is found using the relation $\sigma_i = \partial F/\partial H_i$, where $\sigma_i$ are the components of the total magnetization, and $\mathbf{T} = \mathbf{H} \times \boldsymbol{\sigma}$.

The ferromagnetic component $\boldsymbol{\sigma}(0)$, the susceptibility $\chi$, and cubic coefficients $C_{ijk}$ were calculated assuming a two sublattice system for the anisotropic exchange model and also for the single ion magnetocrystalline anisotropy model. The comparison of the measured coefficients to those calculated for the two models indicates that the anisotropic exchange mechanism is predominant in causing weak ferromagnetism in $YFeO_3$.

THE static magnetic properties of a crystal can be described by the various tensors which appear when the magnetic energy $F$ is expanded in a series in the components of an applied field $H$

$$F = \sum \sigma_i(0)H_i + \tfrac{1}{2}\sum \chi_{ij}H_iH_j \\ + \sum C_{ijk}H_iH_jH_k + \cdots, \quad (1)$$

where $i$, $j$, $k$ stand for the coordinate axes which are assumed to coincide with the crystallographic axes. The vector $\sigma(0)$ is the permanent ferromagnetic moment of the crystal and the second order tensor $\chi$ is the usual susceptibility tensor.

The energy expression (1) must be invariant under all the symmetry operations of the magnetic point group of the crystal. $FeYO_3$ belongs to the point group $D_{2h}{}^1$ and has a ferromagnetic component in the $z$ direction.[2] Consequently,[3] the only nonvanishing coefficients are

$$\sigma_z(0), \quad \chi_{xx}, \quad \chi_{yy}, \quad \chi_{zz}, \quad C_{xxz}, \quad C_{yyz}, \quad C_{zzz}. \quad (2)$$

These coefficients were experimentally evaluated by measurements of the total magnetization $\boldsymbol{\sigma}$ and the torque $\mathbf{T}$ as a function of the applied field, using the relations $\sigma_i = \partial F/\partial H_i$[4] and $\mathbf{T} = \mathbf{H} \times \boldsymbol{\sigma}$. The results are:

$$\sigma_z(0) = 1.2 \text{ emu g}^{-1}$$
$$\chi_{xx} = 18 \times 10^{-6} \text{ emu g}^{-1} \text{ oe}^{-1}$$
$$\chi_{yy} = \chi_{zz} = 12 \times 10^{-6} \text{ emu g}^{-1} \text{ oe}^{-1} \quad (3)$$
$$C_{xxz} = 22 \times 10^{-12} \text{ emu g}^{-1} \text{ oe}^{-2}$$
$$C_{yyz}, \quad C_{zzz} < 10^{-12} \text{ emu g}^{-1} \text{ oe}^{-2}.$$

If a microscopic model is assumed for the crystal, one can calculate these coefficients. A comparison between the calculated and measured coefficients will determine which model fits the material best.

$YFeO_3$ exhibits weak ferromagnetism as shown by the low value of $\sigma_z(0)$. The $Y^{+++}$ ion is diamagnetic. The $Fe^{+++}$ ions are all in the same crystalline environment, and therefore will have the same magnetic moment. It is therefore reasonable to assume that the ferromagnetic component is the result of canting an essentially antiferromagnetic iron lattice. The existence of an antiferromagnetic structure in the isomorphous compounds $LaFeO_3$, $HoFeO_3$, $ErFeO_3$ has been proven by neutron diffraction.[5]

$YFeO_3$ has four iron atoms per unit cell. It can be shown[6] that for a ferromagnetic crystal the magnetic unit cell is equal to the crystallographic unit cell. There are therefore four magnetic sublattices. To facilitate the calculation, a two sublattice model was assumed. This does not change the symmetry properties of the magnetic crystal.[3]

For crystals of this type two mechanisms for canting are known[7,8]: (1) single ion magnetocrystalline anisotropy (sima), (2) antisymmetric exchange interaction (ae). In the former, the magnetocrystalline easy direction of magnetization is different for the two sublattices, energetically favoring canting. The latter interaction has the form $\mathbf{D} \cdot \boldsymbol{\sigma}^1 \times \boldsymbol{\sigma}^2$, where $\boldsymbol{\sigma}^1$, $\boldsymbol{\sigma}^2$ are the magnetizations of the two sublattices, and $\mathbf{D}$ is a constant vector. This interaction tends to align the two sublattices perpendicular to each other and to the vector $\mathbf{D}$.

The coefficients (2) were calculated according to the two models representing these interaction mechanisms. Two sublattices in the $Y$ plane were assumed, with

---

* On leave of absence from the Weizmann Institute of Science, Rehovot, Israel.

[1] S. Geller and E. A. Wood, Acta Cryst. **9**, 563 (1956).
[2] R. C. Sherwood, J. P. Remeika, and H. J. Williams, J. Appl. Phys. **30**, 217 (1959).
[3] D. Treves (to be published).
[4] L. D. Landau and E. M. Lifshitz, *Electrodynamics of Continuous Media* (Pergamon Press Ltd., London, 1960), p. 147.

[5] W. C. Koehler, E. O. Wollan, and M. K. Wilkinson, Phys. Rev. **118**, 58 (1960).
[6] S. Alexander and D. Treves (to be published).
[7] I. Dzyaloshinsky, J. Phys. Chem. Solids **4**, 241 (1958).
[8] T. Moriya, Phys. Rev. **120**, 91 (1960).

TABLE I. Calculated values of the coefficients of the energy series expansion according to the two models studied. The expressions are correct within the approximation $\gamma_0 \ll 1$.

| | Single ion anisotropy model | Antisymmetric exchange model |
|---|---|---|
| $\sigma_z(0)$ | $\sigma_0 H_a \sin 2\alpha / 2H_e$ | $\sigma_0 H_d / H_e$ |
| $\chi_{xx}$ | $\sigma_0 H_a \sin^2 2\alpha / 8H_e^2 \cos 2(\alpha - \gamma_0)$ | $\sigma_0 H_d^2 / 2H_b H_e^2$ |
| $\chi_{yy}$ | $\sigma_0 / H_e$ | $\sigma_0 / H_e$ |
| $\chi_{zz}$ | $\sigma_0 / H_e$ | $\sigma_0 / H_e$ |
| $C_{xxz}$ | $\dfrac{\sigma_0 \sin 2\alpha [\cos(2\alpha - \gamma_0) - H_a \sin^2 2\alpha / 16H_e]}{4H_e^2 \cos^2 2(\alpha - \gamma_0)}$ | $\sigma_0 H_d (1 - H_d^2 / 4H_b H_e) / 2H_b H_e^2$ |
| $C_{yyz}$ | $-\sigma_0 H_a \sin 2\alpha / 16H_e^3$ | $-\sigma_0 H_d / 8H_e^3$ |
| $C_{zzz}$ | $-\sigma_0 \sin 2\alpha / 8H_e^2$ | $-\sigma_0 H_d / 8H_e^3$ |

magnetizations pointing in symmetrical directions around the $z$ axis. The molecular field approximation was used and the magnitude of the spins was assumed field independent. These assumptions are common to both models.

In the sima model one assumes that the easy directions of magnetization are in symmetrical positions in the $Y$ plane, at an angle $\alpha \le \pi/4$ to the $x$ axis. This ensures that the ferromagnetic component at $\mathbf{H}=0$ is in the $z$ direction.

In the ae model the vector $\mathbf{D}$ is assumed parallel to the $y$ axis. In addition one assumes a magnetocrystalline easy direction common to both sublattices and parallel to the $x$ axis. This is needed in order to describe the anisotropy in the $Y$ plane.

In order to calculate the coefficients (2), one expresses the energy $F$ as a function of the interactions assumed in the models. Using the relation $\sigma_i = \partial F / \partial H_i$ one finds that the coefficients (2) are given by

$$\chi_{ii} = \frac{d\sigma_i}{dH_i}, \quad C_{xxz} = \frac{1}{2}\frac{d^2\sigma_z}{dH_x^2}, \quad C_{yyz} = \frac{1}{2}\frac{d^2\sigma_z}{dH_y^2},$$

$$C_{zzz} = \frac{1}{6}\frac{d^2\sigma_z}{dH_z^2}, \quad (4)$$

where the derivatives are calculated at $H=0$, and $\sigma = \sigma^1 + \sigma^2$ is the total magnetic moment of the system.

The results of the calculation of the coefficients for the two models are summarized in Table I. In the table $\sigma_0 = |\sigma^1| = |\sigma^2|$, $H_a = 2K_a/\sigma_0$, where $K_a$ is the anisotropy constant in the sima model, $H_e = \lambda\sigma_0$, where $\lambda$ is the antiferromagnetic molecular field constant, $H_b = 2K_b/\sigma_0$, where $K_b$ is the anisotropy constant in the ae model, and $H_d = D\sigma_0$. Finally, $\gamma_0$ is the canting angle at $\mathbf{H}=0$. For $\gamma_0 \ll 1$ it is given by

$$\gamma_0 = H_a \sin 2\alpha / 4H_e \quad (5a)$$

and

$$\gamma_0 = H_d / 2H_e \quad (5b)$$

in the sima and the ae models, respectively.

Using (5) and $\alpha \le \pi/4$, one can see from Table I that

$$\chi_{xx}^k / \chi_\perp \le \tfrac{1}{4} \quad (6a)$$

$$\chi_{xx}^d / \chi_\perp = H_d^2 / 2H_e H_b. \quad (6b)$$

Here $\chi_\perp$ is the calculated susceptibility in the $X$ plane. It is equal (in both models) to $\lambda^{-1}$. The superscripts $k$, $d$, denote expressions calculated according to the sima and ae models, respectively.

The paramagnetic contribution to the susceptibility $\chi_{xx}$ at room temperature ($T_n = 648°$K) is less than $(\frac{1}{2})\chi_\perp$[9]; therefore, according to (6a) the sima model predicts $\chi_{xx} < \chi_\perp$ for any set of interaction parameters. On the other hand, according to (6b) the ae model predicts $\chi_{xx} > \chi_\perp$ for

$$H_d^2 > 2H_e H_b. \quad (7)$$

The experimental results (3) show that $\chi_{xx}/\chi_\perp = 1.5$, which is compatible only with the ae model.

Consider now the constant $C_{xxz}$. Table I shows that the sign of $C_{xxz}^k$ is determined by the sign of the term $S$

$$S = \cos(2\alpha - \gamma_0) - H_a \sin^2 2\alpha / 16H_e. \quad (8)$$

With Eq. (5a) and $\alpha \le \pi/4$ one can see from (8) that $S$, and therefore $C_{xxz}^k$, is always positive.

On the other hand, $C_{xxz}^d$ is negative for $H_d^2 > 4H_e H_b$. The experimental value of $C_{xxz}$ is negative, a result again compatible only with the ae model. These two results point very strongly towards the conclusion that the antisymmetric exchange is the predominant interaction responsible for weak ferromagnetism in YFeO$_3$.

As a concluding remark it should be pointed out that the set of parameters $H_d = 10^5$ oe, $H_e = 6 \cdot 10^6$ oe, $H_b = 370$ oe, and $\sigma_0 = 72$ emu/g, when substituted in the expressions of Table I, yields values for the coefficients which are in good quantitative agreement with all the measured ones.

[9] R. M. Bozorth, Ferromagnetism (D. Van Nostrand Company, Inc., Princeton, New Jersey, 1959), p. 472.

JOURNAL OF APPLIED PHYSICS    SUPPLEMENT TO VOL. 33, NO. 3    MARCH, 1962

# Neutron Diffraction Studies on Europium Metal*

C. E. OLSEN, N. G. NERESON, AND G. P. ARNOLD

*Los Alamos Scientific Laboratory, University of California, Los Alamos, New Mexico*

Neutron diffraction investigations have been made on metallic europium which show the metal to be anti-ferromagnetic with a Néel temperature of $87° \pm 1°$K. The observed magnetic reflections can be interpreted in terms of a helical spin structure having a pitch of $3.60\ a_0$ and a helix axis in the (100) direction or an amplitude modulated structure in the (100) direction having a periodicity of $3.60\ a_0$. Between $20°$ and $87°$K, within experimental error, the intensities of the magnetic reflections are proportional to $(T_N - T)^{\frac{1}{2}}$, where $T_N$ is the Néel temperature. The helix pitch is not a function of temperature.

## INTRODUCTION

BECAUSE of the anomalous physical and chemical properties of europium metal it appeared that neutron diffraction studies would be useful. Magnetic studies reported by various authors[1-3] show an anomaly between $90°$ and $100°$K. In particular, the results of Bozorth and Van Vleck[3] indicate that the metal neither exhibits typical antiferromagnetic behavior nor is it ferromagnetic. Electrical resistivity measurements[4,5] confirm the magnetic anomaly and show a behavior of a type predicted by De Gennes and Friedel[6] for magnetic ordering. Barrett[7] has established by x-ray diffraction measurements that there are no crystallographic transitions to $5°$K.

## EXPERIMENTAL

The metal used in these investigations was obtained from the Research Chemical Company and Lindsay Chemical Company. The metal was rolled into foils ($\sim 6$ mils) which were found to be partially oriented.

Europium, because of its high neutron absorption, exhibits a smaller peak to background ratio than many other substances. However, between $295°$ and $87°$K, nuclear reflections were observed from the following crystal planes: (110), (200), (211), (220), and (310), as shown in Fig. 1.

Below $87°$K additional temperature-dependent magnetic reflections were found; one close to the central beam, one on either side of the (110) reflection, and a weak reflection on the low-angle side of the (220) reflection. Lattice parameter measurements on the magnetic reflections showed that the observed reflections could not be indexed on a super lattice of the normal body-centered cubic cell ($a_0 = 4.582$ A). No magnetic contributions were observed in the nuclear peaks even though

several beam and specimen geometries were tried. The cell parameters as determined from the nuclear reflections are in agreement with those given by Barrett[7] and show that the thermal expansion in europium virtually ceases below $87°$K.

## DISCUSSION OF RESULTS

Theoretical considerations by Yoshimori[8] and Villain[9] lead to the prediction of a screw type of magnetic spin alignment. Neutron diffraction investigations by Koehler et al.[10] and Wilkinson et al.[11] on holmium, and Herpin et al.[12] on $Au_2Mn$, confirm the existence of such helical spin structures. Koehler[13] derives an expression for the neutron scattering from such a spin structure.

Analysis of intensities of the (110+ and 110−) magnetic reflections on several foil samples show them to be equal when corrections are made for geometric factors and suitable factors inserted for the difference in magnetic form factor. The positions of the two magnetic reflections are symmetrically located in reciprocal lattice space with respect to the nuclear (110) reflection.

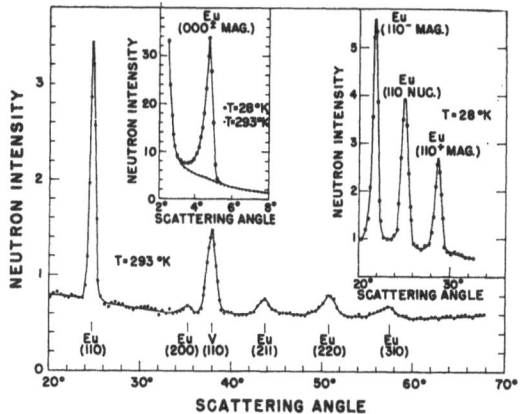

FIG. 1. Neutron diffractometer trace of europium metal at two different temperatures.

---

* Work performed under the auspices of the Atomic Energy Commission.

[1] Klemm and H. Bommer, Z. anorg. u. allgem. Chem. **231**, 138 (1937).

[2] C. Henry La Blanchetais and F. Trombe, Compt. rend. **243**, 707 (1956).

[3] R. M. Bozorth and J. H. Van Vleck, Phys. Rev. **118**, 1493 (1960).

[4] M. A. Curry, S. Legvold, and F. H. Spedding, Phys. Rev. **117**, 971 (1960).

[5] C. E. Olsen, U. S. Atomic Energy Commission LASL-2406, 21 (1960).

[6] P. G. De Gennes and J. Friedel, J. Phys. Chem. Solids **4**, 71 (1958).

[7] C. S. Barrett, J. Chem. Phys. **25**, 1123 (1956).

[8] A. Yoshimori, J. Phys. Soc. Japan **14**, 307 (1959).

[9] J. Villain, J. Phys. Chem. Solids **11**, 303 (1959).

[10] W. C. Koehler, J. W. Cable, E. O. Wollan, and M. K. Wilkinson, Bull. Am. Phys. Soc. **5**, 459 (1960).

[11] M. K. Wilkinson, W. C. Koehler, E. O. Wollan, and J. W. Cable, J. Appl. Phys. **32**, 48S (1961).

[12] A. Herpin, P. Meriel, and J. Villain, Compt. rend. **249**, 1334 (1959).

[13] W. C. Koehler, Acta Cryst. **14**, 535 (1961).

The position and intensities correspond to diffraction from either a helical spin structure or a sinusoidally modulated spin structure[14] having a period $3.60 \pm 0.02$ $a_0$. Assuming a helical structure, the helix axis is parallel to the (100) direction, and the interplanar turn angle between the atoms on the cell face and the body center is calculated to be $50° \pm 0.04°$. The lack of temperature dependence for the scattering angle of the magnetic reflections appears to be associated with the absence of thermal expansion. The absence of the expected satellites on the higher angle reflections appears to arise from the decrease in the magnetic form factor coupled with the high absorption and geometrical factors such as the preferred orientation of the foil. When unoriented powders are used, diffraction theory for helical and sinusoidal spin structures call for additional magnetic contributions nearly at the positions of the nuclear reflections. None were observed indicating the possibility that the rolling of the foil produced, in addition to the preferred orientation, a domain orientation such

that the additional reflections were not observed due to geometric factors. Because of foil orientation it was not possible to calculate a magnetic moment for the europium. Further investigations are being made and will be described in a subsequent report.

Analysis of the temperature dependence of the intensities of the magnetic reflections indicated that they could not be fitted to any reasonable, squared Brillouin function. However, between 87° and 20°K they could be fitted to a function proportional to $(T_N - T)^{\frac{1}{4}}$.

## ACKNOWLEDGMENTS

We would like to thank Karl Gschneidner, CMF-5, for supplying us with a sample of europium metal, Robert W. Keil, CMB-6, for rolling the foil, and John Yarnell, P-2, for the use of his double-scattering neutron spectrometer. We would especially like to thank Dr. W. C. Koehler of the Oak Ridge National Laboratory for his helpful discussions and for pointing out the significance of some of our diffraction results.

[14] W. C. Koehler, J. Appl. Phys. **32**, 20S (1961).

---

JOURNAL OF APPLIED PHYSICS     SUPPLEMENT TO VOL. 33, NO. 3     MARCH, 1962

# Appearance of a Weak Ferromagnetism in Fine Particles of Antiferromagnetic Materials

W. J. Schuele and V. D. Deetscreek
*The Franklin Institute Laboratories, Philadelphia, Pennsylvania*

Several experiments have been carried out on NiO which may be interpreted as confirmation of the suggestion of Néel that antiferromagnetic material in fine particle form should exhibit a weak ferromagnetism and superparamagnetism. The initial susceptibility of the nickel oxide between 22 A and >1000 A increases with decreasing particle size. The susceptibility decreases with decrease in temperature below room temperature for all but the smallest size sample. This sample had a very large susceptibility and an "S"-shaped magnetization curve, characteristic of superparamagnetic material. In the intermediate size range, cooling through the Néel temperature in a field introduces an additional magnetization which is field independent.

## INTRODUCTION

RECENTLY Néel[1-3] has suggested that antiferromagnetic material in fine particle form should exhibit some interesting magnetic properties including superparamagnetism and a weak ferromagnetism. We have carried out several experiments to investigate these suggestions. Only a short account of them will be given here and a more detailed paper will be presented at a later date. The antiferromagnetic state is characterized by an ordered antiparallel arrangement of electron spins. In the simplest case (admittedly oversimplified) an antiferromagnetic material will have two sublattices, $A$ and $B$, which are identical in every respect with the exception that the atomic moments in the $B$ sublattice are antiparallel to the moments in the $A$ sublattice. In

two dimensions this can be represented by rows of spins in which the even numbered rows have all the spins in one direction and the odd rows have them in the reverse direction. When the number of rows is odd, there will be a row of uncompensated spins which (when properly oriented) give rise to a net magnetization. For an even number of rows, there will be complete compensation and no net magnetization. For a large crystal the effect of the uncompensated spins can be neglected. However, as the size of the particle is reduced, the fraction of the total number of uncompensated spins will increase greatly. Néel has also suggested that if the size of the particles is further reduced, the thermal fluctuation of the spins will become an important factor. This effect is analogous to that observed in extremely fine particles of ferromagnetic materials, which gives rise to superparamagnetism. Néel has suggested that the same name

[1] L. Néel, Compt. rend. **252**, 4075 (1961).
[2] L. Néel, Compt. rend. **253**, 9 (1961).
[3] L. Néel, Compt. rend. **253**, 203 (1961).

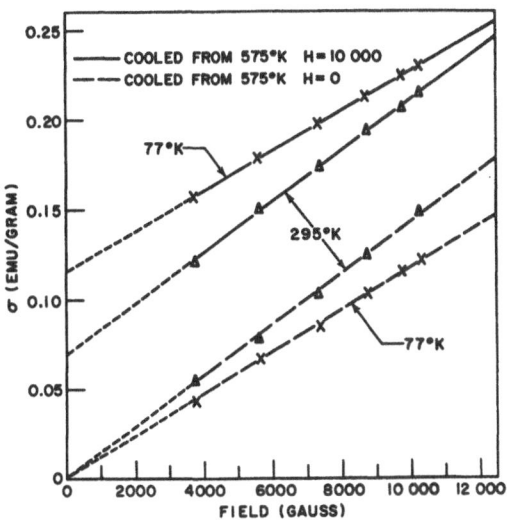

FIG. 1. Magnetization curves for 216-A NiO sample.

TABLE I. Susceptibility and remanent magnetization of NiO particles.

| Particle size (A) | $\chi \times 10^6$ | | emu/g$\times 10^3$ extrapolated to $H=0$ | |
|---|---|---|---|---|
| | 295°K | 77°K | 295°K | 77°K |
| >1000 | 9.7 | 7.6 | 0 | 0 |
| 345 | 11.6 | 8.6 | 4.5 | 6.5 |
| 216 | 14.2 | 11.8 | 69.0 | 115.0 |
| 22 | 140.0[a] | 1170[a] | | |

[a] Initial susceptibility.

be applied to the behavior of very fine antiferromagnetic particles.

## PREPARATIONS AND MEASUREMENTS

Nickel oxide was the first material studied. It was prepared by several techniques used previously to make fine particle ferrites[4,5] and resulted in samples with the following nominal particle sizes: >1000 A, 345 A, 216 A, and 22 A. The nominal particle size for each sample was determined by x-ray line broadening. The x-ray diffraction peaks corresponded to the A.S.T.M. diffraction data card for NiO. No spurious impurity peaks were observed. The susceptibilities of the four nickel oxide samples were determined by measuring the moments of the samples as a function of applied field. The susceptibilities per gram obtained at room temperature and 77°K are given in Table I. To align the uncompensated spins, the samples were now heated above the Néel temperature (530°K) and cooled in a magnetic field of 10 000 gauss. The moment as a function of field was again determined.

## EXPERIMENTAL RESULTS AND DISCUSSIONS

### A. Samples before Cooling in a Field

The susceptibilities of the nickel oxide samples increased as their average particle size decreased (Table I, columns 2 and 3). With the exception of the 22 A particle size (to be discussed below), the susceptibilities showed a decrease with decreasing temperature which is characteristic of antiferromagnetic materials. Our largest particle size sample had a susceptibility in agreement with the literature values.[6] A plot of the moment vs applied field for the >1000 A, 345 A, and 216 A samples resulted in straight lines which extrapolated to zero field through the origin. Figure 1 is typical of the plot obtained and shows the results of the 216-A sample.

### B. Samples Heated above the Néel Temperature and Cooled in a Magnetic Field

The nickel oxide samples were heated above the Néel temperature and cooled in a magnetic field. This treatment had no measurable effect on the largest particle size sample. The magnetization curves obtained for the 345-A and 216-A samples after this treatment have slopes equal to those obtained before cooling through the Néel temperature in a field. The curves were displaced indicating an additional magnetization as the result of this treatment. This additional magnetization is field independent in the region in which measurements were made (3500–10 500 gauss). See solid curves Fig. 1. By extrapolating these lines to zero field we obtain the moment in emu/g shown in Table I. The value is noticeably higher for the 216-A sample than for the 345-A sample.

The 22-A size sample showed a very large initial susceptibility at room temperature, larger than the other three samples by a factor of 10. The complete magnetization curve from −10 500 to +10 500 oe was "S"-shaped passing through the origin, which is characteristic of superparamagnetic material. The initial susceptibility obtained at 77°K was more than 8 times larger than the room temperature susceptibility.

## CONCLUSIONS

It is concluded that fine particles of antiferromagnetic material can exhibit ferromagnetic properties as suggested by Néel. This phenomenon was found to be size dependent, with extremely fine particles behaving essentially like superparamagnetic material.

## ACKNOWLEDGMENT

The authors gratefully acknowledge their indebtness to Dr. S. Shtrikman who suggested the experiment and advised us throughout this investigation.

[4] W. Schuele and V. D. Deetscreek, J. Appl. Phys. 32, 235S (1961).
[5] A. Berkowitz and W. Schuele, J. Appl. Phys. 30, 134S (1959).
[6] G. Foëx, Tables de Constantes et Donnes Numeriques (Masson and Cie, Paris, 1957), Vol. 7, p. 96.

JOURNAL OF APPLIED PHYSICS     SUPPLEMENT TO VOL. 33, NO. 3     MARCH, 1962

# Magnetic Structure of Manganese Chromite*

L. M. CORLISS AND J. M. HASTINGS

*Brookhaven National Laboratory, Upton, New York*

Numerous attempts have been made to explain the anomalous magnetic properties of chromites and of mixed spinels containing chromium on the basis of the theoretical models of Néel and of Yafet and Kittel. Recent calculations by T. A. Kaplan and co-workers, using a classical Heisenberg model, have shown that over an appreciable range of the ratio of the $BB$ to the $AB$ interaction, the ground state is a ferrimagnetic spiral, i.e., a conical spin configuration in which the transverse components progress in spiral fashion along a fixed direction in the crystal. The object of the present study is to determine the suitability of the various proposed theoretical models in the case of manganese chromite, for which rather complex magnetic behavior has been observed by means of neutron diffraction. From diffraction patterns taken at room temperature, it is readily established that $MnCr_2O_4$ is a normal spinel with less than 1% of the $Mn^{2+}$ ions present on $B$ sites. The Curie temperature, as determined from diffraction data is $\sim 43°K$. Below this temperature the magnetic contributions to the fundamental spinel peaks arising from aligned spins increases as the temperature is lowered and is effectively saturated at about 20°K. At 18°K, additional sharp peaks appear at positions which cannot be indexed either on the original unit cell or on any reasonably enlarged cell. These extra reflections persist with unchanged intensities down to 4.2°K. No change in either the positions or intensities of the fundamental lines is observed in going through the transition. Above 18°K a broad, diffuse peak is present in the region where the principal extra lines develop. This diffuse peak decreases with increasing temperature, but is still observable above the Curie point. Application of a magnetic field along the neutron scattering vector decreases the magnetic contributions to the fundamentals, but increases the intensities of the extra reflections. Analysis of the data shows that while little difficulty is encountered in explaining the observed saturation moments, the Néel model and the Yafet–Kittel model fail to account for major qualitative features of the diffraction results. Agreement is not significantly improved by modifications of the Néel model in which reversed spins (either random or ordered) are introduced, or in which some of the spins remain paramagnetic even at low temperatures. On the other hand, all the qualitative aspects of the diffraction results can be explained in terms of the ferrimagnetic spiral model of Kaplan and co-workers. The theoretical calculations give reasonably good quantitative agreement with the positions and intensities of the extra lines as well as the intensities of the fundamentals, while retaining the expected values for the individual ionic moments and the macroscopic saturation moment. Although some discrepancies still remain, there is every reason to believe that these can be removed by further refinement of the theory and by the extension of experimental measurements to single crystals.

* Research performed under the auspices of the U.S. Atomic Energy Commission.

---

JOURNAL OF APPLIED PHYSICS     SUPPLEMENT TO VOL. 33, NO. 3     MARCH, 1962

# Lattice Theory of Spin Configuration

E. F. BERTAUT

*Laboratoire d'Electrostatique et de Physique du Métal, Institut Fourier, et Centre d'Etudes Nucléaires, Grenoble, France*

In the frame of a given crystallographic symmetry a matrix $\zeta(\mathbf{k})$ is constructed, the eigenvectors and eigenvalues of which are directly related to the spin configuration and magnetic exchange energy, respectively. For Bravais lattices the matrix equation reduces to Villain's equation.

When chemical and magnetic unit cells coincide, the eigenvectors of $\zeta(0)$ are shown to be identical to the "basis of irreducible representations" used by Turov and Dzialoshinski in the construction of an invariant Hamiltonian.

The usefulness of both methods is discussed. The matrix theory is more general in the sense that it remains valid for magnetic cells different from the chemical one. Although the theory presented here starts from conventional crystallographic symmetry, it contains all the configurations possible in the so-called magnetic groups and even those not contained there (helical configurations).

The theory includes isotropic as well as anisotropic (crystalline field, dipolar, pseudodipolar, antisymmetric) coupling in the order 2 approximation.

Various examples of magnetic structures in the fields of perovskites, ilmenites, spinels, garnets, etc., are presented.

## INTRODUCTION

THE theory of ferri- and antiferro-magnetism, as developed by Néel,[1] was based on the *a priori* assumption of a decomposition into sublattices of the positions of magnetic ions. Although it was clearly understood that the decomposition into sublattices should minimize the Heisenberg[2] energy, the idea of deriving spin configurations directly from the Heisenberg energy came out much later from the first papers on helical configurations, on chromium by Kaplan,[3] on $MnO_2$ by Yoshimori,[4] and was expressed most clearly

[1] L. Néel, Ann. Physik **3**, 137 (1948).

[2] W. Heisenberg, Z. Physik **49**, 619 (1928).
[3] T. A. Kaplan, Phys. Rev. **116**, 888 (1959).
[4] A. Yoshimori, J. Phys. Soc. Japan **14**, 807 (1959).

by Villain[5] for the case of magnetic atoms on Bravais lattices.

In the meantime, Russian authors, Dzialoshinski[6] and Turov and Nays,[7] were able to derive spin configurations from a phenomenological or macroscopic Hamiltonian which was invariant with respect to spin reversal and symmetry operations of the crystallographic space group. Their very powerful method is, however, limited to the case where magnetic and chemical unit cells coincide.

In the matrix or microscopic method[8] which will be developed here for general lattices, we insist more on the geometrical than on the physical aspect in the sense that particular emphasis is put on the crystallographic symmetry. For the application of the matrix method, no special knowledge of group theory is required. It turns out that the matrix method reduces to Villain's method[5] in the case of Bravais lattices, and that when magnetic and chemical unit cells coincide the results of the microscopic and macroscopic methods are identical. The main advantage of the matrix method is, however, that it directly gives insight into the elementary interactions involved, and that it is not limited to a magnetic cell coinciding with the chemical one. We limit ourselves to the consideration of an Hamiltonian of order 2, following closely the notations of short papers[8] which appeared in the beginning of the year.

## Isotropic Case

The dominant terms in the Hamiltonian $H$ correspond to isotropic exchange (direct or super-exchange). Let the Heisenberg energy be

$$H = -2 \sum_{R,R'} J_{R,R'} \mathbf{S_R} \cdot \mathbf{S_{R'}}$$
$$= -2 \sum_{R,R'} \hat{J}_{R,R'} \sigma_\mathbf{R} \cdot \sigma_{\mathbf{R'}}, \qquad (1)$$

where the summation is over all couples of magnetic atoms in the crystal. Here we introduce unitary spins $\sigma$ by Eq. (2) and new exchange integrals $\hat{J}_{RR'}$ by convention (3):

$$\sigma = \mathbf{S}/S \qquad (2)$$

$$\hat{J}_{RR'} = S_R J_{RR'} S_{R'}. \qquad (3)$$

The motion of spin $\mathbf{S_R}$ at point $\mathbf{R}$ is described by

$$\frac{d\mathbf{S_R}}{dt} = 2 \sum_{R'} J_{RR'} \mathbf{S_{R'}} \wedge \mathbf{S_R} \qquad (4)$$

and at equilibrium $(d\mathbf{S_R}/dt = 0)$, $\mathbf{S_R}$ must be parallel to

the "molecular field"

$$\sum_{R'} J_{RR'} \mathbf{S_{R'}}.$$

This parallelism may be expressed in terms of unitary spins by

$$\lambda_R \sigma_\mathbf{R} = \sum_{R'} \hat{J}_{RR'} \sigma_{\mathbf{R'}}, \qquad (5)$$

where $\lambda_R$ is a real scalar.[9] Equation (5) is not particularly new and may be obtained as well by minimizing Eq. (1) with the condition

$$\sigma_R{}^2 = 1 \qquad (6)$$

("strong" condition of Luttinger-Tisza[10]). $\lambda_R$ appears then as a Lagrange parameter. Substitution of Eq. (5) into Eq. (1) shows that

$$H = -\sum_R \lambda_R. \qquad (7)$$

$-\lambda_R$ represents the contribution of the ion in $\mathbf{R}$ to the exchange energy. Stability requires the positivity of $\lambda_R$ (spin and molecular field must have the same direction).

The only point to which not much attention seems to have been paid in the past is precisely the fact that $\lambda_R$ is a scalar attached to the point $\mathbf{R}$. Transformation properties of scalar quantities are particularly simple. Indeed, scalars are invariant under all crystallographic symmetry operations. Consequently, those $\lambda_R$'s are equal which belong to crystallographically equivalent atoms, i.e., atoms which are related by the conventional symmetry operations of the 230 crystallographic space groups as tabulated in the International Tables for Crystallography.[11] (It is clear that we restrict ourselves to perfectly ordered periodical lattices.) As a consequence, the Hamiltonian is invariant because the $\lambda_R$ are invariant.

Translational symmetry will be taken into account in the following way. We shall number by $i$ (or $j) = 1$, $2, \cdots, n$ the different Bravais lattices of magnetic atoms belonging to the crystallographic unit cell. In the following, $i$ and $j$ will replace the indices $R$ and $R'$ of the foregoing lines. Let us write as many Eq. (5) as there are Bravais lattices, say $n$ and pick out lattice $i$. Multiplying $\lambda_i \sigma_i(\mathbf{R_i})$ by $\exp(2\pi i \mathbf{k} \cdot \mathbf{R_i})$ and summing over all $\mathbf{R_i}$ belonging to lattice $\mathbf{i}$, one obtains using the translational symmetry of the exchange integrals (e.i.)

$$\lambda_\mathbf{i} \mathbf{T_i}(\mathbf{k}) = \sum_i \zeta_{ij}(\mathbf{k}) \mathbf{T_j}(\mathbf{k}). \qquad (8)$$

Here

$$N \mathbf{T_i}(\mathbf{k}) = \sum_{R_i} \sigma_\mathbf{i}(\mathbf{R_i}) \exp(2\pi i \mathbf{k} \cdot \mathbf{R_i}) \qquad (9)$$

[5] J. Villain, J. Phys. Chem. Solids 11, 303 (1959).
[6] I. E. Dzialoshinski, J. Phys. Chem. Solids 4, 241 (1958).
[7] E. A. Turov and V. E. Nays, J. Met. U. S. S. R. 9, 10 (1960).
[8] E. F. Bertaut, Compt. rend. 252, 76, 252, 2032, 2078, 3895 (1961); J. phys. radium 22, 321 (1961).

[9] A quantity which relates 2 vectors is generally a tensor of order 2, but reduces to a scalar when the 2 vectors are strictly parallel.
[10] J. M. Luttinger and L. Tisza, Phys. Rev. 81, 1015 (1952).
[11] International Tables for Crystallography (The Kynoch Press, Birmingham, England, 1952).

and

$$\zeta_{ij}(\mathbf{k}) = \sum_{R_j} \jmath_{R_i R_j} \exp[2\pi i \mathbf{k} \cdot (\mathbf{R}_1 - \mathbf{R}_j)]. \quad (10)$$

$N$ is the number of unit cells in the crystal. $\mathbf{T}_i(\mathbf{k})$ is the Fourier transform of the spins $\sigma_1(\mathbf{R}_i)$. The summation in Eq. (10) is in two parts. One fixes $\mathbf{R}_i$ and sums over all atoms $\mathbf{R}_j$ of lattice $j$ which are first equivalent neighbors of $\mathbf{R}_i$ and belong to the same value of $J_{R_i R_j}$. This procedure is repeated for second neighbors $\mathbf{R}_j$ of $\mathbf{R}_i$ and so on. At least theoretically there is no limitation to the range of magnetic interactions. The advantage is that the infinite number of equations (5) is now replaced by $n$ equations (8) which may be written in the concise form

$$[\zeta(\mathbf{k}) - (\lambda)]\mathbf{T}(\mathbf{k}) = 0. \quad (11)$$

Here $\zeta(\mathbf{k})$ is an Hermitian matrix of order $n$ and elements given by Eq. (10). $(\lambda)$ is the diagonal matrix formed by the elements $\lambda_s \delta_{ij}$. $\mathbf{T}(\mathbf{k})$ is a vector of $n$ components $\mathbf{T}_i(\mathbf{k})$ $(i, j = 1, 2, \cdots, n)$. It is easily recognized that for the case of *one* Bravais lattice $(n=1)$ (10) and (11) reduce to Villain's equation.[5]

## Equivalent Atoms. Magnetic and Chemical Cells Identical

It is perhaps worthwhile to illustrate the manipulation of the foregoing relations by a simple example of equivalent atoms in a magnetic cell identical with the chemical one. One may put $\mathbf{k}=0$. $\mathbf{T}_i(0)$ and $\sigma_1(\mathbf{R}_i)$ become identical (9). We have to do with a simple eigenvalue problem. The number of distinct modes cannot exceed $n$, the number of Bravais lattices. Equation (10) reduces to a sum of products in which an e.i. (exchange integral) multiplies the number of corresponding equivalent atoms. We choose here the simple corindum-type structure as found in $Cr_2O_3$ or $\alpha$-$Fe_2O_3$ where the Bravais lattices have origins in

$$0\,0\,z \; (1); \quad 0\,0,\tfrac{1}{2}+z \; (2); \quad 0\,0\,\bar{z} \; (3); \quad 0\,0\,\tfrac{1}{2}-z \; (4)$$

and are marked by the numbers in parenthesis.

In $\alpha$-$Fe_2O_3$, for instance, atom (1) has 1 neighbor on lattice 3 at 2.88 A and e.i. $J_1$, 3 neighbors on lattice 4 at 2.96 A and e.i. $J_s$, 3 neighbors on lattice 3 at 3.36 A and e.i. $J_s$, 6 neighbors on lattice 2 at 3.69 A and e.i. $J_4$, 6 neighbors on lattice 1 at 5.02 A and e.i. $J_0$. The first line of the matrix equation (11) becomes

$$(A-\lambda)\mathbf{T}_1 + B\mathbf{T}_2 + C\mathbf{T}_3 + D\mathbf{T}_4 = 0, \quad (12)$$

where the following abbreviations are used:

$$\begin{aligned} A &= 6J_0; & B &= 6J_4; \\ C &= J_1 + 3J_s; & D &= 3J_s. \end{aligned} \quad (13)$$

The complete matrix system (11) reads here

$$\begin{bmatrix} A-\lambda & B & C & D \\ B & A-\lambda & D & C \\ C & D & A-\lambda & B \\ D & C & B & A-\lambda \end{bmatrix} \begin{bmatrix} \mathbf{T}_1 \\ \mathbf{T}_2 \\ \mathbf{T}_3 \\ \mathbf{T}_4 \end{bmatrix} = 0. \quad (14)$$

There are 4 solutions with the eigenvectors:

$$\begin{aligned} \mathbf{T}_I &= (1, -1, 1, -1); & \mathbf{T}_{II} &= (1, -1, -1, 1); \\ \mathbf{T}_{III} &= (1, 1, -1, -1); & \mathbf{T}_{IV} &= (1,1,1,1), \end{aligned} \quad (15)$$

and the corresponding eigenvalues:

$$\begin{aligned} \lambda_I &= A-B+C-D; & \lambda_{II} &= A-B-C+D; \\ \lambda_{III} &= A+B-C-D; & \lambda_{IV} &= A+B+C+D. \end{aligned} \quad (16)$$

Modes II and III correspond, respectively, to $\alpha$-$Fe_2O_3$[12] and $Cr_2O_3$.[13] The simple comparison of $\lambda_{II}$ and $\lambda_{III}$ shows that if $J_s > 2J_4$, the spin configuration $\mathbf{T}_{II}$ is more stable than $\mathbf{T}_{III}$. This does not mean that $J_s$ which in $\alpha$-$Fe_2O_3$ relates parallel spins is positive. In fact it can be shown[14] that in $\alpha$-$Fe_2O_3$ all $J_j$ $(j=1$ to $4)$ are negative.

This very simple example shows that we may get at once the spin configuration, its energy, and its stability conditions by a close examination of Eq. (11). As an exercise the reader may try to find out the form of the matrix $\zeta(\mathbf{k})$ in the general case (i.e., for $\mathbf{k} \neq 0$) considered in reference 14.

## Relation with Group Theory

The exchange integral $J_{ij}$ connects spin $\sigma_1$ with spin $\sigma_j$. For instance, in (14) the e.i. $D = 3J_s$ relates the spins 1, 2, 3, 4, respectively, to spins 4, 3, 2, 1. The coefficient of $D$ in the matrix (14) which we denote by the same letter in parenthesis (17) is a substitution, or symmetry, operator.

$$(D) = \begin{bmatrix} \cdot & \cdot & \cdot & 1 \\ \cdot & \cdot & 1 & \cdot \\ \cdot & 1 & \cdot & \cdot \\ 1 & \cdot & \cdot & \cdot \end{bmatrix}. \quad (17)$$

Clearly, geometry shows that $(D)$ corresponds to the operator $\bar{1}$, center of inversion.[15] The reader may check that the matrices $(B)$ and $(C)$ represent binary axes passing through the origin and that $(B) = (C)(D)$.

In the present case the eigenvectors $\mathbf{T}(0)$ of the matrix $\zeta(0)$ appear to be linear combinations of spins which are invariant under the symmetry operations of the crystallographic group. But these invariant linear combinations $\mathbf{T}(0)$ form the "basis of irreducible representations" used by Russian authors[6,7] for the construction of an Hamiltonian, invariant under the symmetry operations of the crystallographic group. The reader may directly check that the vectors $\mathbf{T}_{IV}$, $\mathbf{T}_I$, $\mathbf{T}_{II}$, $\mathbf{T}_{III}$ are respectively identical with the vectors $\mathbf{m}$ and $\mathbf{l}_j$ $(j=1, 2, 3)$ in reference 6. The vector

$$\mathbf{T}_{IV} = \sigma_1 + \sigma_2 + \sigma_3 + \sigma_4 \quad (18)$$

corresponds to the "identity representation," in other words to the ferromagnetic configuration.

[12] C. G. Shull, W. A. Strauser, and E. O. Wollan, Phys. Rev. **83**, 333 (1951).
[13] B. N. Brockhouse, J. Phys. Chem. **21**, 961 (1953).
[14] E. F. Bertaut, J. Phys. Chem. Solids (to be published).
[15] The inversion center is not at the origin, but in $\pm(0\,0\,\tfrac{1}{4})$ in the conventional description of the corindum structure.

More generally, let $\mathbf{C}$ be any symmetry operation which leaves $\zeta(0)$ invariant so that

$$\mathbf{C}^{-1}\zeta(0)\mathbf{C}=\zeta(0). \qquad (19)$$

It is easy to see that if $\mathbf{T}(0)$ is solution, $\mathbf{CT}(0)$ is solution, too. If the corresponding root $\lambda$ is nondegenerate, $\mathbf{T}(0)$ will transform into itself (as above). If the root is $q$ times degenerate, there will be $q$ eigenvectors spanning an invariant subspace of dimension $q$.

As example we consider the so-called $B$- or octahedral sites in spinels situated at points ($\frac{5}{8}\frac{5}{8}\frac{5}{8}$; $\frac{5}{8}\frac{3}{8}\frac{3}{8}$; $\frac{3}{8}\frac{5}{8}\frac{3}{8}$; $\frac{3}{8}\frac{3}{8}\frac{5}{8}$). The matrix $\zeta(0)$ has the same form (14), but with $A=0$ and $B=C=D$. The same vectors $\mathbf{T}(0)$ (15) diagonalize the matrix $\zeta(0)$, but the only nondegenerate root corresponds to the identity representation, the other one is triply degenerate. In fact, the reader may verify that the ternary axis passing through point $\frac{5}{8}\frac{5}{8}\frac{5}{8}$ just interchanges the vectors $\mathbf{T_I}$, $\mathbf{T_{II}}$, $\mathbf{T_{III}}$ which form an invariant subspace of dimension 3.

To summarize the discussion, the eigenvectors of $\zeta(0)$ form the basis of irreducible representations so that they may be found either by the solution of $\zeta(0)$ or *a priori* from group theory.

### Propagation Vector

The vector $\mathbf{k}$ is defined by Eq. (20) and will be noted $[h_1h_2h_3]$

$$\mathbf{k}=h_1\mathbf{b}_1+h_2\mathbf{b}_2+h_3\mathbf{b}_3. \qquad (20)$$

The vectors $\mathbf{b}_j$ are the usual reciprocal vectors related to the unit cell vectors $\mathbf{a}_j$ ($j=1, 2, 3$) by Eq. (21) or the equivalent definition Eq. (22) and circular permutations.

$$\mathbf{a}_j \cdot \mathbf{b}_k = \delta_{jk} \qquad (21)$$

$$\mathbf{b}_1=(\mathbf{a}_2 \wedge \mathbf{a}_3)/(\mathbf{a}_1\mathbf{a}_2\mathbf{a}_3). \qquad (22)$$

A notation like $\mathbf{k}=[\frac{1}{2},\frac{1}{2},0]$ means that spins are propagating in the [110] direction and that the magnetic unit cell is twice the chemical one in the $\mathbf{a}_1$ and $\mathbf{a}_2$ directions. Usually $\mathbf{k}$ is restricted to remain in the first Brillouin zone. However when magnetic and chemical unit cells are identical it is often indicated to use instead of $\mathbf{k}=0$ a vector $\mathbf{k}=[h_1h_2h_3]$ with integer numbers $h_j$, for reasons of geometrical intuition, so that the plane ($h_1h_2h_3$) is a propagation plane in which all spins are "in phase." For instance, the Néel configuration in spinels may be described as a solution of $\zeta(0)$ by an eigenvector $\mathbf{T}(0)=(1, 1, -1, -1, -1, -1)$, where the first two components refer to tetrahedral, the other ones to octahedral sites, but also as a solution of $\zeta(\mathbf{k})$ ($\mathbf{k}\neq0$) with $\mathbf{T}(\mathbf{k})=(1,1,1,1,1,1)$ and $\mathbf{k}=[004]$ or $\mathbf{k}=[444]$. This simply means that in ferrimagnetic spinels spins are parallel in (004) and (444) planes. Physically the two descriptions are equivalent.

From the hermitian nature of $\zeta(\mathbf{k})$ it is evident that to each solution with a propagation vector $\mathbf{k}_0$ corresponds an equivalent solution with a vector $-\mathbf{k}_0$. (This degeneracy is essential; it cannot be removed by any additional anisotropy term.)

One may raise the question of how many $\mathbf{k}$ vectors are needed for the construction of the most general solution, knowing that $\sigma_j(\mathbf{R})$ is given by the Fourier inversion of Eq. (9).

$$\sigma_j(\mathbf{R}_j)=\sum_k \mathbf{T}_j(\mathbf{k}) \exp-2\pi i\mathbf{k}\cdot\mathbf{R}_j. \qquad (23)$$

Let several vectors $\mathbf{k}_1, \mathbf{k}_2, \cdots, \mathbf{k}_p$ give rise to solutions with corresponding matrices $(\lambda)_1, (\lambda)_2, \cdots, (\lambda)_p$. Actually [see Eq. (5)] the $(\lambda)$ matrix does not depend on a particular $\mathbf{k}$ vector. It is practical to think about the different $\lambda_i$ values ($i=1, \cdots, n$) which enter the $(\lambda)$ matrix as coordinates in a $\lambda$ space.[16] The compatibility condition that a solution depends simultaneously on the vectors $\mathbf{k}_1, \mathbf{k}_2, \cdots, \mathbf{k}_p$ is that the matrices $(\lambda)_1, (\lambda)_2, \cdots, (\lambda)_p$ must intersect in the same point $\lambda_i$ ($i=1, \cdots, n$). This condition is only *one* limitation to the existence of such "composed" modes.

Another one is the condition (6) for each $i$. The following three relations may be inferred from the reality of $\sigma_j$ and from Eqs. (6) and (23).

$$\left.\begin{array}{c} \mathbf{T}_j(\mathbf{k}) = \mathbf{T}_j^*(-\mathbf{k}) \\[4pt] \sum_k |\mathbf{T}_j(\mathbf{k})|^2=1 \\[4pt] \sum_{k\neq k'} \mathbf{T}_j^*(\mathbf{k})\mathbf{T}_j(\mathbf{k}') \exp-2\pi i(\mathbf{k}-\mathbf{k}')\cdot\mathbf{R}_j=0 \end{array}\right\}. \qquad (24)$$

Note that the conditions (24) must be satisfied by each component $j$ ($j=1, \cdots, n$). The conditions (24) constitute very severe limitations to the existence of composed modes.

The best proof is of course experiment. The overwhelming mass of neutron diffraction studies shows that in most magnetic configurations only one $\mathbf{k}$ vector is observed.

The author does not mean however that solutions with several $\mathbf{k}$ vectors should be discarded without a careful preliminary investigation nor does he mean that there is always a one-mode solution for an arbitrary $\mathbf{k}$ vector.

Indeed the 2 last conditions (24) read for a one-mode ($\pm\mathbf{k}_0$).

$$|\mathbf{T}_j(\mathbf{k}_0)|^2+|\mathbf{T}_j(-\mathbf{k}:)|^2=1$$
$$\mathbf{T}_j^2(\mathbf{k}) = \mathbf{T}_j^2(-\mathbf{k})=0. \qquad (25)$$

We have shown[14] that these relations are not fulfilled for all modes on $B$ sites of the cubic spinel when $\mathbf{k}=[110]$. Such modes are called incomplete.

A simple solution of Eq. (25) is

$$\mathbf{T}_j(\mathbf{k})=\tfrac{1}{2}(\mathbf{x}+i\mathbf{y}) \exp i\varphi_j, \qquad (26)$$

where $\varphi_j$ is a phase angle, $\mathbf{x}$ and $\mathbf{y}$ are orthogonal unit vectors. We leave it up to the reader to show from

---

[16] M. J. Freiser, Phys. Rev. **123**, 2003 (1961).

relations (23) and (26) that the angle between 2 spins $\sigma_i(\mathbf{R}_i)$ and $\sigma_j(\mathbf{R}_j)$ in a one-mode solution is given by

$$\Theta_{ij}(\mathbf{R}_i, \mathbf{R}_j) = 2\pi\mathbf{k} \cdot (\mathbf{R}_i - \mathbf{R}_j) + (\varphi_i - \varphi_j). \quad (27)$$

## Stability Conditions

Stability conditions for one-mode solutions are found by requiring that the matrix $[(\lambda_0) - \zeta(\mathbf{k}_0 + d\mathbf{k})]$, where $(\lambda_0)$ and $(\mathbf{k}_0)$ correspond to equilibrium, has positive roots for small but arbitrary variations of $d\mathbf{k}$. Examples are found in papers on spinels[3,17] and on the corindum structure.[8,14] The importance of establishing stability conditions is easily recognized. Susceptibility measurements and the determination of $\Theta_p$ and $T_N$ (paramagnetic Curie temperature and Néel temperature), for instance, give some information on linear combinations of exchange integrals. Stability conditions are expressed by inequalities between exchange integrals and limit further their parameter space.

## Occurrence of Ferromagnetism

There cannot be any ferromagnetism if the chemical and magnetic unit cell do not coincide in a one-mode solution. Indeed $\mathbf{T}(0) = 0$ implies $\sum_{R_i} \sigma_i(\mathbf{R}_i) = 0$.

## Coupling of Systems

Let $\{A\}$ be a set of $n_A$ equivalent magnetic atoms and $\{B\}$ another set of $n_B$ equivalent magnetic atoms such that $\{A\}$ is not equivalent to $\{B\}$. (Example: $A$ and $B$ sites in spinels.) The matrix $\zeta(k)$ of the coupled system $\{AB\}$ will have the form

$$\zeta(\mathbf{k}) = \left[ \begin{array}{c|c} \zeta_A(\mathbf{k}) & \zeta_{AB}(\mathbf{k}) \\ \hline \zeta_{BA}(\mathbf{k}) & \zeta_B(\mathbf{k}) \end{array} \right], \quad (28)$$

where $\zeta_A(\mathbf{k})$ and $\zeta_B(\mathbf{k})$ are square matrices having, respectively, eigenvectors $\mathbf{T}_A(\mathbf{k})$ and $\mathbf{T}_B(\mathbf{k})$. $\zeta_{AB}(\mathbf{k})$ is a rectangular $n_A \times n_B$ matrix. When the whole set of the vectors $\mathbf{T}_A(\mathbf{k})$ and $\mathbf{T}_B(\mathbf{k})$ is known, they form a complete system and the solution of Eq. (28) must be a linear combination of the $\mathbf{T}_A(\mathbf{k})$ and $\mathbf{T}_B(\mathbf{k})$. If we call $\phi_A$ and $\phi_B$ the square matrices formed by the eigenvectors of $\zeta_A(\mathbf{k})$ and $\zeta_B(\mathbf{k})$, a simple inspection of the matrix Eq. (29) tells us what modes of $\{A\}$ and $\{B\}$ may be combined.

$$\psi = \tilde{\phi}_A \zeta_{AB}(\mathbf{k})\phi_B. \quad (29)$$

The right combination must diagonalize the perturbation matrix $(p)$ of Eq. (30):

$$(p) = \left[ \begin{array}{c|c} 0 & \zeta_{AB} \\ \hline \zeta_{BA} & 0 \end{array} \right]. \quad (30)$$

Trivial eigenvectors combinations are in the $\zeta(0)$ matrix (28) those of the identity representations

$\mathbf{T}_{Af} = \underbrace{(1, 1, \cdots, 1)}_{n_A}$ and $\mathbf{T}_{Bf} = \underbrace{(1, 1, \cdots, 1)}_{n_B}$ of the separate modes. They occur in the combinations $\mathbf{T}_{Af} + \mathbf{T}_{Bf}$ and $\mathbf{T}_{Af} - \mathbf{T}_{Bf}$. The last combination is the well-known ferrimagnetic mode of Néel where the coupling energy stabilizes the otherwise unstable separate modes.

The method outlined here has been used by the author to derive the 32 modes of $\zeta(0)$ in garnets with 3 magnetic sublattices, labelled $(a)$, $(c)$, and $(d)$ and coupled by exchange forces only.[14] One has

$$\zeta(0) = \begin{bmatrix} \zeta_a & \zeta_{ad} & \zeta_{ac} \\ \zeta_{da} & \zeta_d & \zeta_{dc} \\ \zeta_{ca} & \zeta_{cd} & \zeta_{cc} \end{bmatrix}, \quad (31)$$

where $\zeta_a$ is of order 8, $\zeta_c$ and $\zeta_d$ are of order 12.

## THE ANISOTROPIC CASE

### Coupling of Modes

The most general interaction of order 2 between 2 spins in $\mathbf{R}$ and $\mathbf{R}'$ may be written[18]:

$$W_{RR'} = -2 \sum_{\alpha, \beta} A_{\alpha\beta}(\mathbf{R}, \mathbf{R}')\sigma_\alpha(\mathbf{R})\sigma_\beta'(\mathbf{R}')$$

$$= -2\sigma(\mathbf{R}) \cdot \mathbf{A}_{\mathbf{RR'}} \cdot \sigma'(\mathbf{R}'), \quad (32)$$

where $A_{\alpha\beta}$ is a 9-component tensor ($\alpha, \beta = x, y, z$) and $\mathbf{A}_{\mathbf{RR'}}$ a dyadic. The tensor is readily decomposed[8] into a scalar part which represents the usual isotropic interaction, a vector part which corresponds to the Dzialoshinsky-Moriya[6,19] (D-M) antisymmetric coupling by means of a vector $\mathbf{D}_{\mathbf{RR'}}$ and finally a symmetrical tensor part $\boldsymbol{\phi}_{\mathbf{RR'}}$ of zero trace which includes dipolar and pseudodipolar coupling and, for $\mathbf{R} = \mathbf{R}'$, the action of crystalline field.[8]

In dyadic notation one has

$$\mathbf{A}_{\mathbf{RR'}} = \hat{J}_{RR'}\mathbf{I} + (\mathbf{D}_{RR'} \wedge) + \boldsymbol{\phi}_{RR'} \quad (33)$$

and Eq. (5) reads

$$\lambda_R \sigma_R = \sum_{R'} \mathbf{A}_{\mathbf{RR'}} \cdot \sigma_{\mathbf{R'}}. \quad (34)$$

Equations (10) and (11) are again valid if it is understood that $\hat{J}_{RR'}$ is everywhere replaced by $\mathbf{A}_{RR'}$. It should be noted that in Eq. (34) the term with $\mathbf{R} = \mathbf{R}'$ due to crystalline field may exist too.

Although anisotropy forces are usually small, compared to isotropic exchange forces, they produce at least 2 noticeable effects. First they orient spins into a definite direction, detected by measurements of magneto-crystalline energies and by neutron diffraction. The second effect is the admixture of modes with the same $\mathbf{k}$ vector. The so-called weak ferromagnetism (w.f.) is an admixture of an af and a ferromagnetic mode

[17] T. A. Kaplan, Phys. Rev. 119, 1460 (1960); J. Appl. Phys. Suppl. 32, 13S (1961).

[18] H. Kramers, Physica 1, 182 (1935); J. H. Van Vleck, Phys. Rev. 52, 1178 (1941).

[19] T. Moriya, Phys. Rev. 120, 91 (1960).

TABLE I. Representations and spin modes group $Cmcm$.

|        | $x$ | $y$ | $z$ |
|--------|-----|-----|-----|
| $\Gamma_1$ | $A$ | $C$ | $G$ |
| $\Gamma_2$ | $F$ | $G$ | $C$ |
| $\Gamma_3$ | $G$ | $F$ | $A$ |
| $\Gamma_4$ | $C$ | $A$ | $F$ |

TABLE II. Representations and spin modes group $Pbnm$.

|        | $x$ | $y$ | $z$ |
|--------|-----|-----|-----|
| $\Gamma_1$ | $A$ | $G$ | $C$ |
| $\Gamma_2$ | $F$ | $C$ | $G$ |
| $\Gamma_3$ | $C$ | $F$ | $A$ |
| $\Gamma_4$ | $G$ | $A$ | $F$ |

But admixtures of 2 exclusively af modes may also exist and have been experimentally detected by Frazer[20] in various sulfates.

If we split $\zeta(\mathbf{k})$ into an isotropic and an anisotropic part

$$\zeta(\mathbf{k}) = \zeta_{is}(\mathbf{k})\mathbf{I} + \zeta_{an}(\mathbf{k}), \quad (35)$$

the condition that 2 modes $\mathbf{T}_I$ and $\mathbf{T}_{II}$ may be coupled is that

$$r = \mathbf{T}_I \cdot \zeta_{an}(\mathbf{k}) \cdot \mathbf{T}_{II} \neq 0. \quad (36)$$

More generally if $\phi$ is the matrix which diagonalizes $\zeta_{is}(\mathbf{k})$ ($\phi$ is the eigenvector matrix of the isotropic problem), let us construct the matrix

$$\psi = \tilde{\phi}\zeta_{an}\phi. \quad (37)$$

An off-diagonal term $\psi_{pq}$ indicates coupling of modes $\mathbf{T}_p$ and $\mathbf{T}_q$. The diagonalization of $\zeta(\mathbf{k})$ (35) is carried out by conventional methods.

One may distinguish between an antisymmetrical and a symmetrical w.f. according to whether it is due to the $\mathbf{D}$ vector or to crystalline field. We only indicate some results which may be obtained as well by the matrix as by the group theoretical method when chemical and magnetic unit cells are equal. In ternary crystal classes $(3,\bar{3},6,\bar{6})$ w.f. is unambiguously antisymmetrical and due to the $\mathbf{D}$ vector. In orthorhombic crystals, both effects may be present and can only be distinguished by their different temperature behavior.[21] In tetragonal crystals the nature of w.f. is easy to decide. For instance, in $NiF_2$ the only possible invariant combination of a ferrom. and af mode is symmetrical and necessarily due to crystalline field.

It may also be shown that antisymmetrical w.f. (see $\alpha$-$Fe_2O_3$) does not contribute to directional anisotropy whereas symmetrical w.f. does ($NiF_2$[22] and probably $Cr_2CuO_4$[23] and manganites).

The advantage of the matrix method is that it is able to investigate *separately* the effect of dipolar, pseudo-dipolar, D-M coupling, crystalline field interaction, that symmetry enters the matrix automatically and that the method is not limited to $\mathbf{k}=0$. Explicit expressions of anisotropy tensors and applications are given in reference 23.

We do however not want to minimize the results of the group theoretical method which, by the study of transformation properties, obtains elegant results with very little mathematics.

As an example, let us investigate the sulfates $SO_4M$ belonging to space group $Cmcm$. $M$ is a bivalent transition metal in the positions $0\,0\,0; 0\,0\,\frac{1}{2}; \frac{1}{2}\,\frac{1}{2}\,\frac{1}{2}; \frac{1}{2}\,\frac{1}{2}\,0$ which will be numbered 1, 2, 3, 4. If we note by $\mathbf{F}$ the ferromagnetic vector $\mathbf{T}_{IV}$ and relabel the vectors (15)

$$\mathbf{T}_I = G; \quad \mathbf{T}_{II} = A; \quad \mathbf{T}_{III} = C, \quad (38)$$

according to the nomenclature used by Wollan and Koehler,[24] the following Table I may be constructed. Here vector components on the same line belong to the same one dimensional representation $\Gamma_j$ ($j=1, 2, 3, 4$), in other words transform in the same way in all symmetry operations. For instance, in the second line of Table I, $F_x$, $G_y$, and $C_z$ change sign in the operation of the binary helicoidal axis in $(\frac{1}{2}y\frac{1}{4})$, do not change sign in the inversion operation and so on, so that their products taken 2 by 2 like $F_xG_y$ or $G_yC_z$ are invariants of order 2.

In $CoSO_4$, Frazer[20] finds a superposition of $G$ and $C$. Table I tells that $G$ and $C$ modes can only be associated in the Hamiltonian as products $C_yG_z$ or $G_yC_z$ (in which latter case a w.f. may exist). There is also a high temperature form of some of the sulfates with the same 4-point arrangement of $M$, but belonging to space group $Pbnm$ for which Table II may be constructed.[7,23] (It is easily seen that Table II differs from Table I by replacing $C$ by $G$ and vice versa.) Let us mention that Table II is also valid for rare earth-iron perovskites like $FeErO_3$ and correctly predicts[7,23] the association of w.f. and a.f. modes. It has been found[25] that $G_x$ is associated with $F_z$ at low temperatures and $F_x$ with $G_z$ at high temperatures.

Of course these results may be obtained by the matrix method, too, which gives direct insight into the elementary interactions involved.

As a conclusion, the author would be happy that readers of this paper would recognize at least the great importance of crystallographic symmetry in magnetism whenever they handle the group, theoretical, or the matrix method.

### ACKNOWLEDGMENTS

The author is indebted to Dr. Corliss and Dr. Turov for numerous discussions and to Professor Néel for his encouragement.

[20] B. C. Frazer, Acta Cryst. **13**, 1088 (1960) and B. C. Frazer and P. Y. Brown, Brookhaven Natl. Lab. Rept. No. 5681 (1961).

[21] For instance $\mathbf{D}_{RR'}$ depends on exchange integrals, whereas the crystalline field does not, so that w.f. due to $\mathbf{D}_{RR'}$ will "last" until the Curie-Néel temperature.

[22] T. Moriya, Phys. Rev. **117**, 635 (1960).

[23] E. F. Bertaut (to be published).

[24] E. O. Wollan and W. C. Koehler, Phys. Rev. **100**, 545 (1955).

[25] R. M. Bozorth, V. Kramer, and J. P. Remeika, Phys. Rev. Letters **1**, 3 (1958).

JOURNAL OF APPLIED PHYSICS     SUPPLEMENT TO VOL. 33, NO. 3     MARCH, 1962

# Magnetic Transitions in Cubic Spinels

N. Menyuk, A. Wold, D. Rogers, and K. Dwight

*Lincoln Laboratory,\* Massachusetts Institute of Technology, Lexington 73, Massachusetts*

Magnetization curves of the cubic spinels $MnCr_2O_4$, $CoCr_2O_4$, $MnV_2O_4$, and $CoV_2O_4$ have been obtained between 4.2°K and their Curie points. With the possible exception of $MnV_2O_4$, all the curves indicate the presence of a magnetic transition within this temperature range. Although all these materials have been reported to be normal spinels, inconsistencies between our results and the recent theory of spin configurations in normal cubic spinels led to further experiments which established that $CoV_2O_4$ is not, in general, normal.

Theory suggests the presence of two types of transition in $MnCr_2O_4$ and $CoCr_2O_4$, but only one has been observed in each case. The temperatures at which the observed break in the magnetization curves of these materials occurs can be interpreted by ascribing them to different types of transitions. The interpretation of the $MnCr_2O_4$ magnetization curve is compatible with the neutron diffraction results of Corliss and Hastings.

## INTRODUCTION

A STUDY has been made of the magnetic properties of the cubic spinels $MnCr_2O_4$, $CoCr_2O_4$, $MnV_2O_4$, and $CoV_2O_4$ from the Curie point to 4.2°K. These materials have all been reported to be normal spinels.[1-3] Their magnetic properties are therefore of interest since they should be interpretable in terms of the recent theory of spin configurations in normal spinels in the ground state[4,5] ($T=0$) and at the Curie point ($T=T_c$). This choice of materials also permits a check on the internal consistency of the results, since each substitution (i.e., Co → Mn or Cr → V) is duplicated. Pure chromite samples were prepared by the precursor method as described in detail by Whipple and Wold.[6] The vanadites were prepared by first grinding mixtures of CoO and $V_2O_3$ or MnO and $V_2O_3$ in a nitrogen dry box. The cobalt-vanadium oxide mixture was then heated to 1100°C in an evacuated sealed capsule, while the manganese-vanadium oxide mixture was heated to 1100°C in hydrogen atmosphere. Chemical analysis of the resulting samples indicated that the ratio of $B$-site cations to $A$-site cations was within 0.3% of 2:1.

## THEORETICAL BACKGROUND

The study of the ground state of the classical Heisenberg exchange energy[4,5] in normal spinels indicates that for a range of $B$-$B$ exchange interactions large enough to destabilize the collinear (Néel) spin configuration, the stable configuration is one in which the spin vector of each of the six sublattice sites describes a ferrimagnetic (conical) spiral. This spiral can be precisely defined by a single parameter $u$, which involves the near-neighbor exchange interactions and the atomic spins on each site. (Increasing $u$ corresponds to increasing the relative value of the $B$-$B$ interaction.) Although this spiral can be the ground state only for $u < 1.3$, the striking agreement between this defined spiral and the spin configuration determined experimentally by Corliss and Hastings in manganese chromite,[7] where $u = 1.6$, indicates that the deviation from the defined spiral is fairly small for a finite distance beyond $u = 1.3$.

In addition to the ground state problem, the spin configuration at the Curie temperature has also been considered.[5] It was found that for $u < 2.177$ the spin configuration is collinear. For $u > 2.177$ the material is antiferromagnetic. Thus for some range of $u$ ($1.3 < u < 2.177$), the theory predicts that the spin configuration will change from collinear to the defined spiral to a modified spiral (of much greater complexity) on reducing the temperature from $T_c$ to 4.2°K.

## EXPERIMENTAL RESULTS AND DISCUSSION

The magnetization curves of these materials, as measured on a vibrating coil magnetometer,[8] are shown in Fig. 1. Magnetization curves of $MnV_2O_4$ taken at lower field strengths indicate that the "hump" in the curve is due to a lack of a saturation field. However, in $CoV_2O_4$, the shift between two essentially flat portions of the curve, which occurs between 45° and 70°K, appears to represent a gradual change from one magnetic state to another.

The magnetic moments of these materials at 4.2°K indicate that none of them have a Néel-type spin configuration at low temperature. Furthermore, the value of $u$ computed for these samples from the 4.2°K magnetic moment values[9] indicates that $u > 1.3$ in all

\* Operated with support from the U. S. Army, Navy, and Air Force.

[1] E. J. W. Verwey and E. L. Heilmann, J. Chem. Phys. **15**, 174 (1947).

[2] H. M. Richardson, F. Bell, and G. R. Rigby, Trans. Brit. Ceram. Soc. **53**, 376 (1954).

[3] A. Burdese, Ann. Chim. (Rome) **47**, 827 (1957).

[4] T. A. Kaplan, K. Dwight, D. Lyons, and N. Menyuk, J. Appl. Phys. **32**, 13S (1961).

[5] D. Lyons, T. A. Kaplan, K. Dwight, and N. Menyuk (to be published).

[6] E. Whipple and A. Wold, J. Inorg. and Nuclear Chem. (to be published).

[7] L. Corliss and J. Hastings, J. Appl. Phys. **33**, 1138 (1962), this issue.

[8] K. Dwight, N. Menyuk, and D. Smith, J. Appl. Phys. **29**, 491 (1958).

[9] These values were computed using the spin-only atomic moments of 5, 3, and 3 $\mu_B$, respectively, for $Mn^{2+}$, $Co^{2+}$, and $Cr^{3+}$. However, $S_B$ for $V^{3+}$ was taken as 1.2 $\mu_B$. This value is based on paramagnetic measurements taken with our $MnV_2O_4$ sample. The $V^{3+}$ contribution was determined after subtracting the tetrahedrally-coordinated $Mn^{2+}$ contribution as obtained in spinel structures by P. F. Bongers [Thesis, University of Leiden (1957)].

the samples. Substitution of manganese for cobalt was found to increase the value of $u$ in the vanadites while decreasing it in the chromites. This result is surprising, particularly so in view of the fact that this substitution yields reasonably consistent changes in the Curie temperatures and in the size of the unit cell.[10] This apparent contradiction, coupled with the success of the theory in manganese chromite, led us to question the assumption of normality in these samples, particularly since the experimental data upon which this assumption rests are far from conclusive. On theoretical grounds,[11] the cobalt vanadite (III) sample is the one most likely to be partially inverse. The $CoV_2O_4$ sample was therefore heated to 1100°K and slowly cooled in an evacuated tube. The value of $M_0$ for the annealed sample was found to be about 15% lower than that of the original quenched sample, and the apparent transition between 45° and 70°K was no longer present. These results appear to confirm the fact that $CoV_2O_4$ is not, in general, a normal spinel.

In both $CoCr_2O_4$ and $MnCr_2O_4$, a transition was indicated by a sharp change in the slope of the magnetization curves. The difference in the Curie temperature of $MnCr_2O_4$ and in the value of $M_0$ for $CoCr_2O_4$ between our values and those previously reported[12,13] is probably due to the increased purity of our samples. On warming $MnCr_2O_4$ from 4.2°K, the magnetization curve experienced an anomalous rise in the moment after an initial dip, as shown in Fig. 1. We have no explanation for this effect. However, the cooling curve evidenced a distinct break at 18°K, which is taken as due to a *collinear → defined spiral* transition.[14] This transition temperature ($T_t$) is near the theoretical one predicted by assuming a linear variation from $T_t=0$ at $u=8/9$ to $T=T_c$ at $u=2.177$. This interpretation is compatible with (but *not* definitively verified by) the results of Corliss and Hastings,[7] who observed the onset of a ferrimagnetic spiral below this temperature. The fact that the

[10] R. Arnott, M.I.T. Lincoln Laboratory Quarterly Progress Reports (October 15, 1960) and (July 15, 1961).

[11] A. Miller (to be published).

[12] T. R. McGuire and S. W. Greenwald, *Solid State Physics in Electronics and Telecommunications* (Academic Press Inc., New York, 1960), Vol. 3, Pt. 1, pp. 50–70.

[13] P. L. Edwards, Phys. Rev. **116**, 294 (1959).

[14] P. L. Edwards' failure to observe this transition (reference 13) was probably due to the fact that his measurements were made on a point by point basis. Our magnetization measurements were made as a continuous function of temperature, a procedure which is extremely sensitive to abrupt changes in slope even when the magnitude of the change in magnetization is small. Edwards explained his magnetization curve in $MnCr_2O_4$ on the basis of the Yafet-Kittel triangular spin arrangement. However, it has since been shown that this configuration is not stable in cubic spinels [see T. A. Kaplan, Phys. Rev. **116**, 888 (1959) and reference 4].

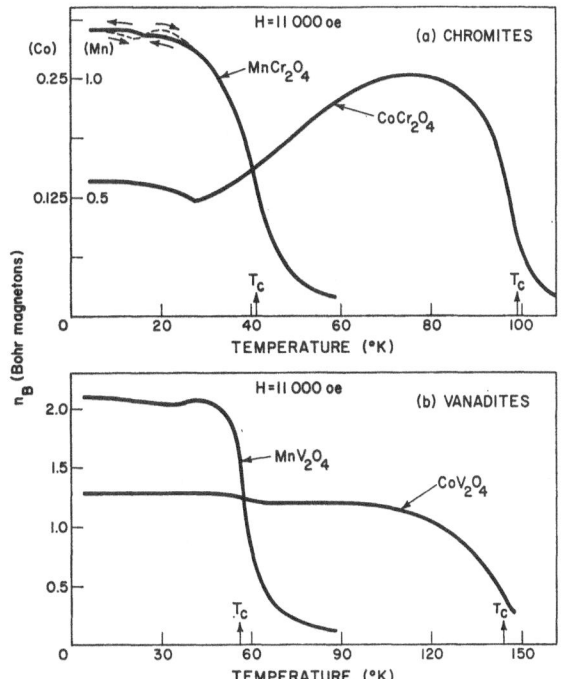

FIG. 1. Magnetization curves taken in field of 11 000 oe. The Curie points shown are based on determinations made with low field values ($H \cong 100$ oe).

normalized transition temperature ($T_t/T_c$) in $CoCr_2O_4$ is lower than in $MnCr_2O_4$, although the former material has the higher value of $u$ ($\sim 2$), indicates that the break in the $CoCr_2O_4$ curve at 27°F is caused by the lower-temperature *defined spiral → modified spiral* transition. Furthermore, on the basis of the curve given in Fig. 1, it seems probable that the Néel mode does not persist down to 27°K, since then the resultant curve, if the spins were constrained to remain collinear, would have to approach the value $n_B=3 \mu_B$ at 4.2°K. This appears unlikely.

However, a note of caution should be expressed regarding the above explanation based on two types of transitions. We have investigated the magnetization curves of a number of other chromites and vanadites, both normal and inverse. In almost all cases, one break in the curve was observed, but we have never observed two transitions in a single sample. There is no *a priori* way at present of predicting whether a given transition will be observable, and it appears unlikely that a molecular field calculation will agree with our data. In view of the failure of magnetization measurements to supply direct evidence of double transitions, it is felt that the resolution of this question will require further neutron diffraction investigations.

JOURNAL OF APPLIED PHYSICS    SUPPLEMENT TO VOL. 33, NO. 3    MARCH, 1962

# Thin Films–2

## H. Rubinstein, *Chairman*

## Ferromagnetic Resonances in Thin Films*

D. Chen and A. H. Morrish

*University of Minnesota, Minneapolis 14, Minnesota*

The ferromagnetic resonance spectra of thin films of Permalloy and other materials have been investigated. Frequently more than one maximum in the resonance line is observed. The evidence from various experiments indicates that films with more than one resonance peak are stratified. A simple model in which two layers of film are coupled via the dipole-dipole interaction is considered.

FERROMAGNETIC resonance absorption in thin films has been studied at 24 kMc with the steady magnetic field applied parallel to the plane of the film. The majority of the investigations have been made on Permalloy films. Some Permalloy films, especially those less than 400 A thick, exhibit one resonance peak. Films of greater thickness frequently possess more than one peak, the usual number being two. The observation of two resonance peaks in some Permalloy films has been reported earlier by van Itterbeek, *et al.*[1] The spin wave model does not account for these extra resonances. When a film with more than one peak is partially etched, some of the peaks disappear. It therefore appears likely that there is a stratification in the films, formed either in the evaporation process, or by oxidation of the film surface.

Measurements have been made on Permalloy films of various thicknesses, prepared under the conditions necessary to produce rotatable anisotropy.[2] Such films are likely to be stratified. The resonance absorption experiment reveals that for the films that are anisotropic (about 1000 A thick), there is a secondary peak on the

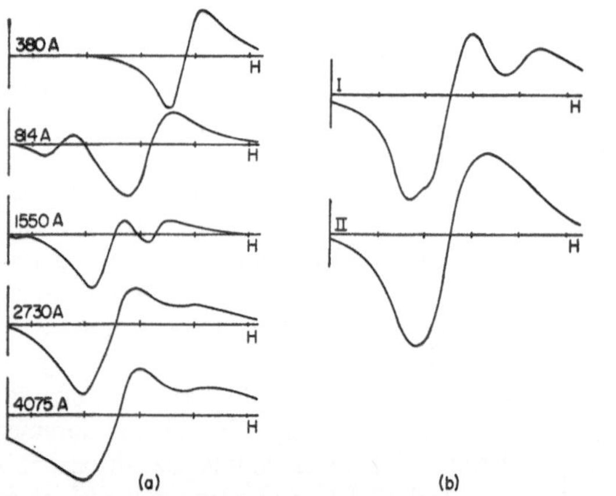

Fig. 1. The differentiated resonance line of some Permalloy films.

low field side of the main peak. For films that possess rotatable anisotropy (>2500 A), the secondary peak is on the high field side of the main peak. This behavior is illustrated in Fig. 1(a). Another set of films with rotatable anisotropy, and with identical initial thicknesses of 3100 A, but etched from the top surface with diluted Mirrofe solution to various thicknesses, has also been studied. Similar results have been obtained. However, the secondary peak shifts from the high field side to the low field side of the main peak at a thickness of around 1800 A. The origin of the shift of the secondary peak is not clearly understood. Nevertheless, these results further confirm that the rotatable films are stratified.[3]

In order to examine this stratification idea, a simple model has been considered. It is assumed that there is a dipole-dipole coupling between two ferromagnetic layers. Without the coupling, the internal magnetic energy per unit area of each layer can be expressed as follows:

$$W = \frac{d}{2}[M_s H_0 l^2 + M_s(H_0 + 4\pi M_s)m^2],$$

where $d$ is the thickness of the layer, $M_s$ is the saturation magnetization, and $H_0$ is the applied dc magnetic field, which is in the $z$ direction. The coordinate system is chosen such that the film is in the $x$-$y$ plane and the direction cosines of the magnetization vector are $l$, $m$, $n$. Since the resonance field is around 5000 oe, the anisotropy energy is neglected and the use of the approximation $n \approx 1 - \frac{1}{2}(l^2 + m^2)$ is justified. With this energy term, the equation of motion yields two resonance fields, one for each layer, given by the relation[4];

$$(\omega/\gamma)^2 = H_0(H_0 + 4\pi M_s).$$

If each layer has the same $M_s$, the resonance fields will be, of course, identical.

Now suppose the two layers are coupled by the dipole-dipole interaction. There will be an added energy term associated with this coupling between the layers of

* This research was supported by the U. S. Air Force through the Air Force Office of Scientific Research (ARDC).

[1] A. van Itterbeek, G. Forrez, J. Smits, and J. Witters, J. phys. radium **21**, 81 (1960); also see P. E. Tannenwald and M. H. Seavy, Jr., Phys. Rev. **105**, 377 (1957), Figs. b and d.
[2] R. J. Prosen, J. O. Holmen, and B. E. Gran, J. Appl. Phys. **32**, 915 (1961).
[3] R. J. Prosen, J. O. Holmen, B. E. Gran, and T. J. Cebulla, J. Appl. Phys. **33**, 1150 (1962).
[4] C. Kittel, Phys. Rev. **71**, 270 (1947).

the form

$$W_c = cl_1l_2 - c'm_1m_2,$$

where $c$ and $c'$ are coupling constants, and depend on the geometry and the magnetic state of the film. When this term is included in the equations of motion, a quartic equation in $H_0$ is obtained, two roots of which are negative. Therefore, in general, for a fixed frequency, there are two resonance fields. In addition, a change in the coupling coefficients will shift the value of the resonance fields.

Measurement of a normal Permalloy film 2000 A thick, gives the differentiated resonance absorption curve shown in Fig. 1(b) I. After 40 sec of etching, the film thickness was reduced to 1000 A. The resonance measurement now gives the curve shown in Fig. 1(b) II. It is clear that the high field resonance peak has disappeared.

In order to study the dipole-dipole coupling mechanism further, sets of films were prepared in which two layers of Permalloy films are separated by a layer of SiO of various thicknesses. Measurements on some two layer films showed that the resonance peaks shifted as a function of the thickness of the SiO layer; data from one set of films are shown in Fig. 2. This result is in accord with our theoretical model. However, the coupling constants obtained experimentally are small, and in order to account for this it is necessary to assume that the magnetization of the film layers is nonuniform. The absorption curve of other two layer films is complicated with four or more peaks being observed. Analysis of such curves is therefore difficult. It is likely that the multiple peaks are the result of stratification within each layer of the two layer films.

Some measurements have also been made at liquid $N_2$ temperature. The peaks are shifted as would be expected in accordance with the temperature dependence of $M_s$.

Films with two layers, one Co and the other Fe in one

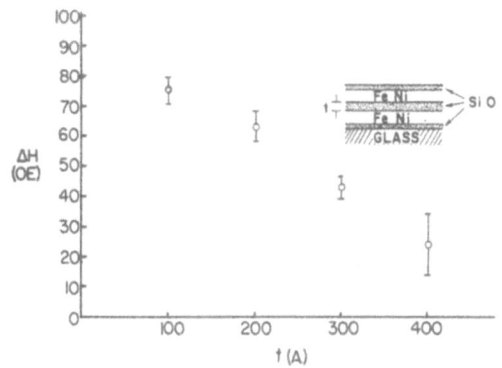

FIG. 2. The shift of one resonance peak as a function of the separation of the spacing between two layers of Permalloy film. The shift is measured relative to the position of the peak in a single layer film.

case and one Ni and the other Fe in another case, have also been investigated. The resonance peaks are found to shift as compared to the peaks for a single-layer film of one of the materials, again in accord with our simple model.

To summarize, the results provide evidence for the occurrence of stratification in many thin films of ferromagnetic materials. The properties of the films are influenced by this stratification. In particular, as a result of coupling, the ferromagnetic resonance line is modified. It appears possible that such films may be useful as a coupling device in circuit applications.

We wish to thank R. J. Prosen and J. O. Holmen of the Minneapolis-Honeywell Research Center, Hopkins, Minnesota, for supplying us with most of the films used in this investigation, as well as for helpful discussions. We also wish to thank S. W. Reubens of Remington-Rand Univac, St. Paul, Minnesota, for manufacturing some additional films for us. We are indebted to W. F. Brown, Jr., for some helpful discussions, and to C. W. Searle for taking some of the resonance measurements.

JOURNAL OF APPLIED PHYSICS    SUPPLEMENT TO VOL. 33, NO. 3    MARCH, 1962

# Stress Effects on the Magnetic Properties of Evaporated Single-Crystal Nickel Films

J. F. Freedman

*IBM Research Center, Yorktown Heights, New York*

Nickel single-crystal films grown epitaxially on NaCl at elevated temperatures shown anomalously large values for the crystalline anisotropy constant, $K_1$, as well as for $K_\perp$, the perpendicular anisotropy (i.e., the anisotropy resulting from rotation out of the plane of the film). These values decrease to normal bulk values when the films are floated off the rocksalt.

It has been shown by x-ray diffraction that the as evaporated films exist in a highly strained condition, resulting in a tetragonal distortion of the cubic symmetry. The parameters of the unit cell change from $a_0 = 3.524$ A for unstrained nickel to $a = b = 3.500$ A in the plane of the film and $c = 3.546$ A normal to the film. It is also shown that the stress causing this strain is elastic in character and is relieved by floating the film off the rocksalt.

A calculation using bulk elasticity data yields a planar compressive stress $\sigma = -1.2 \times 10^{10}$ d/cm². The magnitude and direction of the stress indicates it results from the difference in thermal expansion between the nickel film and the NaCl substrate.

This externally applied stress system influences the magnetic state of the film by contributing to the total energy of the system through a magnetoelastic interaction. Employing the five-constant magnetostrictive equation, it is shown that a planar compressive stress of this magnitude does result in an anomalously high crystalline anisotropy value, as well as an anomalously high value of the anisotropy perpendicular to the plane of the film.

## INTRODUCTION

THERE are three major sources for the introduction of stresses and strains into an evaporated film: (1) a structural contribution arising from any defect distribution caused during deposition or subsequent thermal annealing; (2) an epitaxial misfit resulting during epitaxial growth of a metal film on a substrate of different atom spacings; and (3) strains arising from the constraints imposed by the thermal contraction of the substrate.

Chikazumi[1] has associated the anomalous part of $K_1$ (the value of the fourth-order crystal anisotropy constant exceeding that of the bulk material) in epitaxial single-crystal films to the presence of a tensile stress, possibly arising from the epitaxial misfit between the NaCl substrate and the Ni film. Similarly, Kuriyama et al.[2] indicate the need of a tensile stress to explain the value of the magnetization obtained from ferromagnetic resonance work on films grown on rocksalt. In the case of a metal film evaporated on NaCl at elevated temperatures, however, the stress contribution from the differences in thermal contraction would be quite large, resulting in a residual compressive stress. This work was therefore conducted to investigate the magnitude and the direction of the residual stress in pure single-crystal nickel films evaporated on rocksalt, and the effect of this stress on the magnetic properties.

### EXPERIMENTAL PROCEDURE

Nickel films were prepared by deposition in a vacuum of $10^{-5}$ mm Hg onto the heated (100) cleavage surface of NaCl by the method employed by Chikazumi[1]; namely, preheating of the NaCl at 500°C and evaporation at temperatures in excess of 200°C.

Magnetic anisotropy measurements were made on an automatic torque magnetometer[3] which is capable of detecting 0.002 d-cm. Anisotropy measurements in the plane of the film and perpendicular to the plane of the film were made with the film as evaporated on the rocksalt. The samples were then floated off the rocksalt on water, picked up on glass slides, and the properties remeasured.

Structural measurements were made using electron diffraction and transmission micrograph techniques on specimens floated onto 200 mesh Cu grids. X-ray diffraction measurements were obtained on specimens as evaporated, and after floating.

### EXPERIMENTAL RESULTS

Anomalously large values of $K_1$ (the fourth-order crystal anisotropy constant) and $K_\perp$ (the anisotropy constant resulting from rotation of the magnetization out of the plane of the film) were obtained as has previously been reported by Chikazumi[1]. The values decreased to the theoretical bulk values when the films were floated off and picked up on glass. It was also observed, however, that films exposed to air for a period of time experienced a decay of the anisotropies to normal bulk values. Accompanying the decay was a change in surface morphology in which the film became severely wrinkled. This was interpreted as indication of the fact that the nickel films were under a considerable elastic compressional stress, rather than tension, and this stress is relieved when the film is physically removed from the constraining substrate, or when sufficient water vapor is absorbed during exposure to atmospheric conditions to accomplish the same result.

To detect this stress, x-ray traces were taken using $CoK\alpha$ radiation and a Geiger tube diffractometer. For all nickel lines scanned a NaCl line in the corresponding $\theta$ vicinity was also scanned for purposes of supplying an internal standard for the nickel spacings. This was easily accomplished, since the films were thin enough to allow penetration of the beam through the nickel film and into the substrate. A set of typical results are tabulated in Table I.

[1] S. Chikazumi, J. Appl. Phys. **32**, 81S (1961).
[2] M. Kuriyama, H. Yamanouchi, and S. Hosoya, J. Phys. Soc. Japan **16**, 701 (1961).
[3] E. L. Boyd, IBM J. Research and Develop. **4**, 116 (1960).

TABLE I.

| (hkl) plane[a] | | | $a_0$ | |
|---|---|---|---|---|
| Ni | NaCl | $\theta$ | Ni[b] | NaCl[c] |
| (002) | | 30.29 | 3.5460 | |
| | (004) | 39.56 | | 5.6408 |
| (022) | | 45.92 | 3.5218 | |
| | (333)⎫ | 55.48 | | 5.6404 |
| | (511)⎭ | | | |
| (222) | | 61.68 | 3.5196 | |
| | (044) | 63.76 | | 5.6409 |

[a] The axis of reference has been chosen so that the $x$-$y$ plane represents the plane of the film.
[b] $a_0$ NBS Ni = 3.5238 A at 26°C.
[c] $a_0$ NBS NaCl = 5.6402 A at 26°C.

There is excellent agreement for the cell constant obtained from the various NaCl planes, implying a relative error of 1 part in 10 000. The error for the nickel determinations is somewhat larger because of the considerable peak broadening observed. As a result the estimated relative error in the nickel values is 1 to 2 parts per 1000, which is, however, much smaller than the observed deviations. A reasonable explanation of the observed parameter values is obtained by assuming a tetragonal distortion of the unit cell resulting from an isotropic planar compressive stress and a Poisson's extension normal to the film, resulting in a unit cell $a = b = 3.500$ A, $c = 3.546$ A.

Since this stress must be an elastic stress which is completely removed when the film is relieved of the constraining NaCl substrate, specimens were mounted on the diffractometer and a trace of the (002) planes taken, yielding the values of Table I. Without disturbing the film in its mounting, moist air was blown over the surface. An immediate rescan of the (002) line revealed an increase of $\theta$ to 30.50, resulting in a decrease in cell parameter $a_0$ to 3.5240, the normal bulk nickel parameter. Observation of the surface indicated the typical wrinkling; magnetic anisotropy measurements before and after the x-ray traces yielded a decrease of the values to the normal bulk values.

## DISCUSSION

The stress $\sigma_i$ present in the plane of the film can now be calculated from classical elasticity theory for a cubic system, noting that for a film supported on a substrate $\sigma_z = 0$ (where the $z$ axis is normal to the plane of the film) and the strain in the $z$ direction $\epsilon_z$ is therefore the effective Poisson's contraction set up normal to both $\sigma_x$ and $\sigma_y$. Since $\epsilon_x = \epsilon_y = 0.0057$ cm/cm from the x-ray data, we have

$$\sigma_x = \sigma_y = \epsilon/(S_{11} + S_{12}). \quad (1)$$

Substituting the literature values for the bulk compliances[4] $S_{11} = 0.799 \times 10^{-12}$ cm²/d, $S_{12} = -0.312 \times 10^{-12}$

cm²/d, one obtains for the stress $\sigma = -1.16 \times 10^{10}$ d/cm², a compressive stress in excess of the yield point for normal bulk materials. Such large elastic stresses are, however, in accord with other works in thin films.[5-10] From a consideration of the differences in thermal expansion coefficients of NaCl and Ni, it is found a temperature differential of 250°C is needed to account for the observed strain. This temperature differential is always exceeded when one epitaxially evaporates Ni on NaCl. All films have therefore been deformed past their elastic limit, any further increase in stress being relieved by plastic strain. As a result, $K_1$, $K_\perp$, and the cell parameters appear temperature independent.

The abnormally large values of $K_1$ and $K_\perp$ can now be explained by this isotropic compressive stress acting in the plane of the film. In the case of the fourth-order crystalline anisotropy constant, it can be shown that the contribution to the magnetic anisotropy energy due to the applied stress[11] is

$$U_\sigma = -\sigma\lambda_s = K_{1\sigma}\alpha_i^2\alpha_j^2, \quad (2)$$

where $\lambda_s$ is obtained from the five-constant magnetostrictive equation, and $\alpha_i^2\alpha_j^2$ are the direction cosines between the magnetization and the crystallographic axis. Putting in values for the direction cosines for the stress and crystal symmetry observed yields

$$K_{1\sigma} = \sigma(\tfrac{2}{3}h_4 - 2h_3) \quad (3)$$

but

$$K_{1\sigma} = K_{1(meas)} - K_{1(bulk)} = -5 \times 10^4 \text{ erg/cc.} \quad (4)$$

Employing the values of the magnetostrictive constants of Bozorth and Hamming,[12] namely, $h_3 = -2.8 \times 10^{-6}$, $h_4 = -7.5 \times 10^{-6}$, the necessary value of $\sigma$ required to explain the anomalous value of $K_1$ is $-8 \times 10^{10}$ d/cm². Although this value is larger than the one estimated from elasticity, the agreement is quite reasonable since this calculation is based on the difference of two large terms, both of which are subject to considerable experimental error. Earlier data, in which it was assumed $h_3 = 0$, were used by Chikazumi[1] and led him to the conclusion that a tensile stress was required. Hence, the most important part of this calculation is that a contribution to $K_1$ can result from a planar stress system.

[4] W. Boas and J. K. Mackenzie, Prog. Met. Phys. 2, 90–120 (1950).

[5] J. W. Beams, Structure and Properties of Thin Films (John Wiley & Sons, Inc., New York, 1959), p. 183.
[6] J. W. Menter and D. W. Pashley, Structure and Properties of Thin Films (John Wiley & Sons, Inc., New York, 1959), p. 111.
[7] D. W. Pashley, Nature 182, 296 (1958); Phil. Mag. 4, 316, 324 (1959).
[8] J. E. Gordon, Growth and Perfection of Crystals (John Wiley & Sons, Inc., New York, 1958), p. 219.
[9] D. M. Marsh, J. Sci. Instr. 36, 165 (1959).
[10] C. E. Neugebauer, J. Appl. Phys. 31, 1096 (1960).
[11] P. K. Baltzer, Phys. Rev. 108, 580 (1951).
[12] R. M. Bozorth and R. W. Hamming, Phys. Rev. 89, 865 (1953).

To a first-order approximation, the anomalous part of $K$ ($\Delta K_\perp$) resulting from the planar stress can be shown to be

$$\Delta K_\perp = \sigma h_1.$$

Since $h_1 = -68.8 \times 10^{-6}$, a value of stress $\sigma = -3 \times 10^{10}$

d/cm$^2$ is required to explain the experimental values, $\Delta K_\perp = 2 \times 10^6$ ergs/cc.

## ACKNOWLEDGMENT

The author is indebted to Dr. N. Stemple for assistance in obtaining the x-ray results.

---

JOURNAL OF APPLIED PHYSICS          SUPPLEMENT TO VOL. 33, NO. 3          MARCH, 1962

# Stratification in Thin Permalloy Films

R. J. Prosen, J. O. Holmen, B. E. Gran, and T. J. Cebulla
*Minneapolis-Honeywell Research Center, Hopkins, Minnesota*

Permalloy films have been studied to determine composition variations which exist in a direction normal to the substrate surface. These films are formed by a noninterrupted deposition process. They are classified according to the amount of oxygen present on the substrate prior to deposition. If normal care is taken to produce a clean glass substrate, a thin layer of $\alpha Fe_2O_3$ and $NiFe_2O_4$ is produced at the film-substrate interface. Those films produced on a substrate saturated with oxygen have a thick negative magnetostrictive alloy layer adjacent to the substrate. This nickel-rich alloy presumably results from the diffusion of oxygen through the depositing films and the formation of iron oxide and nickel ferrite. In addition to the chemical analysis of the films, evidence to support the idea of stratification is found in the etch experiments, resonance data, saturation magnetization measurements, and cross section analysis.

THE films studied can be classified into two categories according to the amount of oxygen present on the substrate prior to deposition. The first class, "normal" films, are produced on glass substrates which have been detergent cleaned, vapor degreased, and ion bombarded prior to deposition. Pressure during deposition is in the $10^{-5}$ to $10^{-6}$ torr range. Evaporation at a rate of 1000 A/min takes place from an 83-17 nickel-iron alloy resistance heated in a recrystallized alumina (Morganite) crucible onto a glass substrate held at a temperature of about 200°C.

When normal films are etched with a dilute Mirrofe solution, a residue remains on the substrate surface. A powerful etchant such as aqua regia is required to etch this residue. Interferometric techniques indicate that this residual film is less than 100 A thick. Electron diffraction identifies this residue as $\alpha Fe_2O_3$ and $NiFe_2O_4$. This layer is the likely cause of spin wave pinning at the substrate.[1] Formation of these oxides at the substrate will leave a nickel enriched alloy layer. The final 300 to 400 A etch at a higher rate; this is presumably evidence of stratification.

Polarographic analysis[2] shows the average composition of these normal films to be about 79% Ni even though the melt contained an 83% alloy. Since experimental error is under one percent, this result is difficult to reconcile with other work[3] where an 81% film was reported.

The second class of films, the "high oxygen content"

films, are produced in a manner similar to that described for "normal" films except that after ion bombarding and prior to the deposition, the pressure in the vacuum chamber is increased to $10^{-3}$ torr for a few minutes with dry air or oxygen. The pressure is reduced and the deposition again takes place at $10^{-5}$ to $10^{-6}$ torr. It is assumed that in this process, the substrate becomes saturated with oxygen. These films possess rotatable anisotropy as described previously.[4]

Polarographic analysis shows this class of films in the thickness range of 3000 to 3500 A contains 82 to 83% Ni. In the analysis of thicker films, the average nickel content was found to be less than this. Analysis of an 8930-A film in two parts shows the top 5660 A of this film to be 78% Ni; the bottom 3270 A contains 83% Ni. Torsion magnetometer[5] measurements of saturation magnetization agree with these results. The $B_s$ value for the lower layer is less, indicating a nickel-rich alloy and/or a higher oxide content.

Presumably, the nickel-rich alloy results from the formation of $\alpha Fe_2O_3$ and $NiFe_2O_4$. These oxides are not detected polarographically since they do not dissolve in $0.3M$ nitric acid. In agreement with this assumption, the total weight of the high oxygen content films obtained by direct microbalance weighing is invariably greater than the total weight determined polarographically. In the case of normal films, the total weights obtained by the two methods do agree very well.

Other evidence to support the presence of oxides in

[1] C. F. Kooi, International Conference on Magnetism and Crystallography, September, 1961, Kyoto, Japan.
[2] S. L. Phillips and E. Morgan, Anal. Chem. **31**, 1467 (1959).
[3] D. O. Smith, J. Appl. Phys. **30**, 264S (1959).

[4] R. J. Prosen, J. O. Holmen, and B. E. Gran, J. Appl. Phys. **32**, 91S (1961).
[5] C. A. Neugebauer, Phys. Rev. **116**, 1441 (1959).

the nickel-rich layer of high oxygen films is found in cross section studies performed with the aid of an electron microscope. The films were stripped from their substrates, mounted in Bakelite, polished, and etched with acid ferric chloride to bring out details. These studies show a large concentration of etch pits near the substrate and very few near the film surface. This means there exists a higher concentration of high energy areas in this lower portion of the film and could be interpreted to mean a higher concentration of oxides near the substrate surface.

A further indication of the role of substrate oxygen is well exemplified in the following experiment. An aluminum film is deposited on a glass substrate at about 330°C. The white bloom common to evaporation of Al under these conditions is formed on the surface and is presumably aluminum oxide. A Permalloy film, deposited on top of this layer, also has a surface which appears white. This white material (approximately 200 A thick) identified by electron diffraction has $\alpha Al_2O_3$ as its major phase with $\beta Al_2O_3$ and $\alpha Fe_2O_3$ as minor constituents.

The exact nature and position of the oxides in the high oxygen content films will be known as a result of current experiments which combine film etching with electron diffraction analysis. Electron diffraction on a 3000-A film has shown the presence of a NiO surface layer.[4] On the other hand, no residual oxide film remains on the substrate when the film is dissolved in dilute Mirrofe. While the absence of this interface oxide may account for the poor adhesion of the film to the substrate it does not negate the possibility of oxygen or oxides moving into the film.

In studies already completed, differing etch rates are obtained at distinct thickness levels; these changes in etch rates are accompanied by changes in the 60-cycle hysteresis loop characteristics. Figure 1 shows the loss in weight and change in coercive force as a function of etch time. The property termed rotatable anisotropy[4] refers to the minor loop with a coercive force $h_c$. The changes in the coercive force for the major loop $H_c$, the coercive force for the minor loop $h_c$, and the field necessary to establish an easy direction $H_a$, are all shown. These three meet in a common point when the film loses rotatability.

Throughout the next region, the film has isotropic hysteresis loop characteristics. The coercive force decreases monotonically until the film is about 1500 A thick; here an anisotropic loop appears with the anisotropy field slightly larger than the coercive force. The anisotropy field becomes slightly smaller than the coercive force at about 1000 A at which time cross-tie

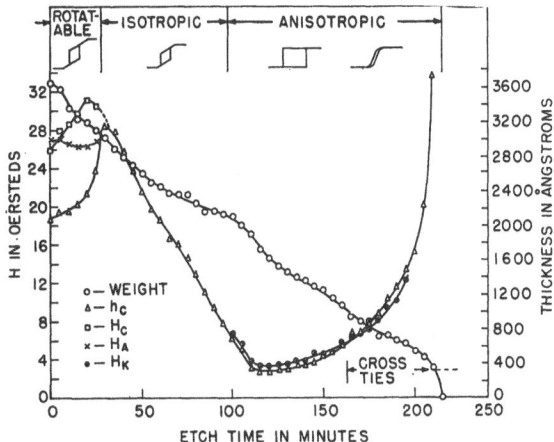

FIG. 1. Etch study of a 3540-A high oxygen content film. Hysteresis loop characteristics and thickness are plotted as a function of etch time. Thickness is estimated from the microbalance weight; one microgram is approximately 10.8 A.

walls are first observed. In the region below 1500 A, the coercive force appears to follow the $T^{-\frac{1}{2}}$ law predicted by Néel.

Throughout the first two regions, the hysteresis loop is skewed as pictured. The origin of this is not strain between the film and the substrate as in the case of mottled films.[6] Removal of the films from the substrate does not affect this loop characteristic. Although we have shown the presence of a negative magnetostrictive layer, any strain present must be contained within the film itself.

Resonance studies conducted along with the etch experiments show the presence of multiple resonance lines which vary in a consistent way with thickness during the etch. These multiple resonances may result from the interactions between layers with differing saturation magnetization values.

Although this study has not yet clarified the origin of anisotropy or rotatable anisotropy, it has provided conclusive evidence for the existence of stratification in permalloy films produced on glass. It is further concluded that this layer-type structure can be enhanced by introducing oxygen onto the substrate prior to deposition.

## ACKNOWLEDGMENTS

It is a pleasure to acknowledge the assistance of K. H. Olsen, J. Lund, and R. A. Neenan in making many of the measurements reported here. Grateful acknowledgement is made to Professor A. H. Morrish and Dr. D. Chen for many stimulating discussions.

[6] E. E. Huber, Jr., and D. O. Smith, J. Appl. Phys. 30, 267S (1959).

JOURNAL OF APPLIED PHYSICS    SUPPLEMENT TO VOL. 33, NO. 3    MARCH, 1962

# Electrodeposition of Magnetic Materials

I. W. Wolf

*The General Electric Company, Syracuse, New York*

A discussion of the electrodeposition of alloys of the Ni-Fe-Co system is presented. Solutions from which these metals are deposited are described and the role of their constituents discussed. The various applications for these deposited materials is reviewed pointing out their advantages and shortcomings. A detailed description of the electrodeposited thin film is given.

## INTRODUCTION

UNLIKE many centenarians,[1] the field of electro-deposited magnetic materials exhibited a great deal more virility in its second fifty years than in its first. Indeed within the last fifteen years the intensity of activity and applications have reached a peak.

Magnetic-materials electrodeposition involves the reduction of nickel, cobalt, and iron ions from solution to a free metal at the cathode solution interface. The reduction is, of course, brought about by supplying the necessary energy in electrical form. Electrodeposition of each of the three metals has been studied and applications found for their products, but in addition, the three binary alloys and the tenary alloy have also been examined and applied. The Ni-Fe-Co system happens to be a remarkably advantageous choice of systems for alloy preparations for two reasons.[2] First, the standard potentials of the three metals are close together (iron $-0.44$ v, cobalt $-0.277$ v, nickel $-0.250$ v) and second, the metals deposit with a high polarization. Thus it has been quite common to mix two of the single metal plating baths together in varying proportions—the anions must be acceptable—and obtain controllable alloys over a wide range of compositions. The operating conditions for the alloy depositions are similar to those of the single constituents.

In the first part of this report headed Deposition, factors and problems in plate preparation common to all phases of this field, are discussed. In the second headed Properties, composition and thickness effects are described with data presented from one system only, and finally in a brief discussion under Applications, the present status of availability for utilization is reviewed.

## DEPOSITION

### A. Baths

An electroplating aqueous solution contains cations of the metal to be plated, anions for the most advantageous deposition conditions and additives for special purposes. Nonreducible cations, such as sodium or calcium ions may be present as well but they act in the same way as additives and thus are considered here as such.

### *Cations*

The cations may exist in the bath as solvated ions or as complexes if there are complexing materials present. In general, they occur in relatively high concentrations. At sufficiently high cathode current density, a cation depletion region occurs in the vicinity of the cathode which is replenished by ion diffusion from the main body of the solution. The diffusion rate is, of course, dependent on the difference in concentration between the solution and depletion zone and thus the former is maintained high for practical plating.

In alloy plating, the composition of the plate depends on the ratio of reducible cations in the depletion region. Thus the plate composition often varies as the depletion layer first forms after the initiation of plating. It is this layer which *agitation* and *pulse plating*[3,4] influence. Normally these actions increase the concentration of the more readily reducible, and thus most depleted, ion in the region of the cathode.

Of the three cations of main concern in this work, cobalt and iron (ferrous) ions are more readily reduced than nickel in the systems generally used for magnetic codeposition. Thus the ion ratios of Co or Fe to Ni is kept small relative to the atom ratio desired in the plate. This fact is, upon first glance, rather surprising since, as can be seen from the potential values quoted in the Introduction, nickel is the most noble metal and hence should be most easily reduced. However, such inverse phenomena are not uncommon in alloy plating and have been referred to as "anomolous codeposition." Possible explanations for this behavior are discussed by Brenner[2] and Korovin.[5]

### *Anions*

The number of possible anions which might be used in the magnetic material plating solution is extremely large. Fortunately for any reviewer of this subject, the number which has been found to offer practical advantages are relatively small. The most important are sulfate, chloride, fluoborate, sulfamate, hypophosphite, and cyanide.

The selection of the anion to be used depends on the specific properties desired for the bath. For instance, the

[1] W. Beetz, Poggendorf's Annalender Physik Chem. (2) III, 107 (1860).

[2] A. Brenner, *Monogram on Alloy Plating*, Chap. 31 (to be published).

[3] G. H. Cockett and E. S. Spenser–Timms, J. Electrochem. Soc. **108**, 906 (1961).

[4] P. Kuttner, and Clemson D., ECS Electrodeposition Symposium, Detroit, 1961.

[5] N. V. Korovin, Zhur. Neorg. Khim. **2**, 2259–63 (1957).

sulfamate is reported to plate with high current efficiency[6] for nickel, cobalt, and iron and thus less hydrogen is evolved at the cathode. Long[7] found that such a bath was useful in depositing permalloy on a moving wire cathode at higher current densities (up to 400 ma/cm²).

The chloride ion improves the anode dissolution problem which sometimes occurs with high nickel content anodes. Chloride is often used in addition to other anions such as sulfate and sulfamate. Thus by using the two anions advantages are obtained without introducing some disadvantages such as higher internal stress that occurs when the all-chloride bath is used.

In many cases, comparable results have been achieved in obtaining specific magnetic properties with different anion compositions. For the case of Permalloy thin films (1000 A) on a sputtered-gold, flat substrate recent experiments with a sulfamate bath[9] gave results almost the same as those obtained with the sulfate-chloride[10] bath through the range of plate compositions 82–100% nickel. Although differences may be found in some cases when different anions are incorporated in the bath if these anions affect physical or metallurgical properties of the deposit such as internal stress grain size or orientation, most of these effects can be countered by including the proper additives. Thus, it is believed that many different baths can yield deposits with very similar magnetic properties if properly developed, and therefore, the choice of anions will not usually be critical.

One glaring exception to the above statement is the case where the anion enters into the reaction at the cathode. An example of this is the inclusion of hypophosphite in the bath. This anion's presence results in the inclusion of phosphorous in the deposit, which, interestingly enough, has led to an increase of $H_c$ for cobalt-nickel[11–12] and cobalt[13] while a decrease of $H_k$ and $H_c$ is achieved for thin nickel-iron.[14,15] The operative mechanisms for the two results are in all probability, entirely different and thus the discrepancy is only apparent for the two effects. In the case of the high coercive force materials, a lamellar structure in which the platelets grow perpendicular to the substrate has been shown and a model in which each lamella is a single domain particle has been proposed.[16] The role that the phosphorous plays in the case of nickel-iron has not been explained.

*Additives*

Into the category of additives falls all of the materials which perform special roles for the plating operation. In this field goes much scientific and artistic ability of the bath fabricator.

As was indicated in the introduction to this section, *nonreducible cations* often appear in the bath. A known contribution that sodium or calcium makes is the increase of solution conductivity and the salts from which they derive are known as "supporting electrolytes." Potassium ion reputedly causes lower stress in the deposition in addition to a conductivity increase. Amonium ion has a complicated and often surreptitious role in the bath since it may act as a complexing agent. Thus it may create entirely different conditions at the cathode, but it is freqeuntly used in nickel plating solutions since its presence is reported in some instances to improve bath throwing power—the ability to plate recesses, etc. Its effect on magnetic properties has not been carefully studied, though it is known to affect the nickel-iron ratio in Permalloy deposition[10] as well as coercivity for cobalt-nickel phosphorous deposits. Magnesium or manganese salts also have been used in the role of "supporting electrolytes." In addition to this effect many of these ions act as "brighteners" of the deposit.

A second type of additive that frequently appears is the *inorganic compound*. An example is the buffering agent such as boric acid or sometimes citric acid—*in this case* the latter behaves like any weak acid though it is organic. The buffering action or *p*H control is especially important in acid baths where hydrogen is competing with the metal ions for reduction and in those baths where plating mechanisms are affected by *p*H. The buffering takes place at the cathode and thus a much greater range of solution *p*H is tolerable when it is present.

Another role of inorganic compound additions is that of the oxidizing agent, added to reduce hydrogen bubble formation and thus pitting at the cathode. Hydrogen peroxide is an example.

The third type of additive is the *organic compound*. Its roles include wetting agents for better cathode and anode behavior, leveling, brightening, complexing and stress reduction.

A great deal of study has been made of the activity of agents which level and brighten. These effects have not been of great direct concern in magnetic material plating, and thus will not be taken up in this report. The reader who might wish to pursue this area further is referred to the extensive work in the last two years directed toward explaining the mechanisms involved.[17]

Wetting agents have been found to be most beneficial in reducing cathode pitting by increasing the tendency of the bath to wet the cathode thus removing hydrogen

[6] R. C. Barrett, Am. Electroplaters Soc., 47th Ann. Tech. Proc., 1960.
[7] T. R. Long, J. Appl. Phys. **31**, 123S (1960).
[8] I. W. Wolf, H. W. Katz, and A. E. Brain, Proc. 1959 Electronic Computers Conference.
[9] A mixture of Barrett Company sulfamate iron and nickel.
[10] I. W. Wolf and V. P. McConnell, Proc. Am. Electroplaters Soc. (43rd), 215 (1956).
[11] T. H. Bonn and D. C. Wendell, Jr., U. S. Patent 2,655,787.
[12] H. Koretzky (to be published).
[13] J. S. Sallo and J. M. Carr, ACS Electrodeposition Symposium, 1961.
[14] P. Kuttner and D. Clemson, ACS Electrodeposition Symposium, 1961.
[15] J. Mathias (private correspondence).
[16] J. S. Sallo and K. H. Olsen, J. Appl. Phys. **32**, 2035 (1961).

[17] See, for instance, papers at 1960, 1961 ECS Electrodeposition Symposium.

bubbles when they form. A similar action at the anode, especially if oxygen is formed there, is beneficial. Common wetting agents such as sodium lauryl sulfate have been used to perform this task. Long[7] implies a warning of possible harmful impurities which may accompany the less pure forms of this agent.

Many additives combine with the ions to form complexes, thus yielding an entity with properties all its own. Citrate, besides being a buffer as mentioned previously, is also a complexing agent and has been used in iron-containing baths to form a soluble complex with ferric ions, which are formed when the ferrous ions are oxidized. Thus the formation of an undesirable precipitate is prevented. As was mentioned with regard to ammonia, the use of complexing agents can lead to great changes in alloy compositions. This fact was well illustrated by Tsu[18] who was attempting to study magnetic characteristics of thin permalloys of less than fifty percent nickel. Thiourea was added as a complexing agent and caused vast changes in plate composition and structure. It is likely that, besides having a complexing action, the thiourea is also partially reduced or irreversibly adsorbed at the cathode.

Of the additive actions, probably the one of most concern in magnetic plating is that of the stress reducers. Because these materials not only reduce stress but simultaneously change crystal size and orientation, brighten and sometimes level, they are of interest to both the fundamental and applied worker in magnetic films.

Stress itself has been a strong contender in accounting for some of the yet unexplained behavior in both low and high coercivity materials.

In the case of the former, an investigation of stress reducers in the permalloy thin film, plated from a sulfate-chloride, bath was conducted.[19] Indications were that of a wide variety of types of organic compounds, the sulfon-groups on resonant structures act as effective stress reducers almost exclusively.[20] The study included compounds with a variety of active groups and the results with the sulfon-groups compared with others was striking. Indeed the effect of the sulfon-group was diminished if a competing group was even attached somewhere else on the ring such as the OH in a naphthol sulfonic acid.

A great deal of evidence has now been gathered[21-23] supporting the concept that the carbon-sulfur bond is broken after the sulfonic is adsorbed at the cathode. Before, during, and/or after the adsorption, the sulfon group is reduced to a sulfide and included in the plate as such. The rest of the molecule detaches from the cathode

and is carried away. Thus the stress reducing action is of a very specific nature and requires particular conditions. The presence of sulfide ion is not sufficient, in itself, for stress reduction nor is that of compounds in general reducible to sulfides.

The mechanism by which a competing attached group, as discussed above, is believed to decrease stress reducing action is by itself acting as the adsorbtion site for the molecule rather than the sulfon-site. Thus OH being more likely to adsorb than $SO_3H$ reduces the stress reducing ability of the molecule in the case of the naphthol sulfonics.

In carrying out these studies with Permalloy thin films, it was observed that a correlation existed between the amount of measurable tensile stress in the film and $H_k$. Stress was measured by the bending of a Mylar strip due to plating and subsequent straightening with a weight and was qualitatively correlated with the amount of plate necessary to cause peeling from a glass substrate.

Lest premature acceptance of the correlation of $H_k$ with stress be generated, it is only fair to point out that very recently reported findings by Kendrick,[24] working with no knowledge of the above study, indicate that the same agents which were found to lower $H_k$ and stress, as mentioned, also act to increase total reflectivity in an all nickel deposit. Moreover, the plate deposited from a no-additive bath (giving high stress deposits) showed "fairly strong" (100) crystal orientation parallel to the substrate with a slight (111), while all "good" stress reducers when added to the bath resulted in plates with "almost" or "completely random" orientation. Thus it may be supposed that along with stress reduction there has been brought about significant crystal changes. Further light may be thrown on this subject in the future with more quantitative correlations of the dependent variables as a function of amount of agent.

The results of the above experiments with stress-reducing suggested a much more crucial experiment with anisotropy in thin films.[25] It is possible to obtain plated films with compressive stress by adding sufficient "stress reducing agent" to the solution. Thus one can compare "tensile" and "compressive" films and observe whether the field induced anisotropy is different in the two cases as might be expected if this anisotropy is due to magnetoelastic internal stress effects alone.

The results of this experiment were negative. When Permalloy of approximately 82:18 was deposited from a solution containing 5 g per liter of saccharin, the easy axis was maintained in the direction of the field.

Since there is always a question of a contribution of "short-range ordering anisotropy"[26-28] due to iron pairs

[18] I. Tsu, Plating **47**, 632 (1960).

[19] I. W. Wolf, ECS Electrodeposition Symposium, 1961.

[20] An exception was *B*-naphthol, which, while not very effective, *did* reduce stress.

[21] H. Brown, 1961 ECS Electrodeposition Symposium.

[22] J. J. Hoekstra and D. Trivich 1961 ECS Electrodeposition Symposium.

[23] B. J. Riley and S. E. Beacon, 1961 ECS Electrodeposition Symposium.

[24] R. J. Kendrick, Plating **48**, 1099 (1961).

[25] The author is indebted to D. Smith for this suggestion.

[26] L. Néel, Compt. rend. **237**, 1468, 1613 (1958); J. phys. radium, **15**, 225 (1954).

[27] S. Taniguchi, Sci. Repts. Research Inst., Tohoku Univ. **A7**, 269 (1955).

[28] D. Smith, J. Appl. Phys. **30**, 264S (1959).

in a nickel matrix for Permalloys, the experiment was also carried out with pure nickel from both sulfate-chloride and a sulfamate baths. In these cases the results again appeared to be negative, but because of the easy axis dispersion created when this very large amount of stress reducer[29] was added to the bath the film appeared to be almost isotropic.[30]

Fisher[31] has measured "macrostress" with a stressometer of the Kushner[32] type for cobalt and nickel films using varying amounts of saccharin in a chloride bath. He found good correlation between stress and coercive force for the latter but for cobalt concluded that the change in crystal structure was the dominant influence.

## B. Plating Conditions

In addition to the bath itself, there are several other factors which affect the characteristics of an electrodeposit, involving the way in which the plate is deposited. Included among the independent variables normally investigated are temperature, current density, substrate, acidity, time varying currents and agitation. The principles involved in the last of these was discussed previously with regard to the ion depletion layer at the cathode. Each has been demonstrated to have an effect on magnetic properties of the deposit, if varied sufficiently, however, the effects are often complex and thus are not easily generalized. A brief view of the maze of results obtained with these variables and comments on each follows.

### Temperature of Solution

For the case of Permalloy deposition from the sulfate-chloride bath at low current densities, the magnetic characteristics of thin films deposited in baths with temperatures ranging from 10°C to 60°C did not vary appreciably. For cobalt-nickel deposits from a sulfate-chloride bath, Kaznachei and Zhogina[33] report that both composition and coercive force are very slightly temperature dependent over approximately this range changing more rapidly above 70°C.

On the other hand, Polukarov[34] reports an appreciable bath temperature dependency for the coercive force of cobalt which also varies in crystalline orientation in this temperature range. White and Kolk[35] report "poor" physical characteristics below 65°C for 97 Fe-3Ni thin films deposited from an all chloride bath.

With so many temperature dependent phenomena possible in such systems, it is no wonder that every experimenter in this field has investigated or has planned to investigate this variable.

### Current Density

As for temperature, optimization for current density cannot be simply formulated. Densities from 0.25 to 500 ma/cm² have been successfully worked with. Alloy composition, crystal size and orientation,[34] stress, and, of course, magnetic properties all are influenced by current density. As was indicated previously in discussing bath compositions, it is possible to alter some undesirable high-current-density effects by changing bath compositions or plating conditions.

### Substrate

The subject of substrates, while of only minor interest to the worker in high coercivity materials where deposits generally are of the order of $10^{-3}$ cm, are of greatest concern to the low-coercivity thin film plater. As many an unfortunate "wrong-guesser" can testify once having decided on the wrong substrate, all subsequent effort toward improving film characteristics was futile.

Surfaces which have been purposely roughened have been shown[36] to yield films of higher coercivity. Many workers have demonstrated that one can achieve anisotropy in deposited Permalloys by depositing on a surface scratched predominantly in one direction. The easy direction in this case will be along the scratches. Those experimenters that have attempted to use rolled metals as substrates find that the easy direction will appear perpendicular to the roll direction if no field is applied during deposition even though the metal appears to have a bright unscratched finish. Orientation in this case is believed due to the slip planes[37] formed in the rolling process.

Probably the most dramatic evidence of the extent to which small substrate differences can effect characteristics is given by a comparison of a film deposited on a sputtered gold substrate on a smooth glass base with a film identically prepared except for a brief heat treatment of the substrate prior to plating.

For the substrate of gold sputtered on glass and plated with Permalloy the following behavior was observed. If the substrate is heated at 120°C for 15 min, $H_c$ and dispersion increase but $H_k$ remains constant. Heating for longer periods causes an increase in the $H_k$ and finally such a degree of degradation that the film appears to be isotropic from hysteresigraph measurements.

Thus by varying the degree of aglommeration of the gold substrate only, a wide variety of film characteristics were obtainable.

The ideal substrate for almost all low coercivity thin film applications should have *minimum roughness* and

[29] Saccharin and SNSR (Barret Chemical Company) were used in the two cases.
[30] The easy and transverse directions *are* distinguishable for low drive hysteresis loops.
[31] R. D. Fisher, ECS Electrodeposition Symposium, 1961.
[32] J. B. Kushner, Proc. Am. Electroplaters Soc. 41, 188 (1954).
[33] B. Yu. Kaznachei and V. M. Zhogina, *Metallography and Working of Non-Ferrous Metals* (U.S.S.R., 1957), pp. 77–90.
[34] Yu. M. Polukarov, Zhur Fiz. Khim. 34, 150–156 (1960).
[35] H. J. White and A. J. Kolk, ECS Electrodeposition Symposium, Detroit, 1961.

[36] J. C. Lloyd and R. S. Smith, J. Appl. Phys. 30, 274S (1959).
[37] This explanation was first offered by M. Cohen, Lincoln Laboratory.

*maximum randomness of orientation* of crystals. Work on substrates in high coercivity *very thin* electrodeposited films has not been reported, however, for thicker plates, mechanical or chemical polishing of a metal substrate has been found adequate.

### Acidity

The upper limit of hydrogen ion concentration in the acid baths is dictated by the tendency of this ion to compete for the electrons at the cathode. Thus, at some minimum $pH$ a sufficient concentration of hydrogen ions are present so that the cathode current efficiency, i.e., the rate of metal reduced divided by the theoretical rate if every electron was captured by a metal ion-begins to drop noticeably with further $pH$ decrease. For the Permalloy deposition from sulfate-chloride this critical $pH$ is approximately 2.5. Data for $pH$ reported by several investigators significantly cut off in the region 2-2.5 for nickel-iron, and nickel-cobalt, however, Polukarov[38] reports data on cobalt from a sulfate bath down to 1.3. For very iron rich iron-nickel electrodeposition (97Fe-3Ni), White and Kolk[35] found the $pH$ range 0.4—2.0 best.

The upper limit of $pH$ is more difficult to define. In depositing iron alloys, unless complexing agents are added, the oxidation of ferrous ions is increased, ferric hydroxide precipitates and finally ferrous hydroxide precipitates as the $pH$ is increased. As for the cobalt-nickel system, many investigators have deposited without complexing agents up to a $pH$ of 7 from "acid" baths. Sree and Rama Char[39] report alloy deposition from an *alkaline* ($pH > 7$) pyrophosphate bath using citrate as an additional complexing agent.

Within the range of $pH$ allowable for iron deposition, workers with Permalloy have not found any large $pH$ dependencies for magnetic characteristics. However, in the cobalt-nickel system the situation is different. The changes observed with $pH$ change include alloy composition, crystal orientation and size, hydroxide inclusion and stress. It is thus not surprising that coercive force and remnant magnetization of deposits are affected by solution acidity. White and Kolk[35] also report a dependency of coercive force on $pH$ for 97Fe-3Ni deposited from their all-chloride bath.

### Time Varying Currents

By far the greatest amount of attention to the effects of time varying deposition current on magnetic properties has been concentrated in the high-coercivity-materials deposition and indeed has been devoted to the task of increasing coercive force and optimizing squareness. Because of its general acceptance as an important variable, it is rarely overlooked. For the

chloride-sulfate bath Kornei[40] and Zapponi[41] found that a superimposed ac on the direct current yielded higher coercive forces and Scheer and York[42] and Koretzky[12] report improved squareness with the use of additives in the same system still utilizing the ac. The effect is $pH$ dependent, however, becoming negligible above $pH$ 5. Quinn, Sulich, and Manley[43] have demonstrated an effect of these mutual variables on crystal structure which correlates with the magnetic effects.

Little has been recorded of such effects on low coercivity materials. Cockett and Spenser[3] recommend the use of a pulsed plating current to eliminate composition gradients through the thickness of Permalloy films, but such factors as proper agitation and low current density, can reduce the need for this.

### PROPERTIES
### Effect of Composition

In the preceding portion of this work, information was presented indicating that an electrodeposited material of a given composition might be found to have any of a very large number of magnetic characteristics depending on its preparation and other factors previously mentioned. Since this is so, any presentation of data relating magnetic effects with composition should be justified.

In the following case, the results are believed to be of interest because of several factors. The deposition conditions were purposely chosen to provide a simple reproducible system. The substrate-sputtered gold on glass-was chosen to meet the smooth, random-orientation requirements for low coercivity and anisotropy discussed above. The bath is of a well known variety and the internal stress kept quite low, but slightly tensile. Similar results for the nickel-iron system have been also achieved by other workers in low coercivity

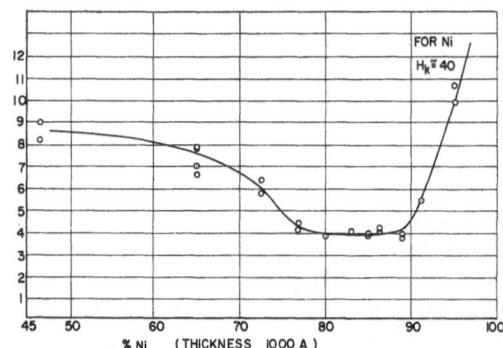

FIG. 1. $H_k$ vs % Ni for 1000 A Ni–Fe films.

[38] Yu. M. Polukarov, Zhur. Fiz. Khim. **32**, No. 5, 1008 (1958).
[39] V. Sree and T. L. Rama Char, Bulletin India Sec. Electrochem. Soc. **7**, No. 3, 72-5 (1958).

[40] Kornei, Office of Science Research & Development, Report 5325.
[41] P. P. Zapponi, U. S. Patent 2,619,454, 1952 (to Brush Development Corporation).
[42] H. C. Scheer, and E. R. York, U. S. Patent 2,834,725, 1958 (to IBM Corporation).
[43] H. F. Quinn, M. Sulich, and G. W. Manley, ECS Electrodeposition Symposium, 1961.

materials using different baths and substrates,[44,45] so that some degree of limited generalizations can be made concerning the results.

The films reported were all plated from a sulfate-chloride bath[10] at 3 ma/cm² current density without agitation, superimposed ac or any known disruptive influences. Reproducibility was good for the systems checked—the Ne–Fe and Ni–Fe–Co reported here were repeated throughout—and hysteresis measurement results were corroborated for these same two systems.[46] Results are for 1000-A samples unless otherwise stated.

In Fig. 1,[47] the anisotropy field is plotted versus composition for the nickel-iron system. As can be seen, $H_k$ goes through a rather flat minimum in the neighborhood of zero magnetostriction. The nickel-rich—magnetostrictively—samples showed poorer linearity of the transverse loop than the iron rich.

In Fig. 2,[47] the coercive force $H_c$ and easy axis disturb $H_{CD}$ are plotted as functions of composition. Here the minima are approximately in the same composition neighborhood but are less well defined on the iron-rich side.

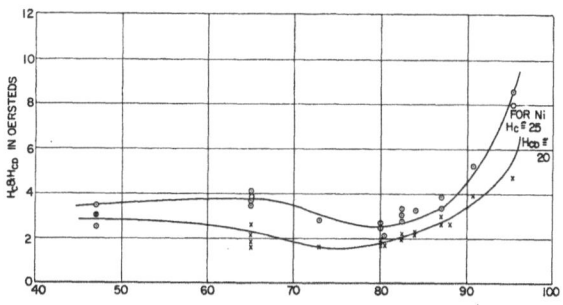

FIG. 2. $H_c$ and $H_{CD}$ (disturb level) vs % Ni for 1000-A Ni–Fe films.

From the results obtained studying composition dependencies, it is now clear that the anisotropy in thin magnetic films is far from a specific composition phenomena. Indeed one is tempted to generalize and say that any thin ferromagnetic metal or alloy will have a field-oriented preferred direction if properly deposited. Two examples of other deposition systems which exhibit the well-defined anisotropy are Co–Ni and Co–Ni–Fe.

Hysteresis loop squareness for the easy axis and linearity of transverse loop were obtained throughout the region of composition investigated, except for the 99% nickel neighborhood where a slight deterioration of squareness ratio is observed.[48]

[44] T. Long (private communication).

[45] J. Mathias (private communication).

[46] The author is indebted to T. Crowther of Linclon Laboratory for this work and for the dispersion measurements and magnetostriction results.

[47] These figures are reprinted from the October, 1961 issue of the Journal of the Electrochem. Soc. (108, No. 10) with permission of the Editor.

[48] In addition, a slight "imperfection" in the corners of some of the Co–Ni loops appeared, possibly due to composition variation through the thickness. This effect resulted in lower $H_{CD}$ readings.

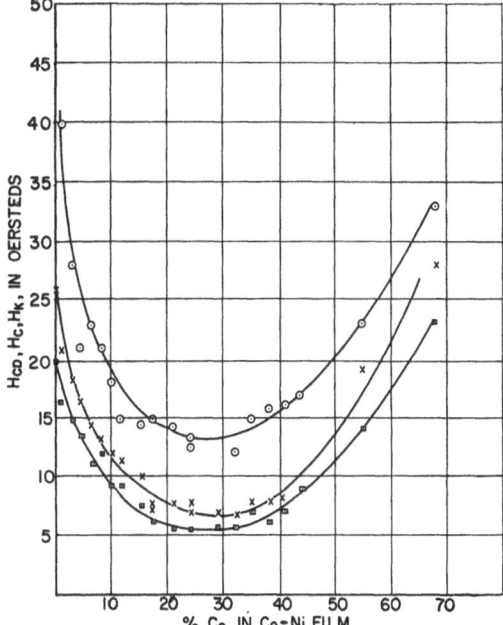

FIG. 3. $H_c$, $H_{CD}$, and $H_k$ vs % Co in 1000-A Co-(Ni:Fe) films.

In Fig. 3, the $H_k$, $H_c$, and $H_{CD}$ are plotted for the composition range 0–68% Co. The minimum value for all these curves occurs in the neighborhood of just under 30%. Although these values *are* thickness dependent, the general shape is also maintained for the 2000-A samples which were tested as well.

In Fig. 4,[49] $H_k$, $H_c$, and $H_{CD}$ values versus cobalt composition are given for an almost constant Ni:Fe ratio of 82:18 in the Ni–Fe–Co system. It was established in this work that the zero magnetostriction composition depends only (within accuracy of measurement) on the Ni:Fe ratios. Thus the results should be less dependent on the magnetoelastic effects which

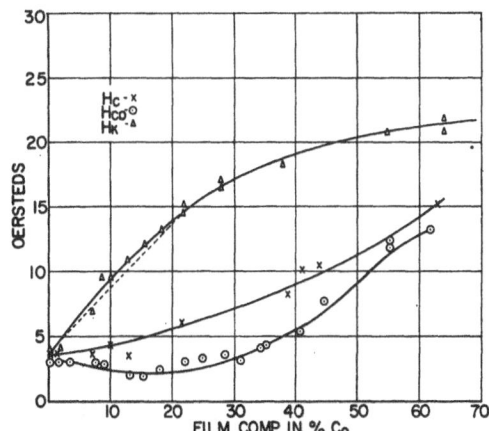

FIG. 4. $H_c$, $H_{CD}$, and $H_k$ vs % Co in 1000-A Co-Ni films.

[49] These figures reported from the collection of papers presented at the Symposium on the Electric and Magnetic Properties of Thin Metallic Layers, Leuven, Belgium, 1961.

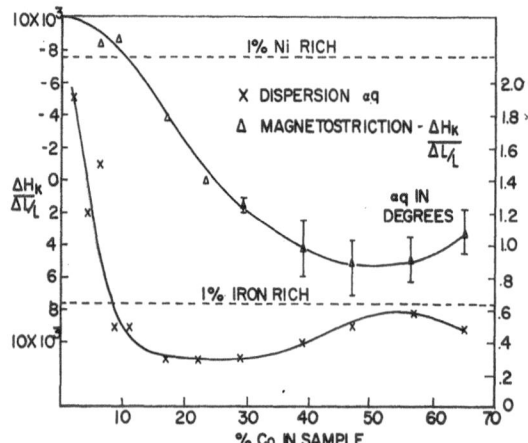

FIG. 5. Dispersion of preferred axis and magnetostriction vs % Co in 1000-A Co-(Ni:Fe) films.

composition data of the other systems inherently incorporate.

If the anisotropy which remains after the magneto-elastic effects are removed, is attributed to short range ordering, then the near linearity of the $H_k$ dependency on cobalt composition which is shown in Fig. 4 for the 0–20% Co range suggests a pairing of Co with Fe.

In Fig. 5,[49] the magnetostriction sensitivity $\Delta H_k/\Delta L/L^{27}$ is plotted as a function of percent cobalt. It can be seen that the parameter does change with cobalt through the entire range from that of about 1% Ni rich to about $\frac{1}{2}$% Fe rich of the zero value, as expressed for Permalloy. However, x-ray fluorescence analysis of the films indicated that the Ni:Fe ratio had varied in roughly the same way as the pseudocomposition film suggested. Thus, accidentally, the belief that the magnetostriction coefficient depended only on the Ni:Fe ratio was tested and found to be correct within the error of measurement.

Results of easy axis dispersion[46] measurements of these films indicated that the introduction of cobalt brings about a sharp drop in this parameter. Though some of the effect might be due to the decrease of the magnetostriction constant, the major part cannot be attributed to this factor. Thus, it can be seen that the Co-Ni-Fe system provides an interesting area for further probing, with many practical advantages for thin film applications possibly to be gained.

## Thickness

The thickness dependency originally predicted[50] for thin films was that of domain wall coercive force and was of an exponential form:

$$H_c = At^{-\frac{1}{3}},$$

where $A$ is a constant and $t$ is the film thickness. Tiller and Clark[51] found that the dependency of $H_c$ on thick-

ness agreed with the theoretical form for Ni–Fe films which were vapor deposited. Lloyd and Smith,[36] however, reported that for films electrodeposited on copper, although the exponential form was realized, the $-\frac{4}{3}$ exponent itself did not represent the results. Instead the exponent was found to be a variable.

In Fig. 6, the $H_c$ versus thickness plots are given for Ni–Fe films of three compositions. As can be seen, even the sign of the exponent changes. In this case the negative of the exponent has been designated as $B$. A plot of $B$, as well as the comparable exponent $D$ for the $H_{CD}$ thickness correlation, are presented in Fig. 7.

In addition to the wall-coercive-force dependency, the anisotropy field, $H_k$, appears to be thickness dependent. This dependency is not of the simple exponential type as is that of $H_c$ and $H_{CD}$. The effect, itself, is composition dependent. The value $H_k$, for a distinctly nickel-rich Permalloy increases with thickness and for a definitely iron-rich film decreases with thickness. The effect is minimum around the zero magnetostriction composition. In Fig. 8, curves for samples of compositions in the three regions are given.

Thus another bit of evidence is provided which links the behavior of $H_k$ with magnetoelastic effects, but any proof of a true casual relationship is lacking.

## APPLICATIONS

Though electrodeposited metals and alloys have found some limited use as magnetic shields[10] and have been investigated as magnetic core material for inductance coils[52] having special properties, it is in the areas of high coercivity magnets, magnetic recording, and thin film memories that most attention has been paid to application.

Iron and iron-cobalt when properly electrodeposited on a mercury cathode result in elongated single domain (ESD) particles. When these particles are compacted, very high coercivity magnets are thereby attained. The

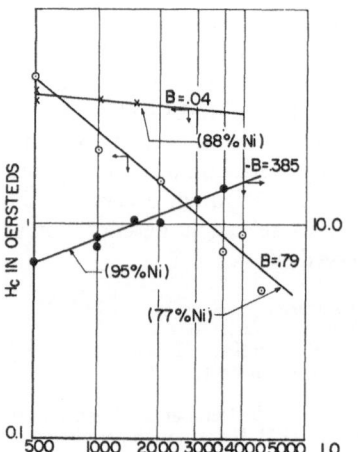

FIG. 6. $H_c$ vs thickness of Ni-Fe films at several compositions.

[50] L. Néel, J. phys. radium **17**, 3 (1956).
[51] C. O. Tiller and G. W. Clark, Phys. Rev. **110**, 2 (1958).

[52] I. W. Wolf, Proc. Am. Electroplaters Soc. **44**, 121 (1957).

phenomena has been extensively treated in the literature and reviewed recently by Luborsky.[53]

## Magnetic Recording

While magnetic tapes with nonmetallic powder coatings are the most frequently used medium for permanent information storage by far, there are properties offered by a deposited metal which make it advantageous for certain applications. Kaznachei and Zhogina[33] conclude their paper on nickel-cobalt electrodeposition with a statement to the effect that the electrodeposited alloy was shown to be superior to powder coatings as far as "reproduction fidelity and signal," "mechanical stability" "noise level" and "resulting" amplitude." While it is true that generalizations in such comparisons between dynamic techniques are often found to be premature, it is clear that where close mechanical tolerances are desired or very good adhesion is necessary, such as high speed rotating drums or disks, there is an advantage in employing a deposited metal. Moreover, the recent developments in electrodeposited films in which coercivities of 2000 oe[13] are obtained justifies at least temporarily, contentions about the high storage densities which can be achieved with deposited metals. Indeed film development is advanced far beyond that of present recording heads,[12] which are limited by mechanical considerations.

## Thin Films

The chief practical use of electrodeposited magnetic thin films has been for high speed random access information storage. Considering the short life of the development efforts the results to date are quite impressive. Memories have been built using the plated wire,[54] plated

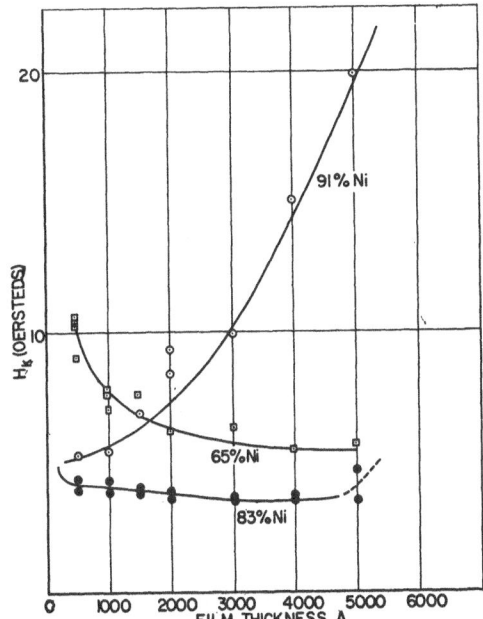

FIG. 8. $H_k$ vs thickness for Ni-Fe films at several compositions.

rod,[35] and plated flat film,[55] and each has been operated successfully. In all cases there is evidence for good reproducibility of magnetic film properties when the process parameters are controlled.

It is difficult to judge at this time the eventual place that electrodeposited thin films will have in magnetic thin film technology, however, at present, the process does offer good reproducibility of acceptable characteristics for memory application.

## ACKNOWLEDGMENTS

The author wishes to express his gratitude to H. Koretzky of the IBM Corporation for his advice in the area of high coercivity plating, to D. Smith, J. Raffel, and T. Crowther of Lincoln Laboratories for their aid and comments on the thin film work described, and to Mrs. B. Harrington and R. Schultz of the General Electric Electronics Laboratory for their active participation in the experiments.

The work carried out by the author as described above was conducted under the sponsorship of Lincoln Laboratory, a center for research operated by Massachusetts Institute of Technology with the joint support of the U. S. Army, Navy, and Air Force.

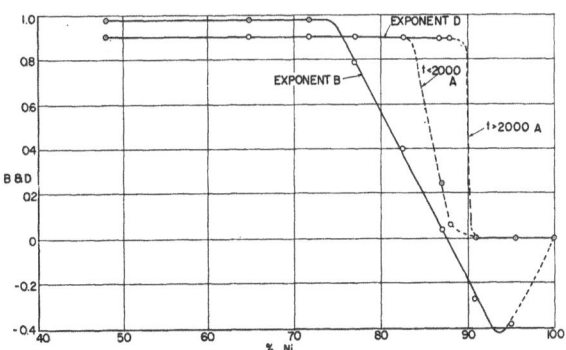

FIG. 7. Exponent coefficients $B$ and $D$ vs % Ni for Ni-Fe films.

[53] F. E. Luborsky, J. Appl. Phys. **32**, 1715 (1961).
[54] T. R. Long, Electrodeposition Symposium, Electrochem. Soc. Conference, 1961.

[55] R. Bogusch and E. A. Fisch, National Electronics Conference, 1960.

JOURNAL OF APPLIED PHYSICS    SUPPLEMENT TO VOL. 33, NO. 3    MARCH, 1962

# Rotatable Anisotropy in Composite Films

J. M. Lommel and C. D. Graham, Jr.*

*General Electric Research Laboratory, Schenectady, New York*

The anisotropy in the plane of the film of thin films of Ni evaporated on Mo-covered substrates and of oxidized Ni films has been measured as a function of field and temperature. At intermediate fields, 200–1200 oe, a large rotatable anisotropy is found, about an order of magnitude larger than the usual magnetic annealing anisotropy. At higher fields the rotatable anisotropy disappears and only the magnetic annealing anisotropy remains. The temperature dependence of the rotatable anisotropy indicates a connection with the antiferromagnetic character of NiO.

NICKEL films have been deposited on Mo-covered and plain glass substrates. All of the Ni on Mo films possessed the property of having a rotatable anisotropy, whereas Ni films on glass showed it only after they were allowed to oxidize. A rotatable anisotropy, following the notation of Prosen, Holmen, and Gran[1] is one whose easy axis can be rotated by application of a large field. The rotatable anisotropy differs from the usual anisotropy in thin films in that with the latter, called the magnetic annealing anisotropy, the easy axis can be rotated only by heating and cooling in a magnetic field. It has been shown[2-4] that the magnetic annealing easy axis can be rotated at room temperature, but it requires orders of magnitude more time than it does to move the rotatable anisotropy.

The composite Ni on Mo films was prepared by depositing Mo from a hot Mo wire on glass substrates in a vacuum of $10^{-5}$ mm Hg. The substrates were exposed to air and then Ni was deposited from a molten source held in an alundum crucible at pressures of $10^{-10}$ or $10^{-5}$ mm Hg. Ni films were deposited on glass substrates in a similar fashion. In all cases, the substrate was baked-out at approximately 400°C in vacuum before the next film was evaporated. This was done to assure removal of adsorbed films of $H_2O$ on the glass or Mo-covered substrates.

Torque curves were measured in a torque magnetometer similar to the one described earlier.[2] In the case of Ni films evaporated at $10^{-10}$–$10^{-9}$ mm Hg, the torque measurements were made in the vacuum system before the films were exposed to air. Films made in the bell jar vacuum system were measured in a separate system at a pressure of $10^{-7}$ mm Hg. The temperature dependence of the torque was measured by heating the outside walls of the vacuum system. The film was assumed to reach the wall temperature, introducing an uncertainty of the order of 10°C; the film was cooler than the indicated temperature.

The field dependence of the torque for Ni on Mo

films was different from that reported for pure Ni films.[4] The behavior is illustrated in Fig. 1 for a Ni film deposited on Mo at a pressure of $2\times10^{-10}$ mm Hg. At low fields, the torque is unidirectional, as was found for pure Ni films, and is of the form $\mathbf{M}_r\times\mathbf{H}$, where $\mathbf{M}_r$ is the remanent magnetization. At higher fields, e.g. 750 oe, a large anisotropy is found which is absent in pure Ni films. It is an order of magnitude larger than the magnetic annealing anisotropy and its easy direction is the direction of the last high field applied to the film. Fields of about 2500 oe were used to rotate this anisotropy. At higher measuring fields, 1500 oe for this film, the rotatable anisotropy disappears and a large rotational hysteresis is found. At the highest field available, 2500 oe, only the small magnetic annealing anisotropy remains.

In the region where the rotatable anisotropy is present, the torque curves are sheared-over because the field is not great enough to keep the magnetization aligned with it. The rotational hysteresis is greatest in the hard directions and is zero in the easy directions. The magnitude of the rotatable anisotropy was determined by plotting the reciprocal of the easy direction slope of the torque curve as a function of $(1/H)$ follow-

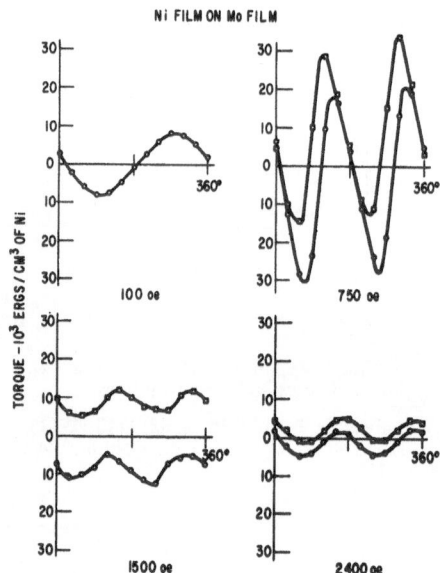

FIG. 1. Torque curves for a Ni film deposited on Mo. The thickness of Ni was 3000 Å and the Ni-Mo interface had an apparent surface area of 0.8 cm².

---

* Temporarily at the Institute for Solid State Physics, University of Tokyo.

[1] R. J. Prosen, J. O. Holmen, and B. E. Gran, J. Appl. Phys. 32, 91S (1961).

[2] C. D. Graham, Jr., and J. M. Lommel, *Proceedings of the International Conference on Magnetism, Kyoto, 1961* (to be published).

[3] Z. Málek, W. Schüppel, O. Stemme, and W. Andrä, Ann. Physik 5, 211 (1960).

[4] C. D. Graham, Jr., and J. M. Lommel, J. Appl. Phys. 32, 83S (1961).

ing Kouvel and Graham.[5] Over a portion of such a plot, before the rotatable anisotropy disappeared, the data fell on a straight line which was extrapolated to infinite field. The temperature dependence of the rotatable anisotropy was determined by the technique of measuring torque curve slopes as a function of temperature. The results for a Ni on Mo film are shown in Fig. 2. The precision of the technique fell off at high temperature because the rotatable anisotropy reached a maximum at lower and lower fields as the temperature increased. The temperature at which the rotatable anisotropy disappeared was about 275°C. The rotatable anisotropy reappeared when the film was cooled; its magnitude and the field at which it was observed decreased from the as-deposited values.

The rotational hysteresis for these composite films is different from that of pure Ni films. Whereas pure Ni films show only one peak in the rotational hysteresis versus field curve[4] just above the coercive force, the composite films exhibit two peaks. The low field peak occurs when the remanent magnetization begins to rotate and the high field peak occurs when the rotatable anisotropy disappears. At higher fields the rotational hysteresis approaches zero. The temperature dependence of the rotational hysteresis associated with the disappearance of the rotatable anisotropy was measured. The rotational hysteresis versus field curves decreased in magnitude and the peak shifted to lower measuring fields as the temperature was increased after a maximum was reached at 100°C. The rotational hysteresis disappeared at the same temperature that the rotatable anisotropy disappeared, a hundred degrees below the Curie temperature for pure Ni.

It was found that pure Ni films on glass made in a bell jar vacuum system or the ultrahigh vacuum system did not show any rotatable anisotropy. Films which had been exposed to the atmosphere for several weeks did show the rotatable anisotropy. What was said about the behavior of Ni on Mo films above could be repeated qualitatively for oxidized Ni films. While the temperature dependence of the rotatable anisotropy is not exactly the same for both films (Fig. 2), the temperature at which the anisotropy disappears is the same.

Because of this we conclude that the rotatable anisotropy in Ni on Mo films is due to the presence of a NiO layer between the Ni and Mo. Oxygen adsorbed on Mo or $MoO_3$ has reacted with Ni when it was first deposited. The rotatable anisotropy in these films is associated with the antiferromagnetic NiO. The Néel temperature of

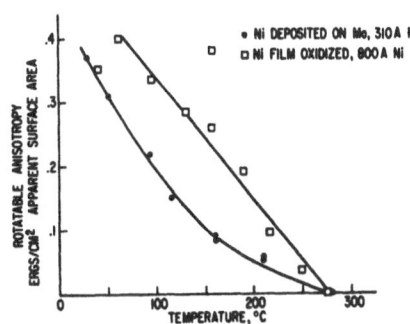

FIG. 2. Temperature dependence of the rotatable anisotropy constant. The apparent surface area was 0.8 cm².

NiO is 250°C[6] which is in good agreement with the experimentally determined temperature at which the rotatable anisotropy disappears. A Ni on Mo film was deposited on NaCl and floated off. X-ray and electron diffraction analysis did not indicate the presence of any intermetallic compounds of the Ni-Mo system.

The forms of two models for the rotatable anisotropy have been suggested.[1] In the first, a uniaxial exchange coupling between ferromagnetic Ni and antiferromagnetic NiO could lead to rotatable anisotropy of the form observed here. Schmid[7] has reported the existence of such a coupling in oxidized Ni powder; he did not find a shifted hysteresis loop, an effect consistent with a uniaxial rather than a unidirectional coupling. The rotatable anisotropy could also be explained by a model in which the distortion[8] in the NiO caused by antiferromagnetic ordering is transmitted to the Ni. The negative magnetostriction of Ni would then give rise to a uniaxial anisotropy with the correct direction for the easy axis. Both of these models require that the high field which aligns the easy axis of the rotatable anisotropy must rotate the spins in the antiferromagnet, which the data of Roth and Slack[9] indicate can be done in the fields used here. The present experiments appear to be inadequate to distinguish between these two models, although detailed calculations have not yet been made.

## ACKNOWLEDGMENTS

The authors are grateful for the stimulating discussions and encouragement of I. S. Jacobs, W. H. Meiklejohn, and J. S. Kouvel.

[5] J. S. Kouvel and C. D. Graham, Jr., J. Appl. Phys. **28**, 340 (1957).

[6] C. Kittel, *Introduction to Solid State Physics* (John Wiley & Sons, Inc., New York, 1953), p. 189.
[7] H. Schmid, Cobalt **6**, 8 (1960).
[8] S. Greenwald and J. S. Smart, Nature **166**, 523 (1950).
[9] W. L. Roth and G. A. Slack, J. Appl. Phys. **31**, 352S (1960).

JOURNAL OF APPLIED PHYSICS    SUPPLEMENT TO VOL. 33, NO. 3    MARCH, 1962

# Unidirectional Hysteresis in Thin Permalloy Films*

W. D. Doyle, J. E. Rudisill,† and S. Shtrikman‡

*The Franklin Institute Laboratories, Philadelphia, Pennsylvania*

A study of torque curves in thin Permalloy films with $h_c > 0.5$, where $h_c = H_c/\bar{H}_k$ is reported. The behavior of a 1400 A, 77% Ni film with $h_c = 0.9$ is discussed as a typical example. For applied fields $h > 0.5$, the torque curves are generally complex and the discrepancies with the Stoner–Wohlfarth (S–W) model wide-spread. For $h < 0.5$, if the film is previously saturated, the torque has a period of $2\pi$ in agreement with the S–W model. However, when $0.3 < h < 0.5$, it is irreversible and the hysteresis is unidirectional, occurring around the easy direction of the bulk of the film. A model is proposed to explain this effect in which it is assumed that the film contains small regions with negative anisotropy $-K_2$, where $K_2$ is of the order of $K_1$, the anisotropy of the bulk of the film. The torque curves calculated for this model are found to agree qualitatively very well with the experimental results.

A STUDY of torque curves in thin Permalloy films, and the field dependence of the rotational hysteresis[1] in particular, has revealed that the results are dependent on the value of the reduced coercive force $h_c = H_c/\bar{H}_k$, where $H_c$ is the coercive force in the easy direction and $\bar{H}_k$ is the average anisotropy field. Further, the films can be classified according to their torque curves in one of two categories, depending on whether $h_c < 0.5$ or $> 0.5$.

A previous paper[2] has discussed the case of $h_c < 0.5$ where both theoretically and experimentally the situation appears relatively straightforward. In that region, the results agree very well with formulas derived by Shtrikman and Treves[3] (S–T) for the magnetization reversal in an infinite cylinder.

The extension of the work to the case $h_c > 0.5$, has shown that the coherent rotation theory of Stoner and Wohlfarth[4] (S–W) to which the model reduces in this range, does not adequately describe the experimental results. This is illustrated by a typical example: a 1400 A, 77% Permalloy film with $h_c = 0.9$. For applied fields $h > 1.5$, using the reduced notation, the torque, as expected from (S–W), is sinusoidal with a period of $\pi$ and a field-independent amplitude. However it is irreversible, as found also by Mayfield[5] and Takahashi *et al.*[6] For $0.7 < h < 1.5$, a well-defined uniaxial hysteresis[7] is observed, associated with the anisotropy of the bulk of the film. Here the behavior is characteristic of the

(S–W) model. For $0.5 < h < 0.7$, the torque is irreversible and rather irregular. For $h < 0.5$, if the film is previously saturated, the torque has a period of $2\pi$ in agreement with the (S–W) model. However, when $0.3 < h < 0.5$, it is irreversible and the hysteresis is unidirectional[7], occurring around the easy direction of the bulk of the film. The experimental curve in this case is shown in Fig. 1(a).

The present discussion will be limited to the origin of this unidirectional hysteresis, found at low fields. This effect can be understood qualitatively very well if it is assumed that the film contains small regions with negative anisotropy $-K_2$ where $K_2$ is of the order of $K_1$, the anisotropy of the bulk of the film. The observations of

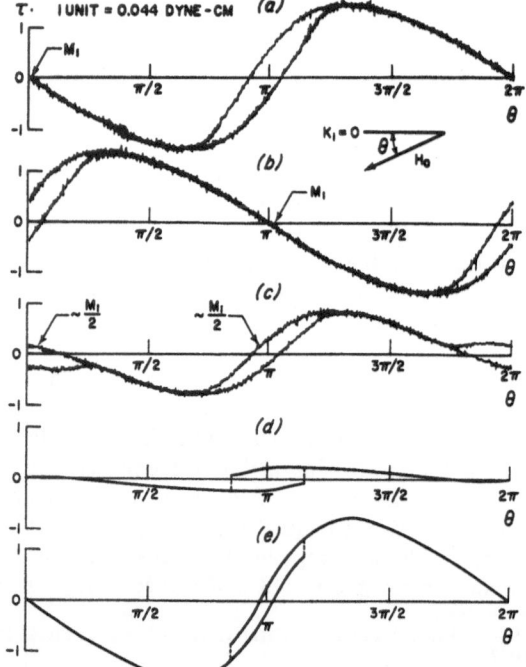

FIG. 1. Experimental and theoretical torque curves. (a) shows the result if the film is first saturated in the $\theta = 0$ direction, (b) if the film is saturated in the $\theta = \pi$ direction, and (c) if the film is demagnetized. In each case, $h = 0.4$ and the torque is recorded for both clockwise and counterclockwise rotations. In (d), the torque calculated for an N. A. region is shown. (e) includes the interaction with $M_1$.

* This research was supported by the United States Air Force under contract, monitored by the Aeronautical Systems Division, Air Force Systems Command.

† Now at Philco Corporation, Philadelphia, Pennsylvania.

‡ On leave of absence from the Weizmann Institute of Science, Rehovoth, Israel.

[1] E. M. Bozorth, *Ferromagnetism* (D. Van Nostrand Company, Inc., Princeton, New Jersey, 1951), p. 514.

[2] W. D. Doyle, J. E. Rudisill, and S. Shtrikman, J. Appl. Phys. 32, 1785 (1961); also, a paper presented at the International Conference on Magnetism, Japan, 1961.

[3] S. Shtrikman and D. Treves, J. phys. radium 20, 286 (1959).

[4] E. C. Stoner and E. P. Wohlfarth, Phil. Trans. Roy. Soc. (London) 240A, 599 (1948).

[5] J. R. Mayfield, J. Appl. Phys. 30, 256S (1959).

[6] M. Takahashi, D. Watanabe, T. Kono, and S. Ogawa, J. Phys. Soc. Japan 15, 1351 (1960).

[7] Uniaxial hysteresis is defined as two irreversible jumps of the magnetization, 180° apart in every 360° rotation of $h$. Unidirectional hysteresis is defined as one irreversible jump in every 360° rotation.

Smith[8] indicated the existence of such regions which have been suggested by him to account for films with $h_c > 1$.

Consider a rotational hysteresis experiment in a negative anisotropy (N. A.) region, described in terms of the (S–W) asteroid,[9] (Fig. 2). Let the easy axis of the bulk of the film be the $y$ axis. Then, the $x$ axis defines the easy axis of the N. A. regions. If the total volume fraction of the N. A. region is small the first-order effect of the bulk of the film, saturated in the $+y$ direction, can be represented by an effective field $\mathbf{H}_{dc}$, also in the $+y$ direction, along the hard axis of the N. A. region. Obviously $|\mathbf{H}_{dc}|$ is determined by the size and shape of the N. A. region. If the applied field $\mathbf{H}_0$, is then rotated by $360°$, the circle it describes in the $(H_x, H_y)$ plane will be shifted a distance $|\mathbf{H}_{dc}|$ towards the cusp in the $+y$ direction and will not, as in the normal case, be symmetric with respect to the asteroid.[10]

Figure 1(d) shows the angular dependence of the torque $\mathbf{M}_2 \times \mathbf{H}_0$, where $\mathbf{M}_2$ is the magnetization of the N. A. region, for clockwise and counterclockwise rotation of $\mathbf{H}_0$, for the particular case $K_1 = 2K_2$, $|\mathbf{H}_0| = 0.4 H_{k_1}$, and $|\mathbf{H}_{dc}| = 1.1 H_{k_2}$. The calculation was made by determining $\mathbf{H}_e$, where $\mathbf{H}_e = \mathbf{H}_0 + \mathbf{H}_{dc}$ for various orientations of $\mathbf{H}_0$. Then, using $\mathbf{H}_e$, the position of $\mathbf{M}_2$ was found from the tables provided by Stoner and Wohlfarth.[4]

The general features of the curve can, however, be deduced by examining Fig. 2. As $\mathbf{H}_0$ is rotated clockwise, $\mathbf{M}_2$ will be moved reversibly, as long as $\mathbf{H}_0$ is in either the first, third, or fourth quadrants. In these areas, $\mathbf{H}_e$ either lies outside the asteroid or $\mathbf{H}_e$ and $\mathbf{M}_2$ already lie in the same quadrant. Only in the second quadrant, when $\mathbf{H}_0$ crosses the asteroid at A does an irreversible jump occur, the only one in a complete cycle of $360°$. For counterclockwise rotations, $\mathbf{M}_2$ is moved reversibly, when $\mathbf{H}_0$ is in any quadrant except the third. There, $\mathbf{M}_2$ jumps when $\mathbf{H}_0$ crosses the asteroid at B. Thus, in this picture, the only part of the torque which is irreversible occurs between the points A and B, straddling an easy direction of the bulk of the film.

If the torque, $\mathbf{M}_1 \times \mathbf{H}_0$, due to the interaction of $\mathbf{H}_0$ with the bulk of the film is superimposed, assuming that the volume fraction of the total N. A. regions is $< 0.2$, the result [Fig. 1(e)], is seen to be qualitatively in agreement with experiment [Fig. 1(a)].

According to the proposed model, the direction about which the loss occurs should be opposite to the direction of $\mathbf{H}_{dc}$, and therefore, of $\mathbf{M}_1$. This has in fact, been verified experimentally. Figure 1(a) shows the unidirectional loss, occurring around $\theta = \pi$, after the film had

[8] D. O. Smith, J. Appl. Phys. **32**, 70S (1961).

[9] L. D. Landau and E. M. Liftshitz, *Electrodynamics of Continuous Media* (Addison–Wesley Publishing Company, Inc., Reading, Massachusetts, 1960).

[10] It is assumed that $H_0 < 0.5 H_{k_1}$, so that the bulk of the film behaves essentially as a permanent magnet.

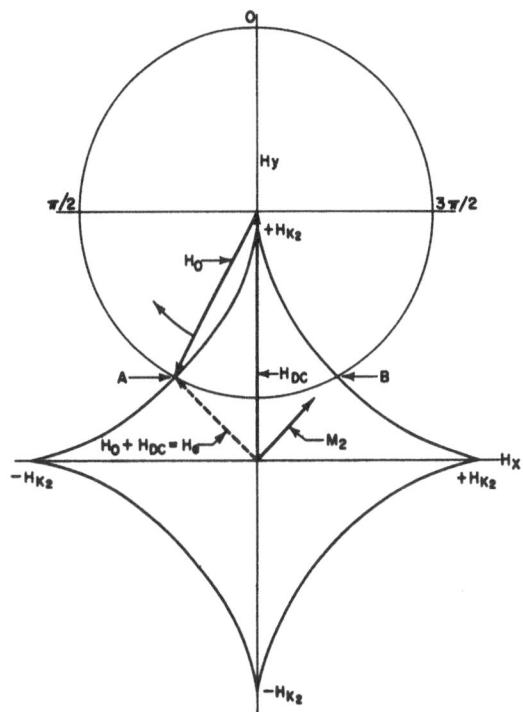

FIG. 2. The (S–W) critical switching curve applied to an N. A. region. The effect of the bulk film saturated in the $+y$ direction is represented by the field $\mathbf{H}_{dc}$. The $x$ axis defines the easy axis of the N. A. region.

been first saturated at $\theta = 0$. In Fig. 1(b), the loss shifted to $\theta = 0$, after the film was saturated at $\theta = \pi$. When the film was demagnetized, by splitting it into domains, half the N. A. regions experienced an interaction field in the $\theta = 0$ direction, and half in the $\theta = \pi$ direction. The loss then appeared as expected, around both directions [Fig. 1(c)].

The magnitude of the unidirectional loss (observed in a series of films[11] whose $h_c$ varied from 0.4 to 1.5) has been found to increase sharply above $h_c = 0.7$. This is indicative of a similar increase in the total volume fraction of N. A. regions, in agreement with Smith's[8] results.

It is concluded that the field dependence of the rotational hysteresis becomes quite complex when $h_c$ is not $\ll 1$. The unidirectional hysteresis observed at low fields in films with $h_c > 0.5$, can be understood qualitatively if the existence of regions with negative anisotropy is assumed. Results interpreted in terms of this model indicate that the total volume fraction of negative anisotropy regions increase sharply above $h_c = 0.7$. An analytic treatment of the model, which will allow a quantitative interpretation of the experimental data, is presently being considered.

[11] Several films were kindly provided by A. Noreika, Philco Corporation, Philadelphia, Pennsylvania; D. O. Smith, Lincoln Laboratory, Lexington, Massachusetts; and M. Prutton, International Tabulator and Computer, Limited, Stevenage, England.

# Ferromagnetic Resonance in Single-Crystal Nickel Films

M. POMERANTZ, J. F. FREEDMAN, AND J. C. SUITS

*International Business Machines Corporation, Thomas J. Watson Research Center, Yorktown Heights, New York*

Ferromagnetic resonance measurements have been carried out on single-crystal nickel films obtained by epitaxial evaporation on NaCl under various evaporation conditions. Resonance measurements on thin films by previous investigators have yielded anomalous values of $4\pi M$ which have tentatively been attributed to stress in the films. In this paper it is shown that there is a second factor which must be taken in consideration to explain the shift in resonance peak.

Films that were grown at high temperatures (400°C) and in good vacuum ($10^{-10}$ mm Hg) exhibited an apparent saturation magnetization only 60–70% of the bulk value. This is in accord with the results of ferromagnetic resonance experiments of Kuriyama *et al.*, who attempted to explain their results by the presence of large tensile stresses in the films. However, electron microscope studies of films grown under the above conditions indicated that the films formed discrete, well-separated islands of nickel. The observed particle sizes are such that the demagnetizing factor perpendicular to the film is 60–70% of the value for a continuous planar film. This implies that the shift in the resonance peak was caused not by stresses but by a lowering of the demagnetizing field resulting from the discrete particle growth.

To further substantiate this, continuous films grown at lower temperatures (300°C) and in technical vacuum ($10^{-5}$ mm Hg) exhibited a magnetization approximately 60% *higher* than the bulk material values. X-ray diffraction indicated the presence of large compressive stresses in these films. That the increased magnetization was induced by these compressive stresses was convincingly shown by the decrease of the saturation magnetization to bulk value when the stress was relieved by the introduction of water vapor into the cavity. No such change in the external field at resonance was observed for the discontinuous films under a similar exposure to water vapor.

## INTRODUCTION

IN recent years there has been considerable experimental work on the ferromagnetic resonance absorption of evaporated ferromagnetic films. In general, these experiments yield a shift in the resonance peak as compared to unstrained bulk materials. This shift yields anomalous values for the saturation magnetization when calculated from the Kittel formula[1]

$$\nu = \frac{\gamma}{2\pi}\{[H+(N_y-N_z)M][H+(N_x-N_z)M]\}^{\frac{1}{2}}, \quad (1)$$

where $\nu$ is the microwave frequency at resonance, $\gamma$ is the gyromagnetic ratio for an electron spin, $H$ is the externally applied magnetic field in the $z$ direction, $M$ is the saturation magnetization, and $N_i$ is the demagnetizing factor in the $i$th direction (for thin planar films $N_x=0$, $N_y=0$, $N_z=4\pi$), where the $Z$ axis is chosen perpendicular to the plane of the film.

It has been suggested by MacDonald[2] that the shift in resonance can sometimes be accounted for by the presence of stresses in the film. In the case of a planar film, stresses can be included in the Kittel formula in the form

$$\nu = \frac{\gamma}{2\pi}\left[H-\left(4\pi M+\frac{3\lambda\sigma}{M}\right)\right] \quad \text{(for } H \perp \text{film)}$$

$$\nu = \frac{\gamma}{2\pi}\left[\left(H+4\pi M+\frac{3\lambda\sigma}{M}\right)H\right]^{\frac{1}{2}} \quad \text{(for } H \parallel \text{film)},$$

$$(2)$$

where $\sigma$ is an isotropic stress (if $\sigma$ is positive, the stress is tensile) acting in the plane of the film, and $\lambda$ is the magnetostriction.

This paper reports on ferromagnetic resonance absorption by single-crystal nickel films grown under widely varying conditions of vacuum ($10^{-5}$ to $10^{-10}$ mm Hg) and substrate temperature (300° to 500°C). These measurements show that a second factor must be taken into consideration when explaining shifts in the resonance peak.

## EXPERIMENTAL TECHNIQUES

Single-crystal nickel films were obtained by epitaxial growth on freshly cleaved (100) faces of NaCl. After preheating one hour at 500°C the substrate temperature was maintained at a temperature between 300° and 500°C. Evaporations were made in two separate vacuum systems, one a conventional oil diffusion pumped system capable of $10^{-5}$ mm Hg during evaporation, and the second a bakeable all glass system with a mercury diffusion pump capable of $10^{-10}$ mm Hg during evaporation. Thicknesses ranged from 250 to 1000 A.

The microwave spectrometer was a Varian 4500 ESR model with modified microwave circuitry. The operating frequency for all measurements was 9.023 kMc, and the external field $H$ was varied to obtain resonance for the various samples. Magnetic fields were measured with a Numar nuclear magnetic resonance gaussmeter. Magnetic field modulation was employed and the derivative of the absorption was recorded. All measurements were made at room temperature.

A section of each film examined in the resonance experiments was floated off the rocksalt substrate in water, picked up on 200 mesh copper grids, and examined by electron transmission diffraction and micrograph techniques. For support, a carbon backing was first evaporated onto the thinner films.

## EXPERIMENTAL RESULTS AND DISCUSSION

If one applies the Kittel formula [Eq. (1)] to an infinite sheet with external magnetic field perpendicular

[1] C. Kittel, Phys. Rev. **71**, 270 (1947); **73**, 155 (1948).
[2] J. R. MacDonald, Proc. Phys. Soc. (London) **A64**, 968 (1951).

to the surface, i.e., $N_z = 4\pi$, $N_x = N_y = 0$, with our value of the microwave frequency ($\nu = 9$ kMc) and the accepted value[3] of $M = 485$ gauss for nickel, the external field at resonance is calculated to be $H = 9200$ oe.

It was found that in films grown at high substrate temperatures ($>400°C$) and in ultrahigh vacuum ($<10^{-9}$ mm Hg) the ferromagnetic resonance occurred at 6900 oe. If this is interpreted according to Eq. (1), it gives an apparently low value of $M = 310$ gauss, about 65% of the accepted value. This is in agreement with the observations of Kuriyama et al.,[4] who further suggested applying Eqs. (2) to this case. Their explanation requires that the stresses $\sigma$ be tensile ($\sigma > 0$) because for nickel the magnetostriction is negative. This is in disagreement with the work of Freedman[5] who showed that for nickel films grown on rocksalt at elevated temperatures the difference in thermal contraction produced a residual compressive stress.

Electron microscope studies of those films for which the field at resonance was low indicated that the films were in the form of discrete, well-separated islands of nickel (Fig. 1). The ratio of particle diameter to thickness is approximately 3:1, and appeared to be relatively independent of thickness. Ellipsoids of this axial ratio have demagnetizing factors of $N_x \approx N_y \approx \pi/2$, $N_z \approx 3\pi$. Such a change in the demagnetizing factor would account for the observed downward shift in resonance peak. The films used by Kuriyama et al. were deposited[4] at high temperatures ($>430°C$) and in high vacuum ($3\times10^{-6}$ mm Hg) so that it is possible that their observations of low magnetization were the result of demagnetization effects rather than stress.

To further substantiate this, continuous films were obtained by growth at lower substrate temperatures ($<400°C$) and in a vacuum of $10^{-5}$ mm Hg. The ferromagnetic resonance in these films occurred at $H = 13\,000$ oe. Applying Eq. (1), one finds an apparent value of $M = 800$ gauss, approximately 60% higher than the accepted value. Electron microscopy showed that these films were continuous, so that the pertinent demagnetizing factors were $N_x = N_y = 0$, $N_z = 4\pi$.

The upward shift of the resonance field in the continuous films can be accounted for by the presence of large compressive stresses. Using the revised Kittel formula [Eq. (2)] and the value[3] of $\lambda_{100} = -46\times10^{-6}$ one obtains $\sigma = -1.2\times10^{10}$ d/cm², in excellent agreement with the elasticity calculation of Freedman[5] based on x-ray data. That the apparent increase in $M$ was induced by compressive stresses was convincingly shown when water vapor was introduced into the cavity. This had the effect of releasing the film from the NaCl substrate and relieving any elastic stresses. It was

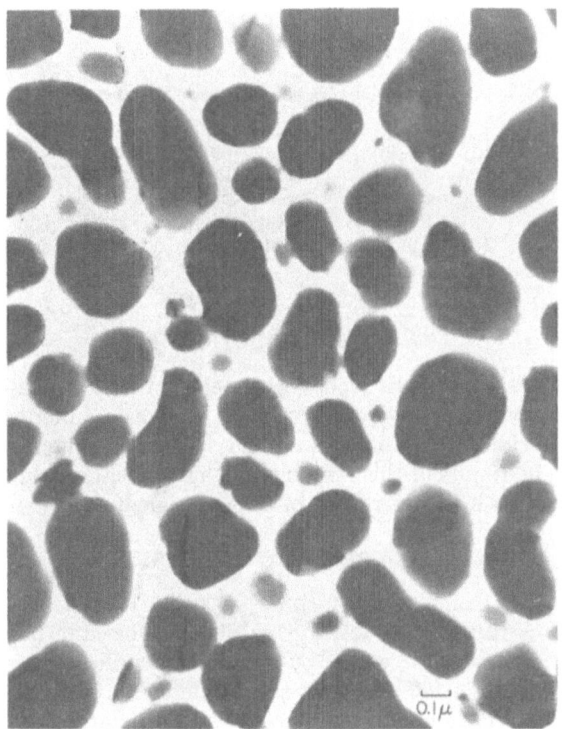

FIG. 1. Electron micrograph of a nickel single crystal grown on NaCl at 400°C and in a vacuum of $(2)\times10^{-10}$ mm Hg. Average film thickness is 1000 A.

observed that the ferromagnetic resonance shifted downward to very nearly the normal value of 9200 oe. after the water vapor treatment. No such change of the field at resonance was observed for the discontinuous films under a similar exposure to water vapor.

## CONCLUSION

In general, it was found that the external magnetic field at ferromagnetic resonance in nickel single-crystal films was strongly influenced by the vacuum and substrate temperature during deposition. The external field at resonance for unstrained, uniform films of nickel is $H = 9200$ oe. for a resonance frequency of 9 kMc. It was found that films grown at high substrate temperatures and ultrahigh vacuum had low values of resonance field, about 7000 oe. These films were observed to be in the form of discrete islands, and we attribute the low resonance field to the fact that the demagnetizing factors for such samples are $N_x \approx N_y \approx \pi/2$, $N_z \approx 3\pi$.

On the other hand, films grown at lower substrate temperatures and technical vacuum had fields at resonance of about $H = 13\,000$ oe. This upward shift of the resonance is caused by compressive stresses in the film; the films were continuous and the demagnetizing factors were $N_x = N_y = 0$, $N_z = 4\pi$. When the stresses were relieved the resonance field in the continuous films dropped to very nearly the normal value.

[3] R. M. Bozorth, Ferromagnetism (D. Van Nostrand Company, Inc., Princeton, New Jersey, 1951).
[4] M. Kuriyama, H. Yamanouchi, and S. Hosoya, J. Phys. Soc. Japan 16, 701 (1961).
[5] J. F. Freedman, J. Appl. Phys. 33, 1148 (1962), this issue.

JOURNAL OF APPLIED PHYSICS     SUPPLEMENT TO VOL. 33, NO. 3     MARCH, 1962

# Isotropic Stress Measurements in Permalloy Films

G. P. WEISS AND D. O. SMITH

*Lincoln Laboratory,\* Massachusetts Institute of Technology, Lexington 73, Massachusetts*

Isotropic stress measurements in Permalloy films have been carried out as a function of thickness, rate of deposition, and substrate temperature. The measurements were made by clamping one end of a substrate consisting of a thin strip of glass or mica and observing the deflection of the free end during deposition. Results indicate that the stress is independent of thickness in the range 100–2000 A. However, depending upon the values of the other deposition parameters, one may observe either tensile, compressive, or zero stress.

## INTRODUCTION

AN investigation has been made of the state of stress in evaporated Permalloy films as a function of thickness, deposition rate, and substrate temperature. The results of this investigation are presented together with a brief description of the experiment.

## EXPERIMENTAL PROCEDURE

The calculation of stress in a thin film from the curvature of the substrate is a well-recognized technique which was first worked out for electrodeposited films by Stoney[1] in 1909. Brenner and Senderoff[2] have given the problem a more rigorous analysis and state that the limits in which Stoney's calculations can be used without considerable error are for films whose thicknesses are no more than a few percent of the substrate thickness.

In this experiment one end of a thin glass or mica strip which was used as the substrate was clamped and the deflection of the free end from its equilibrium position was measured with a traveling microscope whose resolution is 0.0004 in.

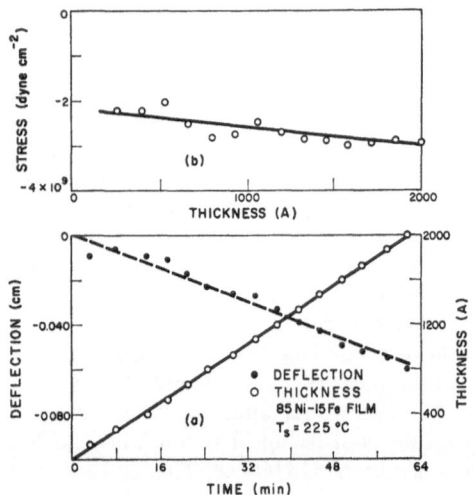

FIG. 1. (a) Experimental plot of film thickness and substrate deflection vs time. (b) Stress vs thickness as calculated from (a).

\* Operated with support from the U. S. Army, Navy, and Air Force.
[1] G. G. Stoney, Proc. Roy. Soc. (London) **A82**, 172 (1909).
[2] A. Brenner and S. Senderoff, J. Research Natl. Bur. Standards **42**, 105 (1949).

In order to study the stress as a function of film thickness, measurements must be made during deposition and the film must be kept at the desired temperature during the evaporation.

A furnace has been designed which consists of a simple box with a built-in heater. A clamp is provided inside the furnace for mounting the substrate. A shield has been installed between the heater and the substrate to prevent direct radiation from the heater upon the substrate. A pair of small openings have been provided for illumination and observation of the free end of the substrate. Provisions were made to mount a resistance monitor on the furnace so that a continuous record of the film thickness and hence the rate of deposition could be had. In addition there are provisions for mounting additional samples for other measurements.

The deposition rates chosen for this experiment varied between 25 and 250 A/min. A number of times during the evaporation, corresponding to predetermined thicknesses as given by the resistance monitor, both the deflection and the time were recorded. Experimental data indicating the constancy of the deposition rate and the deflection of the substrate are shown in Fig. 1(a). At each of the deposition rates selected evaporations were made at a number of different substrate temperatures in the range 25–350°C.

The following experiment was performed to investigate the influence of substrate cleanliness on the measured stress at a given substrate temperature. On several occasions the substrate was maintained at 300°C for one hour and then allowed to cool to a lower temperature just prior to evaporation. No significant differences in the stress were noted between these films and films made on substrates heated directly to the desired temperature. At the completion of the evaporation the sample was coated with silver and the film thickness was determined by interferometry and the stress was then calculated.

## RESULTS

The data as calculated[1,2] [see Fig. 1(b)] show that there is a small thickness dependence of stress in the range 100–2000 A; this dependence is negligible compared to the temperature dependence discussed below. It was found that in the Permalloy range near composition 83% Ni-17% Fe there is no compositional dependence of stress. It was found, however, that there exists a dependence both upon deposition rate and substrate temperature. As shown in Fig. 2 the stress may be either tensile, compressive, or zero depending upon the deposition rate and substrate temperature.

A number of other investigators[3-6] also report tensile stress in metallic films evaporated at pressures of $\sim 10^{-5}$ mm Hg, while Murbach and Wilman[5] report the finding of compressive stress in aluminum evaporated at pressures of $\sim 10^{-4}$ mm Hg upon copper substrates.

It should also be noted in Fig. 2 that there is a shift of the temperature range for compressive stress to the right (higher temperatures) for higher deposition rates.

Hysteresis loop measurements have been made upon a number of samples from each evaporation and it has been observed that for high substrate temperatures (above 200°C) the films are quite normal, while for lower substrate temperatures the films are abnormal and possess a high wall coercive force (30 to 50 oe). The value of the initial permeability of these abnormal films depends upon the values of magnetic fields previously applied. Similar effects have been studied by Cohen.[7]

## DISCUSSION

The mechanism for the generation of the observed stress is unknown. It is speculated that the stress is related to the surface energy of the small crystallites which make up these polycrystalline films. Electron microscopy observations of Permalloy films[7] show that at low substrate temperatures the crystallite size is small (<100 A). It can be imagined that tensile stress is produced by the coalescence of neighboring crystallites which are also firmly bonded to the substrate.[8] At

[3] E. C. Crittenden, Jr., and R. W. Hoffman, Phys. Rev. **78**, 349 (1950).

[4] R. W. Hoffman, F. J. Anders, and E. C. Crittenden, Jr., J. Appl. Phys. **24**, 231 (1953).

[5] H. P. Murbach and H. Wilman, Proc. Phys. Soc. (London) **B66**, 905 (1953).

[6] R. W. Hoffman, R. D. Daniels, and E. C. Crittenden, Jr., Proc. Phys. Soc. (London) **B67**, 497 (1954).

[7] M. S. Cohen (private communication).

[8] D. O. Smith, M. S. Cohen, and G. P. Weiss, J. Appl. Phys. **31**, 1755 (1961).

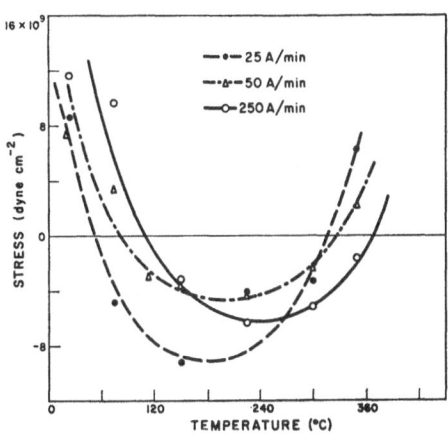

FIG. 2. Stress vs temperature for various deposition rates. Positive values indicate tensile stress while negative values indicate compressive stress.

higher values of substrate temperature the crystallite size is increased and the compressive stress related to surface tension of the individual crystallite predominates.

It should be noted (Fig. 2) that at substrate temperatures in excess of 300°C the state of stress changes from compressive to tensile which is in contradiction to the above discussion. It is possible, however, for both of these mechanisms to be working simultaneously. The state of stress will then depend upon the balance between these two mechanisms.

Crystallite size is also influenced by deposition rate. At low substrate temperatures the crystallite size is observed[7] to increase as the deposition rate is decreased. Thus it is expected that the transition from tensile to compressive stress would occur at a higher temperature as the deposition rate is increased. That this is the case is seen in the experimental data of Fig. 2.

## ACKNOWLEDGMENT

It is a pleasure to acknowledge the assistance of C. Westcott in making all the evaporations.

JOURNAL OF APPLIED PHYSICS   SUPPLEMENT TO VOL. 33, NO. 3   MARCH, 1962

# Hall Effect Determination of Planar Stress in Ferromagnetic Films

R. L. COREN

*Philco Research Division, Blue Bell, Pennsylvania*

Hall effect measurements on a ferromagnetic film can be used to examine the normal anisotropy and the distribution of stresses in the plane of the film. The measured Hall voltage exhibits a smooth transition from low to high field regions rather than the sharp crossover, at technical saturation, which is expected of a single domain. Since the ferromagnetic Hall effect depends upon the component of magnetization perpendicular to the plane one can attribute this difference of behavior to a domain structure describable by a distribution of normal anisotropies. The difference between the single and multi-domain Hall plots depends upon the nature of this distribution and from certain features of the experimentally obtained difference curve one can infer the type and extent of the distribution. If the anisotropy distribution is due to a corresponding range of strains superimposed on the geometric anisotropy of the film one can also estimate the form and extent of the function describing the planar stresses. It is shown that the existence of the anisotropy distribution results in a difference between the field at which the Hall measurements indicate technical saturation and that at which the normal anisotropy distribution is peaked.

## INTRODUCTION

SMALL departures from purity, crystal perfection, stoichiometry, etc., affect the magnetic properties of ferromagnetic films, primarily by inducing strains; evaporated films are known[1] to be stressed in their plane to the extent of about $10^9$ d/cm². In any sample the nonuniformity of these factors leads to a distribution of magnetic properties and, since electrical measurements on ferromagnetic materials reflect their magnetic behavior,[2] this distribution can be studied electrically. From the Hall effect one can obtain a measure of the distribution of magnetic anisotropies normal to the film plane and an estimate can be made of the distribution of planar stress.[3]

In a single domain film the magnetization dependent Hall resistivity may be written[4]

$$\rho_H = E/J = RM \sin\theta, \qquad (1)$$

where $E$ and $J$ are mutually perpendicular in the film plane, $\theta$ is the angle the magnetization $M$ makes to the plane, and $R$ is the ferromagnetic Hall coefficient. Equation (1) neglects a similar field dependent term which can be subtracted experimentally. Sin$\theta$ can be found by minimizing the magnetic free energy $F$. With a magnetic field $H$ applied in the normal direction, the major terms in $F$ are

$$F = K \sin^2\theta - HM \sin\theta. \qquad (2)$$

While many factors might contribute to the anisotropy constant $K$, it is primarily the sum of a demagnetizing energy $2\pi M^2$ and a term $\frac{3}{2}\sigma\lambda$ ($\sigma$=planar stress, $\lambda$=isotropic magnetostriction), arising because of the high stresses mentioned above. Thus,

$$\sin\theta = H/H_K; \quad H_K = 2K/M, \qquad (3)$$

so that the Hall resistivity of Eq. (1) increases uniformly with $H$ and at $H=H_K$ becomes constant. Experimentally[5,6] one finds a smooth transition between these two linear regions due to the fact that the film posesses a multidomain structure with a range of $H_K$ values.

## ANISOTROPY DISTRIBUTION

If the probability of a particular anisotropy field $P(H_K)$ is zero outside of the range $H_1 < H_K < H_2$, then the measured Hall effect is an average over all domains:

$$\rho(H) = RM \left[ \int_{H_1}^{H} P(H_K)dH_K + H \int_{H}^{H_2} \frac{P(H_K)}{H_K}dH_K \right]. \quad (4)$$

The first integral represents the contribution from regions with $H_K < H$, so that $\sin\theta = 1$, the second is over those remaining ($H < H_K < H_2$). It follows that

$$\begin{aligned} \rho(H) &= RMH/H_s, \quad H < H_1 \\ &= RM, \quad H > H_2. \end{aligned} \qquad (5)$$

These lines intersect at the field $H = H_s$ where

$$\frac{1}{H_s} = \int_{H_1}^{H_2} \frac{P(H_K)}{H_K}dH_K. \qquad (6)$$

To obtain a measure of $P$, at each field subtract the experimental multidomain curve [Eq. (4)] from the high and low field lines [Eqs. (5) and (6)]. The difference $\Delta$ is given by

$$\begin{aligned} \Delta(H) &= RM \int_{H_1}^{H} P(H_K)\left(\frac{H}{H_K} - 1\right)dH_K; \quad H_1 < H < H_s \\ &= RM \int_{H}^{H_2} P(H_K)\left(1 - \frac{H}{H_K}\right)dH_K; \quad H_s < H < H_2 \end{aligned} \qquad (7)$$

and is zero outside the region $H_1 - H_2$.

[1] R. W. Hoffman, R. D. Daniels, and E. C. Crittenden, Proc. Phys. Soc. (London) **B67**, 497 (1954).

[2] H. J. Juretschke in *Structure and Properties of Thin Films*, edited by C. A. Neugebauer, J. B. Newkirk, and D. A. Vermilyea (John Wiley & Sons, Inc., New York, 1959), p. 410; R. L. Coren and H. J. Juretschke, J. Appl. Phys. **32**, 292S (1961).

[3] W. Hellenthal, Z. Physik **156**, 573 (1959).

[4] E. M. Pugh, Phys. Rev. **36**, 1503 (1930); E. M. Pugh and T. W. Lippert, Phys. Rev. **42**, 709 (1932).

[5] A. Colombani and G. Goureaux in *Structure and Properties of Thin Films*, edited by C. A. Neugebauer, J. B. Newkirk, and D. A. Vermilyea (John Wiley & Sons, Inc., New York, 1959), p. 393.

[6] R. Coren and H. J. Juretschke, J. Appl. Phys. **28**, 806 (1957).

In principle, $P(H)$ can be obtained by taking the second slope of $\Delta$ but as $\Delta$ is generally a small quantity and is obtained by graphical subtraction it is not sufficiently well defined to allow such detailed operations. Consequently, in studying $\Delta$ it is desirable to use those of its features which do not depend sensitively upon detailed shape and which can be measured without actually constructing $\Delta$. The crossover field $H_s$, the field values $H_1$ and $H_2$ at which $\Delta$ goes to zero, and $\Delta(H_s)$, its maximum value satisfy these criteria. From $H_1$, $H_2$, $H_s$ we determine the experimental fractional limits

$$f_1' = (H_s - H_1)/H_s; \quad f_2' = (H_2 - H_s)/H_s. \quad (8)$$

For any function $P$ Eqs. (6) and (7) yield $H_s$ and $\Delta(H_s)$ as functions of the parameters describing $P$ so that by comparing experimental and computed values of $\Delta(H_s)/RM$ for the same $f_1'$, $f_2'$ these parameters can be chosen.

To illustrate this scheme for analyzing the distribution we refer to Table I which contains data, on evaporated nickel films, from Hall effect measurements which have been reported elsewhere.[6] These data can be analyzed in terms of a distribution with a fraction $n$ of the domains at anisotropy field $H_0$ and the remaining $1-n$ distributed uniformly between $H_1$ and $H_2$.

$$P(H_K) = n\delta(H_0) + \frac{1-n}{(f_1+f_2)H_0}, \quad H_1 < H_K < H_2, \quad (9a)$$

$$f_1 = (H_0 - H_1)/H_0, \quad f_2 = (H_2 - H_0)/H_0. \quad (9b)$$

$f_1$ and $f_2$ are the theoretical fractional limits and $\delta$ is the Dirac point function. While other distributions may also be in accord with the measurements, this simple form can be used to estimate relevant properties. $n$ may be taken as a measure of the distribution sharpness. For the films under consideration the parameters of Eq. (9) are shown in Table I. Since $f_1 = 0$ it follows that $H_0 = H_1$.

### DISCUSSION

For both films $H_0 \doteq 2.7$ koe, which is somewhat lower than the value calculated from their known magnetization,[6] i.e., $H_K = 4\pi M \approx 3.8$ koe. The difference can be attributed to a tensile strain approximately equal to that which should develop as a result of the difference in thermal expansion coefficients of the nickel and glass substrate, upon cooling the sample to room temperature after annealing at 275°C.

Since only a relatively small part of the domains ($n = 0.15$, $0.38$ at 100 A, 60 A, respectively) have the anisotropy value $H_0$, it appears that in most parts of the film the tension described above is not the only factor influencing the anisotropy. As the entire distribution is at values of $H_K$ greater than $H_0$, local factors which

TABLE I. Analysis of normal anisotropy distribution.

| | 100 A | 60 A | | 100 A | 60 A |
|---|---|---|---|---|---|
| $H_1$ | 2.69 koe | 2.70 koe | $n$ | 0.15 | 0.38 |
| $H_s$ | 2.90 koe | 3.20 koe | | | |
| | | | $f_1$ | 0 | 0 |
| $H_2$ | 3.05 koe | 4.65 koe | | | |
| | | | $f_2$ | 0.14 | 0.72 |
| $f_1'$ | 0.072 | 0.156 | | | |
| | | | $H_s/H_0$ | 1.07 | 1.18 |
| $f_2'$ | 0.052 | 0.453 | | | |
| $\Delta(H_s)/RM$ | 2.17% | 8.27% | | | |

vary in magnitude at different points of the film must act to relieve the tensile stress, i.e., they are entirely compressive in nature. This must be characteristic of the particular mechanism of strain development, e.g., the inclusion of interstitial impurity atoms such as oxygen or nitrogen and the formation of oxides at crystal boundaries would produce compressive stresses in the magnetic material. This mechanism is reasonable in view of the fact that these films were grown quite slowly at $10^{-6}$ torr.[6]

The anisotropy distribution arising from the planar stresses results in a separation of the fields at technical saturation and at the distribution maximum. This difference, which may be quite large (18% in the case of the 100-A film), can be adjusted through Eq. (6). In view of the unresolved situation with regard to quantitative theories of the extraordinary Hall effect[7] such corrections may be quite significant.

It should be noted that a slight misalignment of the applied field from the normal direction to the plane results in a gradual approach to saturation and, even in a single domain film, this produces a rounding of the Hall curve similar to that produced by the anisotropy distribution.[8] For this reason extreme care must be taken to assure exact field alignment when performing measurements to be analyzed in this way. With the films discussed here, proper geometry was assured by adjusting the film position, in a field near $H_s$, until the Hall signal was maximum.

While no attempt has been made to analyze the sensitivity of this method to the choice of anisotropy distribution, it may be noted that the data for the films considered here could not be fitted to a distribution involving a power law dependence or to a uniform distribution with two superimposed delta functions, except in the limit that these cases reduced to Eq. (9). Within this model the limiting factor in defining the theoretical parameters is experimental accuracy.

[7] J. M. Luttinger, Phys. Rev. 112, 739 (1958); J. Smit, Physica 24, 39 (1958).

[8] K. M. Koch, W. Rindner, and K. Strnat, Z. Naturforsch. 13a, 113 (1958).

JOURNAL OF APPLIED PHYSICS    SUPPLEMENT TO VOL. 33, NO. 3    MARCH, 1962

# Magnetoelastic Behavior of Thin Ferromagnetic Films as a Function of Composition*

E. N. MITCHELL AND G. I. LYKKEN
*University of North Dakota, Grand Forks, North Dakota*

A series of films of composition varying from 72% Ni, 28% Fe to 90% Ni, 10% Fe were evaporated at a pressure of the order of $10^{-5}$ mm Hg onto a heated glass substrate (250°C) in the presence of a magnetic orienting field in the plane of the film. The composition was controlled by placing a charge of the selected composition on the filament in an amount just sufficient to make the desired 3000 A film when all of the material was evaporated.

The magnetoelastic behavior of the films was studied by measuring $\eta = \Delta H_k/(\Delta l/l)$, here $\Delta H_k$ is the change in the anisotropy field upon the application of a stress sufficient to strain the film an amount $\Delta l/l$ (typically $4 \times 10^{-5}$). $\eta$ was found to decrease linearly with composition to a first approximation with a slope of $-26.5 \times 10^3$ oe per percent nickel, with $\eta$ being zero for a film composed of 81.5% Ni, 18.5% Fe.

The value of $\eta$ was compared with that of $\Lambda_m$ for bulk material where $\Lambda_m$ is the change in the magnetization divided by the applied stress in the case where the stress approaches zero and where the bulk material is biased by an applied field sufficient to yield the maximum change in $\Lambda$. Using bulk data for the saturation magnetization and Youngs modulus the value of $\eta$ measured here yields a value for $\Lambda_m$ which is too large by an order of magnitude.

## INTRODUCTION

THE subject of this paper is a study of the response of anisotropy field to mechanical strain of a particular set of iron nickel films ranging in composition from 72% Ni–28% Fe to 90% Ni–10% Fe. This study established that within the framework of certain restrictions there exists a one-to-one correlation between the magnetoelastic behavior of these films and their composition. A comparison with comparable data for bulk material indicates that the response differs from that for bulk material by an order of magnitude.

## EXPERIMENTAL PROCEDURE AND RESULTS

The films studied were made by evaporation of iron and nickel wire from a tungsten filament onto a glass substrate heated to about 250°C. The deposition was made in the presence of a dc magnetic field oriented parallel to the plane of the glass and about 20 oe in

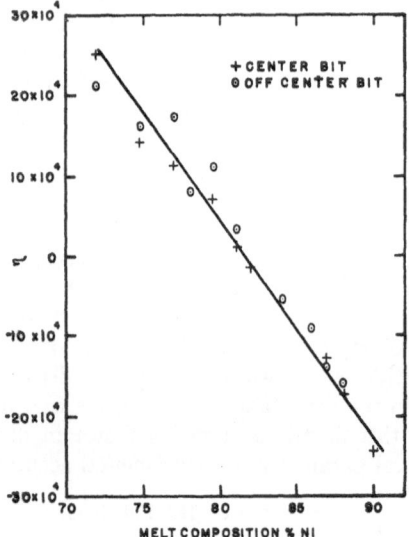

FIG. 1. Magnetoelastic response of iron nickel films of specified compositions in terms of $\eta$.

* This research was supported in part by a grant from the National Science Foundation.

magnitude. The residual pressure in the system during deposition was of the order of $10^{-5}$ mm Hg and was free of organic material.

The composition was predetermined by placing a metal charge on the evaporation filament of the end composition desired for the particular film in an amount sufficient to make a film about 3000 A thick. During the deposition, the filament was rapidly raised to a temperature sufficient to evaporate all the iron and nickel metal in about 30 sec. The filament assemblies were weighed before and after typical evaporations and it was determined that the weight loss of the assembly was less than the weight of the iron and nickel placed on the filament. The minimum amount of material left unevaporated on the filament was typically 20% of the metal placed on the filament for evaporation. This did not appear to be a severe limitation on the method in as much as isolated cases occurred in which as much as 40% of the metal remained whose magnetoelastic properties did not deviate from the monotonic variation of the magnetoelastic properties with composition. In addition, no systematic variation between the amount of unevaporated metal and the composition placed on the filament could be found, indicating that the mechanism was not preferential with respect to composition. The films had values of anisotropy field ($H_k = Ms/X_0$) ranging from 1.5 to about 6 oe, were about 3000 A thick and were 8 mm in diameter. The measurement of the initial magnetic susceptibility ($X_0$) was made in the difficult direction at low applied field.

The parameter chosen for measurement was designated $\eta$ and is defined as $\eta = \Delta H_k/(\Delta l/l)$, where $\Delta H_k$ is the change in anisotropy field in oersteds when the strain of the film is $\Delta l/l$. Measurements of $\eta$ were made with a drive field of the order of tenths of oersteds. The films were extended or compressed along the difficult magnetic direction by supporting the microscope substrates on fulcrums near the ends of slides and deflecting the centers of the slides down to a stop placed a predetermined distance below the undeflected plane of the slide. The strain was computed by assuming that the median surface of the slide for small deflections was a cylindrical section and that this median surface had not been extended or compressed in the bending process. The strain unless otherwise noted was $4 \times 10^{-5}$. The error introduced by the above simplification would be about equal to the precision with which the strain can be measured.

$\eta$ was measured on 22 different samples made in a total of 15 evaporations ranging in composition from 72% Ni–28% Fe to 90% Ni–10% Fe. These data are plotted in Fig. 1. A straight line is drawn on the graph to fit the data with a slope of $-26.5 \times 10^3$ oe/% Ni. On the basis of this straight line, zero magnetoelastic effect is seen to occur for a film of 81.5% Ni–18.5% Fe and is in substantial agreement with data for bulk material.[1] The error in $\eta$ introduced by limitations of precision in the measurement of $H_k$ is $7 \times 10^3$ oe. No allowance for a constant error which could result from an error in the measurement of $\Delta l/l$ was made.[2] No effort was made to estimate the precision with which the composition was determined.

The results reported here can be compared to measurements made on bulk material[1] of a parameter designated $\Lambda_m$ where $\Lambda_m = B_\sigma/\triangle$ and where $B_\sigma$ is the change in magnetization upon the application of a stress $\sigma$ to the material. The measurement is made in the limit as $\sigma$ approaches zero and with an applied magnetic field such that $\Lambda_m$ is a maximum. The general shape of the two sets of data are similar but they do not agree in detail. In particular the bulk data show an approach towards a maximum for data corresponding to the lower percentages of nickel reported here and a tendency toward a minimum for the higher percentages.

It can be shown that quantitative comparison between the two sets of data can be made by comparing $B_s\eta/H_k$ for films with $Y\Lambda_m$ for the bulk material where $B_\sigma$ is the saturation magnetization of the film, and $Y$ is the Youngs modulus of the bulk material. The former quantity is found to be an order of magnitude larger than the latter. In order for the comparison to be valid, the quantities should be measured for situations of similar values of stress (or strain) and applied magnetic field. However, the value of $\eta$ is found to be essentially independent of the magnitude of the magnetizing field and varies too slowly with $\Delta l/l$ to account for the differences.

The large difference in the behavior between the thin film and bulk forms seems to be due to the difference in form. The film in its initial state probably is under approximately isotropic strain in the plane of the film and it exhibits an anisotropy in its magnetic properties which the bulk material does not exhibit.

The question of uniqueness of the results here stated arises. In an effort to answer this and also help answer questions concerning the difference between bulk and film data, films have been made at different substrate temperatures (higher) than those reported here and it appears that the value of $\eta$ may be a function of substrate temperature. This is currently under further investigation.

## SUMMARY

In conclusion, one may state that there appears to exist a linear relation between $\eta$ and composition in the interval reported here, that the slope of this line is $-26.5 \times 10^3$ oe/% nickel, that for small strain $\eta$ varies little with strain, that for a magnetizing field $H < H_k, \eta$ is independent of the driving field and that the results reported here are at variance with bulk data by an order of magnitude. Finally, preliminary investigation indicates that $\eta$ may be a function of substrate temperature.

[1] R. M. Bozorth and H. J. Williams Revs. Modern Phys. 17, 72 (1945).

[2] *Note added in proof.* Measurements made with a new more accurate instrument indicate that $\Delta l/l$ was subject to appreciable systematic error and that $\Delta l/l$ was larger than reported.

# Magnetization

## S. Foner, *Chairman*

## Statistical Mechanics of Ferromagnetism*

Herbert B. Callen

*Laboratory for Research on the Structure of Matter, University of Pennsylvania, Philadelphia 4, Pennsylvania*

The fact that reliable statistical mechanical theories of ferromagnetism existed only in the very low temperature region (spin wave theory) and in the very high temperature region (Kramers and Opechowski) has seriously hampered the development of magnetic theory. The Weiss molecular field theory neglects all spin-spin correlations and therefore gives too high a Curie temperature, misrepresents the details of the phase transition, and is unreliable as a basis for analyzing any effect depending on spin interactions. Cluster methods attempt to introduce correlation, but they are both inadequate and sometimes run into severe difficulties, such as the appearance of an anti-Curie temperature (in the Bethe-Peierls-Weiss theory). The source of such difficulties is demonstrated by the formalism of P. W. Kasteleijn and J. Van Kranendonck [Physica **22**, 317 (1956)] who showed that the cluster theories correspond to the choice of a self-inconsistent two-particle density matrix. Furthermore the spin wave theory indicates that the interactions which are dominant in establishing spin-spin correlations are iterated interactions through enormously many long and complicated paths connecting the spins. Cluster methods are unable to include these long paths, and stress instead the direct short-path interactions. The need for a many-body theory is therefore indicated.

One such many-body theory is the Green function treatment of Bogoliubov and Tyablikov.[1] That treatment is equivalent to the random phase approximation. The simplest formulation starts with the spin wave theory of Dyson.[2] The destruction operator for a Dyson spin wave is

$$a_k \equiv \frac{1}{2NS^{\frac{1}{2}}} \sum_j S(\mathbf{R}_j) \exp(i\mathbf{k} \cdot \mathbf{R}_j).$$

Similarly Dyson introduces the Fourier components of the $z$ components of spin

$$S_k{}^z \equiv \sum_j S_z(\mathbf{R}_j) \exp(i\mathbf{k} \cdot \mathbf{R}_j).$$

Unfortunately the spin waves defined in terms of these operators are neither orthogonal nor are they eigenstates of the Hamiltonian. In fact the commutation relations are

$$[S_k{}^z, a_\lambda] = a_{k+\lambda}, \quad [a_k, a_\lambda{}^+] = (1/NS)S_{k-\lambda}{}^z$$

and

$$[H, a_k] = -\mu H a_k - (2SN)^{-1} \sum_\lambda (\epsilon_{\lambda \cdot k} - \epsilon_\lambda)(S_\lambda{}^z a_{k-\lambda} - a_\lambda S_{k-\lambda}{}^z),$$

where $\epsilon_k$ is the energy of a simple spin wave, expressed as a function of $\mathbf{k}$; for small $k$ it is proportional to $k^2$. The random phase approximation "decouples" the awkward higher-order terms in the right-hand members of these commutation relations. In lowest order this decoupling consists of replacing $S_\lambda{}^z$ by $N\langle S_z \rangle \delta_{\lambda 0}$, where $\langle S_z \rangle$ is the average value of the $z$ component of spin, to be evalu-

ated self-consistently. Then the commutation relations become

$$[a_k, a_\lambda{}^+] = (1/S)\langle S_z \rangle \delta_{\lambda k}$$

and

$$[H, a_k] = -[\mu H + \epsilon_k \langle S_z \rangle / S] a_k.$$

The first of these states that the quasi-particles (magnons) produced by the creation operators $a_k{}^+$ are bosons. The second commutation-relation states that the energy of such a quasi-particle is $\mu H + \epsilon_k \langle S_z \rangle / S$. Thus the average number of such magnons, of wave vector $k$, is

$$\langle n_k \rangle = \langle S_z \rangle / S [\exp \beta (\mu H + \epsilon_k \langle S_z \rangle S^{-1}) - 1]^{-1},$$

which is the result given by Tyablikov[1] and by Englert.[3] It differs from the simple spin wave theory by reducing the energy of each spin wave proportionally to the average magnetization.

The Green function method (or RPA) gives a result which forms a useful interpolation throughout the entire temperature range. It agrees to second order in $1/T$ with the Kramers-Opechowski series in the high temperature (paramagnetic) region. Near the Curie temperature the theory reduces to the Weiss theory, and in the very low temperature region it agrees with the simple spin wave theory. The corrections to the simple spin wave theory at slightly higher temperatures are incorrectly given, however, as the RPA approximation predicts a correction to the magnetization varying as $T^3$, whereas Dyson has shown that the lowest order correction varies as $T^4$. This error in the theory arises from the decoupling procedure, which ignores spin-wave correlation effects in replacing $S_\lambda{}^z$ by zero (if $\lambda \neq 0$).

A direct approach to a statistical theory consists in a series expansion of the partition sum, a diagrammatic representation of the terms, and a partial summation of those terms corresponding to the long, circuitous paths connecting and correlating two spins. Such a theory has been carried out by Brout[4] and co-workers and by Horwitz and Callen.[5] The results have been published, or are in publication in detailed form elsewhere for the Ising model, and the extension to the Heisenberg model will be presented in a separate more extensive publication. Again the theory provides a useful interpolation between the low and high temperature region. Evaluation of the theory at low temperatures, by R. Stinchcomb,[6] indicates that it properly gives the Dyson dynamical ($T^4$) correction.

---

* Supported by the Office of Naval Research.
[1] S. V. Tyablikov, Ukrain. Mat. Zhur. **11**, 287 (1959); see also D. N. Zubarev, Soviet Phys.—Uspekhi **3**, 320 (1960).
[2] F. J. Dyson, Phys. Rev. **102**, 1217, 1230, (1956).
[3] F. Englert, Phys. Rev. Letters **5**, 102 (1960).
[4] R. Brout, Phys. Rev. **115**, 824 (1959); **118**, 1009 (1960).
[5] G. Horwitz and H. B. Callen, Phys. Rev. **124**, 1757 (1961).
[6] R. Stinchcomb (private communication).

JOURNAL OF APPLIED PHYSICS     SUPPLEMENT TO VOL. 33, NO. 3     MARCH, 1962

# Magnetization of Localized States in Metals

P. A. Wolff, P. W. Anderson, A. M. Clogston, B. T. Matthias, M. Peter, and H. J. Williams

*Bell Telephone Laboratories, Murray Hill, New Jersey*

A study has been made of the magnetic properties of dilute solutions of iron in various nonmagnetic $4d$ series elements and alloys. In some cases the iron atoms possess a localized magnetic moment which manifests itself as an inverse temperature dependence in the susceptibility of the solution. The occurrence of the moment is determined by the valence electron concentration in the solute element (or alloy). As a function of this quantity one finds portions of the $4d$ series in which localized moments are strictly absent, interspersed by regions—one centered near Mo, the other in the vicinity of Rh and Pd—in which magnetization occurs and at whose edges the moments appear almost discontinuously.

This behavior may be understood in terms of a theoretical model in which the magnetization is ascribed to a virtual level, of the type proposed by Friedel, of the iron atom. Polarization occurs when the virtual level lies near the Fermi level and is sufficiently sharp in energy. A self-consistent Hartree-Fock calculation indicates that under these circumstances the impurity atom develops an exchange potential which splits the level—causing it to have different energies for spin-up and spin-down electrons and thus giving rise to magnetization. The model gives a qualitative description of the experiments and, in particular, is able to account for the fact that the impurity atoms sometimes carry a moment that is a fraction of a Bohr magneton.

## INTRODUCTION

THE aim of the work described below is to investigate the magnetic behavior of iron atoms when dissolved in nonmagnetic matrices. For this purpose we have measured[1] the susceptibility, as a function of temperature, of dilute solutions of Fe in an extensive series of alloys of the $4d$ metals. In some of the alloys the susceptibility is essentially temperature independent. Other cases show a Curie-Weiss type temperature dependence, indicating that in these solvents the Fe atoms have localized magnetic moments. Each solvent alloy is characterized by an electron concentration $N$ obtained by counting all electrons outside the closed $4p$ shell. The experiments show that the magnetic behavior of the dissolved Fe is principally determined by $N$. As a function of $N$ one finds portions of the $4d$ series within which localized moments are strictly absent, interspersed by regions (one centered near Mo, the other in the vicinity of Rh and Pd) where magnetization occurs, and at whose edges the moments appear almost discontinuously. In most cases the moments are of the order of two Bohr magnetons or less, but exceptions occur for alloys containing large percentages of Pd and for the Mo-Pd system in which the Fe atoms acquire very large moments.

To interpret these observations we must understand how electrons can localize and polarize in the vicinity of an Fe atom. In a metal it is unlikely that a single electron can be confined to the vicinity of the impurity. However, a form of electron localization can occur if there is a sharp resonance—corresponding to a virtual level—in the impurity scattering cross section. Under these circumstances much of the electron density in the vicinity of the scatterer comes from electrons of a rather well-defined energy, and the situation is like that which obtains with a truly localized state. Such a virtual level will magnetize if it lies near the Fermi

level. It then becomes energetically favorable to split the level—placing the resonance above the Fermi level for one sign of spin, and below for the other. In so doing one reduces the Coulomb energy—which mainly arises from interactions between electrons with parallel spin—of the localized state. A self-consistent Hartree-Fock calculation indicates that the conditions for magnetization are that the Coulomb self-energy of the localized charge be larger than both the level width, and its displacement from the Fermi level.

Because the Coulomb energy is rather large, the conditions for polarization are not particularly stringent. Thus the theory suggests—as is observed experimentally—that Fe atoms should often have a moment. It also predicts correctly the shape of the moment vs $N$ curve at the onset of magnetization, and provides a very natural explanation for the occurrence of moments that are a small fraction of a Bohr magneton. The model is not, however, sufficiently detailed to give the complete form of the moment vs $N$ curve which depends in a sensitive way on the level width and position, as well as on the band structure of the solvent alloy.

## EXPERIMENT

In the work described below the susceptibility of various alloys was measured at a field of 14000 gauss over the temperature range $1.4°K$ to room temperature. The results of these experiments, which mainly involved alloys of adjacent members of the second long transition period, are presented in Tables I, II, and III and in Figs. 1 and 2. In the three tables are listed the various alloys studied along with the average number of electrons $N$ outside the closed $4p$ shell. Each alloy is to be understood as containing one atomic percent of iron. In Table I, covering the alloys from Zr to Ru, the first four and last four entries had essentially temperature independent susceptibilities. In the remaining cases the susceptibility $\chi$ increased markedly at low temperatures, though that of the background matrix (without dissolved Fe) was nearly temperature inde-

[1] B. T. Matthias, M. Peter, H. J. Williams, A. M. Clogston, E. Corenzwit, and R. C. Sherwood, Phys. Rev. Letters **5**, 542 (1960); A. M. Clogston, B. T. Matthias, M. Peter, H. J. Williams, E. Corenzwit, and R. C. Sherwood, Phys. Rev. (to be published).

TABLE I. Magnetic moment and Curie temperature for 1% solutions of Fe in 4*d* alloys from Zr to Ru.

| Alloy | Structure | $N$ | $\dfrac{\mu}{\mu_B}$(Curie-Weiss) | $\theta(°K)$ |
|---|---|---|---|---|
| Zr | hcp | 4 | 0.0 | |
| Nb | bcc | 5 | 0.0 | |
| $Nb_{0.8}Mo_{0.2}$ | bcc | 5.2 | 0.0 | |
| $Nb_{0.6}Mo_{0.4}$ | bcc | 5.4 | 0.0 | |
| $Nb_{0.4}Mo_{0.6}$ | bcc | 5.6 | 0.3 | $-9\pm3$ |
| $Nb_{0.3}Mo_{0.7}$ | bcc | 5.7 | 0.6 | $-4$ |
| $Nb_{0.2}Mo_{0.8}$ | bcc | 5.8 | 1.1 | $-1$ |
| $Nb_{0.1}Mo_{0.9}$ | bcc | 5.9 | 1.9 | $-3$ |
| Mo | bcc | 6.0 | 2.1 | $-4$ |
| $Mo_{0.8}Re_{0.2}$ | bcc | 6.2 | 2.1 | $-6$ |
| $Mo_{0.6}Re_{0.4}$ | bcc | 6.4 | 2.2 | $-5$ |
| Re | hcp | 7.0 | 0.0 | |
| Tc | hcp | 7.0 | 0.0 | |
| $Re_{0.5}Ru_{0.5}$ | hcp | 7.5 | 0.0 | |
| Ru | hcp | 8.0 | 0.0 | |

pendent. The data are fit very well by a Curie-Weiss law of the form

$$\chi = \chi_0 + [n p^2 \mu_B^2 / 3k(T-\theta)], \qquad (1)$$

where $\chi_0$ is the susceptibility of the alloy without dissolved Fe, $n$ is the number of magnetic centers, $\theta$ the Curie temperature and $p$ the effective magnetic moment of the iron atoms given by $g[S(S+1)]^{\frac{1}{2}}$. In Table I we have listed, for each alloy, the quantities $\theta$ and $\mu/\mu_B = gS$, which is obtained from $p$ by assuming that $g=2$.

For alloys beyond Ru (listed in Table II) the susceptibility of the matrix was sufficiently temperature dependent that a separate measurement of $\chi_0$ as a function of $T$ was made in each case. The corrected susceptibility $(\chi-\chi_0)$ was then used as before to obtain $\mu/\mu_B$ and $\theta$. This procedure gives reasonable results up to $Rh_{0.7}Pd_{0.3}$, yielding the values of the parameters listed in Table II. Beyond $Rh_{0.7}Pd_{0.3}$, however, this subtractive correction is no longer adequate. The plots of $1/(\chi-\chi_0)$ vs $T$ begin to deviate appreciably from

TABLE II. Magnetic moment and Curie temperature for 1% solutions of Fe in 4*d* alloys from Ru to Pd.

| Alloy | Structure | $N$ | $\dfrac{\mu}{\mu_B}$(Curie-Weiss) | $\theta(°K)$ | $\dfrac{\mu}{\mu_B}$(Sat. Mag.) | $T_c$(°K) |
|---|---|---|---|---|---|---|
| Ru | hcp | 8.0 | 0.0 | | | |
| $Ru_{0.75}Rh_{0.25}$ | hcp | 8.25 | 0.0 | | | |
| $Ru_{0.63}Rh_{0.37}$ | hcp | 8.37 | 0.8 | $-21\pm2$ | | |
| $Ru_{0.5}Rh_{0.5}$ | hcp | 8.5 | 1.3 | $-13\pm2$ | | |
| $Ru_{0.25}Rh_{0.75}$ | fcc | 8.75 | 1.7 | $-17\pm2$ | | |
| Rh | fcc | 9.0 | 2.2 | $-14\pm2$ | | |
| $Rh_{0.7}Pd_{0.3}$ | fcc | 9.3 | 4.5 | $-2$ | | |
| $Rh_{0.55}Pd_{0.45}$ | fcc | 9.45 | 5.9(100°K) | $-2$ | | |
| $Rh_{0.4}Pd_{0.6}$ | fcc | 9.6 | 7.1(100°K) | 1 | | |
| $Rh_{0.2}Pd_{0.8}$ | fcc | 9.8 | 9.6(100°K) | 14 | 7.1 | 11 |
| $Rh_{0.1}Pd_{0.9}$ | fcc | 9.9 | 11.4(100°K) | $33\pm2$ | 9.5 | 27 |
| $Rh_{0.05}Pd_{0.95}$ | fcc | 9.95 | 12.7(100°K) | $49\pm6$ | 10.8 | 39 |
| Pd | fcc | 10.0 | 11.3(100°K) | $55\pm3$ | 9.7 | 39 |
| $Pd_{0.75}Ag_{0.25}$ | fcc | 10.25 | 8.3(100°K) | 12 | 6.3 | 11 |

straight lines, and it is impossible to make a meaningful fit to a simple Curie-Weiss law. In this range of the 4*d* series the Fe atoms have quite large moments and it seems likely that their spins polarize the surrounding matrix. Assuming that this is correct, one may proceed in the simplest possible way by supposing that the moment associated with each iron atom is proportional to the susceptibility $\chi_0$ of the matrix. If these moments are large, as is true in the present alloys, one may approximately write

$$P = P_r(\chi_0/\chi_{0r}), \qquad (2)$$

where $P_r$ and $\chi_{0r}$ designate the moment and susceptibility at some reference temperature $T_r$. With this assumption Eq. (1) takes the form

$$\frac{(\chi_0/\chi_{0r})^2}{(\chi-\chi_0)} = \frac{3k(T-\theta)}{n p_r^2 \mu_B^2}. \qquad (3)$$

Plots of the left-hand side of Eq. (3) against temperature give quite satisfactory straight lines which enable

TABLE III. Magnetic moment and Curie temperature for 1% solutions of Fe in various alloys.

| Alloy | Structure | $N$ | $\dfrac{\mu}{\mu_B}$(Curie-Weiss) | $\theta(°K)$ |
|---|---|---|---|---|
| $V_{0.67}Ru_{0.33}$ | CsCl | 6.0 | 1.1 | $-3$ |
| $V_{0.6}Ru_{0.4}$ | | 6.2 | 2.4 | $-4$ |
| $V_{0.5}Ru_{0.5}$ | Tetra. | 6.5 | 2.2 | $-2$ |
| $Mo_{0.75}Rh_{0.25}$ | hcp | 6.75 | 1.9 | $-27\pm2$ |
| $Mo_{0.72}Rh_{0.28}$ | hcp | 6.84 | 0.7 | $-4$ |
| $Mo_{0.9}Pd_{0.1}$ | bcc | 6.4 | 11.2 | $-33$ |
| $Mo_{0.8}Pd_{0.2}$ | bcc | 6.8 | 6.0 | $-10$ |
| $Mo_{0.6}Pd_{0.4}$ | bcc | 7.6 | 2.6 | $-5$ |
| $Mo_{0.45}Pd_{0.55}$ | bbc+fcc | 8.2 | 3.0 | $-15$ |
| $Mo_{0.4}Pd_{0.6}$ | fcc | 8.4 | 4.1 | $-9$ |
| $Mo_{0.2}Pd_{0.8}$ | fcc | 9.2 | 6.1 | 0 |

one to evaluate $\theta$ and $p(100°K)$. In this way we have determined $\theta$ and $\mu(100°K)$ for the alloys beyond $Rh_{0.6}Pd_{0.4}$, obtaining the values listed in Table II and presented in Fig. 1. These alloys are ferromagnetic at low temperatures so we have also included in Table II the observed saturation moment per atom of iron, and the Curie temperature $T_c$ obtained by the method of plotting $H/\sigma$ vs $\sigma_g^2$, where $\sigma_g$ is the magnetization per gram. The fact that the results for $\theta$ and $T_c$, and for $\mu/\mu_B$ determined in the ferromagnetic and paramagnetic regions, are in qualitative agreement gives support to our method of analyzing the paramagnetic susceptibility.

In addition to the main sequence of alloys presented in Tables I and II we have investigated the behavior of various other systems, partly to fill in gaps in the picture, and partly to test to what extent alloys with the same electron concentration but different composition will correspond to Fig. 1. These results are collected in Table III and shown superimposed on the curve of

Fig. 1 in Fig. 2. It is noteworthy that the vanadium-ruthenium and molybdenum-rhodium alloys—composed of elements whose $N$ values differ by three—taken together trace out a curve similar in shape to the peak defined by the Nb−Mo−Re alloys in Fig. 1. This result suggests that the band structures of all these alloys are similar, and that the rigid band model is a fair approximation. The data for the Mo−Pd dilute iron alloys do not fit on the curve of Fig. 1, though there is a deep minimum corresponding to the region where no local moment exists in the Re−Ru system. This failure is not surprising, however, since Mo and Pd are so far apart in the periodic table that the rigid band model can have little meaning for their alloys.

## INTERPRETATION

The results described above show that, in some portions of the periodic table, an iron atom present as a dilute impurity in a host lattice carries with it a localized magnetic moment. This experimental fact has two implications: First, there must be wave functions associated with the iron atoms which, in some sense, are local in character; and secondly, these local wave functions have a net spin and show a magnetic moment that may vary between zero and several Bohr magnetons. In this section we would like to discuss the nature of these local states in metals, and then consider the circumstances under which they can magnetize.

The simplest sort of localized state one can conceive is that in which a single electron moves in a closed orbit in the vicinity of an impurity. Such a model is known to provide an excellent description of impurity states in semiconductors and insulators. In metals, on the other hand, it is much less likely to be correct. There one has large conduction band widths and it is probable that the energy of a localized level will coincide with that of some conduction band states. Under these circumstances one-electron theory does not permit localization of such a state—there will always be some coupling of the impurity wave function to those of the conduction band so that an electron, initially localized in the former, will eventually leak away to the distant parts of the crystal. This is almost certainly the situation in the alloys described above. Thus we are forced to examine the problem of localization for a situation in which the energy of the impurity level lies *within* the conduction band. This question has been considered by Friedel[2] who points out that a sharp resonance in the impurity scattering cross section can be responsible for a form of electron localization that is very much akin to that occurring in the semiconductor case. The scattering cross section is proportional to the square of the electron wave function at the impurity. Thus a resonance in the former implies a maximum, at that energy, of the electron density in

[2] J. Friedel, Suppl. Nuovo cimento **7**, 287 (1958); A. Blandin and J. Friedel, J. phys. radium **20**, 160 (1959).

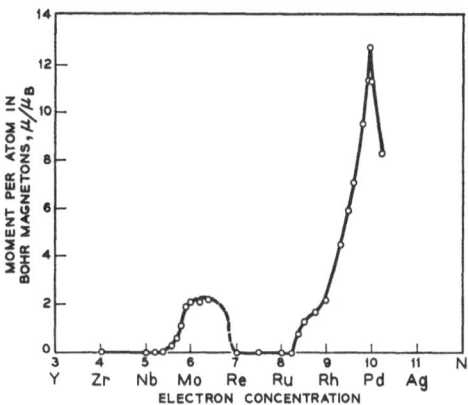

FIG. 1. Moment per iron atom vs electron concentration in alloys made up of adjacent members of the 4*d* series.

the vicinity of the scatterer. With a sharp resonance one may attain a situation—usually called a virtual level—in which most of the electron density at the impurity is due to conduction electrons lying in a narrow energy range. This state of affairs is similar to that which obtains in the semiconductor case, the principal difference being that in the metal the electrons which form the localized state have a finite (though small) energy spread. We believe that states of this nature are responsible for electron localization and polarization in the Fe solutions discussed above.

With our present knowledge of band theory, it is not feasible to attempt a detailed calculation of the wave functions associated with iron impurities in the alloys discussed above. Instead, we have investigated[3] a number of simple models to gain some insight into the general features of virtual levels in metals. They all indicate that, near resonance, the electron density in the orbital centered at the impurity, $|U_k(R_0)|^2$,

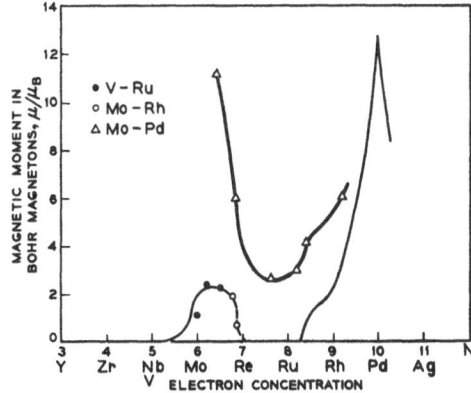

FIG. 2. Moment per iron atom vs electron concentration in alloys composed of nonadjacent members of the 4*d* series.

[3] P. W. Anderson, Phys. Rev. **124**, 41 (1961); P. A. Wolff, Phys. Rev. (to be published). The model discussed by Anderson can easily be generalized to describe a virtual level containing more than one electron.

should vary with energy as

$$|U_k(R_0)|^2 = \frac{(\Delta/\pi\rho)}{[(E(k)-E_0)^2+\Delta^2]}, \qquad (4)$$

where $E_0$ is the position of the resonance, $\Delta$ its width, $\rho$ the density of states in the conduction band at energy $E_0$, $k$ the wave vector of the incoming electron, and $E(k)$ its energy. Note that

$$\sum_k |U_k(R_0)|^2 = \int \rho(E)|U(R_0)|^2 = 1 \qquad (5)$$

indicating that the virtual level contains a single electron. More sophisticated models would be required to provide for localization of more than one electron.

We will use Eq. (4) to discuss the behavior—in particular the magnetization—of virtual levels of the dilute iron alloys. The quantities $E_0$ and $\Delta$ will be left as adjustable parameters. Presumably, the change of these parameters with $N$ is responsible for the variation of moment illustrated in Fig. 1. Unfortunately it is not possible, at present, to calculate them from first principles. All the models suggest, however, that $\Delta$ should be proportional to the density of states $\rho$ in the conduction band. Furthermore, in situations such as that of Fe in Nb, where the atomic wave functions of the impurity have a rather different character from those that form the conduction states of the background matrix the models indicate that the virtual levels should have an energy close to that of the iron orbital, with a width that is proportional to both $\rho$ and the matrix element of the Hamiltonian between the Fe orbital and a nearest neighbor orbital of the atom that forms the background. Crude estimates in this case give a value of $\Delta$ in the range 2–5 ev.

The next problem is to investigate the magnetization of a state such as that described by Eq. (4). Here it should be recognized that in a virtual level the electron density in the vicinity of the impurity is due to *many* electrons whose individual wave functions extend throughout the crystal. We cannot obtain a localized moment by polarizing one (or even many) of these. Rather, one must assume that there are spin dependent terms in the impurity potential which attract electrons with one sign of spin and repel those of the other, thus giving rise to a net spin density. Such a spin density will, in turn, create an exchange field of the type required to produce the initial spin pile-up. The crucial question we must ask, therefore, is whether or not such fields can be produced in a *self-consistent way* (within the framework of the Hartree-Fock theory). It is at this point that the virtual level plays a crucial role. Under ordinary circumstances a reasonable sized spin dependent term in the impurity potential will only produce a small concentration of spin at the impurity. The resultant exchange field is not then large enough to

produce self-consistency. However, if a fairly sharp virtual level happens to lie near the Fermi level a rather small spin dependent potential can split it sufficiently to place the resonance *above* the Fermi level for one sign of spin, and *below* for the other. Since the virtual level contributes much of the electron density in the vicinity of the impurity it is then possible to obtain a large spin polarization and an exchange field big enough to be self-consistent.

To make these statements more quantitative we investigate the effect of adding a spin dependent term to the impurity potential. The energies of the impurity are then shifted by amounts $\delta V_\uparrow$ and $\delta V_\downarrow$, giving rise to virtual levels at $E_0+\delta V_\uparrow$ and $E_0+\delta V_\downarrow$. To make the problem self-consistent we calculate the expectation value of the Hartree-Fock field with respect to the impurity orbital $\varphi$. One finds

$$(\varphi_\sigma, V_{HF}\varphi_\sigma)$$

$$= \sum_{\substack{k'\sigma' \\ \text{(filled states)}}} \left\{ \int \varphi_\sigma^*(r)\psi_{k'\sigma'}^*(r') \frac{e^2}{|r-r'|} \psi_{k'\sigma'}(r') \right.$$

$$\times \varphi_\sigma(r)d^3r d^3r' - \delta_{\sigma\sigma'} \int \varphi_\sigma^*(r)\psi_{k'\sigma'}^*(r')$$

$$\left. \times \frac{e^2}{|r-r'|}\psi_{k'\sigma'}(r)\varphi_\sigma(r')d^3r d^3r' \right\}, \quad (6)$$

where the $\psi_{k'\sigma'}$'s are the wave functions, including scattering from the Fe impurity, of the conduction electrons. The biggest contributions to Eq. (6) come from those terms in $\psi_{k'}$ and $\psi_{k'}^*$, that involve the orbital $\varphi$ at the impurity atom. There will, of course, be matrix elements to other orbitals but they are smaller, and change relatively less as $E_0$ goes to $E_0+\delta V$. Thus we will ignore all contributions to Eq. (6) except those involving the orbital $\varphi$. With this approximation the direct and exchange integrals in Eq. (6) are the same, and of the form

$$J = \int |\varphi(r)|^2 \frac{e^2}{|r-r'|}|\varphi(r')|^2 d^3r d^3r'. \qquad (7)$$

Thus, the parallel spin Coulomb integral is cancelled by the exchange integral and Eq. (6) takes the form

$$(\varphi_\uparrow, V_{HF}\varphi_\uparrow) = J \sum_{\substack{k' \\ \text{(filled states)}}} [|U_{k,\downarrow}(R_0)|^2] \qquad (8)$$

with a corresponding expression for the down spin matrix element. We now make the problem self-consistent by requiring that the *change* in the matrix element of $V_{HF}$ (as compared with the unpolarized state) be $\delta V$. With the aid of Eqs. (4) and (5) we find the formula

$$\delta V_\uparrow = \frac{J}{\pi} \sum_{\substack{k \\ (\text{filled states})}} \left\{ \frac{\Delta/\rho}{[E(k) - E_0 - \delta V_\downarrow]^2 + \Delta^2} \right.$$

$$\left. - \frac{\Delta/\rho}{[E(k) - E_0]^2 + \Delta^2} \right\}$$

$$= \frac{J\Delta}{\pi} \int_{-\infty}^{E_F} dE \left\{ \frac{1}{[E - E_0 - \delta V_\downarrow]^2 + \Delta^2} - \frac{1}{(E - E_0)^2 + \Delta^2} \right\}$$

$$= \frac{J}{\pi} \left\{ \tan^{-1}\left[ \frac{E_F - E_0 - \delta V_\downarrow}{\Delta} \right] - \tan^{-1}\left[ \frac{E_F - E_0}{\Delta} \right] \right\}. \quad (9)$$

Here $E_F$ is the Fermi energy, and there is a corresponding equation, with $\delta V_\uparrow$ and $\delta V_\downarrow$ interchanged. These equations have been rather extensively investigated so we will only discuss the simplest possible case here. That arises when the virtual level lies at the Fermi level. The coupled equations that determine $\delta V_\uparrow$ and $\delta V_\downarrow$ then take the form

$$\delta V_\uparrow = -\frac{J}{\pi} \tan^{-1}\left( \frac{\delta V_\downarrow}{\Delta} \right)$$

$$\delta V_\downarrow = -\frac{J}{\pi} \tan^{-1}\left( \frac{\delta V_\uparrow}{\Delta} \right). \quad (10)$$

Their solutions are most easily studied graphically. In Fig. 3 we illustrate plots of $\delta V_\uparrow$ vs $\delta V_\downarrow$ and $\delta V_\downarrow$ vs $\delta V_\uparrow$ for a magnetic case ($J/\Delta = 5$) and a nonmagnetic case ($J/\Delta = 1$). In the former there are three sets of roots of Eqs. (10)—an unstable one with $\delta V_\uparrow = \delta V_\downarrow = 0$; and two stable ones, corresponding to a local moment, with $\delta V_\uparrow = -\delta V_\downarrow$. On the other hand, in the latter case there is but a single nonmagnetic solution with $\delta V_\uparrow = \delta V_\downarrow = 0$. From these graphs it is clear that the criterion for magnetization is

$$\left. \frac{\partial(\delta V_\uparrow)}{\partial(\delta V_\downarrow)} \right|_{\delta V_\downarrow = 0} \geq 1. \quad (11)$$

This result is valid even when the virtual level is not at the Fermi level and leads to the condition

$$\frac{J}{\pi} \left[ \frac{\Delta}{\Delta^2 + (E_F - E_0)^2} \right] \geq 1, \quad (12)$$

for the existence of magnetization.

Since Eqs. (9) have been studied in some detail in the literature,[3] we will not attempt to analyze them further here, but merely mention some of their consequences. It is clear, first of all, that even in our simple

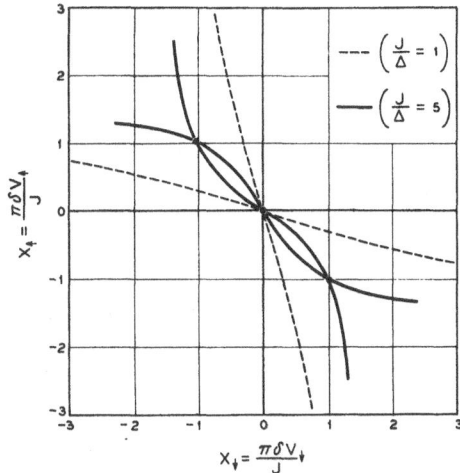

FIG. 3. Plots of $\delta V_\uparrow$ vs $\delta V_\downarrow$ for magnetic and nonmagnetic cases.

model the occurrence of localized moments is determined by a rather complicated interplay of virtual level width and position. This complexity will be further compounded in real metals by the presence of degenerate and overlapping bands. Thus it seems hopeless at the present to attempt to calculate where, in the moment vs $N$ curve, the Fe atoms are magnetized. In other respects, however, the virtual level description of localized moments is rather satisfactory. For Fe the Coulomb energy $J$ is about 10 ev. This figure, in combination with Eq. (12) and our previous rough estimates of $\Delta$, suggests that polarization should be fairly common in dilute iron alloys. It is important to realize, however, that the numerics are favorable here only because it is the Coulomb energy, which is relatively large, that appears in Eq. (12) rather than a true exchange integral which is of the order 1–2 ev in Fe. In this respect our point of view differs significantly from that of Friedel and his co-workers.

Detailed calculations of moment vs level width or position show that at the onset of polarization the moment curve has a vertical tangent, and thereafter rises continuously to a value of one Bohr magneton. A more complicated model would be required to give the values greater than unity observed experimentally. However, the behavior near threshold is in fair accord with the experimental curve. The latter approaches zero quite abruptly with some tailing at small values of the moment which may be due to the fact that the background matrix is an alloy, rather than a perfect crystal as has been assumed above. Finally, it is noteworthy that the theory provides an explanation for the occurrence of moments that are a small fraction of a Bohr magneton. This behavior arises quite naturally in our model, whereas it would be very difficult to understand in terms of the simple one-electron localized state of semiconductor physics.

# Pyromagnetic Test of Spin Wave Theory in Metallic Nickel

Emerson W. Pugh and Bernell E. Argyle
*IBM Research Center, Yorktown Heights, New York*

The temperature dependence of the spontaneous magnetization of metallic nickel has been studied between 4.2° and 120°K by a pyromagnetic technique developed by the authors. Fractional changes in magnetization as small as a few parts per million could be detected near 4.2°K. The resultant data were fitted by the method of least squares to a theoretical equation containing terms descriptive of thermal excitation of spin waves in the presence of an effective magnetic field plus a $T^2$ term descriptive of collective electron behavior. The best fit of the data to this equation is obtained using the spin wave terms alone, provided an intrinsic energy gap is assumed in the spin wave dispersion law of 2.7°K for magnetization parallel to the [111] axis and 1.9°K parallel to the [100] axis. Enhancement of this gap by an externally applied field follows theoretical predictions. It is noted that the measured difference between the gap temperature along the two principal axes has the value theoretically predicted from previous measurements of magnetic anisotropy energy, however the isotropic contribution observed in this experiment has not been theoretically anticipated. A possible origin for the isotropic gap is proposed in terms of interaction of polarized **s** and **d** electrons. It is also pointed out, however, that the "isotropic effective field" may be a spurious result, originating in thermal expansion effects not included in the theoretical equation to which the data were fitted. Finally, a new type of pyromagnetic measurement is described which can be used to determine the temperature dependence of the magnetic anisotropy.

**I**N spite of the broad advances that have been made in understanding ferromagnetic phenomena, there is considerable lack of agreement regarding the theoretical description for spontaneous magnetization in metals. The major difficulty appears to be in determining the degree to which the ferromagnetically aligned electrons are localized or free to wander through the lattice. Complete localization leads to the Heitler-London-Heisenberg model of ferromagnetism and to the Bloch-Dyson spin wave theory, while complete nonlocalization leads to the Stoner-Wohlfarth collective electron model. To the first approximation, the decrease in spontaneous magnetization with increasing temperature is predicted to follow a $T^{\frac{3}{2}}$ law by spin wave theory and a $T^2$ law by the collective electron model. A more exact spin wave treatment[1] gives weak terms of order $T^{\frac{5}{2}}$ and higher, as well as a correction for an effective internal field which may be expressed by lower order terms beginning with $T$ for temperatures sufficiently above the spin wave gap temperature.[2]

The possibility that these two models are not mutually exclusive was first discussed by Herring and Kittel,[3] who proposed that the low lying stationary states of a metal are of two types: states distributed in energy as described by the itinerant electron model with no spin waves excited, and states derivable from these by excitation of spin waves. The result of this proposal would be an $M$ versus $T$ curve containing both a $T^{\frac{3}{2}}$ and a $T^2$ term. Edwards and Wohlfarth[4] have reiterated this concept on the basis that electron correlation effects in metallic ferromagnetism could be treated from the view of plasma theory and thus introduce some feature of spin wave theory into the itinerant electron model.

Experimentally it had been difficult to make an unambiguous selection between the $T^{\frac{3}{2}}$ and the $T^2$ terms, let alone resolve the admixture of one in the other, or to find higher or lower order terms. The over-all problem is further complicated by the fact that spin wave theory is not expected to be valid except at relatively low temperatures. Furthermore, thermal expansion of the crystal may change the magnitudes of various interactions and thus add an additional temperature dependence. These later problems can be somewhat lessened by working only at low temperatures, but this increases the sensitivity required for meaningful measurements.

In order to provide more accurate experimental data in the important low temperature range, work was begun to develop a new pyromagnetic measuring technique. The first measurements,[5] published a year ago, showed that there is no anomaly near 10°K as previously reported[6] and served to establish the capability of the measuring technique. In the present paper a complete set of magnetization curves, obtained by somewhat improved techniques of a single crystal of nickel, are presented and shown to be well described by spin wave theory.

## I. THE PYROMAGNETIC TECHNIQUE

The basic method of the pyromagnetometer is to detect the change in sample magnetization when it is heated from some reference temperature, typically chosen as 4.2°K, to a new temperature in a fixed external field. The change in magnetization is detected inductively while the change in temperature is sensed by a thermocouple referred to a liquid helium bath. The heart of the system is the cryostat portion illustrated in Fig. 1. A half-inch diameter spherical sample is centered within the lower Helmholtz coil and thermally insulated from it by two vertical quartz cylinders, the bottom one supporting the sample while the top one is spring loaded and holds the sample in place. A 99.999% pure copper heater cap containing a $\frac{1}{4}$-w, 200-ohm carbon resistor is attached to each side of the sample as shown on the front side. The thermo-

[1] F. J. Dyson, Phys. Rev. **102**, 1217 and 1230 (1956).
[2] S. H. Charap, Phys. Rev. **119**, 1538 (1960).
[3] C. Herring and C. Kittel, Phys. Rev. **31**, 869 (1951).
[4] D. M. Edwards and E. P. Wohlfarth, J. phys. radium **20**, 136 (1959).
[5] E. W. Pugh and B. E. Argyle, J. Appl. Phys. **32**, 334S (1961).
[6] S. Foner and E. D. Thompson, J. Appl. Phys. **30**, 229S (1959).

couple wires (not illustrated) are wrapped twice around the sample and the bead soldered to it with Wood's metal to assure good thermal contact. The entire assembly illustrated in Fig. 1 is contained in a copper can immersed in liquid helium. The cylindrical copper plug at the bottom of the lower coil form assembly is a continuous part of it, and is soldered into a hole in the base of the can to assure constant temperature of the coils. Approximately 4 cm Hg pressure of helium gas is maintained in the can to facilitate the return of the sample temperature to 4.2°K after the heater power is turned off. A measurement is made by applying a predetermined voltage to the heaters. The change in flux and temperature are recorded simultaneously. At a point in the pulse, when thermal equilibrium is nearly reached and power input to the sample is only slightly greater than the rate of heat loss to the surroundings, the power is turned off and the flux and temperature return to their original values. A Helmholtz geometry for the pickup coil was chosen to minimize changes in inductive coupling resulting from thermally induced sample motion. Its 4710 turns are bucked against the same number wound on the matching coil to minimize any field fluctuation effects. Changes in sample magnetization of 1 part in $10^7$ are easily detected and changes as small as 1 part in $10^8$ could be detected with minor modifications. However, various sources of possible consistent error currently reduce our over-all accuracy to a few parts in $10^6$ near 4.2°K.

Some important aspects of the system, relating to sources of consistent error, are discussed below:

*(1) Thermometry.* Glass fiber insulated, 5-mil.-diam gold-cobalt versus normal silver thermocouples referred to a liquid helium bath were used throughout this work. They were calibrated by us against a platinum resistance thermometer and found to agree excellently with the curves of Powell *et al.*,[7] using the ratio method to bring the two curves into coincidence. The bead of the thermocouple is soldered to the sample to assure good thermal contact and several centimeters of the wire are wrapped around the sample to avoid rapid conduction of heat away from the contact point. Temperatures are read to the nearest 0.1° and should be accurate to at least ±0.2° over the entire range.

*(2) Thermal equilibrium of sample.* Heat is fed into the sample through the heater caps during thermal pulsing and simultaneously conducted away by the helium transfer gas. The constancy of temperature from one point to another in the sample must therefore be carefully examined. Calculations, based on the thermal input required to maintain a given temperature, indicate temperature differences in the sample should be less than 1 to 2 deg at 100°K and considerably smaller at lower temperatures. In addition, two important experiments have been performed. First, one thermocouple was soldered to the sample surface in the usual manner

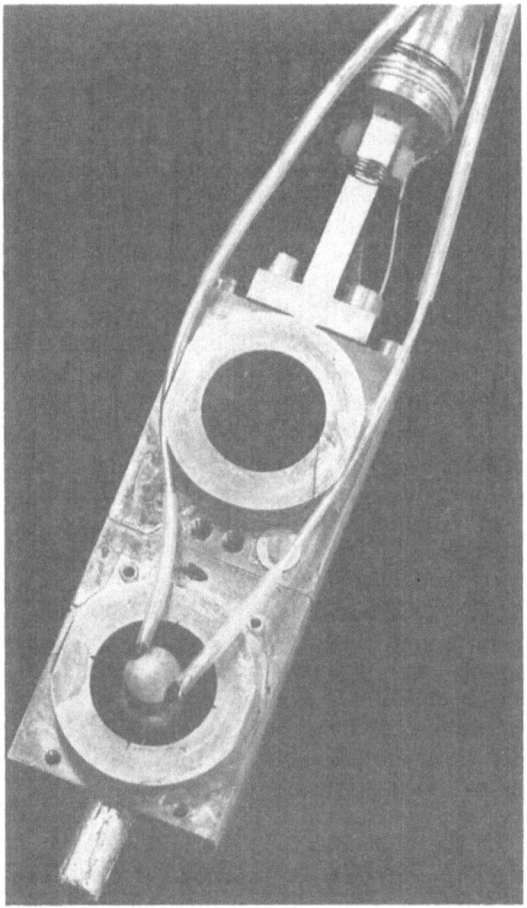

FIG. 1. Photograph of cryostat portion of pyromagnetometer. One half inch diameter nickel sample is shown centered in the lower Helmholtz coil with heater cap attached. Upper Helmholtz coil bucks out fluctuations in applied field. Assembly shown is contained by copper can immersed in liquid helium.

and the second was imbedded in the center. Less than 0.5° difference was recorded at 100°K. Second, one heater cap was removed and the temperature difference between the hot side and the cold side was found to be less than 2° during thermal pulsing to 100°K. Under normal conditions in which the sample is heated from both sides, the temperature error should be less than 0.5° to 1° at 100°K, which corresponds to a 1 to 2% error in $\Delta M/M$. This error becomes relatively smaller at lower temperatures.

*(3) Sample motion.* Change in flux coupling due to sample motion resulting from thermal expansion has been minimized by using Helmholtz pickup coils designed to satisfy the optimum conditions determined by Ruark and Peters.[8] There is no way simultaneously to minimize the effects of the motion of image poles relative to the coils, but their total coupling, by virtue of their greater distance to the coils, is only 0.01 of that of the sample. Taking account of the expansion of the quartz support cylinder and the sample itself, an error

[7] R. L. Powell, M. D. Bunch, and R. J. Curruccini, Cryogenics 1, 139 (March, 1961).

[8] A. E. Ruark and M. F. Peters, J. Opt. Soc. Am. and Rev. Sci. Instr. 13, 205 (1961).

in the measured $\Delta M/M$ of only $10^{-6}$ should occur at $100°$K. This corresponds to a fractional error of $10^{-4}$ in the output signal, and allows ample safety factor for possible tilting of the quartz support cylinder.

(4) *Thermal expansion of the coils.* The copper coils never change temperature by more than $8°$ during thermal pulsing. The resultant thermal expansion of the area of the coil in the presence of the applied field should give negligible flux pickup. Even if some signal were detected, it would be eliminated with other background signals as described under 5.

(5) *Background signal.* The major source of background signal results from the paramagnetism of the carbon heaters. This signal has been measured under a variety of conditions and found to be reproducible for any given power input. Compensation for this and any other background signal (e.g., changing thermal emfs induced in the pickup coil leads during pulsing) has been achieved by measurements on a 99.999% pure copper sphere[9] as a function of field and temperature. The correction to the data is less than 2.5% above $30°$K and increases to 30% at $6°$K. This background signal contributes the major source of possible consistent error at low temperatures since corrections for it can probably be relied on only to about ±30%, giving possible consistent errors in $\Delta M/M$ of 1% above $30°$K and increasing to 10% for temperatures near $6°$K.

(6) *Integrator drift.* The integrator used is of the Cioffi type[10] as modified by B. A. Calhoun of this laboratory. Its drift during a thermal pulse is roughly proportional to the length of time the sample is maintained at an elevated temperature. It is negligible for small pulses and held to a maximum of about 2% for large pulses.

Summarizing the above *possible* sources of consistent error in the measured change in magnetization $\Delta M$, it can be estimated that such errors decrease rapidly from 10% near $6°$K to 1% near $30°$K. They then may increase to as much as 2 or 3% for pulses in excess of $100°$K.

The absolute calibration of the pickup coil has been achieved by plotting the measured flux $\phi$ versus $H$ for the nickel sample at liquid nitrogen temperature. From the knee of the curve, which must occur at $H = 4\pi/3M$, a value of $M_0 = 510 \pm 5$ gauss has been obtained. Extrapolation of the straight line curve above the knee back to $H = 0$ gives a value for $\phi_0$, which corresponds to $M_0$ of the sample. The total estimated error in determining the calibration ratio of $M_0/\phi_0$ is ±2%.

## II. THEORY OF SPONTANEOUS MAGNETIZATION
### A. Spin Wave Theory

Using a general theory of the exchange interactions of ferromagnetic spin waves in a rigid cubic lattice,

Dyson[1] has derived the following expression for the spontaneous magnetization per unit mass of a ferromagnet in zero effective magnetic field:

$$(M-M_0)/M_0 = -CT^{3/2} - DT^{5/2} - ET^{7/2} - FT^4 + \cdots. \quad (1)$$

In this equation $M$ and $M_0$ are, respectively, the spontaneous magnetizations at temperature $T$ and absolute zero, while $CT^{3/2}$ is the dominant Bloch term, $DT^{5/2}$ and $ET^{7/2}$ arise from the discreteness of the lattice, and $FT^4$ is the lowest order term arising from interactions between spin waves.

Equation (1) has also been obtained by Oguchi[11] using an expansion of the spin wave operator method of Holstein and Primakoff[12] and even more recently by Keffer and Loudon[13] using easily understood semi-classical arguments. Further elaboration of this equation is therefore unnecessary here.

In an actual measurement, the spin waves are acted on by an effective magnetic field which is the sum of the intrinsic field, resulting from internal interactions, and the applied field (in our case 1000 to 12 000 oe after correcting for the demagnetizing field). These fields create an energy gap in the spin wave spectra which must be considered in any detailed processing of low temperature magnetization data. The calculation of this effect will be presented here for the low temperature long wave length approximation corresponding to the $CT^{\frac{3}{2}}$ term of Eq. (1).

The fractional decrease in magnetization of an assembly of atoms is represented in simple theory by

$$(M-M_0)/M_0 = \frac{g\beta}{M_0} \sum_k \langle N_k \rangle, \quad (2)$$

where $g\beta$ is the moment associated with a unit of spin excitation (spectrographic splitting factor times the Bohr magneton), $M_0$ is the maximum moment of the assembly of atoms, and $\sum_k \langle N_k \rangle$ is the sum over-all $\mathbf{k}$ values of the thermally excited spin wave occupation numbers. The right-hand side of this equation may readily be placed in the integral form

$$\frac{g\beta}{M_0} \left(\frac{1}{2\pi}\right)^3 \int_0^\infty \frac{4\pi k^2 dk}{\exp(E/k_BT) - 1}$$

by integrating the Bose-Einstein distribution function $[\exp(E/k_BT) - 1]^{-1}$, characteristic of spin waves, over all of $\mathbf{k}$ space and multiplying by the $(1/2\pi)^3$ states available per unit volume of $\mathbf{k}$ space in a unit volume of material. Boltzman's constant is indicated by $k_B$ to distinguish it from $\mathbf{k}$ space. The spin wave energy in the long wave length approximation is given by

$$E = 2SJl^2k^2 + g\beta H_e,$$

[9] Heat treated to remove any trace of temperature dependent magnetic properties in a manner previously described by E. W. Pugh, Rev. Sci. Instr. **29**, 1118 (1958).
[10] P. P. Cioffi, Rev. Sci. Instr. **21**, 624 (1950).

[11] T. Oguchi, Phys. Rev. **117**, 117 (1960).
[12] T. Holstein and H. Primakoff, Phys. Rev. **58**, 1098 (1940).
[13] F. Keffer and R. Loudon, J. Appl. Phys. **32**, 2S (1961).

where $S$ is the spin per atom, $J$ the exchange term, and $l$ the cube edge length of the unit cell, and $H_e$ is an effective field, defined by

$$H_e = H_0 + H_i.$$

Here $H_0$ is the applied field, after correction for demagnetizing fields, and $H_i$ is the intrinsic internal field which may be divided into an isotropic portion $H_i'$ and an anisotropic portion $H_i''$. The anisotropic field has been calculated theoretically to be

$$H_i'' = (nK_{10}/M_0)(\Gamma - \tfrac{1}{5}),$$

where $n$ is a constant equal to 10 and $\Gamma = \alpha_1^2\alpha_2^2 + \alpha_2^2\alpha_3^2 + \alpha_3^2\alpha_1^2$—the $\alpha_i$'s being the cosines between the magnetization direction and the cube edges. Thus $\Gamma$ is zero along the [100] axis and $\tfrac{1}{3}$ along the [111] axis. No theoretical calculation is available for the isotropic field $H_i'$. Such a field has been postulated here to facilitate interpretation of our experimental results. It may be thought of, for example, as originating in interactions between polarized **s** and **d** electrons.

Making the substitution $\mathcal{K}^2 = 2SJl^2k^2/k_BT$ and $T_g/T = g\beta H_e/k_BT$ in the preceding expressions, Eq. (2) becomes

$$\frac{M - M_0}{M_0} = \frac{g\beta}{M_0}\left(\frac{1}{2\pi}\right)^3\left(\frac{k_BT}{2JSl^2}\right)^{\frac{3}{2}}$$
$$\times 4\pi\int_0^\infty \frac{\mathcal{K}^2 d\mathcal{K}}{\exp(\mathcal{K}^2 + T_g/T) - 1}. \quad (3)$$

Performing the indicated division, the integral becomes

$$\int_0^\infty \mathcal{K}^2 d\mathcal{K}\sum_{j=1}^\infty \exp[-j(\mathcal{K}^2 + T_g/T)]$$

which may be rewritten

$$\sum_{j=1}^\infty \exp(-jT_g/T)\int_0^\infty \exp(-j\mathcal{K}^2)\mathcal{K}^2 d\mathcal{K} = \frac{\sqrt{\pi}}{4}\zeta\left(\frac{3}{2}, \frac{T_g}{T}\right),$$

where

$$\zeta\left(\frac{3}{2}, \frac{T_g}{T}\right) \equiv \sum_{j=1}^\infty \left(\frac{1}{j}\right)^{\frac{3}{2}}\exp(-jT_g/T).$$

Equation (3) may now be rewritten

$$\frac{M - M_0}{M_0} = \frac{1}{\nu S}\left(\frac{k_BT}{2JS}\right)^{\frac{3}{2}}\left(\frac{1}{4\pi}\right)^{\frac{3}{2}}\zeta\left(\frac{3}{2}, \frac{T_g}{T}\right). \quad (4)$$

Here use has been made of the relation $M_0 = g\beta NS$, where $N$ is the number of atoms in a unit volume, and the identity $\nu \equiv 1/Nl^3$, where $\nu$ is 1, 2, and 4 for sc, bcc, and fcc structures. For the special case of $T_g = 0$,

$$\zeta\left(\frac{3}{2}, \frac{T_g}{T}\right)$$

reduces to the Riemann zeta function of $\tfrac{3}{2}$, equal to 2.612, and the right-hand side of Eq. (4) reduces to the $CT^{\frac{3}{2}}$ term of Eq. (1). In a similar manner it can be shown that the $T^{\frac{5}{2}}$ coefficient of Eq. (1) should be modified by multiplying it by

$$\zeta\left(\frac{5}{2}, \frac{T_g}{T}\right)\Big/1.341,$$

where

$$\zeta\left(\frac{5}{2}, \frac{T_g}{T}\right) = \sum_{j=1}^\infty \left(\frac{1}{j}\right)^{\frac{5}{2}}\exp(-jT_g/T)$$

reduces to 1.341 when $T_g = 0$. Dyson has shown that $D$ is about $2.75 \times 10^{-4}$ as large as $C$, and the other terms are smaller yet. So that for measurements below $100°$K all terms may be neglected except the $C$ and $D$ terms and even the $D$ term is expected to contribute less than $3\%$ to the magnetization change at $T = 100°$K.

To a good approximation, Eq. (1) may be rewritten to include the temperature gap with only two terms

$$(M - M_0)/M_0 = -\frac{C}{2.612}\zeta\left(\frac{3}{2}, \frac{T_g}{T}\right)T^{\frac{3}{2}}$$
$$-\frac{D}{1.341}\zeta\left(\frac{5}{2}, \frac{T_g}{T}\right)T^{\frac{5}{2}} + \cdots. \quad (5)$$

As pointed out by Charap,[2] this complicated expression for the magnetization may be expanded in a simple power series in temperature by using Robinson's[14] expansions for the modified Riemann zeta functions which converge for $|\alpha| \leq 2\pi$:

$$\zeta(\tfrac{3}{2}, \alpha) = -3.54\alpha^{\frac{1}{2}} + 2.61 + 1.46\alpha - 0.104\alpha^2 + \cdots$$
$$\zeta(\tfrac{5}{2}, \alpha) = 2.36\alpha^{\frac{3}{2}} + 1.34 - 2.61\alpha - 0.730\alpha^2 + \cdots.$$

Substituting these into Eq. (5) and using the fact that $T_gD \ll C$ gives

$$(M - M_0)/M_0 \cong \cdots - 0.56T_gCT^{\frac{1}{2}}$$
$$+ 1.35T_g^{\frac{3}{2}}CT - CT^{\frac{3}{2}} - DT^{\frac{5}{2}} + \cdots. \quad (6)$$

The theoretical expression for magnetization versus temperature is sufficiently complicated that one should not expect to find a linear plot for $\Delta M$ versus $T$ on log-log paper except over a limited temperature range. This is illustrated in Fig. 2 where theoretical plots of $(M_0 - M)/M_0C$ versus $T$ are shown for various values of $T_g$. The curve for $T_g = 0$ would be a straight line of slope 1.5 if $D/C$ were set equal to zero instead of $2.75 \times 10^{-4}$, the theoretically predicted value.

### B. Effects of Thermal Expansion

The constants $J$, $g$, and $S$ of the spin wave expression of Eq. (3) are dependent on the overlap of electronic wave functions and therefore may vary with thermal expansion of the crystal. Any attempt to discuss these

[14] J. E. Robinson, Phys. Rev. **83**, 678 (1951).

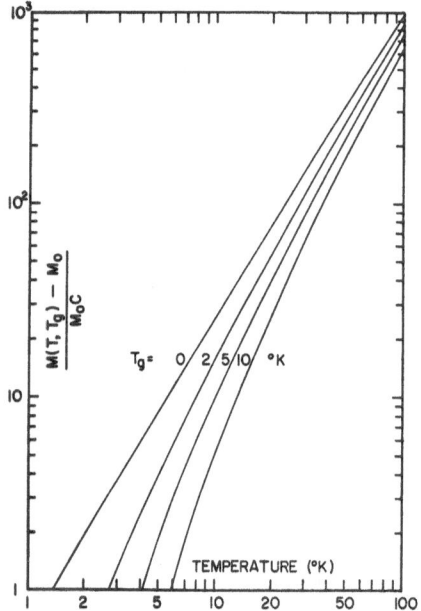

FIG. 2. Theoretical curves for $(M_0-M)/M_0 C$ versus $T$ illustrated on a log-log graph. As the assumed gap temperature is increased from $T_g=0$ to 15, the slope increases more and more, especially at low temperatures. A value $D/C=2.75\times10^{-4}$ is assumed for all curves. If $D=0$ were assumed, the curve for $T_g=0$ would have a slope of 1.5 and run from corner to corner in the graph.

effects in detail becomes involved. Let it suffice to note that $J$ and $S$ variations with temperature can be estimated by combining the pressure dependence of the Curie temperature measured by Patrick[15] with thermal expansion data of Nix and MacNair.[16] An upper limit on the variation of the $g$ factor with temperature can be determined from the resonance data of Standley and Reich,[17] assuming $\partial g/\partial V$ is constant over the temperature range of interest.

Our approximations indicate that variations in $J$ should contribute a fractional change in magnetization of less than 2 parts in $10^4$ below 100°K. The other effects may be larger and have a dominant $T^2$ term in a temperature range near 100°K, the coefficient of which is the order of magnitude of $10^{-7}$. These contributions are therefore expected to be small enough to act only as a perturbation on spin wave phenomena below 100°K.

## C. Collective Electron Model

A theoretical relation for the decrease of spontaneous magnetization with increasing temperature was determined by Stoner[18] for a parabolic band shape and subsequently by Wohlfarth[19] for a more general case in which the density of states is proportional to $E^n$, where

[15] L. Patrick, Phys. Rev. **93**, 384 (1954).
[16] F. C. Nix and D. MacNair, Phys. Rev. **60**, 597 (1941).
[17] K. J. Standley and K. H. Reich, Proc. Phys. Soc. (London) **68B**, 713 (1955).
[18] E. C. Stoner, Proc. Roy. Soc. (London) **A165**, 372 (1938).
[19] E. P. Wohlfarth, Phil. Mag. **42**, 374 (1951).

$n$ is any positive or negative constant. If all the states for spins aligned parallel to the magnetization are of lower energy than the Fermi level, then an exponential temperature dependence occurs. If, however, unfilled states for spins both parallel and antiparallel to the magnetization direction occur at absolute zero, then the temperature dependence is of the form

$$(M-M_0)/M_0 \cong ST^2.$$

The Stoner-Wohlfarth coefficient $S$ depends critically on the details of the assumed band shape and exchange energy and there are no good theoretical estimates of its magnitude for nickel. A purpose of this experiment is therefore to determine the magnitude of this term, or at least to set an upper limit for it.

## III. EXPERIMENTAL RESULTS—EASY AXIS

Experimentally obtained decreases in magnetization for a nickel sample heated to various temperatures above 4.2°K are plotted in Fig. 3 for three different fields applied parallel to a [111] axis. The larger the applied field, the smaller the magnetization decrease. This can be attributed to field suppression of spin wave excitation, or equivalently to an increase in the spin wave energy gap. Quick comparison with theory can be made using the highest order term in $T$, containing $T_g$, of Eq. (6). However, exact analysis of the effect is more satisfying, and has been achieved by a numerical least

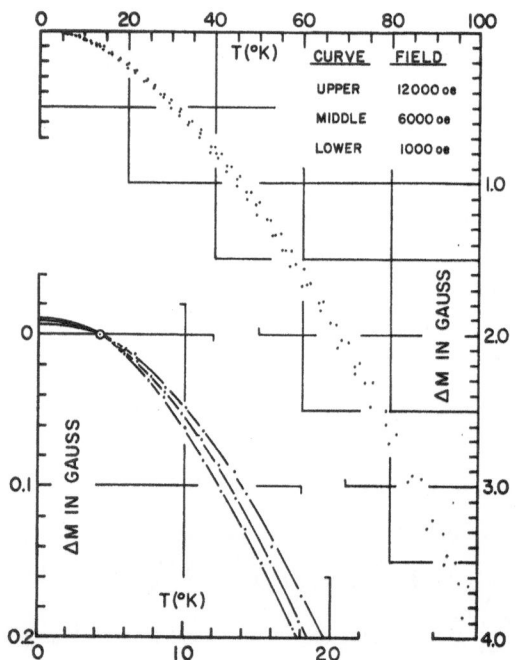

FIG. 3. Experimentally observed decreases in magnetization parallel to the [111] axis for various temperatures above 4.2°K ($\Delta M \equiv 0$ at 4.2°K). The indicated applied fields have been corrected for demagnetizing effects and thus represent the applied field inside the sample. The curves in the lower left hand corner are the low temperature portions of the larger curves. Note the different scales for the two sets of curves.

squares fitting[20] of the equation

$$\frac{\Delta M}{M} = \frac{C}{2.612}\zeta\left(\frac{3}{2}, \frac{T_g}{T}\right)T^{\frac{3}{2}}\Big]_{4.2}^{T}$$

$$+ \frac{D}{1.341}\zeta\left(\frac{5}{2}, \frac{T_g}{T}\right)T^{\frac{5}{2}}\Big]_{4.2}^{T} + ST^2\Big]_{4.2}^{T}. \quad (7)$$

The first two terms come from Eq. (5), while the $ST^2$ term is inserted to permit analysis of contributions from the Stoner-Wohlfarth collective electron model.

From this analysis it has been learned that the best statistical fit to the data occurs for a purely spin wave model with the intrinsic field approximately nine times the 2000-oe field predicted from anisotropy measurements. The statistical analysis leading to this conclusion is indicated below and numerical values for the coefficients given. Representation of the data by $\log M$ versus $\log T$ is discussed, and a comparison of our results and those of other experimenters is given.

## A. Curve Fitting by Least Squares Method

A least squares fit to Eq. (7) of the magnetization versus temperature data for each of the four applied fields was obtained using an IBM 7090 computer over a range of assumed gap temperatures under three conditions (1) $C$ the only independent fitting parameter, $D = 3 \times 10^{-9}$, and $S = 0$; (2) $C$ and $S$ both independent and $D = 3 \times 10^{-9}$; and finally, (3) $C$ and $D$ independent and $S = 0$. The standard deviations $\sigma$ of the experimental data about the calculated curves were then plotted versus the gap temperature. This analysis was performed for each applied field for three temperature ranges: 4.2°K up to 40°, 70°, and 120°K. The results for data up to 70° are shown in Fig. 4. For the case in which $C$ alone was allowed to vary, a pronounced minimum in the standard deviation versus $T_g$ was observed for each set of data. It is important to note that $T_g$ corresponding to the minimum $\sigma$ increases with the applied field as expected from theory. Converting these values of $T_g$ to effective fields and substracting from them the applied fields in each case yields nearly the same intrinsic field. The values in kilooersteds are 15.3, 17.3, 17.7, and 18.5 for the four sets of data, taken in order of increasing applied field. Similarly, values of 24.8, 20.0, 16.4, and 17.2 were obtained for data up to 40°K. An average value of $H_i = 18.4 \pm 3$ koe is obtained from these results.

The excellence of the fit to pure spin wave theory with large internal field included, for data up to 70°K, is attested to by the fact that allowing either the $D$ or $S$ term to vary independently of $C$ produces a negligible decrease in the standard deviation at that field—in

[20] A similar least squares fitting to the spin wave equation has been used recently to analyze NMR results in CrBr$_3$ by A. C. Gossard, V. Jaccarino, and J. P. Remeika, Phys. Rev. Letters 7, 122 (1961).

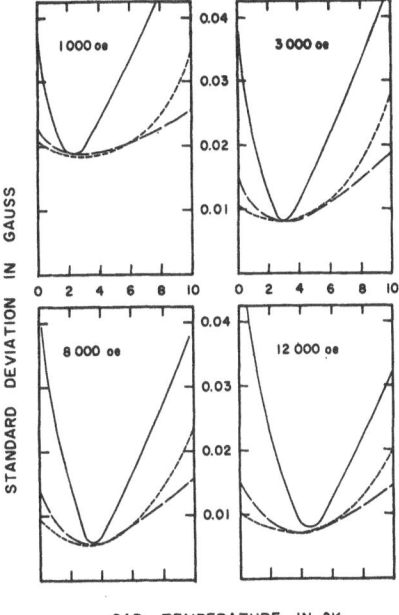

FIG. 4. Standard deviation of the experimental data from Eq. (7) is plotted versus assumed values of the gap temperature $T_g$ for four different applied fields. The solid curve in each plot corresponds to the condition $C$ determined by method of least squares while $D = 3 \times 10^{-9}$ and $S = 0$, the long dashes correspond to $C$ and $D$ determined with $S = 0$, and the short dashes correspond to $C$ and $S$ determined with $D = 3 \times 10^{-9}$. The minimum in the solid curve occurs at the best experimental value for the gap temperature.

spite of the quite general fact that addition of a second independent term must always produce a better fit.

In Table I we have tabulated results for $C$, $S$, $D$, and $\sigma$ obtained by least squares analysis for two values of intrinsic field: the value $H_i = 18.4$ koe indicated by the results illustrated in Fig. 4 and also the value $H_i = 2$ koe as expected from anisotropy measurements. When $H_i = 18.4$ koe as opposed to 2 koe, the value of $C$ is nearly the same for the four different applied fields; and it is only slightly affected by letting either $S$ or $D$ vary independently. Also the weighted average value obtained for $D$ is $2.5 \times 10^{-9}$ which results in $D/C = 2.7 \times 10^{-4}$ as compared to the theoretically computed ratio of $2.75 \times 10^{-4}$. This agreement is better than anticipated from our experimental accuracy, but indicates the excellent fit to pure spin wave theory which has been obtained. In the $C$, $S$ row of Table I, $D$ is again set equal to $3 \times 10^{-9}$ while $C$ and $S$ vary independently. For $H_i = 18.4$ koe, a mixed set of very small values is obtained for $S$ which are best described by $|S| < 10^{-7}$. Thus to this accuracy there is no evidence of a $T^2$ term from any of the possible sources suggested in Sec. II.

In the second column of Table I, with $H_i = 2000$ oe as theoretically anticipated, it will be noted that when only $C$ varies, a very high standard deviation results. With both $C$ and $D$ or $C$ and $S$ varying, a substantially smaller standard deviation is achieved; however, $\sigma$ is still larger in this case, with two independent coefficients,

TABLE I. Least squares fit of various combinations of coefficients for [111] data up to 70°K.

| Independent coefficient | $H\times10^{-3}$ | $H_i = 18\,400$ oe | | | | $H_i = 2000$ oe | | | |
|---|---|---|---|---|---|---|---|---|---|
| | | $C\times10^6$ | $D\times10^9$ | $S\times10^7$ | $\sigma\times10^3$ | $C\times10^6$ | $D\times10^9$ | $S\times10^7$ | $\sigma\times10^3$ |
| $C$ | 1 | 9.58 | | | 19.8 | 7.85 | | | 30.5 |
| $D\equiv3\times10^{-9}$ | 3 | 9.58 | | | 8.5 | 8.04 | | | 26.2 |
| | 6 | 9.57 | | | 5.7 | 8.15 | | | 22.0 |
| | 12 | 9.62 | | | 7.7 | 8.39 | | | 20.8 |
| $C, D$ | 1 | 9.90 | −1.6 | | 19.3 | 6.92 | 18.6 | | 20.9 |
| | 3 | 9.64 | 2.3 | | 8.4 | 6.98 | 19.3 | | 11.5 |
| | 6 | 9.62 | 2.4 | | 5.7 | 7.18 | 17.9 | | ⎸8.7 |
| | 12 | 9.58 | 3.6 | | 7.7 | 7.45 | 16.5 | | 8.9 |
| $C, S$ | 1 | 10.40 | | −0.8 | 19.3 | 5.94 | | 2.3 | 20.8 |
| $D\equiv3\times10^{-9}$ | 3 | 9.74 | | −0.1 | 8.4 | 5.86 | | 2.5 | 10.0 |
| | 6 | 9.70 | | −0.1 | 5.7 | 6.01 | | 2.3 | 7.3 |
| | 12 | 9.52 | | 0.9 | 7.7 | 6.15 | | 2.3 | 8.2 |

than it was for $H_i = 18.4$ koe and only one independent coefficient $C$. Furthermore, the fit of the data to the theoretical curve, when examined point by point, is not nearly as good for the low field case and two variables as it is for the higher field and one variable. This is particularly true at low temperatures where the spin wave energy gap becomes important.

Tabulated results similar to those of Table I have been obtained for all data up to 120°K. All coefficients agree within a few percent with those of Table I, although there is a consistent trend toward larger coefficients for the higher order terms in $T$. These deviations are within the estimated limits of possible consistent experimental error for large thermal pulses, and it is therefore dangerous to base any conclusion on the apparent trend.

## B. Analysis by $\mathrm{Log}\,M$ versus $\mathrm{Log}\,T$ Plots

It has already been pointed out in Sec. II that a simple 1.5 slope should not be expected on a log $M$ versus log $T$ plot and several theoretical curves were illustrated. Such curves do tend to be reasonably straight, however, over a given temperature range, and this fact can be used to make more graphical the results already determined by the method of least squares. In Fig. 5 average theoretical slopes as determined between 25° and 100°K are plotted as a function of gap temperature. The dashed curves are for $\Delta M$ measured from 0°K while the solid curves are for $\Delta M$ measured from 4.2°K. Aside from facilitating comparison with our experimental results, the solid curves illustrate the erroneous results that can be obtained by log-log plots if the extrapolation of $M$ to absolute zero has even a small error. Slopes obtained experimentally from our data are plotted on the curve in a position corresponding to the applied field energy gap. Shifting the points to the right by the intrinsic gap temperature, $T_g = 2.7$°K, places them all on the theoretically predicted curve within the limits of accuracy to which the slope of the data could be determined.

## C. Comparison With Other Experimental Results

Since our data cover only the range up to 120°K, it is instructive to compare them with data obtained by other experimenters to gain a better over-all picture. Such a comparison is provided in Table II. Extrapolation of results to absolute zero is dangerous, especially for the Foner-Thompson[6] data where large anomalies were recorded near 10°K and for the Budnick[21] nuclear magnetic resonance data which are available only down to liquid nitrogen temperature. All data have therefore been normalized to $\Delta M/M_0 = 0$ at $T = 70$°K. A smooth curve through the data was used to facilitate comparison at identical temperature points. Little if any agreement is found between our results and those of Foner and

TABLE II. Tabulated values for $\Delta M/M_0\times10^3$ at various temperatures as obtained by Pugh and Argyle (present work, $H_0 = 5$ koe), Budnick ($H_0 = 0$),[21] Fallot ($H_0$ extrapolated to zero),[22] and Foner and Thompson ($H_0 = 15$ koe).[6] In order to avoid a questionable extrapolation to absolute zero, all data has been normalized to $\Delta M/M\equiv0$ at $T = 70$°K. The P. and A. value for $T = 0$°K is based on a theoretical spin wave extrapolation.

| $T$ (°K) | P. and A. | Budnick | Fallot | F. and T. |
|---|---|---|---|---|
| 0.0 | 4.23 | | | |
| 4.2 | 4.21 | | | 2.06 |
| 10 | 4.11 | | | 0.39 |
| 20 | 3.73 | | | 2.45 |
| 30 | 3.21 | | | 2.20 |
| 40 | 2.50 | | | 1.75 |
| 50 | 1.85 | | 3.0 | 1.20 |
| 60 | 0.96 | | | 0.55 |
| 70 | 0 | 0 | 0 | 0 |
| 80 | −1.11 | −1.15 | | −0.70 |
| 90 | −2.31 | −2.38 | −1.0 | −1.40 |
| 100 | −3.71 | −3.87 | | −2.10 |
| 120 | −6.69 | −7.13 | | −3.70 |
| 140 | | −10.8 | −8.2 | −6.70 |
| 160 | | −15.8 | | −10.5 |
| 180 | | −21.3 | | −14.8 |
| 200 | | −27.5 | −19.4 | −20.1 |
| 250 | | −47.8 | −33.8 | −35.7 |
| 300 | | −76.3 | −50.3 | −56.5 |

[21] J. I. Budnick, unpublished nuclear magnetic resonance measurements.

Thompson or of Fallot[22] below 70°K. As higher temperatures are reached, the fractional differences between all measurements become smaller. The best over-all agreement appears to be between the pyromagnetic data and the nuclear magnetic resonance data.

## IV. EXPERIMENTAL RESULTS—HARD AXIS

Effects associated with thermal excitation of spin waves are expected to be the same in the hard [100] direction as in the easy direction except for a possibly smaller intrinsic energy gap in the spin wave spectrum. Technical anisotropy effects can, however, dominate the spin wave effects. It can be shown, for example, that when the single crystal is slightly misaligned with the applied field not exactly parallel to the [100] axis, increasing temperatures actually give rise to an apparent *increase* in magnetization. This is because our experimental equipment does not measure $\mathbf{M}$ but rather $M_H$, the component of $\mathbf{M}$ parallel to the applied field. Thus, as the temperature is increased, the anisotropy decreases and permits the magnetization to swing more nearly parallel to the applied field.

This effect is illustrated in Fig. 6 for magnetization versus temperature curves obtained with $H$ applied at $1.5°\pm0.3°$ from the [100] axis, the angle having been determined by x-ray analysis. As the applied field is

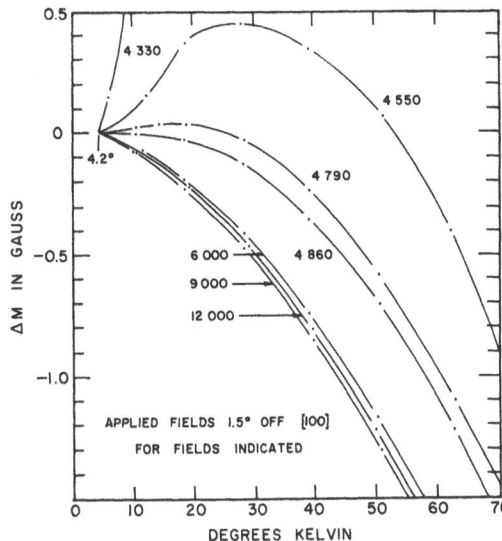

FIG. 6. Experimentally observed changes in the component of $\mathbf{M}$ parallel to the applied field with the field 1.5° off of the [100] axis. Increases in $M_H$ are observed for increasing temperature when $H_0$ is sufficiently small, as described in the text.

decreased from 12 000 to 9000 oe, the decrease in $T_g$ dominates, while from 9000 to 6000 oe the anisotropy effect dominates. As the applied field is further reduced, large anisotropy effects occur, until at a field of approximately 4900 oe $dM_H/dT=0$ for small $T$. If the data for $M_H$ versus $T$ are assumed to comply with the equation

$$K_1/K_{10}=(M/M_0)^n,$$

and the anisotropy constant at absolute zero, $K_{10}$ is assumed to be $8\times10^5$ ergs/cc,[23] then a detailed analysis of the data[24] illustrated in Fig. 6 reveals that

$$n\cong100$$

at $T=4.2°K$ instead of 10 as is theoretically predicted.[25]

Our major interest here is, however, the high field region where spin wave excitation effects dominate the anisotropy effects. An analysis of these effects indicates that reliable spin wave information should be obtainable from all data for which $H_0>6000$ oe. Accordingly, a least squares fit of the [100] results to Eq. (7) was obtained for $H_0=7, 9,$ and 12 koe in the same manner as previously described for the [111] results. Curves of $\sigma$ versus $T_g$ similar to those of Fig. 4 were plotted and indicate the existence of an intrinsic positive field parallel to the [100] axis of $13\pm5$ koe, equivalent to $T_g=1.9°K$. The results of the least squares analysis were found to conform well with the [111] results tabulated in Table I, when the [100] data was processed for $H_e=13$ koe as indicated by experiment and $H_e=-3$

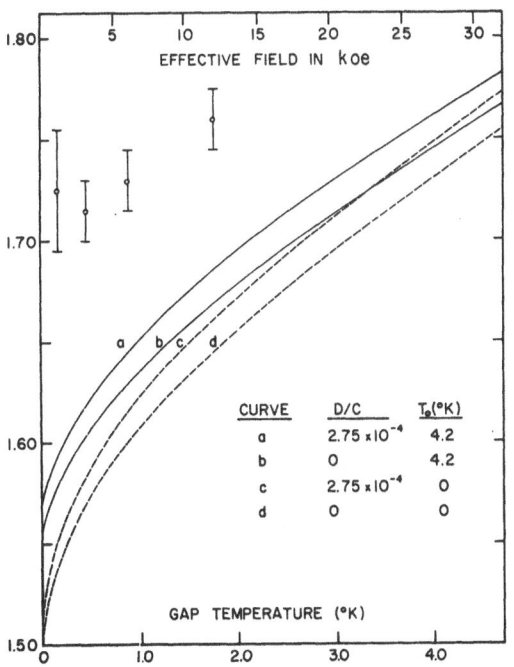

FIG. 5. The straightline character of $\log (M_0-M)/M_0C$ versus $\log T$ curves in the high temperature region (see Fig. 2) has been used in this figure to plot average slopes between 25° and 100°K against an assumed gap temperature. Experimentally measured slopes for [111] data are indicated at a gap temperature corresponding to the applied field. The intrinsic gap temperature is the distance by which these points must be shifted to the right to fall on the theoretical curve a.

CURVE table:

| CURVE | D/C | $T_g(°K)$ |
|---|---|---|
| a | $2.75\times10^{-4}$ | 4.2 |
| b | 0 | 4.2 |
| c | $2.75\times10^{-4}$ | 0 |
| d | 0 | 0 |

[22] M. Fallot, Ann. phys. **6**, 305 (1936).

[23] $K_{10}=7.5\times10^5$ ergs/cc by H. J. Williams and R. M. Bozorth, Phys. Rev. **55**, 673 (1939); $K_{10}=8.3\times10^5$ ergs/cc by K. H. Reich, Phys. Rev. **101**, 1647 (1956).
[24] The analysis of these data as well as additional experimental results will be presented in a subsequent paper by the authors.
[25] J. H. Van Vleck, J. phys. radium **20**, 124 (1959).

koe as suggested by theory. The coefficients for Eq. (7) were the same within a few percent, although the standard deviations were slightly larger as would be expected from the limited number of experimental points available. The same conclusion is again reached; namely, the best fit is achieved with the higher intrinsic field and no $T^2$ term.

## V. CONCLUDING DISCUSSION

Combining all of the experimental results obtained along both crystallographic axes gives the following best values for the coefficients to Eq. (7):

$$C = (9.65 \pm 0.25) \times 10^{-6}$$
$$D = (2.6 \pm 1.5) \times 10^{-9}$$
$$D/C = (2.7 \pm 1.5) \times 10^{-4}$$
$$T_g[111] = 2.7 \pm 0.5°K \quad H_i[111] = 18.4 \pm 3 \text{ koe}$$
$$T_g[100] = 1.9 \pm 0.7°K \quad H_i[100] = 13 \pm 5 \text{ koe}$$
$$|S| < 10^{-7}.$$

It is interesting to note that $H_i[111] - H_i[100] = 5 \pm 6$ koe. This is just the anisotropic field predicted from theory. The isotropic field or energy gap is therefore $H_i' = 16 \pm 4$ koe or $T_g' = 2.4 \pm 0.6°K$.

The origin of the experimentally observed isotropic intrinsic field is somewhat puzzling. Magnetostatic fields associated with spin waves, and also magnetostrictive effects, both have the character of an isotropic field to spin waves, but the magnitudes are too small. Sufficiently large magnitudes are conceivable, however, for interactions between polarized s and d electrons.[26] A model may be envisioned in which the s·s and d·d

[26] J. C. Slonczewski (private discussion).

interactions are both larger than s·d interactions. In this case polarized s electrons would not be expected to follow the details of the spin wave excitation of the d electrons but instead would always contribute an average torque to the d spins tending to align them parallel to the direction of M. This model appears to be satisfactory except that the resultant effective field has not been seen in resonance experiments. The shorter wave length of thermally excited spin waves as compared to resonance waves may be the solution to this dilemma, since s electrons may be able to follow in detail the long wavelength d spin waves but not the shorter ones.

The difficulties of envisioning a satisfactory model for the origin of the intrinsic field suggests that other effects, e.g., thermal expansion of the lattice should be examined in more detail. Because the thermal expansion of nickel is negligible below 20°K and increases rapidly until about 140°K, $\Delta M$ resulting from expansion effects should be negligible below 20°K and then increase very rapidly. If treated as a perturbation on spin wave effects, this would give the appearance of suppressing magnetization changes in the low temperature range in a similar manner as the effective field—where as in reality it would actually be enhancing the rate of change of $M$ with increasing temperature. This explanation of the "intrinsic field" as a spurious thermal expansion effect is currently under study.

## ACKNOWLEDGMENTS

The authors are indebted to S. H. Charap for pointing out the principles involved in the field dependence of the spin wave analysis. They are also indebted to W. J. Doherty for programming the least squares analysis.

JOURNAL OF APPLIED PHYSICS    SUPPLEMENT TO VOL. 33, NO. 3    MARCH, 1962

# NMR in Domains and Walls in Ferromagnetic CrBr₃

A. C. Gossard, V. Jaccarino, and J. P. Remeika
*Bell Telephone Laboratories, Inc., Murray Hill, New Jersey*

Nuclear magnetic resonance absorptions have been observed in zero applied field in the insulating ferromagnet CrBr₃. Three lines are shown to be a quadrupole split triplet from nuclei in the bulk of the domains. A fourth, with a resonance frequency of 56.58 Mc at 1.34°K and a 250-kc linewidth, is ascribed to nuclei in the domain walls. The nuclear resonance frequencies are proportional to the local magnetization, and thus measure the temperature dependence of the magnetization both in domain walls and in domains. This is the first measurement of the difference between wall and domain magnetizations, and shows that $(\Delta M/M)_{\text{wall}} - (\Delta M/M)_{\text{domain}} = T/200$ in the spin wave region.

NUCLEAR resonant absorption of rf power by $Cr^{53}$ nuclei in the insulating ferromagnet $(T_c = 37°K)$[1] CrBr₃ has been observed both for nuclei lying in the bulk of the ferromagnetic domains and for nuclei in the domain walls. Since the magnetic field at the nuclei and hence the resonance frequency are proportional to the electronic magnetization, it is possible to obtain directly the temperature dependence of the magnetization both in the domains and in the walls.

The results in CrBr₃ are of special interest because CrBr₃ more closely approximates the Heisenberg model of ferromagnetism on which spin wave theory is based than do previously known ferromagnets, which universally are either electrical conductors or involve substantial ferrimagnetic interactions. A detailed comparison between the observed temperature dependence of the magnetization $M(T)$ in the domains and the predictions of spin wave theory has been reported elsewhere[2] and shows that the observed results for the domain $M(T)$ are entirely consistent with spin wave theory. $M(T)$ in the domain walls, on the other hand, is found to be distinctly different from $M(T)$ in the domains. We believe that these are the first measurements of a difference in magnetization between domains and walls.

The sample consisted of two grams of compressed polycrystalline flakes of CrBr₃. The CrBr₃ was formed by passing bromine gas over hot (800°C) chromium powder. The sample was placed in a cryostat within the resonant coil of a modified frequency swept[3] Robinson rf oscillator, whose rf voltage level versus radio frequency could be displayed either on an oscilloscope or recorded with a strip chart recorder.

On searching the frequency range from 30 to 200 Mc four prominent absorption lines were found. Three of the lines are a quadrupole split triplet attributed to nuclei in the bulk of the domains. The spacing between these lines was nearly 295 kc, independent of temperature, and the frequency of the central line was extrapolated to 58.096 Mc at $T = 0°K$. The fourth line was 250 kc wide (more than 10 times as broad as the other three) but was quite intense, with about 30 times the

integrated intensity of the other lines. Lying at a lower frequency than the domain lines, it also behaved differently as the temperature was varied, as shown in Fig. 1.

By assuming that the three narrow lines come from nuclei in the bulk of the domains and that the fourth line comes from nuclei in the walls, we can consistently explain these observations. In either domains or walls, the largest contribution to the field at the nuclei arises from the exchange polarization of inner closed-shell $s$ electrons[4] by the three $3d$ electrons of the orbital singlet $^4A_2$ ground state configuration of the aligned $Cr^{3+}$ ions in the octahedral field of the surrounding $Br^-$ ions. In addition, however, anisotropic quadrupolar and dipolar contributions to the $Cr^{53}$ nuclear energy, which will be different in domains and walls, are expected in CrBr₃. The crystal structure of CrBr₃ is the highly anisotropic hexagonal layer structure of bismuth tri-iodide: $DO_5$[5]. Layers of $Cr^{3+}$ ions with

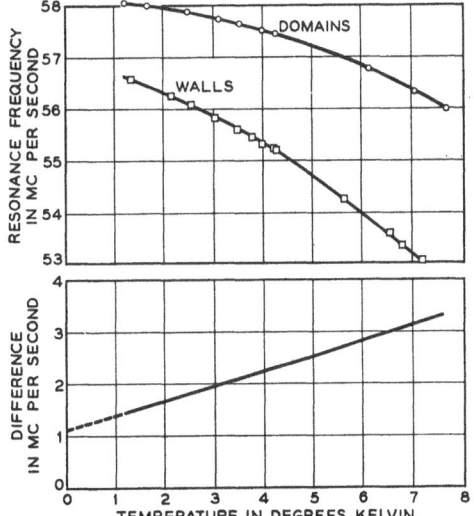

FIG. 1. Top: $Cr^{53}$ nuclear magnetic resonance frequency vs temperature in CrBr₃ in zero applied field for nuclei in the bulk of the domains and in the walls. NMR frequency for $Cr^{53}$ is 0.2406 Mc per kilogauss. Bottom: Difference between nuclear resonance frequency in domains and in walls vs temperature.

[1] Ichiro Tsubokawa, J. Phys. Soc. Japan **15**, 1664 (1960).
[2] A. C. Gossard, V. Jaccarino, and J. P. Remeika, Phys. Rev. Letters **7**, 122 (1961).
[3] F. N. H. Robinson, J. Sci. Instr. **36**, 481 (1959); R. J. Blume (to be published).
[4] G. W. Pratt and J. H. Wood, Phys. Rev. **107**, 995 (1957); V. Heine, Phys. Rev. **107**, 1002 (1957).
[5] A. Braekken, Kgl. Norske Videnskab. Selskabs, Forh. **5**, No. 11 (1932).

3.61-A interionic distances are separated by 6.07 A. A saturation magnetization of three Bohr magnetons per $Cr^{3+}$ ion with the easy direction of magnetization the $c$ axis, perpendicular to the layers of $Cr^{3+}$ ions, was found by Tsubokawa.[1] The magnetization within the domains, as well as the dipolar magnetic field and principal axis of the electric field gradient will be uniformly aligned in the $\pm c$ directions at the nuclei within the domains. Interaction of the $Cr^{53}$ quadrupole moments with the electric field gradient will split the $Cr^{53}$ $(I=\frac{3}{2})$ nuclear magnetic resonance absorption into three lines within the domains. Within the domain walls, on the other hand, the magnetization makes varying angles with respect to the $c$ axis. Dipolar fields and the angle between the magnetization and principal field gradient axis thus vary from site to site in the wall, and one expects a wide distribution of resonance frequencies and a single broad line of complex shape from nuclei within the walls. Differences in the temperature dependence of the resonance frequencies and resonance intensities are also predicted between nuclei in walls and in domains. As the temperature is raised above $0°K$, we expect that there will be wall excitations which will be thermally driven, resulting in a faster decrease of the magnetization in the walls than in the domains.[6] This will result in a faster decline of the nuclear resonance frequency with temperature in walls than in domains. Finally, it is likely that the rf susceptibility is due to a combination of wall motion and domain rotation. Both of these processes will excite the nuclear resonance,[7] but generally nuclear resonant losses from the wall motion mechanism are greater than those from the domain rotation mechanism. The observed quadrupole splittings, linewidths, temperature effects, and intensities of the resonances all agree with these predictions based on assigning three of the observed resonances to nuclei in domains and the fourth to nuclei in walls.

In Fig. 1, the difference between the nuclear resonance frequency for the wall line and the central line of the three domain lines is plotted against temperature. This difference in frequencies indicates that the difference in nuclear fields can be expressed as $H_N^{\text{domain}} - H_N^{\text{wall}} = 4450 + 1200T$. We may take $H_N^{\text{wall}}$ to be the field at the center of the wall since the number of nuclei per unit change in $H_N$ will have a sharp maximum at the wall center and since the enhanced rf field at the nuclei should also be maximum at the center of the wall. For the magnetization deviations $\Delta M$ we have $(\Delta M/M)_{\text{wall}} - (\Delta M/M)_{\text{domain}} = T/200$. One can account for part of the 4450-oe difference at $T=0°K$ by the difference in dipolar fields between $M$ parallel and perpendicular to

the $c$ axis. We have calculated this difference at $Cr^{3+}$ sites in $CrBr_3$ with a moment of $3\ \mu_B$ on each $Cr^{3+}$ site and find a difference of 2420 oe between the two directions. The remaining 2030 gauss can perhaps be accounted for by an anisotropic orbital contribution to the hyperfine field. The presence of substantial orbital effects, in spite of the fact that the $Cr^{3+}$ ground state is nominally an orbital singlet, is indicated by the observed 6850 oe[8] of magnetocrystalline anisotropy with the easy direction perpendicular to the $Cr^{3+}$ layers, in contradiction of what is expected from dipolar anisotropy. Abragam and Pryce[9] have proposed that states higher in energy in the crystalline field can be mixed into the singlet ground state by spin orbit coupling, and that in noncubic symmetry this will then lead to a single ion anisotropy (the $DS_z^2$ term in the usual spin Hamiltonian) as well as an anisotropic hyperfine interaction. The difference in orbital fields parallel and perpendicular to the $c$ axis will be related to the first-order anisotropy constant $K_1$ by $\Delta H_{\text{orbital}} \approx 4K_1\mu_B\langle 1/r^3\rangle/\lambda S$. Taking $K_1$ to be 0.24 $cm^{-1}$ and $\lambda$, the spin orbit coupling, to be 300 $cm^{-1}$, $\mu_B$ the Bohr magneton, $S=\frac{3}{2}$ and $\langle 1/r^3\rangle$ to be $[0.36 \times 10^{-8}\ cm]^{-3}$ we predict 420 oe from this mechanism, still too small to explain the observed 2030 oe. Sugano and Peter[10] have pointed out that configuration mixing of higher excited states is also an important process. This may lead to a different relationship between $\Delta H_{\text{orbital}}$ and $K_1$ and thus to a different anisotropy of the orbital hyperfine interaction.

The term in $H_N^{\text{domain}} - H_N^{\text{wall}}$ proportional to $T$ remains. It is evidently caused by the more rapid decrease of magnetization with temperature in the wall than in the domains. Effects important in the wall magnetization are (1) scattering of spin waves by walls, (2) propagation of spin waves entirely within the walls, and (3) thermal excitations of wall translations. Calculations by Suhl[6] and Winter[11] have indicated a linear temperature dependence of the difference between the wall and domain magnetizations due to these effects. One expects the greatest differences to occur in materials with the narrowest walls. Since wall thickness is proportional to $(J/K)^{\frac{1}{2}}$ and $CrBr_3$ has a low exchange (5.44°K between nearest neighbors[2]) and high anisotropy, we are not surprised that this effect can be observed in $CrBr_3$, but not in iron, cobalt, and nickel.

### ACKNOWLEDGMENTS

We wish to acknowledge many helpful discussions with Dr. L. R. Walker, Dr. M. Peter, Dr. J. F. Dillon, and Dr. S. Geschwind, and able technical assistance from J. L. Davis.

[6] H. Suhl, Bull. Am. Phys. Soc. 5, 175 (1960).
[7] A. M. Portis in *Magnetism: A Treatise on Modern Theory and Materials*, edited by G. T. Rado and H. Suhl (Academic Press Inc., New York, to be published), Chap. 19.

[8] J. F. Dillon, J. Appl. Phys. 33, 1191 (1962), this issue.
[9] A. Abragam and M. H. L. Pryce, Proc. Roy. Soc. A205, 135 (1951).
[10] S. Sugano and M. Peter, Phys. Rev. 122, 381 (1961).
[11] J. M. Winter, Phys. Rev. (to be published).

JOURNAL OF APPLIED PHYSICS     SUPPLEMENT TO VOL. 33, NO. 3     MARCH, 1962

# Temperature Dependence of YIG Magnetization

Irvin H. Solt, Jr.*

*Research and Development Laboratories, Fairchild Semiconductor, Palo Alto, California*

The temperature dependence of the saturation magnetization, $4\pi M$, has been measured for pure, polished YIG spheres in the temperature range from 4–50°K. The technique used involved measuring the magnetic field separation of the 210 and 220 magnetostatic modes as a function of temperature. The mode spacing is equal simply to $4\pi M/5$. This method offers a high degree of precision since the linewidths are less than 1 gauss and the separation of the two modes is the order of 500 gauss. Data were taken for several samples and with the external magnetic field oriented in both the [100] and [111] directions. The experimental results have been fitted to the theoretical expression of Dyson. Only the first-order correction term to Bloch's equation was necessary, the result being

$$M = M_0(1 - 8.19 \times 10^{-6} T^{\frac{3}{2}} - 1.03 \times 10^{-7} T^{\frac{5}{2}}).$$

A fit to the $T^{\frac{3}{2}}$ term alone is definitely excluded.

## I. INTRODUCTION

IN recent years, numerous attempts[1–3] have been made to refine Bloch's[4] original spin wave description of a ferromagnet. These refinements have attempted, with varying degrees of success, to take into account (a) deviations of the energy spectrum from the (wavelength)$^{-2}$ law due to the discreteness of the lattice, and (b) various interaction mechanisms of the spin waves with each other. In particular, Dyson[1] has derived the equation

$$M = M_0(1 - a_0 T^{\frac{3}{2}} - a_1 T^{\frac{5}{2}} - a_2 T^{7/2} - a_3 T^4) \quad (1)$$

for the temperature dependence of the saturation magnetization, where $T$ is the absolute temperature and $M_0$ is $M$ at $T = 0$. The terms $a_1$, $a_2$, $a_3$ give the deviation from the original Bloch $T^{\frac{3}{2}}$ dependence. The $T^{\frac{5}{2}}$ and $T^{7/2}$ terms result from the discreteness of the lattice and the $T^4$ term is the lowest-order spin wave interaction term. Recently there have appeared two measurements of the magnetization of a ferromagnet. One of these was for nickel[5] and the other for the nonconductor $CrBr_3$.[6] Both of these experiments required the higher-order terms to fit the data points.

Concurrently, it also seemed to us to be of considerable interest to measure the magnetization of a ferrimagnet.[7] It had been pointed out by Mercereau and Feynman[8] that one could determine the magnetization from the separation of the magnetostatic modes. This method is potentially quite precise for materials with narrow linewidths. One can, in principle, use any pair of modes, but certain of the modes do have distinct advantages. For the following reasons the 210 and 220 modes seemed best suited:

(1) The magnetization is linearly related to the separation of these modes: $H_{210} = \omega/\gamma + 8\pi M/15$ and $H_{220} = \omega/\gamma - 4\pi M/15$, where $H$ is the resonant magnetic field, $\omega$ is the angular frequency, and $\gamma$ the gyromagnetic ratio. Hence $4\pi M/5 = H_{210} - H_{220}$.

(2) The low-order modes have stronger absorptions and thus give better signal-to-noise ratios than the higher-order modes.

(3) The shift in resonant field due to propagation effects is the same for both modes.[9]

Yttrium iron garnet (YIG) appeared to be the best choice of material since it is available in a high degree of purity and since the linewidth is quite narrow (<1 gauss).

There are several precautions which one has to take. First, the frequency must be chosen so that none of the higher-order interacting modes are in the vicinity of either the 220 or 210 mode[10,11]; otherwise, considerable displacements of the line positions may occur. Secondly, one must operate with the magnetic field applied in either the [100] or [111] crystal direction, since in all other orientations changes in the line separation would be due not only to changes in the magnetization but also to changes in the anisotropy.[12]

## II. EXPERIMENTAL METHOD

The experimental measurements were made in the temperature range from 4.2–50°K. The sample, microwave cavity, and Dewar were initially cooled with liquid helium. The helium was then blown out and the

* The experimental work was done while the author was with Hughes Research Laboratories, Malibu, California.
[1] F. J. Dyson, Phys. Rev. **102**, 1217, 1230 (1956).
[2] F. Keffer and R. Loudon, J. Appl. Phys. **32**, 2S (1961).
[3] R. Brout and H. Haken, Bull. Am. Phys. Soc. **5**, 148 (1960).
[4] F. Bloch, Z. Physik **61**, 206 (1930).
[5] B. E. Argyle and E. W. Pugh, Bull. Am. Phys. Soc. **6**, 125 (1961).
[6] A. C. Gossard, V. Jaccarino, and J. P. Remeika, Phys. Rev. Letters **7**, 122 (1961).
[7] There is some question whether Eq. (1) is valid for a ferrimagnet. Kaplan has shown that the ferrimagnetic spin wave excitations are analagous to the acoustical and optical modes of lattice vibrations. In the low temperature region only the long wave acoustical type spin waves are important and hence a complicated ferrimagnetic arrangement of spins is equivalent to that of a ferromagnet having the same spin moment/unit cell.
[8] J. Mercereau and R. P. Feynman, Phys. Rev. **104**, 63 (1956).
[9] M. Rene Plumier, Compt. rend. **251**, 1356 (1960).
[10] P. C. Fletcher, I. H. Solt, Jr., and R. O. Bell, Phys. Rev. **114**, 739 (1959).
[11] P. C. Fletcher and I. H. Solt, Jr., J. Appl. Phys. **30**, 181S (1959).
[12] I. H. Solt, Jr., and P. C. Fletcher, J. Appl. Phys. **31**, 100S (1960).

measurements made as the sample warmed up. A typical run took several hours. The temperature was monitored simultaneously with a carbon resistance thermometer and a copper-Constantan thermocouple referenced to liquid helium. The resistance and emf were continuously recorded on chart recorders. Since it was impossible to measure the position of the 210 and 220 lines at the same temperature (due to the slowly increasing temperature), they were measured alternately by switching the magnetic field back and forth from the one line to the other. For each magnetic field measurement, a marker was put on the charts. The line positions of both modes were then plotted as a function of temperature. The separation of the two lines at the desired temperature was taken from the difference in plots of the two line positions.

The magnetic fields were measured with a proton resonance detector, and with the narrow YIG linewidths a measurement precision of about 0.01 gauss was obtained. The sample used was a pure, polished YIG sphere. The microwave frequency was 9092 Mc at 4.2°K and rose slightly with temperature. The sample was mounted on an insulating standoff and positioned in the microwave cavity so that the absorptions by the 210 and 220 modes were approximately equal in intensity.

### III. DISCUSSION OF EXPERIMENTAL RESULTS

A typical experimental run is shown in Fig. 1, where $4\pi M/5$ has been plotted vs $T^{3/2}$. One can clearly see that the data points do not lie on a straight line and that one needs terms of higher order than $T^{3/2}$ to obtain a proper fit.[13] The method of least squares was used to evaluate the constants in Eq. (1), with the result:

$$M = M_0(1 - 8.19 \times 10^{-6} T^{3/2} - 1.03 \times 10^{-7} T^{5/2}). \quad (2)$$

Since the fit of this equation to the experimental points was within the experimental scatter $\pm 0.03$ gauss), the addition of the next highest term seemed unnecessary and meaningless. If one normalizes the temperature to the Curie temperature $T_c$, one obtains

$$M = M_0(1 - 0.105\,\theta^{3/2} - 0.722\,\theta^{5/2}), \quad (3)$$

where $\theta = T/T_c$ and $T_c$ is assumed to be 545°K for YIG.

The standard deviation for the constants in Eq. (2) and (3) is $\pm 20\%$. This is obtained from the average of the constants determined for the different runs. Although the constants for one run in the [111] direction differed by more than the error, it is felt that the true

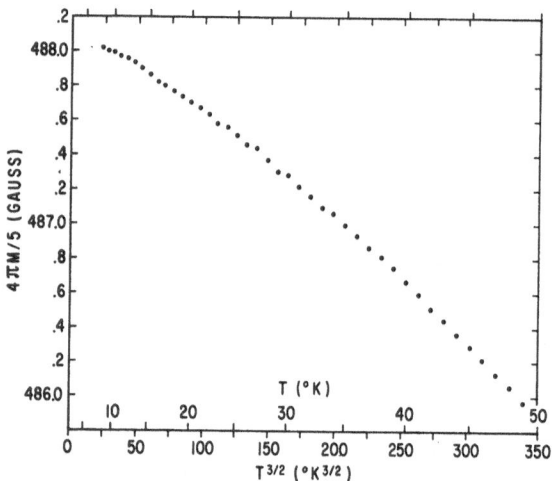

FIG. 1. Variation in $4\pi M/5$ vs $T^{3/2}$ for an 0.712-mm pure, polished YIG sphere. The microwave frequency was 9092 Mc and the applied field was along the [100] crystal direction.

value lies somewhere within the above confidence region. It is interesting to compare the coefficient of the $T^{3/2}$ term with the theoretical value.[2] Using the experimental values for exchange constant and magnetization in YIG, one obtains $10.1 \times 10^{-6}$. This is a difference of 23% from the experimental average and just outside of the confidence region.

There are several possible sources of systematic errors which are worth discussing. First, measurements were made under a dynamic temperature condition rather than a static condition. Hence, it is possible that the temperature of the sample might lag the temperature at the thermocouple and resistance thermometers. If this were the case, the true curve would depart still further from the $T^{3/2}$ dependence. However, it is felt that the temperature rise was sufficiently slow that the errors could not be larger than a fraction of a degree. The second potential source of error was in the temperature measurement. A Leeds and Northrup precision copper-Constantan thermocouple wire was used and the data were based on the NBS curves for the emf versus temperature with a liquid helium reference temperature. The agreement was within 0.1°K at the temperature of liquid nitrogen and hence was assumed to be good over most of the range. The third source of error might arise from the asphericity of the sample. The sample was aspherical since the measured value of $M$ varied with crystal orientation, in excess of the expected anisotropy variation. However, Fletcher[14] has shown that for prolate and oblate spheroids, the separation of the two lines in question is still linearly proportional to M. Hence, although the measured value of magnetization may be incorrect, the temperature dependence is identical with that of a sphere.

---

[13] Due to a slight asphericity of the sample the values of magnetization given on the ordinate of Fig. 1 are not the true values of the magnetization for YIG. Hence it must be emphasized that the data represent a precision measurement of the temperature dependence of the magnetization and not a precision measurement of the magnetization itself.

[14] P. C. Fletcher (private communication).

## IV. CONCLUSIONS

The magnetization of YIG has been measured very precisely over the temperature range from 4° to 50°K ($T \leq 0.9 T_c$). The experimental results show quite conclusively that the simple $T^{\frac{3}{2}}$ temperature dependence cannot fit the data. Several potential errors have been discussed. These are not sufficient to account for the observed deviation from the $T^{\frac{3}{2}}$ dependence.

## ACKNOWLEDGMENTS

The author gratefully acknowledges many helpful discussions with R. L. White and P. C. Fletcher and the invaluable technical assistance of E. H. Gregory, R. Morrison, and Judith Osmer. The author particularly wishes to thank R. A. Lefever for supplying him with the pure YIG single crystals used in the experiment.

---

JOURNAL OF APPLIED PHYSICS    SUPPLEMENT TO VOL. 33, NO. 3    MARCH, 1962

# Ferromagnetic Resonance in CrBr₃

J. F. DILLON, JR.
*Bell Telephone Laboratories, Incorporated, Murray Hill, New Jersey*

Ferromagnetic resonance has been observed in single crystals of anhydrous chromium tribromide using frequencies of 20 to 27 kMc. At 1.5°K a $g$ value of 2.006 is found along with a uniaxial anisotropy, $K = 9.4 \times 10^5$ ergs/cm³. Linewidths as narrow as 3.5 oe have been observed at 1.5°K.

CHROMIUM tribromide is one of the very small number of ionic crystals which are now known to become ferromagnetic at some temperature.[1] As such, its magnetic properties are of particular interest. Recently, Tsubokawa[2] has published the results of measurements of magnetization, susceptibility, and anisotropy. Gossard, Jaccarino, and Remeika[3] have reported on the variation of the nuclear magnetic resonance frequency of the chromium ion with temperature, and the relation of these data to spin wave theory. In this paper we report the observation of ferromagnetic resonance in single crystals of CrBr₃, the variation of the field for resonance with temperature and crystal direction, the $g$ value, and the linewidth.

## SAMPLES

Our samples were grown by J. P. Remeika who used a procedure very similar to that described by Tsubokawa. The best linewidth data were obtained on a nearly rectangular flake about 0.035 by 0.035 by 0.0007 in. In many other samples somewhat wider lines were obtained. In thinner samples the magnetostatic modes tended to overlap one another so seriously that it was not possible to get a linewidth readily.

## FIELD FOR RESONANCE

The experimental apparatus has been described in some detail previously.[4] The frequencies used here were 20.24 and 23.24 kMc. At a temperature of 1.5°K the sample was rotated about an axis in the hexagonal plane, and traces made of the absorption versus field at ten degrees intervals. From the $H_{res}$ versus angle curves we located the hexagonal axis. Recordings were then made of the line shape over a temperature range up to 300°K. Figure 1 shows the course of the field for resonance over this range for the field parallel and normal to the hexagonal plane. At 1.5°K a very thin section was rotated with the field in its major plane. Though the absorption structure observed as it was rotated through 180° consisted of several magnetostatic modes, it was clear that there was no significant anisotropy in the plane. Thus the anisotropy is uniaxial, and in agreement with Tsubokawa[3] we find the hexagonal axis to be the easy direction.

H. J. Williams and R. C. Sherwood measured the magnetization of a single crystal of CrBr₃ over the range from 1.5° to 120°K, using applied fields of 5000 to 14 000 oe. By interpolation in their data we found the magnetization in the field required for resonance at each temperature. Rather than assuming the sample to be an infinitely thin disk, an estimate of the appropriate demagnetizing factors ($N_x = N_y = 0.05$, $N_z = 0.90$) was obtained from the spacings of the magnetostatic modes.[5] The displacement of the field for resonance was computed from $M$ at $H_{res}$ and these demagnetizing factors. Curve (d) of Fig. 1 represents the field for resonance with this displacement removed; it is the field at which resonance would occur for a spherical sample. The field interval between $\omega/\gamma$ and (d) at any temperatures is equal to $2K/M$, the anisotropy field. Multiplication by $M/2$ yields the anisotropy constant $K$ as a function of temperature. This is given in Fig. 2.

It is to be noted that our resonance value for $K$ is almost twice the torque $K$ published by Tsubokawa.[2] The reason for this discrepancy is not clear.

[1] W. N. Hansen, J. Appl. Phys. **30**, 304S (1959).
[2] I. Tsubokawa, J. Phys. Soc. Japan **15**, 1664 (1960).
[3] A. C. Gossard, V. Jaccarino, and J. P. Remeika, Phys. Rev. Letters **7**, 122 (1961); J. Appl. Phys. **33**, 1187 (1962), this issue.
[4] J. F. Dillon, Jr., and J. W. Nielsen, Phys. Rev. **120**, 105 (1960).

[5] J. F. Dillon, Jr., J. Appl. Phys. **31**, 1605 (1960).

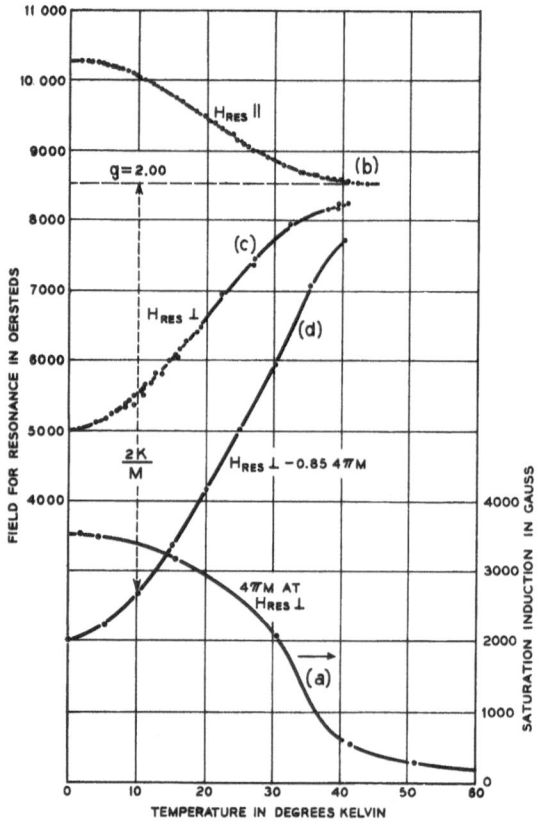

FIG. 1(a) Magnetization versus temperature of CrBr₃ at the applied field for resonance perpendicular to the hexagonal plane, $f$=23.4 kMc as measured by H. J. Williams and R. C. Sherwood. (b) $H_{RES}$ ($T$) with the field in the hexagonal plane for sample 0.035 by 0.035 by 0.0007 in. (c) $H_{RES}$ ($T$) with the field normal to the hexagonal plane. (d) $H_{RES}^\perp$ less estimated demagnetizing field.

## g VALUE

The spectroscopic splitting factor $g$ used above was seen to be close to 2.00 in the ferromagnetic region by a comparison of the field for resonance curves taken at 20.3 and 23.4 kMc. In addition, for 1.5°K, very careful measurements were made at three frequencies, ∼20, ∼23, ∼27 kMc. With the field parallel to the hexagonal axis, $g|_{1.5°}=2.007\pm0.001$. Finally, measurements in the paramagnetic range gave $g|_{181°}=2.014\pm0.003$, and $g|_{295°}=2.019\pm0.003$. The ground state of Cr⁺⁺⁺ is ⁴$F$, but in an octahedral field the lowest lying level is effectively an $S$ state. For this we would expect a value of $g$ very close to the free-electron value.

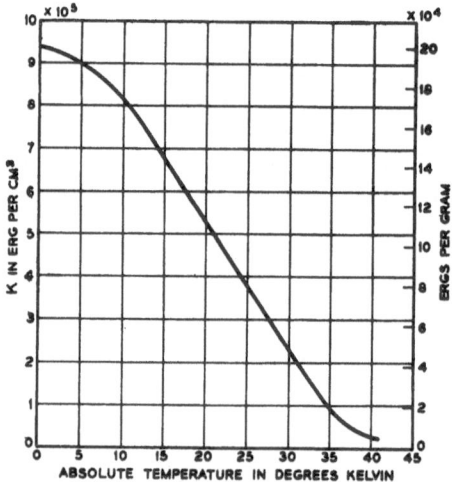

FIG. 2. Anisotropy constant $K(T)$ for CrBr₃ as determined from resonance measurements at 20.2 kMc.

## LINEWIDTH

Linewidths were measured on semiautomatic plots of absorption against field for the range 1.5° to 300°K. For brevity, the detailed data must be presented in a later publication. The sample was the 0.035- by 0.035- by 0.0007-in. piece mentioned above. The linewidth is about 10 oe in the liquid helium range increasing monotonically with temperature up to at least 300°K. Values are 130 oe at 37°K, the Curie temperature, and 450 oe at 300°K. At low temperatures there is a considerable anisotropy in linewidth, certainly not unexpected for a uniaxial crystal such as this.

At the lower end of the temperature range, linewidth is very clearly tied in with the physical condition of the sample. The first time this crystal was cooled to liquid helium temperatures, we measured linewidths of about 3.5 oe with the field parallel to the sixfold axis. On the following day the best value obtainable under apparently identical conditions was twice as large. These crystals are very soft and it seems to be very difficult to perform any manipulation with them at all without degrading the ferromagnetic resonance linewidth.

## ACKNOWLEDGMENTS

The author is indebted to J. P. Remeika for the crystals. He wishes to thank A. C. Gossard, V. Jaccarino, and F. Keffer for many discussions. The data were taken with the assistance of H. E. Earl.

# Oxides–1

## L. M. CORLISS, *Chairman*

## Acoustic Losses in Ferromagnetic Insulators

R. C. LECRAW, E. G. SPENCER, AND E. I. GORDON

*Bell Telephone Laboratories, Inc., Murray Hill, New Jersey*

Studies of acoustic losses in single crystal garnets and ferrites using magnetostrictively driven spherical resonators are described. Resonant modes have been observed in spheres of yttrium iron garnet (YIG) at $\sim$9 Mc and room temperature with $Q$'s of $\sim$10$^7$. This is approximately six times greater than any other known material at the same frequency and temperature. These high $Q$'s have now been observed in both shear and compressional modes. [R. C. LeCraw, E. G. Spencer, and E. I. Gordon, Phys. Rev. Letters **6**, 620 (1961). In this reference only a certain compressional mode showed the unusually high $Q$.] It should be noted that the above $Q$ in YIG may be only a lower bound. Contact losses have been eliminated by levitation and by using higher order modes. However, due to complicating factors such as the relatively large and variable dislocation density (revealed by etching) and surface defects, we are not yet certain whether the measured $Q$'s are characteristic of perfect YIG. The intrinsic $Q$ could be higher than the above figure and still be consistent with the present results.

The dependence of the internal friction $Q^{-1}$ in pure YIG on frequency and temperature are given. The frequency dependence at 300°K is shown to be $Q^{-1} \sim f$ in the megacycle range with possibly a slower dependence on $f$ in the microwave range. Data are given on $Q^{-1}$ at $\sim$20 Mc from 300° to 20.3°K. $Q^{-1}$ is relatively independent of temperature down to $\sim$50°K. Below 50°K, $Q^{-1}$ increases rapidly down to 20.3°K, indicating a large absorption maximum below 20.3°K. The origin of this maximum is at present unknown. Data are also presented on $Co_{1.1}Fe_{1.9}O_4$ at $\sim$20 Mc from 300° to 20.3°K. This material has a very large maximum ($Q^{-1} \sim 10^{-2}$) at 290°K. Below 50°K, however, it has quite remarkable properties, with $Q^{-1}$ being lower than YIG. Instead of $Q^{-1}$ increasing rapidly below 50°K, as in YIG, it decreases rapidly from 50° to 20.3°K, indicating very low losses in the helium range. Investigation of all the above effects is continuing in detail to obtain an understanding of the many questions raised by these results about acoustic losses in ferromagnetic insulators.

---

## X-Ray and Magnetic Studies of CrO$_2$ Single Crystals

W. H. CLOUD, D. S. SCHREIBER,* AND K. R. BABCOCK

*Central Research Department, E. I. du Pont de Nemours and Company, Wilmington, Delaware*

The oxygen parameter of CrO$_2$, which has the rutile structure, has been determined to be $u = 0.301 \pm 0.004$. Single-crystal measurements of cell constants over the temperature range $-200°$ to $+200°$C show the unusual decrease in the $c$ axis with increasing temperature in agreement with x-ray powder results reported by other workers. An estimate of the magnetocrystalline anisotropy has been obtained from measured magnetization curves of a small single-crystal sphere of CrO$_2$. The directions of easy magnetization lie in (100) planes at an angle of approximately 40° to the tetragonal axis.

### CRYSTAL STRUCTURE STUDIES

CHROMIUM dioxide has the rutile structure, space group $P4/mnm$, with $a = 4.421$ A, $c = 2.917$ A.[1,2] The chromium ions lie at the corners and body-center of the unit cell. Oxygen ions lie at $u$, $u$, 0; $\bar{u}$, $\bar{u}$, 0; $u-\frac{1}{2}$, $\frac{1}{2}-u$, $\frac{1}{2}$; $\frac{1}{2}-u$, $u-\frac{1}{2}$, $\frac{1}{2}$. A value of $u = 0.294$ has been reported[3] from x-ray powder measurements of the intensities of the 111 and 210 reflections.

The chromium ions do not contribute to the x-ray reflections for which the sum of the indices is odd. Since the $z$ position of oxygen is fixed at $\frac{1}{2}$, the intensities of $hk0$ reflections for which $h+k = 2N+1$ should determine

oxygen positions. Most of these are too weak to be measured in powder patterns. Six of these reflections were observed using a single-crystal sphere 0.3 mm in diameter and Mo $K\alpha$ radiation. The single crystals were prepared under hydrothermal conditions similar to those used to prepare powder specimens for which chemical analyses gave a stoichiometry equal to that for pure CrO$_2$ within the accuracy of the measurement.[1,2] From the six reflections the oxygen parameter was determined to be $u = 0.301 \pm 0.004$ by a least squares fit between observed and calculated structure factors. In the temperature factor, $\exp(-B \sin^2\theta/\lambda^2)$, $B$ was determined to be 0.84 A$^2$. The discrepancy factor, $\sum |\Delta F|/\sum |F|$, was 0.030. The data are shown in Table I. The observed intensities are an average of data taken from two quadrants of the single-crystal sphere.

During the course of the single-crystal x-ray studies,

* Present address: Cornell University, Ithaca, New York.
[1] T. J. Swoboda, P. Arthur, N. L. Cox, J. N. Ingraham, A. L. Oppegard, and M. S. Sadler, J. Appl. Phys. **32**, 374S (1961).
[2] T. J. Swoboda *et al.*, J. Phys. Chem. Solids (unpublished).
[3] K. Siratori and S. Iida, J. Phys. Soc. Japan **15**, 210 (1960).

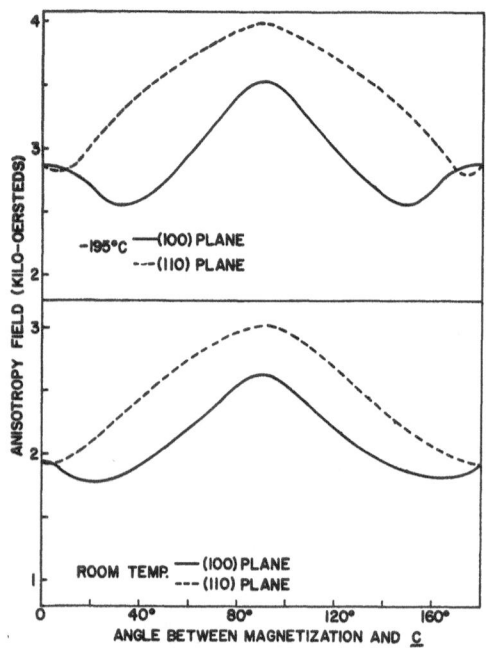

FIG. 1. Anisotropy field measured on a CrO₂ single-crystal sphere.

TABLE I. X-ray intensities for a CrO₂ single-crystal sphere.

| hkl | Relative intensity | $F_{obs}$ | $F_{calc}$ |
|-----|-----|-----|-----|
| 120 | 3878 | 10.05 | 9.90 |
| 140 | 882 | 6.68 | 6.98 |
| 160 | 180 | 3.83 | 3.59 |
| 320 | 230 | 3.17 | 3.15 |
| 340 | 148 | 3.07 | 3.18 |
| 360 | 46 | 2.02 | 1.99 |

the cell constants were measured as a function of temperature. The data are in excellent agreement with powder data reported by Siratori and Iida[3] which show that $c$ increases with increasing temperature from −196°C to approximately −100°C and then decreases with temperature up to 200°C, the highest temperature used in the measurement. The $a$ axes and volume of the unit cell increase monotonically with temperature. Siratori and Iida have proposed that this unusual behavior of $c$ versus temperature may be due to Jahn-Teller distortion of the $(3d)^2$ orbitals of $Cr^{4+}$. The single-crystal measurements give room temperature values of $a = 4.4218$ A, $c = 2.9182$ A.

## MAGNETOCRYSTALLINE ANISOTROPY

Single-crystal magnetization measurements were made on an apparatus in which the force on the specimen in a magnetic field gradient was measured by observing the deflection of a pendulum. The magnetization versus field curves of an 80-microgram single-crystal sphere of CrO₂ were measured for fields applied in various crystallographic directions. Although hysteresis

of the pole caps of the electromagnet restricted accuracy for fields less than 1500 oe, significant differences in the shapes of magnetization curves could be found for applied fields in different crystallographic directions. The energy $K$ required to magnetize the sphere was taken as

$$K = \int_0^{M_s} H dM = M_s H_s - \int_0^{H_s} M dH, \qquad (1)$$

where $M_s$ is the saturation magnetization and $H_s$ is a field large enough to produce saturation. The saturation magnetization of CrO₂ is 640 gauss at −196°C and 490 gauss at 25°C.[1,2] The anisotropy field $H_A = 2K/M_s$.

Values of $H_A$ were determined for applied fields at various angles to $c$ in the (100) and (110) planes at room temperature and liquid nitrogen temperature. The results are shown in Fig. 1. Although the individual values of $H_A$ were not accurate to more than 500 oe, the existence of the minimum in the (100) plane at an angle of approximately 40° to the $c$ axis was verified by repeated measurements. Measurements with the applied field in the (001) plane at various angles to an $a$ axis showed that it required more energy to magnetize in a [110] direction than in a [100] direction.

The data of Fig. 1 show that the directions of easy magnetization lie in (100) planes at an angle of approximately 40° to the $c$ axis. This result is supported by neutron diffraction measurements on CrO₂ powders[4] which show that the angle between the atomic moments and the $c$ axis is approximately 40°.

## ACKNOWLEDGMENT

The single crystal of CrO₂ was prepared by N. L. Cox.

[4] A. E. Austin, E. Adelson, and C. M. Schwartz (private communication).

JOURNAL OF APPLIED PHYSICS          SUPPLEMENT TO VOL. 33, NO. 3          MARCH, 1962

# Substitutions of Divalent Transition Metal Ions in Yttrium Iron Garnet

S. Geller, H. J. Williams, R. C. Sherwood, and G. P. Espinosa
*Bell Telephone Laboratories, Inc., Murray Hill, New Jersey*

Various amounts of the divalent ions of manganese, iron, cobalt, and nickel have been substituted for trivalent iron in yttrium iron garnet. The effect of substitution of $Fe^{2+}$ or $Mn^{2+}$ ions in dodecahedral sites has also been studied. Electrical balance was accomplished by the simultaneous substitution of tetravalent silicon or germanium. Under these conditions, it is concluded that the divalent ions prefer octahedral sites.

The great affinity of silicon for the tetrahedral sites in the garnets results in the reduction of trivalent to divalent iron at high temperatures even in air atmosphere when appropriate amounts of finely divided silica are present. In $Y_3Fe_x^{2+}Fe_{5-2x}^{3+}Si_xO_{12}$, the maximum attainable value of $x$ (under the conditions of our experiments) appears to be 0.45. In attempting to put $Fe^{2+}$ ions into dodecahedral sites, greatest success was attained when there was also some in the octahedral sites as in $\{Y_{2.9}Fe_{0.1}^{2+}\}$ $[Fe_{0.3}^{2+}Fe_{1.7}^{3+}](Fe_{2.6}^{3+}Si_{0.4})O_{12}$. The 0.1 was the maximum amount that could be put into the dodecahedral sites.

Although, it appears that the $Co^{2+}$ ion is the easiest to substitute into yttrium iron garnet, it is perhaps the most difficult to understand, and only a few of the results on experiments with this ion are included.

In all substitutions of divalent magnetic ions not involving the dodecahedral sites, the 0°K moments are smaller than those calculated on the basis of the simplest interaction model. Several possible explanations are explored; a plausible one is the random canting of some tetrahedral ion moments resulting from weaker interaction with the octahedral divalent ions. The possibility that some of the divalent ions enter tetrahedral sites is not entirely excluded especially in the case of the $Co^{2+}$ ion.

The divalent magnetic ions do not behave like the trivalent magnetic lanthanide ions when substituted in the dodecahedral sites. They appear not to contribute to the net spontaneous moment appreciably.

## INTRODUCTION

THERE is now some history associated with the incorporation of divalent iron in the iron garnets but little is known of the crystalchemical and associated magnetic behavior of this ion in the garnets. Furthermore although it has been shown that divalent transition metal ions enter the garnets in substantial amount,[1] almost nothing has been reported on their behavior in the magnetic garnets. The present investigation involves the substitution of the divalent ions of Mn, Fe, Co and Ni in YIG, with electrical compensation mainly by tetravalent $Si^{4+}$ ions.

TABLE I. Crystallographic and magnetic data from selected specimens.

| Specimen | $a$, A | $n_B(\infty, 0), \mu_B$ obs | calc[a] |
|---|---|---|---|
| 1. $\{Y_3\}[Fe_2](Fe_3)O_{12}$ | 12.376 | 4.96[b] | 5.00 |
| 2. $\{Y_{2.60}Ca_{0.40}\}[Fe_2](Fe_{2.60}Si_{0.40})O_{12}$ | 12.344 | 2.98 | 3.00 |
| 3. $Y_3Fe_{0.30}^{2+}Fe_{4.40}Si_{0.30}O_{12}$ | 12.356 | 3.63 | 3.74 |
| 4. $Y_3Fe_{0.40}^{2+}Fe_{4.20}Si_{0.40}O_{12}$ | 12.350 | 3.20 | 3.32 |
| 5. $Y_3Fe_{0.45}^{2+}Fe_{4.10}Si_{0.45}O_{12}$ | 12.345 | 2.98 | 3.11 |
| 6. $\{Y_{2.98}Fe_{0.02}^{2+}\}[Fe_{0.32}^{2+}Fe_{4.34}Si_{0.34}O_{12}$ | 12.350 | 3.45 | 3.47 |
| 7. $\{Y_{2.90}Fe_{0.10}^{2+}\}[Fe_{0.30}^{2+}Fe_{4.30}Si_{0.40}O_{12}$ | 12.340 | 3.15 | 2.82 |
| 8. $Y_3Ni_{0.40}Fe_{4.20}Si_{0.40}O_{12}$ | 12.341 | 3.95 | 4.08 |
| 9. $Y_3Ni_{0.50}Fe_{4.00}Si_{0.50}O_{12}$ | 12.335 | 3.78 | 3.88 |
| 10. $Y_3Mn_{0.40}Fe_{4.20}Si_{0.40}O_{12}$ | 12.359 | 2.89 | 3.00 |
| 11. $\{Y_{2.90}Mn_{0.10}\}Fe_{4.90}Si_{0.10}O_{12}$ | 12.368 | 4.39 | 4.00 |
| 12. $\{Y_{2.75}Mn_{0.25}\}Fe_{4.75}Ge_{0.25}O_{12}$ | 12.375 | 3.68 | 2.50 |
| 13. $Y_3Co_{0.10}Fe_{4.80}Si_{0.10}O_{12}$ | 12.371 | 4.70 | 4.63 |
| 14. $Y_3Co_{0.40}Fe_{4.20}Si_{0.40}O_{12}$ | 12.349 | 3.05 | 3.52 |
| 15. $Y_3Co_{0.60}Fe_{3.80}Si_{0.60}O_{12}$ | 12.333 | 2.32 | 2.78 |
| 16. $Y_3CoFe_3SiO_{12}$ | 12.296 | 0.67 | 1.30 |

[a] The calculated values are based on the simplest model of interaction in which (1) all divalent ions not specifically indicated to be in dodecahedral sites are assumed to be in octahedral sites; (2) the moments of the divalent ions are assumed to be 4.2, 2.3, 5.0, and 3.7 for $Fe^{2+}$, $Ni^{2+}$, $Mn^{2+}$, and $Co^{2+}$ respectively, as in the ferrospinels [see E. W. Gorter, Philips Research Repts. **9**, 321 (1954) and pertinent references therein]; (3) there is no kind of canting of any of the moments; (4) moments of magnetic ions in dodecahedral sites are assumed to be antiparallel to the resultant moment from the octahedral and tetrahedral ions.
[b] From M. A. Gilleo and S. Geller, Phys. Rev. **110**, 73 (1958).

[1] S. Geller, J. Appl. Phys. **31**, 30S (1960).

## DISCUSSION

In the last column of Table I, the values of the 0°K moments, calculated under the conditions listed as a footnote to the table, do not give completely satisfactory results for any of the garnets containing divalent ions. However, the second specimen listed, which does not contain any magnetic divalent ions, gives the theoretically expected moment. Thus the deviations in the remainder of the cases have perhaps a more subtle origin. There are unfortunately a number of possibilities or combinations thereof for explaining the results.

If we plot moment extrapolated to infinite field and 0°K of the specimens containing $Fe^{2+}$ ions mainly, we hope, in the octahedral sites, a rather good straight line connects these points; this line could be reproduced by assigning a moment of $4.6\,\mu_B$ to the $Fe^{2+}$ ion, a value which is apparently high. Similarly higher than expected moments would be implied for the other divalent ions involved: 2.5 for $Ni^{2+}$ and 5.3 for $Mn^{2+}$. For cobalt, higher values than 3.7 are implied by all results but that from the 0.10 Co specimen.

Another possibility is that of distribution of the divalent ions over both tetrahedral and octahedral sites. In the cases of $Fe^{2+}$, $Co^{2+}$, and $Ni^{2+}$ the measured moments would imply that higher amounts of over-all substitution result in smaller percentages of these ions to tetrahedral sites, the larger percentage, of course, always being in the octahedral sites. This also takes into consideration possible differences in moments in the two different sites. In the case of the $Mn^{2+}$ ion, if its moment really is $5\,\mu_B$, it should make no difference to the net moment whether the $Mn^{2+}$ ions are in octahedral or tetrahedral sites. Also, we believe the size of the $Mn^{2+}$ ions really precludes their being in tetrahedral sites in the garnets. Thus the possible distribution of the divalent ions over the octahedral and tetrahedral sites is not in itself an explanation of the results although there is indication from other experiments that some $Co^{2+}$ ions do enter tetrahedral sites.

The possibility that the results were caused by multiphase material cannot be completely eliminated. The self-consistency of both the crystallographic and magnetic data, their reproducibility, and certain other indications requiring lengthy discussion, however, do give us some confidence that the observed deviations especially of the $Co^{2+}$ substituted specimens are indeed from compositions of single phases (except perhaps for specimens 7, 12, and 13). A plausible tentative explanation might be the following. We can expect that the $Me^{2+}-O^{2-}-Fe^{3+}$ interactions are all weaker than $Fe^{3+}-O^{2-}-Fe^{3+}$ interactions. The substitution of nonmagnetic ions in octahedral sites has a larger effect than a proportional substitution of nonmagnetic ions in tetrahedral sites.[2-3] It may well be that the introduction of the divalent magnetic ions randomly in the octahedral sites results in a random canting of moments of tetrahedral $Fe^{3+}$ ions. This is reasonable because the *percentage* deviation of net moment increases with increasing substitution of divalent ions. It is probable that the tetrahedral ions most affected are those which are linked to less than two octahedral $Fe^{3+}$ ions as in the case of nonmagnetic ion substitution. It is also possible that the cause is a combination of the random canting, the entry of some divalent ions into the tetrahedral sites and different $g$ values from those assumed.

It is now rather well known that when magnetic rare-earth ions are substituted for Y in YIG, they interact with the magnetic ions in the $Fe^{3+}$ ion sublattices, and that the net moment of such a garnet is the difference between the resultant moment of the iron sublattices and the net moment of the $c$ position ions. But we note that a similar result is not obtained for the $Fe^{2+}$ and $Mn^{2+}$ ion substitutions for $Y^{3+}$ ion. Considering that for specimens 3, 4, and 5, there is a difference of about $0.12 \mu_B$ between observed and calculated, the observed values for specimens 6 and 7 should be about 3.35 and $2.70 \mu_B$, respectively. Even if the

result for specimen 6 were not considered to be significant, the result for specimen 7 does appear to be so. In fact, if we were to ignore the $Fe^{2+}$ ion in the $c$ sites, the calculated moment would be $3.24 \mu_B$ and subtracting the $0.12 \mu_B$ from this, we would expect $3.1 \mu_B$ in good agreement with the observed value. Also note that the lattice constant of specimen 7 indicates that $Fe^{2+}$ ions are really *in* the dodecahedral sites; it is 0.016 A less than that of specimen 3 and 0.003 A less than that of specimen 5 which contains more silicon. Even if the extrapolation of $n_B$ is made to $H=0$, a value of $3.0 \mu_B$ is obtained for specimen 7,[4] and this is still too high but could indicate a canting or partial paramagnetism of the $Fe^{2+}$ ions in the $c$ sites.

In the case of specimen 11, if the $Mn^{2+}$ ions in $c$ sites are ignored, a moment of $4.5 \mu_B$ would be expected. At 1.4°K, this specimen saturates at 8000 oe. Thus here again the interaction of the $c$-site magnetic ions if any is unlike that of the rare earths.

Specimen 12 was an attempt to put more $Mn^{2+}$ ion into the $c$ sites; 0.10 seemed to be very near the maximum that would enter when $Si^{4+}$ ion was used for electrical balance. We find that 0.25 is the maximum substitution we can make in the dodecahedral positions. If we ignore the $Mn^{2+}$ ion in specimen 12, the calculated moment would be $3.75 \mu_B$. If we extrapolate moments to $H=0$ for specimen 12, we obtain $3.4 \mu_B$. The difference in the $H=\infty$ and $H=0$ values could result from the presence of a slight amount of impurity phase which is indicated by the x-ray photograph. Nevertheless, the indications are very strong indeed that the divalent magnetic ions in $c$ sites do not behave like the magnetic rare earths in the iron garnets.

It has been suggested to us by M. Peter that the divalent ions in the $c$ sites may be forming antiferromagnetic pairs similarly to the manner in which divalent $Mn^{2+}$ ions form such pairs when these ions are substituted for $Mg^{2+}$ ion MgO.[5] This would imply that such interactions must be stronger than those between $c$-site and $d$- or $a$-site ions.

[2] M. A. Gilleo, J. Phys. Chem. Solids 13, 33 (1960).

[3] Results, not yet published, that we have obtained on other systems give definitive corroboration to this statement.

[4] When this specimen is cooled in a field of 14 240-oe saturation at 1.4°K is attained at a field of 11 600 oe and the moment is $3.14 \mu_B$.

[5] B. A. Coles, J. W. Orton, and J. Owen, Phys. Rev. Letters 4, 116 (1960).

JOURNAL OF APPLIED PHYSICS    SUPPLEMENT TO VOL. 33, NO. 3    MARCH, 1962

# Cation-Cation Three-Membered Ring Formation

JOHN B. GOODENOUGH

*Lincoln Laboratory,* Massachusetts Institute of Technology, Lexington 73, Massachusetts*

The stoichiometric compound FeS undergoes an electron-ordering transition at $T_\alpha \approx 140°C$ on heating, $T_\alpha \approx 100°C$ on cooling. It is pointed out that the various magnetic, crystallographic, and electric properties associated with this transition are compatible with the formation of three-electron $Fe^{2+}$—$Fe^{2+}$ bonds within the basal planes. This rather unexpected effect brings to nine the number of different electron-ordering transitions that can be identified. It is also noted that $Mn_2Mo_3O_8$ probably represents a closely related type of electron ordering.

STOICHIOMETRIC (presumably) FeS, which has a NiAs structure, exhibits three transition temperatures: antiferromagnetic ordering below $T_N \approx 320°C$; a spin-flip transition with spins parallel to the $c$ axis below $T_s$ and perpendicular to the $c$ axis for $T_s < T < T_N$; a three-cation-clustering transition at $T_\alpha$, where $T_\alpha = T_s \approx 140°C$ on heating and $T_\alpha \approx 100°C < T_s \approx 120°C$ on cooling.[1,2] The perpendicular susceptibility $\chi_\perp$ is smallest below $T_\alpha$, largest in the small temperature interval $T_\alpha < T < T_s$ found in the cooling cycle, and intermediate for $T > T_s$. There is a sharp discontinuity in the electrical conductivity parallel to the $c$ axis at $T_\alpha$ (not at $T_s$), but only a small discontinuity perpendicular to this axis. Neutron-diffraction data[3,4] give ferromagnetic basal planes coupled antiferromagnetically to one another for all $T < T_N$ and a low-temperature $Fe^{2+}$ moment corresponding roughly to $4\mu_B$, or its spin-only value. High-temperature susceptibilities[5] obey a Curie-Weiss law with $\theta_p \approx -875°K$ and $\mu_{eff} \approx 5.25\mu_B$.

Since the melting point $T_{mp}$ of a compound is primarily determined by ordering of the outer $s$ and $p$ electrons, which have greater radial extension than $d$ electrons of partially filled shells of lower principal quantum number, transition-metal compounds frequently exhibit $d$-electron-ordering transitions at a $T_t < T_{mp}$. It is reasonable to assume that interpretation of the magnetic and electric properties of stoichiometric FeS depends primarily upon a proper description of the $3d$ electrons of $Fe^{2+}$. However, it must be realized that in the NiAs structure the energy required to transfer a cation from its octahedral site to an interstitial, bipyramidal site is relatively small, so that even for stoichiometric samples the number of interstitial $Fe^{2+}$, and therefore the physical properties of a given sample, may depend sensitively on the heat treatment of the specimen.

There are two approximations that are used in the description of outer electrons in solids: (1) the molecular-orbital or band model, which in its simple form is only applicable to outer $s$ and $p$ electrons, and (2) the Heitler–London, ligand-field model, which is successful for the case of tightly bound $4f$ electrons or for $3d$ electrons in ionic compounds with relatively large separation $R$ of the cations. For intermediate cation separations, a collective-electron model that is modified to include strong electron-lattice interactions, ligand-field splittings, and electron correlations must be devised. It is convenient to define a critical cation-cation separation $R_c$ such that for $R < R_c$ the cation-sublattice $d$ electrons are "collective" and for $R > R_c$ they are "localized." Although the transition is probably not sharp at $R_c$, a semiempirical critical separation for cation $3d$ electrons in oxides has been shown to be[6]

$$R_c(3d) \approx [3.00 - 0.05(Z - Z_{Ti})] \text{ A}, \quad (1)$$

where $Z_{Ti}$ and $Z$ are the atomic numbers of Ti and of the transition element in question. In the presence of more polarizable anions, such as $S^{2-}$, $R_c(3d)$ is somewhat ($\lesssim 0.4$ A) larger. Whereas the $Fe^{2+}$ ions of ideal, stoichiometric FeS are separated by about $3.5$ A $> R_c$ in the basal plane for $T > T_\alpha$, they form triangular clusters with $R = 3.00$ A $\approx R_c$ within a cluster, but $3.73$ A between clusters.[2] Clustering is identical in alternate pairs of basal-plane layers. Within a pair of basal planes, $R(c \text{ axis}) = 2.94$ A; between pairs the layer separation is $2.97$ A and $R = 3.04$ A. These intercation distances are compatible with a ligand-field model for $T > T_\alpha$, but with relatively strong cation-cation interactions along the $c$ axis that broaden the $c$-axis-oriented states into a narrow band of collective-electron states. At $T < T_\alpha$ they are compatible with collective-electron orbitals within the basal planes for the "triangular molecule" of a cluster. However, the band width for the collective-electron states may be assumed small relative to the ligand-field splittings. Therefore a qualitative energy-level scheme may be obtained from the ligand-field model with Hamiltonian

$$H = H_0 + V_{el} + V_c + V_t + V_{LS} + V_\lambda + \sum_{ij} J_{ij} \mathbf{S}_i \cdot \mathbf{S}_j, \quad (2)$$

where $V_{el}$ is the correction to the spherical approximation $H_0$ for the true electrostatic interaction between outer electrons, $V_c$ and $V_t$ are the cubic and noncubic

* Operated with support from the U. S. Army, Navy, and Air Force.

[1] E. Hirahara and M. Murakami, J. Phys. Chem. Solids **7**, 281 (1958); J. Phys. Soc. Japan **13**, 1407 (1958).

[2] E. F. Bertaut, Bull. Soc. franç. minéral et crist. **79**, 276 (1956).

[3] S. S. Sidhu and D. Meneghetti, Phys. Rev. **91**, 436 (1953).

[4] A. F. Andresen, Acta Chem. Scand. **14**, 919 (1960).

[5] R. Benoit, J. chim. phys. **52**, 119 (1955); Compt. rend. **234**, 2174 (1952).

[6] J. B. Goodenough, "Magnetism and the chemical bond," to be published as Supplement to Vol. III, *Progress in Inorganic Chemistry,* edited by F. A. Cotton (Interscience Publishers, Inc., New York).

components of the ligand fields, $V_{LS}$ is the spin-orbit coupling and has the form $\lambda \mathbf{L} \cdot \mathbf{S}$ for Russell-Saunders coupling, $V_\lambda$ is the elastic-coupling energy associated with interstice distortions that accompany electron ordering at neighboring cations, and the last term is the magnetic-exchange coupling. The first three terms are large relative to the last four terms, which may all be of comparable magnitude. Since Hund's rule applies, $V_{el} > V_c$ and a qualitative energy-level diagram can be obtained from the one-electron model. However, the ground state is a six-electron state.

The cubic portion of the octahedral-site field splits the fivefold-degenerate $d$ level into an upper, doubly degenerate $\Gamma_3(d_{z^2}, d_{x^2-y^2})$ and a lower, triply degenerate $\Gamma_5(d_{xy}, d_{yz}, d_{zx})$. Whereas the trigonal-field component $V_t$ favors stabilization from the $\Gamma_5$ of a nondegenerate $\Gamma_{T1}$, which quenches the orbital angular momentum and concentrates the "extra-electron" charge along the $c$ axis, spin-orbit coupling would stabilize the twofold degenerate $\Gamma_{T3}$, which concentrates this charge in the basal plane. Therefore the ground state at temperature $T$ of the NiAs phase is $\alpha\Gamma_{T1} + \beta\Gamma_{T3}$, where $\alpha + \beta = 1$ and $V_t$ favors $\alpha$, $V_{LS}$ favors $\beta$. In addition, the magnetic-exchange term consists of two parts, an antiferromagnetic cation-anion-cation interaction between basal planes, which is independent of the magnitudes of $\alpha$ and $\beta$, and a cation-cation interaction along the $c$ axis (between basal planes) that is ferromagnetic for $\alpha \neq 0$, is antiferromagnetic for $\alpha = 0$.[7] Since the magnetic-exchange energy is optimum if the two effects are cooperative, this term favors $\alpha = 0$. Further, the cation-cation interactions are especially sensitive to $R(c$ axis), increasing exponentially with decreasing $c$. (The relative sensitivity to lattice-parameter changes of cation-cation to cation-anion-cation interactions gives an exchange-inversion temperature in $Mn_{2-x}Cr_xSb_{0.95}In_{0.05}$.) Since the $c$ parameter varies as the temperature, this means that so long as $\alpha \neq 0$, both $\alpha$ and the coupling between magnetic sublattices (which is measured by $n$, where $\chi_\perp = 1/n$) must decrease with temperature. However, once $\alpha = 0$ the two magnetic-exchange contributions add, and $n$ increases with decreasing $c$ parameter, or temperature. These considerations lead to the following qualitative interpretation of the properties of stoichiometric FeS.

(1) The paramagnetic susceptibility should obey a Curie-Weiss law, but with $\mu_{eff}$ (obs.) $> \mu_{eff}$ (spin-only) $= 4.90\mu_B$ both because of the presence of an orbital-momentum contribution ($\beta \neq 0$) and $\theta_p = \theta_{p0} + aT$.

(2) In the interval $T_\alpha \leq T \leq T_N$, $\chi_\perp = 1/n$ increases with decreasing temperature because $\alpha \neq 0$. The discontinuous increase in $\chi_\perp$ on cooling through $T_s$ reflects an added lowering of $\alpha$ and $c$ due to strong spin-orbit coupling below $T_s$, where $\mathbf{L}$ and $\mathbf{S}$ are parallel.

(3) For $T < T_\alpha$, $\alpha = 0$ and a discontinuous decrease in $\chi_\perp = 1/n$ on cooling through $T_\alpha$ reflects the fact that

below $T_\alpha$ the two magnetic-exchange terms add, above they subtract.

(4) There are two contributions to the magnetic anisotropy: dipole-dipole interactions, which stabilize the spins in the basal plane, and spin-orbit coupling, which stabilizes the spins along the $c$ axis. The spin-orbit coupling increases with decreasing $\alpha$, so that the dipole-dipole term predominates for $T > T_s$, the spin-orbit term for $T < T_s$. Since $\alpha = 0$ below $T_\alpha$, it follows that $T_\alpha \leq T_s$.

(5) Below $T_\alpha$, where $\alpha = 0$, the four atomic $\Gamma_{T3}$ orbitals are three-fourths filled, which is compatible with ferromagnetic cation-cation exchange within the basal planes.[7] For $R$ (basal plane) $\approx R_c$, there exists the possibility of stabilizing the system via the formation of cation-cation molecular orbitals to give $R < R_c$ for bonded cations and $R > R_c$ for nonbonded cations.[7] Such a bonding distorts the structure; if this distortion lowers the crystalline symmetry, the elastic coupling $V_\lambda$ will induce a cooperative phenomenon. The most stable cation-cation bond would be a homopolar bond, the $\Gamma_{T3}$ hole of neighboring cations being ordered into the same ligand. Such ordering would require antiferromagnetic coupling between bonded cations, or a new magnetic order, and it would give rise to a sharp decrease in the symmetry from hexagonal to orthorhombic or monoclinic. Less cation-cation bonding energy is realized by three-membered ring formation (three-membered ozone is less stable than $O_2$), which contains three-electron bonds compatible with ferromagnetic basal planes, but the induced crystallographic distortion is more compatible with the close-packed-hexagonal anion matrix. The fact that three-membered ring formation is observed below $T_\alpha$ indicates that the sum of the elastic and cation-cation binding energies stabilizes the structure with the smaller binding energy. Three-membered ring formation in $Mn_2Mo_3O_8$[8] is probably due to similar elastic constraints. (With no localized $4d$ electrons simultaneously present at $Mo^{4+}$ there is no induced atomic moment.) This brings to nine the number of types of electron ordering that can be distinguished in transition-metal compounds.[6]

(6) The angular momentum associated with the $\Gamma_{T3}$ "hole" is larger for the molecular orbital, or large radius, than for the atomic orbital, so that the $c$ axis remains the easy axis. However, a discontinuous change in the orbital contribution to the moment can be expected at $T_\alpha$. Further, the spin contribution to the moment remains the same for the $\Gamma_3$ and $\Gamma_{T3}$ electrons, but may be slightly reduced for $\Gamma_{T1}$ below $T_\alpha$ due to some tendency to homopolar bonding between pairs of basal planes. The net result would give an atomic moment roughly the same as the spin-only value, but noticeable changes in the $g$ factor in the temperature interval $T_\alpha \leq T \leq T_s$.

[7] J. B. Goodenough, Phys. Rev. **117**, 1442 (1960); **120**, 67 (1960).

[8] W. H. McCarroll, L. Katz, and R. Ward, J. Am. Chem. Soc. **79**, 5410 (1957).

(7) Interpretation of the electrical conductivity vs temperature is dangerous without a knowledge of the number of interstitial $Fe^{2+}$ that are present. However, ordering of the electrons below $T_\alpha$ to give $\alpha=0$ is qualitatively compatible with the observed $c$ axis resistivity changes, since the bonding and antibonding $\Gamma_{T1}$ states are split and with $\alpha=0$ the bonding states are full, the antibonding states empty. (With the above model, conductivity within the basal plane can only be sustained below $T_\alpha$ if interstitial $Fe^{2+}$ are present to provide bridges between the "triangular" molecules. The possible influence of interstitials on the form of the low-temperature phase has not been included in the present qualitative discussion.)

(8) The data are compatible with $R_c$ ($Fe^{2+}$ in sulfur lattice) $\approx 3.00$ A.

---

JOURNAL OF APPLIED PHYSICS     SUPPLEMENT TO VOL. 33, NO. 3     MARCH, 1962

# Perminvar Characteristics of Nickel-Cadmium Ferrites with Small Additions of Cobalt and Molybdenum

H. Lessoff and A. P. Greifer*

*Radio Corporation of America, Memory Products Operation, Needham Heights 94, Massachusetts*

During an investigation of low-loss ferrites, it was found that magnetically annealed nickel-cadmium ferrites containing minor additions of cobalt and molybdenum oxide exhibited rectangualr hysteresis loops with squareness ratios approaching unity at temperatures up to 400°C. Data showing the effect of formulation and temperature on squareness ratio and coercive force are presented.

## INTRODUCTION

IN the development of low-loss nickel zinc ferrites for use in a variety of devices that operate at high frequencies up to 1000 Mc and low amplitudes, the inclusion of small amounts of cobalt ferrite has been found to be of definite advantage towards increasing the useful frequency range, the magnetic $Q$, and in some cases, the initial permeability over a restricted temperature interval.[1-4] Minor additions of oxide of molybdenum[3] further improve the initial permeability and the magnetic $Q$ without adversely affecting the high-frequency response.

In many of the compositions prepared with cobalt oxide, especially those containing amounts of ferric oxide beyond that required for stoichiometry, "wasp-waisted" loops, similar to those originally found in Perminvar,[5] have been observed.[6-8] The good high-frequency properties frequently found for these compositions are presumably associated with a reduction of the residual losses[9-11] and a near absence of hysteresis at low fields, resulting in a linear relation between the applied field and the induction.[2] It has been known for some time that a square hysteresis loop is obtained after such compositions are heated in a magnetic field.[6] Thus, it is not surprising that these unusual "Perminvar" properties have been used with some success in our laboratory as a criterion for compositions that might have low losses and relatively high permeabilities at higher frequencies.

During the development of high-frequency materials along the above lines,[3,4,12] it was decided to investigate the effect of substitution of cadmium for zinc in nickel zinc ferrites containing minor additions of cobalt and molybdenum. Although cadmium ion behaves in the same general manner as zinc in ferrites, it was anticipated that the cadmium ion, being some 25% larger than the zinc ion and of approximately the maximum size that can be accommodated within an oxidic spinel lattice, would have considerable effect on the high-frequency properties of the system. It was found that the properties obtained are not as desirable as the compositions made with zinc; however, the square loop properties are of considerable interest. An investigation has shown the system to have constricted hysteresis loops which, upon magnetic annealing, develop loops of high rectangularity. For certain compositions, a high squareness ratio is maintained to 400°C.

## EXPERIMENTAL

Using standard ferrite processing techniques, compositions prepared were based on the following for-

* Present address: Clevite Corporation, Electronic Research Division, Cleveland, Ohio.

[1] C. M. van der Burgt, Philips Research Repts. 12, 97 (1957).

[2] O. Eckert, Proc. Inst. Elec. Engrs. 104B, 428 (1957).

[3] H. Lessoff, W. Croft, and J. McCusker, Proc. 1959 Electronic Components Conf. 1959 132.

[4] A. P. Greifer, Y. Nakada, and H. Lessoff, J. Appl. Phys. 32 382S (1961).

[5] G. W. Elmen, J. Franklin Inst. 206, 317 (1928); 207, 583 (1929).

[6] Y. Kato and T. Takei, J. Inst. Elec. Engrs. Japan 53, N. 538, 408 (1933).

[7] M. Kornetzki, J. Brochman, and J. Frey, Naturwissenschaften 42, 482 (1955).

[8] O. Eckert, German patent No. 1,057,255 (1959).

[9] M. Kornetzki, Elektrotech. Z. A80 No. 17, 605 (1959).

[10] C. M. van der Burgt, J. Inst. Elect. Engrs. (London) 104B, 550 (1957).

[11] C. Guillaud, G. Villers, A. Marais, and M. Paulus, *Solid State Physics* (Proc. of International Conference in Brussels, June 2–7, 1958) (Acedemic Press, Inc., New York 1960), Vol. 3, Part 1, p. 71.

[12] N. A. Smoskov and S. A. Gushchino, Fiz. Metal i Metalloved., Akad. Nauk S. S. S. R. 8, 557-61 (1959).

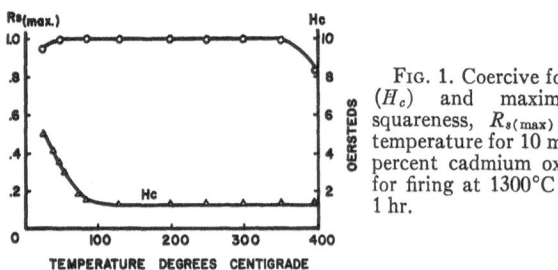

FIG. 1. Coercive force ($H_c$) and maximum squareness, $R_{s(max)}$ vs temperature for 10 mole percent cadmium oxide for firing at 1300°C for 1 hr.

mulations:   $NiO_{(0.4875-x)}CdO_{(x)}Fe_2O_{3(0.4975)}Co_2O_{3(0.01)}$ $MoO_{3(0.005)}$, where "$x$" was varied from 0 to 0.30 in steps of 0, 0.06, 0.08, 0.10, 0.12, 0.14, 0.20, 0.25, and 0.30.

The magnetic properties were measured on a 60-cycle hysteresis-loop tester which provided visual display of the loops on an oscilloscope. In general, magnetic annealing was performed by heating the sample above the Curie temperature and then cooling it in an ac field of 50 amp-turns. However, no differences in the final properties were observed when the material was annealed in a dc field.

All the formulations containing cadmium exhibited constricted loops prior to magnetic annealing and rectangular loops after magnetic annealing. The coercive force decreases with increasing cadmium content and the maximum squareness ratio increases from about 0.6 at 0 mole percent cadmium oxide to approximately 0.8 at 14 mole percent cadmium oxide and then decreases.

All the formulations showed an increase in the squareness ratio and a decrease in the coercive force as the ambient temperature was increased (Fig. 1). By use of proper firing conditions, a squareness ratio of approximately unity can be achieved after annealing. The Curie temperature decreases orderly with increasing cadmium content, which is similar to the effect of zinc additions.

Complete or partial substitution of zinc oxide for cadmium oxide in these compositions destroys both the constricted character of the loop and the tendency to squareness after magnetic annealing. In addition, formulations made without molybdenum oxide did not exhibit constricted hysteresis loops nor high rectangularity after annealing. All compositions are spinel phase only; for both microscopic examination and x-ray diffraction did not reveal the presence of a second phase.

## DISCUSSION

The directional order model developed independently by Néel[13] and Taniguchi[14] and amended by Penoyer and Bickford[15] to apply to cobalt bearing ferrites adequately

explains many of the properties of the nickel-cadmium ferrites described above. In ferrites containing small amounts of cobalt, the uniaxial anisotropy responsible for the associated Perminvar characteristics is created during slow cooling by the migration of cobalt ions to preferred octahedral sites that establish an axis of symmetry along a particular (111) easy direction.[15] In the absence of an external field, the preferred (111) axis in each domain is determined by its direction of spontaneous magnetization. Likewise, within the domain wall itself, the spin directions in the wall determine the migration sites.

Gorter and Esveldt[16] noted that compositions of nickel cobalt, magnesium cobalt, and manganese cobalt ferrites with low ferrous iron contents did not possess either the constricted loop after slow cooling or the rectangular loop after magnetic annealing as contrasted to the presence of these characteristics when the ferrous iron content is increased. They suggest that the presence of $Fe^{++}$ promotes the diffusion of $Co^{++}$. However, ferrous ion, itself, may be responsible for the easier diffusion of cobalt but only be indicative of the presence of cation vacancies. We have found that the "Permivar" characteristic is lost in the compositions investigated when cadmium is not present and when zinc is substituted for cadmium. Since divalent cadmium with an ionic radius of 1.03 A probably has to "squeeze" into the lattice, the consequent "stretching" of the lattice may permit the easier diffusion and ordering of cobalt. We do not exclude the possibility that some cadmium oxide is volatilized at the high firing temperatures, thus, in effect, creating cation ion vacancies by increasing the mole percent ferric oxide in the fired piece; however, we had not found evidence of extensive loss of cadmium.

The role of molybdenum is difficult to define; the Perminvar characteristics in the compositions studied are absent when molybdenum is omitted. The presence of molybdenum trioxide enables the ferrite reaction to proceed at much lower firing temperatures,[4] and apparently facilitates cobalt diffusion. It is our opinion that both the above results are accomplished by molybdenum by the creation of cation vacancies. Such behavior may be expected of molybdenum which can assume valencies of 3,4,5, and 6. We have pointed out that in regard to losses, a dense, small-grained structure appears desirable.[4] The addition of molybdenum trioxide makes it possible to fire a ferrite to this "prescription."

## ACKNOWLEDGMENTS

The writers wish to acknowledge their indebtedness to J. A. McCusker for his helpful suggestions and to R. Bouchard, A. Boulanger, and R. Gravel for the sample preparations and measurements.

[13] L. Néel, J. phys. radium 15, 225 (1954); J. Appl. Phys. 30, 35 (1959).
[14] S. Taniguchi and M. Yamamoto, Sci. Repts. Research Insts. Tohoku Univ. A6, 330 (1954); A7, 269 (1955); A8, 173 (1956).
[15] R. F. Penoyer and L. R. Bickford, Jr., Phys. Rev. 108, 271 (1957).
[16] E. W. Gorter and C. J. Esveldt, Proc. Inst. Elec. Engrs. 104B, 418 (1957).

JOURNAL OF APPLIED PHYSICS    SUPPLEMENT TO VOL. 33, NO. 3    MARCH, 1962

# Low Temperature Anisotropy of Manganese-Iron Ferrites

WILFRED PALMER

*IBM Research Center, Yorktown Heights, New York*

The anisotropy of the composition system $Mn_xFe_{3-x}O_4$, $0.40 \leq x \leq 1.80$, has been measured between 4.2° and 77°K by the torque method, supplementing previous measurements on this system by Penoyer and Shafer between 77° and 313°K. The anisotropy curves for $0.4 \leq x \leq 1.25$ show a minimum below 110°K and appear to be the summation of a number of components of different sign and different temperature dependence—negative components which decrease in magnitude relatively slowly with increasing temperature and positive components which decrease more rapidly. Over most of the composition range studied $K_1$ is negative at 4.2°K, but at $x=0.80$ the magnitude of the positive, low temperature contribution is sufficient to change the sign of $K_1$ from negative to positive at 34°K. In the region $x \leq 1.0$, this contribution can be attributed to the presence of $Fe^{2+}$ ions, but such an explanation is not applicable to the positive contributions observed in compositions for which $x \geq 1.25$ and in which there is no $Fe^{2+}$. For $1.25 \leq x \leq 1.80$, negative values of $K_1$ at 4.2°K of unexpectedly large magnitude ($< -3.7 \times 10^5$ergs/cm³) are observed. Unlike values at 77°K in this composition region, these values of $K_1$ are of too large a magnitude to be attributed to the splitting of the $3d$ levels of $Fe^{3+}$ ions in a cubic crystalline field.

## INTRODUCTION

THE magnetic anisotropy of the cubic spinel system $Mn_xFe_{3-x}O_4$, $x \leq 1.80$, has been measured by the torque method between 77° and 320°K by Penoyer and Shafer.[1] This investigation revealed an anomalous composition dependence of the anisotropy characterized by a pronounced dip in the magnitude of $K_1$ in the region $0.4 < x < 1.0$. An unexpectedly small temperature dependence of $K_1$ at low temperature was also found in this region. The anisotropy in the region $1.00 \leq x \leq 1.80$ was satisfactorily explained by attributing it to the spiltting of the $3d$ levels of $Fe^{3+}$ ions in a cubic crystalline field.

In the present work, determinations of $K_1$ by the torque method for $0.40 \leq x \leq 1.80$ have been extended to 4.2°K.

## EXPERIMENTAL PROCEDURE AND RESULTS

The samples were single-crystal spheres for which values of $x$ were 0.40, 0.60, 0.80, 1.00, 1.25, 1.55, and 1.80. They were the same samples used by Penoyer and Shafer, and the method of preparation and chemical analyses are discussed in reference 1.

Torque measurements were made in a field of 22 500 gauss with a vacuum-enclosed self-balancing torque magnetometer[2] whose output was automatically recorded as a function of field direction on an X-Y recorder. Liquid helium was used as a refrigerant, and temperatures between 4.2° and 77°K were maintained by means of a resistance heater attached to the sample holder.

Values of $K_1$ over the temperature range 4.2° to 300°K are shown in Fig. 1. Anisotropy values above 77°K were provided by R. F. Penoyer.

## IONIC DISTRIBUTION AND ANISOTROPY CONTRIBUTIONS

In the past few years, neutron diffraction,[3] electrical conductivity,[4] and magnetic moment[5] data have been used to resolve uncertainties as to the valence and anion coordination of Mn ions in manganese-iron ferrites. Neutron diffraction studies by Hastings and Corliss have shown that in $MnFe_2O_4$ approximately 80% of the Mn ions occupy tetrahedral sites (A sites). A sharp increase in the activation energy of the electrical conductivity has been found by Záveta to occur in stoichiometric samples between $x=1.06$ and $x=1.12$. Since the

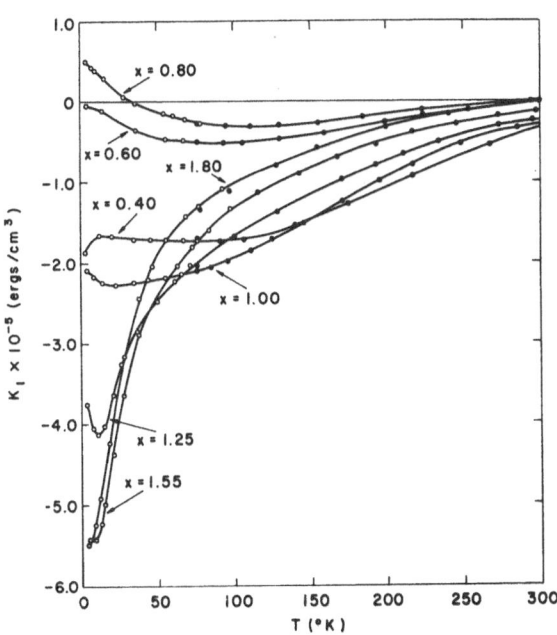

FIG. 1. $K_1$ as a function of absolute temperature for various ferrites of the composition series $Mn_xFe_{3-x}O_4$. Open circles denote values obtained by the author. Solid dots denote values provided by R. F. Penoyer.

[1] R. F. Penoyer and M. W. Shafer, J. Appl. Phys. **30**, 315S (1959).
[2] R. F. Penoyer, Rev. Sci. Instr. **30**, 711 (1959).
[3] J. M. Hastings and L. M. Corliss, Phys. Rev. **104**, 328 (1956).
[4] K. Záveta, Czechoslov. J. Phys. **9**, 748 (1959).
[5] A. H. Eschenfelder, J. Appl. Phys. **29**, 378 (1958).

small activation energies ($\sim$0.1 ev) observed for $x \leq 1.06$ indicate that, as in magnetite, the mechanism of conduction in the low manganese region is an electron exchange between $Fe^{2+}$ and $Fe^{3+}$ ions on octahedral sites ($B$ sites), the increase of activation energy near $x = 1.1$ indicates an almost complete replacement of $Fe^{2+}$ ions by $Mn^{2+}$ ions at this manganese concentration. Moreover, the failure of this increase to occur at $x = 1.0$ indicates the substitution of a certain amount of manganese with a valence larger than 2 in the region $x \leq 1.0$.

The foregoing results are most easily rationalized by assuming that, for small values of $x$, the major fraction of Mn ions occupies $A$ sites in a divalent state, with a smaller fraction ($<0.2$) occupying $B$ sites in a trivalent state, and that this distribution of ions prevails until complete occupancy of the $A$ sites by $Mn^{2+}$ ions is achieved. Such a distribution of ions is in accord with the expected tendencies of $Mn^{2+}$ ions to form partially covalent bonds with tetrahedral anion coordination and of $Mn^{3+}$ ions to form bonds with octahedral coordination.[6] Further additions of manganese occupy $B$ sites, probably as $Mn^{3+}$,[7] and the distortion of the surrounding oxygen octahedron by the $Mn^{3+}$ ion ultimately produces a tetragonal distortion of the crystal at $x = 1.85$.[8]

The anisotropy curves of compositions for which $0.40 \leq x \leq 1.00$ are characterized by a minimum at low temperature and appear to be the summation of a number of components of different sign and different temperature dependence—negative components which decrease in magnitude relatively slowly with increasing temperature and positive components which decrease more rapidly. Over most of the composition range studied, $K_1$ is negative at 4.2°K, but at $x = 0.80$ the magnitude of the positive contribution is sufficient to change the sign of $K_1$ from negative to positive at 34°K and to produce a minimum at the relatively high tem-

perature of 110°K, thus accounting for the small temperature dependence above 77°K previously observed in this composition region. This positive component has been attributed by Slonczewski[9] to $Fe^{2+}$ ions whose contribution per ion increases with decreasing $Fe^{2+}$ concentration because of simultaneous decreases in the separation of the two lowest energy levels of the $Fe^{2+}$ ion and in the effective exchange field.

Positive low temperature contributions are also indicated at $x = 1.25$ by a minimum in the anisotropy curve and in $x = 1.55$ by a flattening of the anisotropy curve below 10°K. These components cannot be explained in the same way as those in the region $x \leq 1.0$, since, on the basis of conductivity data, the presence of appreciable amounts of $Fe^{2+}$ can be regarded as unlikely at such large manganese concentrations.

In the region $1.25 \leq x \leq 1.80$, negative values of $K_1$ at 4.2°K of unexpectedly large magnitude ($< -3.7 \times 10^5$ ergs/cm³) are observed. The reasonable magnitude and regular decrease of $K_1$ above 77°K with increasing $x$ for $1.00 \leq x \leq 1.80$ can be explained in the manner proposed by Penoyer and Shafer. However, this explanation cannot account for the large negative anisotropy at low temperature, since the splitting parameter required would be unreasonably large.

It is doubtful that the anisotropy of the manganese-iron ferrite system at low temperature can be explained on the basis of any simple model. However, the large differences in anisotropy values at 4.2° and 77°K observed in the composition region $1.25 \leq x \leq 1.80$ point out the need for low temperature anisotropy data over a wide range of compositions in this and probably other composition systems before a reliable assignment of anisotropy sources can be made.

### ACKNOWLEDGMENTS

The author wishes to express his thanks to R. A. Fiorio for assistance in making the torque measurements; to M. W. Shafer for preparation of the single crystals; to R. F. Penoyer for the use of unpublished data; and to J. C. Slonczewski and B. A. Calhoun for many helpful discussions.

[6] J. B. Goodenough and A. L. Loeb, Phys. Rev. **98**, 391 (1955).

[7] Magnetic moment values for $x > 1.0$ are significantly lower than would be expected from the substitution of $Mn^{3+}$ with an effective moment of $4\mu_B$. However, this result can be attributed to a noncolinear spin arrangement on $B$ sites [cf., V. L. Moruzzi, J. Appl. Phys. **32**, 59S (1961)].

[8] A microwave resonance determination of the anisotropy for the composition $x = 1.85$ between 77° and 390°K has been made by J. Overmeyer J. Appl. Phys. **32**, 142S (1961).

[9] J. C. Slonczewski, J. Appl. Phys. **32**, 253S (1961).

JOURNAL OF APPLIED PHYSICS     SUPPLEMENT TO VOL. 33, NO. 3     MARCH, 1962

# Some Superparamagnetic Properties of Fine Particle δFeOOH

A. W. SIMPSON

*The Plessey Company, Ltd., Caswell Research Laboratories, Towcester, Northamptonshire, England*

Various samples of δFeOOH have been prepared all with substantially the same particle shape and size but with differing low-temperature saturation magnetizations. The particles were found to be hexagonal platelets with a mean diameter of 200 A and a thickness of 30 A. The magnetization of the samples was varied by varying the preparation conditions and also by partially decomposing the material to αFeOOH and αFe₂O₃. It has been shown that at a little above room temperature the powders were all superparamagnetic and that the variations of magnetization with preparation conditions qualitatively support a model proposed by Francombe and Rooksby. The variation of the Blocking temperature of a particular preparation as its magnetization was reduced by successive decomposition is also discussed. The results are tentatively interpreted in terms of a constant term corresponding to the normal anisotropy energy and a variable one due to the Lorentz internal field.

## 1. INTRODUCTION

IF a freshly prepared alkaline precipitate of ferrous hydroxide is rapidly oxidized, a strongly magnetic, finely divided black powder is formed. This material was first described by Glemser and Gwinner in 1939[1] and was shown to have a hexagonal x-ray structure very similar to that for Heamatite ($\alpha Fe_2O_3$) and denoted $\delta Fe_2O_3$. Recently Rooksby and Francombe[2] and simultaneously Bernal, Dasgupta, and MacKay[3] have shown that the x-ray structure can best be interpreted in terms of $Fe_2O_3 \cdot H_2O$ or FeOOH, denoted delta ferric oxyhydroxide ($\delta FeOOH$), with a crystallographic structure based on ferrous hydroxide $Fe(OH)_2$ from which it is formed by oxidation. They have shown that on slowly warming δFeOOH a gradual crystallographic change occurs above 100°C, resulting in the formation of Goethite ($\alpha FeOOH$) which is followed by dehydration to Heamatite ($\alpha Fe_2O_3$) at temperatures between 150° and 250°C. The interpretation of the x-ray results showed that δFeOOH consists of a hexagonal, close-packed array of oxygen and hydroxyl ions with the iron ions randomly distributed among the interstital sites. Careful investigation of the x-ray line intensities by Rooksby and Francombe showed that approximately 20% of the iron ions were in tetrahedral sites and the rest in the octahedral sites. It was suggested that the material was ferrimagnetic with superexchange occurring between the $Fe^{3+}$ ions on octahedral and tetrahedral sites via a colinear oxygen ion in an analogous manner to the exchange interaction in cubic $\gamma Fe_2O_3$. The transition from δFeOOH to αFeOOH only involves the movement of all the iron ions onto the octahedral sites, therefore, on warming, the transition should cause the magnetization to slowly fall to zero.

## 2. PREPARATION AND PHYSICAL PROPERTIES

An alkaline precipitate of ferrous hydroxide was prepared at room temperature by slowly adding dilute sodium hydroxide to a solution of approximately 10 wt % of ferrous sulphate until the $p$H rose to 9. In order to obtain various speeds of oxidation the precipitate was warmed up to a particular temperature and then oxidized by adding a considerable excess of 15% hydrogen peroxide. The reaction evolved a little heat, and the oxidation temperature was taken as the mean of the temperature immediately before and after the reaction. Finally the precipitate was filtered, washed, and dried in a vacuum desiccator for a few weeks. After preparation particular samples were held for 15 hr at successively higher temperatures up to 250°C, in order to partially decompose the material by conversion to αFeOOH and αFe₂O₃. The water content and hence the composition were estimated after each decomposition temperature by finding the loss in weight.

Electron microscope and x-ray investigations showed that the δFeOOH particles were hexagonal platelets with a mean diameter of 200 A and a mean thickness of 30 A with the plane of the particles normal to the $C$ axis. Thermal decomposition had no effect on the mean dimensions of the particles, and there were only nominal variations in the particle size from one preparation to another. This was because the precipitation conditions for the ferrous hydroxide formation were maintained constant throughout the experiments, and the oxidation process had no effect on the particle size.

Samples for magnetic measurements were prepared by adding a few percent of organic binder to the dry powder and pressing into cylindrical compacts.

## 3. MAGNETIC MEASUREMENTS AND RESULTS

The magnetization of the samples in fields up to 9000 oe was measured between room temperature and liquid nitrogen temperature by means of a modified Curie-Weiss method. Measurements of coercivity and remanence were also made in the same apparatus.

The room temperature coercivity for most of the samples was exactly zero, indicating that the materials were superparamagnetic, but at 77°K the values lay between 100 and 1400 oe. Superparamagnetism was confirmed by plotting the room temperature and liquid nitrogen magnetization measurements against the ap-

[1] O. Glemser and E. Gwinner, Z. anorg. u. allgem. Chem. **240**, 163 (1939).
[2] M. H. Francombe and H. P. Rooksby, Clay Minerals Bull. **4**, 1 (1959).
[3] J. D. Bernal, D. R. Dasgupta, and A. L. MacKay, Clay Minerals Bull. **4**, 15 (1959).

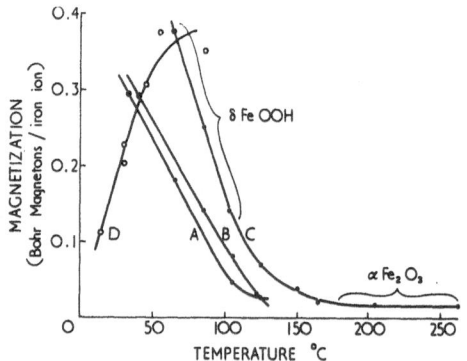

FIG. 1. Dependance of magnetization on preparation conditions. Curves A, B, and C show the variation with decomposition temperature and curve D with initial oxidation temperature.

plied field divided by the absolute temperature when, except in the low field regions, the points lay on a single curve.

The magnetization per iron ion at 77°K was obtained by extrapolating the magnetization to infinite field by means of the Frolich approach to saturation formula modified to suit the Langevin curve. The number of iron ions present in the sample was obtained by completely decomposing the material to $\alpha Fe_2O_3$ and weighing.

Some of the results for the variation of saturation magnetization per iron ion as a function of oxidation and decomposition temperatures are shown in Fig. 1. Curve D shows the increase in magnetization per iron ion with increasing oxidation temperature, while curves A, B, and C show, for three separate preparations, how the magnetization falls steadily with increasing decomposition temperature due to progressive ordering of the iron ions onto the octahedral sites.

Measurements were also made of the temperature above which the material becomes superparamagnetic, i.e., the Blocking temperature. This was achieved by measuring the remanent magnetization from 77°K to room temperature and defining the Blocking temperature as that at which the remanent magnetization had fallen to 1% of the saturation magnetization at 77°K. Some definition of this nature is essential because the variation of remanence with temperature shows a pronounced tail, presumably due to the range of particle size within each sample.

Blocking temperatures as high as 330°K were observed for the sample with the highest magnetization; also the Blocking temperature fell steadily with increasing decomposition temperature.

## 4. DISCUSSION AND CONCLUSIONS

The variation of magnetization per iron ion shown in Fig. 1 qualitatively supports the model suggested by Francombe and Rooksby, firstly in that the magnetization increases with increasing oxidation temperature (curve D). This is because the more rapidly ferrous

hydroxide is oxidized the more disordered will be the material, and therefore the greater the probability of an iron ion remaining on a tetrahedral site and hence the greater the magnetization. Secondly the fall in magnetization with increasing decomposition temperature shown in curves A, B, and C also supports the idea of gradual ordering of the iron ions onto the octahedral site. This occurred while still maintaining a chemical formula FeOOH and an x-ray structure similar to the undecomposed sample (up to 105°C. See Fig. 1.) It was hoped to correlate the magnetization with the fraction of iron ions on the tetrahedral site, but it proved impossible to interpret the x-ray line intensities in terms of the model proposed by Francombe and Rooksby.

For normal superparamagnetic materials the particle shape and size may be varied, but the magnetization remains constant. With $\delta FeOOH$, however, it is possible to keep the particle shape and size constant and vary the magnetization successively. Figure 2 indicates the consequence of this and shows the variation of Blocking temperature with magnetization for two samples corresponding to curves A and B of Fig. 1.

The Blocking temperature obtained on extrapolation to zero magnetization is finite, and although this may be a function of an independent variable, it is tempting to interpret the results in terms of a proposal made by Bean and Livingston.[4] They suggested that the Lorentz internal field due to the magnetization itself could produce a finite blocking temperature. Then the slope of the curve in Fig. 2 corresponds to the variation of Lorentz field with magnetization, and the intercept corresponds to the Blocking temperature due to anisotropy alone.

Finally it is interesting to note that the completely decomposed $\delta FeOOH$ in the form of platelets of $\alpha Fe_2O_3$ (curve C, Fig. 1) is not strongly superparamagnetic. Néel[5] has shown that $\alpha Fe_2O_3$ is antiferromagnetic with the iron ions lying on parallel planes normal to the $C$ axis, with successive layers magnetized alternately in opposite directions. Also he has shown that below 260°K the direction of the antiferromagnetic magnetization

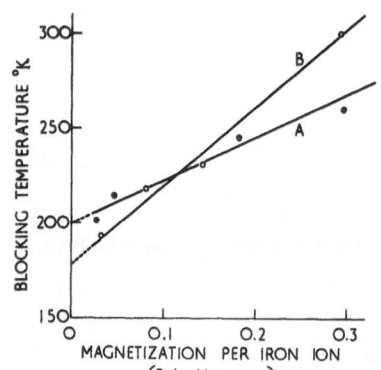

FIG. 2. Dependence of Blocking temperature on sample magnetization for the samples A and B of Fig. 1.

[4] C. P. Bean and J. D. Livingston, J. Appl. Phys. **30**, 120S (1959).
[5] L. Néel, Revs. Modern Phys. **25**, 58 (1953).

vectors is parallel to the $C$ axis. Therefore, since the platelets are thin in the $C$ direction, a platelet with an odd number of layers either over the whole or part of the surface should have an over-all magnetic moment below 260°K. A 30-A thick film would have, on an average, thirteen layers of iron ions resulting in a moment of 0.2 Bohr magnetons per iron ion, compared to a measured value of only 0.017 Bohr magnetons.

The low magnetization of the platelets of $\alpha Fe_2O_3$ may occur because the powder is formed well below its Néel temperature, and it is perhaps energetically more favorable to have no over-all magnetic moment.

---

JOURNAL OF APPLIED PHYSICS    SUPPLEMENT TO VOL. 33, NO. 3    MARCH, 1962

# Rare-Earth Ruthenates

R. ALÉONARD, E. F. BERTAUT, M. C. MONTMORY, AND R. PAUTHENET

*Laboratoire d'Electrostatique et de Physique du Métal, Institut Fourier, Grenoble, France*

In the pyrochlore compounds $Ru_2M_2O_7$ (where M=Pr, Nd, Gd, Tb, Dy, Ho, and Y), $Ru^{IV}$ contributes to the Curie constant 1 instead of the value 3 which would be expected for a $d^4$ state with $g=2$ and $j=2$.

THE pyrochlore structure $A_2B_2O_7$ can be derived from the $UO_2$ or $U_4O_8$ structure ($CaF_2$ type) by omitting one oxygen atom over eight and by ordering the cations A and B in a doubled unit cell.[1]

Roth[2] has prepared pyrochlores of formula $2TiO_2 \cdot M_2O_3$ with M=Sm, Gd, Yb, Y. Actually $RuO_2$ and $IrO_2$ belong to the rutile-type structure $TiO_2$ and their cations have similar radii (Goldschmidt radii are respectively 0, 64; 0, 65 and 0, 66 A for $Ti^{IV}$, $Ru^{IV}$, and $Ir^{IV}$). As the study of $4d$ and $5d$ cations and their possible interactions with rare earths might be interesting, we have synthesized the rare-earth ruthenates $Ru_2M_2O_7$ by decomposition of a mixture of nitrates and of the corresponding series $Ir_2M_2O_7$ by heating the oxide mixture up to 800°C in sealed silica tubs.

Both series of compounds show the pyrochlore structure. The parameter measurements show the well-known phenomenon of the lanthanide contraction, i.e., the decrease of the ionic radius with the atomic number.[3,4]

The susceptibilities have been measured in the $Ru_2M_2O_7$ compounds with M=Pr, Nd, Gd, Tb, Dy, Ho, and Y from 2° to 1300°K. The general behavior is paramagnetic with a saturation effect at low temperatures. In the Gd and Tb compound (see Fig. 1) there is

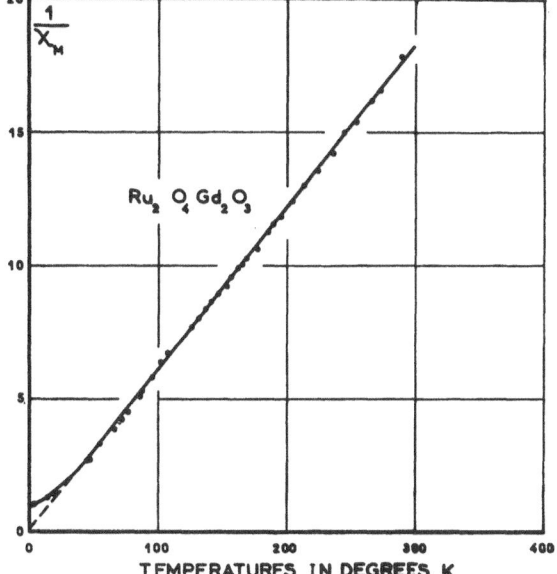

FIG. 1. Inverse of the magnetic susceptibility vs temperature in $Gd_2Ru_2O_7$.

a marked curvature in the $1/\chi$ plot vs $T$ at low temperatures which may be indicative of antiferromagnetic ordering.

Above 500°K, the inverse of the susceptibility is linear in all compounds and allows the definition of a Curie constant $C_{exp}$ which is reproduced in Table I. After deduction of the theoretical contribution $C(M)$ of the rare-earth ion M, one obtains a definite contribution $C(Ru)$ for the $Ru^{IV}$ ion (see last line of Table I). This contribution is near to $C(Ru) \sim 1$, which is quite different from the value 3 which would be expected for a $d^4$ state with $g=2$ and $j=2$.

Crystalline field effects may explain the low spin state of $j=1$, i.e., a $(t_0)^4$ state. Calculations have not yet been carried out.

TABLE I. Curie constants of $Ru_2M_2O_7$. M=rare earth and Y.

| M | Pr | Nd | Gd | Tb | Dy | Ho | Y |
|---|---|---|---|---|---|---|---|
| $C_{exp}$[a] | 5.0 | 5.7 | 17.8 | 25.4 | 30.8 | 30.2 | 2.2 |
| $C(M)$ | 1.6 | 1.65 | 7.875 | 11.8 | 14.2 | 14.1 | ... |
| $C(Ru)$ | 0.9 | 1.2 | 1.03 | 0.9 | 1.2 | 1.0 | 1.1 |

[a] $C_{exp} = 2[C(M) + C(Ru)]$.

[1] C. Hermann, O. Lohrmann, and H. Philipp, *Strukturbericht 1928–1932* (Akademische Verlagsgesellschaft, Leipzig, 1937), Vol. II, p. 58.
[2] R. S. Roth, J. Research Natl. Bur. Standards **56**, 17 (1956).
[3] E. F. Bertaut, F. Forrat, and M. C. Montmory, Compt. rend. **249**, 829 (1959).
[4] M. C. Montmory and F. Bertaut, Compt. rend. **252**, 4171 (1961); cf. Acta Cryst. **13**, 1015 (1960).

JOURNAL OF APPLIED PHYSICS     SUPPLEMENT TO VOL. 33, NO. 3     MARCH, 1962

# Fine-Grained Ferrites. II. $Ni_{1-x}Zn_xFe_2O_4$

W. W. Malinofsky, R. W. Babbitt, and G. C. Sands

*U. S. Army Signal Research and Development Laboratory, Fort Monmouth, New Jersey*

A series of fine-grained single-phase ferrites was prepared according to the formula $Ni_{1-x}Zn_xFe_2O_4$, where $x$ took the values 0, 0.10, 0.33, 0.50, and 0.67. A previously-described process combining the so-called flame-spraying and hot-pressing techniques was used to obtain grain sizes of approximately 0.1 micron in the densified bodies. $\mu'$ and $\mu''$ were measured to 1000 Mc, with grain size and composition as parameters. Measurements were also made to 3800 Mc on a ferrite annealed through the critical size for multidomains, confirming previously-reported theory that magnetic poles on the domain walls are the source of the microwave peak at about 2000 Mc. Wall displacement appears to contribute to the static $\mu_0$ in large-grained ferrites when $x>0$.

In addition, the temperature dependence of initial permeability $\mu_0$ was studied. The temperature coefficient of $\mu_0$ was found to increase with grain size. While the $\mu_0$ vs temperature curves were of various shapes generally, still a set of ferrites made up of differing compositions and grain sizes were obtained having linear temperature dependences. Of practical interest are those with slopes of $-220$ ppm/°C, 0 ppm/°C (NPO), and 220 ppm/°C.

Microwave properties were studied also. Because zero-field ($H_{dc}=0$) measurements showed that fine-grains (i.e., below the critical size for multidomains) eliminated the so-called microwave loss peak in the rf dispersion, a decrease in the low-field loss follows. Also, microwave measurements of the main resonance loss susceptibility vs rf power were made at $X$ band using a cavity technique. Although normal critical fields $h_c$ were generally found, one sample with $x=0.67$ and $\Delta H=655$ oe showed an anomalous rise in $\chi''/\chi_0''$ at high powers, significantly expanding its power-handling capability.

## INTRODUCTION

A PROCESS was previously described[1] for the preparation of up to 98% dense $NiFe_2O_4$ (theoretical density $d=5.38$ g/cm³) with average grain sizes as low as $0.06\,\mu$, i.e., below the critical size ($\approx0.1$ micron) for domain wall formation. This permitted a study of the magnetic properties, including the high-frequency dispersion of the complex permeability with and without the presence of domain walls, from which it was concluded that domain rotation is the source of both the rf and microwave resonances observed in ordinary-grain-sized $NiFe_2O_4$. This study has now been extended to nickel ferrites substituted with zinc.

## PREPARATION

The same combination of flame-spraying and hot-pressing processes was used to prepare a series of single-phase $Ni_{1-x}Zn_xFe_2O_4$ compositions, except that hot-pressing could generally be done about 200°F lower to obtain comparable densities to $NiFe_2O_4$.

TABLE I. Dependence of temperature coefficient of $\mu_0$ on grain size in $Ni_{1-x}Zn_xFe_2O_4$.

| $x$ | $t$ microns | T.C. ppm/°C | $d$ g/cm³ |
|---|---|---|---|
| 0 | 0.095 | 220 | 4.37 |
| 0 | 0.16 | 370 | 4.37 |
| 0 | 0.20 | 480 | 4.43 |
| 0 | 0.23 | 540 | 4.47 |
| 0 | 0.10 | 290 | 5.28 |
| 0 | $\approx15$ | 2300 | 4.89 |
| 0.50 | 0.03 | $-220$ | $\approx4$ |
| 0.50 | 0.11 | 620 | $\approx5.2$ |
| 0.67 | 0.07 | $-990$ | 4.38 |
| 0.67 | 0.12 | 1110 | 5.22 |

X-ray diffraction techniques were again used to obtain average crystallite size and crystal structure.

## RADIO FREQUENCY PROPERTIES

*Temperature dependence.* Table I shows the dependence of the temperature coefficient of initial permeability (T.C.) on grain size ($t$) for $x=0$, 0.50, and 0.67. Whereas in conventionally-prepared ferrites $t$ and $d$ are very closely tied together, the first section of Table I shows the results of an attempt to maintain $d$ constant (within 1.5%) as $t$ was doubled by a series of anneals. For $x=0$ the average T.C. for $-50°$ to 85°C becomes increasingly more positive, increasing linearly with grain size up to 0.23 micron. Also, a nonlinearity in T.C. is found near or above the critical size for domain wall formation. For both $x=0.50$ and 0.67 the finer-grained materials had a negative T.C., becoming positive at larger grain sizes. Judging from the $x=0$ series, it seems that as far as T.C. is concerned density is of only secondary importance compared to gain size (at least, for smaller grain sizes). Of practical importance are some fine-grained ferrites (shown in Table II) with low T.C.'s, and which are approximately linear. The first and third can be used to compensate each other. T.C., given in parts per million per °C, and $\mu_0$ were measured at 4 Mc. The maximum quality $Q_{max}$ obtainable is also given, along with its corresponding frequency $f_Q$.

*High-frequency dispersion.* The complex permeability was measured to 3800 Mc for the $x=0$ series mentioned above. A weak microwave resonance peak near 2000 Mc was observed only for grain sizes above approximately 0.2 micron, the density having increased only 1.5% to 4.43 g/cm³. This is below the critical density (4.85 g/cm³) given by Pippin and Hogan[2] for observation of this peak. Previously,[1] it was reported for another sample having a greater density (5.28 g/cm³) that again

[1] W. W. Malinofsky and R. W. Babbitt, J. Appl. Phys. 32, 237S (1961).

[2] J. E. Pippin and C. L. Hogan, Sci. Rept. No. 1, Gordon McKay Laboratory of Applied Science, Harvard University (April, 1959), p. 22.

TABLE II. Fine-grained $Ni_{1-x}Zn_xFe_2O_4$ with linear temperature coefficients of $\mu_0$.

| T.C. ppm/°C | x | t microns | $\mu_0$ | $Q_{max}$ | $f_Q$ Mc |
|---|---|---|---|---|---|
| −220 | 0.5 | 0.03 | 9.0 | 95 | 25 |
| 0±50 | 0.6 | 0.03 | 13 | 60 | 10 |
| 220 | 0 | 0.095 | 8.2 | >300 | 55 |

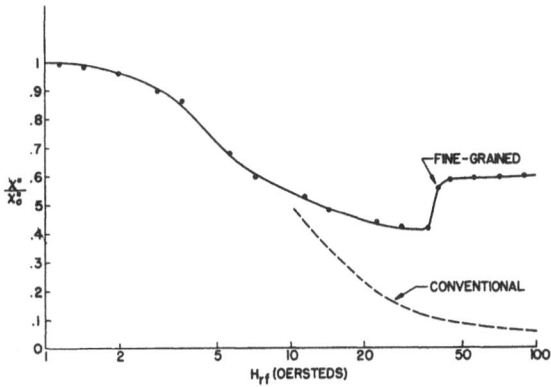

FIG. 1. Microwave resonance loss at high power in $Ni_{0.33}Zn_{0.67}Fe_2O_4$.

this peak became evident only when the grain size became larger than about 0.2 micron. The similar results thus tend to confirm the theory[1] that the microwave resonance is caused by magnetic poles forming on domain walls.

Table III lists some observed rf magnetic properties of fine-grained $Ni_{1-x}Zn_xFe_2O_4$. The last column gives the ratio of the observed rf resonance frequency to that calculated on the basis of domain rotation, in the manner of Smit and Wijn.[3] For all the samples with $x=0$, this ratio is 0.8, or closer to the theoretical value than the 0.5 shown in their table for normal size grains. Among the other values of $x$, only the last sample showed a very low ratio (0.2). Whereas it had previously[1] been found that in fine-grained $NiFe_2O_4$ the static $\mu_0$ of very dense material was nearly equal to that of the large-grained ferrites of similar density, it has now been found that in dense NiZn-ferrites the $\mu_0$ is far less than in the corresponding conventional materials. This indicates that besides domain rotation another mechanism, probably wall displacement, is operative in conventional NiZn-ferrites.

## MICROWAVE PROPERTIES

*Low field loss.* The previously mentioned microwave peak in the dispersion curve[1] occurs for large grains at 2000 Mc for $x=0$, and at about 1500 Mc for $x=0.1$. This peak began to vanish for grain sizes below about

TABLE III. Theoretical rotational resonance frequency compared to experimental value.

| x | t microns | $\mu_0$ | $f_{rexp}$ Mc | d g/cm³ | $M_s$ gauss | $f_{rexp}/f_{rtheor}$ |
|---|---|---|---|---|---|---|
| 0 | 0.095 | 7.1 | 630 | 4.37 | 219 | 0.8 |
| 0 | 0.16 | 9.4 | 500 | 4.37 | 219 | 0.8 |
| 0 | 0.20 | 12.8 | 330 | 4.43 | 222 | 0.8 |
| 0 | 0.23 | 15.5 | 290 | 4.47 | 224 | 0.8 |
| 0.10 | 0.12 | 12.1 | 500 | 4.87 | 293 | 0.7 |
| 0.33 | 0.06 | 9.1 | 750 | 3.52 | 275 | 0.9 |
| 0.33 | ≈0.1 | 38 | 150 | 4.94 | 386 | 0.6 |
| 0.50 | 0.06 | 49 | 125 | 4.91 | 368 | 0.7 |
| 0.50 | 0.11 | 76 | 90 | ≈5.3 | 396 | 0.7 |
| 0.67 | 0.07 | 24 | 125 | 4.38 | 211 | 0.6 |
| 0.67 | 0.12 | 83 | 15 | 5.22 | 251 | 0.2 |

0.2 micron, reducing the loss over the range 1 to 5000 Mc. For example, the reduction in the former at 3000 Mc is by a factor of about 2.5. Since the dispersion curve represents the limit of $H_{dc}=0$ in a plot of absorption due to domain rotation vs dc magnetic field,[4] the above result indicates a decrease in low field losses over this range. This is of practical interest for resonance isolators and microwave phase shifters intended to be operated in this low microwave frequency range.

*High-power microwave resonance loss.* In general, the fine-grained ferrites had critical fields[5] $h_c$ and spin-wave linewidths about the same as ordinary-grain-size ferrites. However, the last sample in Table III (with $\Delta H=655$ oe) gave an anomalous result. Figure 1 shows its normalized loss susceptibility $\chi''/\chi_0''$ vs $H_{rf}$, the amplitude of the applied rf field, as measured in a cavity at $X$ band. Whereas the conventionally-prepared[6] large-grained composition ($x=0.6$) shows a continual decrease in loss with $H_{rf}$, the fine-grained ferrite shows a reversal in the degradation of the main resonance followed by an improving trend out to high power levels, until arcing begins in the pressurized waveguide. This effect did not occur in a smaller grain-size sample of the same composition and lower density (next-to-last sample in Table III), indicating the possible existence of an optimum grain size. This effect is of interest in maintaining the figure of merit of resonance isolators out to very high power levels.

## ACKNOWLEDGMENTS

The authors wish to thank Dr. J. J. Green of Raytheon Company for the high-power microwave resonance measurements, I. Bady for advice on measurement techniques, and Dr. E. Both for support of this work.

[3] J. Smit and H. P. J. Wijn, *Ferrites* (John Wiley & Sons, Inc., New York, 1959), Table 50.I, p. 270.

[4] R. F. Soohoo, *Theory and Application of Ferrites* (Prentice-Hall, Inc., Englewood Cliffs, New Jersey, 1960), p. 88.
[5] H. Suhl, Proc. Inst. Radio Engrs. 44, 1270 (1956).
[6] J. J. Green, Sci. Rept. No. 2 (Series 2), Gordon McKay Laboratory of Applied Science, Harvard University (December, 1959), Fig. 5–22.

JOURNAL OF APPLIED PHYSICS    SUPPLEMENT TO VOL. 33, NO. 3    MARCH, 1962

# Vanadium Iron Oxides

A. WOLD, D. ROGERS, R. J. ARNOTT, AND N. MENYUK

*Lincoln Laboratory,\* Massachusetts Institute of Technology, Lexington 73, Massachusetts*

An investigation in the ternary system FeVO has been made of the chemical, crystallographic, and magnetic properties of the corundum $FeVO_3$, and the spinels $FeV_2O_4$ and $Fe_2VO_4$. $FeVO_3$ has been reported in the literature as $Fe^{2+}V^{4+}O_3$ and antiferromagnetic at room temperature. However, the study reported in this paper shows that these results are incorrect. Oxidation of the spinel $FeV_2O_4$ results in the formation of some $FeVO_3$. Conversely, reduction of $FeVO_3$ ultimately forms $FeV_2O_4+Fe$. Therefore both the iron and vanadium are trivalent in the compound $Fe^{3+}V^{3+}O_3$. Furthermore, $FeVO_3$ has a small magnetic moment at room temperature which persists up to approximately 445°K as well as an anomalous rise in the magnetic moment below 150°K. A change in the relative intensity of several lines in the x-ray pattern of the material, which may indicate a shift in the position of the metal atoms was observed down to liquid nitrogen temperatures. However, no change in the symmetry was observed. The magnetic properties of $FeV_2O_4$ and $Fe_2VO_4$ are also studied. The magnetization curves of both these materials indicate the existence of a magnetic transition at 275°K for $Fe_2VO_4$ and 70°K for $FeV_2O_4$.

## INTRODUCTION

THE ternary system Fe–V–O contains several compounds which possess the spinel, ilmenite, or corundum structure. These compounds show interesting magnetic properties and give further insight into the phenomenon of direct cation-cation interactions. This paper reports the preparation, crystallographic, and magnetic properties of the compounds $FeVO_3$, $Fe_2VO_4$, and $FeV_2O_4$.

The spinels $Fe_2VO_4$ and $FeV_2O_4$ have been prepared previously by several investigators.[1–4] However, no adequate chemical analyses were reported and neither the magnetic nor electrical properties were studied. Berry and Combs[5] have reported $FeVO_3$ to be antiferromagnetic and assigned the valencies $Fe^{2+}-V^{4+}$ to the metal constituents. Their choice was made because the *d* spacings of $FeVO_3$ and $\alpha Fe_2O_3$ are similar. They therefore assigned the valencies $Fe^{2+}-V^{4+}$ on the basis that the ionic radius of $V^{3+}$ is substantially larger than that of $Fe^{3+}$. Actually, according to the radii obtained by Geller[6] from perovskite data, the ionic radius of $V^{3+}$ is slightly smaller than that of $Fe^{3+}$. Thus it would appear that ionic radii do not present a sufficient argument for the assignment of valencies, and the existence of $Fe^{3+}V^{3+}O_3$ is still possible. The study reported in this paper shows that $FeVO_3$ is not antiferromagnetic and both the iron and vanadium are present as trivalent ions.

## CHEMICAL RESULTS AND DISCUSSION

Pure $FeVO_3$ was prepared by prolonged grinding of $Fe_2O_3$ and $V_2O_3$ under a nitrogen atmosphere and then heating the sample in an evacuated silica tube at 1100°C for 48 hr. Unfortunately our numerous attempts to prepare the spinel $FeV_2O_4$ by the reaction of FeO and $V_2O_3$ in evacuated silica tubes failed. Despite the precautions taken to prevent oxidation, all the products became slightly oxidized and contained impurity lines corresponding to the phase $FeVO_3$. However, the use of certain $CO—CO_2$ gas mixtures resulted in the preparation of the pure spinels $FeV_2O_4$ and $Fe_2VO_4$ free of $M_2O_3$ contamination.

A study of the system $Fe_2O_3—V_2O_3$, under various $CO—CO_2$ atmospheres, is summarized in Table I. It can be seen that the initial reduction of iron to $Fe^{2+}$ resulted in the formation of iron-vanadium spinel phases and further reduction with higher concentrations of carbon monoxide gave products which contained metallic iron. Attempts to prepare pure $FeV_2O_4$ under a $CO—CO_2$ atmosphere, rich in $CO_2$ (1:5), showed the presence of $FeVO_3$ in the diffraction patterns, which continued to grow at the expense of the spinel. An increase of carbon monoxide in the gas mixture formed products which had considerably less $FeVO_3$ present, and at a concentration of $CO:CO_2$ of 3:1 only $FeV_2O_4$ was formed. Where the ratio of Fe:V is 1:2, a $CO—CO_2$ gas mixture of 3:1 provides a sufficient reducing condition to maintain vanadium in the 3+ state without the reduction of $Fe^{2+}$ to the metal. This can be accomplished because

\* Operated with support from the U. S. Army, Navy, and Air Force.

[1] J. Andrieux and H. Bozon, Compt. rend. **228**, 565 (1949).

[2] A. Burdese, Ann. Chim. (Rome) **47**, 817 (1957).

[3] G. B. H. Lovell, Trans. Brit. Ceram. Soc. **50**, 315 (1951).

[4] H. M. Richardson, F. Ball, and G. R. Rigby, Trans. Brit. Ceram. Soc. **53**, 376 (1954).

[5] C. R. Berry and C. M. J. Combs, J. Appl. Phys. **131**, 1130 (1960).

[6] S. Geller, Acta Cryst. **10**, 248 (1957).

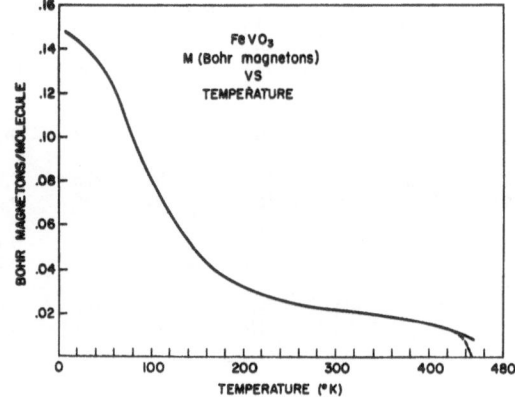

FIG. 1. $FeVO_3$ *M* (Bohr magnetons) vs temperature.

TABLE I. Summary of experimental conditions and resulting products.

| Fe:V | Atm. conditions | Products | % oxid. state | Remarks |
|------|-----------------|----------|---------------|---------|
| 1:1 | vacuum | pure $FeVO_3$ | 100% $Fe^{3+}-V^{3+}$ | chem. anal. does not differentiate between $Fe^{3+}V^{3+}$ and $Fe^{2+}-V^{4+}$ |
| 1:1 | pure $CO_2$ | unident. phase | | |
| 1:1 | $1CO:4CO_2$ | $Fe-V$ spinel phases only | | |
| 1:1 | $1CO:1CO_2$ | $FeV_2O_4+Fe$ | | |
| 1:2 | $1CO:5CO_2$ | $FeV_2O_4+$corundum | | $FeVO_3$ lines grow at the expense of spinel |
| 1:2 | $1CO:1CO_2$ | $FeV_2O_4+$corundum | | $FeVO_3$ considerably decreased |
| 1:2 | $3CO:1CO_2$ | $FeV_2O_4$ | 100% $Fe^{2+}-V^{3+}$ | no foreign lines |
| 2:1 | $1CO:3CO_2$ | $Fe_2VO_4$ | 99.8% $Fe^{2+}-V^{4+}$ | no foreign lines |

of the stabilization of $Fe^{2+}$ by the spinel lattice, and the pure spinel $FeV_2O_4$ is formed. It can be concluded that as long as iron is present as $Fe^{2+}$, only the spinels $FeV_2O_4$ or $Fe_2VO_4$ will form. The valencies of the metal constituents in $FeVO_3$, therefore, are assigned the values $Fe^{3+}V^{3+}O_3$.

## PHYSICAL MEASUREMENTS AND DISCUSSION

The Curie points and the saturation moments at 4.2°K and 11 000 gauss were determined for $FeVO_3$, $Fe_2VO_4$, and $FeV_2O_4$ by means of a vibrating coil magnetometer. Their crystallographic parameters were obtained using a Norelco diffractometer and vanadium filtered chromium $K\alpha$ radiation. Low-temperature observations were made using the apparatus described by Schwartz et al.[7] All cell edges agree with those reported previously.

The magnetization curve for $FeVO_3$ is plotted on Fig. 1. It can be seen that pure $FeVO_3$ has a moment of 0.15 Bohr magnetons per molecule measured at 4.2°K and a field of 11 000 gauss. On warming, there is initially a rapid decrease in the magnetic moment with increasing temperature, but a small spontaneous net moment persists up to approximately 445°K. Additional field vs moment measurements have shown that the materials have a spontaneous net moment up to this temperature. X-ray studies between room temperature and liquid

nitrogen temperature showed no crystallographic transition. However, a gradual change in the relative intensity of the lines was noted down to liquid nitrogen temperature. This change may indicate a shift in the position of the metal atoms. From our crystallographic measurements, there appears to be no evidence of cation ordering in $FeVO_3$. This is in agreement with the work of Cox et al.[8] who have reported that no cation ordering can be detected by x-ray diffraction techniques. In addition the line positions in the Mössbauer spectrum of $FeVO_3$ indicate the presence of $Fe^{3+}$.

The Curie point and saturation moment at 4.2°K for $FeV_2O_4$ are 109°K and 1.95 Bohr magnetons/molecule. Those for $Fe_2VO_4$ and 440°K and 0.72 Bohr magnetons/ molecule at 4.2°K. The magnetization curves of $FeV_2O_4$ and $Fe_2VO_4$ were also measured between the Curie points and 4.2°K for both of these materials. Both of these samples indicated the presence of a magnetic transition as evidenced by an anomalous change in the slope of the magnetization curves. These transitions occurred at 275°K for $Fe_2VO_4$ and 70°K for $FeV_2O_4$. Since these samples are pure single-phase materials, the transitions probably correspond to a change from a simple Néel configuration to a more complex canted-spin model.[9]

[7] R. Schwartz, B. Post, and I. Fankuchen, Rev. Sci. Instr. **22**, 281 (1951).

[8] D. E. Cox, W. J. Takei, G. Shirane, and S. L. Ruby (private communication).

[9] N. Menyuk, A. Wold, D. Rogers, and K. Dwight, J. Appl. Phys. **33**, 1144 (1962), this issue.

JOURNAL OF APPLIED PHYSICS       SUPPLEMENT TO VOL. 33, NO. 3       MARCH, 1962

# Preparation and Properties of Ferrospinels Containing Ni³⁺

M. W. Shafer

*IBM Research Center, Yorktown Heights, New York*

Nickel-iron spinels with compositions between $NiFe_2O_4$ and $Ni_2FeO_4$ have been prepared under oxygen pressures up to 2000 psi and at temperatures between 250° and 1000°C. Good agreement was found between the measured lattice parameters of the resulting spinel phases and the values obtained by extrapolating from the previously determined spinel series between $Fe_3O_4$ and $NiFe_2O_4$. Spinel compositions with lattice parameters as low as 8.3005 A were obtained; this compares with 8.3885 for $Fe_3O_4$ and 8.3394 for $NiFe_2O_4$. As is the case of the spinel solid-solution series between $Fe_3O_4$ and $NiFe_2O_4$, the composition of these nickel-rich spinels was found to depend strongly on both the oxygen pressure and the temperature. These relationships have been determined and are briefly discussed. Since it had been shown that there is essentially no measurable deviation from Végard's law between $Fe_3O_4$ and $Ni_2FeO_4$, lattice parameter measurements were a convenient method of determining the compositions of the spinel phase. The fact that single phase nickel-iron spinels can be prepared with nickel concentrations greater than $NiFe_2O_4$ indicates that $Ni^{3+}$ replaces $Fe^{3+}$ in the spinel lattice. Of the several possible configurations for a composition such as $Ni_2FeO_4$, there is strong evidence from magnetic moment measurements that on octahedrally ligated spinel sites the preferred arrangement is the $[Ni^{2+} Ni^{3+}]$ ion pair. The corresponding decrease in moment as $Ni^{3+}$ substitutes for $Fe^{3+}$ between $NiFe_2O_4$ and $Ni_2FeO_4$ agrees very well with that predicted by the Néel antiparallel array theory. A lowering of the Curie temperature was also observed in this compositional range.

## INTRODUCTION

THE preparation and properties of ferrites in the spinel system $Ni_xFe_{3-x}O_4$ have been discussed quite extensively in recent years,[1-3] especially for those compositions between $x=0$ and $x=1$. However, this previous work was done only up to one atmosphere oxygen pressure where it was evident that the maximum amount of nickel that the spinel structure could tolerate was a value of about $x=1$. The purpose of this paper is to present the results of an investigation in which high oxygen pressures were used to prepare nickel ferrites with $1 \leq x \leq 2$.

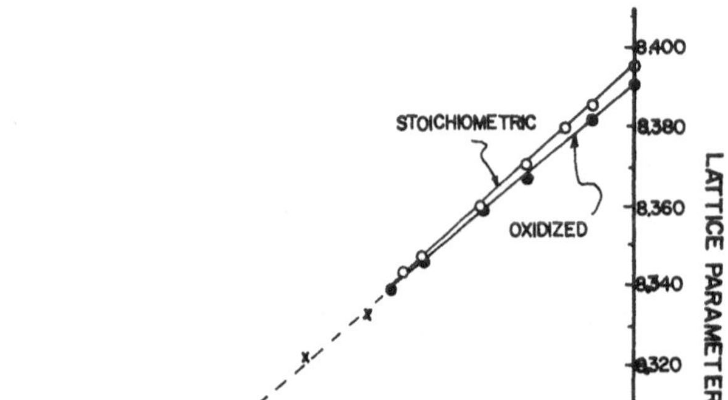

## MATERIALS AND EXPERIMENTAL TECHNIQUES

The starting materials were prepared by either co-precipitating the hydroxides of iron and nickel or by forming a gel from an alcoholic solution and then vacuum drying at low temperatures. The resulting material was an amorphous homogeneous mixture of "iron and nickel oxide" which was analyzed to determine the exact iron-nickel ratio.

This material was then placed in gold containers and reacted at the desired temperature and pressure. The pressure apparatus consisted of four externally heated, stellite-pressure vessels which were connected with capillary tubing to a source of high pressure oxygen. The pressures were read by Bourdon gauges to $\pm 200$ psi. After the mixtures were reacted for sufficient time to reach an equilibrium condition, they were quenched and examined by x ray to determine the composition of the spine phase.

## RESULTS AND DISCUSSION

In Fig. 1 the lower of the two solid lines shows nickel-iron spinels containing the maximum number of cation vacancies as determined by the phase diagram. The upper curve represents stoichiometric spinels where the metal to oxygen ratio is exactly three to four.[4] The smaller lattice constants in the oxidized samples are due to an increased $Fe^{3+}/Fe^{2+}$ ratio. As there is no width to the spinel field in this region,[1,3] the curves converge around $x=1$. The lattice parameters for spinel compositions with $x>1$ continue to fall on the straight line, i. e., the broken line in Fig. 1, indicating that Végard's law is still followed between $x=1$ and $x=2$. Since the spinels prepared in this region were reacted at relatively low temperatures (up to 1000°C), resulting in very fine-grained materials, the lattice parameters could not be

FIG. 1. Variation of cell dimensions with composition $x$ for the $Ni_xFe_{3-x}O_4$ system.

[1] A. Paladino, J. Am. Ceram. Soc. **42**, 168 (1959).
[2] C. Okazaki, J. Phys. Soc. Japan **15**, 2013 (1960).
[3] M. W. Shafer, J. Phys. Chem. **65**, 2055 (1961).
[4] J. E. Weidenborner and M. W. Shafer (to be published).

measured with a degree of precision which would allow one to distinguish whether the extension of the oxidized or stoichiometric curve would be more closely followed. As expected, the lattice parameters, and hence the composition of the spinel phase, in this region were a function of both temperature and oxygen pressure. Although the phase diagrams showing these compositional changes cannot be shown here, it should be mentioned that the general phase relations that were determined for compositions where $x>1$ were followed—that is, at high temperatures (in this case 500°C for some compositions) the spinel loses oxygen causing NiO to precipitate with the formation of an iron-rich spinel. Likewise, at the lower oxygen pressures used the reaction also went in this direction. By isothermally studying the variation of the lattice parameters as a function of oxygen pressures, the conditions under which single-phase stoichiometric spinels were formed could be established by noting the pressure at which no further change was observed. A temperature of about 370° and an oxygen pressure of 2000 psi were the conditions at which the maximum amount of nickel was observed to go into the spinel structure. This was a single-phase composition of $Ni_{1.71}Fe_{1.29}O_4$ and a composition containing some nickel oxide at $Ni_{1.8}Fe_{1.2}O_4$. When the reaction temperature dropped below 370° nickel oxide was usually present, indicating a nonequilibrium condition.

For the existence of single-phase nickel-iron spinels with nickel concentrations greater than $x=1$, it is necessary for either some of the iron to assume an oxidation state of four or some of the nickel three. Magnetic moment measurements were made in an attempt to resolve which of the two cases is preferred, and the results are shown in Fig. 2. If we consider the model where the $[Ni^{3+}Ni^{2+}]$ ion pair occupy the octahedral sites, it is seen from Fig. 2 that there is excellent agreement between the measured values and the calculated moments as predicted by the Néel antiparallel array theory. Of course, a model where $[Fe^{4+}]$ is present on the tetrahedral sites would give identical values for the moments, but such an arrangement is unlikely for several reasons. In the first place, the formation of $Ni^{3+}$ should be favored over $Fe^{4+}$ as the fourth ionization potential of iron is about 20 ev larger than the third of nickel[5]; and if the $Fe^{4+}$ ion should be present, it probably would prefer the octahedral positions since it is a $3d^4$ ion.[6] Rather crude dc conductivity measurements also indicated that the resistivity of these nickel-rich spinels was several orders of magnitude lower than that for nickel ferrite ($NiFe_2O_4$) prepared in the same manner—a probable indication that $Ni^{3+}$ and $Ni^{2+}$ occupy identical lattice

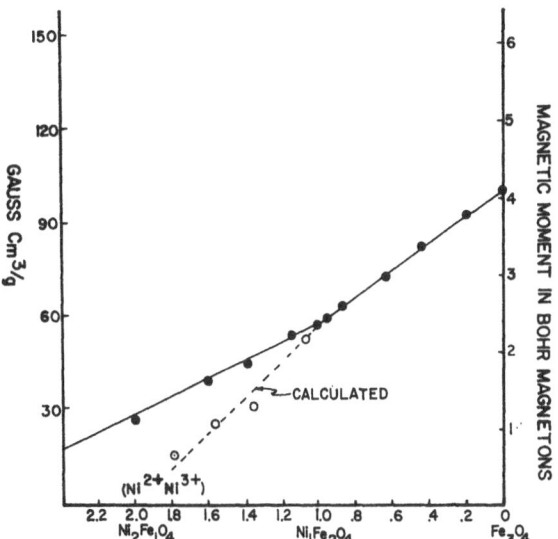

FIG. 2. Variation of magnetic moment with composition for the $Ni_xFe_{3-x}O_4$ system. Closed circles represent measurements on high-temperature preparations. Open circles represent samples prepared at high oxygen pressures. The dashed curve is calculated from the Néel model for a composition of $x=2$ and a distribution of $Fe^{3+}[Ni^{2+}Ni^{3+}]O_4$.

positions.[7] In addition to the calculated curve (dashed) shown in Fig. 2, the solid line shows the moments as a function of the spinel composition. For nickel concentrations greater than $x=1$, this curve represents a dilution of $NiFe_2O_4$ with nonmagnetic nickel oxide.

Curie temperature measurements showed that as the nickel content of the spinel increases above $x=1$, there is a decrease from 595° for $NiFe_2O_4$ to 525° for $Ni_{1.4}Fe_{1.6}O_4$.

## SUMMARY

Single-phase nickel-iron spinels can be prepared with nickel concentrations greater than nickel ferrite ($NiFe_2O_4$). However, the amount of nickel that can be substituted in the spinel lattice is strongly dependent on the oxygen pressure used. For example, at 2000 psi the maximum pressure used in this investigation, a spinel corresponding to a composition of $Ni_{1.8}Fe_{1.2}O_4$, was obtained.

There is good evidence from magnetic moment measurements that these spinels contain $Ni^{3+}$ and that it prefers the octahedral sites of the spinel lattice.

## ACKNOWLEDGMENTS

The author gratefully acknowledges the help of H. G. Schaefer who helped in the preparation of the samples, of J. E. Weidenborner who assisted with the x-ray work, and of H. R. Lilienthal who made the magnetization measurements.

[5] R. Parsons, *Handbook of Electrochemical Constants* (Academic Press Inc., New York, 1959).

[6] J. D. Dunitz and L. E. Orgel, J. Phys. Chem. Solids **3**, 318 (1957).

[7] H. Lord, Nature **188**, 929 (1960).

# Soft Magnetic Materials

P. A. ALBERT, *Chairman*

## Two Effects of Changes in Tension

OSAMU YAMADA

*Laboratoire d'Electrostatique et de Physique du Métal, Institut Fourier, Place Doyen Gosse, Grenoble, France*

In the Rayleigh region of the initial magnetization, increasing or decreasing of a tension leads to an irreversible increase in magnetization. This phenomenon can be attributed to two effects, namely the first and the second effect. For the first effect, the magnitude of the change of magnetization is directly proportional to the value of magnetization itself and is also directly proportional to the change of tension but independent of the sign of the tension change. For the second effect, the change of magnetization is also directly proportional to the value of magnetization but is not proportional to the change in tension. The second effect has, in general, a different value for increase and decrease of tension and is accompanied by the appearance of unsymmetrical hysteresis loops. The first effect is caused by the displacement of 180° walls, in contrast to the second effect which is associated with 90° walls.

## I. INTRODUCTION

IN previous papers,[1,2] the change of magnetization due to the change of temperature occurring in a wire specimen of carbon steel was investigated. It has been shown that heating or cooling always leads to an irreversible increase in magnetization and that this phenomenon can be attributed to two effects, namely the first and the second effect. From the results of experiments it is concluded that these two effects are caused by the change of the state of internal stress due to the change of temperature.

On the other hand, many studies[3-6] have been done about the effects of stresses on magnetic properties and the experimental data are explained by the assumption that only the displacement of 90° walls occurs through the magnetostriction by application of tension or compression. But in the region of weak magnetic field, the phenomenon is so complex that the experimental values of changes in magnetization do not agree with the theoretical predictions. For instance, Lliboutry[6] has observed an increase of magnetization for application of a tension and also for removal of a tension in the Rayleigh region of carbon steel specimen. With the analogy of the effects caused by a change of temperature, the complex characters of the effects of a change in tension will be analyzed in the present paper.

## II. SUPERPOSITION OF THE TWO EFFECTS

If a wire specimen of carbon steel ($H_c = 16$ oe) initially demagnetized under tension of 15.6 kg/mm² is sub-

jected to a weak magnetic field $H_0 = 2.2$ oe (the magnetization is A in Fig. 1), then removed the tension of 15.6 kg/mm² at the same field 2.2 oe, one observes an increase of magnetization from A to C in Fig. 1. Afterwards, if one increases the field under zero tension, one obtains the magnetization curve which is designated by $J_{2.2}$ oe ($\bar{\sigma}_{15.6} \sigma_0$) in Fig. 1. This curve is not joined to $J(\bar{\sigma}_0)$, namely the magnetization curve measured without tension after demagnetization without tension but joined rapidly to $J(\bar{\sigma}_{15.6} \bar{\sigma}_0)$ which is the magnetization curve traced without tension after demagnetization under tension of 15.6 kg/mm² and after removal of this tension. It is the same when $H_0$ takes another value of magnetic field.

As the general formula we shall designate by $J(\bar{\sigma}_a \bar{\sigma}_b \bar{\sigma}_c \ldots \bar{\sigma}_i, \sigma_j)$ the magnetization obtained under the last tension of $\sigma_j$, applying the magnetic field at the tension $\bar{\sigma}_i$, after successive changes in tension from $\sigma_a$ to $\sigma_i$. The specimen is always demagnetized under the first tension $\sigma_a$.

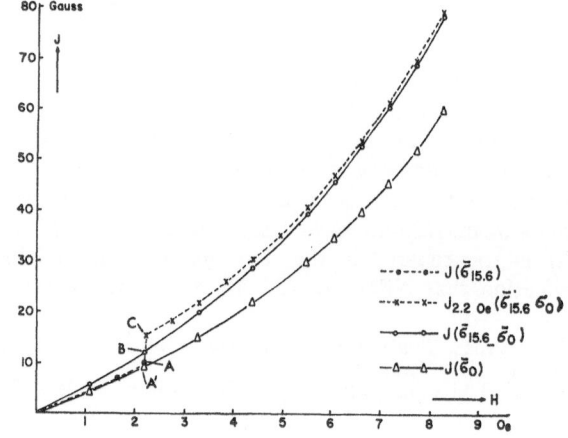

FIG. 1. Magnetization curves of carbon steel wire ($H_c = 16$ oe); BC corresponds to the first effect; A'B to the second effect.

[1] O. Yamada, J. Phys. radium **21**, 116S (1960); Compt. rend. **250**, 4313; **251**, 662, 860 (1960); **253**, 403, 629 (1961).
[2] O. Yamada, International Conference on Magnetism and Crystallography, Kyoto (1961).
[3] R. M. Bozorth, *Ferromagnetism* (D. Van Nostrand Company, Inc., Princeton, New Jersey, 1951), p. 595.
[4] R. M. Bozorth and H. J. Williams, Revs. Modern Phys. **17**, 72 (1945).
[5] W. F. Brown, Phys. Rev. **75**, 147 (1949).
[6] L. Lliboutry, Ann. Phys. **6**, 731 (1951).

From the above-mentioned experimental results, one can deduce that the change of the magnetization due to a change of tension comprises two parts, the first of which is designated $X_\sigma(\sigma_{15.6}\,\sigma_0)$ and the second proceeds from the difference between $J(\bar{\sigma}_{15.6}\,\bar{\sigma}_0)$ and $J(\bar{\sigma}_{15.6})$ namely AB in Fig. 1. The value of $X_\sigma(\sigma_{15.6}\,\sigma_0)$ is given by the expression

$$X_\sigma(\sigma_{15.6}\,\sigma_0) = J(\bar{\sigma}_{15.6}\,\sigma_0) - J(\bar{\sigma}_{15.6}\,\bar{\sigma}_0). \qquad (1)$$

For the case of application of tension, one can obtain in the same way the expression

$$X_\sigma(\sigma_0\sigma_{15.6}) = J(\bar{\sigma}_0\sigma_{15.6}) - J(\bar{\sigma}_0\bar{\sigma}_{15.6}). \qquad (2)$$

We shall designate the first part $X_\sigma$ by the name of the "first effect due to a change of tension." Actually, the two curves $J(\bar{\sigma}_{15.6}\,\bar{\sigma}_0)$ and $J(\bar{\sigma}_0)$ are very different, although they are plotted under the same tension of zero, so the parameter $Y_{\sigma d}$ or $Y_{\sigma e}$ which is given, for example, by

$$Y_{\sigma d}(\sigma_{15.6}\sigma_0) = J(\bar{\sigma}_{15.6}\bar{\sigma}_0) - J(\bar{\sigma}_0) \qquad (3)$$

or

$$Y_{\sigma e}(\sigma_0\sigma_{15.6}) = J(\bar{\sigma}_0\bar{\sigma}_{15.6}) - J(\bar{\sigma}_{15.6}) \qquad (4)$$

may be called the "second effect due to a change of tension."

From Eq. (1) and (3) one obtains

$$X_\sigma(\sigma_{15.6}\sigma_0) + Y_{\sigma d}(\sigma_{15.6}\sigma_0) = J(\bar{\sigma}_{15.6}\sigma_0) - J(\bar{\sigma}_0) \qquad (5)$$

or, from Eq. (2) and (4),

$$X_\sigma(\sigma_0\sigma_{15.6}) + Y_{\sigma e}(\sigma_0\sigma_{15.6}) = J(\bar{\sigma}_0\sigma_{15.6}) - J(\bar{\sigma}_{15.6}). \qquad (6)$$

In other words, it takes the place of a superposition of the two effects.

### III. FIRST EFFECT DUE TO A CHANGE OF TENSION

The results of experiments show that, for the first effect, the magnitude of the change of magnetization is directly proportional to the value of magnetization itself, and is also directly proportional to the change of tension but independent of the sign of the tension change that is,

$$X_\sigma(\sigma_a\sigma_b) = X_\sigma(\sigma_b\sigma_a) = k_\sigma J\sigma, \qquad (7)$$

where $\sigma$ is a difference between two tensions, $\sigma_a$ and $\sigma_b$, and $k_\sigma$ is a constant. For a wire specimen of carbon steel ($H_c = 16$ oe), we have obtained $k_\sigma = 0.013$ (kg/mm²)⁻¹. It is deduced by experiments that the first effect is caused by the displacement of 180° walls.

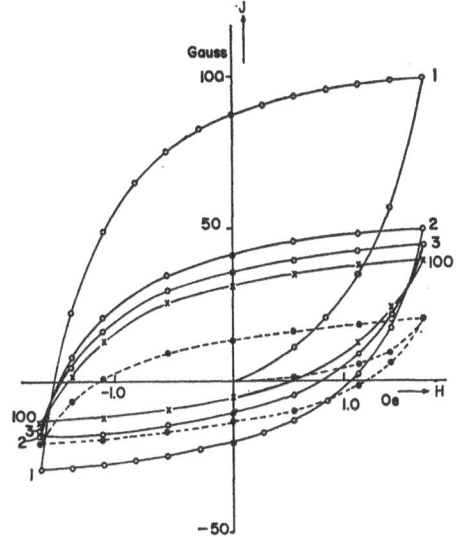

Fig. 2. Comparison of unsymmetrical hysteresis loops accompanied by the second effect with the ordinary hysteresis loop for 60% Ni Permalloy. All loops are measured under the same tension of 16.5 kg/mm². —O—O— $J(\bar{\sigma}_0\bar{\sigma}_{16.5})$, —x—x— the 100th cycle of $J(\bar{\sigma}_0\bar{\sigma}_{16.5})$, — —●— —●— — $J(\bar{\sigma}_{16.5})$.

### IV. SECOND EFFECT DUE TO A CHANGE OF TENSION

It is shown by experiments that, for the second effect, the change of magnetization is also directly proportional to the value of magnetization but is not proportional to the change of tension. The second effect has, in general, a different value for increase and decrease of tension and is accompanied by the appearance of unsymmetrical hysteresis loops which are shown in Fig. 2 for a specimen of 60% Ni Permalloy. The second effect is written by

$$Y_{\sigma d}(\sigma_b\sigma_a) = k_{yd}J(\bar{\sigma}_a)\sigma^n \qquad (8)$$

or

$$Y_{\sigma e}(\sigma_a\sigma_b) = k_{ye}J(\bar{\sigma}_b)\sigma^n, \qquad (9)$$

where $\sigma$ is a difference of two tensions $\sigma_a$ and $\sigma_b\,(\sigma_a<\sigma_b)$ and $k_{yd}$ or $k_{ye}$ a constant. We have obtained for the same specimen of carbon steel, $n=1.525$, $k_{yd}=4.87\times10^{-3}$, and $k_{ye}=3.62\times10^{-3}$.

The results of experiments indicate that the second effect is associated with 90° walls.

A more detailed report will be published.

# Recent Developments in Soft Magnetic Alloys

EDMOND ADAMS

*U. S. Naval Ordnance Laboratory, Silver Spring, Maryland*

The present status of soft magnetic alloys used in power generation, communication, and special devices applications will be reviewed with emphasis on future prospects for their improvement. Although the study of microscopic structure of magnetic materials has led to important advances, present and future improvements require an understanding of the fundamental processes of magnetization and a study of submicroscopic and atomic influences.

The bulk of present day magnetic materials is based on alloys of iron with silicon, aluminum, nickel, and cobalt in varying proportions. Their quality has been constantly upgraded by better control of purity and composition through improvements in melting, processing, and annealing practices. Some examples of how each of these factors affects the ultimate magnetic properties of soft magnetic alloys will be given. Brief summaries will also be presented on the magnetic properties of current soft magnetic materials and how they relate to modern devices. Some interesting properties of magnetic alloys based on iron, silicon and/or aluminum will be reported. The development of a hardenable, corrosion and abrasion resistant nonmagnetic alloy based on a ductile intermetallic compound TiNi will also be disclosed.

## INTRODUCTION

MODERN civilization is largely dependent on electric power for its mobility, communication, and various control devices. These, in turn, depend on the phenomenon of ferromagnetism, which is found chiefly in four elements—iron, nickel, cobalt, and the rare-earth element gadolinium. These few elements, and their various alloys, provide magnetic materials having a wide range of properties.

In most engineering applications the most desired magnetic properties are: (1) High saturation-induction (to provide greatest output per volume); (2) high permeability to reduce magnetizing currents to a minimum; and (3) low hysteresis and eddy current losses for efficiency. Most of our present high permeability magnetic alloys possess low magnetostriction values and either low anisotropy values or well-defined crystallographic orientation in the plane of the sheet material.

The most common soft magnetic materials in use today for power applications are the silicon-iron alloys from 1-4% silicon which are used in huge tonnage quantities. It is the most economical material from the standpoint of its high induction value and resistivity. The aluminum-iron alloys also have these same advantages and furthermore remain ductile up to about 10% aluminum, while the silicon-iron alloys containing over 5% silicon are virtually unworkable. However, the exploitation of the 0–10% aluminum-iron alloys has been neglected.

The second class of magnetic materials, usually alloys of iron with nickel and cobalt, are those which have the best magnetic characteristics for a specific application notwithstanding any premium cost. The nickel alloys exhibit very high permeabilities and low losses, but because of low saturation-induction values are used in low level applications. The cobalt alloys have a high saturation but are difficult to process and are expensive.

## SILICON-IRON ALLOYS

The magnetic properties of silicon-iron alloys have been over the years upgraded by increasing their purity through improved melting and annealing practices. However, economic considerations dictate the extent to which these modern techniques can be adopted. The chief improvement in silicon-iron alloys came when Goss[1] developed a procedure for the alignment of the [100] directions of the grains in the rolled sheet material. This is known as cube-on-edge texture, since only one easy direction of magnetization lies in the plane of the rolled strip. Recently another improved texture has been announced, first by Assmus *et al.*[2] and others[3,4] where two orientations of the [100] directions lie in the plane of the rolled strip. This type of texture provides "easy" directions in the rolling direction, transverse to the rolling direction, and perpendicular to the sheet surface. The commercial production of this type of texture is imminent and promises lower losses for small transformers at high induction values; however, in large transformers smaller gains are to be expected, because present designs efficiently circumvent unfavorable orientations. In small or medium transformers "cube" textured silicon-iron alloys may in many cases be equivalent or even superior to most nickel-iron alloys. Besides the reduction in core size, exciting current, and copper windings, cube-textured silicon-iron sheet lends itself to use in laminated configurations such as square, *U*-cores, or *E-I* shaped cores. This is important in many military applications because these configurations resist degradation in stress and shock environments better than the usual tape-wound cores.

Another silicon-iron alloy which has received considerable attention is the one containing 6.4% silicon. This alloy has the lowest anisotropy constant of all the silicon irons, along with a very low magnetostriction value. However, alloys with this high silicon content are difficult to hot work and virtually impossible to cold work.

[1] N. P. Goss, Trans. Am. Soc. Metals **23**, 515–531 (1935).

[2] F. Assmus, K. Detert, and G. Ibe, Z. Metallkunde **48**, 344–349 (1957).

[3] J. L. Walter, W. R. Hibbard, H. C. Fiedler, H. E. Grenoble, R. H. Pry, and P. G. Frischmann, J. Appl. Phys. **29**, 363 (1958).

[4] G. Wiener, P. A. Albert, R. H. Trapp, and M. F. Littman, J. Appl. Phys. **29**, 366 (1958).

## ALUMINUM-IRON ALLOYS

The aluminum-iron alloys from 0–16% aluminum have been under limited investigation during the past one-half century. The most important alloys for engineering applications were found at approximately 12–16% aluminum content. Because of their high resistivities, strain insensitivity, and hardness, these materials are used chiefly in the synchro and recording industry. Another series of aluminum-iron alloys from 0–10% aluminum also have been investigated but not as thoroughly as the higher percentages.

The major differences found between low percentage aluminum-iron alloys and equivalent percentages of silicon-iron alloys are: (1) Aluminum is a more powerful deoxidizer than silicon, and (2) the aluminum-iron alloys show improved ductility. Whereas even small percentages of cold reduction are difficult to obtain on silicon-iron alloys containing over 4% silicon, it is possible to cold reduce aluminum-iron alloys containing up to 6% aluminum by as much as 90%, and those alloys containing up to 12% aluminum are capable of standing considerable cold reduction without fracture. Commercial production of aluminum-iron alloys was limited in their early stages of development because of the relative high cost of aluminum versus silicon and because of the greater difficulty in preventing segregation of aluminum during the melting process. However, the cost picture has equalized in recent years, and with modern induction melting equipment the segregation problem has practically been eliminated.

Recent work by Sugihara[5] in Japan and Helms[6] of the Naval Ordnance Laboratory has illustrated that magnetic properties of isotropic aluminum-iron alloys are comparable to those of equivalent silicon-iron alloys. Table I illustrates properties of equivalent alloying compositions after standard annealing procedures. It may be seen that properties of the lower percentage additions are more favorable in the aluminum-iron

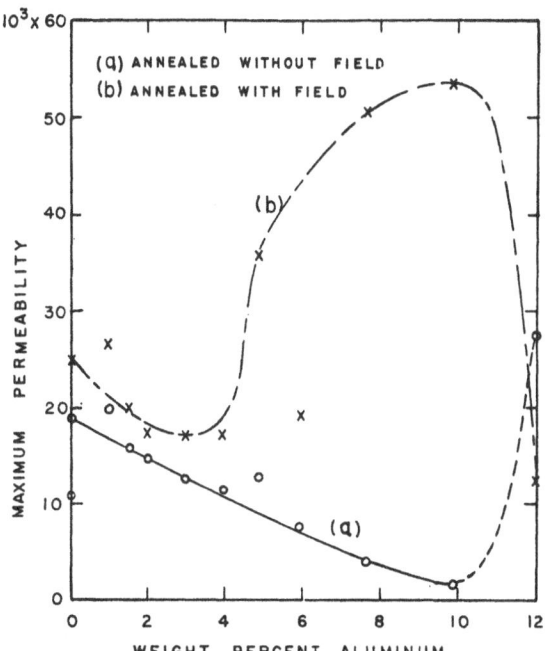

FIG. 1. Response of iron-aluminum alloys to annealing in a magnetic field of 1.7 oe.

system; however, when the alloy addition is increased the advantages tend toward the silicon-iron alloys. Just as in the case of silicon-iron alloys, the low percentage aluminum-iron alloys are very responsive to magnetic annealing. Figure 1 illustrates effects of magnetic annealing versus standard annealing on 0.014-in. thick laminated cores in the aluminum-iron series. Helms has obtained maximum permeabilities in the range of 53 000 by magnetic annealing alloys containing 8% to 10% aluminum. Of course, properties of this magnitude can be obtained in the higher 12% to 16% aluminum-iron range without magnetic annealing, but the loss in ductility in this range precludes a wide acceptance for most applications.

## SILICON-ALUMINUM-IRON ALLOYS

The excellent dc magnetic characteristics of the mixed silicon-aluminum-iron alloys as announced by Masumoto[7] have long been known. The brittleness of the 9.5% silicon and 5.6% aluminum, known as Sendust, has precluded its use for ac applications. Recent work by Adams and Hubbard[8] at the Naval Ordnance Laboratory has produced sheet by various powder metallurgy techniques such as slip casting, liquid-phase sintering, and sintering of loose powder, which promise to have some use in several special applications. The magnetic characteristics of Sendust type alloy by various processes are listed in Table II.

TABLE I. Comparison of magnetic properties of silicon-iron vs aluminum-iron alloys.

| Percent silicon or aluminum[a] | dc magnetic properties | | | | | 60 cps core loss (watts/lb) 10 kgauss |
|---|---|---|---|---|---|---|
| | $\mu_{20}$ | $\mu_{max}$ | $H_c$ oe | $B_r$ gauss | $B_m{}^b$ gauss | |
| 1.0 Al | 2041 | 20 214 | 0.329 | 14 101 | 16 348 | 0.70 |
| 0.94 Si | 860 | 14 780 | 0.442 | 14 140 | 16 400 | 0.69 |
| 3.0 Al | 1923 | 12 815 | 0.450 | 13 741 | 16 278 | 0.70 |
| 2.81 Si | 1720 | 17 440 | 0.192 | ... | 15 340 | 0.58 |
| 5.0 Al | 2381 | 12 625 | 0.329 | 10 153 | 15 446 | 0.64 |
| 4.8 Si | 1410 | 17 160 | 0.201 | 7410 | 15 230 | 0.43 |
| 8.0 Al | 513 | 3461 | 0.550 | 3242 | 13 392 | 1.07 |
| 6.4 Si | 1720 | 28 860 | 0.147 | 11 310 | 14 050 | 0.40 |
| 8.0 Al[c] | 1140 | 50 900 | 0.269 | 10 900 | 12 800 | 0.45 |
| 6.4 Si[c] | 1390 | 66 940 | 0.197 | 12 100 | 14 600 | 0.37 |

[a] Annealed at 1000°C in hydrogen.
[b] $H = 30$ oe.
[c] Magnetic anneal at 1.7 oe.

[5] M. Sugihara, J. Phys. Soc. Japan 15, 8 (1960).
[6] H. H. Helms (to be published).

[7] H. Masumoto, Sci. Repts. Tôhoku Imp. Univ. 25, 388–402 (1936).
[8] E. Adams and W. M. Hubbard, U. S. Patent 2,992,474 dated July 18, 1961; U. S. Patent 2,988,806 dated June 20, 1961.

TABLE II. Characteristics of Sendust alloys by various fabricating processes.

| Process | Thickness in. | Density g/cc | dc permeability $\mu_{20}$ | dc permeability $\mu_{max}$ | $B_r$ gauss | $B_{max}$ $H=30$ oe | $H_c$ oe |
|---|---|---|---|---|---|---|---|
| Cast Ring[a] | 0.250 | 6.96 | 15 000 | 110 000 | 7000 | 11 000 | 0.04 |
| Sintered[b] | | | | | | | |
|   a. liquid phase | 0.350 | 6.50 | 6200 | 25 000 | 5000 | 8200 | 0.10 |
|   b. liquid mix. | 0.260 | 6.51 | 10 000 | 33 700 | 4500 | 9500 | 0.07 |
|   c. slip cast[c] | 0.022 | 5.50 | 2000 | 15 000 | 4400 | 7900 | 0.23 |
|   d. loose powder | 0.400 | 6.30 | 11 500 | 46 000 | 5200 | 8800 | 0.08 |
|   e. hot pressed | 0.250 | 6.86 | . . . | . . . | . . . | . . . | . . . |

[a] Cast alloy: 1000°C $H_2$ atm. anneal.
[b] Sintered alloys: a. (FeSi-AlSi-Fe) mix.-30 K psi 1175°C-He atm. b. 20% (FeSi-AlSi-Fe) +80% Sendust powder mix 30 K psi. c. (FeSi-AlSi-Fe) mix.-water-alginate slurry. d. Sendust powder-no pressure. e. Sendust powder-2 K psi-1260°C.
[c] Losses =0.2 w/lb.-60 cycles.

Because present magnetic materials may be exposed to unusual and extreme environmental conditions of high levels of nuclear radiation and temperature (up to 500°C) various comprehensive studies have been made. In the area of soft magnetic materials it was found that silicon-iron alloys suffered the least degradation in these environments. However, very little materials research has been concerned with developing new alloys with improved environmental resistance. One such alloy consisting of 3% silicon, 1% aluminum and balance iron showed excellent oxidation resistance, and its rectangular hysteresis loop remained essentially un-remained essentially undamaged after nuclear radiation (see Fig. 2).

### NICKEL-IRON ALLOYS

A recitation of the specific applications of the high permeability nickel-iron alloys would be superfluous.

Needless to say, our electronic industries involving communication, computers, and control are tremendously dependent on their availability. The chief problem in these applications is reproducibility. Many of the magnetic characteristics are structure-sensitive and vary tremendously with impurity level, processing details, and heat treatment. Frequently, this structure-sensitive peculiarity is used to develop a square hysteresis loop, low remanence, high initial or maximum permeability, or any other desired property. An example of the development of an improved magnetic material by understanding the fundamental processes of magnetization is given by Nesbitt and Gyorgy[9] in their work on a gold Permalloy tape for high speed switching. The usual Permalloy composition tape core depends upon residual stresses arising from cold working for its rectangular hysteresis loop and coercive force. These investigators introduced a gold rich phase in a matrix of strain-free Permalloy composition which allowed switching by domain rotation yet had the proper coercive force from the precipitated gold phase. The switching cores of this material were found to be four times faster than standard Permalloy compositions.

One way to control the purity of nickel-iron magnetic alloys is to start from constituent powders of known impurity content. Besides purity, the composition limits can also be held to closer tolerance than with cast alloys. In general, however, Scholefield and Richardson[10] found that, although reproducible alloys could be made in this manner, carefully vacuum-cast alloys yielded a superior product and with considerably more economy. Recent work at the Naval Ordnance Laboratory[11] confirms this observation; however, some nickel-iron alloys were found to have a somewhat higher saturation-flux density than cores manufactured from the cast alloy, probably due to absence of any diluent manganese and other impurities. The cores also exhibited a superior thermal stability than the cast material, perhaps due to the presence of a microscopic

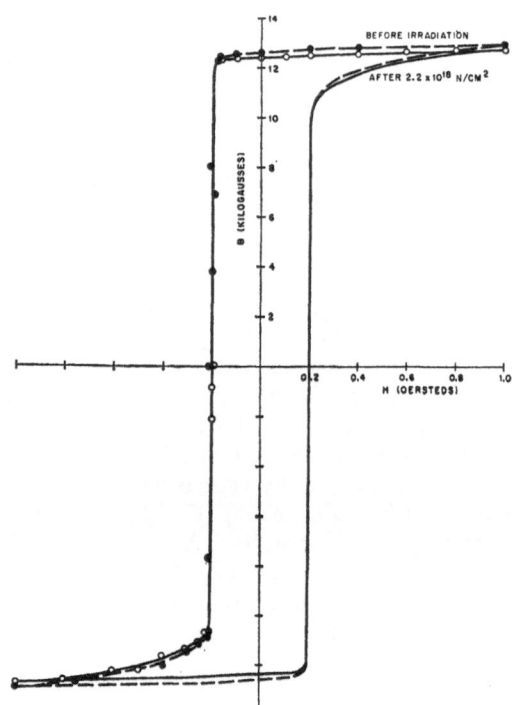

FIG. 2. Stability of a 3% silicon, 1% aluminum-balance-iron alloy to nuclear radiation of 2.2×10^18 fast neutrons/cm².

[9] E. A. Nesbitt and F. M. Gyorgy, J. Appl. Phys. 32, 1305 (1961).
[10] H. H. Scholefield and S. G. G. Richardson, Powder Met. Bull. 4, 44 (1959).
[11] W. M. Hubbard, J. F. Haben, and E. Adams, Navy Ordnance Rept. 6257, U. S. Naval Ordnance Laboratory, Silver Spring, Maryland (1959).

TABLE III. Comparison of USA and USSR commercial magnetic materials.

| USA and USSR alloy[a] | Type | dc permeability | | $B_r$ gauss | $B_m$ gauss | $H_c$ oe |
|---|---|---|---|---|---|---|
| | | $\mu_0$ | $\mu_{max}$ | | | |
| 3.3 silicon-iron | O[b] | ... | 55 000 | 9500 | 15 500 | 0.08 |
| E330 | O | 800 | 45 000 | ... | 14 000 | 0.10 |
| 50 Ni-50 Fe | I[b] | 4000 | 40 000 | 10 700 | 15 600 | 0.13 |
| 50 NP | I | 2600 | 35 500 | 11 100 | 14 250 | 0.16 |
| None | ... | ... | ... | ... | ... | ... |
| 80NKhS[c] | ... | 28 000 | 113 000 | ... | 6650 | 0.016 |
| 4-79 mo-Permalloy | I | 20 000 | 100 000 | 6000 | 8700 | 0.030 |
| 4-79 mo-Permalloy (vac. melt) | I | 60 000 | 400 000 | 5500 | 8000 | 0.010 |
| 79 NM | I | 24 000 | 275 000 | 5750 | 8050 | 0.013 |
| 16 Alfenol | I | 3000 | 100 000 | 4200 | 7600 | 0.024 |
| IU 16 | I | ... | 67 000 | 3500 | 5650 | 0.022 |
| 65 Permalloy[d] | MA[b] | 5000 | 1 500 000 | $B_r/B_s$ 0.98 | 13 900 | 0.011 |
| 65NP | MA | 1300 | 412 000 | $B_r/B_s$ 0.98 | 13 300 | 0.036 |
| 50 Ni-50 Fe[c] | O | 1300 | 100 000 | $B_r/B_s$ 0.95 | 14 980 | 0.100 |
| 50 NP[c] | O | 700 | 62 000 | $B_r/B_s$ 0.92 | 14 800 | 0.185 |

[a] 0.014-in. laminations unless noted otherwise.
[b] Symbols: O =grain-oriented, I =isotropic, and MA =magnetic anneal.
[c] 80 Ni (Cr-Si add.) bal. Fe.
[d] 0.002-in. tape wound cores.

porosity whicn provided this stability by "built-in air-gaps." However, except for specialty items it is expected that the bulk of the nickel-irons will be manufactured by induction melting and upgraded by improved melting techniques. A new alloy of 80% nickel and balance iron with undetermined amounts of chromium and silicon is reported by the Russians.[12] Table III shows this alloy and other USSR alloys compared with equivalent US commercial alloys.

## SPECIAL APPLICATIONS

The existence of an improved magnetic core material or alloy in itself will not solve a particular problem. In many instances the utilization may take considerable engineering effort and perserverence. For example, a core material was required for a high-power, broadband, variable-reluctance sonic transducer. The core material should have a high flux density at low bias magnetizing forces with a high incremental permeability and a low incremental core loss. At present the theoretical efficiency of the variable-reluctance transducer using grain-oriented silicon-iron "C" cut cores is about 80%. Since the size of the power plant for the range and frequencies involved is quite large, any increase in the efficiency of the transducer would permit a more compact design or increase its output.

The improvement at the Naval Ordnance Laboratory[13] was accomplished by developing a technique to cut a strain-sensitive, domain-oriented 49 Co, 49 Fe, 2V magnetic core material known commercially as Supermendur. Up to this time it had been considered impossible to retain the excellent toroidal magnetic

characteristics after bonding and cutting by usual commercial techniques. The method developed consisted of the rigid encapsulation of a Supermendur core in an aluminum box and cutting it by electrolytic erosion. The resulting degraded cut surfaces were then carefully lapped until no further improvement of its magnetic characteristics occur. The magnetic evaluation of this cut core showed a 43% greater residual induction and a 15% greater maximum induction than the presently used silicon-iron core material. This is shown in Table IV.

## RECORDING HEAD MATERIALS

Since the introduction of magnetic tape recording, there has been a continuing search for an ideal magnetic transducer head material. Ideally, it should have a high initial permeability, high resistivity, and be physically hard. Tape heads wear because of the abrasive action of the moving tape upon the surface of the head. The intimacy of the contact between tape and head at the

TABLE IV. Magnetic properties of transducer core materials (60 cps).

| Material | $H_c$ (oe) | $B_r/B_m$ | $B_m$ (gauss) at 11 oe |
|---|---|---|---|
| Grain-oriented silicon iron tape wound core, uncut | 0.76 | 0.86 | 18 500 |
| Grain-oriented silicon iron tape wound core, cut | 0.70 | 0.49 | 18 500 |
| Supermendur tape wound core, uncut | 1.0 | 0.92 | 23 000 |
| Supermendur tape wound core, impregnated and cut | 0.81 | 0.46 | 19 200 |
| Supermendur tape wound core, aluminum boxes and cut | 0.68 | 0.61 | 21 200 |
| Supermendur tape wound core, polyurethane elastomer encapsulated, and cut | 0.84 | 0.49 | 21 800 |

[12] O. N. Al'tgauzen, N. A. Semenova, and A. N. Stepanova, Elektrichestvo 1, 51–55 (1961).
[13] J. F. Haben, Naval Weapons Rept. 7337, U. S. Naval Ordnance Laboratory, Silver Spring, Maryland (1960).

TABLE V. Comparison of various magnetic recording head materials.

| Material | Permeability | | Coercive force $H_c$ | Saturation-induction | Resistivity micro-ohm cm | Hardness annealed (Rockwell) |
|---|---|---|---|---|---|---|
| | $\mu_0$ | $\mu_{max}$ | | | | |
| Mumetal | 30 000 | 100 000 | 0.05 | 7000 | 62 | 35 $R_B$ |
| Permalloy | 25 000 | 100 000 | 0.03 | 9000 | 55 | 35 $R_B$ |
| Sinimax | 5000 | 40 000 | 0.10 | 11 000 | 55 | 76 $R_B$ |
| 16 Alfenol | 5000 | 80 000 | 0.04 | 8000 | 140 | 29 $R_C$ |
| Si-Al-Fe alloy | 6500 | 25 000 | 0.10 | 9000 | ~200 | 48 $R_C$ |
| Ferrite | 2000 | 4000 | 1.00 | 4000 | $>10^8$ | 54 $R_C$ |

gap surface determines the ultimate high frequency output from the head. For this reason pressure is generally applied to the tape to force it against the head gap, thereby insuring good contact and also increasing the wear.

Because of the physical design of the tape head, wear causes the gap width to continually increase. Since the width of the gap determines the ultimate high frequency response, this continual increase in gap width due to wear degrades the high frequency response of the head.

Another problem associated with tape recorder heads is the loss of applied signal due to eddy currents in the head material. These eddy currents are dependent upon the square of the frequency of the applied signal and the resistivity of the material comprising the head. In some cases involving high frequency erase, the eddy current losses are so great that, if ordinary head materials such as mumetal or Permalloy are used, not enough flux can be generated in the gap to erase the tape; multiple gap erasing is then resorted to.

Because standard composition Permalloy head materials are strain-sensitive, they must be carefully heat-treated and very carefully handled after heat treatment to avoid introducing strains which degrade the magnetic permeability severely.

Ferrites and aluminum-iron alloys are some of the newer materials recently introduced to solve some of the above mentioned deficiencies. Ferrites, however, being brittle, chip away at the gap edges thus degrading the resolving power at the gap. Recently Duinker[14] of Netherlands solved this deficiency by introducing a thin coating of glass as the spacer in the gap, thus cementing the ferrite. The aluminum-irons have a high hardness and resistivity but suffer from a low permeability. A summary of various transducer materials for recording heads is shown in Table V. Adams and Hubbard[8] suggest the use of a sintered silicon-aluminum-iron alloy of approximately Sendust composition as an excellent transducer recording head material. It possesses a small-grain size, high resistivity, fair initial permeability, low losses, and a hardness in excess of presently used materials. Its low loss characteristics are shown in Fig. 3. Tada[15] reports the use of an alloy of similar composition in video transducer heads manufactured in Japan. The excellent wear characteristic of this head as reported by Tada is shown in Fig. 4.

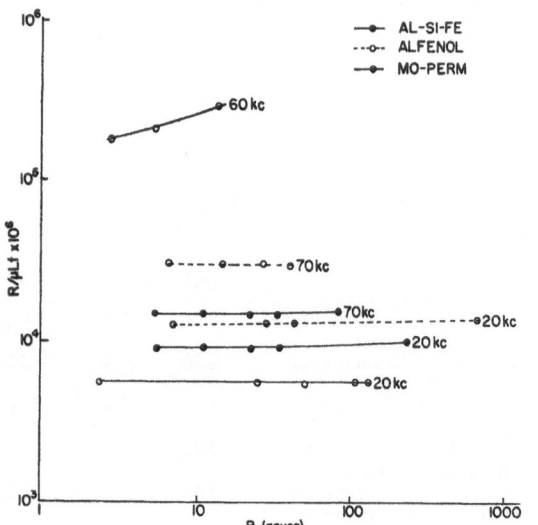

FIG. 3. Loss characteristics (ohms per henry×permeability and frequency) of standard recording heads at various flux levels.

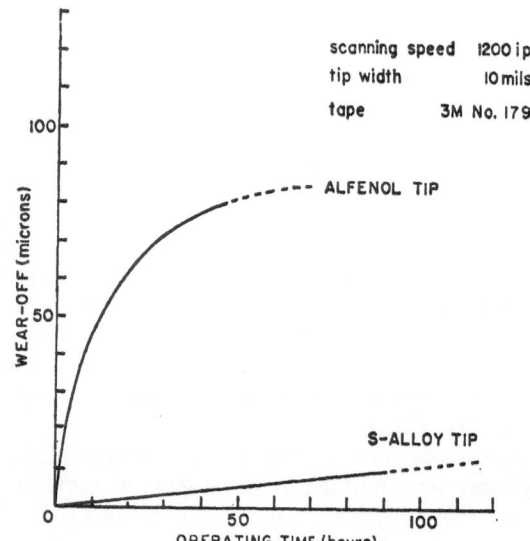

FIG. 4. Wear characteristics of a silicon-aluminum-iron alloy ($S$ alloy) compared with an aluminum-iron alloy (Alfenol) used in a video transducer head (after Tada).[15]

[14] S. Duinker, Philips Research Repts. **15**, 342–367 (1960).
[15] M. Tada, Tech. Rept., Sony Corporation No. 1, Japan (1961).

## NONMAGNETIC ALLOYS

It is just as difficult to find a completely nonmagnetic alloy with required physical characteristics as it is to find an ideal magnetic alloy. Recently the Naval Ordnance Laboratory reported[16] a new class of nonmagnetic alloys composed of nickel and titanium and based on the ductile intermetallic compound NiTi and associated phases NiTi$_2$ and Ni$_3$Ti. These alloys are nonmagnetic, extremely corrosion resistant, and Fig. 5 shows that their hardening capabilities are comparable to tool steel (Rockwell "C" 62).

Investigations into the TiNi alloy containing 54.5% Ni revealed some unusual mechanical vibration damping properties. At room temperature the alloy had a very

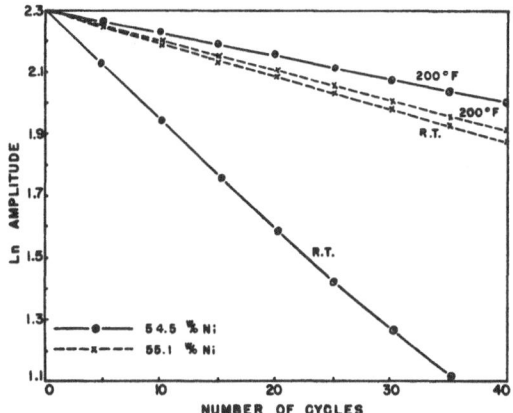

FIG. 6. Decay in torsional vibration of 54.5 and 55.1 weight% nickel-titanium alloys at room temperature and 200°F.

corrosion and abrasion resistance are a problem. They appear particularly useful as a material for nonmagnetic tools in mine disposal applications and in various components of magnetometers, mine-laying and servicing craft where the above characteristics are required. Their corrosion and abrasion resistance suggests their use in food and chemical processing industries. The marked changes in damping characteristics with temperature indicates a possible application of NiTi in temperature sensing devices.

## SUMMARY

The present trend of progress for the improvement of commercial soft magnetic alloys has been chiefly concerned with attaining reproducible and ideal magnetic characteristics. Modern melting techniques, such as consumable arc and vacuum melting, have upgraded present magnetic alloys to quality levels previously limited to laboratory specimens.

Greater progress has been made in developing specialized magnetic materials to solve a specific problem. Tremendous strides have been made towards the solution of magnetic material problems in the area of re-

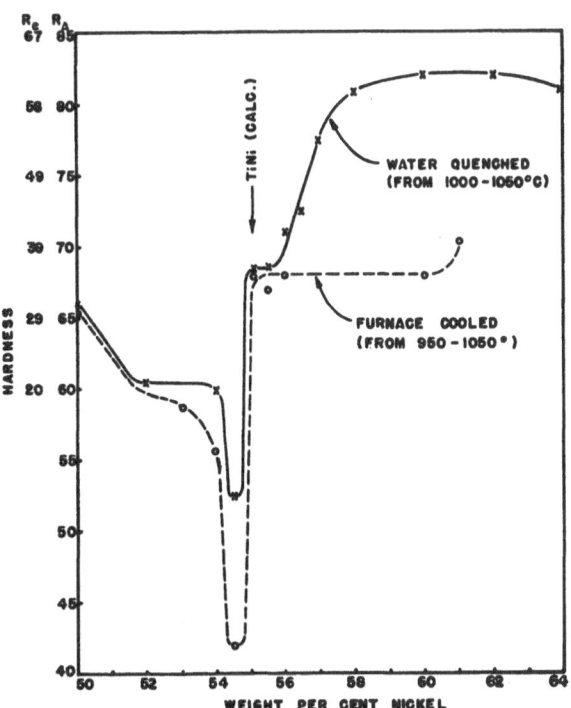

FIG. 5. Hardness as a function of nickel content for furnace-cooled and water-quenched NiTi alloys.

high damping capacity, upon heating to temperatures slightly in excess of room temperature the damping capacity markedly decreases. This phenomenon is illustrated by torsion studies in Fig. 6.

Magnetic susceptibility measurements over a wide temperature range as shown in Fig. 7 indicates that the TiNi base materials are paramagnetic with a permeability value approaching unity (<1.002). The permeability of this class of materials was also found to be stable with variations in working and heat treatment.

These alloys offer a potential solution for many troublesome nonmagnetic material applications where low permeability, strength, hardness, fabricability, and

16 W. J. Buehler, NOLTR 61–75, U. S. Naval Ordnance Laboratory, Silver Spring, Maryland (1961).

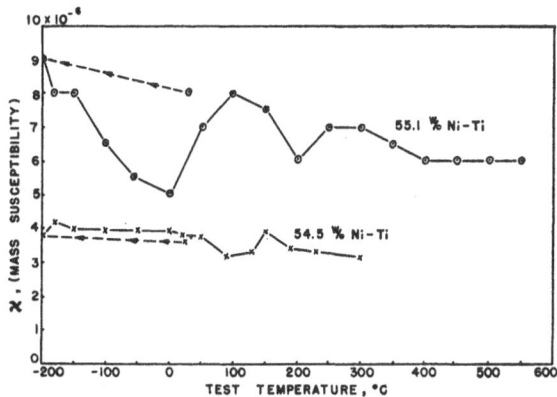

FIG. 7. Magnetic susceptibility of 54.5 and 55.1 weight% nickel-titanium alloys as a function of temperature.

cording heads, switching cores for computors, pulse transformers, etc. This suggests that further improvement in other areas can be expected by increased effort.

Since the variety of magnetic materials commercially available today is so great, one should study the magnetic properties carefully and select the proper material for the application. The use of the correct material may result in better performance or even prove more economical. In many cases design is more important than an improved magnetic material.

If commercially available, the use of newly developed magnetic materials should be encouraged consistent with economic and performance benefits. This stimulates larger volume for the suppliers and reduces future material costs.

### ACKNOWLEDGMENT

The author wishes to acknowledge the assistance of the staff members of the Magnetism and Metallurgy Division of this Laboratory who contributed to the preparation of this manuscript.

---

JOURNAL OF APPLIED PHYSICS　　　SUPPLEMENT TO VOL. 33, NO. 3　　　MARCH, 1962

# Textured 6.5% Silicon-Iron Alloy

G. Facaros
*Westinghouse Materials Manufacturing Department, Blairsville, Pennsylvania*

AND

R. G. Aspden
*Westinghouse Research Laboratories, Pittsburgh, Pennsylvania*

Sheet of an iron base alloy containing 6.5% silicon and second phase inclusions was prepared by hot rolling and pack rolling. Two structures were obtained by varying the rate of heating to the final annealing temperature. A slow rate of heating yielded {110} grains in a secondary recrystallization structure, and a fast rate, the primary recrystallization and normal grain growth structure. Only during the slow rate of heating were inclusions effective in suppressing normal grain growth and in promoting growth of {110} secondaries.

The secondary recrystallization texture was of the {110} ⟨001⟩ type. The normal grain growth texture had a similar degree of alignment of ⟨100⟩ directions with the rolling direction and a nearly random distribution of planes of the ⟨100⟩ zones parallel to the rolling plane. Similar dc and ac magnetic characteristics were observed for these two structures since they had the same degree of alignment of ⟨100⟩ directions (easy directions of magnetization) with the rolling direction.

## INTRODUCTION

IRON base alloys containing about 6.5% silicon have a potential application in magnetic devices as a result of low losses, high permeabilities and low magnetostriction.[1] The main factor limiting their commercial application has been the lack of ductility at room temperature. Ruder[2] measured the total ac losses of sheet from alloys containing up to 8% silicon and found a minimum near 6.5% silicon. Goertz[3] observed that magnetic field annealing increased the maximum permeability of alloys containing from 2 to 10% silicon. The highest maximum permeability was obtained for a 6.5% silicon single crystal slowly cooled with the magnetic field parallel to the ⟨100⟩ directions. Magnetic annealing of polycrystalline cast rings[3,4] and of sheet[5] improved the maximum permeability to a lesser extent.

In the above work on polycrystalline sheet no relationship between texture and magnetic characteristics was reported. When a material has a positive crystal anisotropy constant, one would expect the magnetic properties to be influenced by texture. Since this alloy has a positive anisotropy,[6] different textures were developed by normal grain growth and by secondary recrystallization. Magnetic characteristics of these textures were measured.

## EXPERIMENTAL PROCEDURE

An iron base alloy containing 6.49% silicon, 0.16% manganese, and 0.022% sulfur was prepared by air melting commercial grades of alloying elements. Manganese and sulfur additions were made since they have been found to promote the formation of the {110} ⟨001⟩ texture in iron-3% silicon alloys.[7,8]

The melt was poured into a square ingot mold. One hour after pouring, the ingot was stripped hot, placed in a furnace at 700°C, and heated slowly to 1000°C. Subsequently, the ingot was forged to a thickness of

[1] W. J. Carr in *Ferromagnetism*, edited by R. M. Bozorth (D. Van Nostrand Company, Inc., Princeton, New Jersey, 1951), p. 649.
[2] W. E. Ruder, Proc. Inst. Radio Engrs. **30**, 437 (1942).
[3] M. Goertz, J. Appl. Phys. **22**, 964 (1951).
[4] P. Albert, J. Appl. Phys. **29**, 351 (1958).
[5] J. F. Nachman and W. J. Buehler, NAVORD Report 4304, PB 121545 (May 10, 1956).
[6] L. P. Tarasov, Phys. Rev. **56**, 1231 (1939).
[7] J. E. May and D. Turnbull, Trans. Am. Inst. Mining, Met., Petrol. Engrs. **212**, 769 (1958).
[8] J. E. May and D. Turnbull, J. Appl. Phys. **30**, 210S (1959).

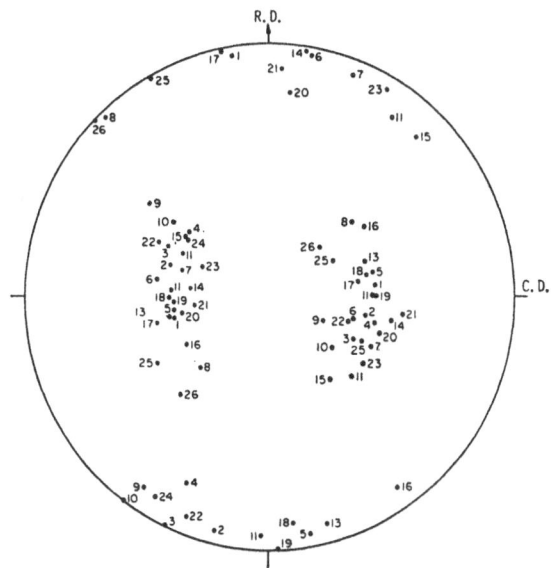

FIG. 1. {200} pole figure of grains in the secondary recrystallization structure. Rolling direction and cross direction are designated R. D. and C. D.

TABLE I. dc and ac magnetic properties.

| Structure | dc magnetic properties from 10 oe | | Core losses in w/lb —60 cycle | |
|---|---|---|---|---|
| | Residual induction, gauss | Coercive force, oe | 10 kgauss | 12 kgauss |
| Normal grain growth | 6750 | 0.092 | 0.38 | 0.61 |
| Secondary recrystallization | 4450 | 0.094 | 0.40 | 0.66 |

magnetic data were obtained by a ballistic method. Core loss values were measured by a wattmeter method.

Macroscopic examinations were made on samples etched at 70°C in a solution containing 241 g of ferric ammonium sulfate, 40 cc of sulfuric acid, and 1000 cc of water. The individual grains were large enough for orientation determinations by the conventional Laue back reflection method. The {200} poles of these grains were used to construct pole figures.

## RESULTS AND DISCUSSION

Strip was prepared with two structures: one resulting from primary recrystallization and normal grain growth and the other from {110} secondary recrystallization. A normal grain growth structure (grain diameter of 2 to 3 times sheet thickness) was obtained by rapid heating to the annealing temperature. The samples were inserted in a furnace at 1200°C and held at 1200°C for 16 hr in a reducing atmosphere. Secondary recrystallization (grain diameter of about 1 in.) was effected by inserting decarburized specimens in a furnace at 750°C, heating at 60°C per hr to 1100°C, holding 16 hr at 1100°C, heating at 25°C per hr to 1200°C, and holding 2 hr at 1200°C. Only in the latter anneal were manganese sulfide inclusions effective in suppressing normal grain growth and in promoting growth of {110} secondaries.

In the {200} pole figure, Fig. 1, a number is included beside the {200} poles of each x-rayed grain. The secondaries tended to have a {110} ⟨001⟩ type texture. The normal grain growth structure had a similar distribution of ⟨100⟩ directions parallel to the rolling direction and a near random distribution of planes of the ⟨100⟩ zones parallel to the rolling plane. The magnetic characteristics of these two structures are given in Fig. 2 and in Table I. Similar dc and ac properties were observed for both structures since they had the same degree of alignment of ⟨100⟩ directions (easy direction of magnetization) with the rolling direction.

## ACKNOWLEDGMENTS

We would like to acknowledge the helpful discussions with G. Wiener and E. C. Bishop.

3 in. reheated to 1000°C and hot rolled to a thickness of 0.140 in. The hot-rolled sheet was pack-rolled (as suggested by Nachman[6] et al.) at temperatures ranging from 750° to 350°C to a thickness of about 0.025 in., 86% reduction in thickness. The sheet was pickled or surface ground to remove oxides, and final sheet thickness was 0.022 in.

Epstein specimens ($1\frac{3}{16} \times 12$ in.) were sheared from the sheet at 400°C with the long axes parallel to the rolling direction. A variation in the textures was obtained by heat treatment. Magnetic measurements were made in an Epstein frame. Calculations were based on an assumed density of 7.48 g per cm³. Direct current

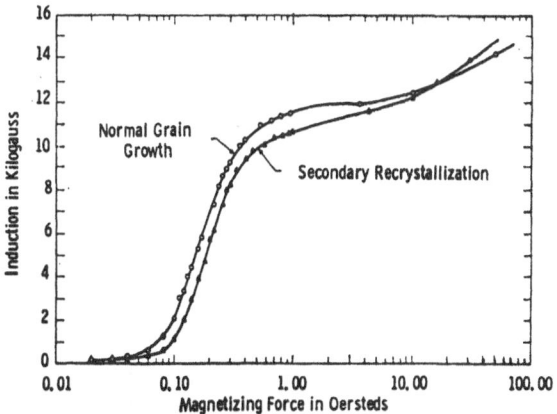

FIG. 2. dc magnetization curves for secondary recrystallization and normal grain growth structures.

JOURNAL OF APPLIED PHYSICS     SUPPLEMENT TO VOL. 33, NO. 3     MARCH, 1962

# Anisotropy in Rolled Permalloy Tape

E. M. GYORGY AND D. TREVES*

*Bell Telephone Laboratories, Incorporated, Murray Hill, New Jersey*

The anisotropy induced in gold-doped Permalloy by cold rolling and subsequent heat treatments has been investigated using a recording torque magnetometer. The material was cold rolled from 0.014 in. to 0.000125 in. and then heat treated for 2 hr in a dry hydrogen atmosphere. The torque curves obtained could be decomposed into a hard and soft component. For fields below 200 oe, the torque curves due to the hard component have a tangent-like shape with the peak a few degrees from the hard direction, while the torque curves due to the soft component have the regular shape corresponding to uniaxial magnetocrystalline anisotropy. Both components are uniaxial with the easy direction along the roll direction.

The hard component is independent of the heat treatments up to about 900°C and has an equivalent anisotropy field of the order of several hundred oersteds. The high apparent anisotropy of the hard component and its independence of the heat treatment suggest that this component arises from a magnetostatic effect. By etching the tape it has been shown that the hard component is due to the surface structure of the tape. The surface structure is probably the result of the rolling process.

The soft component is strongly dependent upon the heat treatment subsequent to the cold work and almost vanishes at an annealing temperature of about 650°C. This decrease is consistent with the ordered pair mechanism suggested by Chikazumi. The small residual soft component at high annealing temperatures is very sensitive to the rate of quenching. This behavior may indicate the presence of partial ordering and some degree of crystal orientation. The value of the anisotropy field due to the soft component is approximately equal to the measured value of the threshold for rapid rotational flux reversal.

## INTRODUCTION

THE precipitation of gold in Permalloy has made it possible to control the coercive force and, to some degree, the threshold for rotational flux reversal by appropriate heat treatments.[1] In order to gain further understanding of the magnetic properties of this alloy, the effect of heat treatments on the roll induced anisotropy and on the threshold for nonuniform rotational flux reversal has been investigated. The alloy investigated contained 75% Ni, 18% Fe, and 7% Au. At a thickness of 0.014 in. the material was heat treated at 900°C for 2 hr in dry hydrogen. The material was then cold-rolled to a $\frac{1}{8}$ mil thickness without further anneal. For the flux reversal studies, the tape was insulated and wrapped on a standard ceramic bobbin. Flat circular samples 18 mm in diameter were used for the torque measurements. Both types of samples were then heat treated for 2 hr in dry hydrogen. The effect of this final heat treatment will be the subject of this paper. The anisotropy was measured using a recording torque magnetometer,[2] and the thresholds for the rapid rotational flux reversal were obtained using standard magnetic flux switching techniques.[3]

The torque curves obtained as a function of the angle $\theta$ between the roll direction and the applied field could be decomposed into two parts. One, which we shall call the soft component, is dependent on the thermal history; the other, which we shall call the hard component, is independent of heat treatments up to 900°C. The presence of the two components is illustrated in Fig. 1 which gives the torque curves for samples annealed at different temperatures. The curves in Fig. 1 can be quite accurately described as the sum of the 900°C curve and a sin2$\theta$ component of the appropriate amplitude. The sin2$\theta$, or soft component, decreases with the temperature of anneal and is negligible in the 900°C curve. This curve may be taken to represent the hard component.

## HARD COMPONENT

The value of the anisotropy field of the hard component was evaluated by applying a field near the hard direction and observing the field at which the torque

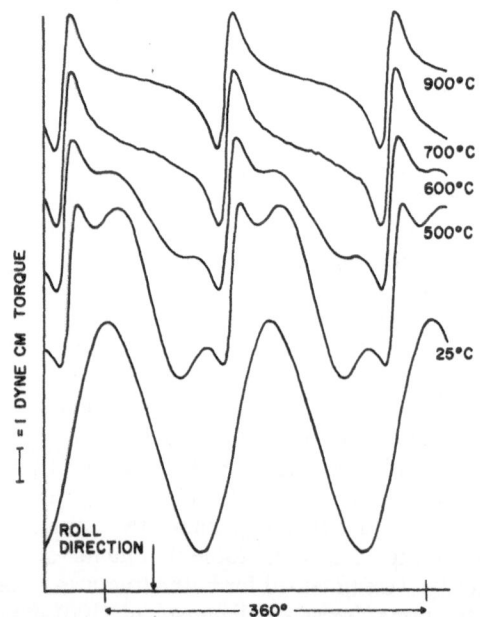

FIG. 1. Torque curves for samples annealed for 2 hr at various temperatures. For the 25°C curve, the sensitivity of the recorder has been decreased by a factor of 5. The weight of the samples was approximately 7 mg. The field was 200 oe for all curves and the rotational hysteresis was negligible. The curves are displaced vertically for presentation.

* On leave of absence from the Weizmann Institute of Science, Rehovot, Israel.

[1] E. A. Nesbitt and E. M. Gyorgy, J. Appl. Phys. **32**, 1305 (1961).

[2] G. T. Croft, F. J. Donahoe, and W. F. Love, Rev. Sci. Instr. **26**, 360 (1955).

[3] E. M. Gyorgy and F. B. Hagedorn, J. Appl. Phys. **30**, 1368 (1959).

reaches a maximum. In analogy to the Stoner-Wohlfarth model,[4] this field (350 oe) gives a measure of the anisotropy of the hard component.

At low fields the amplitude of the hard component torque increases linearly with the applied field. Above 200 oe the peak begins to decrease slightly and the shape approaches a $\sin 2\theta$ curve. The independence of the hard component on the annealing temperature points to the fact that it is due to some kind of shape anisotropy. The high value of the equivalent anisotropy field and the slight decrease in peak torque observed at high fields also suggest this conclusion. The shape anisotropy is thought to arise from imperfections on the surface of the tape. A microscopic observation of the samples revealed striations parallel to the roll (easy) direction and of the order of $1~\mu$ in width. In order to verify that the hard component was really due to the surface structure, the samples were etched to 60% of their original weight, which is equivalent to removing less than $1~\mu$ per side. Any kind of etch will not appreciably alter the surface structure unless the layer removed is at least several times thicker that the surface inhomogeneities. Furthermore, microscopic observation did not reveal any change in the appearance of the surface after the etch. After the etch, the torque due to the hard component is unchanged while the reduction of the torque due to the soft component is approximately proportional to the weight loss. The fact that the soft component is reduced while the hard component is practically independent of the weight loss during etching indicates that the hard component is due to the shape anisotropy connected with the surface structure.

The rapid rise with $\theta$ of the hard component torque is inconsistent with the Stoner-Wohlfarth model. The observed shape can be somewhat better explained if one considers closure domains, or spikes, of the Néel type[5] around the striations. In the spikes, the magnetization is parallel to the striations. At low fields the torque, due to these closure domains, will be given by

$$T = HMV \sin\alpha,$$

where $M$ is the saturation magnetization, $H$ is the applied field, $\alpha$ is the angle between the applied field and the roll direction, and $V$ is the volume of the closure domains. The volume of these domains is obviously zero for $\alpha=0$ and increases with increasing $\alpha$ yielding a torque which rises faster than $\sin\alpha$. This more rapid increase of the torque with $\alpha$ is a better approximation of the experimental low field curves than a usual torque curve obtained from crystalline anisotropy.

## SOFT COMPONENT

The soft component is, as was mentioned earlier, dependent on the heat treatment. The value of the anisotropy field $H_k$ due to the soft component decreases monotonically from 60 oe for no anneal to a value between zero and 1.5 oe for the 900°C anneal. This spread in values for the 900°C anneal is the result of different rates of cooling. In fact, for all annealing temperatures greater than 550°C, the value of $H_k$ for samples slowly cooled is approximately 1.5 oe higher than the value obtained for sample quenched in cold hydrogen. We may interpret this difference if we assume that the anisotropy of the quenched samples is due primarily to the directional ordering mechanism suggested by Chikazumi.[6] For temperatures above 550°C quenched Permalloy is in the disordered state and has a very low crystalline anisotropy.[7] As a result, the contribution to $H_k$ from oriented crystallites will be negligible. It is experimentally found that samples quenched from above 650°C have a negligible soft component.

The slowly cooled samples may have some degree of ordering, and as a result, a substantially higher crystalline anisotropy.[7] The observed [100] fiber axis perpendicular to the roll direction and in the plane of the tape[8] will then lead to an additional uniaxial component along the roll direction. The anisotropy for the samples slowly cooled from above 650°C is then primarily due to the crystalline anisotropy.

In thin evaporated Permalloy films the threshold for nonuniform rotational flux reversal is equal to the measured uniaxial anisotropy.[9-11] This agreement between the threshold for rotational flux reversal ($H_0$) and the anisotropy field of the soft component ($H_k$) is also observed for the rolled materials studied here. Experimental limitations restricted this comparison to values of $H_0$ less than about 7 oe. Since the hard component is due to surface imperfections, it presumably does not have a significant effect on the reversal process of the bulk of the material.

[4] E. C. Stoner and E. P. Wohlfarth, Trans. Roy. Soc. (London) A240, 599 (1948).

[5] L. Néel, Cahiers phys. 25, 21 (1944).

[6] S. Chikazumi, J. Appl. Phys. 29S, 346 (1958).

[7] R. Bozorth, Ferromagnetism (D. Van Nostrand Company, Inc., Princeton, New Jersey, 1959), p. 571.

[8] We wish to thank Mrs. M. H. Read for the x-ray analysis.

[9] D. Olson and A. V. Pohm, J. Appl. Phys. 29, 274 (1958).

[10] F. B. Humphrey and E. M. Gyorgy, J. Appl. Phys. 30, 935 (1959).

[11] E. M. Gyorgy, J. Appl. Phys. 31, 110S (1960).

# Small Angle X-Ray Scattering from MnS in Silicon Steels*

Bani R. Banerjee, R. E. Lenhart, and W. H. Robinson†

*Research Division, Crucible Steel Company of America, Pittsburgh 13, Pennsylvania*

In silicon steels, a dispersed MnS impurity phase is believed to pin grain boundaries, inhibit grain growth, and permit the desired secondary recrystallization for optimum orientation, grain size, and magnetic properties of the finished material. However, no experimental evidence has yet been obtained to establish the size distribution of this impurity phase. Therefore, the size distribution of such a precipitate was explored by a small-angle x-ray scattering technique, with a spectrometer using a Johansson-focusing monochromator; spurious surface effects were carefully avoided. Laboratory heats of 3.2% silicon steel compositions with varying sulfur, were prepared by vacuum induction melting of high-purity components; these materials were carefully processed to 0.0005-in. thick foils. Using a sulfur-free sample as reference standard for residual surface scattering, a plot of $\ln \tau (2\theta)$ vs $(2\theta)^2$ showed the presumed MnS particles to be 20 A to 150 A in diameter. These results agreed with preliminary observations by transmission-electron microscopy.

* Abstract only; full version to be submitted to the Journal of Applied Physics.
† Metals Research Laboratory, Carnegie Institute of Technology, Pittsburgh 13, Pennsylvania.

---

# A Method of Magnetic Annealing of Vanadium Permendur-Type Alloys

R. E. Burket* and D. M. Stewart

*Allegheny Ludlum Steel Corporation Research Center, Brackenridge, Pennsylvania*

A method of magnetic annealing of 2 V Permendur (2% V, 49% Co, 49% Fe) stamped rings and laminations in zero externally applied magnetic field is discussed. It is shown that an internal magnetic field biasing technique yields highly improved magnetic properties.

This method involves magnetizing the samples when they are still in a hard condition, i.e., as stamped or low temperature (500° to 600°C) annealed, and then subsequently annealing them at the normal time and temperature employed for the development of optimum magnetic properties (800–900°C). No external magnetic field is applied during any of the anneals.

This technique utilizes the internal magnetic field which remains after magnetizing the samples at room temperature. It is shown that a marked improvement may be obtained in 1-DU type laminations although the samples are unstacked after the initial magnetization (before annealing). Apparently, although the samples are partially self-demagnetized, there is still a sufficiently high internal field remaining to produce a self-magnetic annealing effect during the anneal. It is also shown that the cooling rate employed in this anneal is important—just as it is in conventional magnetic annealing.

Normal dc and 60 cycle magnetic properties are presented for 0.014-in. ring and 1-DU samples from a number of heats. The data indicate that the effect is manifested by a two to threefold increase in the dc $\mu_{max}$ and a 35 to 50% increase in the rectangularity of the dc hysteresis loop ($Br/Bm$). It does not necessarily cause a decrease in the dc coercive force or the 60-cycle watt loss.

I T is a well-established fact that for certain iron-nickel, iron-silicon, iron-cobalt, and iron-cobalt-nickel alloys annealing in the presence of an externally applied magnetic field can cause marked improvements in certain properties.

One of these alloys, a high purity 2% V, 49% Co, 49% Fe known as Supermendur[1] or its less pure version known as vanadium Permendur, is probably the best known soft magnetic alloy which illustrates the beneficial effects.

After being processed to 0.014 in. in the normal manner for this alloy, vanadium Permendur exhibits the magnetic properties which are indicated in Tables I and II.

This same alloy can be annealed to develop a high coercive force, high $Br$ hysteresis loop suitable for use in self-biasing, magnetostriction type transducers as illustrated by the data in Table III.

It was thought that it may be possible to take advantage of this self-biased conditioned of the material as illustrated by the high $Br$, high $Hc$ and obtain an effect similar to a magnetic anneal by a sequence of annealing, magnetizing and reannealing. It was thought that if the material were annealed at 1022°F, magnetized and left at remanence, and then reannealed at 1350° to 1550°F that during the second anneal the residual induction in the strip may act in a manner similar to an externally applied field. It was believed that this effect would be possible in vanadium Permendur because the normal annealing temperatures are 150° to 250°F below the Curie temperature.

To check this hypothesis two of the samples which have been annealed at 1022°F as previously reported were magnetized and left at remanence. Then they were reannealed at 1550°F and both fast cooled by withdrawing the annealing box at temperature and slow cooled to 1100°F before withdrawing. Previously

* Present address: Westinghouse Electric Corporation, Blairsville, Pennsylvania. Previously employed at Allegheny Ludlum Steel Corporation Research Center.

[1] H. L. B. Gould and D. H. Wenny, in the Proceedings Conference on Magnetism and Magnetic Materials, AIEE, 1956, p. 675 or Elec. Eng. **76**, 208 (March, 1957).

unannealed samples were annealed with the reannealed samples as cross checks. The magnetic results on these samples are tabulated in Tables IV and V.

Both of the reannealed samples had improved squareness ratios and greatly improved maximum permeabilities. The controlled cooling rate caused a greater increase in the $\mu_{max}$ and lowered the $Hc$ to the vicinity of that obtained on the regular magnetic annealed samples.

Subsequent to this work additional heats and other sample configurations were studied. Many applications of this type of material require the use of certain types of laminations which have air gaps in the circuit. Since air gaps have the effect of shearing hysteresis loops, it was believed that although the technique appeared to work very well in gapless structures, that it may not work in laminated structures containing gaps.

To check the effect on laminations, 0.014-in. 1-DU laminations were stamped from eight heats and the samples were treated in the same manner as stamped Rowland rings. The 1-DU laminations are double backed laminations and are usually stacked in a 1×1 interleaved fashion which to a great extent eliminates the gap effect. The properties obtained on these stacks then are often equivalent to those obtained on rings or toroidal cores.

In this particular instance, however, the 1-DU's were annealed at 1022°F, stacked, tested, separated, restacked, reannealed at 1550°F, restacked, and tested.

TABLE I. 0.014-in. vanadium Permendur rings annealed at 1550°F for 4 hr, furnace cooled to 1100°F, and withdrawn.

| | | dc | | | |
|---|---|---|---|---|---|
| | | $Br$ from | $Hc$ from | | |
| $B$ at $10H$ | $B$ at $100H$ | $100H$ | $100H$ | $\mu_{max}$ | $Br/B_{100}$ |
| 20 800 | 23 000 | 9700 | 0.45 | ... | 0.42 |

TABLE II. 0.014-in. vanadium Permendur rings annealed at 1550°F for 4 hr in PdH$_2$ in the presence of an externally applied circumferential magnetic field of 5 to 10 oe, furnace cooled to 1100°F in the presence of the field, withdrawn and cooled to room temperature in the presence of the magnetic field.

| | | dc | | | |
|---|---|---|---|---|---|
| | | $Br$ from | $Hc$ from | | |
| $B$ at $10H$ | $B$ at $100H$ | $100H$ | $100H$ | $\mu_{max}$ | $Br/B_{100}$ |
| 21 800 | 23 100 | 18 700 | 0.205 | 45 000 | 0.81 |
| 21 400 | 23 000 | 19 400 | 0.192 | 60 000 | 0.85 |

TABLE III. Annealed at 1022°F for 4 hr in pure dry hydrogen and withdrawn from the furnace (0.014-in. stamped rings).

| | | dc | | | |
|---|---|---|---|---|---|
| | | $Br$ from | $Hc$ from | | |
| $B$ at $10H$ | $B$ at $100H$ | $100H$ | $100H$ | $\mu_{max}$ | $Br/B_{100H}$ |
| 1650 | 21 400 | 18 100 | 20.6 | 735 | 0.845 |
| 1700 | 21 400 | 17 900 | 20.1 | 717 | 0.837 |
| 1650 | 21 300 | 17 900 | 20.2 | 726 | 0.837 |

TABLE IV. Annealed at 1550°F 4 hr and fast cooled.

| | | | dc | | | |
|---|---|---|---|---|---|---|
| | | $B$ at | $Br$ from | | | |
| Condition | $B$ at $10H$ | $100H$ | $100H$ | $Hc$ | $Br/Bm$ | $\mu_{max}$ |
| Reannealed | 21 100 | 23 000 | 12 600 | 0.412 | 0.55 | 24 400 |
| Not previously annealed | 21 050 | 23 100 | 8600 | 0.460 | 0.37 | 10 300 |

TABLE V. Annealed at 1550°F 4 hr, furnace cooled to 1100°F, and withdrawn.

| | | | dc | | | |
|---|---|---|---|---|---|---|
| | | $B$ at | | $Hc$ from | | |
| Condition | $B$ at $10H$ | $100H$ | $Br$ | $100H$ | $Br/Bm$ | $\mu_{max}$ |
| Reannealed | 20 700 | 23 100 | 14 500 | 0.276 | 0.63 | 34 900 |
| Not previously annealed | 21 800 | 23 000 | 9700 | 0.452 | 0.42 | 10 800 |

TABLE VI. Average magnetic properties on eight heats of 0.014-in. vanadium Permendur.

| Annealed at 1550°F, 4 hr, cooled to 1100°F, and withdrawn | | | | | |
|---|---|---|---|---|---|
| | | dc | | | 60~wpp |
| Sample | $\mu_{max}$ | $B$ at $100H$ | $Br/B_{100}$ | $Hc$ | at 20kb |
| Ring | 10 090 | 22 950 | 0.50 | 0.68 | 2.18 |
| 1-DU | 11 375 | 22 890 | 0.45 | 0.49 | 1.43 |
| Annealed at 1022°F, magnetized, reannealed at 1550°F, slow cooled to 1100°F, and withdrawn | | | | | |
| Ring | 23 550 | 22 830 | 0.77 | 0.63 | 2.88 |
| 1-DU | 25 460 | 22 900 | 0.74 | 0.43 | 1.98 |

It was thought that taking the structure apart after the initial magnetizing may permit demagnetization to occur to the extent that the residual induction would not be effective in subsequent annealing.

The data presented in Table VI indicate that 1-DU laminations can effectively be treated in the same manner as rings and that apparently demagnetization of the separated laminations does not preclude the use of the residual induction to effect magnetic annealing.

These average data indicate that the dc $\mu_{max}$ is very markedly increased and that the coercive force is not necessarily changed as a result of the self-bias magnetic anneal treatment.

Some additional work has been done on magnetizing as cold-rolled (as stamped) samples and subsequently annealing them with the 1550°F slow cool cycle and it has been established that this technique also results in improvements similar to those obtained by the anneal, bias, anneal technique.

It is believed that the data presented illustrate that the residual induction in samples previously magnetized can be effectively used to magnetically anneal material. It is believed that this technique may have application in cases in which it is not practical or possible to magnetic anneal in the conventional manner.

JOURNAL OF APPLIED PHYSICS · SUPPLEMENT TO VOL. 33, NO. 3 · MARCH, 1962

# Magnetic Anisotropy of the Demagnetized State

Jean-Claude Barbier and Bernadette Ferlin-Guion

*Laboratoire d'Electrostatique et de Physique du Métal, Grenoble, France*

An isotropic polycrystalline substance, demagnetized in a slowly decreasing alternating field, is in general magnetically anisotropic. The magnetization curve $J(H)$ in a given direction will depend on the angle between $H$ and the direction of demagnetization. We have compared the two magnetization curves for $\varphi = 0$ and $\varphi = \pi/2$. Their difference is a maximum for a field of 0.75 $H_c$, and can be more than 60% of $J$. This effect does not exist in substances with uniaxial magnetic anisotropy like hexagonal Co, or in samples with large internal stresses. We have also investigated the variation of the anisotropy of the demagnetized state of a sample which was made progressively more uniaxial by an increasing stress. Furthermore, if a complete demagnetization is followed by a partial demagnetization in the perpendicular direction, this partial demagnetization only modifies the initial state if a certain critical field is exceeded. We have determined some characteristics of this critical field.

IN many magnetic studies, a reference state results from demagnetization by an alternating field of constant direction and slowly decreasing amplitude. In a study of transverse susceptibilities, Covo[1] has shown that this reproducible reference state, even if it is well characterized by a zero resultant magnetization, is nevertheless magnetically anisotropic. This means that the magnetization curve $J(H)$ for a given direction depends upon the angle between that direction and the direction of the demagnetizing field. This phenomenon is often undetected since, for reasons of experimental expediency, it is customary to measure magnetization curves in the same direction as that of the demagnetizing field. We have studied this anisotropy of demagnetization in polycrystalline samples in which other sources of anisotropy are absent.

## DEMONSTRATION OF THE ANISOTROPY OF DEMAGNETIZATION

We have compared the magnetization curves $J(H)$ for angles $\varphi = 0$ and $\varphi = \pi/2$ between the direction of measurement of the magnetization curve and the direction of the demagnetizing field. To distinguish the different cases of demagnetization we shall denote by $J_k$ the magnetization obtained in a measuring field $H_m$, where $k$ represents the particular demagnetization process. Thus we shall denote by:

$k = x$ the demagnetization for $\varphi = 0$,
$k = y$ the demagnetization for $\varphi = \pi/2$,
$k = x + y$ a demagnetization of type $x$ followed, before the measurement, by a demagnetization of type $y$.
$k = y + x$ a demagnetization of type $y$ followed by a demagnetization of type $x$.

We assume for the present that the demagnetizations are complete, that is to say, that the maximum demagnetizing field is much higher than the coercive force of the sample.

In Fig. 1 we have shown results for a sample of Fe–Ni, annealed to relieve strains, and characterized by a coercive force $H_c = 0.27$ oe. In weak magnetic fields the two magnetization curves $J_x(H)$ and $J_y(H)$ are significantly different but they tend to coincide with each other as saturation is approached. The ratio $J_x/J_y$ changes with $H$ and goes through a maximum for $H = 0.75 H_c$. The maximum value of the ratio is approximately 1.6. The influence of the direction of the demagnetizing field is thus very important.

## RELATION BETWEEN THE ANISOTROPY OF DEMAGNETIZATION AND THE STRUCTURE OF THE SAMPLE

We have observed this anisotropy of demagnetization in many cubic substances which have three easy axes of magnetization. It may be supposed that in the course of the demagnetization these axes are not equivalent and that their respective orientations with respect to the demagnetizing field are of importance in the establishment of the initial domain structure. We are now studying the different distributions of domains

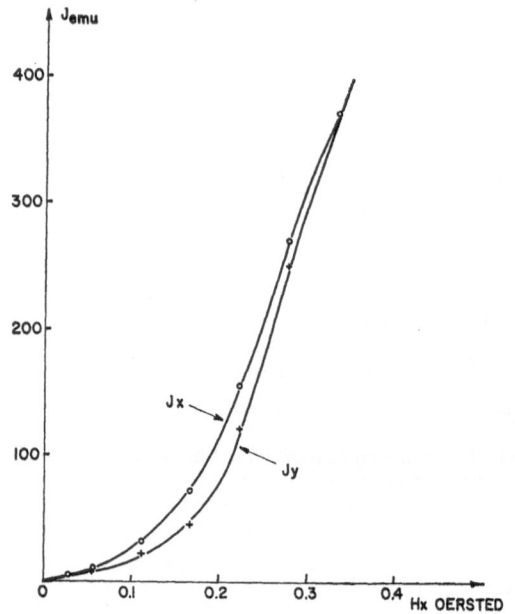

Fig. 1. First magnetization $J_x$ and $J_y$ versus field for Fe–Ni wire.

[1] J. Covo, Thèse de 3ème cycle, Grenoble (October, 1960).

associated with different types of demagnetizations.

If one chooses a uniaxial substance, such as hexagonal cobalt, instead of a substance of cubic structure, the direction of demagnetization should no longer have any influence. This had been verified experimentally. We have found that $J_x/J_y$ was equal to unity within the precision of the measurements.

We have as well made measurements on an as-extruded, unannealed sample of Fe-Ni in which there were internal strains. The anisotropy of demagnetization disappeared. The internal strains created favored directions of the magnetization and the influence of the demagnetizing field had disappeared.

We have rerun the annealed sample of Fe–Ni, without strain, and we have progressively introduced in it uniaxial character by submitting it to progressively increasing stress. The coercive force varies with increasing stress but as long as the anisotropy of demagnetization persists, the ratio $J_x/J_y$ is a maximum for $H_m$ of the order of 0.75 $H_c$. In order to study the variation of the ratio $J_x/J_y$ with the applied stress, we worked at $H_m/H_c$ equal to a constant, where $H_m$ designates the measuring field for $J_x$ and $J_y$. For small stresses (elastic range) $J_x/J_y$ decreases linearly with $\sigma$. Because of plastic deformation it is difficult to study the disappearance of the anisotropy of demagnetization for high stresses.

### THRESHOLD FIELD OF DEMAGNETIZATION

We have considered up to now total demagnetizations in which the maximum demagnetizing field is well above the coercive force. We have noted that in such a case the demagnetized state $y+x$ is equivalent to the state $x$.

To analyze the process of reorientation of the demagnetized state we have made a total demagnetization of the type $y$ followed by a partial demagnetization of the type $x$, the latter by application of an alternating field of decreasing amplitude, according to the classical technique, but starting from an initial maximum amplitude $H_{Dx}$ which may be varied. In varying $H_{Dx}$ we have been able to trace the curve $J_{y+x} = f(H_{Dx})$.

According to what we have said $J_{y+x} = J_x$ when $H_{Dx}$ becomes very large as in the case of total demagnetization of the type $x$. In contrast we have observed that $J_{y+x}$ remains equal to $J_y$ as long as $H_{Dx}$ is below a threshold field $H_S$. In other words, for $H_{Dx} < H_S$ the demagnetization of type $x$ does not modify the preceding demagnetization of type $y$.

Having characterized the anisotropy of the demagnetization by the value of the ratio $J_x/J_y$ as a function of the measuring field $H_m$ we have studied the variation of the threshold field $H_S$ with $H_m$. For a measuring field smaller than 0.22 oe (sample of coercive force $H_c = 0.27$ oe) we have noticed that $H_S$ varies proportionally with $H_m$. If this variation is essentially linear it appears to us that when $H_m$ tends toward zero, $H_S$ tends toward a

value of the order of 0.025 oe for the sample cited above. For $H_m = 0.11$ oe, $H_S$ is little different from 0.09 oe.

We have also noticed that if one changes the state of stress of the sample, the threshold field remains constant, for a fixed value of the measuring field, and it is therefore independent of the state of stress.

### MODIFICATION OF THE DEMAGNETIZED STATE BY FACTORS OTHER THAN THE APPLIED FIELD

Yamada[2] has shown that for certain ferromagnetic samples one obtains different magnetizations, in a given applied field, if the temperature is changed between the demagnetization and the application of the measuring field.

We have inquired whether a change of the state of stress between the demagnetization and the measurement could produce a change in the magnetization for a given applied field. Lliboutry[3] has already noted that if, after having demagnetized under zero stress, there is applied a stress, and then the field, the magnetization obtained is always greater than that obtained when the field is applied to the sample without stress. Moreover, the magnetization increases with tension or compression.

In a more general way we have obtained curves $J(H)$ for which the demagnetizations were carried out in a state of stress $\sigma$ and the measurements made in a state of stress $\sigma'$. We have changed the stress from $\sigma$ to $\sigma'$ after the demagnetization. We observed that for each value of the applied field, the smallest value of the magneti-

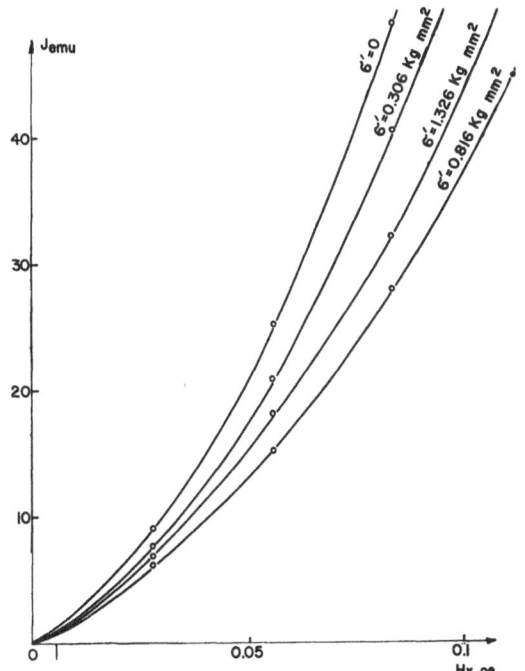

FIG. 2. Magnetization at stress $\sigma' = 0.816$ kg/mm². Demagnetization at various stresses $\sigma'$. $0 \leqslant \sigma' \leqslant 1.32$ kg/mm².

[2] O. Yamada, Compt. rend. **251**, 860 (1960).
[3] L. Lliboutry, Ann. Phys. **6**, 731 (1951).

zation corresponded to $\sigma = \sigma'$. For $\sigma < \sigma'$ or $\sigma > \sigma'$ a larger magnetization was observed. Figure 2 gives an indication of the results obtained. In comparing these results with those of Yamada, we have been led to suppose that

there is a great similarity between the effect due to a change in temperature and that due to a change in the state of stress. Yamada has undertaken a detailed study of this similarity between the two effects.

---

JOURNAL OF APPLIED PHYSICS          SUPPLEMENT TO VOL. 33, NO. 3          MARCH, 1962

# Grain-Size Effects in Oriented 48% Nickel-Iron Cores at 400 Cycles

M. F. LITTMANN, E. S. HARRIS, AND C. E. WARD

*Research Center, Armco Steel Corporation, Middletown, Ohio*

The constant-current, flux-reset test at 400 cycles has been used widely to evaluate core material for magnetic amplifier applications. An important criterion by this test is the "gain" or ratio of flux change to change in dc control magnetizing force. It is well known that gain is related to the squareness ratio $Br/Bm$ of the hysteresis loop.

This study shows that gain is also very sensitive to the grain structure as well as the crystal orientation of the core material. Material having a homogeneous structure has high gain, whereas if *partial* secondary grain growth has occurred, resulting in a large difference in grain size for the material, the gain can be reduced by a factor 1.5 to 3 times. The effect was artificially simulated by combining two cores of different grain size. Individually, the cores had high gain but together they had very low gain. The cause is ascribed to the difference in coercive force between the large and small grains. The conclusion is reached that high gain is obtainable from material consisting entirely of either large or small grains of proper crystal orientation.

## INTRODUCTION

IN a previous paper,[1] the relation of dc magnetic properties to dynamic tests related to magnetic-amplifier performance for oriented 48% nickel-iron was studied. Since then the CCFR test has gained wide usage as a method of grading cores and evaluating core material.

Knowledge of the structural changes in core material and their relation to the CCFR tests has been very helpful in making better materials. For example, when crystal orientation changes take place during annealing which destroy the cubic or (100) [001] texture, there is an accompanying decrease in squareness ratio and coercive force together with deterioration in "gain." On the other hand, it was found possible to obtain cores with only a little less squareness when annealed at higher temperatures to obtain lower coercive force. However, the "gain" characteristics could be worse for some of these cores with high squareness and low coercive force than for cores with unacceptable squareness.

## EXPERIMENTAL PROCEDURE AND RESULTS

To learn the cause for this behavior, studies of dc magnetic properties, grain structure, and 400-cycle CCFR tests were made on $1\text{-}\times 1\frac{1}{4}\text{-}\times\frac{1}{4}$-in. cores prepared from 2- and 4-mil 48 Orthonik. Dc measurements were in accord with ASTM method A341-55 covering ballistic tests for normal induction and hysteresis properties. The CCFR tests followed the procedure of

AIEE Standard No. 432 for oriented 50% nickel-iron alloys.[2]

Detailed investigation of the full range of annealing

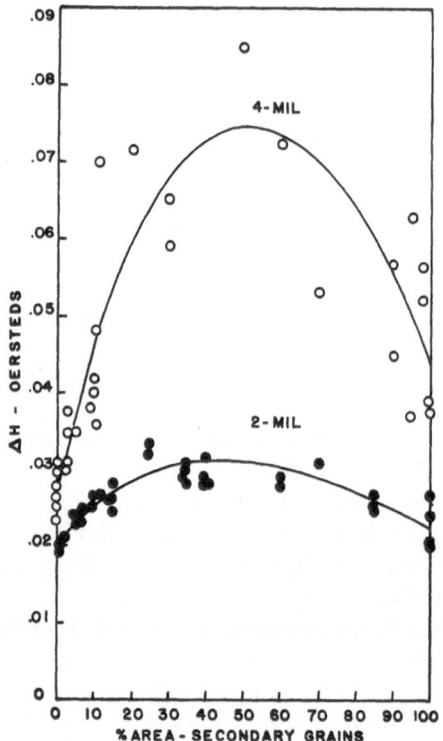

FIG. 1. Effect of grain structure on $\Delta H$ at 400 cycles by CCFR test.

---

[1] C. E. Ward and M. F. Littmann, Trans. Am. Inst. Elec. Engrs. **74**, Part I, 422 (1955).

[2] AIEE Magnetic Amplifier Comittee Report, Trans. Am. Inst. Elec. Engrs. **77**, Part I, 524 (1959).

temperature for certain lots of core material showed that the value of $\Delta H$,[3] which is inversely proportional to gain, appeared to go through a maximum and then return to a low value similar to that for the lower anneal temperatures. In such cases, we found there was a change in macro-grain structure which could be a possible cause. It was observed that the high $\Delta H$ (low gain) was accompanied by a heterogeneous structure consisting of a mixture of primary (100) [001] grains with a grain size of about ASTM 8 at 100× and (120) [001] secondary grains with a grain size of about ASTM 5 to 8 at 1×.

Figure 1 shows the relation between $\Delta H$ and the extent of secondary growth. For both 2- and 4-mil cores it appears that the greatest change in gain corresponds to about 40 to 50% secondary growth, although the shape of the curves is distinctly different as will be discussed later.

The reason for the maximum is believed to lie in the different ease of magnetization for the large vs the small grains, the smaller grains having higher coercive force. To test this hypothesis, a simple experiment was performed using two 2-mil cores. One core contained only large secondary grains with low coercive force and the other core small primary grains with relatively high coercive force. Both cores had a squareness ratio of 0.940 or higher. Tested separately by CCFR at 400 cycles, each core exhibited low $\Delta H$ values. When stacked and retested as one core with a single set of windings, the squareness value assumed the average value of the individual cores, but the $\Delta H$ increased by a factor of 3 as shown in Table I.

The test is a clear demonstration of the effect of the duplex grain structure on magnetic properties. In studying the results in greater detail, it appears that the size of the secondary grains also affects the results as does also the amount of twinning present in the secondary grains. The effect of the twins is considered to be a texture effect rather than grain size.

It is thought that the stronger effect of grain structure, in the case of the 4-mil cores compared to 2-mil cores, is a combined result of the larger secondary grains together with a higher proportion of twins. Also, the $\Delta H$ for the 4-mil material returned to a higher value at 100% growth (compared to 0%) than the 2-mil mate-

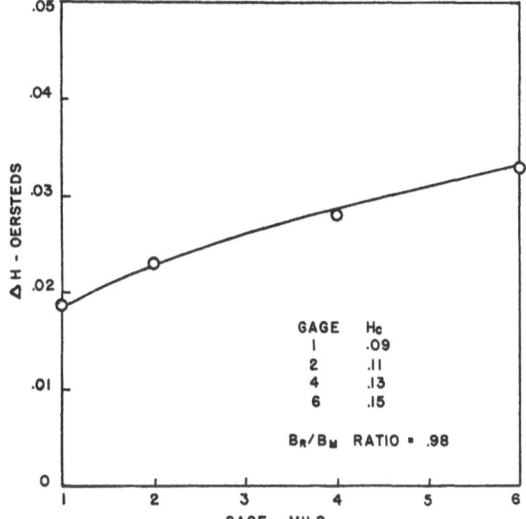

FIG. 2. Effect of thickness on $\Delta H$ at 400 cycles by CCFR test for (100) [001] primary grain structure.

rial because the squareness ratio was reduced more by the presence of annealing twins with undesirable orientation.

From the standpoint of materials, it is advantageous either to retain a fine-grained primary structure with the cubic texture or to achieve *complete* (120) [001] secondary growth free of twins. The latter texture would be useful, of course, only for tape-wound cores where the favorable [001] crystallographic direction can be utilized. For the transverse direction, the squareness ratio would be poor because the [001] direction is about 20° from the applied field.

The necessity of retaining a primary grain structure relatively free of secondary growth tends to place a lower limit on the coercive force attainable. In practice, this coercive force tends to decrease with decreasing thickness. Because of eddy current effects noted earlier by Lord,[4] and to a much smaller extent, because of lower coercive force, the gain obtainable from 48 Orthonik improves with decreasing thickness as shown in Fig. 2.

## SUMMARY

The gain characteristic of oriented 48% nickel-iron cores by CCFR tests at 400 cycles is strongly influenced by grain structure. For high gain (low $\Delta H$), the grain structure should be homogeneous and may consist either of very small (100) [001] primary grains or large (120) [001] secondary grains free of annealing twins. Mixtures of large and small grains cause low gain because of the difference in hysteresis characteristics between the large and small grains.

TABLE I. Effect of combining cores $-400\sim$ CCFR tests.

| Core | $Br/Bm$ | $H_1$/oe | $\Delta H$/oe | Grain structure | dc $H_c$/oe |
|------|---------|----------|---------------|-----------------|-------------|
| 2510 | 0.937 | 0.100 | 0.018 | 100% secondary grains | 0.028 |
| 2508 | 0.977 | 0.215 | 0.024 | 100% primary grains | 0.139 |
| Combined | 0.950 | 0.117 | 0.068 | ... | 0.108 |

[3] $\Delta H$ is the change in dc magnetizing force to reset core flux between 10 and 20 kgauss; see reference 2.

[4] H. W. Lord, Trans. Am. Inst. Elec. Engrs. **73**, Part I, 721 (1954).

JOURNAL OF APPLIED PHYSICS    SUPPLEMENT TO VOL. 33, NO. 3    MARCH, 1962

# Magnetic Properties of Tape-Wound Cores of High-Purity 3% Silicon-Iron with the (110)[001] Texture

J. L. WALTER

*General Electric Research Laboratory, Schenectady, New York*

Tapes, 1 mil thick, with sharp (110)[001] textures have been prepared from high-purity 3% silicon-iron. Surface energy driving forces were utilized to obtain the texture in the impurity-free silicon-iron.

Magnetic properties of wound cores were measured at 60, 400, and 1000 cps. Exceptionally high values of peak induction, residual induction, and squareness, compared to ordinary silicon-iron, were obtained. Losses and excitation were comparable to 1-mil Deltamax properties.

In a half-wave magnetic amplifier, the silicon-iron cores produced as much as 22% higher current output and 40% higher gain than did equivalent Deltamax cores.

## INTRODUCTION

A NEW type of driving force for grain growth has been utilized for the control of crystal orientations through secondary recrystallization in high-purity silicon-iron.[1-4] It derives from the differences in gas-metal interfacial energies of various crystallographic surfaces. Utilization and control of this driving force has made it possible to obtain exceptionally sharp (110) [001] textures in high-purity silicon-iron sheet[4] and provides several advantages in terms of magnetic quality over conventional means for controlling textures in silicon-iron.

First, the formation of the texture does not depend upon the presence of impurities in the material. In fact, the driving force is most effective if the material is free of impurities. Thus, residual impurities can be eliminated by using starting material which is free of impurities.

Second, the nature of the driving force is such that only grains with (110) planes within about 5 deg of the surface act as nuclei for secondary recrystallization.[4] This range of deviation of the (110) planes is about half that obtained by conventional (impurity controlled) secondary recrystallization.

Third, the perfection of the texture is not a function of the thickness of the material as is the case with impurity controlled grain growth.

This combination of conditions, i.e., lack of residual impurities, sharply improved crystal orientation, and the ability to produce these textures in thin laminations results in improved magnetic properties when compared to commercial silicon-iron, particularly in the case of material less than about 4 mils thick.

This report describes the ac magnetic properties of wound cores made from 1-mil (0.001 in.) thick tapes of high-purity 3% silicon-iron with the (110) [001] texture. The properties are compared to those obtained from identical cores of 1-mil Deltamax-type material.[5] In addition, control characteristics of both types of cores in a magnetic amplifier have been obtained and are compared.

## EXPERIMENTAL PROCEDURE

### Preparation of Tapes

The high-purity silicon-iron was prepared by melting and casting electrolytic iron and silicon *in vacuo*. The silicon content was 3.1%. The impurity content totaled less than 0.005% by weight. The ingots were hot- and cold-rolled to tape, 0.001 in. thick, with appropriate intermediate anneals. The tape was slit to 0.250-in. widths, cleaned, and given the final anneal at 1200°C in vacuum to develop the (110) [001] texture. Pressure during the anneal was $10^{-5}$ mm Hg.

The tapes were then coated with MgO and wound into toroids having an i.d. of 1.0 in. and an o.d. of 1.375 in., weighing about 19 g. After a stress-relief anneal at 900°C, the toroids were placed in a plastic case for protection during coil winding and handling.

The silicon-iron cores and the Deltamax-type cores were tested under identical conditions.

### Magnetic Tests

Major dynamic hysteresis loops to peak magnetizing forces of 2 and 10 oe were obtained at 60, 400, and 1000 cps. Input voltage was sinusoidal.

Values of peak magnetizing force and core loss were obtained at inductions from 10–18 kgauss for the same frequencies.

Control characteristics at 60, 400, and 1000 cps were obtained by placing the cores in a half-wave magnetic amplifier with resistive load.

The data to be presented are typical of the test results of many samples.

## EXPERIMENTAL RESULTS

Core loss values, obtained from high-purity silicon-iron core 15 and the Deltamax core (labeled 49 Squaremu # 1), are plotted versus induction in Fig. 1.

Data from the dynamic hysteresis loops obtained from high-purity silicon-iron core 23 are given in Table I. The bracketed figures are data from Squaremu core # 1, tested under identical conditions.

[1] J. L. Walter and C. G. Dunn, Trans. AIME **215**, 465 (1959).
[2] J. L. Walter, Acta Met. **7**, 424 (1959).
[3] J. L. Walter and C. G. Dunn, Acta Met. **8**, 497 (1960).
[4] J. L. Walter and C. G. Dunn, Trans. AIME **218**, 1033 (1960).
[5] Obtained from Magnetic Metals Company, Camden, New Jersey. Material called "49 Squaremu."

## DISCUSSION

As is evident from the data, the magnetic properties of the high-purity silicon-iron rival the properties of Deltamax in most respects while exceeding Deltamax in useful flux density. Use of this new silicon-iron in transformers would allow an increase in voltage level by about 25% compared to Deltamax. In addition, at 60 cps, core losses and core weight would be lower. At higher frequencies, the core loss advantage would disappear but the advantage of higher output would be retained.

The exceptionally high values of induction and squareness of the high-purity silicon-iron are reflected in performance of core 23 in the magnetic amplifier. Output current at 400 cps was 22% greater than that provided by the Deltamax core. Core 23 also provided about 40% higher current gain than did the Deltamax core. Similar results were obtained for control characteristics at 1000 cps. At 60 cps, the gain was almost identical to the gain provided by the Deltamax core as was the control current for minimum output. At higher frequencies, core 23 required higher control currents for minimum output, compared to the Deltamax core; this would be expected on the basis of the higher coercive force of core 23, compared to the Deltamax core, at higher frequencies.

Comparing the properties of the high-purity silicon-iron to the properties of ordinary oriented-silicon-iron shows the marked improvement in properties effected by the combination of improved texture and high-purity. For example, 4-mil Silectron has $B_r$ of 14 kgauss and $B_p$ of 15.5 kgauss and $H_c$ of 0.55 oe at 60 cps.[6] The high-purity silicon-iron core 23 would provide at least 15% greater output and less than half the losses of 4-mil Silectron.

An even greater difference in properties is obtained if the properties of core 15 are compared to the properties of 1-mil commercial silicon-iron. At 14 kgauss, the 400-cps losses of 1-mil "Oriented T"[7] are a factor of 2.5 times the losses obtained from core 15 at the same induction. The 1000-cps losses of this material are higher

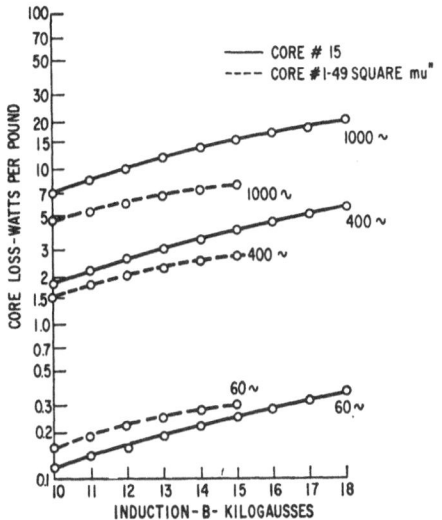

FIG. 1. Plot of ac losses vs induction.

by a factor of 1.6 than the losses of core 15 at 1000 cps. In addition to lower losses, much smaller exciting fields are required to reach high values of induction than is the case for Oriented T; under dc excitation, core 15 required a magnetizing force of only 0.20 oe to reach 15 kgauss whereas 1-mil Oriented T required 8.5 oe to reach the same induction.[8]

Along with markedly superior room temperature magnetic properties, core 15 also exhibited much lower temperature dependence of $B_r$ and $B_p$ and better stability at elevated temperatures compared to commercial silicon-iron. At 500°C, the percent decrease of $B_p$ and $B_r$ was only about half that obtained for Silectron.[6] The properties did not change after several cycles to 500°C in marked contrast to the degradation of properties of Silectron after only one cycle. The Deltamax core suffered much greater changes in magnetic properties at elevated temperatures because of its low Curie temperature. At 300°C, $B_r$ of the Deltamax core was reduced by 40% and the hysteresis loop was no longer square.

The stability at elevated temperatures and the superior magnetic properties at room temperature are directly related to both the improved crystal texture and the low residual impurity content of the high-purity silicon-iron.

### ACKNOWLEDGMENTS

The author is indebted to Miss E. A. Freeman for material preparation. O. A. Vitillo, E. G. Siwek, A. E. Kettner, H. F. Storm, and D. L. Watrous of G. E.'s General Engineering Lab provided magnetic properties data and also gave valuable comments concerning test procedures.

TABLE I. Dynamic loop data; silicon-iron core 23 and Deltamax core (brackets).

| Frequency (cps) | Peak $H$ (oe) | $H_c$ coercive force (oe) | $B_r$ res. ind. (kgauss) | $B_p$ peak ind. (kgauss) | $\dfrac{B_r}{B_p}$ |
|---|---|---|---|---|---|
| 60 | 2 | 0.25(0.21) | 16.8(14.2) | 17.9(14.5) | 0.94(0.98) |
|  | 10 | 0.38(0.22) | 17.3(14.2) | 18.8(14.8) | 0.92(0.95) |
| 400 | 2 | 0.40(0.26) | 17.6(14.8) | 18.2(15.1) | 0.97(0.98) |
|  | 10 | 0.50(0.27) | 17.9(14.8) | 19.0(15.1) | 0.94(0.96) |
| 1000 | 2 | 0.57(0.33) | 18.1(14.9) | 18.5(15.1) | 0.98(0.99) |
|  | 10 | 0.78(0.36) | 18.1(14.9) | 19.0(15.5) | 0.95(0.96) |

[6] M. Pasnak and R. Lundsten, Trans. AIEE 78, 1033 (1960).
[7] Produced by Armco Steel Corporation, Middletown, Ohio.

[8] Catalog, "Thin electrical steels," Armco Steel Corporation, 14 (1957).

JOURNAL OF APPLIED PHYSICS    SUPPLEMENT TO VOL. 33, NO. 3    MARCH, 1962

# Cooling Rate Effect on Initial Permeability of 4-79 Molybdenum Permalloy

Armand A. Lykens

*The Carpenter Steel Company, Research Laboratory, Reading, Pennsylvania*

This paper shows the combined effects of Ni and Mo on the optimum cooling rate of 4-79 molybdenum Permalloy and provides an empirical equation for calculating this rate ($CR_{opt}$) for highest 60 cycle initial permeability at $B=40$ gauss ($u_0$). Cold-rolled strip was obtained from heats of the Carpenter Steel Company's HyMu 80 with Ni between 79.29 to 80.63% and Mo between 4.08 to 4.44%. Before cooling rate studies were run, duplicate ring lamination samples from each heat were thoroughly purified by annealing in hydrogen at 2050°F to remove carbon and other nonmetallics. This was done to obtain the true effects of Ni and Mo on the $u_0$ —cooling rate relationship. Magnetic measurements were made after each four hours at temperature until the $u_0$ of each heat stabilized. An average cooling rate of 150°F/hr through the critical ordering temperature range was employed. As annealing time increased, $u_0$ gradually increased and then stabilized for each heat at various values between 30 500 and 43 250. The stabilized samples, when repeatedly heated to 2050°F in hydrogen and cooled at different rates between 60° and 810°F/hr, displayed values of $CR_{opt}$ between 100° to 600°F/hr. As the $CR_{opt}$ decreased, the peak of the $u_0$—cooling rate curve became more abrupt. The following regression equation was calculated: $CR_{opt}$ (°F/hr) $= -26201.2 + 365.0$ (%Ni) $-633.3$ (%Mo). The standard deviation of the difference between the observed and the predicted values was calculated to be 45°F/hr. The Mo has nearly twice the influence as Ni on the $CR_{opt}$ but in an opposing direction.

## INTRODUCTION

IT is believed that a critical degree of atomic ordering is necessary to obtain the highest initial permeability in 4-79 molybdenum Permalloy. It occurs when the alloy is cooled at some optimum rate through the ordering temperature range (1112° to 572°F) during the final hydrogen anneal. Work by Chegwidden and Ashworth[1] shows the separate effects of Ni and Mo upon the optimum cooling rate ($CR_{opt}$) when one element is held constant and the other varied. This paper shows the joint effects of Ni and Mo and provides an empirical equation for calculating the $CR_{opt}$ for highest initial permeability at $B=40$ gauss ($u_0$).

## EXPERIMENTAL PROCEDURE AND RESULTS

For this study, 13 electric arc-melted heats of the Carpenter Steel Company's HyMu 80 alloy were employed with Ni ranging from 79.29 to 80.63% and Mo from 4.08 to 4.44%. A typical heat analysis is:

| C | Mn | Si | P | S | Cr | Ni | Mo | Fe |
|---|---|---|---|---|---|---|---|---|
| 0.030 | 0.75 | 0.40 | 0.005 | 0.005 | 0.08 | 80.00 | 4.25 | Bal. |

An ingot of each heat was mill processed into 0.014-in. thick strip by hot forging to slab and hot rolling the slab to strip 0.100 to 0.200 in. in thickness. The acid-cleaned hot-rolled strip was then cold rolled to 0.014 in. using appropriate intermediate anneals. Ring laminations of $1\frac{1}{2}$ in. o. d. and 1 in. i.d. were blanked from these strips and stacked into $\frac{1}{2}$-in.-high samples for annealing and testing purposes.

The annealing equipment consisted of a resistance wound furnace and a sealed Inconel muffle through which pure hydrogen dried to a minimum dew point of −90°F was circulated. The seal was a Neoprene gasket inserted between the end plate and the muffle. To cool the furnace, a variable speed fan was attached to an opening in the rear of the furnace directly behind the muffle. Cooling rate was controlled by setting the fan speed at a predetermined rate when the furnace temperature reached 1200°F. An iron-Constantan thermocouple, connected to a recorder, was located inside the muffle just above the samples. With this apparatus, the average cooling rate was calculated over the ordering temperature range (1112° to 572°F) by dividing the temperature difference of 540°F by the time in hours required to cool the samples through this range. Because forced air cooling was used, the rate of cooling was not constant but decreased with decreasing temperature.

The hydrogen purification system used was typical of most installations. It consisted of an activated charcoal filter for gas degreasing, a commercial palladium catalyst (De-Oxo unit) for converting uncombined oxygen to water, and finally an activated alumina electrodryer.

The testing equipment consisted of a voltmeter-ammeter, sine current tester, and a sample holding jig with a 20-turn secondary coil and a one-turn primary. Tests were conducted according to ASTM Standard A346–58 for 60-cycle ac initial permeability at $B=40$ gauss.

Before the cooling rate studies, duplicate samples from each heat were thoroughly purified in hydrogen to remove carbon and other nonmetallics. Purification was performed to obtain the true effects of Ni and Mo on the $u_0$—cooling rate relationship. The purification treatment involved annealing at 2050°F for 4 hr in dry hydrogen with a flow rate of 5 cu ft/hr. An average cooling rate of 150°F/hr was employed through the critical temperature range. Then, the $u_0$'s were measured. This annealing and testing cycle was repeated on the same samples until the $u_0$ of each heat stabilized. With increasing annealing time, the $u_0$ gradually increases and then stabilizes. The values range from 30 500 to 43 250. This variance between heats is attributed to differences in $CR_{opt}$; e.g., the heat with the highest permeability has an $CR_{opt}$ close to the 150°F/hr used in the purification anneal. Other experiments have indicated that the permeability increase with annealing time is caused by the removal of carbon. An inverse linear relationship

[1] R. Chegwidden and J. Ashworth in *Ferromagnetism*, edited by R. M. Bozorth (D. Van Nostrand Company, Inc., Princeton, New Jersey, 1951), fourth printing, Chap. 5, p. 137.

was found to exist between carbon and permeability; the lower the carbon the greater the permeability. Magnetic stabilization occurred when carbon was reduced to its lowest level, approximately 0.005%. The minimum annealing time required for magnetic stabilization varied between 10 and 20 hr depending upon the initial carbon level.

These purified samples were then reheated to 2050°F and immediately cooled at a selected rate through the critical range. This procedure was repeated numerous times on the same samples in order to investigate cooling rates ranging from 60 to 810°F/hr. After each annealing cycle, the $u_0$ of each sample was measured and the average value plotted against cooling rate, as shown in Fig. 1, for representative heats with different Ni and Mo contents. The $u_0$ of the purified samples ranged from 17 700 to 54 900 with the cooling rates investigated. Each analysis displayed an $CR_{opt}$ that was between approximately 100° and 600°F/hr. As the $CR_{opt}$ decreased, the peak of the $u_0$—cooling rate curve became more abrupt; thus making it more difficult to cool within their optimum range the heats with slow optimum cooling rates.

The influence of Ni and Mo on the $CR_{opt}$ is also apparent in Fig. 1. By comparing curves A, B, D, and E, where the Mo was between 4.08 to 4.19%, one observes that increasing Ni from 79.29 to 80.33% raises the optimum rate from approximately 100° to 550°F/hr. The curves suggest that a sharp decrease in $u_0$ occurs when Ni falls below approximately 79.50%. Curves B and C show that when Ni is held constant, an increase in Mo from 4.13 to 4.35% causes a decrease in $CR_{opt}$ from approximately 400° to 250°F/hr.

Based upon the empirical data, the following equation to predict the $CR_{opt}$ for maximum $u_0$ for these two elements was calculated by means of a multiple regression program performed on an IBM 650 computer:

$$CR_{opt}(°F/hr) = -26\ 201.2 + 365.0\ (\%\ Ni)$$

$$-633.3\ (\%\ Mo).$$

The standard deviation of the difference between the observed and predicted values was calculated to be 45°F/hr. This means that 95% of all predicted values of

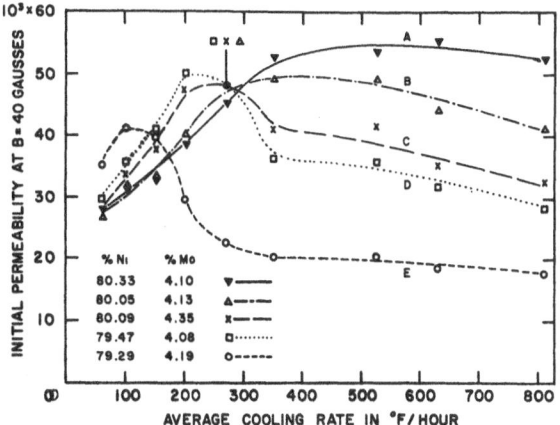

FIG. 1. Initial permeability at $B=40$ gauss for a 0.014-in. strip of representative Carpenter HyMu 80 heats versus average cooling rate in °F/hr through the critical ordering range (1112° to 572°F) from the final hydrogen anneal.

$CR_{opt}$ will be within plus or minus 90°F/hr of the actual values. The coefficient of determination ($r^2$) was calculated to be 93%. The formula indicates that Mo has nearly twice as much influence as Ni on the $CR_{opt}$ but in an opposing direction.

## SUMMARY

Cooling rate studies on hydrogen purified 4–79 molybdenum Permalloy show the strong influence of cooling rate on $u_0$. Each analysis displays an $CR_{opt}$ for maximum $u_0$ of approximately 50 000 that depends upon Ni and Mo. From the experimental data, a regression equation was calculated for predicting the $CR_{opt}$ from this alloy with Ni between 79.29 to 80.63% and Mo between 4.08 to 4.44%.

## ACKNOWLEDGMENTS

The author wishes to express his appreciation to E. L. Frantz for performing the magnetic tests and anneals, to F. P. Orlando for statistical analysis assistance, and to the Carpenter Steel Company for permission to publish this paper.

JOURNAL OF APPLIED PHYSICS    SUPPLEMENT TO VOL. 33, NO. 3    MARCH, 1962

# Rare Earths

J. F. DILLON, *Chairman*

## Spinwave and Uniform Precession Linewidths in Rare-Earth Substituted Yttrium Iron Garnet

P. E. SEIDEN

*IBM Research Center, Yorktown Heights, New York*

The temperature dependence of the spinwave and uniform precession linewidths of yttrium iron garnet substituted with the order of one atomic percent of holmium, erbium, and europium was measured from 4.2° to 300°K. The data are compared to the theory of de Gennes, Kittel, and Portis which attributes the linewidth to the rapidly relaxing rare-earth ions. Although the theory qualitatively predicts the temperature dependence of the linewidth, the detailed behavior varies considerably in these materials and remains to be accounted for. A new technique is briefly described that is used for obtaining $\Delta H_k$ from the main resonance nonlinear behavior of the susceptibility which avoids the detailed point measurements previously necessary.

## INTRODUCTION

THE temperature dependence of both the spinwave and uniform precession linewidths of rare-earth substituted yttrium iron garnets has been studied in the temperature range from 4.2° to 300°K. The spin-wave linewidth measured is that of the $z$-directed degenerate spinwave associated with the premature saturation of the uniform precession resonance line.[1] The data have been compared with the predictions of the theory of de Gennes, Kittel, and Portis[2] which attributes the linewidths of rare-earth garnets, where the rare-earth ion is not in an $S$ state, to the fact that the rare-earth ion has a very short relaxation time which relaxes the iron sublattice magnetization through the iron-rare earth exchange field. Due to space limitations we will describe here results for only three rare earths. The three samples considered are single crystals of yttrium iron garnet containing 0.5% of holmium or 1% of erbium or europium. Small percentage substitutions are used since one cannot saturate the 100% rare-earth garnets as is necessary in order to measure $\Delta H_k$.

## EXPERIMENTAL TECHNIQUE

The usual technique[3] for the measurement of the linewidth of the spinwave associated with the main resonance saturation is to plot a point by point curve of the susceptibility as a function of the inverse rf magnetic field and then extrapolate the linear portion of the curve back to the small signal susceptibility in order to find the critical field necessary to calculate $\Delta H_k$. This technique involves a large amount of experimental and computational work, especially if many measurements are to be taken.

In the present technique we use an experimental arrangement containing a reflection cavity which is matched to the system when the sample is off ferrimagnetic resonance. Analysis of this cavity shows that in the region where the susceptibility is proportional to the inverse rf magnetic field the power reflected from the cavity will be a constant independent of incident power. Therefore, we may record reflected power as a function of incident power continuously by an X–Y recorder and perform an extrapolation of this region of constant reflected power directly on the data curves to find the critical power level. This procedure eliminates the extensive point by point measurements and calculations. The details of this technique will be the subject of a separate publication.[4]

## EXPERIMENTAL RESULTS

The shape of the curves of spinwave and uniform precession linewidths versus temperature for the samples measured here are similar to those measured previously

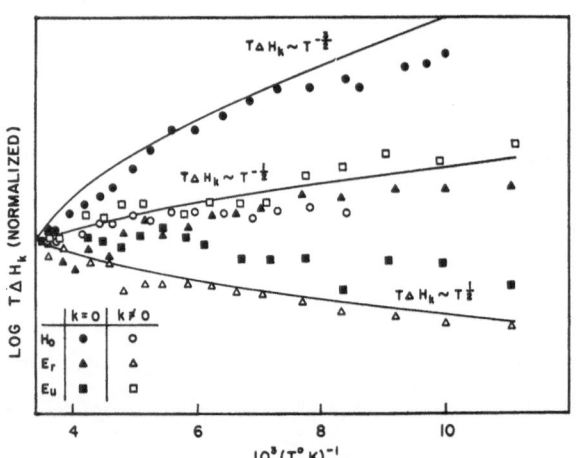

FIG. 1. Uniform precession and spinwave linewidths for temperatures *above* the linewidth peak plotted so as to give the temperature dependence of the rare-earth ion relaxation time.

[1] H. Suhl, J. Phys. Chem. Solids 1, 209 (1957).

[2] P. -G. de Gennes, C. Kittel, and A. M. Portis, Phys. Rev. 116, 323 (1959).

[3] P. E. Seiden and H. J. Shaw, J. Appl. Phys. 31, 225S (1960).

[4] P. E. Seiden, IRE Trans. on Microwave Theory Tech. (to be published.)

for yttrium iron garnet of low purity and 100% rare-earth garnets.[5] These curves were explained qualitatively by the theory of de Gennes $et\ al.$[2] The theory predicts that at temperatures above that at which the linewidth maximum occurs (where the rare-earth ion relaxation frequency is $greater$ than the iron-rare earth exchange frequency) the temperature dependence of the linewidth is given by (assuming that the exchange frequency is independent of temperature),

$$\Delta H \sim \tau/T, \qquad (1)$$

where $\tau$ is the rare-earth ion relaxation frequency. Some recent work[6] on relaxation mechanisms indicate that in some rare earths $\tau$ may vary as,

$$\tau \sim e^{\Delta/T}, \qquad (2)$$

where $\Delta$ is the crystal field splitting parameter of the rare-earth ion. In Fig. 1 we give normalized log $T\Delta H_k$ [essentially log $\tau$ using equation (1)] as a function of $T^{-1}$ for both the uniform precession ($k=0$) and spinwave ($k\neq0$) linewidths. On this plot the exponential behavior of Eq. (2) will be a straight line. Also shown on the same figure are the power laws of $\tau \sim T^{-\frac{3}{2}}$, $T^{-\frac{1}{2}}$, $T^{\frac{1}{2}}$. It is seen that the exponential law does not supply a good fit to the data and that the power laws provide only a fair fit to some of the data. On the basis of Eq. (1) and the data for $\Delta H$ for Eu and, $\Delta H_k$ for Er we would find that $\tau$ $increased$ as a function of temperature; this behavior is quite unlikely.

The theory of de Gennes $et\ al.$ predicts the following temperature dependence for temperatures below the linewidth peak (rare-earth ion relaxation frequency $less$

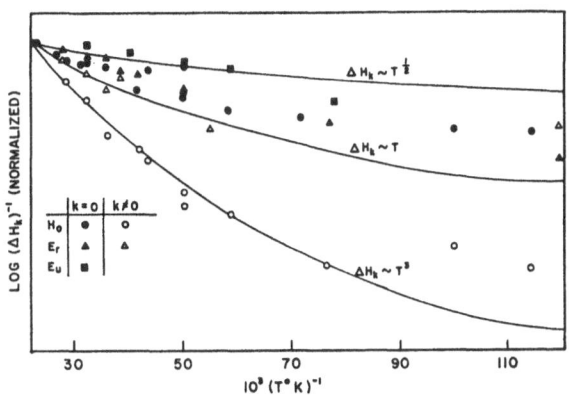

FIG. 2. Uniform precession and spinwave linewidths for temperatures $below$ the linewidth peak plotted so as to give the temperature dependence of the rare-earth ion relaxation time.

than the iron-rare earth exchange frequency),

$$\Delta H \sim 1/\tau. \qquad (3)$$

Figure 2 shows normalized log $(\Delta H_k)^{-1}$ (again essentially log $\tau$) as a function of $T^{-1}$ for this region. Here again neither the exponential law nor the power laws provide a good fit to the data.

From the data presented here, it is seen that although the theory of de Gennes $et\ al.$ provides a good qualitative fit over the whole temperature range, it is not the whole story, and the detailed behavior of the differences among the rare earths and between the spinwave and uniform precession linewidths remains to be accounted for.

### ACKNOWLEDGMENTS

The author would like to thank E. A. Giess and C. E. Hallas for growing the crystals and H. R. Lilienthal and R. J. Bennett for sample preparation.

[5] G. P. Rodrigue, H. Meyer, and R. V. Jones, J. Appl. Phys. **31**, 376S (1960).

[6] C. B. P. Finn, R. Orbach, and W. P. Wolf, Proc. Phys. Soc. (London) **77**, 261 (1961); R. Orbach, Proc. Phys. Soc. (London) **77**, 821 (1961).

JOURNAL OF APPLIED PHYSICS     SUPPLEMENT TO VOL. 33, NO. 3     MARCH, 1962

# Magnetocrystalline Anisotropy of Rare-Earth Iron Garnets

R. F. PEARSON

*Mullard Research Laboratories, Redhill, Surrey, England*

This paper will review recent anisotropy data, deduced from static and resonance measurements, on the rare-earth iron garnets of formula $5Fe_2O_3 \cdot 3M_2O_3$, where M is Y, Gd, Yb, Er, Sm, Tb, Dy, and Ho, respectively. Both the high temperature behavior (up to 300°K) which can be expressed in terms of the anisotropy constants $K_1$ and $K_2$ and the anomalous anisotropy properties which occur at liquid helium temperatures are discussed. The results are interpreted in terms of the single ion model which considers the anisotropy resulting from the effect of the combined crystal and exchange fields of the garnet on the energy levels of the individual magnetic ions, and the extent to which this model can describe the anisotropy of rare-earth iron garnets will be briefly discussed.

## INTRODUCTION

SINCE the rare-earth iron garnets were first discovered by Bertaut and Forrat (1956)[1] and independently by Geller and Gilleo,[2] their magnetic properties have been the subject of considerable attention. Considering first their structure, the garnet unit cell, which is cubic with space group $O_h^{10}-Ia3d$, contains 8 molecules of formula $R_3Fe_5O_{12}$, where R is yttrium or a rare-earth ion of electronic configuration $4f^x(5s)^2(5p)^6$ where $x$ varies from 5 to 14. The $Fe^{3+}$ ions are situated on the $16a$ (octahedral) sites and the $24d$ (tetrahedral) sites while the ions $X^{3+}$ lie on the $24c$ sites around which the nearest oxygen ions are situated on the corners of a rather distorted cube. This rare-earth ion site, in which we shall be mainly interested, possesses orthorhombic symmetry and has 3 twofold axes oriented in turn along the two [110] axes and one [100] axis of the unit cell, thus giving rise to six magnetically inequivalent sites in the unit cell. The spontaneous magnetization of the iron garnets containing Y, Sm, Eu, Gd, Tb, Dy, Ho, Er, Tm, and Yb has been measured by Pauthenet[3] as a function of temperature and he concluded that in most cases the results could be interpreted in terms of a ferrimagnetic arrangement of the $Fe^{3+}$ ions on the $a$ and $d$ sublattices giving a resultant magnetization which is aligned antiparallel to the moments of the rare earth ions on the $c$ sublattice. As the rare-earth ions contributed increasingly to the ferrimagnetic moment at low temperatures, magnetic compensation points are possible in the system and were observed in the case of Gd, Ho, Dy, Er, and Tb iron garnets.

If we now consider the magnetocrystalline anisotropy of the garnets, it seems reasonable in view of the localized nature of the magnetic moments, to use the "one ion model" which has been successfully used to interpret the anisotropy of the spinel ferrites.[4-6] We therefore need to know the effect on the individual magnetic ions produced by the crystal field, spin orbit interaction, and exchange interactions, respectively. In the case of the rare-earth ions where the magnetic moment is due to the $4f$ electrons, the crystal fields are much smaller than those in the transition metal ferrites, while the spin orbit interaction is much larger. Approximate values of these parameters for the rare-earth ions are shown in Table I together with the values of a transition metal ion $Co^{2+}$ for comparison. Figure 1, due to Smit, illustrates how the effective anisotropy is determined by the weakest interaction which for the cobalt is the spin orbit interaction $\lambda$, whereas for the rare-earth ion it is probably the exchange interaction. Of course, an explicit expression for the anisotropy can only be gained from an exact knowledge of the lowest energy levels of the individual ions, from which we can calculate the free energy and the anisotropy of the whole crystal by summing up the contributions from ions on all the possible sites. As the ground states of the rare-earth ion as shown in Table I vary so widely from, for example, an $S$ state in $Gd^{3+}$ to an $I$ state for $Ho^{3+}$, we may expect quite a wide range of anisotropy behavior as we go through the iron garnet series. In the following section we shall describe the anisotropy measurements and results in detail together with their possible theoretical interpretation in terms of the one ion model.

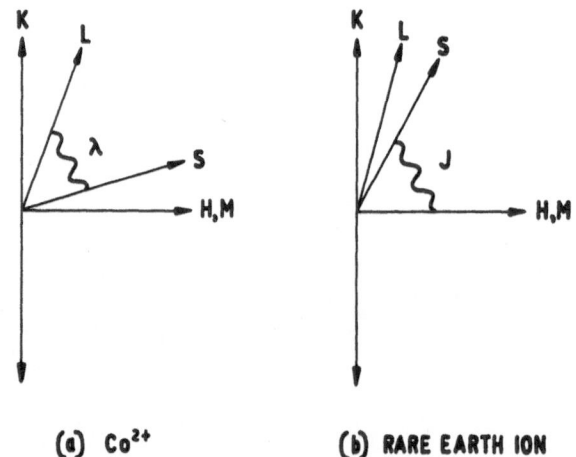

(a) $Co^{2+}$     (b) RARE EARTH ION

FIG. 1. Schematic representation of the orientation of the spin and angular momentum vectors with respect to the axis of a uniaxial potential for a $Co^{2+}$ ion and a rare-earth ion in an applied magnetic field.

[1] F. Bertaut and F. Forrat, Compt. rend. **242**, 382 (1956).
[2] M. A. Gilleo and S. Geller, J. Phys. Chem. Solids **3**, 30 (1957).
[3] R. Pauthenet, Ann. Phys. **13**, 1, 424 (1958).
[4] W. P. Wolf, Phys. Rev. **108**, 1152 (1957).
[5] V. J. Folen and G. T. Rado, J. Appl. Phys. **29**, 438 (1958).
[6] K. Yosida and M. Tachiki, Prog. Theoret. Phys. (Kyoto) **17**, 331 (1957).

Table I. Properties of rare-earth ions.

| | Electron configuration | Ground state of free ion | Spin orbit coupling[a] | Crystal field splitting[b] |
|---|---|---|---|---|
| $Sm^{3+}$ | $4f^5$ | $^6H_{5/2}$ | 1200 cm$^{-1}$ | 245 cm$^{-1}$ |
| $Gd^{3+}$ | $4f^7$ | $^8S_{7/2}$ | ... | 0 |
| $Tb^{3+}$ | $4f^8$ | $^7F_6$ | 1770 cm$^{-1}$ | 130 cm$^{-1}$ |
| $Dy^{3+}$ | $4f^9$ | $^6H_{15/2}$ | 1860 cm$^{-1}$ | 115 cm$^{-1}$ |
| $Ho^{3+}$ | $4f^{10}$ | $^5I_8$ | 2000 cm$^{-1}$ | 100 cm$^{-1}$ |
| $Er^{3+}$ | $4f^{11}$ | $^4I_{15/2}$ | 2350 cm$^{-1}$ | 90 cm$^{-1}$ |
| $Tm^{3+}$ | $4f^{12}$ | $^3H_6$ | 2660 cm$^{-1}$ | 80 cm$^{-1}$ |
| $Yb^{3+}$ | $4f^{13}$ | $^2F_{7/2}$ | 2940 cm$^{-1}$ | 70 cm$^{-1}$ |
| $Co^{2+}$ | $3d^7$ | $^4F_{3/2}$ | 540 cm$^{-1}$ | 10 000 cm$^{-1}$ |

[a] B. Bleaney, Proc. Phys. Soc. (London) **A68**, 937 (1955).
[b] R. L. White and P. J. Andelin, Phys. Rev. **115**, 1435 (1959).

## MEASUREMENT OF ANISOTROPY

With the production of single crystals as described by Nielsen and Dearborn[7] more detailed measurements on the garnets became possible and these led immediately to microwave studies of the line widths, $g$ values and anisotropy of these compounds. The resonance line widths of the rare-earth iron garnets are found to increase rapidly at low temperatures and consequently the measurement of anisotropy by this method is not possible at temperatures lower than approximately 200°K. Microwave resonance studies at lower temperatures have been carried out on crystals of yttrium iron garnet containing small percentages of rare-earth ions. The work of Dillon[8] in this respect, showed the very large effects on anisotropy fields that small amounts of rare-earth impurities cause at temperatures near 4.2°K.

If the anisotropy energy can be represented by an expansion of the following form,

$$E = K_1(\alpha_1^2\alpha_2^2 + \alpha_2^2\alpha_3^2 + \alpha_3^2\alpha_1^2) + K_2\alpha_1^2\alpha_2^2\alpha_3^2,$$

where $\alpha_1$, $\alpha_2$, $\alpha_3$, are the direction cosines of the magnetization, then the field for resonance at principal directions in the (110) plane may be expressed in terms of the resonance frequency, the $g$ factor and $K_1/M$, $K_2/M$, where $K_1$, $K_2$ are the usual anisotropy constants and $M$ the magnetization. Further experimental details of this method are described by Rodrigue[9] and Dillon.[8]

Static measurements of the anisotropy may be derived from either magnetization curve or torque measurements in the principal crystallographic planes. Simple expressions for the torque in terms of the usual anisotropy constants are given by the derivative of the magnetocrystalline energy surface in these principal planes, although in cases where this simple expression is not applicable it is more convenient to discuss the experimental results in terms of the energy surface obtained by integration of the torque curve. Experi-

mental details of the torque method are described by Penoyer[10] and Pearson.[11] Torque measurements at helium temperatures, 1.5–4.2°K, may be made using apparatus very similar to that described in these latter references. Very few static measurements of anisotropy in garnets have been reported in the literature and most of the values reported in this paper have not been previously published.

## DISCUSSION OF ANISOTROPY DATA

### Yttrium Iron Garnet Y₃Fe₅O₁₂

The experimental variation of $K_1$ with temperature is shown in Fig. 2. The resonance results, taken from Rodrigue and Jones[9] are essentially the same within the experimental errors of the static values obtained from torque measurements. $K_2$ was found to make a negligible contribution to the anisotropy energy. If we use the single ion model, the anisotropy is given by the sum of the contributions from the $Fe^{3+}$ ions (which are the only magnetic ions present as $Y^{3+}$ is nonmagnetic) on the tetrahedral and octahedral sites. The ground state of the $Fe^{3+}$ ion, which has electronic configuration $3d^5$ is $^6S_{5/2}$ and therefore we can directly apply the results of Yosida and Tachiki[6] and Wolf[4] who calculated the energy levels of $S$-state ions as a function of exchange field (i.e., sublattice magnetization) using a general spin Hamiltonian of the form:

$$H = \frac{a}{6}(S_x^4 + S_y^4 + S_z^4) + DS_\alpha^2 + FS_\alpha^4$$

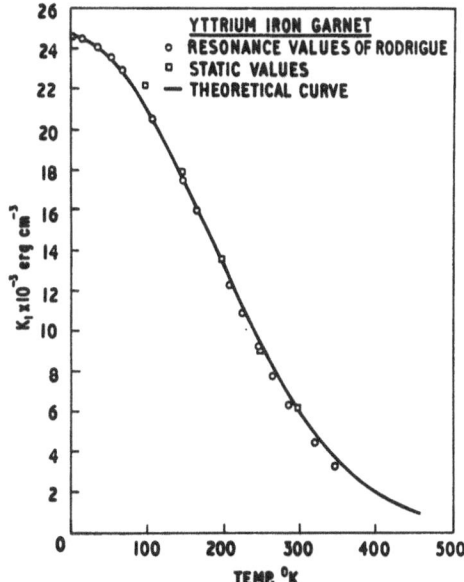

Fig. 2. Comparison of static, resonance, and theoretical values of $K_1$ for yttrium iron garnet as a function of temperature.

[7] J. W. Nielsen and E. F. Dearborn, J. Phys. Chem. Solids **5**, 202 (1958).
[8] J. F. Dillon and J. W. Nielsen, Phys. Rev. **120**, 105 (1960).
[9] G. P. Rodrigue, H. Meyer, and R. V. Jones, J. Appl. Phys. **31**, 376S (1960).

[10] R. F. Penoyer, Proc. A.I.E.E. Conference on Magnetism, p. 365 (1956).
[11] R. F. Pearson, Proc. Phys. Soc. (London) **74**, 505 (1959).

which assumes the crystal field environment of the ion is mainly cubic with a small axial distortion in direction $\alpha$. Values of $a$, $D$, $F$ corresponding to the energy level splitting in the octahedral and tetrahedral sites can in principle be obtained from paramagnetic resonance measurements of $Fe^{3+}$ in nonmagnetic garnet structures. Using these energy levels, the partition function and hence the free energy of the system is derived. The anisotropy constants are then found by comparing the directional dependence of the free energy with the usual expression for cubic crystals. The theoretical curve derived in this way for YIG by Rodrigue[9] is shown in Fig. 2. Although the agreement with experiment seems quite good, the values of the crystal field parameters $a$ and $F$ required to fit the experimental results differ from those found by paramagnetic resonance on $Fe^{3+}$ in yttrium gallium garnet by Geschwind[12] which led to a theoretical value of $K_1$ at $0°K$ of $0.37$ cm$^{-1}$ per unit cell compared to the experimental value of $0.24$ cm$^{-1}$. Possible reasons for this discrepancy were discussed by Geschwind,[12] but it is still not quite certain whether it can be accounted for by a difference in the crystal field parameters between the nonmagnetic and the magnetic garnet or due to the neglect of another anisotropy mechanism. A possible mechanism to explain the discrepancy was put forward by Walker[13] in terms of the zero point motion of spin waves in the ferrimagnetic ground state.

### Gadolinium Iron Garnet $Gd_3Fe_5O_{12}$

If we substract the appropriate values of $K_1$ for YIG from those corresponding to GdIG at a particular temperature, the remainder is the effective anisotropy contribution from the $Gd^{3+}$ ions. The experimental values of this contribution[9] are plotted in Fig. 3 together

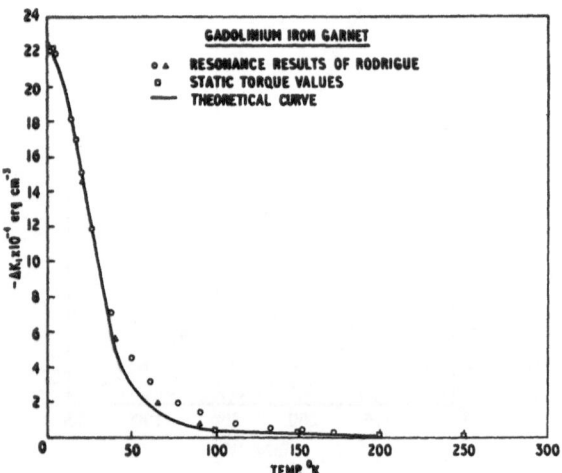

FIG. 3. Comparison of resonance, static, and theoretical values of the contribution to $K_1$ from $Gd^{3+}$ ions in GdIG.

[12] S. Geschwind, Phys. Rev. **121**, 363 (1961).
[13] L. R. Walker, J. Appl. Phys. **32**, 264S (1961).

with values deduced from torque experiments. The static values of $K_1$ agree reasonable well with the microwave results; the small discrepancies at lower temperatures can probably be ascribed to small impurities in the GdIG crystals used in the earlier resonance experiments. As the ground state of the free $Gd^{3+}$ ion is $^8S_{7/2}$ we can again use the theory of anisotropy for $S$-state ions used in the case of YIG. The theoretical curve deduced using this theory by Rodrigue and Jones[9] is plotted in Fig. 3. In the theoretical analysis the term in $D$ which was neglected in the YIG analysis produces a much larger effect on the anisotropy of the $Gd^{3+}$ sublattice and was therefore taken into account. It would be interesting to see whether comparison of the results with paramagnetic resonance data on $Gd^3$ in YGaG reveal a similar discrepancy to that found in yttrium iron garnet between the theoretical and experimental values of $K_1$ at $0°K$. The experimental values of $K_2$ the second-order anisotropy constant are small compared

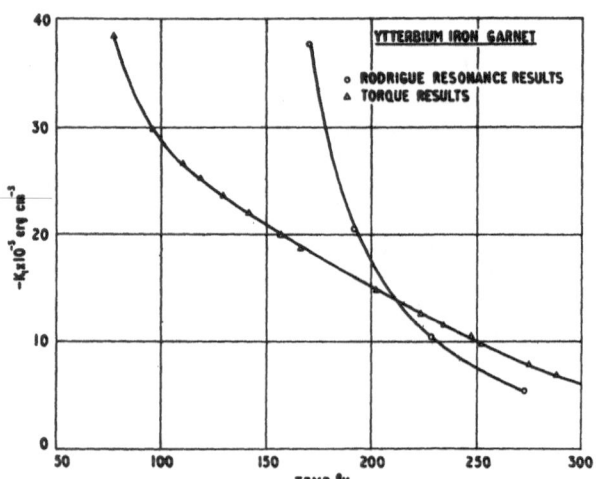

FIG. 4. Static and resonance values of $K_1$ for YbIG between $80°$ and $300°K$.

to $K_1$ as would be expected from the theoretical treatment as $K_2$ would correspond to higher order terms in the crystal field becoming important.

### Ytterbium Iron Garnet $Yb_3Fe_5O_{12}$

In Fig. 4 are plotted the values of $K_1$ between $80°$ and $300°K$ deduced from torque experiments and resonance measurements of Rodrigue.[9] As the temperature is reduced below $220°K$ the resonance results indicate a rapid rise in $K_1$ which is not shown by the static measurements. This discrepancy has been investigated in greater detail by Teale.[14,15] It appears that the anisotropy in the field for resonance is not wholly due to magnetocrystalline anisotropy; an anomalous mech-

[14] R. W. Teale and R. F. Pearson, Proc. Phys. Soc. (London) **76**, 308 (1960).
[15] R. W. Teale, R. F. Pearson, and M. J. Hight, J. Appl. Phys. **32**, 150S (1961).

anism which may be of a relaxation nature also contributes. The [100] direction is unique in that this anomalous mechanism does not have affect on the field for resonance until much lower temperatures are reached.

At 80°K the contribution to $K_1$ from the $Yb^{3+}$ is still quite small ($\sim 10\,000$ ergs/cm$^{-3}$) but on cooling to 20°K the value of $K_1$ increases to $-450\,000$ ergs cm$^{-3}$. Further torque measurements on the YbIG are impracticable at lower temperatures as owing to the continuing decrease of magnetization the applied fields required to saturate the crystal become prohibitively large. Further information on the anisotropy of the $Yb^{3+}$ ion at low temperatures may only be gained from experiments on samples of yttrium iron garnet containing small percentages of $Yb^{3+}$. The values of $K_1$ for YbIG at 4.2°K extrapolated from measurements on a 10% Yb doped YIG crystal was $6.7 \times 10^6$ ergs cm$^{-3}$. Measurements on samples containing substitutions of 1%, 2.1%, and 10%

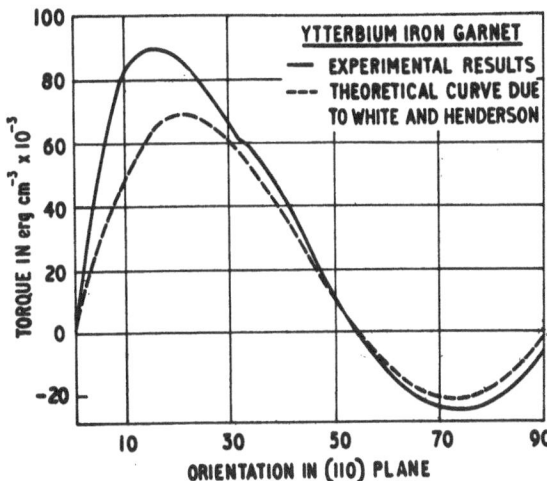

Fig. 6. Theoretical and experimental torque curves for YbIG at 1.5°K.

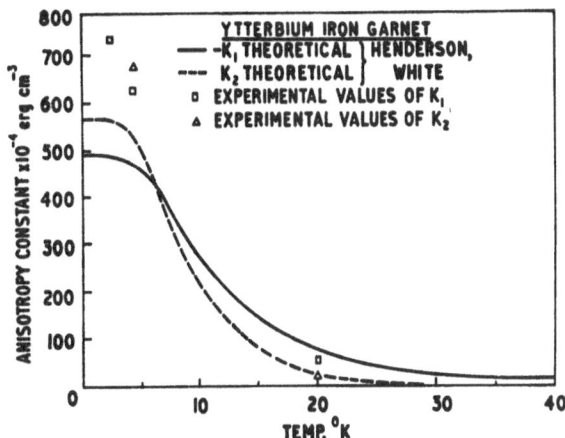

Fig. 5. Theoretical and experimental values of $K_1$, $K_2$ for YbIG plotted against temperature.

Yb in YIG gave values of $K_1$ which were approximately linearly dependent on composition, within the experimental error of the chemical analysis of the ytterbium concentration.

The energy levels of the $Yb^{3+}$ ion in the garnet structure were investigated theoretically by Ayant and Thomas[16] and later paramagnetic resonance measurements by Wolf[17] and White[18] on $Yb^{3+}$ in YGaG confirmed that the ground state is a degenerate doublet which splits in a magnetic field like a spin of $S=\frac{1}{2}$ with a nearly isotropic $g$ value of 24/7. The anisotropic splitting of this doublet by the exchange field of the $Fe^{3+}$ ions in the rare-earth iron garnet was clearly demonstrated by the optical measurements of Wickersheim[19] who found that absorption spectra corre-

sponding to transitions between this ground state doublet and higher excited states were dependent on the orientation of the resultant magnetization of the YbIG, and the values of these doublet splittings for the inequivalent sites occupied by the rare-earth ion in the garnet lattice could be expressed in a form similar to that found for the $g$ tensor found in corresponding paramagnetic resonance experiments on $Yb^{3+}$ in YGaG. Assuming only the ground state doublet is occupied at low temperatures (the next excited state is 550 cm$^{-1}$ higher[16]), the free energy of the system was derived from the spectroscopic splittings by Henderson and White[20] and from this the anisotropy was directly deduced. Figure 5 shows the theoretical values of $K_1$ and $K_2$ plotted as a function of temperature with the corresponding experimental data deduced from measurements on the doped samples. Figure 6 also shows the theoretical and experimental torque curve at 1.5°K. The agreement between the results is very reasonable, when we consider that the ground state splittings in the pure YbIG may be slightly different from those in the doped Yb in YIG crystal on which the torque measurements were made. However, a significant difference between theory and experiment is the absence in the theoretical curve of the anomaly at about 30° from [100] which is shown by experimental torque curves and was first observed in microwave resonance experiments by Dillon.[8] According to Kittel[21] this type of anomaly may be interpreted in terms of crossovers or near crossovers of the low lying energy levels, but the spectroscopic evidence indicates that in YbIG the separation of the ground state doublet is always at least 15 cm$^{-1}$, and therefore should not provide such an anomaly.

[16] Y. Ayant and J. Thomas, Compte rend. **248**, 387 (1959).
[17] W. P. Wolf, D. Boakes, G. Garton, and D. Ryan, Proc. Phys. Soc. (London) **74**, 663 (1959).
[18] R. L. White and J. W. Carson, J. Appl. Phys. **31**, 53S (1960).
[19] K. A. Wickersheim, Phys. Rev. **122**, 1376 (1961).

[20] R. L. White and J. W. Henderson (to be published).
[21] C. Kittel, Phys. Rev. **117**, 681 (1960).

### Terbium Iron Garnet $Tb_3Fe_5O_{12}$

The static values of $K_1$ deduced from torque measurements at temperatures between 100° and 300°K are plotted in Fig. 7. At 80°K torque measurements in a (110) plane gave values of $K_1 = -760 \times 10^3$ and $K_2 = -7.6 \times 10^6$ ergs cm$^{-3}$, respectively. The Tb$^{3+}$ ion possesses such a large anisotropy that accurate measurements at lower temperatures may only be deduced from experiments on samples of YIG containing small percentages of terbium. Torque measurements on a sample containing 0.19% Tb in YIG were made at 4.2° and 1.5°K. The energy surface deduced from these measurements, shown in Fig. 8, has minima in the [100] and [111] directions, while the [110] direction is a maximum. This type of behavior is consistent with $K_1$ being positive and $K_2$ being larger and of opposite sign although the energy surface cannot be fitted exactly with these constants. The energy levels for the terbium ion Tb$^{3+}$ in the garnet structure have been calculated by Walker,[22] assuming that the crystal field on the rare-earth ion site has orthorhombic symmetry. The crystal field parameters in the Hamiltonian used to calculate the energy levels were adjusted to give agreement between the giant resonance peaks found experimentally and the crossover position of the low lying energy levels. The energy surface derived from this energy level data of Walker[22] is compared with the experimental surface derived from the torque results as shown in Fig. 8. Although one might at first think that the reasonable agreement between the energy level crossovers and anomalous resonance peaks (and indeed the anomalies found in the torque curves[23]) indicated the theory to be correct, it seems difficult to reconcile with this the discrepancy between the experimental and theoretical energy surfaces. However, it was pointed out by Walker

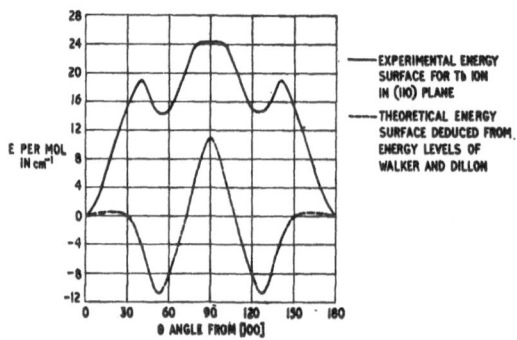

FIG. 8. Experimental energy surface for Tb iron garnet at 1.5°K extrapolated from doped garnet data and compared with the theoretical curve due to Walker and Dillon (1961).

that by adjusting the crystal field parameters to fit the shape of the energy surface rather than to the crossover positions (which seem to be more insensitive to variation of these parameters) much more satisfactory agreement might be obtained.

### Erbium Iron Garnet $Er_3Fe_5O_{12}$

The static values of $K_1$ between 100° and 300°K are plotted in Fig. 9 together with the resonance result of Rodrigue.[9] Here again we notice as in the case of YbIG a difference between the static and resonance results. It remains to be seen whether the situation here resembles closely that already described for YbIG. Therefore assuming that the static results give the true anisotropy, the contribution to $K_1$ from the Er$^{3+}$ (obtained by subtracting $K_1$ for YIG) is very small and in fact is less than the experimental error between 100° and 300°K. Torque measurements have been made on a sample of ErIG at 4.2°K which on integration give an energy surface with minima at the [100] and [111] directions indicating that $K_1$ is much less than $K_2$ and of opposite sign. Further measurements on a crystal containing 2% erbium in YIG showed similar behavior. The energy surface was approximately fitted with values of $K_1 = +100 \times 10^3$ and $K_2 = -630 \times 10^3$ ergs cm$^{-3}$ at 4.2°K. These results show that the contribution to $K_1$ from the erbium ions is approximately zero at 100°K and rises to a large positive value of approximately $6 \times 10^6$ at 4.2°K. Theoretical interpretation of this data in terms of the one ion model is still difficult in view of the uncertainty as to the ground state energy levels. In an orthorhombic crystal field, the $^4I_{15/2}$ ground state of the free ion is split into eight doublet levels. Paramagnetic resonance on the lowest level by Ball[24] indicated that the Er$^{3+}$ ion possesses large anisotropy which is very sensitive to small changes in the environment of the ion. This latter behavior is not borne out by the magnetic garnets which show essentially the same type

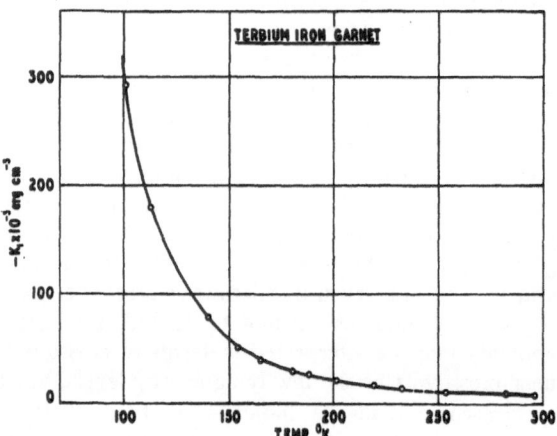

FIG. 7. Static values of $K_1$ for TbIG between 100° and 300°K.

[22] L. R. Walker, J. Appl. Phys. **33**, 1243 (1962); J. F. Dillon, J. Appl. Phys. **33**, 1191 (1962), this issue.
[23] R. F. Pearson and R. W. Cooper, J. Appl. Phys. **32**, 265S (1961).

[24] M. Ball, G. Garton, M. J. M. Leask, D. Ryan, and W. P. Wolf, J. Appl. Phys. **32**, 267S (1961).

of anisotropy in 2% and 100% concentrations of erbium in YIG.

## Samarium Iron Garnet Sm₃Fe₅O₁₂

The static values of $K_1$ between 80° and 300°K are plotted in Fig. 10. Resonance results of Rodrigue[9] between 180° and 300°K are in reasonable agreement with the static values. In order to deduce values of $K_1$ at lower temperatures measurements were made on a sample containing 1% Sm in YIG. At 80°K there is an interesting discrepancy between the properties of doped and pure samples. For pure SmIG $K_1$ is $-1.2\times10^6$ ergs cm⁻³ and $K_2$ is $+1.0\times10^6$ ergs cm⁻³ while for the corresponding sample containing 1% Sm $K_1$ is negative (extrapolated to 100% is $-1.55\times10^6$) and $K_2$ is also negative (very small at 80°K). At 4.2°K the results for SmIG indicated that $K_2$ has increased much more than $K_1$ as the energy surface has minima at [110], and maxima at [111] and [100] directions compared to the expected minimum at the [111] shown by the 1% Sm in YIG results, which give a value of $K_1$ approximately equal to $-400\,000$ ergs cm⁻³. The energy levels of the Sm³ ion are obviously quite sensitive to crystal field changes resulting from the replacement of Sm by yttrium in the doped crystal. The value of $K_1$ deduced at 0°K from the energy levels for a cubic crystal field given by White and Andelin (see reference b of Table I) is about $3\times10^6$ ergs cm⁻³ compared with the value of $-4\times10^7$ ergs cm⁻³ extrapolated from the doped crystal measurements.

## Holmium Iron Garnet Ho₃Fe₅O₁₂

The values of $K_1$ between 80° and 300°K are plotted in Fig. 10. At 80°K the values of $K_1$ and $K_2$ are approximately $-800\times10^3$ ergs cm⁻³ and $-270\times10^3$ ergs cm⁻³, respectively. At 4.2°K, $K_1$ and $K_2$ are still negative and $K_1$ is at least equal to $12\times10^6$ ergs cm⁻³ and could possibly be larger. Measurements on a sample containing 0.19% Ho in YIG gave a value of $K_1$ (extrapolated to 100% concentration) of about $-3\times10^7$ at 4.2°K. Assuming the crystal field on the rare-earth ion is cubic,

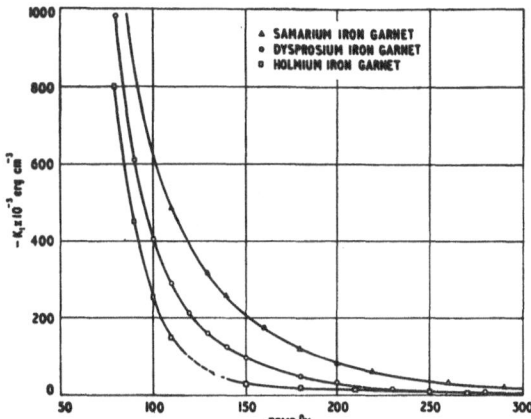

FIG. 10. Static values of $K_1$ for SmIG, HoIG, and DyIG between 80° and 300°K.

the value of $K_1$ at 0°K derived from the energy levels given by White and Andelin is about $-3\times10^6$ ergs cm⁻³. It seems again necessary to calculate the energy levels for the case of an orthorhombic crystal field to see whether a larger value of anisotropy is obtained.

## Dysprosium Iron Garnet Dy₃Fe₅O₁₂

The static values of $K_1$ between 80° and 300°K are shown in Fig. 10. Although the values of $K_1$ appear to be not quite continuous through the compensation point at 221°K, this may be due to lack of saturation of the torque curves in the region near this point. Further measurements on a sample containing 1% Dy in YIG give values which are in good agreement with those for DyIG at 80°K, where $K_1=-970\times10^3$ and $K_2=+214\times10^3$ ergs cm⁻³. At 4.2°K, the energy surfaces shown by the two crystals are of similar shape with minima at 36°, 70° from [100], and maxima at [100], [111], and [110] directions though the energy difference between the [100] and [111] directions is approximately $-43\,000$ for the 1% Dy in YIG sample compared to $-775\,000$ for the DyIG which was probably not completely saturated at 4.2°K. Note that here the qualitative agreement between the anisotropy surface in the pure and doped garnets at low temperatures is contrary to that expected from the magnetic susceptibility measurements on Dy³⁺ in yttrium aluminum or gallium garnets by Wolf et al.[24]

## CONCLUSIONS

The results reviewed in this paper confirm as was expected that a very wide range of anisotropy properties are found in the rare-earth iron garnets. Owing to the high anisotropy developed in the garnets containing Tb, Ho, Dy, Sm, Yb, and Er, accurate measurements of their anisotropy at temperatures below 20°K will require measurements to be carried out in much higher static fields (possibly up to 500 000 oe) than are avail-

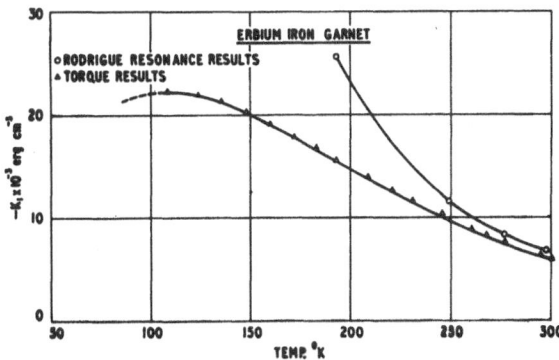

FIG. 9. Static and microwave values of $K_1$ for ErIG between 100° and 300°K.

able at the moment. However, the results here indicate that experiments on crystals of YIG containing small percentages of appropriate rare-earth ions will give a reliable indication of the properties of the pure rare-earth iron garnets except in the case of Tb and Sm which seem to show different results in doped and 100% crystals. The single ion model of anisotropy gives reasonable agreement with experiment for the case of YIG, GdIG, and YbIG and there seems no reason to doubt that if the ground state energy levels of the other rare-earth ions could be determined similar agreement could be obtained. However, for the case of TbIG and SmIG, either these ions are particularly sensitive to their local crystal field environment and therefore the anisotropy would tend to be not a linear function of concentration, or the single ion model seems to break down. Further experiments on crystals containing larger percentages of Tb and Sm might resolve this anomaly.

## ACKNOWLEDGMENTS

The author is grateful to many of the persons mentioned in the references for access to their papers before publication, to Dr. R. W. Teale, Mr. K. Tweedale, and Dr. F. W. Harrison for their valuable comments, Mr. R. W. Cooper who made the torque measurements, Mr. J. L. Page for crystal preparation and finally, Dr. K. Hoselitz under whose direction this work was carried out. The paper is published by kind permission of Mr. P. E. Trier, Director of the Mullard Research Laboratories.

JOURNAL OF APPLIED PHYSICS      SUPPLEMENT TO VOL. 33, NO. 3      MARCH, 1962

# Neutron Magnetic Scattering from Rare-Earth Ions

M. BLUME

*Theoretical Physics Division, AERE, Harwell, Berkshire, England*

AND

A. J. FREEMAN

*Materials Research Laboratory, Ordnance Materials Research Office, Watertown, Massachusetts*

AND

R. E. WATSON[*][†]

*Avco, RAD, Wilmington, Massachusetts*

The theory of neutron scattering by paramagnetic rare-earth ions has been given by Trammell and more recently by Odiot and Saint-James. Unlike the case for the iron series transition metals, for which the orbital angular momentum of the $3d$ electrons is almost completely quenched, the magnetic scattering of neutrons from the rare earths (except for gadolinium and europium) arises from both the spin and orbital contributions to the magnetization. Using recently determined Hartree-Fock wave functions for some of the trivalent rare-earth ions we have determined the functions $\langle j_n \rangle$ and $\langle g_n \rangle$ (for $n=0, 2, 4,$ and 6) which are necessary for the theoretical evaluation of the spin and orbital contributions respectively to the magnetic form factor. A comparison of these theoretical results with the experiments of Koehler and Wollan for $Nd^{3+}$ and $Er^{3+}$ and of Koehler, Wollan and Wilkinson for $Ho^{3+}$ is presented. Differences are found between theory and experiment and some possible reasons for this are discussed.

A paper on these matters is being submitted to the Physical Review.

[*] Part of the work of this author was supported by the Ordnance Materials Research Office, Watertown, Massachusetts.
[†] Present address: Theoretical Physics Division, AERE, Harwell, Berkshire, England.

JOURNAL OF APPLIED PHYSICS    SUPPLEMENT TO VOL. 33, NO. 3    MARCH, 1962

# Ferromagnetic Resonance in Terbium–Doped Yttrium Iron Garnet

L. R. WALKER

*Bell Telephone Laboratories, Murray Hill, New Jersey*

Calculations have been made on a specific model of a terbium-doped yttrium iron garnet system to determine its behavior in ferrimagnetic resonance. Good qualitative agreement has been found with the experiments of Dillon. Quantitative agreement is probably as good as can be expected in view of the fact that the crystal field has to be determined empirically in the course of the calculations.

ABOUT two and a half years ago J. F. Dillon[1] made measurements on the ferrimagnetic resonance of YIG samples in which a fraction of a percent of the yttrium ions had been replaced by those of a rare earth. He found that at temperatures in the liquid helium range the field for resonance might vary by as much as several kilo-oersteds when the angle between the applied dc magnetic field and the crystal axes was changed by a few degrees. This extremely anisotropic behavior progressively disappeared as the temperature was raised. In Fig. 1 the effect is shown for a nominal 0.2% atomic substitution of terbium. There are, with the applied field in the (110) plane, four sharp peaks in the resonant field as a function of angle, and also a considerable depression of the resonant field between the peaks below the value for a undoped sample. The latter is also shown in Fig. 1 for comparison.

An attempt has been made to account for these results in a quantitative way by analyzing numerically a specific model for the system. The calculation, while not entirely successful, does appear to reflect most of the observed facts.

The most significant factor determining the properties of the rare-earth garnets and the doped garnets, aside from the ferrimagnetic alignment of the sublattices, is the approximate equality of the exchange and crystal fields felt by the rare-earth ions. As one consequence of this, the magnetization of the latter has no reason to lie along the direction of the iron exchange field; it will, in general, be canted at any particular site and the over-all moment of the system will not be that for mutually opposed sublattices. Also, one must expect the energy levels of the rare earth to be radically shifted as the angle between the exchange field and the crystal axes is varied; this will be reflected in a complicated anisotropy for most measured quantities. In particular, for ferrimagnetic resonance at low temperatures, when most of the ions are occupying their lowest state, Kittel[2] pointed out that the total energy of the system will be strongly angularly dependent. The conventional formula for ferrimagnetic resonance contains the curvature of the total energy surface, and Kittel thus associated the sharp peaks seen by Dillon with such sharp curvatures as might

arise if the two lowest energy levels of the rare earth were to have a near-crossing in some directions.

The model which has been analyzed here divides the doped sample into two subsystems. One of these consists of the two very tightly coupled iron sublattices. We assume that the terbium ions produce no local distortion of the order of the sublattices. We also have to suppose that the iron sublattices lie along the applied dc magnetic field. This means that the analysis is, in general, only good for locally hard or easy directions. To include the realignment of the iron system in the anisotropy field of the terbium would lead to a self-consistency problem of impossible complexity. Clearly the analysis will be better the lower the terbium concentration. The second subsystem consists of the terbium ions in the crystal field and the exchange field of the iron. The concentrations are sufficiently low for interactions between terbium ions to be ignored.

The interaction between the two subsystems may be described as follows. The iron subsystem, with magnetic moment $\mathbf{M}_1$, will be governed by the usual

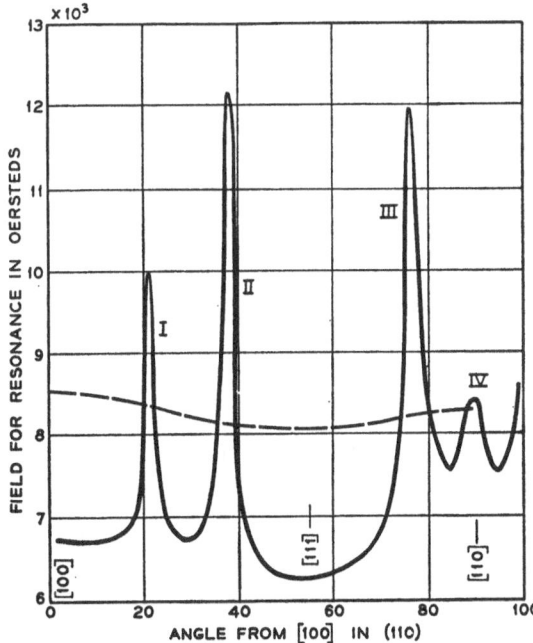

FIG. 1. $H_{res}$ in (110) at 1.5°K in YIG (0.2% Tb). The frequency was 22 989 Mc. The peaks as seen in this plane are designated by roman numerals. The dashed curve is $H_{res}$ for the purest YIG with which we have worked.

[1] J. F. Dillon, Jr., and J. W. Nielsen, Phys. Rev. Letters **3**, 30 (1959); Phys. Rev. **120**, 105 (1960).
[2] C. Kittel, Phys. Rev. Letters **3**, 169 (1959); Phys. Rev. **117**, 681 (1959).

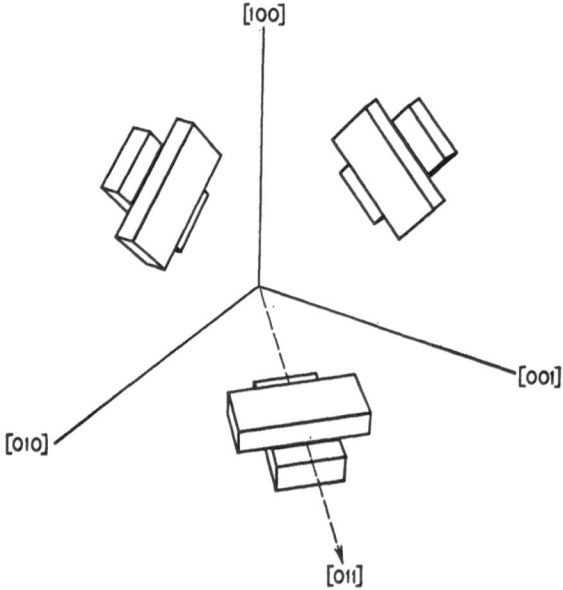

FIG. 2. The relative orientation of the six inequivalent types of dodecahedral site in YIG. Each parallelepiped represents the local orthorhombic symmetry of the sites. The indices refer to the crystallographic axes of the YIG crystal.

equation

$$\mathbf{M}_1 = \gamma(\mathbf{M}_1 \times \mathbf{H}_{\text{eff}}),$$

where $\mathbf{H}_{\text{eff}}$ is the total effective magnetic field. $\mathbf{M}_1$ will be resolved into a dc part $\mathbf{M}_{10}$ along the applied field $H_0\mathbf{n}$ and an rf part $\mathbf{m}_{11}e^{i\omega t}$ at right angles to $\mathbf{n}$. The linearized equation of motion is now

$$i(\omega \mathbf{m}_{11}/\gamma) = (\mathbf{M}_{10} \times \mathbf{h}_1) + (\mathbf{m}_{11} \times \mathbf{H}).$$

The dc part $\mathbf{H}$ of $\mathbf{H}_{\text{eff}}$ also lies along $\mathbf{n}$ and consists of the applied field $H_0$ and the part of the exchange field $\kappa(\mathbf{M}_{20} \cdot \mathbf{n})$ from the terbium ions which lies along $\mathbf{n}$, $\kappa$ being a molecular field constant and $\mathbf{M}_{20}$ the dc magnetization of the terbium. Similarly the rf part $\mathbf{h}$ of $\mathbf{H}_{\text{eff}}$ will be of the form $\kappa m_{21}$, where $m_{21}$ is the component of the rf magnetization of the terbium lying in the plane at right angles to $\mathbf{n}$. The terbium ions themselves are acted upon by an rf field $\kappa \mathbf{m}_{11}$, the rf exchange field of the iron. This rf field will produce a magnetic moment $\mathbf{m}_{21}$ related to the rf field by an rf susceptibility tensor $\|\chi\|$ so that

$$\mathbf{m}_{21} = \|\chi\| \kappa \mathbf{m}_{11}.$$

The moment $\mathbf{m}_{21}$ being known, $m_{21}$ can be evaluated and substituted into the iron equation of motion. One now has two equations for the two components of $\mathbf{m}$; after eliminating these, an equation is found which $H_0$ must satisfy. This has the form

$$\{H_0 - \kappa(M_{20} \cdot n) - \kappa^2 M_{10}\chi_{xx}\}$$
$$\times \{H_0 - \kappa(M_{20} \cdot n) - \kappa^2 M_{20}\chi_{yy}\}$$
$$= |i(\omega/\gamma) + \kappa^2 M_{10}\chi_{xy}|^2,$$

where the subscripts $x$ and $y$ on the susceptibilities refer to two mutually perpendicular directions in the

plane normal to $\mathbf{n}$. We may note here that $H_0$ is also properly involved in $\|\chi\|$, but we ignore it because it is small compared to the dc exchange field of the iron.

To proceed it is necessary to find expressions for the dc magnetic moment and of susceptibility of the terbium system. This involves finding the energy levels and wave functions of the terbium ion in the combined exchange and crystal fields. Because the ground multiplet $J=6$ of the terbium lies about 2000 cm⁻¹ below any other multiplet, we can reasonably operate within this multiplet. The dc moment will then be proportional to the thermal average of the operator $\mathbf{J}$, and the rf susceptibility, which can be calculated by the usual methods of time-dependent perturbation theory, is given by the expression:

$$\chi_{\alpha\beta} = C \sum_{n,m} (\rho_{mm}{}^0 - \rho_{nn}{}^0) \left[ \frac{(J^\alpha)_{nm}(J^\beta)_{mn}}{\hbar\omega + E_n - E_m} - \frac{(J^\alpha)_{mn}(J^\beta)_{nm}}{\hbar\omega + E_m - E_n} \right],$$

where $C$ is a constant; $\rho^0 = \exp(-\beta\mathfrak{IC})/\text{trace}\exp(-\beta\mathfrak{IC})$ the normalized density matrix with $\mathfrak{IC}$ the Hamiltonian for the iron in the exchange and crystal fields; $E_m$, $m=1$ to 13 for the case of terbium are the eigenvalues of $\mathfrak{IC}$; $(J^\alpha)_{nm}$ is the matrix element of the $\alpha$th component of $\mathbf{J}$ between the states $n$ and $m$. It should be mentioned at this point that there are six inequivalent sites for the terbium ions; the crystal field at each is the same in form, with orthorhombic symmetry, but the local symmetry axes differ in orientation. Fig. 2 shows the orientation of these sites. This means that for any direction of the applied dc field six eigenvalue problems have to be solved. The six dc magnetic moments and rf moments have then to be added before putting them into the iron equations. Before leaving the subject of the field for resonance it is of some interest to see qualitatively how the observed effects arise. The origin of the peaks is fairly clear; they will occur when the susceptibility of the terbium changes rapidly with angle and this can only be due to a sudden change in one of the energy denominators—or, in fact, to a near crossing of two energy levels; those at low temperatures will have to be the two lowest ones. It is also apparent that if the level separation becomes comparable with $\hbar\omega$ there will be very marked effects of frequency upon peak height. One is dealing here with paramagnetic resonance in the rare-earth subsystem. Dillon has observed such a drastic effect of frequency upon one of the peaks shown in Fig. 1 at 50 kMc. The origin of the depression of the field for resonance is slightly more subtle. Qualitatively it arises because the equations of motion of one sublattice contain the dc magnetization and the rf susceptibility of the other. If the usual kind of two sublattice system with very small anisotropy compared to exchange were analyzed, it would be found that the leading terms of these quantities which are linear in the exchange would exactly cancel and the anisotropies would appear directly in the field for resonance. When the crystal

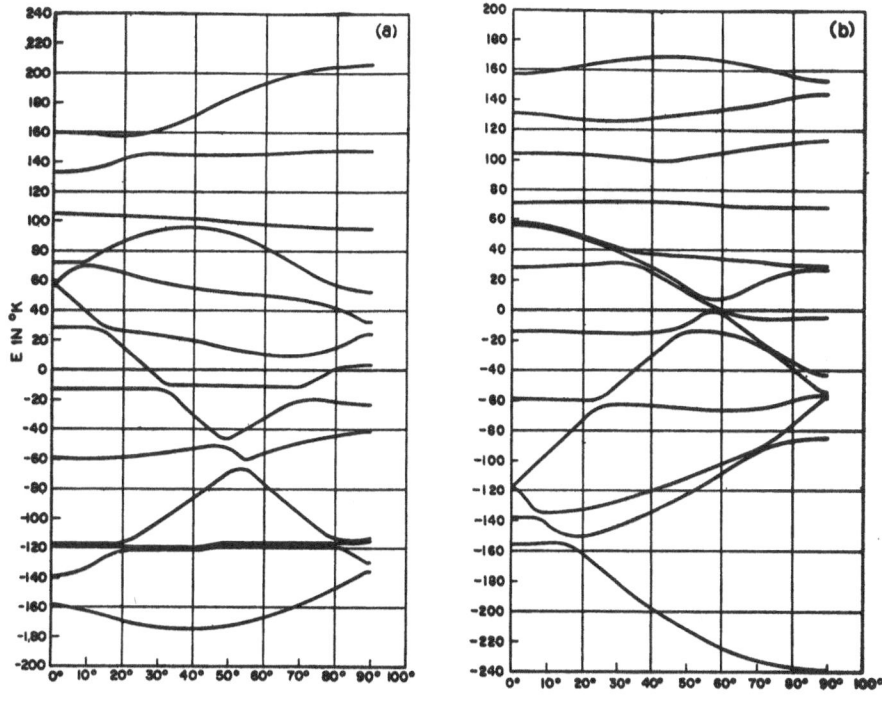

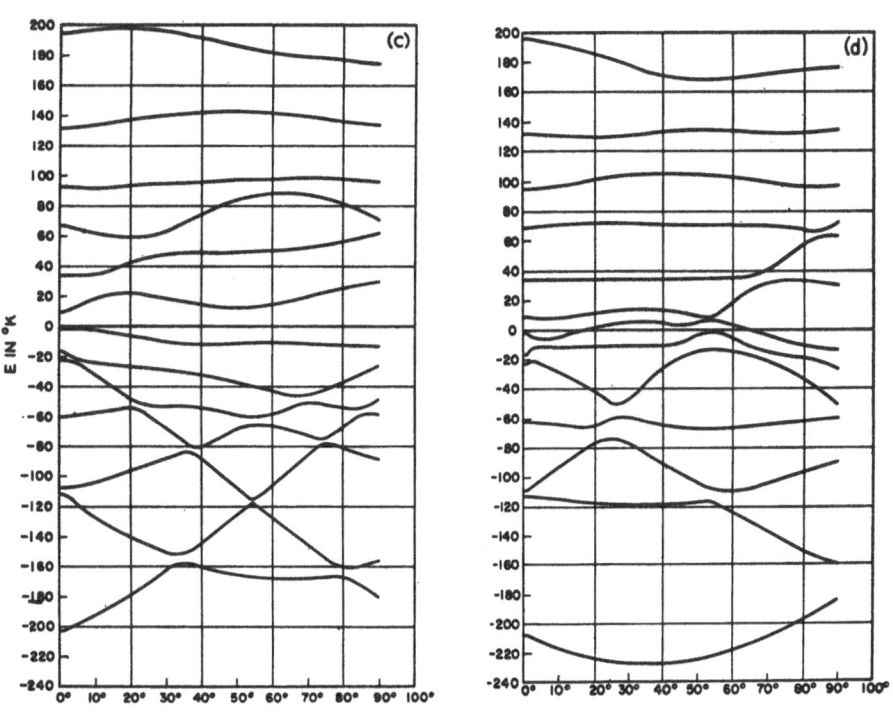

ANGLE IN (1̄10)-PLANE

FIG. 3. Energy levels for the four inequivalent sites as a function of the angle in the (110) plane between the [100] axis and the direction of magnetization for an acceptable set of parameters.

field is comparable with the exchange this cancellation has no reason to take place, and so quite large shifts of the order of the terbium dc exchange field with little angular dependence may arise even for small terbium concentrations.

The program outlined above would be feasible enough if the crystal field in which the terbium ions move were known, but this is unfortunately not the case. One, therefore, has to find a plausible crystal field which will lead to the observed results. As a starting point

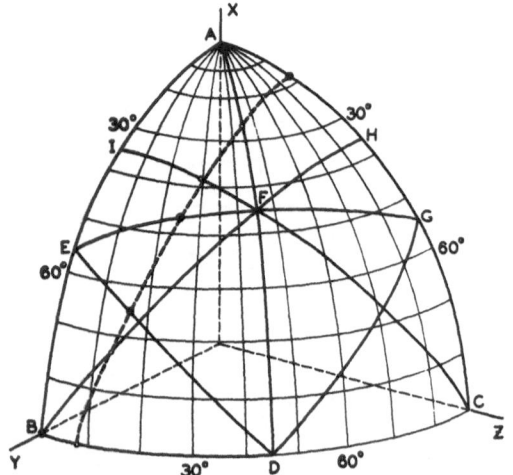

FIG. 4. Locus of near-crossings of the two lowest energy levels as a function of the relative directions of the magnetization and the local orthorhombic axes of a site. $x$, $y$, and $z$ are the site axes. When the magnetic field is turned in the (110) plane, the arcs AB, AC, DEF, and DGZ are explored; when it is turned in the (100) plane, arcs BC and AD are searched; finally, the (111) crystal plane corresponds to arcs BH and CI. The four solid circles represent the four near-crossings found experimentally in the (110) plane; their assignment to sites is based upon computation. The four open circles represent near-crossings found in other crystallographic planes and the assignment to sites is based upon plausibility.

the point charge field of the surrounding oxygen ions has been used. For $f$ electrons in an orthorhombic field one needs the second-, fourth-, and sixth-order spherical harmonics in the potential, a total of $2+3+4 = 9$ terms in all. From these the usual operator equivalents in the $J=6$ manifold are constructed. To do this one has to know the average values of $r^2$, $r^4$, and $r^6$, where $r$ is the radial coordinate of a $4f$ electron, for the terbium ion. Unfortunately, terbium wave functions have to be found by scaling from gold or thallium. The shielding constant used in this scaling is rather large and not too well known. The result is that the required averages are rather uncertain. The following procedure is therefore adopted. It is assumed that the relative sizes of terms in the potential of a given order, two, four, or six, are given correctly by the point charge calculation, but the over-all magnitude of second-, fourth-, and sixth-order terms are left as three adjustable parameters. For finding the relative positions of the energy levels as a function of angle this is all that is needed, for we can use the as yet unchosen size of the exchange field as the energy unit. When the field for resonance is calculated, the exchange must be fixed and this has been done by fitting as well as possible the magnetization curve of terbium iron garnet. Thus, basically three parameters in the crystal field are at one's disposal.

It was found that for applied fields in the (110) plane, for which, incidentally, the number of inequivalent sites is reduced to four, the required four near-crossings could be found at the correct angles for a

rather wide choice of the three parameters. In Fig. 3 are shown the energy levels for the four sites as a function of angle in the (110) plane for a typical set of parameters. More exactly there appears to be a one parameter family of values for the three crystal field constants which give a reasonable fit over a certain range of this parameter. It will be seen that two of the sites contribute one near-crossing each, one yields two of them, and one none. If it is recalled that a fixed field direction for four variously oriented sites is equivalent to four field directions for one site, one can represent the data as in Fig. 4. Here the axes are the symmetry axes of any one of the sites. When the field is moved in the crystallographic (110) plane the arcs AB, AC, DEF and DGZ are explored by the four sites. If the crossings are assigned to particular sites, as is now possible, one finds they occur at the indicated points. It is reasonable to suppose that three of these lie on the curve shown and that the other one is isolated—which, since it occurs at a symmetry point, is allowed to be. If the line which has been drawn is actually the locus of near-crossings it is possible to predict where peaks should be observed for motions of the applied field in other crystallographic planes. This has been done and all the peaks observed by Dillon in the (100), (111), and (112) planes have been assigned to sites and are shown as dots in Fig. 4. Thus, the local situation is not really very complex. The two lowest-energy surfaces appear to "cut" along the indicated line and there is a local minimum in their separation along one of the local symmetry axes.

We now come finally to the field for resonance. It was hoped originally that some of the ambiguity in the crystal field parameters would be removed when the field for resonance was calculated. This is true to some extent, but much remains. Since there is some question about the terbium concentration in the sample for which a fit was to be found, calculations were done for two values of the concentration. It was never possible to achieve a really excellent fit to all the peaks at the same time; summarizing it can be said that there is a range of values for the three crystal field parameters for which one can get all peaks within $\pm 3°$ of the right position and the heights of three of the four peaks rather well—the fourth will then probably be off by a factor of 2 in height. It seems likely that to do better one would need to vary more of the crystal field parameters, rather than three out of nine. It is also true that in fitting we concentrated primarily upon the peaks, while actually there is a considerable difference also in the extent to which the calculation predicts the correct background depression of field. In Figs. 5(a) and (b) are seen some calculated resonance field curves—the first for a concentration of 0.1%, and the second 0.19%. The values of the crystal field parameters required for the first case are shifted very extensively from the starting values based upon the

FIG. 5(a). Comparison between the experimental and computed fields for resonance in the (110) plane for a concentration of 0.1%. The smooth broken curve shows the field for resonance without doping. The iron ion anisotropy deduced from this has been included in the computed fields. (b). Fields for resonance in the (110) plane for a concentration of 0.19%. A complete field for resonance curve was not computed; the points for 0° and 90° are shown by small axes. The crystal field parameters were chosen to give the best fit for each concentration individually.

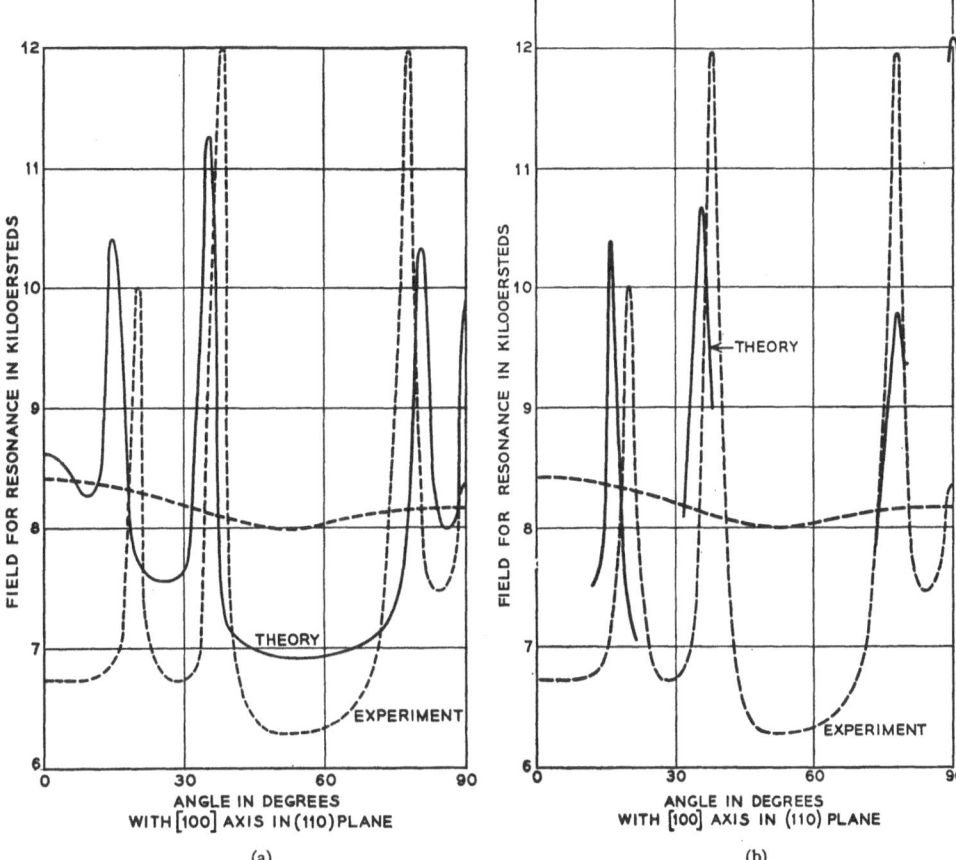

(a)

(b)

estimated moments $r^2$, $r^4$, and $r^6$ and would require the $4f$ wave function to be abnormally large. The values in the second case are somewhat more reasonable. The second case also reproduces the background depression better than the first. A detailed attempt to reproduce the temperature variation of the peaks has not been made in the absence of good quantitative agreement at absolute zero, but it is clear that the level separations are such as to give approximately the right rate of decrease.

It is somewhat difficult to assess the "success" of this calculation. It is true that the main qualitative features of the data are explained and its complexities shown to arise from the multiplicity of sites combined with a rather simple feature of the ionic energy sur-

faces. It is rather surprising and unfortunate that the experimental data does not force a rather narrow choice of crystal fields. Two sets of experimental data which one might attempt to correlate with these calculations are the measurements of dc anisotropy of rare-earth doped garnets and the peaks in line width seen by Dillon in some of these materials. He has advanced the suggestion that one may here be seeing near-crossings of higher levels. It might be noted that it is quite difficult to make use of the information found for terbium doping in analyzing any other rare earth. One is again confronted with a lack of information about the $4f$ wave functions and a consequent uncertainty about the changes to be made in the spin Hamiltonian.

JOURNAL OF APPLIED PHYSICS    SUPPLEMENT TO VOL. 33, NO. 3    MARCH, 1962

# Far Infrared Spectra of Magnetic Materials*

M. TINKHAM

*Department of Physics, University of California, Berkeley, California*

The far infrared spectral region ($\sim$10–100 cm$^{-1}$) corresponds to $kT$ for $T = 15$–150°K or $\beta H$ for $H = 10^5$–$10^6$ oe. Materials with characteristic temperatures or fields of these orders of magnitude may have interesting far infrared spectra. We have studied far infrared resonance spectra in antiferromagnetic and in ferromagnetic rare-earth iron garnets.

Antiferromagnets have a resonance frequency which depends on the exchange and anisotropy fields, $H_E$ and $H_A$. If $H_E$ is found from $\chi_\perp$, $H_A$ can be found from $\omega_0$, and compared with theory. We have done this with FeF$_2$, MnO, and NiO, which have resonances for $T \approx 0$ at frequencies of 52.7 cm$^{-1}$, 27.5 cm$^{-1}$, and 36.6 cm$^{-1}$, respectively. As the temperature is raised, these frequencies fall, reaching zero at $T_N$. When a magnetic field is applied along the easy axis of FeF$_2$, there is a first-order Zeeman splitting, from which a $g$ value of 2.25 was determined. In MnO and NiO, however, there is an easy (111) *plane*. In this case the resonant mode is nondegenerate and hence the Zeeman effect is second-order and unobservably small.

The exchange coupling between the rare earth and iron ions in garnets is typically of order 20 cm$^{-1}$ (30°K), while the iron ions are coupled strongly together, corresponding to their Curie temperature of about 550°K. Because anisotropy breaks down

selection rules, transitions in which a single rare-earth ion spin flips in the exchange field of the iron may be observed at far infrared frequencies. We also observe Kaplan-Kittel exchange resonances, in which the entire sublattice of rare-earth ions precess together and induce a corresponding precession of the iron sublattice. Since the frequency of such a resonance depends on the rare earth sublattice magnetization, it is quite temperature dependent, in contrast to the single ion absorptions. To obtain quantitative agreement with experiment, the Kaplan-Kittel theory must be generalized to take account of anisotropy energy. In YbIG at 2°K, with the magnetization along the [111] easy direction, we find single ion resonances at 23.4 and 26.4 cm$^{-1}$ and an exchange resonance at 14.1 cm$^{-1}$. The latter rises in frequency as the temperature is raised, whereas the former frequencies are nearly constant up to 60°K, where the intensity becomes too low for observation. The corresponding frequencies at low temperatures in ErIG are 18.2, 21.6, and 10.0 cm$^{-1}$, whereas in SmIG the exchange resonance occurs at 33.5 cm$^{-1}$ and *decreases* with increasing temperature. The small temperature dependence of the single ion exchange splittings suggests a rare earth-rare earth coupling (perhaps via spin waves in the iron sublattice) of magnitude $\sim$4% of the iron-rare earth coupling.

## I. INTRODUCTION

THE far infrared spectral region is roughly the decade from 10–100 cm$^{-1}$, corresponding to wavelengths of 0.1–1 mm or 100–1000 $\mu$. In temperature units, this corresponds to about 15–150°K, and in terms of an effective magnetic field acting on a Bohr magneton, it is the region of $10^5$–$10^6$ oe. Since many magnetic materials order below Neél temperatures in the range 50–1000°K, they can be expected to have characteristic exchange energies near or in the far infrared range. A well-known selection rule prevents one from observing transitions in which one spin in an isotropic ferromagnet or antiferromagnet turns over against its exchange field. However, when at least two distinct sublattices are present, even if there is no anisotropy, one can observe an exchange resonance of the sort predicted by Kaplan and Kittel[1] in 1953, which depends directly on the exchange constant. If anisotropy is present, additional single ion transitions in the exchange field *can* be observed. If the material orders into two equal antiparallel sublattices to form an antiferromagnet, antiferromagnetic resonance is possible, as described by Kittel[2] and Nagamiya[3] in 1951. In this resonance, it is essentially a geometric mean of exchange and anisotropy fields which determines the resonance frequency. Since $H_A$ is usually less than $H_E$, this resonance will usually occur for frequencies considerably less than that given by $\hbar\omega \sim kT_N$. These frequencies often

lie in the far infrared. In the following sections, these various types of far infrared resonances will be discussed in more detail.

## II. EXPERIMENTAL TECHNIQUE

At present, the best source of energy in this far infrared region is the long wavelength tail of the radiation from a mercury arc. The desired radiation in a 10% bandwidth is only $\sim$10$^{-10}$ w, and it must be separated from several hundred watts of background at shorter wavelengths. This is done with the monochromator shown in Fig. 1, and described previously.[4] Gratings ruled on dural or solder form the dispersive element, but transmission filters of quartz, black polyethylene, and rocksalt are also essential for eliminating shorter wavelength harmonic radiation diffracting in higher orders. The radiation is chopped at 9 cps to allow a lockin amplifier scheme to be used. The detector is a carbon bolometer, made from an ordinary radio resistor, with evaporated indium electrodes. These bolometers operate at $\sim$1.5°K, where their sensitivity is $\sim$10$^{-12}$ w with a 10-sec electronic time constant. This high sensitivity allows measurements to be made even with the extremely low source power mentioned above, although a further improvement of either source or detector would greatly facilitate further progress in this field of research.

The desired spectra are studied by measurements of transmission through disk samples placed in the brass light pipe used to convey the radiation from monochromator to bolometer. Since this pipe must be of at least $\frac{1}{2}$ in. in diameter (to reduce attenuation by the

* Research supported in part by the Office of Naval Research, The National Science Foundation, and the Alfred P. Sloan Foundation.

[1] J. Kaplan and C. Kittel, J. Chem. Phys. **21**, 760 (1953).
[2] C. Kittel, Phys. Rev. **82**, 565 (1951).
[3] T. Nagamiya, Prog. Theoret. Phys. (Kyoto) **6**, 342 (1951).
[4] R. C. Ohlmann and M. Tinkham, Phys. Rev. **123**, 425 (1961).

light pipe to an acceptable level), the samples must have a similarly large size to avoid loss of energy. Thickness of sample is adjusted so that electric absorption by the tails of the lattice modes at higher frequencies is not excessive. This usually limits the sample thickness to about 1 mm. If the magnetic absorption is then of the same order or larger than this background electric absorption, resonance lines may be observed upon sweeping the frequency by rotating the grating in the monochromator. Because of the variation with frequency of grating efficiency, source intensity, background absorption, and interference effects, it is necessary to make sample-in and sample-out runs, and to take the ratio of signals on a point-by-point basis to obtain reliable information about the existence, frequency, and strength of an absorption line. In view of the low signal-to-noise ratio and the amount of numerical analysis involved, this is a rather tedious process, and data are accumulated rather slowly. Still, interesting results have now been obtained on quite a few systems, and some of these will be described below.

### III. ANTIFERROMAGNETIC RESONANCE

The first of these high-frequency field-free resonances to be predicted and observed was antiferromagnetic resonance. As shown by Kittel[2] and Nagamiya,[3] the resonance frequency in the case of uniaxial symmetry about the sublattice polarization axis should be

$$\omega_0 = \gamma[H_A(2H_E + H_A)]^{\frac{1}{2}} \quad (1)$$
$$= \gamma(2K/\chi_\perp)^{\frac{1}{2}},$$

where $\gamma$ is the gyromagnetic ratio (assumed isotropic here), $H_E = \lambda M$ is the exchange field, $H_A = K/M$ is the anisotropy field, $K$ is the anisotropy constant, and $\chi_\perp$ is the static susceptibility perpendicular to the easy axis. [With lower symmetry, (1) must be generalized.] When $H_A$ arises from purely dipolar anisotropy, $H_A \ll H_E$ (for usual cases), and the frequency is seen to be determined by a geometric mean of $H_A$ and $H_E$ which is much less than $H_E$. Thus in MnF$_2$, $\hbar\omega_0 \sim \frac{1}{5}kT_N$, which falls near 1-mm radiation. However, in FeF$_2$, where there is a large anisotropy arising from crystal field effects acting through the spin-orbit coupling, $\hbar\omega_0 \sim kT_N$, which falls near 50 cm$^{-1}$ or 0.2 mm. Since $H_E$ can be estimated quite well from $\chi_\perp$ or $T_N$, the principal information obtained from measurements of $\omega_0$ concerns the anisotropy energy. This information is of value, because the absence of any net moment in an antiferromagnet makes the anisotropy relatively inaccessible to static measurements.

### FeF$_2$

The first observation of far infrared antiferromagnetic resonance was by Ohlmann and Tinkham[5] on FeF$_2$. This material was chosen because data[6] on

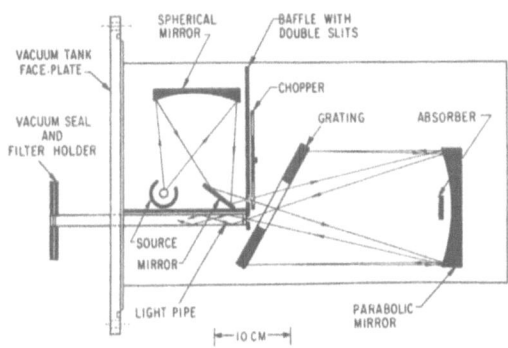

FIG. 1. Schematic diagram of far-infrared monochromator. The diagonal mirror is usually replaced by a reststrahl plate or zero-order filter grating to improve radiation purity.

paramagnetic resonance and anisotropy in the static susceptibility on dilute FeF$_2$ in ZnF$_2$ had shown the existence of a large term $DS_z^2$ in the spin Hamiltonian of each Fe$^{++}$ ion, due to crystal field effects. Since $D$ was estimated to be $-7.3$ cm$^{-1}$, this anisotropy was an order of magnitude larger than the dipolar anisotropy, and the frequency of the resonance was estimated to lie near 50 cm$^{-1}$ (200$\mu$) rather than in the microwave range. After considerable development of far infrared technique, the resonance was observed, the frequency being

$$52.7 \pm 0.2 \text{ cm}^{-1},$$

at $T = 0$. As the temperature was raised, the resonance frequency decreased and the linewidth increased markedly.

Both of these changes were of the sort to be expected qualitatively, but a reliable quantitative theory of these changes is not yet available. Nevertheless, we were able to account for these properties quite well by a simple model in which the macroscopic anisotropy constant $K$ is calculated as a statistical average over the various $S_z$ sublevels of the $S = 2$ manifold for each ion. The energy

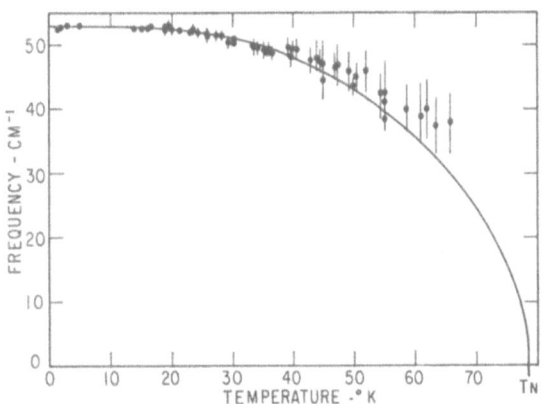

FIG. 2. Temperature dependence of the antiferromagnetic resonance frequency in FeF$_2$, measured by Ohlmann and Tinkham (reference 4). The experimental points represent the frequencies at the transmission minima, which will be slightly displaced from $\omega_0(T)$ when the linewidth is great. The solid curve is the result of the theory sketched in the text, with both $\omega_0(0)$ and $T_N$ fitted to experiment.

[5] R. C. Ohlmann and M. Tinkham, Bull. Am. Phys. Soc. **3**, 416 (1958); Phys. Rev. **123**, 425 (1961).
[6] M. Tinkham, Proc. Roy. Soc. (London) **A236**, 535, 549 (1956).

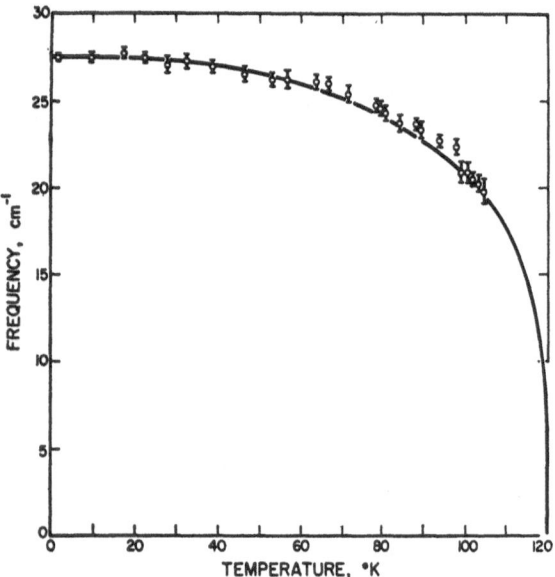

FIG. 3. Temperature dependence of the antiferromagnetic resonance frequency in MnO, as measured by Sievers and Tinkham. The solid curve is the square root of the Brillouin function for $S = \frac{5}{2}$.

expression includes both a molecular-field exchange term linear in $S_z$ and the crystal field $DS_z^2$ term. In this way we calculated $M(T) \sim \langle S_z \rangle(T)$, $\bar{\kappa}(T) = \langle \kappa \rangle(T)$, and $\langle (\kappa - \bar{\kappa})^2 \rangle(T)$, where $\langle \kappa \rangle$ here refers to a value of $K$ per atom averaged over the ensemble.[7] Using these results, we could calculate $\omega_0(T)$ from $K(T)$ and the measured value of $\chi_\perp(T)$. We could also calculate $\Delta\omega(T)$ from $\langle (\kappa - \bar{\kappa})^2 \rangle(T)$ and higher moments. Both of these results are in quite good agreement with experiment. However, $\omega_0(T)$ falls significantly more slowly than the theoretical prediction, as may be seen in Fig. 2. An improved theory for $K(T)$ would be highly desirable here, and for the other materials to be discussed later. The large experimental errors indicated in Fig. 2 arise because of the extremely great linewidth at the higher temperatures. The intrinsic linewidth is about 1 cm$^{-1}$ at 30°K, and it increases approximately as $T^4$. The apparent linewidth is even greater because there is almost 100% absorption even in a sample only 0.5 mm thick until one is well out into the wings of the line.

When a uniaxial antiferromagnet is placed in a magnetic field along the easy axis, the two circularly polarized modes are split apart by a term $\pm\gamma H_0$. By performing this experiment on FeF$_2$, we were able to split the resonance into two components separated by 3.9 cm$^{-1}$ at 18 700 oe. From this result, it was inferred that

$$g_{\parallel} = 2.25 \pm 0.05,$$

in good agreement with the value found by paramagnetic resonance[6] in the dilute salt. In a perpendicular field, the modes are shifted only by an amount of order $(\gamma H_0)^2/\omega_0$, which is unobservably small in this case.

## MnO and NiO

The next materials to be studied were MnO and NiO. These have higher $T_N$ (120° and 523°K) but predominantly only dipolar anisotopry. Accordingly, it turns out that the resonant frequencies at $T = 0$ are 27.5 and 36.6 cm$^{-1}$, respectively, which are considerably less than that of FeF$_2$, although FeF$_2$ has a Neél temperature of only 78°K. NiO has been previously studied by Kondoh,[8] and our data confirm his results and extend them to lower temperatures. The work on MnO by Keffer, Sievers, and Tinkham was reported briefly at this conference last year.[9] These two materials differ from FeF$_2$ in that the dipolar anisotropy produces an easy (111) *plane*, as has been discussed by Keffer and O'Sullivan.[10] Within this plane, there is only a much weaker anisotropy. Keffer and O'Sullivan worked out the normal modes, and found them split into a high-frequency mode involving the strong out-of-plane anisotropy and a low-frequency mode involving only the weak in-plane anisotropy. Only the high-frequency mode lies in the far infrared, and its frequency was predicted within 6% of the observed value by their theory, as corrected in reference 9.

Because the degeneracy of the two circularly polarized modes is completely destroyed in this (easy-plane) case, there can be no first-order Zeeman effect for any orientation of an external field. This was demonstrated experimentally by the absence of any observable shift or broadening of the line when a field of 10 000 oe was applied to a powder sample.

The temperature dependence of the resonant frequency was measured for both MnO and NiO. As shown in Fig. 3, the experimental points for MnO fall very close to the square root of the Brillouin function for $S = \frac{5}{2}$. The experimental points for NiO (not shown) fall almost exactly on top of those for MnO, when $\omega_0(T)/\omega_0(0)$ is plotted against $T/T_N$. However, these points lie somewhat below the square root of the Brillouin function for $S = 1$, which is appropriate for Ni, although they lie well above the Brillouin function itself. Simple

TABLE I. Far infrared antiferromagnetic resonance data.

| Material | $T_N$(°K) | $\omega_0(0)$ (cm$^{-1}$) | Dominant anisotropy |
|---|---|---|---|
| FeF$_2$ | 78 | 52.7 | easy axis |
| MnO | 120 | 27.5 | easy plane |
| NiO | 523 | 36.6 | easy plane |

[7] In this connection it should be noted that Kanamori has shown recently (in unpublished work) that the second order terms $\sim D^2/J$ found in reference 4 are canceled by another effect, leading to the simple relation $K = -6D$ at $T = 0$. This new result considerably improves the agreement between the antiferromagnetic resonance value of anisotropy and that found in reference 6 for the dilute salt.

[8] H. Kondoh, J. Phys. Soc. Japan 15, 1970 (1960).
[9] F. Keffer, A. J. Sievers, III, and M. Tinkham, J. Appl. Phys. 32, 65S (1961).
[10] F. Keffer and W. O'Sullivan, Phys. Rev. 108, 637 (1957); 110, 1484 (1958).

spin-wave and molecular field theories predict a dependence of $\omega_0(T)$ on $M^{\frac{3}{2}}(T)$ and $M(T)$, respectively. Thus there is no simple explanation of these experimental temperature dependences at present. A possible explanation is that $M(T)$ falls more slowly than the Brillouin function. Such behavior would be expected from the lattice distortion theory of Bean and Rodbell reported earlier in this conference.

Some resonance data are summarized in Table I.

## IV. EXCHANGE RESONANCES AND SPLITTINGS IN RARE-EARTH IRON GARNETS

The rare-earth iron garnets have the formula

$$5Fe_2O_3 \cdot 3R_2O_3.$$

The ferric ions are strongly coupled ferrimagnetically, with 6 ions having spin-up and 4 having spin-down. This leaves a net moment of $2 \times (5/2) \times 2 = 10$ Bohr magnetons at $T = 0$. Since the Curie point is $\sim 550°K$, the iron sublattices are nearly completely ordered for the temperatures of interest here, which are below $\sim 70°K$.

The rare-earth ions are coupled to the iron lattice much more weakly, with characteristic exchange energies of $\sim 30°K$. The rare earth-rare earth coupling is weaker still, being of the order of only $1°K$. Thus, at low temperatures, the important excitations are those in which the iron acts as a unit with $M_1 = 10\beta$ per formula unit, and the important exchange constant is that between iron and rare-earth ions, which we denote by $\lambda$. This coupling is of a sense such as to align the spin of the rare earth opposite to that of the iron. In addition to this exchange coupling, the rare-earth ions are also subjected to a crystalline electric field which at least partially lifts the $(2J+1)$-fold degeneracy of the ground state. In fact, these crystal field splittings are usually somewhat larger than the exchange splittings. Therefore, at low temperatures only the lowest Kramers doublet left by the crystal field, but split by the exchange field, will be occupied. (The crystal field must leave at least the two-fold Kramers degeneracy provided only that $J$ is half integral, as it is for the ions—Yb, Er, and Sm—which we have studied. For integral $J$, the following discussion would need modification.) Given these large crystal field effects, the $g$ value and magnetic moment of the rare earth ion will in general differ from that of the free ion, and they will usually be highly anisotropic. In this case, as shown by Van Vleck,[11] it is usually more appropriate to use an "effective" angular momentum $j'$ based on the number of levels (e.g., $j' = \frac{1}{2}$ for a doublet) than an unobservable "true" angular momentum. These points have also been discussed by the author in a paper[12] dealing with the theory of the excitations in the garnets.

In the situation described here, there are three types

TABLE II. Exchange frequencies for rare-earth iron garnets for $T \approx 0$.

| Material | Exchange splitting of ground doublet, $\omega_2(cm^{-1})$ | Exchange resonance frequency, $\omega_e(cm^{-1})$ | $2/5\bar{\omega}_2(cm^{-1})$ |
|---|---|---|---|
| YbIG | 23.4, 26.4 | 14.1 | 10.0(9.8) |
| ErIG | 18.2, 21.6 | 10.0 | 8.2(7.6) |
| SmIG | | 33.5 | |

of excitations which we can observe. The first is the splitting of the ground Kramers doublet by the exchange field $H_E = \lambda M_1$ of the iron. This frequency we denote $\omega_2$, and express as

$$\omega_2 = \gamma_2 \lambda M_1. \qquad (2)$$

It should be remembered that $\gamma_2$ is in general anisotropic, and the exchange constant $\lambda$ itself also appears to be anisotropic (based on resonance data of Boakes et al.,[13] and of Carson and White,[13] and infrared data of Wickersheim[14]). Thus, $\omega_2$ will be different for ions which are inequivalently oriented with respect to the iron exchange field. In our usual experiments, no external field is used, so the iron magnetization lies along the easy direction, which is the [111] direction in YbIG, for example. With the iron magnetization along this symmetry direction, there are only two inequivalent types of sites, and a pair of lines arising from the exchange splitting is observed. The frequencies (for $T \approx 0$) are given in the second column of Table II.

The second type of excitation is a transition to a higher crystal field level, as split by the exchange field. In YbIG, the lowest excited level above the ground doublet is at $\sim 550$ cm$^{-1}$, as shown by Pappalardo and Wood.[15] Since this is well outside the far infrared region, we found no evidence of such transitions. However, in ErIG, there appear to be three excited doublets which lie below 100 cm$^{-1}$ in the pure crystal field. These have been observed in the homologous compound ErGaG, where the exchange field is absent. In the iron garnet, each of these doublets leads to 4 lines, since the double degeneracy is lifted by different amounts in the two types of sites. The resulting spectrum is rather complicated, and we have not yet been able to make any definite assignment of the observed transitions. The problem of assignment is aggravated by the fact that the exchange splittings are comparable to the crystal field splittings. In this case, not only do magnetic sublevels arising from different crystal field levels intermingle, but also second order effects due to off-diagonal matrix elements of the exchange energy are significant. Because of these second order effects, the exchange-split levels need not be exactly symmetric about the unperturbed crystal field level. Moreover, the crystal field

[11] J. H. Van Vleck, Phys. Rev. 123, 58 (1961).
[12] M. Tinkham, Phys. Rev. 124, 311 (1961).

[13] D. Boakes, G. Garton, D. Ryan, and W. P. Wolf, Proc. Phys. Soc. (London) 74, 663 (1959); J. W. Carson and R. L. White, J. Appl. Phys. 31, 53S (1960).
[14] K. A. Wickersheim, Phys. Rev. 122, 1376 (1961).
[15] R. Pappalardo and D. L. Wood, J. Chem. Phys. 33, 1734 (1960).

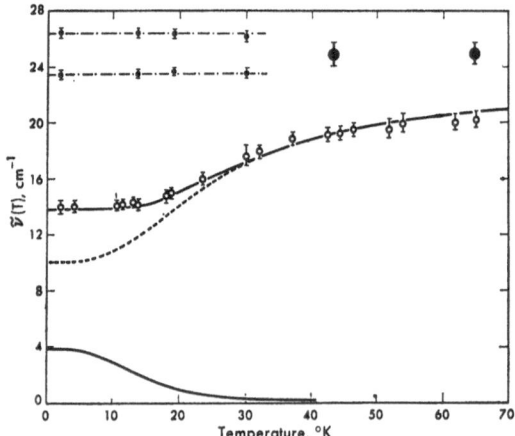

FIG. 4. Temperature dependence of the resonance frequencies in YbIG, as measured by Sievers and Tinkham (reference 16). The two highest frequencies are nearly independent of temperature. Above 30°K only the center of gravity of the doublet could be measured. The dashed curve gives the calculated temperature dependence of the exchange resonance frequency in the isotropic two-sublattice approximation. The solid curve demonstrates the good agreement between theory and experiment when anisotropy energy is introduced. The low-frequency curve is the predicted "ferrimagnetic" resonance with the same model.

splittings would be expected to be somewhat different in ErIG than in ErGaG, in any case.

The third type of excitation is the exchange resonance predicted by Kaplan and Kittel.[1] This is a collective mode, in which the entire iron and rare-earth sublattices undergo a mutual precession reminiscent of that in antiferromagnetic resonance. However, whereas anisotropy is essential for a nonzero antiferromagnetic resonance frequency, the exchange resonance has a nonzero frequency proportional to the exchange constant, provided only that the two sublattices are inequivalent. The theoretical frequency for the isotropic case treated by Kaplan and Kittel is

$$\omega_e = \lambda(\gamma_2 M_1 - \gamma_1 M_2), \qquad (3)$$

provided that the two magnetizations are antiparallel. (If they are parallel, the minus sign becomes a plus sign if magnitudes are used throughout.) If we compare (2) and (3), we see that

$$\omega_e = \omega_2(1 - \gamma_1 M_2/\gamma_2 M_1)$$
$$= \omega_2(1 - J_2/J_1). \qquad (4)$$

In applying this to the garnets, where there is anisotropy, we make a first approximation by simply averaging the two observed values of $\omega_2$, which typically differ by only 15%. Then we note that $J_1 = 5\hbar$ per formula unit (since it arises from 2 ferric ions), whereas $J_2 = 3\hbar$ (since it comes from 6 rare-earth ions, each with a single Kramers doublet entering, with effective angular momentum $j'$ of $\frac{1}{2}$). Therefore (4) becomes

$$\omega_e = \bar{\omega}_2\left[1 - \frac{3}{5}\frac{M_2(T)}{M_2(0)}\right], \qquad (5)$$

where we have allowed for the decrease of $J_2$ upon

disordering of the rare-earth ions at higher temperatures. From this result, we see that at $T=0$, $\omega_e$ should equal $(2/5)\bar{\omega}_2$ if this isotropic approximation were completely accurate. The values given in the last two columns of Table II show that there is only approximate agreement, the experimental frequencies being about 30% greater than predicted.

To eliminate this discrepancy, anisotropy must be introduced into the theory. Since the actual source of the macroscopic anisotropy energy is the anisotropic exchange splitting of the rare earth energy levels, averaged over all sites, it seemed natural to approximate this near an easy axis by an axially symmetric exchange coupling between the iron sublattice and a composite rare earth sublattice. In the theory developed in reference 12, this is expressed by use of an anisotropic $\gamma_2$, with $\gamma_{\parallel} \neq \gamma_{\perp}$. The theory would go through in the same way if one had taken $\lambda_{\parallel} \neq \lambda_{\perp}$, or a combination of the two, which is probably the true situation. In any case, one can evaluate these parameters in terms of the observable macroscopic anisotropy energy $E(\theta)$ by the relation

$$\left[-\frac{1}{E}\frac{\partial^2 E}{\partial \theta^2}\right]_0 = \frac{\lambda_{\parallel}^2 \gamma_{\parallel}^2 - \lambda_{\perp}^2 \gamma_{\perp}^2}{\lambda_{\parallel}^2 \gamma_{\parallel}^2}, \qquad (6)$$

where the derivative is evaluated at the easy axis. Using (6) to eliminate the model parameter $\lambda_{\perp}\gamma_{\perp}$, we find for the resonant frequencies

$$\omega_0 = \frac{\omega_e}{2}\left\{1 \pm \left[1 + \frac{4\gamma_{\perp}\gamma_{\parallel}\lambda_{\parallel}}{\omega_e^2}\left(\frac{\partial^2 E}{\partial \theta^2}\right)_0\right]^{\frac{1}{2}}\right\}, \qquad (7)$$

where $\omega_e = \lambda_{\parallel}(\gamma_{\parallel}M_1 - \gamma_1 M_2)$. In the absence of anisotropy, this reduces correctly to give $\omega_0 = 0$, $\omega_e$. When $\partial^2 E/\partial \theta^2 > 0$, as it must be at an easy axis, the resonance near $\omega_e$ is increased in frequency above $\omega_e$, as is observed. Also, this expression predicts the frequency of the low frequency ferrimagnetic resonance mode in the presence of the anisotropy "field," but no external field.

Returning to the comparison with experiment, the only detailed check of this theory has been carried out on YbIG, and reported recently by Sievers and Tinkham.[16] For YbIG, values of $M_2(T)$ and $(\partial^2 E/\partial \theta^2)$ $(T)$ have been calculated by Henderson and White[17] from the spectroscopic results of Wickersheim,[14] and the microwave resonance[18] g values. These computed values are in quite good agreement with the magnetization measured by Pauthenet[18] and with the torque measurements of anisotropy by the Mullard group[19] and the resonance measurements of anisotropy by

[16] A. J. Sievers, III, and M. Tinkham, Phys. Rev. 124, 321 (1961).

[17] J. W. Henderson and R. L. White, Phys. Rev. 123, 1627 (1961).

[18] R. Pauthenet, Ann. Phys. 3, 424 (1958).

[19] R. F. Pearson and R. W. Cooper, J. Appl. Phys. 32, 265S (1961); Proc. Int. Conf. on Magnetism and Crystallography, Kyoto, 1961 (to be published).

Dillon and Nielson.[20] Since the two latter types of measurements were of necessity made on dilute samples of YbIG in YIG, we used the computed dependences of Henderson and White for our comparison. As shown in Fig. 4, the agreement between theory and experiment is highly satisfactory in view of the fact that no free parameters have been used. Note the characteristic approach of the exchange resonance frequency to the single ion splitting frequency $\bar{\omega}_2$ as the rare-earth sublattice disorders with increasing temperature.

The experimental data on ErIG also show the exchange resonance moving up in frequency as the temperature is increased. Lacking precise quantitative information on sublattice magnetization and torques, we have not made a detailed fit to the theory. However, there seems to be little doubt that the agreement would be similar to that found for YbIG. For example, the required anisotropy correction from $2/5\bar{\omega}_2$ to $\omega_0(0)$, as given by Table II, is of the same order of magnitude as in YbIG, and results of Dillon and Nielson[20] suggest that the anisotropy energy in these material is, in fact, also roughly the same.

A new result arising from analysis of the ErIG spectrum is that there is a measurable shift of the frequency of the single ion exchange splittings with temperature. This shift appears to be fit by an expression of the form

$$\omega_2(T) = \gamma_2[\lambda M_1 + \lambda' M_2(T)], \qquad (8)$$

where $\lambda'$ represents a ferromagnetic coupling between rare-earth ions. For the higher frequency component of the doublet, the magnitude of $\lambda'$ is about 4% of that of $\lambda$. Since $M_2(0) = 3.3 M_1$ in ErIG, this results in a 13% correction. Experimentally, this spectrum can be followed to a temperature where $M_2(T)/M_2(0)$ has fallen to about $\frac{1}{2}$, with a resulting frequency shift of about 1.3 cm$^{-1}$. Curiously, the shift seems to be much less for the low-frequency component of the doublet. The reason for this difference is unknown at present. Naturally the values of $\omega_2$ appearing in equations (4) and (5) should be corrected for this shift. This correction lowers the value of $\bar{\omega}_2$ entering the last column of Table II for ErIG by about 0.6 cm$^{-1}$ to 7.6 cm$^{-1}$. This value is given parenthetically in the table. Having discovered this effect in ErIG, we re-examined the YbIG spectrum and found a small shift of the upper component there also, reaching about 0.4 cm$^{-1}$ at 21°K. This also fits (8) with $\lambda'/\lambda \approx 0.04$. The change is harder to observe because $M_2(Yb)$ is only $\frac{1}{3}$ as large as $M_2(Er)$.

The strength of this rare earth-rare earth coupling is surprisingly large in view of the susceptibility measurements of Wolf and co-workers[21] on YbGaG and ErGaG. They found ferromagnetic coupling in the former, but antiferromagnetic coupling in the latter. Moreover, the Curie-Weiss $\theta$ value was only $\sim 0.1$°K for Yb. The interaction was also small in the case of Er, although no $\theta$ value could be assigned. A possible explanation for our observed ferromagnetic coupling in the *iron* garnets has been given by Suhl.[22] This is based on his previous theory[23] of nuclear spin interactions in ferromagnets, and it utilizes the second order coupling via the polarization of the iron spin wave spectrum by exchange coupling to the rare-earth ions. The order of magnitude of the coupling is correctly predicted to be

$$\lambda' \sim \lambda^2/c,$$

where $c$ is the iron-iron exchange constant. The original theory leads only to a coupling of transverse components, but given the large anisotropy of the rare-earth exchange coupling, the longitudinal component should be coupled only somewhat less effectively. It is conceivable that a difference in this reduction factor between the two types of sites would account for the different magnitude of shifts noted above.

Finally, we mention our preliminary results on SmIG. Here, the only strong line is the exchange resonance, found at 33.5 cm$^{-1}$ at $T = 2$°K. As the temperature is increased, this line moves rapidly toward *lower* frequencies, in contrast to the behavior found for YbIG and ErIG. The reason for this behavior is simply that in Sm, spin and orbital angular momentum are antiparallel, so that the *antiferromagnetic* coupling between the spin and that of the iron tends to align $J$ *ferromagnetically*. Therefore, a plus sign enters into (4) and (5). In that case, the exchange resonance lies *above* the exchange splitting, and it drops down toward the latter as the temperature is raised. It will be interesting to see if we are able to achieve the same degree of success in interpreting the spectrum of SmIG as has been possible in the cases of YbIG and ErIG. In any case, the general outline seems clear, and it appears that we are well on the way to a detailed understanding of these interesting rare-earth iron garnet systems.

## ACKNOWLEDGMENTS

It is a pleasure to acknowledge explicitly the contributions of R. C. Ohlmann in design and construction of the monochromator and in his pioneer experiments on FeF$_2$; of A. J. Sievers, III, for his tireless and efficient work in measuring the spectra of MnO, NiO, YbIG, ErIG, SmIG, ErGaG, and ErAlG; and of Dr. R. L. White, Dr. K. A. Wickersheim, Dr. W. P. Wolf, Professor F. Keffer, and Professor H. Suhl in helpful discussions. Thanks are also due to Professor J. W. Stout for providing the samples of FeF$_2$, to Professor R. V. Jones for polycrystalline garnet samples, and to Dr. R. A. Lefever and co-workers for single-crystal garnet samples.

[20] J. F. Dillon, Jr., and J. W. Nielson, Phys. Rev. **120**, 105 (1960).
[21] M. Ball, G. Garton, M. J. M. Leask, D. Ryan, and W. P. Wolf, J. Appl. Phys. **32**, 267S (1961); *Proceedings of the VIIth International Conference on Low Temperature Physics* (University of Toronto Press, Toronto, 1961), p. 129.
[22] H. Suhl, unpublished comment at Osaka Magnetic Resonance Symposium, October, 1961.
[23] H. Suhl, Phys. Rev. **109**, 606 (1958).

JOURNAL OF APPLIED PHYSICS    SUPPLEMENT TO VOL. 33, NO. 3    MARCH, 1962

# Electron Paramagnetic Resonance of Trivalent Gadolinium in the Yttrium Gallium and Yttrium Aluminum Garnets

L. Rimai and G. A. deMars

*Raytheon Company, Research Division, Waltham, Massachusetts*

The spin Hamiltonian parameters for $Gd^{+++}$ when substituted for $Y^{+++}$ in two diamagnetic garnets, YAlG and YaG, have been determined by electron paramagnetic resonance techniques. From these results the contribution of the independent $Gd^{+++}$ ions to the anisotropy of gadolinium iron garnets may be estimated. The fourth-order constants in the spin Hamiltonian are very similar in YAlG and YGaG and their contribution to the anisotropy constant in GdIG shows good agreement with the experimental data. The contribution of the second-order constants falls short of the values needed to fit experimental data by an order of magnitude. These constants, however, differ markedly in the two garnets and thus their use to predict anisotropy in GdIG is not justified. The present results, therefore, are not in disagreement with a model that ascribes the origin of the crystalline anisotropy in magnetic garnets to the anisotropy energy of the individual ions in the presence of the crystalline electrostatic field.

PARAMAGNETIC ions with $S$ ground states usually exhibit, in the diamagnetic environment of a host crystal, splittings of the order of 1 cm$^{-1}$.[1] They are thus accessible to direct measurement by EPR techniques with conventionally available magnetic fields and microwave frequencies. When such ions are coupled into a magnetic sublattice by relatively strong exchange interactions, as in magnetic garnets, these splittings increase to values beyond the range where determinations by simple, direct methods are at present feasible. This increase comes from the large Zeeman splitting in the presence of the molecular field which represents the net effect of the isotropic exchange interactions of the ion with its magnetic neighbors. The over-all splittings of the ground states of individual ions will, however, still exhibit the small contribution from the crystal field. This contribution will make itself felt as an anisotropy in the total magnetic energy of the sublattice to which they are coupled. As a consequence this energy will depend on the orientation of the sublattice magnetization with respect to the crystal axes.[2]

The interaction of the ion ground state with the crystal field and the effective magnetic field **H** may be described phenomenologically by a spin Hamiltonian[1] which we write as

$$\mathcal{H}(\mathbf{S}) = g\beta \mathbf{H} \cdot \mathbf{S} + \sum_{l,m} b_{l,m} Y_{l,m}(\mathbf{S}), \qquad (1)$$

$g$ is the spectroscopic splitting tensor, which for $S$ state ions may be taken as isotropic.[1] $Y_{l,m}(\mathbf{S})$ are polynomials in the components of the ground state spin vector **S**, that transform under rotation as corresponding $l,m$th spherical harmonics.[3] The parameters $b_{l,m}$ describe the strength and symmetry of the crystal field interaction at the site occupied by the ion. This local symmetry at the various ion positions is not necessarily the same as the over-all symmetry of the unit cell. The anisotropy

of the sublattice, however, is that given by the cell symmetry. Due to this fact and also to the thermal disorder in the magnetic lattice, the energy levels obtained from (1) have to be properly averaged over the magnetically inequivalent sites and over the various $|m_s\rangle$ ground substates according to their thermal population, in order to yield the contribution to the total anisotropy. Such average calculations[2,4] have been performed for the three sublattices in garnets and as a result expressions exist for the contribution to the total magnetic anisotropy from the individual ion anisotropy as represented by the parameter $b_{l,m}$. Since these parameters are not available for the ions in the magnetic garnet, we have to use parameters determined for these ions when included substitutionally in isomorphous diamagnetic crystals. Such results, however, have to be considered with reservations as we shall discuss later. For the $Fe^{+++}$ sublattices in YIG this has been carried out by using parameters obtained from EPR in yttrium gallium garnet YGaG.[5] No detailed agreement was found between the experimental fourth-order anisotropy constants $K_1$ in YIG and the value calculated from the EPR data. Nevertheless, magnitude-wise the agreement was sufficiently good as to show that this independent ion mechanism is a major contributor to the total anisotropy of the magnetic crystal.[6]

It seemed, therefore, of interest to carry out a similar investigation for the $Gd^{+++}$ sublattice in gadolinium iron garnet (GdIG). We performed the EPR measurements at a frequency of 36 kMc on two isomorphous crystals, YGaG[7] and YAlG (yttrium aluminum garnet).

For an arbitrary orientation of the external dc magnetic field **H**, there are six inequivalent yttrium sites, each having three mutually perpendicular axes (of two-fold symmetry). The analysis of the angular dependence of the spectra, in agreement with results of structural

[1] W. Low, *Paramagnetic Resonance in Solids*, Solid State Physics Series, Supplement 2 (Academic Press, Inc., New York, 1960).
[2] W. P. Wolf, Phys. Rev. **108**, 1152 (1957).
[3] M. E. Rose, *Elementary Theory of Angular Momentum* (John Wiley & Sons, Inc., New York, 1957).
[4] K. Yosida and M. Tachiki, Progr. Theoret. Phys. (Kyoto) **17**, 331 (1957).
[5] S. Geschwind, Phys. Rev. **121**, 363 (1961).
[6] EPR data of $Fe^{+++}$ in YAlG further justify this statement [L. Rimai and G. deMars (to be published)].
[7] The YGaG crystals were grown by Mr. C. Quadros of Harvard University.

TABLE I. Spin Hamiltonian parameter of $Gd^{+++}$ in the diamagnetic garnets (in $cm^{-1}$).

| Crystal $T(°K)$ | YGaG 300°K | YGaG 4.2°K | YAlG 300°K | YAlG 4.2°K |
|---|---|---|---|---|
| $g$ | $1.992\pm0.003$ | $1.992\pm0.003$ | $1.990\pm0.003$ | $1.990\pm0.003$ |
| $b_{2,0}$ | $(441.1\pm0.5)\times10^{-4}$ cm$^{-1}$ | $(448.5\pm0.5)\times10^{-4}$ | $(777.7\pm0.5)\times10^{-4}$ | $(776.4\pm0.5)\times10^{-4}$ |
| $b_{2,2}$ | $(-269.7\pm0.5)\times10^{-4}$ cm$^{-1}$ | $(-283.8\pm0.5)\times10^{-4}$ | $(-85.1\pm0.5)\times10^{-4}$ | $(-96.9\pm0.5)\times10^{-4}$ |
| $b_{4,0}$ | $(-42.2\pm0.5)\times10^{-4}$ | $(-45.2\pm0.5)\times10^{-4}$ | $(-46.9\pm0.5)\times10^{-4}$ | $(-48.6\pm0.5)\times10^{-4}$ |
| $b_{4,4}$ | $(21.1\pm0.5)\times10^{-4}$ | $(22.1\pm0.5)\times10^{-4}$ | $(23.2\pm0.5)\times10^{-4}$ | $(23.5\pm0.5)\times10^{-4}$ |
| $b_{6,0}$ | $(+0.6\pm0.5)\times10^{-4}$ | $(+4.15\pm0.5)\times10^{-4}$ | $(+3.77\pm0.5)\times10^{-4}$ | $(4.62\pm0.5)\times10^{-4}$ |
| $b_{6,4}$ | $(0.1\pm0.5)\times10^{-4}$ | $(1.28\pm0.5)\times10^{-4}$ | $(2.43\pm0.5)\times10^{-4}$ | $(4.29\pm0.5)\times10^{-4}$ |

determinations,[8] indicates that two of these axes lie in a (100) plane. The third one is, by necessity, the [100] direction perpendicular to this plane. The two axes in the plane are found to be parallel to the two [110] directions in this (100) plane (within the experimental accuracy of $\pm0.5°$). The various possibilities of combining three such directions yield indeed six inequivalent sites.

The $O^{--}$ ions which are the nearest neighbors of the $Y^{+++}$ lie on the vertices of a distorted cube. If the cube were perfect, one of the [110] from the above axes would be parallel to the edge, and the two others (the [100] and the other [110]) parallel to two face diagonals. The spin Hamiltonian that describes the splitting of the $S$ ground state of $Gd^{+++}$ in the crystal field of the $Y^{+++}$ sites will assume its simplest form in a coordinate system for which the $z$ axis is parallel to the "cube edge," and $x$ and $y$ are the two other two-fold axes. Due to the distortion of the $O^{--}$ cube, $z$ will not be rigorously fourfold; it will be, however, among the three axes, the one closest to it. This means that when we examine the spectrum obtained with $\mathbf{H}$ parallel to these three axes, the axis we should label $z$ is that which exhibits the smallest rhombic term and the largest fourth-order splittings. In this system of coordinates, we rewrite (1) as

$$\mathcal{3C}(\mathbf{S}) = g\beta\mathbf{H}\cdot\mathbf{S} + b_{20}Y_{20} + b_{2,2}(Y_{2,2}+Y_{2,-2}) + b_{40}Y_{40}$$
$$+ b_{4,4}(Y_{4,4}+Y_{4,-4})b_{60}Y_{60} + B_{6,4}(Y_{6,4}+Y_{6,-4}). \quad (2)$$

The conventions in (2) are such that the matrix elements $\langle S,M_s' | Y_{l,m} | S,M_s \rangle$ are proportional to the corresponding Wigner coefficient $C(S,l,S; M_s',mM_s)$ with a constant of proportionality independent of $m$ chosen in such a way that the $b_{l0}$ have their usual meaning.[1]

From (2) the energy levels, as a function of the direction and magnitude of $\mathbf{H}$ and thus the spectrum, may be calculated. This was done by rotating the coordinates to new axes such that $z || \mathbf{H}$, and then the crystal field terms were treated by perturbation theory to second order in $1=2$ terms and first order in the others. The parameters were determined by fitting the spectrum when $z$ is parallel to the two 110 axes $z$ and $x$. Figure 1 shows the angular dependence of a $Gd^{+++}$ ion when $\mathbf{H}$ is rotated in a (100) plane. $\theta=0°$ corresponds to $H||z$. The parameters so obtained are listed for two tempera-

tures on Table I. The signs are absolute and were determined by observing the temperature dependence of the relative intensities of the various lines.[5]

Using the results of Wolf's calculation,[2] Rodrique et al.[9] (RMJ) have fitted the temperature dependence of the Gd contribution to the anisotropy of GdIG and have determined two parameters that relate to the $b_{l,m}$ in the spin Hamiltonian.

They write the first-order anisotropy constant per ion of the $Gd^{+++}$ sublattice as

$$k_1 = k_1 r(T) + k_2(T)/T.$$

The expressions for $k_1$ and $k_2$ in terms of the $b_{l,m}$ in (2) are

$$k_1 = -7b_{40} - \frac{3(70)^{\frac{1}{2}}}{5}b_{4,4}$$

$$k_2 = \frac{T}{k}\left(\frac{1}{6}b_{20}^2 - \frac{5(6)^{\frac{1}{2}}}{9}b_{20}b_{2,2} - \frac{7}{9}b_{2,2}^2\right).$$

For this case we had to generalize slightly $k_2$ from the expression obtained by Wolf to include $b_{2,2}$ term. $r(T)$

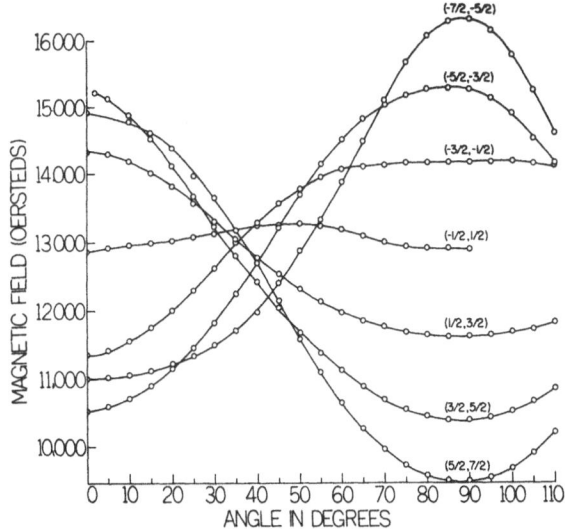

FIG. 1. Angular dependence of the spectrum of $Gd^{+++}$ in YGaG. $H$ is in the (100) plane forming an angle $\theta$ with the [110] direction.

[8] F. Euler (private communication).

[9] P. Rodrique, H. Meyer, and R. V. Jones, J. Appl. Phys. 31, 376S (1960). This reference will be hereafter called RMJ.

TABLE II. Comparisons of Gd contributions to anisotropy in GdIG as calculated from EPR data and from a fit to directly measured values (in cm$^{-1}$).

|  | YGaG 300°K | YGaG 4.2°K | YAlG 300°K | YAlG 4.2°K | RMJ (fit to experimental data in GdIG)[a] |
|---|---|---|---|---|---|
| $k_1$ | $189.48 \times 10^{-4}$ | $205.46 \times 10^{-4}$ | $211.84 \times 10^{-4}$ | $222.23 \times 10^{-4}$ | $209 \times 10^{-4}$ cm$^{-1}$ |
| $k_2$ | $11.24 \times 10^{-4}$ | $11.76 \times 10^{-4}$ | $15.13 \times 10^{-4}$ | $15.97 \times 10^{-4}$ | $504 \times 10^{-4}$ cm$^{-1}$ |

[a] See reference 9.

and $t(T)$ are the temperature dependent factors determined by Wolf. Table II shows a comparison between values for $k_1$ and $k_2$ obtained by RMJ[9] and from our EPR results.

The reasonable agreement obtained for $k_1$ seems to be meaningful since the fourth-order parameters on which it depends do not change appreciably from YAlG to YGaG. This lends some justification to the procedure whereby RMJ calculated the Gd contribution to the anisotropy of GdIG. They subtracted from the total anisotropy constant for this crystal that measured for YIG. This in turn lends support to the assumption that contributions to anisotropy from interactions between the Gd$^{+++}$ and the Fe sublattices is small since if it were not so the various contributions would not be additive.

The lack of agreement for $k_2$ seems, on the other hand, to go along with the fact that the second-order terms are quite different in the two diamagnetic garnets. The fourth-order terms not being too different for the two crystals might be related to the fact that the total fourth-order term in (2) corresponds to a field of practically cubic symmetry, with one axis parallel to the local $z$ axis (110), the other two rotated from $x$, $y$, by an angle $\alpha = 36.7°$. The cubic field around Gd in a dodeca-

hedral oxygen arrangement is probably quite independent of the host crystal. The influence of the host, however, will show itself markedly in the distortions of the cubic field, either by the distortion of the oxygen cube itself or by the effect of further neighbors. A similar situation seems to prevail for Fe$^{+++}$ in an octahedral oxygen environment.[5] The perturbation mechanisms that give rise to the ground state splittings might be such as to enhance the effect of the distortions represented by the $l=2$ terms.

In conclusion, we feel that the present results lend further support to the independent ion theory of magnetocrystalline anisotropy in rare-earth iron garnets. More detailed agreement with experiment can only be hoped for if either direct measurements of the $b_{l,m}$ parameters in YIG become available, or if the theory of these parameters for $S$ state ions is developed to a point permitting their calculation from first principles.

### ACKNOWLEDGMENTS

We would like to acknowledge Dr. P. B. Nutter and Professor R. V. Jones for many helpful discussions, and to thank Dr. S. Warshaw of this laboratory for growing the YAlG crystals.

JOURNAL OF APPLIED PHYSICS    SUPPLEMENT TO VOL. 33, NO. 3    MARCH, 1962

# High Temperature Susceptibility of Garnets : Exchange Interactions in YIG and LuIG

Peter J. Wojtowicz

*RCA Laboratories, Princeton, New Jersey*

Power series expansions of the high temperature susceptibility and its inverse in ascending powers of the reciprocal temperature have been obtained through the use of an extension of the method of Rushbrooke and Wood. The Heisenberg form of exchange is adopted and interactions between neighboring spins from different sublattices (a-d exchange) only are included. The coefficients in the series are derived for arbitrary spins on the two sublattices. The calculations have been carried out to terms including the fifth power of the exchange divided by the temperature; the molecular field theory by contrast is rigorously valid only to the first power term of its expansion. The inverse susceptibility series have been used to determine the magnitude of the a-d exchange interactions in the ferrimagnetic garnets, YIG and LuIG. The intra-sublattice interactions are assumed to be zero, in keeping with the increasing suspicion on the part of numerous investigators that these interactions have been seriously overestimated by molecular field considerations. The data of Aléonard was analyzed by a least-squaring method, and the following values for the a-d exchange were obtained: $J/k = -35.0°K$ for YIG, and $J/k = -34.5°K$ for LuIG. The deviation of theory from experiment is about 1%. In addition, the value for YIG is in satisfactory agreement with the Landau-Lifshitz constant determined from the heat capacity and spin-wave spectrum. It is concluded that these results support the hypothesis of relatively weak intra-sublattice exchange in the garnets.

## I. THEORY OF THE SUSCEPTIBILITY

EXACT series expansions of the high temperature susceptibility and its inverse in ascending powers of the reciprocal temperature have been obtained for the general two sublattice ferrimagnet through an extension[1] of the method of Rushbrooke and Wood.[2] The Heisenberg form of exchange is adopted and interactions between nearest neighboring spins from different sublattices only are considered. The coefficients in the series are derived for general spins and arbitrary $g$ factors on the two sublattices. Unfortunately, space limitations permit the presentation of the results on the garnets alone; the detailed mathematical treatment of the general ferrimagnetic susceptibility will be published elsewhere.

The physical system of interest is one formula weight of garnet, $C_3A_2D_3O_{12}$ containing $5N$ transition metal cations distributed among two nonequivalent sublattices. The $A$ sublattice contains $2N$ octahedrally coordinated sites occupied by cations with spin $S_A$, while the $D$ sublattice has $3N$ tetrahedrally coordinated sites occupied by cations with spin $S_D$. The $C$ sublattice is occupied by nonmagnetic cations of zero spin. Each $A$ site cation has six nearest neighbor $D$ site cations with which strong indirect exchange interactions occur (commonly called a-d exchange). Each $D$ site cation has four nearest neighbor $A$ sites with which it interacts. The intra-sublattice exchange interactions (a-a and d-d exchange) are assumed to be negligible. This is in keeping however, with an increasing suspicion on the part of numerous investigators that these interactions play a minor role in the garnets and have been seriously overestimated by molecular field considerations. The spin Hamiltonian of this system in the presence of an external magnetic field $H_z$ has the form,

$$\mathcal{H} = -2J\mathbf{P} - g\mu H_z \mathbf{Q},$$
$$\mathbf{P} = \sum_{ij} \mathbf{S}_i \cdot \mathbf{S}_j, \quad \mathbf{Q} = \sum_i S_{iz} + \sum_j S_{jz}, \quad (1)$$

where $\mu$ is the Bohr magneton and $J$ is the magnitude of the a-d exchange interactions ($J < 0$ for ferrimagnetism). The operator $\mathbf{P}$ is the sum of Heisenberg exchange operators for all nearest neighbor $A$-$D$ pairs in the lattice, while $-g\mu H_z\mathbf{Q}$ is the Zeeman energy operator for the entire system ($i$ and $j$ are summed over $A$ and $D$ sublattices, respectively).

The molar zero field susceptibility is expanded in a power series in the variable $J/kT$:

$$X = \frac{C_M}{T}\{1 + \sum_n a_n (J/kT)^n\}, \quad (2)$$

where $k$ is the Boltzmann constant and $C_M$ the molar Curie constant. The first term is Curie's law for noninteracting spins, while succeeding terms represent increasing orders of the statistical mechanical perturbation of the exchange on the free ion paramagnetism. According to a straight-forward statistical mechanical development analogous to that reported in reference 2, the coefficients $a_n$ are given by

$$a_n = \frac{3 \cdot 2^n \Gamma_N \langle \mathbf{P}^n \mathbf{Q}^2 \rangle}{n! N (2\bar{A} + 3\bar{D})}, \quad (3)$$

where $\langle \mathbf{X} \rangle$ means $(2S_A + 1)^{-2N}(2S_D + 1)^{-3N}$ times the trace of the matrix representation of the operator $\mathbf{X}$, and where $\Gamma_N \langle \mathbf{X} \rangle$ means "that part of $\langle \mathbf{X} \rangle$ which is proportional to $N$." The quantity $\bar{A}$ is $S_A(S_A + 1)$ while $\bar{D}$ is $S_D(S_D + 1)$. The most difficult and tedious part of the calculation is the evaluation of the $\Gamma_N \langle \mathbf{P}^n \mathbf{Q}^2 \rangle$. This has been carried out to $n = 5$ by using the diagramatic analysis described in reference 2. The resulting $a_n$ are (for the same $g$ factor on both sublattices)

$$a_1 = \Omega, \quad a_2 = \Omega[6\bar{A} + 10\bar{D} - 3]/6,$$
$$a_3 = \Omega[288\bar{A}\bar{D} - 51\bar{A} - 81\bar{D} + 12]/45,$$
$$a_4 = \Omega[1576\bar{A}^2\bar{D} + 2680\bar{A}\bar{D}^2 - 2312\bar{A}\bar{D} - 192\bar{A}^2$$
$$- 520\bar{D}^2 + 324\bar{A} + 510\bar{D} - 45]/270, \quad (4)$$
$$a_5 = \Omega[506\,560\bar{A}^2\bar{D}^2 - 239\,160\bar{A}^2\bar{D} - 278\,360\bar{A}\bar{D}^2$$
$$+ 160\,662\bar{A}\bar{D} + 22\,320\bar{A}^2 + 58\,020\bar{D}^2$$
$$- 16\,524\bar{A} - 25\,974\bar{D} + 1728]/14\,175,$$

[1] P. J. Wojtowicz, J. Appl. Phys. **31**, 265S (1960).
[2] G. S. Rushbrooke and P. J. Wood, Molecular Phys. **1**, 257 (1958).

where $\Omega = 16\bar{A}\bar{D}/(2\bar{A}+3\bar{D})$. Considerable care was taken to ensure the reliability of these results; at several stages in the computation the results were checked against those of Rushbrooke and Wood by removing the restriction that the spins and sublattices be nonequivalent. The inverse susceptibility may also be expanded in a power series,

$$X^{-1} = \frac{T}{C_M}\{1 + \sum_{n=1} b_n(J/kT)^n\}. \qquad (5)$$

The $b_n$ are too complex to reproduce here, but may be computed from the formula

$$b_n = -\sum_{r=1}^{n} a_r b_{n-r}, \quad b_0 = 1. \qquad (6)$$

For the special case of $S_A = S_D = 5/2$ to be considered in the next section, the numerical values of $a_1$ through $a_5$ are 28, 693$\frac{1}{3}$, 13 008.800, 272 420.81, and 5 216 287.1, respectively.

## II. EXCHANGE INTERACTIONS IN YIG AND LuIG

Proceeding under the assumption that the intra-sublattice interactions are indeed negligible in garnets, the derived susceptibility series were used to determine the magnitudes of the $a$-$d$ exchange interactions in the important ferrimagnets, yttrium iron garnet (YIG) and lutetium iron garnet (LuIG). The recent experimental data of Aléonard[3] was treated in two separate ways. First, the molar Curie constant was taken to be fixed at the spin-only value, $C_M = 21.89$. A least-squaring method then gave the following for the $a$-$d$ exchange: $J/k = -35.7°$K for YIG and $J/k = -37.2°$K for LuIG. Secondly, both $C_M$ and $J$ were adjustable and a similar least-squaring procedure yielded $C_M = 21.7$, $J/k = -35.0°$K for YIG and $C_M = 21.3$, $J/k = -34.5°$K for LuIG. In both methods of analysis precautions were taken to include only those experimental points which fell within the range of the practi-

cal convergence of the truncated series. For the present materials this corresponded to including in the analysis only those measurements made above 700°K. The agreement between theory and experiment was quite satisfactory in this range, the rms deviation being on the order of 1% in both cases.

A more stringent test of the validity of the derived values of $J$ is the comparison of theory with experiments other than those from which the values were obtained. For this purpose it is convenient to convert the exchange integral $J$ into the equivalent Landau-Lifschitz stiffness constant $A$ and the spin-wave dispersion relation constant $D$. The required formulas are[4]

$$A = 4SD/a^3 = (8J_{aa}+3J_{dd}-5J_{ad})S^2/2a, \qquad (7)$$

where the $J_{kl}$ are the indicated interactions and $a$ is the cubic lattice constant. For YIG, the representative value $J_{ad}/k = -35.0°$K (with $J_{aa} = J_{dd} = 0$) corresponds to $A = 6.1 \times 10^{-7}$ ergs/cm and to $D = 1.15 \times 10^{-28}$ ergs cm$^2$. A few typical values of these quantities derived from experiments are as follows: $A = 4.4 \times 10^{-7}$ ergs/cm and $D = 0.85 \times 10^{-28}$ ergs cm$^2$ from measurements of the heat capacity in the low temperature spin-wave region[5]; $A = 5.3 \times 10^{-7}$ ergs/cm and $D = 0.99 \times 10^{-28}$ ergs cm$^2$ from microwave studies of the spin-wave spectrum.[6] The agreement is certainly good; by contrast, the corresponding values derived from a molecular field treatment[3] of the susceptibility (adjusting all three interaction constants) are $A = 2.4 \times 10^{-7}$ ergs/cm and $D = 0.44 \times 10^{-28}$ ergs cm$^2$.

In summary, the susceptibilities of YIG and LuIG have been analyzed with the aid of newly derived and highly accurate series expansions under the assumption of negligible intra-sublattice exchange. The subsequent agreement between theory and a variety of experiments was found satisfactory. It is thus concluded that these results support the hypothesis of relatively weak $a$-$a$ and $d$-$d$ interactions in the garnets.

[3] R. Aléonard, J. Phys. Chem. Solids 15, 167 (1960) and private communication.

[4] R. L. Douglass, Phys. Rev. 120, 1612 (1960).
[5] J. E. Kunzler, L. R. Walker, and J. K. Galt, Phys. Rev. 119, 1609 (1960); S. S. Shinozaki, Phys. Rev. 122, 388 (1961).
[6] E. H. Turner, Phys. Rev. Letters 5, 100 (1960); R. C. LeCraw and L. R. Walker, J. Appl. Phys. 32, 167S (1961).

JOURNAL OF APPLIED PHYSICS     SUPPLEMENT TO VOL. 33, NO. 3     MARCH, 1962

# The Contribution of Rare-Earth Ions to the Anisotropy of Iron Garnets

B. A. Calhoun, M. J. Freiser, and R. F. Penoyer

*IBM Research Center, Yorktown Heights, New York*

The low-temperature properties, in iron garnets, of rare-earth ions with an odd number of electrons can be described by a simple model. This model represents the energy levels of the rare-earth ion by an isolated doublet which is subjected to an anisotropic exchange interaction. The complex shapes of the torque curves, predicted by the model and observed at low temperature, are represented by Fourier expansions. A small torque contribution proportional to the applied magnetic field is predicted, and evidence of its existence is observed in (YbY) and (YbGd) iron garnets. A comparison of the behavior of Yb and Dy in YIG and GdIG indicates a moderate influence of the host lattice on Yb and a drastic influence on Dy. The anisotropy anomaly observed for Yb in YIG [at 32° from [001] in the (110) plane] is not present when Yb is substituted into GdIG.

MICROWAVE resonance studies of rare-earth-doped YIG crystals[1] have shown that the anisotropy energy contributed by the rare-earth ions usually has a very complex angular dependence. A detailed study of Tb by Dillon and Walker[2] has shown that the anomalies, in this case, are due to cross-overs or near cross-overs of the energy levels of the Tb ions, as suggested by Kittel.[3] Dillon and Nielsen[1] have pointed out that the anisotropy anomalies can be broadly divided into two groups: sharp peaks associated with ions with an even number of electrons, and broad peaks with ions with an odd number of electrons. In this paper, we describe a simple model applicable to this latter group and present torque data for Dy and Yb ions in YIG and GdIG which agrees with calculated results.

The crystal field splitting of the ground state of the rare earth ion is assumed to be much larger than that produced by the exchange interactions. We include only the iron-rare-earth exchange interaction, neglecting both the weaker rare-earth–rare-earth interactions and the small influence of the rare earth ions on the behavior of iron ions. At sufficiently low temperatures, we can consider only the lowest doublet of an individual rare-earth ion and can write an "effective" Hamiltonian

$$\mathcal{K} = -\mu(g_J-1)/(g_J)\mathbf{s}\cdot\mathbf{g}\mathbf{G}\cdot\mathbf{M}_{Fe} - \mu\mathbf{s}\cdot\mathbf{g}\cdot\mathbf{H}. \quad (1)$$

$\mu$ is the Bohr magneton, and $g_J$ is the Lande $g$ factor of the rare-earth ion. $\mathbf{g}$ is the usual $g$ tensor, $\mathbf{s}$ is the effective spin, and $\mathbf{H}$ is the applied magnetic field. The second term represents effect of an external magnetic field. The first term differs in the form of the "exchange field" $\mathbf{G}\cdot\mathbf{M}_{Fe}$ and in the presence of the factor $(g_J-1)/g_J$ which enters because the "exchange field" acts only on the spin and not on the total magnetic moment of the rare earth ion.[4] It can be shown that the tensor $\mathbf{G}$ will have the same symmetry and

the same principal axes as $\mathbf{g}$.[5] In this local frame $\mathbf{G}$ and $\mathbf{g}$ are both diagonal, and therefore these tensors commute. In the samples we consider where the concentration of anisotropic rare-earth ions is small, the magnetization of the iron ions $\mathbf{M}_{Fe}$ will be nearly aligned with the external field.

Treating the second term in Eq. (1) as a perturbation, we can write the partition function

$$Z = 2\cosh\beta\epsilon + \beta\mu H \sum_i g_i^2 G_i l_i^2$$
$$\times \left[\sum_i g_i^2 G_i^2 l_i^2\right]^{-\frac{1}{2}} \sinh\beta\epsilon, \quad (2)$$

where

$$\epsilon = \mu(g_J-1)g_J^{-1}M_{Fe}\left[\sum_i g_i^2 G_i^2 l_i^2\right]^{\frac{1}{2}}/2 \quad (3)$$

is one-half of the exchange splitting of the doublet and $\beta=1/kT$. $g_i$ and $G_i$ are the diagonal elements of the tensors $\mathbf{g}$ and $\mathbf{G}$ along the symmetry axes $i$ of the particular rare earth site, and $l_i$ are the direction cosines of $\mathbf{H}$ and $\mathbf{M}_{Fe}$ relative to these axes. Since both terms in Eq. (2) depend on the orientation of $\mathbf{M}_{Fe}$ and $\mathbf{H}$, it is clear that there will be two contributions to the torque: a large one from the first term corresponds to the usual anisotropy, and a much smaller one from the second term will be proportional to the magnetic field. This field-dependent torque will be closely related to the angular variation of the magnetization of the rare earth ions since both are derived from the field-dependent term in the partition function. When torque or magnetization is calculated from Eq. (2), it is necessary to sum over the differently oriented sites in the crystal. At sufficiently low temperatures we have $\epsilon > 2kT$ for all orientations, and the field-independent part of the torque becomes simply $L = \sum_{sites} d\epsilon/d\theta$. In a (110) plane we can derive the following Fourier series for $L$ ($\theta$ is the angle from [001] to $\mathbf{H}$):

$$L = \sin2\theta\{-(T_1\alpha+T_2\beta+2T_3\gamma)-\cdots\}$$
$$+\sin4\theta\{(1/4)(T_1\alpha^2+T_2\beta^2+2T_3(\gamma^2-\delta^2))+\cdots\}$$
$$+\sin6\theta\{-(3/32)(T_1\alpha^3+T_2\beta^3+2T_3\gamma(\gamma^2-3\delta^2))-\cdots\}$$
$$+\text{higher terms.} \quad (4)$$

[1] J. F. Dillon and J. W. Nielsen, Phys. Rev. **120**, 105 (1960).
[2] J. F. Dillon and L. R. Walker, Phys. Rev. **124**, 1401 (1961).
[3] C. Kittel, Phys. Rev. **117**, 681 (1960).
[4] P.-G. de Gennes, Compt. rend. **247**, 1836 (1958).

[5] W. P. Wolf, Proc. Phys. Soc. (London) **74**, 665 (1959).

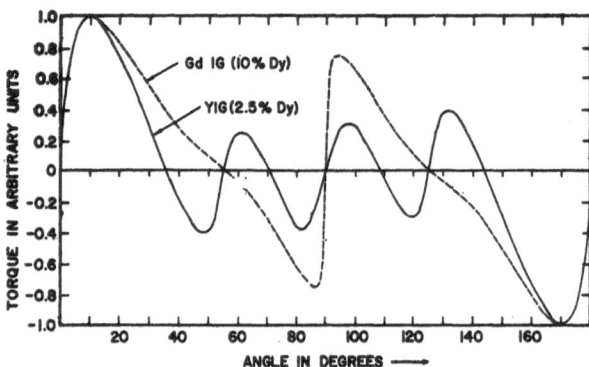

FIG. 1. Torque curves at 1.6°K in field of 24 700 oe for crystals indicated. The curves have been adjusted to same height to show the difference in shape. The torque per Dy ion in GdIG is approximately twice as large as for Dy in YIG and an order of magnitude larger than the torque per Yb ion.

TABLE I. Fourier coefficients of the torques due to Yb ions in YIG and GdIG, measured at 1.6°K with a field of 24 700 oe. The torques are in units of $10^4$ ergs/cc.

|  | $(Yb_{0.3}Gd_{2.7})IG$ | $(Yb_{0.3}Y_{2.7})IG$ | Calculated (see text) |
|---|---|---|---|
| $\sin 2\theta$ | 13.3 | 14.3 | 10.7 |
| $\sin 4\theta$ | 19.0 | 17.8 | 14.5 |
| $\sin 6\theta$ | 6.7 | 5.7 | 3.6 |
| $\sin 8\theta$ | 4.6 | 5.6 | 1.2 |

Here

$$T_1 = [1/2(g_x{}^2 G_x{}^2 + g_y{}^2 G_y{}^2)]^{\frac{1}{2}};$$

$$T_2 = [1/2(g_x{}^2 G_x{}^2 + g_z{}^2 G_z{}^2)]^{\frac{1}{2}};$$

$$T_3 = [1/2(2g_x{}^2 G_x{}^2 + 3g_y{}^2 G_y{}^2 + 3g_z{}^2 G_z{}^2)]^{\frac{1}{2}};$$

$$\alpha = (g_x{}^2 G_x{}^2 - g_y{}^2 G_y{}^2)/2T_1{}^2;$$

$$\beta = (g_x{}^2 G_x{}^2 - g_z{}^2 G_z{}^2)/2T_2{}^2;$$

$$\gamma = (g_y{}^2 G_y{}^2 + g_z{}^2 G_z{}^2 - 2g_x{}^2 G_x{}^2)/2T_3{}^2;$$

$$\delta = 8^{\frac{1}{2}}(g_y{}^2 G_y{}^2 - g_z{}^2 G_z{}^2)/2T_3{}^2.$$

It is clear that, unless $\alpha$, $\beta$, $\gamma$, and $\delta$ are all much smaller than unity, the series will contain substantial high-order harmonics. If the splitting of the doublet is appreciably anisotropic, due to anisotropy in the $g$ factor or the exchange interaction or both, we expect an anisotropy energy which will be an appreciable fraction of the exchange splitting and which will have a complex shape due to the presence of many harmonics. There will be, however, no sharp discontinuities or anomalies on the basis of this model.

Torque curves were obtained on crystals of YIG and GdIG containing from 2.5% to 10% of Dy and Yb. The crystals were grown by the flux technique, and the specimens used were (110) disks cut parallel to natural faces of the crystals. The torque balance was similar to an earlier one[6] except that the bearings were replaced by two fibers, and the balance was enclosed in a vacuum system for low-temperature operation. The magnet used provided fields up to 24 700 oe in a $2\frac{1}{4}$-in. gap. For comparison with Eq. (4), the experimental curves were analyzed to obtain the Fourier coefficients of both sine and cosine terms. The cosine terms arise from a variety of experimental errors, and their magnitude provides an estimate of the influence of these errors.

The Fourier coefficients for $(Yb_{0.3}Gd_{2.7})$ and $(Yb_{0.3}Y_{2.7})$ iron garnets, corrected for the contributions of Fe and Gd, are listed in Table I along with values calculated from Eq. (4) using Wickersheim's optical measurement of the exchange splitting of the Yb doublet in YbIG.[7] Henderson and White[8] have made the same calculation and compared their result with measurements by Pearson on a $(Yb_{0.06}Y_{2.94})IG$ crystal.[9] The difference between the torques due to Yb in YIG and GdIG is nearly as large as the difference between the calculated value and that observed in YIG. Two other features were apparent in the torque curves of these crystals. The curves for the Yb-Y garnets exhibit a step, at 32° from the [001] direction, which corresponds to the small anomalous peak observed in resonance studies.[1] Such a step was not observed with the Yb-Gd crystals and, of course, cannot be calculated from Eq. (2). Small field-dependent torques were observed with all these crystals. These torques differed in shape, but were of the same magnitude as the effects due to the incomplete alignment of the magnetization and the field. The torque curve for Dy-YIG in Fig. 1 has the shape expected from resonance measurements. The marked differences in the torque due to Dy in YIG and in GdIG are not surprising in view of the large differences in the $g$ values of Dy in different nonmagnetic garnets.[10]

We expect that this model of an isolated doublet subjected to anisotropic exchange will also be applicable to Er and Nd ions in garnets. Studies of iron garnets containing these ions and further work on the behavior of Dy ions are planned.

### ACKNOWLEDGMENTS

The authors are indebted to E. A. Giess for growing the crystals used in this work and to R. A. Fiorio for his assistance with the measurements.

6 R. F. Penoyer, Rev. Sci. Instr. **30**, 711 (1959).

7 K. A. Wickersheim, Phys. Rev. **122**, 1376 (1961).
8 J. W. Henderson and R. L. White, Phys. Rev. **123**, 1627 (1961).
9 R. F. Pearson (unpublished data quoted in reference 8).
10 M. Ball, G. Garton, M. Leask, D. Ryan, and W. Wolf, J. Appl. Phys. **32**, 267S (1961).

JOURNAL OF APPLIED PHYSICS    SUPPLEMENT TO VOL. 33, NO. 3    MARCH, 1962

# Devices and Phenomena

## C. L. Hogan, *Chairman*

## Properties of Reset Cores in Radar Pulse Transformers

Reuben Lee

*Westinghouse Electric Corporation, Baltimore, Maryland*

This paper outlines permeability and loss measurements on large uncut toroidal cores of Hipersil and Supermendur alloys. With forced cooling, large increments of core induction can be utilized; with a reset core, pulse permeability is much higher than it is in cut cores, and alters pulse transformer design radically. At inductions up to 24 kgauss, pulse permeability of fully oriented Hipersil and Supermendur alloys were about 6000, compared to approximately 1500 for cut cores at lower flux densities. It appears practicable from these tests to operate Hipersil alloy with excursions of induction up to 30 kgauss and Supermendur alloy up to 40 kgauss, and to reduce pulse transformer weight by 80%.

EXPLORATORY measurements reported in this paper comprise permeability and loss at large flux density excursions up to 45 kgauss, with 2 and 6.5 $\mu$sec repetitive unidirectional pulses. If proper precautions are observed, these large flux density excursions may be used in reducing size of pulse transformers. An experimental line-type modulator used to test sample transformers had the following rating:

Pulse duration 6.5 $\mu$sec
Primary voltage 11 kv
Peak power 7.5 megawatts
Pulse repetition frequency 300 cps
Average power 15 kw.

Tests were made on cores of 2-mil Hipersil, 2-mil Supermendur, and 4-mil Supermendur alloys. The cores were wound in the form of toroids, biased to negative saturation with resetting $H_{do}=1.4$ oe. Preliminary calculations[1] indicated that during a 6.5-$\mu$sec pulse the flux penetration was better for Hipersil alloy than for Supermendur, because of the higher resistivity (50 $\mu$-ohm cm compared to 26 for Supermendur). Cacullated pulse permeability was higher for Hipersil alloy than for either thickness of Supermendur. Measurements confirmed these calculations for flux density excursions below saturation.

Model transformers for 6.5-$\mu$sec pulses were wound on 12-lb. toroidal cores and tested in the experimental modulator. Pulse leading edge risetime was 0.67 $\mu$sec, pulse top was virtually flat, and losses were reasonably close to calculated values.

The flat pulse top was due to high open-circuit inductance, which in turn showed that the high core permeability predicted by calculations was actually exceeded on test. For 2-mil Hipersil and Supermendur alloys this permeability in the the range of 5000 to 6000 compared with 600–1500 for non-reset cut cores,

depending upon $\Delta B$, the flux density. Such permeability at these inductions means improved radar performance, especially at high output power.

Measurements of pulse permeability and loss were made on transformers using circuits described on pages 639 and 647 of reference 1. Oscillograms of hysteresis loops, exciting current and volt-time integral for 2-mil Hipersil and 2-mil Supermendur alloys were taken for several values of $\Delta B$. Typical pulse permeability $\mu_e$ and equivalent loss resistance $R_e$ derived from the oscillograms are given in Table I. Comparing the pulse permeability measured for these two materials, it will be noted that at 24 kgauss the Hipersil alloy permeability is the greater.

Limitations on large increments of induction are saturation and effective means of cooling the core. With

TABLE I. Values of permeability and loss resistance determined from oscillograms.

| | 4-mil Supermendur alloy | 2-mil Supermendur alloy | 2-mil Hipersil alloy |
|---|---|---|---|
| $R_2$ (ohms) | 100 000 | 84 200 | 84 200 |
| $C_2$ (mf) | 0.01 | 0.0161 | 0.0161 |
| $V_2$ (volts) | 50 | 40 | 39 |
| $\Delta B=\dfrac{R_2 C_2 V_2 \times 10^8}{N A_c \times 6.45}$ (gauss) | 22 100 | 24 000 | 23 400 |
| $i_m = V_1/R_1$ (amp) | 23.3 | 10 | 9.35 |
| $\Delta H=\dfrac{0.5\,N i_m}{1_c}$ (oe) | 10.54 | 4.52 | 4.23 |
| $\mu_e = \Delta B/\Delta H$ | 2100 | 5300 | 5540 |
| $E_1 = 10\,800 \Delta B/B_m$ (volts) | 8140 | 8850 | 8630 |
| $i_e = V/R_1$ (amp)[a] | 6.67 | 4.0 | 6.2 |
| $R_e = \dfrac{E_1}{i_e}$ (ohms) | 1220 | 2210 | 1390 |
| Core loss (watts av) | 200 | 110 | 175 |

[a] $V$ is measured graphically as in reference 1, Fig. 15.21.

[1] G. N. Glasoe and J. V. Lebacqz, *Pulse Generators* (McGraw-Hill Book Company, Inc., New York, 1948), Radiation Lab. Series, Vol. 5, p. 631.

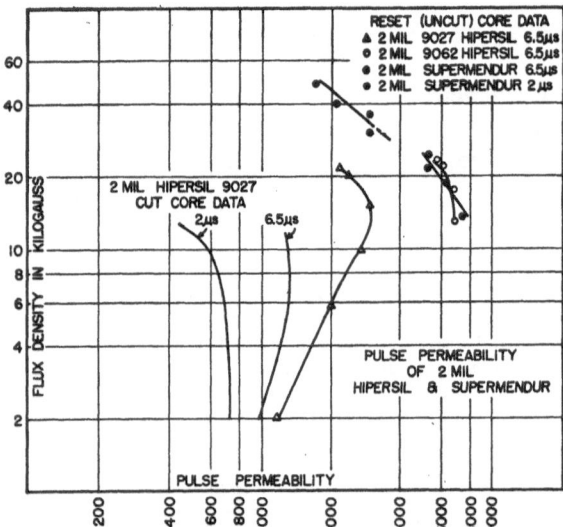

FIG. 1. Pulse permeability of 2-mil Hipersil and Supermendur alloys.

natural convection, 11 to 12 kgauss is the largest practicable value of $\Delta B$ that can be used safely. At flux density excursions of the order of those reported here, some method of artificial cooling should be used. Then the high loss associated with large $\Delta B$ is dissipated safely, and the increase in $\Delta B$ afforded by the newer materials may be fully exploited. These exploratory tests indicate that substantial increase in pulse permeability, as compared with that obtained in cut cores, and in older grades of material, may be obtained by using reset bias. Figure 1 affords an idea of the increase thus obtained. Data for the uncut cores of 9027 grade Hipersil alloy agree with those in reference 2. Since

[2] P. Fenoglio, C. W. Peck, F. R. Richardson, H. W. Lord, and A. Boyajian, "High power-high voltage pulse transformer design criteria and data," dated February 1, 1953 (ASTIA Report AD-21236), Fig. 9–74.

$B_{10} = 23.2$ kgauss for Supermendur and 18.5 kgauss for Hipersil alloy, it appears that operating inductions of 40 and 30 kgauss, respectively, would be practicable for these materials, provided the strip thickness is appropriate to the pulse duration.

Weights of experimental models were as follows:

| Pulse transformer | 18 lb |
|---|---|
| Bias choke | 4 lb |
| Total | 22 lb. |

A transformer of the same rating with cut core weighed 105 lb. These are dry weights with no mounting or cooling means included. Choice of cooling means depends on the amount and kind of insulation, and hence on the voltage. A large pump or fan would easily cancel the advantages of toroidal construction. Therefore it is important to use the same cooling means needed in the radar equipment to dissipate losses in tubes and other components, and thus retain the weight advantage. Likewise, a source of resetting direct current already available in the radar equipment eliminates the need for a separate bias source.

Heat runs were made to ascertain the need for external cooling. These heat runs confirmed the fact that core loss, as represented by $R_e$ in Table I, could be removed by circulating oil, water, or other coolant in close contact with the core. Water-cooling washers, for example, were found to reduce core temperature rise from 180° to 5°C with 1 gal per min of water flow.

### ACKNOWLEDGMENTS

Assistance on the core tests at 6.5 $\mu$sec was rendered by R. A. Hill and Roy Johnson. Supermendur test data at 2 $\mu$sec were taken on smaller (0.13 oz) cores by Ray E. Lee.

# Coaxial Ferrite Phase Shifter for High Power Applications*

A. S. BOXER AND R. S. McCARTER

*Bell Telephone Laboratories, Inc., Whippany, New Jersey*

The low- and high-power behavior of a coaxial ferrite-loaded transmission line has been studied experimentally at microwave frequencies. Longitudinal magnetic biasing fields both above and below the cutoff range ($\omega/\gamma - 4\pi M_s \leq H_{dc} \leq \omega/\gamma$) were explored, the two regions yielding devices with markedly different characteristics. The mechanical structure consisted of a rigid $\frac{7}{8}$-in. coaxial line fully loaded with Trans-Tech 1–103 ferrite together with appropriate quarter-wave dielectric transformers.

(1) Operation below cutoff ($0 \leq H_{dc} < \omega/\gamma - 4\pi M_s$), $f = 3080$ Mc: These parameters permit construction of a compact, light-weight phase shifter producing 360° of phase shift in 10 in. of active length upon application of only 100 oe. Approximate values for $h_{crit}$ are determined from the measured threshold powers and plotted versus applied field.

(2) Operation above resonance ($H_{dc} > \omega/\gamma$), $f = 1350$ Mc: With these parameters, a device can be built that is capable of handling at least several hundred kilowatts of rf peak power while maintaining a figure of merit of 1000 or more. The loaded coaxial line provides 360° of phase shift in 24 in. of active length when biased from 950 to 1350 oe. The high-power and loss characteristics of the latter device permit application to electronic steering of the transmitter beam in a linear array antenna where relatively few phase shifters suffice. The low-field device is suited for beam control in a planar array where large numbers of relatively small phase shifters of medium peak-power capability are needed.

T HIS paper reports the results of an inquiry into the suitability of a longitudinally magnetized, ferrite-filled coaxial line as a high-power microwave phase shifter. Such a line, with appropriately chosen dimensions for the inner and outer conductors, is suited for the transmission of high-power microwave signals since it avoids the severe local field concentrations associated with many other geometries. One outstanding characteristic of this structure is that there are two distinct and widely separated ranges of applied dc magnetic field over which practical operation is possible. The rapid variation of phase shift with applied field in the low-field operating range points to a compact device requiring relatively little control power. High-power operation in this region is, however, hampered by the presence of the subsidiary absorption. By increasing the dc magnetic bias to a point well beyond ferromagnetic resonance, the device characteristics undergo a marked change. Although the phase shift is considerably less sensitive to applied field than in the low-field case, the absorption loss is so reduced that the figure of merit (the ratio of the phase shift in degrees between two values of magnetic field to the peak-insertion loss in the same range) is actually increased. In addition, the absence of the subsidiary absorption makes this an inherently high-power device.

If it can be assumed that the propagation through a coaxial line of sufficiently narrow annular spacing and sufficiently large radius is not unlike that between a pair of parallel planes, the latter structure may be used as a model for the phase shifter. Suhl and Walker[1] have given an approximate treatment of the propagation characteristics and mode configurations of the parallel-plane transmission line filled with lossless ferrite. Propagation is assumed to be in the direction of the applied magnetic field, and the microwave fields are assumed to be independent of the transverse coordinate parallel to the planes. For a sufficiently thin layer of ferrite, the propagation constant is given by[1]

$$\beta^2 = \beta_1^2 [1 - (\sigma + p)^2]/[1 - \sigma(\sigma + p)] \quad \sigma \neq 1, \quad (1)$$

where $\sigma = \gamma H/\omega$, $p = 4\pi\gamma M_s/\omega$, and $\beta_1$ is the propagation constant in the unbounded, unmagnetized medium.

The behavior of $\beta$ as a function of $\sigma$ is as follows: From its value at $\sigma = 0$, $\beta$ decreases monotonically until cutoff is reached at $\sigma = 1 - p$. Propagation is forbidden for $1 - p < \sigma < 1$. Above resonance, $\beta$ again decreases monotonically and approaches $\beta_1$ for large $\sigma$. Although $\beta$ varies most rapidly with $\sigma$ in the neighborhoods of $\sigma = 1 - p$ and $\sigma = 1$, these are, as expected, regions of high loss. Thus there are two ranges of magnetic field well removed from a cutoff zone, which warrant consideration for phase shifting applications.

A series of measurements of phase shift and insertion loss were made on a $\frac{7}{8}$-in. coaxial line (the i.d. of the outer conductor was 0.812 in. and the diameter of the inner conductor 0.375 in.) fully loaded with TTl-103[2] ferrite ($4\pi M_s \sim 700$ gauss) and appropriate Stycast[3] dielectric quarter-wave transformers. During these tests both the dc biasing field and the microwave power level were continuously variable.

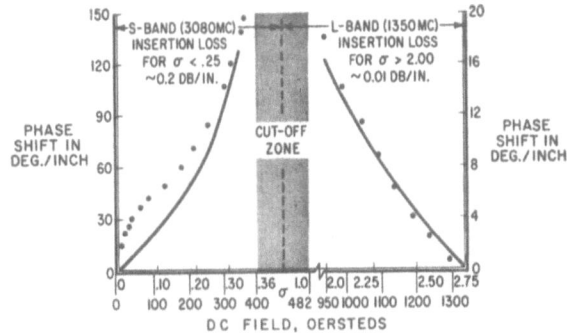

FIG. 1. Phase shift as a function of applied dc magnetic field. These measurements were made using a $\frac{7}{8}$-in. coaxial line, filled with Trans-Tech 1–103 ferrite, at 3080 and 1350 Mc.

* Supported by the Bureau of Ships, Department of the Navy.
[1] H. Suhl and L. R. Walker, Bell System Tech. J. **33**, 1133 (1954).
[2] Trans-Tech, Inc., Gaithersburg, Maryland.
[3] Emerson and Cuming ,Inc., Canton, Massachusetts.

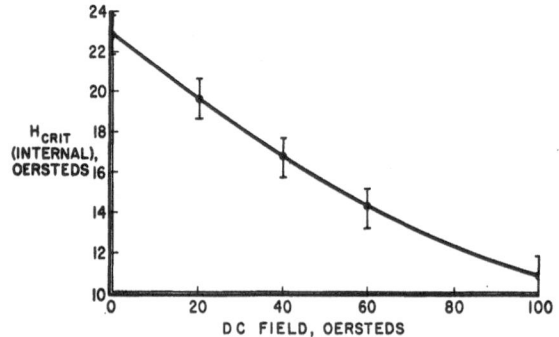

FIG. 2. Approximate internal threshold magnetic field at 3080 Mc as a function of applied dc magnetic field for Trans-Tech 1–103. This measurement was made with the ferrite filling a $\frac{7}{8}$-in. coaxial line.

Since the measurements were made as part of a program to develop ferrite phase shifters for use at $S$ and $L$ bands, the work was guided in part by certain practical considerations. First, for reasons not germane to the present discussion, a compact phase shifter was desired for use at $S$ band. Since, in addition, the magnetic fields required to bias the device above resonance are inconveniently large, it was natural to confine the measurements at 3.08 Gc to the low-field region. Second, most materials display such excessive low-field losses at room temperature in the $L$-band frequency range that the low-field region becomes unattractive. Thus, the high-field measurements were made most conveniently at 1.35 Gc.

Figure 1 shows the experimental phase shift in degrees per inch as determined in each region plotted against the normalized field $\sigma$. Since $\sigma$ is defined differently above and below the cutoff zone, a second abscissa giving the actual field values is shown to facilitate direct comparison. The reader's attention is also directed to the different ordinate scales used for the two curves. The solid curves give the phase shifts predicted by Suhl and Walker. The phase shifts for two branches are calculated independently, without regard to sign, and zero phase shift is defined at $H=0$ and $H=1350$ oe.

The most striking feature of the below-cutoff branch is the rapid variation of phase shift with field below saturation. Taking advantage of this phenomenon, a phase shifter effecting economies in both length and bias field requirements can be constructed. The high-field branch of the curve, while showing no rapid variation of phase shift with applied dc field, does display a

remarkably low insertion loss. The figure of merit can assume values of 1000 or more using ordinary commercial materials. It is also interesting to note that raising the operating temperature improves the performance of both devices; the insertion loss at low fields and the bias required for efficient operation above resonance are each reduced.

Turning now to the high-power stability, it should first be noted that the power limitations arise primarily through the subsidiary resonance,[4] since both operating regions are far removed from the main resonance so that the latter's saturation has a negligible effect on the insertion loss.

In the range below cutoff, the peak power at which nonlinearities first appeared varied from 19 kw at zero applied field to 4 kw at $H=100$ oe. These numbers, however, depend on the geometry, and it would be more appropriate to record the peak internal rf field. Figure 2 gives the values of $h_{\mathrm{crit}}$ (actually the maximum transverse component of the rf magnetic field parallel to the planes) determined using Suhl and Walker's expressions and the experimentally determined threshold powers.

Above resonance no effect due to the subsidiary absorption is expected.[5] The structure was tested at power levels up to 300-kw peak with no apparent deterioration of the figure of merit.[6]

In summary, the fully loaded coaxial line may be operated over two distinct ranges of magnetic field with the low range suited for $S$-band and the high range more appropriate for $L$ band use. The two devices thus obtained have characteristics that are, in a sense, complementary. The first is a compact low-field device capable of controlling a moderate amount of rf power (of the order of a few tens of kilowatts). A device with these characteristics would be appropriate, for example, for use in a planar phased-array antenna. The second, a low-loss, high-power, but rather massive structure would more closely fulfill the demands of a linear array.

The authors would like to express their appreciation to W. H. von Aulock, J. F. Ollom, and I. Jacobs, for many helpful discussions. In addition they would like to thank S. Hershenov, E. F. Landry, and H. E. Noffke, for their help with the instrumentation and measurements.

[4] H. Suhl, J. Phys. Chem. Solids 1, 209 (1957).
[5] P. C. Fletcher and N. Silence, J. Appl. Phys. 32, 706 (1961).
[6] A. S. Boxer, S. Hershenov, and E. F. Landry, IRE Trans. on Microwave Theory Tech. MTT-9, 577 (1961).

JOURNAL OF APPLIED PHYSICS    SUPPLEMENT TO VOL. 33, NO. 3    MARCH, 1962

# Slow-Wave uhf Ferrite Phase Shifters

N. G. SAKIOTIS AND D. E. ALLEN

*Solid State Electronics Department, Motorola, Incorporated, Phoenix, Arizona*

A brief general discussion of the loss mechanisms of ferrites and garnets is given. An idealized representation of the absorption loss as a function of frequency and the applied magnetic field is utilized to define five distinct regions within which devices using ferrimagnetic materials may be operated with low absorption loss.

Slow-wave transmission lines for uhf frequencies are discussed, and the observed phase shifts per unit length for these and a conventional transmission line are given.

It is concluded that through the use of slow-wave transmission lines phase shifters can be designed at uhf frequencies which require as little as one-fourth the length required by a conventional transmission line.

PRESENT ferrite phase shifter designs, which operate in the uhf frequency range, have the disadvantages of bulkiness, high absorption loss, and the applied magnetic field intensities are often on the order of 1500 oe. The purpose of this paper is to partially describe the results of an investigation to determine improved techniques for the design of compact phase shifters in the 200–800 Mc frequency range.

In order to appreciate the problems that arise in the determination of an optimum phase shifter design operating in the uhf range, it is important to understand the various loss mechanisms of the ferrimagnetic medium. Figure 1 is a highly idealized sketch of the absorption as a function of frequency and the applied magnetic field intensity. The low frequency absorption loss peak is caused by a resonance in the domain walls.[1] When a magnetic field is applied, this absorption loss decreases and vanishes at magnetic saturation when the material behavior is essentially that of a single domain. The higher frequency absorption loss, which extends from $\omega_\alpha$ equals $\gamma$(Hanis) to $\omega_{max}$ equals $\gamma$(Hanis$+4\pi Ms$) at zero applied field, is caused by gyromagnetic resonance resulting from the material anisotropy field. As the material is subjected to an increasing magnetic field intensity, the frequency bandwidth will narrow and the absorption maximum will occur at an increasing frequency. Finally, upon magnetic saturation, this resonance is the familiar induced gyromagnetic resonance with the frequency bandwidth being a function of the material linewidth.

It is clear that a material suitable for a phase shifting device is required to exhibit a region of low absorption loss and such regions can always exist, in general, for a given material. Region I includes all frequencies greater than $\omega_{max}$ and all values of applied field less than that required for resonance. Region II includes frequencies much greater than $\omega_\alpha$ and less than $\omega_{max}$ with the applied field being less than that of resonance but greater than zero. Region III includes all frequencies greater than $\omega_\alpha$ and all values of the applied field strength greater than that required for resonance. This region is generally referred to as "above resonance".

Region IV includes frequencies between the maximum frequency of the domain wall resonance absorption and less than $\omega_\alpha$. Region V includes frequencies between zero and the minimum frequency affected by the domain wall resonance. The applied field for regions IV and V may have any value including zero.

The measured value of the real part of the initial permeability of typical ferrites is greater in region V than region IV and greater in region IV than region I. Within region IV, the value is approximately one-half of the dc permeability and within region I, the value is nearly one.

The effective permeability and the change in the permeability, which are important in the design of a phase shifter, are clearly controlled and limited by the region the medium falls within at the frequency of interest. The change in the effective permeability, with a change in the applied magnetic field, is caused by a change in the domain parameters within regions IV and V. The change in the effective permeability within regions I, II, III is caused by a change in the permeability tensor components.

Depending upon the actual region of operation, different transmission line geometries may be more advantageous for phase shifter designs.

(1) If region I operation applies, a transmission line ferrimagnetic material geometry should be used that will provide a low $N_z$, where $N_z$ is the demagnetizing factor in the applied field direction.

(2) If region II operation applies, a geometry that

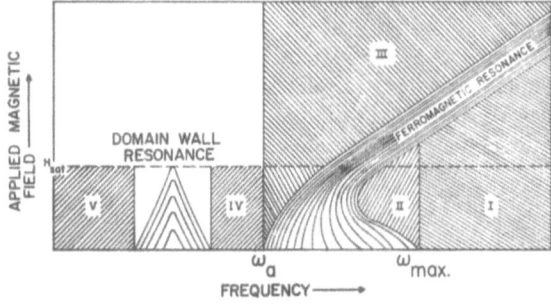

FIG. 1. Definition of regions of operation for an idealized polycrystalline material.

---

[1] J. E. Pippen and C. L. Hogan, "Initial permeability spectra in ferrites and garnets," Scientific Report #1, April 1, 1959.

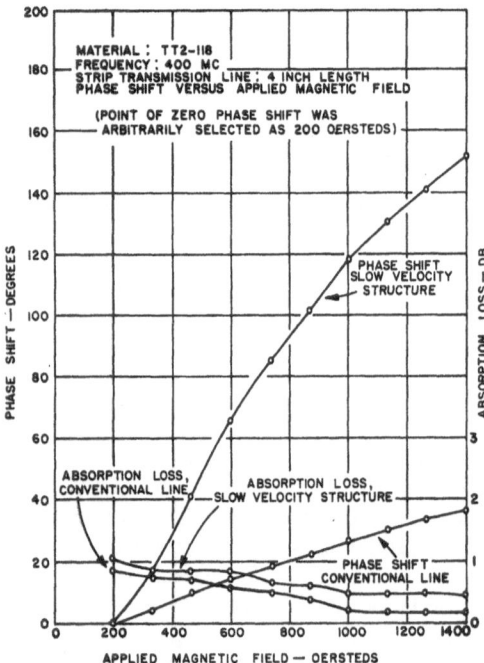

FIG. 2. Comparison of slow-wave stripline with conventional stripline.

maximizes the difference between the magnetic field intensities required for saturation and resonance would result in the best phase shifter performance.

(3) If region III operation applies, it is advantageous to use a geometry that requires the lowest value of external field for resonance.

(4) If region IV or V is applicable, it is best to use a geometry that affords magnetic saturation at low values of external field. In addition, a transverse applied field may be better in these regions.

During this investigation, most of the commercially available materials were evaluated in the 400–1000 Mc frequency range by obtaining the initial permeability spectra. Most of the materials evaluated offered only region III operation between 400–1000 Mc. Only two materials will be discussed further. MO62, a hybrid garnet, could possibly be operated in region II above 700 Mc. Below 700 Mc, operation would be in region III. TT2-118, on the other hand, presented possible operation in region V below 400 Mc.

The types of transmission lines that were evaluated included slow-wave striplines, slow-wave coaxial lines, and a conventional stripline. The purpose of the conventional stripline was only to provide a phase shift per unit length reference for the slow-wave structures. The slow-wave coaxial line, loaded with thin toroids of MO62, was evaluated at 725 Mc. The result was that the absorption loss increased monotonically with the applied field strength until operation was in region III. In other words, region II operation did not occur as predicted. The slow-wave stripline, loaded with TT2-118, was used to study the phase shift per unit length and merit factors as a function of the cutoff frequency of the slow-wave transmission line. To eliminate the effect of the material properties varying with frequency, a series of transmission lines were evaluated. The results were:

| Transmission line cutoff frequency | Phase shift per inch | Merit factor |
|---|---|---|
| 500 Mc | 29 deg | 240 deg/db |
| 700 Mc | 27 deg | 260 deg/db |
| 850 Mc | 19 deg | 190 deg/db |
| 1000 Mc | 18 deg | 170 deg/db. |

Both the phase shift per unit length and the merit factor increase as $\omega \rightarrow \omega_c$. The difference in merit factors of the 500- and 700-Mc transmission lines is smaller than the accuracy of the test equipment. From these data, $\omega/\omega_c$ should be approximately 0.5 for this material (TT2-118) at this frequency.

Figure 2 is a plot of the measured values of phase shift and merit factor for $\omega/\omega_c$ equal to 0.4, and a conventional stripline. The frequency is 400 Mc. The phase shift per unit length of the slow-wave transmission line is approximately 4.5 times that of the conventional line. This means a phase shifter based on a slow-wave structure would be less than $\frac{1}{4}$ the length of a device using conventional transmission line.

Although phase shift per unit length can be increased through the use of slow-wave transmission lines, a better phase shifter will not necessarily result if these transmission lines are used. A major factor may be the control power[2] for the device, and this may be greater for the slow-wave structure because of an increase in the cross-sectional area.

### ACKNOWLEDGMENTS

The authors wish to express their sincere appreciation to Wayne E. Kivett who performed the measurements necessary to obtain the permeability spectra and the measurements on the phase shifter models.

This work was supported by the Air Force Cambridge Research Center under contract.

[2] N. G. Sakiotis and D. E. Allen, "Final report on a uhf ferrite phase shifter," March 15, 1961.

JOURNAL OF APPLIED PHYSICS    SUPPLEMENT TO VOL. 33, NO. 3    MARCH, 1962

# Magnetodynamic Mode Ferrite Amplifier*

Roy W. Roberts
*Melabs, Palo Alto, California*

AND

Bert A. Auld
*Stanford University, Palo Alto, California*

AND

Robert R. Schell
*Melabs, Palo Alto, California*

A longitudinally pumped ferrite amplifier has been operated using two magnetodynamic modes for the signal and idler resonant circuits. This is in contrast to the more usual electromagnetic cavity-type resonance or the ferrite magnetostatic modes. These magnetodynamic modes result from the coupling of a cavity-type resonance with a magnetostatic resonance. The coupling or mixing is strongest when the two unperturbed resonant frequencies are the same.

For the amplifier described in this paper a single crystal yttrium-iron garnet sphere was used. The magnetic bias field applied was 17 kMc/$\gamma$ oe. The resulting magnetodynamic modes were resonant at 21 and 14 kMc, respectively, for the two branches of the mode tuning curve.

Advantages of these modes for ferrite amplification are ease of coupling and lack of interference from other resonant modes.

## GENERAL

A PARAMETRIC amplifier which uses a ferromagnetic medium as a nonlinear coupling element was first proposed by Suhl on a theoretical basis.[1] Realizations of this proposal which employed transverse pumping, that is rf magnetic field perpendicular to the dc magnetic field, were not practical due to excessive pump power requirements.[2] The large pump power requirements were due to spin waves which were pumped to instability and blocked the pumping. This difficulty was overcome by Denton with a longitudinal pumping scheme.[3] With longitudinal pumping, spin waves are also parametrically excited, but are not as effective in inhibiting the pumping as in the transverse case.

As in Denton's treatment of the longitudinally pumped amplifier, two equations for the signal and idle mode amplitudes are derived by making a normal mode expansion

and

$$\left[1+2jQ_1\frac{\omega_s-\omega_1}{\omega_1}\right]A_s-jA_i{}^*F_1=\frac{-Q_1}{Q_{1e}}A_s$$

$$\left[1-2jQ_2\frac{\omega_i-\omega_2}{\omega_2}\right]A_i+jA_s{}^*F_2{}^*=0. \qquad (1)$$

Here $A_s$ and $A_i$ are the signal and idle field amplitudes at frequencies $\omega_s$ and $\omega_i$, respectively. In this case, however, the expansion is in terms of the modes of the ferrite-loaded cavity, using an orthogonality relation

due to Schaug-Pettersen,[4] rather than in terms of magnetostatic modes. This approach permits the implicit inclusion of external coupling, (the $Q_1A_s/Q_{1e}$ term), wall proximity effects, and propagation effects in the sample. From Eq. (1), one may easily show that the threshold of oscillation occurs when the product of the parametric coupling terms

$$F_1F_2{}^*=Q_1Q_2H_p{}^2M_0{}^{-2}(4I_1I_2)^{-1}\int_{\text{cavity}}|M_1\cdot M_2{}^*|^2dv, \quad (2)$$

equals unity. Here, $I_1, I_2$ = total stored energy in modes 1,2; $H_p$ = longitudinal pump field; $M_1, M_2$ = transverse components of magnetization in modes 1,2; $M_0$ = saturation magnetization. Expressions for amplifier gain, bandwidth, and noise figure may also be derived. This

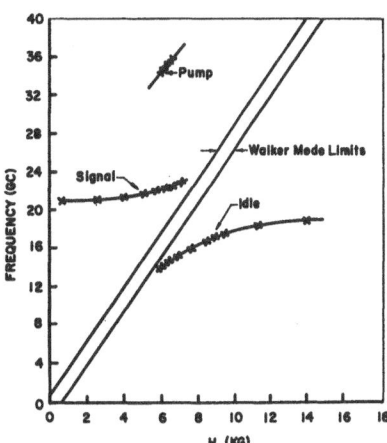

FIG. 1. Experimental magnetodynamic mode tuning curve.

* The work herein reported was supported by the U. S. Army Signal Research and Development Lab, Fort Monmouth, New Jersey.

[1] H. Suhl, J. Appl. Phys. **28**, 1225 (1957).
[2] M. T. Weiss, Phys. Rev. **107**, 317 (1957).
[3] R. T. Denton, Proc. Inst. Radio Engrs. **48**, 937 (1960).
[4] T. Schaug-Pettersen, "Some relations for lossless waveguide and resonators," Internal memorandum, Hansen Laboratory, Stanford University.

approach is completely general and includes magneto-static and magnetodynamic operation as special cases.

## MAGNETODYNAMIC MODES

The term magnetodynamic was first applied to modes in a ferrite cylinder by Coleman and Steier.[5] The same phenomena had been studied earlier by Heller[6] for ferrite loaded cavities. For the results herein reported, the modes appear in a spherical geometry. These modes result when a dielectric sphere resonance is strongly coupled to a magnetostatic mode of similar field distribution. The resulting modes are split from their unperturbed values in a manner analogous to two tuned coupled circuits near the point of degeneracy. The resulting tuning curve is illustrated by Fig. 1 which presents experimental data obtained from measurements made on the actual amplifier configuration. The coupling between the two is strongest in the region of unperturbed resonance crossover. The experimental curves become asymptotic, at zero and infinite magnetic field, to a frequency of 19 Gc. A dielectric sphere with open circuit boundary conditions and a dielectric constant equal to that of the YIG has degenerate $TE_{101}$, $TE_{111}$ and $TE_{1-11}$ resonances at a frequency of about 18.5 Gc, so that the agreement is quite good.

As employed in the amplifier, the upper branch of the tuning curve is used as the signal resonance and the lower branch as the idler. In the region of interest, both branches lie well outside the limits of the Walker modes. Also shown in Fig. 1 is a pump tuning curve. This is simply the sum of the upper and lower tuning curves for various dc fields. It will be noted that this curve has a slope of 2.8 Mc/oe rather than the 5.6 Mc/oe which would be expected for magnetostatic operation.

## EXPERIMENTAL RESULTS

The amplifier under discussion employs a 0.149-in. single crystal YIG sphere which has been x-ray oriented. The sphere is mounted on a polystyrene post through the narrow wall of a section of K-band waveguide. The garnet is positioned with the dc bias field in a (110)

plane. The garnet is rotatable about a (110) axis and has adjustable penetration into the guide. The pump field is applied through an iris in the opposite narrow wall with a double stub tuner for matching. A sliding short in one end of the signal guide provides variable coupling of the magnetodynamic resonance to the external circuit. The signal guide is below cutoff at the idler frequency and provides a reactive termination for the idler. The signal at 21 Gc is fed in the guide and the reflected output signal monitored by a directional coupler.

During operation, the dc bias field is adjusted to about 17 Gc/$\gamma$ oe. The pump frequency is 34.8 Gc. A change of operating point from these values results in degraded amplifier performance which indicates that the coupling is strongest at these points.

This amplifier has provided gains of 50 db with a pump power of about 20 w.[7] The gain saturates at a signal output power of about 1 w. The pump power required for 0-db net gain is about 4 w. This is much larger than predicted by theory and is due to inefficient matching in the pump circuit. This configuration is at a disadvantage, in this respect, over those that have a material resonance at the pump frequency. The performance is also limited by superregenerative operation.

## CONCLUSIONS

The fact that both the signal and idle resonances lie outside the Walker mode limits is advantageous from two standpoints. First, it is quite easy to tightly and selectively couple to the magnetodynamic modes. Secondly, there is no direct coupling of the signal and idle modes to spin modes lying within the Walker mode limits, which might be pumped above thermal equilibrium and contribute excess noise.

It is hoped, by increasing pump circuit efficiency, to operate the amplifier with a klystron pump. Here, longer pump duration would be available to simulate cw operation and also the excess noise now contributed by the pump would be eliminated, allowing noise figure evaluations. This would allow identification of the noise sources and a subsequent prediction of the usefulness of the ferrite amplifier at millimeter wave frequencies.

[5] W. H. Steier and P. D. Coleman, J. Appl. Phys. 30, 1454 (1959).

[6] G. S. Heller, "Ferrite loaded cavity resonators," MIT Lincoln Laboratory Report M82-5, May 6, 1959.

[7] R. W. Roberts, Proc. Inst. Radio Engrs. 49, 963 (1961).

JOURNAL OF APPLIED PHYSICS    SUPPLEMENT TO VOL. 33, NO. 3    MARCH, 1962

# Efficient Frequency Doubling from Ferrites at the 100-Watt Level

A. S. RISLEY AND I. KAUFMAN

*Space Technology Laboratories, Inc., Canoga Park, California*

A ferrite frequency doubler in the microwave range with a conversion efficiency of 5% to an external second harmonic load, for an input power of 300 w, is described. The design of this 8.5 to 17 Gc converter is based upon theoretical and experimental work reported earlier. A design diagram was drawn from the equation determining efficiency as a function of $\chi_0''$ and $h_\omega$ and experimental curves of $\chi_0''$ vs $h_\omega$ for various materials. From this diagram, the optimum material for a given input power was determined. In the case treated, single crystal manganese ferrite was optimum.

THIS paper describes a ferrite frequency doubler whose best efficiency to date is about 5% at an input power of 300 w at 8.5 Gc. This represents a considerable reduction of the input power level required for reasonably efficient frequency doubling.[1]

For frequency doubling with ferrimagnetic materials in a circuit simultaneously resonant to frequencies $\omega$ and $2\omega$, the second harmonic output power $P_{2\omega}$ is[2]

$$P_{2\omega} = P_\omega^2 F_0^2 R_{2\omega}^{-1}. \qquad (1)$$

Here the output power is seen to depend upon the power $P_\omega$ absorbed in the ferrite, the ferrite figure of merit $F_0$, and the microwave circuitry (through the symbol $R_{2\omega}$). For $P_{2\omega}/P_\omega < 0.05$, $F_0$ is given by

$$F_0 = 0.5(1-\epsilon^2)\mu_0 M_0^{-1}\chi_0''. \qquad (2)$$

In (2), mks units with $B = \mu_0 H + M$ are used. $M_0$ is the saturation magnetization, $\epsilon$ is the ratio of the two transverse components of the magnetization and is minimized by a disk that has $h_\omega$ and $H_{dc}$ in its plane, and $\chi_0'' = 2p_\omega/\omega\mu_0 h_\omega^2$. Here $p_\omega$ is the absorbed power per unit volume; $H_{dc}$ and $h_\omega$ are intensities of dc and rf magnetic fields, respectively. We define $R_{2\omega}$ as $R_{2\omega} = 4\omega U_{2\omega}/(Q_{L2\omega}h_{2\omega}^2)$, where $Q_{L2\omega}$ is the loaded $Q$ of the $2\omega$ circuit and $U_{2\omega}$ is the energy stored therein at field intensity $h_{2\omega}$. Equation (1) assumes that the specimen is uniformly excited, that the $2\omega$ circuit is matched to its load, and that operation is at ferrimagnetic resonance at frequency $\omega$.

Combining (1) and (2), we get an expression for the conversion efficiency, $\eta$

$$\eta = 0.25 R_{2\omega}^{-1} M_0^2 (1-\epsilon^2)^2 \mu_0^2 \chi_0''^2 P_\omega. \qquad (3)$$

This equation can be solved for the value of $\chi_0''$ required for particular $\eta$ and $P_\omega$. Moreover, since $p_\omega = P_\omega/V$, for a specified volume of magnetic material $V$, these chosen values of $\eta$ and $P_\omega$ also specify $h_\omega$, by the definition of $\chi_0''$ (above). The relation between $h_\omega$ and $\chi_0''$ at various values of input power $P_\omega$ is given by the two diagonal lines of Fig. 1, for two values of efficiency $\eta$, our $R_{2\omega}$ of $4.77 \times 10^{-5}$ ohm-m², and a 0.200-$\times$0.010-in. disk.

To determine the actual efficiency attainable with a given material, we plot its $\chi_0''$ vs $h_\omega$ characteristic, or, rather, the inverse relation $h_\omega$ vs $\chi_0''$, on the graph of Fig. 1. In general, a particular efficiency can be achieved with a given material if the line corresponding to that efficiency intersects or is tangent to the $h_\omega$ vs $\chi_0''$ curve of that material. The three curves of Fig. 1 correspond to three different materials. From Fig. 1, single crystal $MnFe_2O_4$ became the obvious choice for our generator.

The harmonic generator used is shown in Fig. 2. The $\omega$ fields filled the entire internal volume, whereas the $2\omega$ fields were confined to the left of plane $X$-$X$. By isolating the $2\omega$ fields in this manner, we kept $R_{2\omega}$ small by minimizing $U_{2\omega}$, and we allowed tuning of the $\omega$ circuit without affecting the $2\omega$ mode. The $2\omega$ cavity was matched to its load and had a $Q_{L2\omega}$ of 3800. About

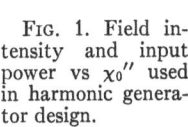
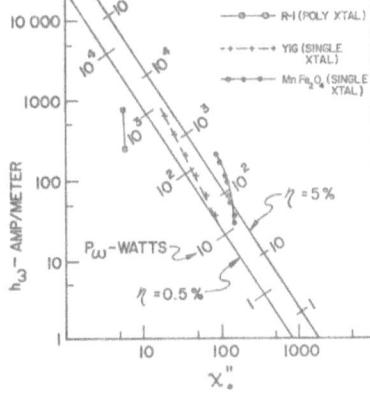

FIG. 1. Field intensity and input power vs $\chi_0''$ used in harmonic generator design.

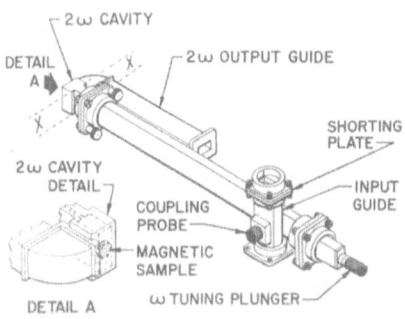

FIG. 2. Doubly-resonant microwave circuit.

[1] J. H. Melchor, W. P. Ayres, and P. H. Vartanian, Proc. Inst. Radio Engrs. **45**, 644 (1957).

[2] D. D. Douthett, I. Kaufman, and A. S. Risley, J. Appl. Phys. **32**, 1905 (1961).

90% of the power entering the input guide was absorbed in the disk. There was no need for tuning the $2\omega$ circuit, since the $\omega$ source was tunable.

The samples were driven with a magnetron (2 $\mu$sec, 40 pps, 8.5 Gc). A number of different disks of single crystal $MnFe_2O_4$ were tested. In the most efficient case, the cavity input power was 334 w and the output 18.4 w. The other disks were somewhat poorer but still of about the same order. From Fig. 1, we could expect higher efficiencies at higher input powers. Actually, we found that above a certain drive level the output became unstable and was not usable.[3]

To compare theory and experiment, $P_\omega$, $P_{2\omega}$, and absorption linewidth $\Delta H$ were measured. For a particular 0.198-$\times$0.009-in. disk, we had $P_\omega = 332$ w, $\Delta H$

[3] See also, M. T. Weiss, J. Appl. Phys. 31, 103S (1960).

(full line) = 3190 amp/m (40 oe). The expected value of $P_{2\omega}$ was then 33.6 w; the measured value was 13.4 w.

Two factors contributing to the disagreement between theory and experiment were (1) $\chi_0''$ was calculated from $\Delta H$; because saturation effects caused a non-Lorentzian line shape, this method is subject to some error. (2) Eq. (1) assumes that the fields at the specimen are uniform and are the unperturbed cavity fields. Although the correction for the perturbation is difficult, that for a cosine distribution is simple. Its application reduces $P_{2\omega\text{theory}}$ from 33.6 to 27.2 w.

A 0.150-$\times$0.010-in. disk of single crystal YIG was also tested. As predicted by Fig. 1, its efficiency was poor. The highest efficiency attainable was 0.1%. Here $P_\omega$ was 2 w. In a separate test with cw power, however, we did obtain 16-$\mu$w output from a 60-mw input.

---

JOURNAL OF APPLIED PHYSICS    SUPPLEMENT TO VOL. 33, NO. 3    MARCH, 1962

# A Miniaturized Ferrimagnetic High Power Coaxial Duplexer-Limiter*

J. CLARK AND J. BROWN

*Applied Physics Section, Sperry Microwave Electronics Company, Clearwater, Florida*

The device to be described in this presentation is a miniaturized high power coaxial component performing within a single package the functions of circulation, power limiting, and preselection. Attention is principally directed to the limiting and preselection functions which are accomplished through the use of a crossed-strip gyromagnetic coupler which differs from those previously reported in that the second-order nonlinear process rather than the first-order process is the basis for its operation. This limiter has been tested independently at power levels as high as 8 kw and found to exhibit a flat leakage output less than 200 mw and a spike leakage energy of about 0.5 erg. The limiter is mechanically tunable over a frequency range greater than 10% in C band. At a fixed setting the 3 db bandwidth is about 20 Mc. The recovery time of the limiter is determined by the kickback of energy from the garnet spin system. Experimental values of this quantity will be presented. The duplexer-limiter package operates at peak powers up to 40 kw into a 2:1 antenna mismatch. Input VSWR at the transmitter port is less than 1.35. Insertion loss from the transmitter port to the antenna port is less than 0.7 db. Low power insertion loss from antenna port through the limiter to the receiver port is 1.0 db.

RECENT advances in ferrimagnetic techniques have been utilized to construct in $C$ band a compact microwave circuit that serves as a combined duplexer, preselector, and power limiter. Principal components of the device are two miniaturized $Y$-junction circulators[1] and an unusual version of the crossed-strip limiter (see Fig. 1). The most significant difference between the previously reported limiters [2–4] and the present crossed-strip limiter is the fact that this limiter utilizes the second-order nonlinear effect[5] as the active limiting process.

The device is mechanically tunable from 5.4 to 5.9 Gc. A summary of electrical characteristics is presented in Table I.

FIG. 1. Breadboard model of the duplexer-limiter.

* This work was supported by the U. S. Army Signal Research and Development Laboratory, Fort Monmouth, New Jersey, under contract.

[1] J. Clark, J. Appl. Phys. 32, 323S (1961).
[2] R. W. DeGrasse, J. Appl. Phys. 30, 155S (1959).
[3] F. J. Sansalone and E. G. Spencer, IRE Trans. or Microwave Theory Tech. MTT-9, 272 (1961).
[4] F. R. Arams, M. Grace, and S. Okwit, Proc. Inst. Radio Engrs. 49, 1308 (1961).
[5] H. Suhl, J. Phys. Chem. Solids 1, 209 (1957).

The reactive nature of the limiting process utilized in this unit requires a 4-port duplexer to circulate the power reflected from the limiter into a load connected to the appropriate duplexer port. Two high power versions of a previously developed 3-port junction-type circulator[1] were joined to form the desired 4-port duplexer. The nature of these components is such that exceptional, high power performance can be obtained with only slight modification to the original structure.

The gyromagnetic limiter-preselector is of more current interest. The coupling element is a single crystal YIG sphere 26 mils in diameter with $\Delta H \approx 0.3$ oe. As is well known, the coincidence of main and subsidiary resonance does not occur in single crystal YIG spheres at $C$ band frequencies. The limiting observed must be attributed to the second-order nonlinear process.

For a given sphere size a "second-order" limiter can be expected to have a substantially higher threshold than a "first-order," or coincidence type of limiter. The highest possible threshold which still provides adequate crystal protection is desirable in that the limiter's usefulness is extended to higher input power levels. The strip to strip isolation with no biasing field (very nearly equivalent to dynamic limiting range) in the limiter reported here is between 45 and 50 db over the 5.4- to 5.9-Gc band. This is adequate to extend the useful range of the limiter to beyond 5 kw of peak input power.

The spike leakage limiter, measured at peak input powers in excess of 5 kw, was approximately 0.5 erg.[6] The shape of the leading edge of the input pulse was such that it reached 90% of full peak power in about 40 nsec.

Recovery time in this type of limiter is determined by the kickback of energy from the garnet spin system. This kickback was observed as an apparent extension of the trailing edge of the limiter pulse beyond the

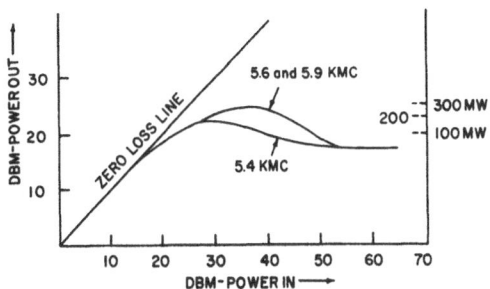

FIG. 2. Typical high power electrical characteristics of the limiter.

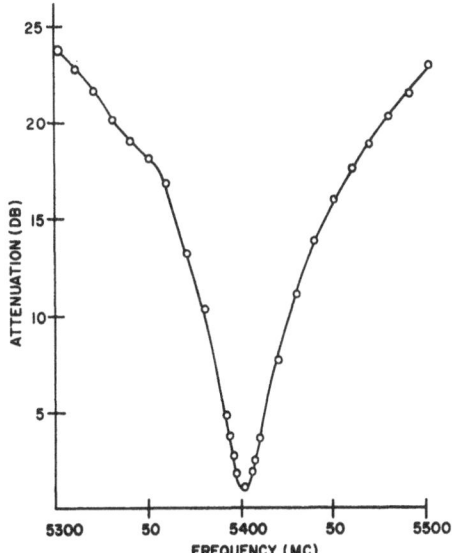

FIG. 3. Low power preselection characteristics of the limiter at 5.4 kMc.

trailing edge of the input pulse. The length of this extension, effectively the recovery time of the limiter, is about 0.2 $\mu$sec.

Flat leakage from the present model of this limiter is such that protection of type 1N23 or similar crystals may be marginal. Improvement in this characteristic should not be difficult to obtain.

Another problem area is found in the critical response of the limiter to temperature change. This is principally due to drift in applied field with temperature. Effects due to anisotropy change with temperature can be minimized by proper orientation of the single-crystal sphere with respect to the applied field.

### ACKNOWLEDGMENTS

The authors wish to thank J. L. Allen for many helpful suggestions and D. E. Tribby and A. C. Setlow for invaluable technical assistance.

TABLE I. Electrical characteristics.

| | |
|---|---|
| Power capacity | 40-kw peak, 40 $w_{av}$ into tran. with 2:1 mismatch at ant. |
| Insertion loss tran. to ant. | 0.7 db or less |
| Low power insertion loss ant. to rec. | 1 db or less |
| Input VSWR | 1.35 or less |
| Preselection | See Fig. 3 |
| Flat leakage | Somewhat power dependent. (Fig. 2) less than 200 mw with 40-kw input to duplexer |
| Spike leakage | Also somewhat power dependent. 0.5 erg with greater than 5-kw peak power into limiter |

[6] For a thorough discussion of spike leakage in a rectangular waveguide second-order limiter see C. N. Patel, Technical Report No. 411-1, Stanford Electronics Laboratories.

JOURNAL OF APPLIED PHYSICS    SUPPLEMENT TO VOL. 33, NO. 3    MARCH, 1962

# Nonlinear Effects in Ferrite-Filled Reduced-Size Waveguide*

F. S. HICKERNELL, B. H. AUTEN, AND N. G. SAKIOTIS

*Solid State Electronics Department Motorola, Inc., Scottsdale, Arizona*

Nonlinear effects were investigated at 9375 Mc in ferrite-filled reduced-size rectangular waveguide with transversely applied magnetic fields. The cross-sectional area of standard X-band waveguide was reduced to as low as one-tenth its original value, maintaining approximately the same aspect ratio. Factors investigated were insertion loss, threshold power level, spike characteristics, dynamic range of limiting, and breakdown power level. These are discussed with relation to waveguide size, frequency, ferrite loading geometry, ferrite material parameters, and temperature. Thresholds as low as 3 w peak were observed using standard polycrystalline ferrite materials. It was possible to realize useable limiter configurations whose low power losses were less than 1 db. The application of these effects in the design of a compact X-band limiter are mentioned, as well as the advantages and limitations of devices utilizing these effects.

THE ferromagnetic nature of ferrite materials makes possible a large variety of modes of propagation in a guided system. Ferrite-filled waveguides of arbitrarily small cross section can support at least two distinct modes of propagation, the normal dominant mode and the so-called birefringent modes, which are peculiar to the ferrite medium.

Seidel[1] has investigated the effect of birefringence in a magnetized gyromagnetic medium with respect to propagation characteristics in guides of arbitrarily small cross section. The presence of these modes depends upon certain relations satisfied by the diagonal component $\mu$ and the off-diagonal component $\kappa$ of the permeability tensor. Seidel has shown that for the birefringent modes to occur, the diagonal term of the permeability tensor must be positive, but less in magnitude than the off-diagonal component $\kappa$; that is, $\mu > 0$, $\mu^2 - \kappa^2 < 0$. The propagation in this region is characterized by an intense concentration of the electromagnetic field to a region of the waveguide near the side wall. The span of applied magnetic fields in the birefringent region lies below the value required for resonance at any given frequency and its width is proportional to the magnetization of the material.

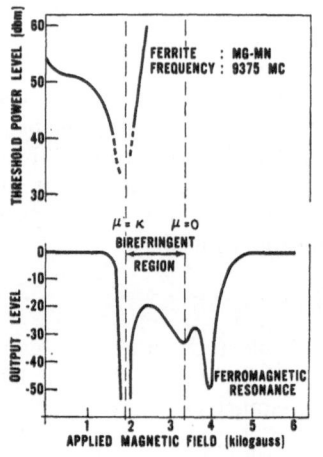

FIG. 1. Transmission characteristics and threshold power level in a ferrite-filled reduced-size waveguide.

In the region $\mu^2 - \kappa^2 > 0$, $\kappa < 0$ of the dominant mode, corresponding to applied field values below those of the birefringent modes, the ferrite material behaves as a low-loss, high index of refraction dielectric. It is characterized by the intense concentration of rf magnetic field as $\mu$ and $\kappa$ approach equivalent numerical values. It is the region of applied magnetic field values just below and just above the $\mu = \kappa$ region that are of particular interest in the investigation of first-order nonlinear subsidiary absorption effects.

Figure 1 shows typical absorption-loss characteristics of a ferrite-filled reduced rectangular waveguide at milliwatt power levels. Operation is at 9375 Mc with the waveguide cross-sectional area reduced by a factor of 10 from standard X-band waveguide maintaining the same aspect ratio. The waveguide is fully loaded with a magnesium-manganese ferrite. The magnetic field is applied across the width dimension. At low biasing field strengths the loss is approximately 0.5 db. A sharp cutoff near 2000 gauss defines the lower boundary of the birefringent region ($\mu = \kappa$) with losses greater than 50 db. The upper boundary was defined by an absorption loss relative maximum at approximately 3250 gauss. Resonance absorption was observed near 4000 gauss. Minimum loss in the birefringent region could be decreased by reducing the ferrite length, increasing the cross-sectional area of the waveguide, or decreasing the temperature of the ferrite. The loss apparently resulted from slow wave propagation, a characteristic of the birefringent region.

Nonlinear effects were investigated at 9375 Mc in the ferrite-filled, reduced-size rectangular waveguide with transversely applied magnetic fields. The cross-sectional area of standard X-band waveguide was reduced to as low as one-tenth its original value, maintaining approximately the same aspect ratio. The factors of primary interest were loss, threshold power level, spike characteristics, dynamic range of limiting, and breakdown power level. These were investigated with respect to waveguide size, frequency, ferrite loading geometry, ferrite material parameters, biasing field, and temperature.

A representative threshold power level response for the onset of nonlinear effects as a function of applied

* Part of this work was supported by the U. S. Navy Department, Bureau of Ships, under contract.
[1] H. Seidel, Proc. Inst. Radio Engrs. 44, 1410 (1956).

field is also shown in Fig. 1. The applied field at which the minimum threshold power level was observed was primarily a function of ferrite length and the $4\pi M$ value for a given waveguide cross section. The field position of minimum threshold could be adjusted to either side of the $\mu = \kappa$ cutoff region by proper choice of these two parameters. Threshold power levels were observed as low as 3 w peak; however, associated low power loss was in excess of 10 db.

The observed nonlinear characteristics as a function of power level for the reduced rectangular waveguide with MgMn ferrite are shown in Fig. 2. Operation is in the low-loss region of the dominant mode. The plateau and spike power levels of the output pulse are shown and the spike energy characteristics are also given. Low-level loss in the ferrite section is 1 db. The threshold power level is at 25 w and the plateau response remains at this level across a dynamic range of 30 db. The spike width decreased and the spike amplitude increased with increasing power in such a way that the spike energy remained essentially constant slightly under 100 ergs. The dynamic range increased with increasing ferrite length. Thus, there existed a means by which the dynamic range could be extended, limited ultimately by the breakdown power level. For the particular configuration of Fig. 2, power levels to 25 kw were utilized without breakdown.

The nonlinear effects are temperature and frequency sensitive through the displacement of the $\mu = \kappa$ region for a given configuration. A temperature decrease or frequency decrease displaces the cutoff region to lower field values. Increasing the frequency or temperature displaces the cutoff region to higher field values.

Similar nonlinear effects to those described were observed in a reduced-size ferrite-filled circular guide with

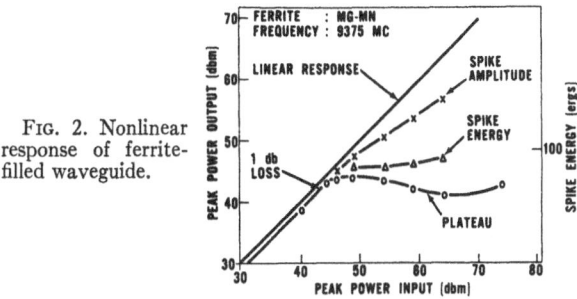

FIG. 2. Nonlinear response of ferrite-filled waveguide.

an applied axial field. No advantages were apparent for the design of a limiter using the circular guide configuration.

The cutoff effects in a ferrite-filled reduced-size waveguide may be used to advantage in the development of a ferrite limiter. Dynamic range may be controlled by the adjustment of ferrite length. Threshold power level may be controlled by adjustment of cross-sectional area subject to the limitation of cutoff of the dominant mode. The field position of minimum threshold power level may be adjusted by varying the geometry or the saturation magnetization of the material. Additional advantages of a reduced-guide configuration are the use of polycrystalline materials, compactness and simplicity of construction. Dielectrically loaded tapered sections are used to match into the ferrite loaded waveguide. Temperature and frequency sensitivity appear to be the only limitations. Motorola has fabricated a small compact limiter which displayed the characteristics shown in Fig. 2.

The authors wish to express appreciation to L. Derting and B. Ziegner for their assistance in making the measurements, and to the Navy Department, Bureau of Ships, for sponsoring part of this work.

JOURNAL OF APPLIED PHYSICS    SUPPLEMENT TO VOL. 33, NO. 3    MARCH, 1962

# Millimeter Wave Parametric Amplification with Antiferromagnetic Materials

R. A. MOORE

*Applied Physics Group, Air Arm Division, Westinghouse Electric Corporation, Baltimore, Maryland*

The use of material with natural magnetic resonances in the millimeter and submillimeter wave bands makes possible electromagnetic interaction without an extremely large applied magnetic field. Antiferromagnetic materials most suitable for microwave application $Cr_2O_3$ and $MnFe_2$ display resonances at approximately 160 and 250 kMc.

In the presence of a magnetic field applied parallel to the sublattice magnetizations, the uniform precessional mode resonates at different frequencies for oppositely directed circularly polarized excitation. A perturbation analysis has been conducted to determine the parametric response of the oppositely directed uniform precessional modes to longitudinal pumping. For materials with a sufficiently narrow linewidth, very close coupling between the uniform precessional mode on an external coaxial, stripline, or waveguide signal circuit is possible. Calculations have been conducted on the response of an antiferromagnetic medium in a silver cavity resonant to the pump frequency. The pump power absorbed by the cavity at the threshold of amplification for $Cr_2O_3$ would be $2(10)^4$ w. Narrower linewidth materials are needed for efficient antiferromagnetic millimeter and submillimeter amplification.

A more efficient utilization of the pump power and a simplified structure for millimeter wave application results when the pump fields are supported by a dielectric rod mode resonance. If the rod is placed across the narrow dimention of rectangular waveguide, the broad metallic wall forms the conducting end plates of the $H_{11}$ mode. The signal $TE_{01}$ waveguide mode couples to the appropriately directed circularly polarized uniform precessional resonance of the antiferromagnetic material on the rod.

## INTRODUCTION

MANY of the familiar ferrite microwave components have been scaled down for millimeter wave devices with only the difficulty of constructing the very small components. The extremely high applied magnetizing fields required for the operation of many components constructed from ferrimagnetic materials would limit their use in many applications. The extremely high magnetic field requirement can be greatly reduced by utilizing the very high internal fields of antiferromagnetic materials. The very large exchange and anisotropy fields result for different materials in natural resonances at a number of millimeter and submillimeter frequencies.[1–3] The possibility of utilizing antiferromagnetic materials for millimeter and submillimeter device application has been suggested by Foner and by Heller *et al.*[4,5] Heller *et al.* demonstrate the use of $Cr_2O_3$ with a resonance linewidth of 350 oe in a magnetic resonance-type isolator.

## SIMPLIFIED ANTIFERROMAGNETIC DYNAMIC RESPONSE

The dynamic response of antiferromagnetic materials has been treated by a number of authors.[6,7] In its simplest form, the antiferromagnetic material can be thought of as consisting of two sets of interpenetrating magnetic moments, equal in magnitude and oppositely directed. Each of the magnetic moments is under the influence of strong exchange and anisotropy fields as well as an externally applied magnetic field. The dynamic response can be expressed in terms of two equations for a two-lattice structure. The response of each magnetic moment system is then obtained by solving the two equations simultaneously. For temperatures very close to $0°K$, the resonant condition becomes $\omega = \omega_C - \omega_H$, where $\omega_H = \gamma H$ and $\omega_C = \gamma(2H_E H_A H_A^2)$ and $H$, $H_E$, and $H_A$ are the applied, exchange, and anisotropy fields, respectively. The plus and minus sign gives the resonant condition for the two directions of circular polarization. Each of the system of magnetic moments responds in both directions of circular polarization. The total response is equal to the sum of the response of the magnetic moments in each direction. This response is clearly illustrated by Heller.

## LONGITUDINAL PUMPING

The possibility of parametrically pumping magnetostatic modes by a longitudinal field was demonstrated by Denton. His amplifier utilized two magnetostatic modes of ferrimagnetic material and showed that amplification with a relatively low pump power is possible. In antiferromagnetic material, the oppositely rotating circularly polarized uniform precessional mode resonances can be used to support the signal and idler fields. The use of uniform precessional modes should lend readily to tight coupling with external electromagnetic fields.

Amplification will be formulated as a perturbation on the individual responses of the antiferromagnetic medium to signal and idler excitation. The responses to both signal and idler excitations are the sum of components of both magnetic moment systems. If the signal, idler, and pump exciting fields are substituted into the equations for the dynamic response, a set of four equations and their complex conjugates results for the circularly polarized signal and idler components of each system of magnetic moment. The set of four equations can be solved simultaneously to yield the response of each of the magnetic moments. The sum of the responses

[1] F. M. Johnson and A. H. Nethercott, Phys. Rev. **114**, 705 (1959).
[2] A. M. Portis and Dale Teaney, Phys. Rev. **116**, 838 (1959).
[3] R. C. Ohlman and M. Tinkham, Phys. Rev. **123**, 425 (1961).
[4] S. Foner, J. phys. radium **20**, 336 (1959).
[5] G. S. Heller, J. J. Stickler, and J. B. Thaxter, J. Appl. Phys. **32**, 307S (1961).
[6] C. Kittel, Phys. Rev. **82**, 565 (1951).
[7] R. K. Wangsness, Phys. Rev. **86**, 146 (1952).

of the two signal components can be expressed as

$$\chi_S^{\pm} = \frac{2[\Delta_i{}^2(\omega_A \mp j1/T_H) - \gamma^2 |h_z|^2(\omega_A \pm j/T_H)\omega_m]}{\Delta^4}, \quad (1)$$

where

$$\Delta^4 = \Delta_i{}^2 \Delta_S{}^2 + \gamma^4 |h_z|^4 - |h_z|^2 \gamma^2$$

$$\times \left[ 2(\omega_C{}^2 + \omega_H[\omega_H \pm \omega_S + \omega_i]) + j\frac{2(\omega_S - \omega_i)}{T} \right],$$

$$\Delta_S{}^2 = \omega_C{}^2 - (\omega_H \pm \omega_S)^2 + j2\omega_S/T_S,$$

$$\Delta_i{}^2 = \omega_C{}^2 - (\omega_H \mp \omega_i)^2 - j2\omega_i/T_i,$$

and the $T$'s are the appropriate relaxation times. The threshold of amplification for the medium occurs when the denominator in Eq. (1) vanishes. If the signal and idler modes are assumed to be in the positive and negative directions of circular polarization, the resonant conditions can be expressed as $\omega_S = \omega_C - \omega_H$ and $\omega_i = \omega_C + \omega_H$, respectively. For this resonant condition the pump field intensity required at the threshold of amplification is given approximately as

$$|h_z|^2 = \frac{\Delta H_i \Delta H_S \omega_S \omega_i}{2[\omega_C{}^2 + \omega_H{}^2 + (\omega_i{}^2 - \omega_S{}^2)]}, \quad (2)$$

where $\Delta H_S$ and $\Delta H_i$ are the unloaded resonant linewidths of the signal and idler modes.

## ANTIFERROMAGNETIC AMPLIFIER

In an amplifier the pump field can best be applied by means of a high $Q$ resonant cavity. The signal can either be coupled to a dynamic transverse cavity mode or to a transmission structure. For example, the high $Q$ $TE_{01}$ mode in a cylindrical cavity might be utilized to couple to the pump field. The signal could then be coupled to the $TE_{11}$ circularly polarized mode. The cavity would be designed to provide magnetic field maxima at the coupling medium. The natural resonant frequency $Cr_2O_3$ is approximately 165 kMc for which a line width of 350 oe has been reported.[5] The pump power absorbed by the cavity can then be calculated by integrating the Poynting vector over its surface. If a silver-coated cavity is assumed, the pump power to the cavity must be $2.1 (10)^4$ w. This does not include any power absorbed by the medium or the increased power which might be required as a result of excitation of unstable spin waves.

It is evident that antiferromagnetic materials with a much narrower linewidth would be required to make this amplifier practical. With narrow linewidth material the signal mode could be coupled directly to its exciting transmission structure rather than by means of a resonant cavity.

A more efficient utilization of the pump power requires that the excitation fields be more tightly coupled to the parametric medium. This can be accomplished along with a much simplified structure for millimeter waves by utilizing an electromagnetically resonant section of propagation rod structures or dielectric rod mode type resonances to support the excitation fields. If the rod is placed across the narrow dimensions of rectangular waveguide, the broad metallic walls form the conducting end planes for the $HE_{11}$ mode. Parris and Moore[8] show that in the presence of a slightly nonuniform longitudinal magnetizing field or by means of a slightly nonuniform structure (such as a dielectric disk at one end) the $HE_{11}$ mode is tightly coupled to the $TE_{01}$ mode of the metallic waveguide. The pump can be supported by means of the mode on the antiferromagnetic material analogous to the $H_{01}$ dielectric rod mode. The pump generator would be coupled from a port at one end of the structure.

## CONCLUSIONS

The use of antiferromagnetic materials with natural magnetic resonance in the millimeter and submillimeter wave bands makes possible electromagnetic interaction without an extremely large applied magnetic field. In the presence of a magnetic field applied parallel to the sublattice magnetization, the uniform precessional mode resonates at different frequencies for oppositely directed circularly polarized excitation. These modes can be utilized for parametric amplification in the presence of a longitudinal pumping field. Calculations conducted on the response of presently available materials for a silver cavity resonant at the pump frequency show that excessive pump power would be required. The use of a dielectric rod mode resonance should result in a more efficient utilization of the pump power. It is felt that the potential usefulness of antiferromagnetic properties for millimeter and submillimeter wave application warrant effort toward narrow resonant linewidth materials.

[8] W. J. Parris and R. A. Moore, "Ferrite post microwave resonant structure," Military Electronics Conference, Los Angeles, California, Winter, 1961.

JOURNAL OF APPLIED PHYSICS    SUPPLEMENT TO VOL. 33, NO. 3    MARCH, 1962

# A Microwave Magnetic Microscope*

R. F. SOOHOO

*California Institute of Technology, Pasadena, California*

A microwave magnetic probe capable of measuring the spatial variation of the magnetic properties of materials has been developed. In contrast to other forms of microscopes, this instrument measures the spatial variation of the important magnetic properties of a sample directly by a selective microwave resonance technique. To date, we have observed clear resonances in portions of Permalloy films (~1000 A thick) with a cross-sectional diameter of 0.010 in. By using a cavity made of superconducting material, it is anticipated that resonances in spots as small as 200 A in cross-sectional diameter may be observed. Some representative experimental results using this microscope are given to indicate its utility.

## INTRODUCTION

A MICROWAVE magnetic probe capable of measuring the spatial variations of the magnetic properties of materials is described below. In contrast to other forms of microscopes, this one measures the spatial variations of the important magnetic properties of a sample directly. By a selective microwave resonance absorption technique, the gyromagnetic ratio, magnetization, linewidth, exchange constant, surface and anisotropy energies, etc. of small portions of a ferromagnet may be determined consecutively.

The principles of operation of the microscope may be briefly described as follows. By placing the magnetic sample against the outer face of the back wall of a cavity which has a small centered hole, only that portion of the material directly opposite the opening is exposed to the microwave radiation from the cavity. By varying the static magnetic field applied to the sample, the resonance spectrum of the radiated portion may be ascertained. The magnetic properties of other portions of the material may likewise be obtained by shifting these portions against the cavity hole successively. In this way, the spatial variation of the magnetic properties of the entire sample may be determined.

Since the radiation eminates from a coupling hole that is quite small compared to the wavelength, there will be a diffraction of the microwaves. This implies that the field at the sample, aside from being nonuniform, will also be not completely confined to the vicinity of the hole. However, experiments to date using a hole as small as 0.001 in. indicate that the uniformity and confinement problem is not severe enough to limit the use of the microscope.

## CAVITY DESIGN PROBLEM

In experiments utilizing the microscope, we wish to maximize the radiation field eminating from the cavity hole to excite the sample to resonance. To do this, the cavity back wall should be made quite thin since the hole there acts like a waveguide beyond cutoff.

The rf magnetic field **h** in a waveguide must obey the wave equation

$$\nabla^2 \mathbf{h} + (\omega/c)^2 \mathbf{h} = 0, \qquad (1)$$

where $\omega$ is the rf frequency. If we assume that $\mathbf{h} \propto e^{-\gamma z}$, where $z$ is the direction of propagation, we find from Eq. (1) that $\nabla_t^2 \mathbf{h} + k_c^2 \mathbf{h} = 0$, where

$$k_c^2 = (\omega/c)^2 + \gamma^2 \qquad (2)$$

and $k_c$ is the cutoff wave number. For the dominant $TE_{11}$ mode of a circular waveguide, it can be easily shown from Eq. (2) that the reactive attenuation $\alpha$ of length $l$ of a waveguide of diameter $d$ is[1]

$$\alpha \simeq -32(l/d) \text{ db} \qquad (3)$$

if the waveguide is far below cutoff. Thus, to obtain a small $\alpha$, $l/d$ must be kept small. If a superconducting cavity is used, $l$ may be made quite small.

By probing the microwave radiation near the hole, we have roughly confirmed the applicability of Eq. (3) to our problem of determining the field intensity at the coupling hole. Furthermore, we found that the radiation-field configuration, being largely confined to the vicinity of the hole, is highly nonuniform. Confinement of the radiation enables exploration of only the selected portion of the sample. That this is so has been demonstrated experimentally. A portion of the sample with and without the rest of the sample attached had resonance intensities that were practically identical when it was placed against the cavity hole. However, if the effect of pinning of the surface spins of a ferromagnet is significant, as it was assumed in our recent calculations,[2] the intensities of the resonance peaks would be dependent upon the location of the excited portion of the spin system.

The complexity of the field configuration at the hole, however, presents a more serious problem of interpretation of the data. Spin wave peaks, which may be only weakly excited in uniform fields, may be highly accentuated by the complex field in the region of the film to be studied. Conversely, the highly nonuniform field may be noninducive to uniform-mode excitation. The ex-

[1] R. F. Soohoo, *Theory and Application of Ferrites* (Prentice-Hall, Inc., Englewood Cliffs, New Jersey, 1960), p. 212.
[2] R. F. Soohoo, "Relaxation of inhomogeneously excited spin systems," presented at the International Conference on Quantum Electronics, March 23–25, 1961, Berkeley, California; Lincoln Laboratory Group Report 53G-0057.

* Part of the research reported in this paper was carried out while the author was at the MIT Lincoln Laboratory (see Lincoln Laboratory Group Report 53G-0064).

perimental evidence for this has been reported previously.[3]

The back wall of the $TE_{10n}$ cavity used in our experiment was made of 0.001-in. thick copper. Because of its thinness, it would vibrate in an acoustical field of rather small intensity. This vibration, coupled with the high $Q$ of the cavity, gave rise to a background signal which may obscure the desired signal due to magnetic resonance. Making the cavity out of a solid piece of dielectric plated by a silver coating would eliminate this difficulty.

## DETECTION SYSTEM

As is well known, whereas the noise of the crystal detector decreases with frequency, the noise of the amplifiers increases with frequency. A broad minimum in fact is centered about 32 Mc. Thus, in principle, a super-heterodyning system whereby the microwave frequency is translated to 32 Mc by means of a local oscillator should constitute the most sensitive detection system. However, due to the unavoidable instability in microwave oscillators, this system is usually impractical at the high-microwave frequencies. The alternative to this approach is to modulate the static magnetic field at as high a frequency as possible, and to detect the resonance signal of the sample at the modulation frequency. In conjunction with a synchronous detector, narrowband amplification may be obtained, resulting in low noise and consequent high sensitivity. In this way, the derivative of the absorption rather than the microwave absorption itself is obtained.

The modulation of the static magnetic field was accomplished by a pair of coils attached to the electromagnet supplying the static field. The reflected signal from the cavity would then have a component at the modulation frequency and may be detected and amplified. As the static magnetic field is varied through the value required for ferromagnetic resonance of the sample, the reflection from the cavity at the modulation frequency would change indicating resonance absorption.

## SCANNING MECHANISM

The sample, a thin Permalloy film, was placed on a platform which could move independently in two orthogonal directions horizontally, the movements being

[3] R. F. Soohoo, Solid State Research Quarterly Progress Report, Lincoln Laboratory, MIT (April 15, 1961), p. 63.

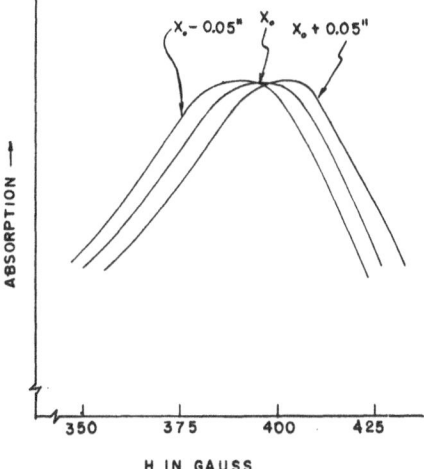

FIG. 1. Resonance absorption vs static magnetic field for portions of a Permalloy film ⅜ in. in diameter and 3500 A thick at 5.5 kMc. $X_0$ represents some arbitrary location on the surface of the film.

controlled by two micrometers with calibrations accurate to within 0.0001 in. The height of the platform was adjusted independently until the film was pressed against the cavity hole. At this point, the static magnetic field was varied through resonance of the portion of sample opposite to the cavity hole. Then, another portion of the sample was placed against the hole by manipulating the micrometers. This process was continued until the entire sample was examined by the probe.

## EXPERIMENTAL RESULTS

To indicate briefly the possible utility of our microscope, we have shown in Fig. 1 the actual resonance curves of three adjacent regions of a Permalloy film obtained by using the microscope described in this paper. The cavity hole was 0.020 in. in diameter and the operating frequency was 5.5 kMc. in these cases. It is seen that there are differences between the resonance fields of the different regions which may be attributed the difference in magnetization or anisotropy field of these regions. In this experiment, no modulation of the static field was applied so that Fig. 1 indicates the absorption rather than the dispersion curve of the sample.

*Note added in proof.* A probe similar to the one reported here has been independently developed by Z. Frait, Czechoslov. J. Phys. **9**, 403 (1959).

JOURNAL OF APPLIED PHYSICS    SUPPLEMENT TO VOL. 33, NO. 3    MARCH, 1962

# Sampling Magnetometer Based on the Hall Effect

H. H. WIEDER

*U. S. Naval Ordnance Laboratory, Corona, California*

A pulsed sampling magnetometer is described, based on the Hall effect in vacuum evaporated films of InSb. Pulsed operation of thin film Hall effect detectors provides an increase in sensitivity over cw operation of conventional Hall devices. This is due, in part, to the improved thermal dissipation of the fim, the reduction in joule heating, and the greatly improved noise discrimination because of stroboscopic sampling. A sensitivity of better than $10^{-4}$ v/gauss, over a frequency range between 100 cps to 5 Mc and an angular orientation sensitivity of $10^{-5}$ v/degree have been obtained in devices built to date.

## INTRODUCTION

THE inherent simplicity of the Hall effect is an obvious advantage in its use for the detection and measurement of magnetic fields. A rectangular plate of length $x$, thickness $y$, and width $z$ is oriented along corresponding cartesian axes. For a current $I_x$ and a magnetic induction $B_y$, a potential difference $V_z$ will arise in a direction orthogonal to both the electric and magnetic vectors. The proportionality between the magnetic induction and the Hall voltage $V_z$ is the basis for the magnetometer applications of the Hall effect. If the sensitivity of such a device is defined as the Hall potential per unit magnetic field ($V_z/H$), then:

$$\frac{V_z}{H} = \mu_e z \left[ \frac{P_m}{y} \mu R_h \right]^{\frac{1}{2}}. \tag{1}$$

Equation (1) presumes that the Hall plate is embedded in a medium with an effective permeability $\mu_e$ and that $P_m = I_x^2 (\sigma z^2 y)^{-1}$ is the peak power density per unit surface area that may be dissipated by heat conduction without materially affecting the Hall coefficient $R_h$. The electron concentration and the electron mobility $\mu$ are assumed to be much greater than the corresponding hole concentration and mobility.

Materials having a high electron mobility and Hall coefficient will increase the sensitivity of Hall detectors. The intermetallic semiconductors InSb and InAs are such materials.[1,2] Indium arsenide has a lower electron mobility than indium antimonide, but a much smaller temperature dependence of $R_h$. Noise effects due to the passage of current are particularly small in InSb in comparison with other semiconductors.[3,4] The sensitivity of Hall detectors may be further increased by increasing their width $z$ and decreasing their thickness $y$. Increasing the width requires that the length of the plate be increased as well in order to avoid electrostatic shorting of the Hall electrodes,[5,6] yet for many applications, it is desirable to keep a Hall probe as small as possible.

Decreasing the thickness of a Hall plate is highly desirable and suggests the use of thin films.[7,8] Such films, because of their large surface to volume ratio, also dissipate heat more efficiently than Hall detectors fabricated from bulk crystalline materials. The magnetometer to be described subsequently, uses thin films of InSb evaporated onto a microscope cover glass substrate. The desired Hall plate contour is obtained either by suitable masking of the substrate or by ultrasonic cutting of the desired pattern from the glass slip.

The peak allowable power density $P_m$ for a particular film geometry and conductivity is determined primarily by the Joule heating. Its magnitude is the steady-state power dissipation in terms of the dc current density $J_x$. If the current is applied in the form of rectangular pulses of duration $\tau$ and repetition rate $\nu$, then for an equivalent dc heating effect, $J_x = J_p (\tau \nu)^{-\frac{1}{2}}$, where $J_p$ is the peak pulse amplitude. An increase in sensitivity may thus be obtained by pulse driving a Hall detector and Eq. (1) should be divided by $(\tau \nu)^{\frac{1}{2}}$. A further increase in sensitivity may be obtained by placing the Hall plate between ferrite or $\mu$-metal field concentrators. The effective permeability $\mu_e$ in the gap between them is determined by the permeability and geometry of the field concentrators as well as the gap spacing. In the case of static magnetic fields, empirical design details of such field concentratiors are given by Hieronymus and Weiss.[9]

The subsequently described results were obtained on an InSb film Hall detector mounted between ferrite field concentrators having a nominal permeability of 500. The gap between the concentrators is 0.035 cm. The dimensions of the Hall plate are: $x = 0.48$ cm, $y = 1.6 \times 10^{-4}$ cm, and $z = 0.24$ cm. At $+25°$C, the conductivity of the film was determined to be $\sigma = 26.6$ (ohm-cm)$^{-1}$ and the Hall coefficient as $R_h = 155.3$ cm$^3$/coul. The effective mobility is then $\mu = 4.13 \times 10^3$ cm$^2$ (volt-sec)$^{-1}$. For a steady-state magnetic field, identical values of $V_z$ are obtained either with a dc or a pulsed current drive up to a peak value of $I = 9$ ma. Above 9 ma, $V_z$ still increases linearly with the pulse current $J_p$. Joule heating affects $R_h$, however, for larger dc currents and results in a nonlinear dependence of $V_z$ upon $I_x$. From the foregoing,

[1] H. Welker, Z. Naturforsch. **7a**, 744 (1952).
[2] H. Weiss, Z. Naturforsch. **8a**, 463 (1953).
[3] D. J. Oliver, Proc. Phys. Soc. (London) **70**, 331 (1957).
[4] G. H. Suits, W. D. Schmitz, and R. W. Terhune, J. Appl. Phys. **27**, 1385 (1956).
[5] R. F. Wick, J. Appl. Phys. **25**, 741 (1954).
[6] F. Kuhrt and H. J. Lippmann, Naturwissenschaften **45**, 156 (1958).

[7] K. G. Gunther, Z. Naturforsch. **13a**, 1081 (1958).
[8] R. F. Potter and G. G. Kretschmar, J. Opt. Soc. Am. **51**, 693 (1961).
[9] H. Hieronymus and H. Weiss, Siemens Z. **31**, 404 (1957).

the peak power density is then $P_m = 0.336$ w/cm². Without the field concentrators $\mu_e = 1$ and for a pulse duration and repetition rate of $\tau = 10^{-7}$ sec and $\nu = 10^3$ pps, the sensitivity of the pulsed Hall plate is seen to be $(V_z/H) = 8.8 \times 10^{-3}$ v/oe. With the field concentrators in place, $V_z$ was found to increase by a factor of six, therefore $\mu_e \simeq 6$. The effective gap permeability was found, however, to be field dependent decreasing to about 2 at a frequency of 5 Mc.

If the magnetic field to be measured is a periodic function of time, then a sampling method may be employed in conjunction with the pulsed Hall detector. The waveform as well as the amplitude and direction of a magnetic field may be determined and the usual advantages of sampling procedures may thus be realized. Figure 1 shows a block diagram of such a sampling magnetometer. The fast sawtooth signal synchronized to the field is compared against a slowly rising ramp in the voltage comparator. Their coincidence triggers a pulse generator which is thus slowly phase modulated with respect to the input signal. The current pulses sample the magnetic field producing a proportional Hall signal in the detector. After suitable integration and amplification of this signal, a low-frequency replica and synthesis of the magnetic field is obtained at the output of the magnetometer. Sinusoidal magnetic fields from 100 to $5 \times 10^6$ cps have been sampled in this fashion with an output frequency between 0.01 and 1 cps.

The maximum sensitivity of this magnetometer may be determined by considering the integrated output of the Hall detector for a train of rectangular pulses. Equation (1) may then be expressed as:

$$\frac{V_z}{H} \simeq \mu_e z \cdot \left(\frac{J_x}{J_p}\right) \left[\frac{P_m \mu R_h}{y}\right]^{\frac{1}{2}}. \tag{2}$$

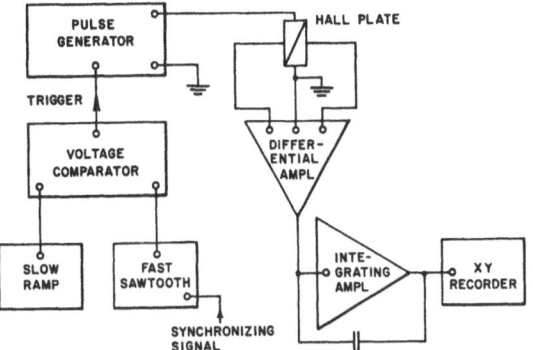

FIG. 1. Block diagram of the apparatus used for the sampling magnetometer.

Taking $J_p = 3J_x$ and $\nu = 10^6$ pps, then with $\tau = 10^{-7}$ sec and $\mu_e = 6$, Eq. (2) yields $(V_z/H) \simeq 5.6 \times 10^{-4}$ v/oe. If the minimum detectible signal above noise is 1 $\mu$v then the minimum field $H_{\min}$ detected by the magnetometer is $H_{\min} = 1.8 \times 10^{-3}$ oe. Experimentally, it is found that $H_{\min} = 4 \times 10^{-3}$ oe for a sinusoidal magnetic field of 1 kc, in fair agreement with the above calculation. This minimum was found, however, to be frequency sensitive rising to about 1 oe at $5 \times 10^6$ cps, probably because of the restricted bandwidth of the integrating amplifier. In any case, the output signal is a linear function to better than 2% of the magnetic field amplitude between $10^{-3}$ and $10^4$ oe. Thus far, the magnetometer has been used to plot magnetic field contours in and around solenoids with and without ferrite cores and to map the steady state fringing field of an electromagnet by pulse driving the Hall detector. The magnetometer may be improved considerably by using thinner films of higher mobility InSb, designing the field concentrators for an optimum $\mu_e$ and bandwidth and improving the thermal heat transfer between the film and its surroundings.

JOURNAL OF APPLIED PHYSICS    SUPPLEMENT TO VOL. 33, NO. 3    MARCH, 1962

# New Type of Flux-Gate Magnetometer

WILLIAM A. GEYGER

*U. S. Naval Ordnance Laboratory, Silver Spring, Maryland*

The use of an ordinary toroidal core without air-gap as a flux-gate magnetometer has been overlooked. A magnetometer using such a core as the field-sensitive element with semicircularly wound and differentially connected second-harmonic detector windings has been developed. The semicircle portions of a nickel-iron-alloy ring core act here like two separate cores (corresponding to the two parallel nickel-iron-alloy strips or scrolls of conventional forms of flux-gate elements), as far as second-harmonic flux components are concerned. By using tape-wound or laminated (washer-type) Supermalloy cores having i.d.-o.d. ratios in the range from 0.85 to 0.98, a sensitivity of 1000 $\mu$amp/oe or 1 v/oe can be achieved. Such "ring-core flux-gate elements" make it possible to: (1) use ultra-thin, $\frac{1}{8}$-mil tape and correspondingly high excitation frequencies, 10–40 kc/sec, (2) make "point measurements" by reducing the core diameter to 0.5 in., or less, (3) minimize the magnetizing-current requirements, (4) obtain linear characteristics, (5) eliminate memory effects, (6) facilitate matching of the magnetic characteristics of the two active parts of the flux-gate element which belong here to the same core, (7) detect very small changes in the earth's magnetic field, and (8) apply multiple detector windings on a common core. The power requirements of a portable magnetometer, operated from a 6-v battery, have been reduced to 50 mw by combining the ring-core flux-gate element with a switching-transistor magnetic-coupled multivibrator in such a way that the oscillation frequency is solely determined by the parameters of the ring core and its excitation windings.

## INTRODUCTION

A RECENT laboratory project required the use of a second-harmonic flux-gate magnetometer of the following characteristics:

(1) The size of the flux-gate element should be within the limits of 0.5–1.5 in. to permit "point measurements" on very small areas of the external magnetic field to be investigated.

(2) Several ranges, e.g., 0. . . 1.5 oe and 0. . .15 oe, should be provided simply by varying a shunt resistor connected with the D'Arsonval type, indicating or recording, dc instrument in the detector system.

(3) The portable magnetometer should be operated from a 6-v battery with a minimum of current drain, not exceeding 50 ma.

It was possible to meet these requirements by using, as the flux-gate element, an ordinary toroidal core (without air gap), as indicated in Fig. 1(a). A particularly favorable method for producing the exciting ac flux is a combination of the ring-core flux-gate element with a switching-transistor magnetic-coupled multi-vibrator (static dc to ac converter) in such a way that

this element determines the oscillation frequency of the multivibrator, as shown in Fig. 1(b).

## OPERATING PRINCIPLE OF RING-CORE FLUX-GATE ELEMENTS

The two semicircle portions of the nickel-iron-alloy ring core, Fig. 1(a), act like two separate cores (corresponding to the two parallel nickel-iron-alloy strips or scrolls of conventional forms of flux-gate elements), as far as second-harmonic flux components are concerned. Consequently, these flux components can be utilized by means of two differentially-connected windings, which are associated with the semicircle portions of the core. Actually, the series-aiding-connected ac excitation (primary) windings $N_P'$, $N_P''$ and the series-opposing-connected second-harmonic detector (secondary) windings $N_S'$, $N_S''$, Fig. 1(b), are uniformly distributed along the semicircle portions of the core.

The graphical symbols of Fig. 1(a) make it evident that the semicircle portions of the core are influenced, during both half-cycles (black and white arrows) of the exciting ac magnetomotive force $H_{ac}$, alternately, by the sum or difference of $H_{ac}$ and the external steady field $H_X$ to be measured. Thus, the total magnetomotive forces in these semicircle portions will be $H_1 = H_{ac} + H_X$, and $H_2 = H_{ac} - H_X$, respectively. The magnetomotive forces $H_1$ and $H_2$ give rise to corresponding fluxes $\Phi_1 = \Phi_{ac} + \Phi_{SH}$ and $\Phi_2 = \Phi_{ac} - \Phi_{SH}$. The first component $\Phi_{ac}$ is common to both semicircle portions of the core and varies with the oscillation frequency of the multivibrator circuit. The second (even-harmonic) component $\Phi_{SH}$ exists because of the presence of $H_X$ combined with $H_{ac}$.

The corresponding phase-reversible second-harmonic voltage $E_{SH}$ across the input terminals of the detector system (polarity-sensitive demodulator circuit D', D'', C, M), Fig. 1(b), is a measure of the external steady field influencing the toroidal core acting as a flux-gate element (FG). In a given field, this voltage $E_{SH}$ varies from a maximum when the axis of the ring-core element and the field are parallel to zero when they are perpendicular (cosine-law directivity).

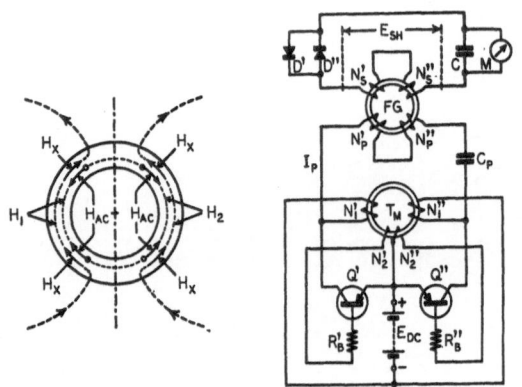

FIG. 1. Ring-core flux-gate magnetometer. (a) Fundamental principle. (b) ac excitation with switching-transistor magnetic-coupled multivibrator.

## ac EXCITATION OF RING-CORE FLUX-GATE ELEMENTS

A particularly favorable method consists in combining the ring-core flux-gate element with a switching-transistor magnetic-coupled multivibrator in such a way that the oscillation frequency is solely determined by the parameters of the toroidal core and its excitation windings.[1]

In the example of Fig. 1(b), the excitation windings $N_P'$, $N_P''$ are supplied from a common-emitter-type multivibrator circuit operating as a static dc to ac converter with an efficiency of about 70–80%. This circuit consists of two *p-n-p* switching transistors $Q'$, $Q''$, a transformer $T_M$ with (unsaturated) Supermalloy 2-mil tape core, two fixed bias resistors $R_B'$, $R_B''$, and the dc power supply (6-v battery).

## CONSTRUCTION OF RING-CORE FLUX-GATE ELEMENTS

In order to obtain high sensitivity of this type of flux-gate element (e.g., 1000 μamp/oe or 1 v/oe), it is necessary to achieve a sufficiently high effective permeability which is a function of i.d.–o.d. ratio and mean diameter of the core as well as of the permeability of the magnetic material itself. Actually, used values of the i.d.–o.d. ratio are in the range from 0.85–0.98; and the mean diameter is about 0.5–1.5 in.

Application of toroidal cores in field-sensitive flux-gate constructions makes it possible to use relatively high excitation frequencies, e.g., 12 000 cps, by taking full advantage of the favorable magnetic properties of $\frac{1}{4}$-mil and $\frac{1}{8}$-mil nickel iron-alloy tapes (considerably reduced eddy-current losses and skin effects). Such ultrathin tapes can not be used successfully for the construction of the conventional forms of small-size flux-gate elements (two parallel strips or scrolls), as required for "point measurements" on very small areas of inhomogenous magnetic fields.

The various circuit configurations and different techniques for extracting the second-harmonic or total even-harmonic output information form ring-core flux-gate elements include operation as a "self-balancing" flux-gate magnetometer.[2] In this case, the dc flux in the semicircle portions of the core is completely balanced by an opposing dc flux. In a modified circuit with two equally rated ring-core flux-gate elements acting as a gradiometer, a similar balance method may be applied for measurement of inhomogeneity of magnetic fields.

[1] W. A. Geyger, AIEE Trans. **79**, 106 (1960).

[2] W. A. Geyger, AIEE Trans. **77**, 213 (1958).

JOURNAL OF APPLIED PHYSICS     SUPPLEMENT TO VOL. 33, NO. 3     MARCH, 1962

# Waveform of the Time Rate of Change of Total Flux for Minimum Core Loss

H. L. Schenk and F. J. Young

*Westinghouse Research Laboratories, Pittsburgh, Pennsylvania*

In testing magnetic materials core losses are measured when the time rate of change of total flux $d\varphi/dt$ in the core varies sinusoidally. Although it may seem intuitive that the core loss for a given flux density is a minimum when $d\varphi/dt$ is sinusoidal, measurements and a simple analysis indicate otherwise. It is shown that the core loss is smaller if the time rate of change of total flux in the core follows a square wave than it would be if the time rate of change of total flux followed a sine wave, with the peak induction being the same in both cases. By a variational method the ratio of square-wave to sine-wave eddy current losses for a given value of peak flux density is found to be $8/\pi^2$.

The ratio of square-wave to sine-wave core losses is measured as a function of induction for a fixed frequency and as a function of frequency for a fixed induction using a Hipersil specimen (cube-on-edge orientation). A feedback circuit is used to prevent distortion in the waveform. When the measurement is made at 60 cps the ratio is about 0.90 for flux densities below 12.5 kgauss and rises with increasing flux density to about 0.96 at 17.5 kgauss. For an induction of 5 kgauss the ratio decreases from about 0.92 at 30 cps to 0.87 at 120 cps.

## INTRODUCTION

IT is the purpose of this paper to show both theoretically and experimentally that the minimum iron loss does not occur when the time rate of change of total flux in the core varies sinusoidally. Although a complete theoretical treatment of iron losses as a function of the waveform of flux would be difficult, it is easy to examine the influence of waveform on the classical eddy current loss. A waveform of the time rate of change of the total flux is calculated which minimizes the classical eddy current losses. Because the classical eddy current loss is only part of the total core loss, one cannot be sure that the waveform calculated here will yield the minimum possible total iron loss but rather a loss smaller than would be obtained using a sinusoidal variation of the time rate of change of total flux.

## SIMPLE THEORY

Since the power loss due to eddy currents is proportional to the square of the eddy currents, which in turn are proportional to the time rate of change of total flux $d\varphi/dt$, then the instantaneous eddy current power loss is proportional to $(d\varphi/dt)^2$. The average eddy current loss $P$ can be found by integrating over one-half cycle giving

$$P = \frac{2K}{T} \int_0^{T/2} \left(\frac{d\varphi}{dt}\right)^2 dt, \tag{1}$$

where $T$ is the period of $d\varphi/dt$ and $K$ is a constant. Hence, to find the $d\varphi/dt$ that causes the least eddy losses, $P$ must be minimized. It is well known that the peak induction $B_m$ is given by

$$B_m = \frac{1}{2A} \int_0^{T/2} \left(\frac{d\varphi}{dt}\right) dt, \tag{2}$$

where $A$ is the cross-sectional area of the core. Whatever $d\varphi/dt$ is used to obtain a particular value of $B_m$ must satisfy Eq. (2). Therefore, a function $[d\varphi(t)/dt]$ is sought which minimizes $P$ given by Eq. (1) and also satisfies Eq. (2). Rewriting Eqs. (1) and (2) in a form more suitable to the calculus of variations and replacing $[d\varphi(t)/dt]$ by $(1/N_s)v(t)$ where $N_s$ is a constant yields

$$P = \frac{2K}{TN_s^2} \int_0^{T/2} F(t,v,v') dt \tag{3}$$

and

$$B_m = \frac{1}{2AN_s} \int_0^{T/2} G(t,v,v') dt, \tag{4}$$

where $v' = dv/dt$, $F = [v(t)]^2$ and $G = v(t)$. The solution is found from Euler's equation which is

$$\frac{\partial F}{\partial v} + \lambda \frac{\partial G}{\partial v} - \frac{d}{dt}\left(\frac{\partial F}{\partial v'} + \lambda \frac{\partial G}{\partial v'}\right) = 0, \tag{5}$$

where $\lambda$ is a numerical parameter of unspecified value. Eq. (5) reduces to $v(t) = -(\lambda/2)$. Substituting this into Eq. (4) and solving for $\lambda$ yields $\lambda = -8fN_sAB_m$, where $f = 1/T$ is the frequency of $d\varphi/dt$. Then, since the $\lambda$ and $v(t)$ just obtained minimize $P$, the time rate of change of flux which minimizes eddy current loss is a square wave of amplitude $4fAB_m$ and frequency $f$.

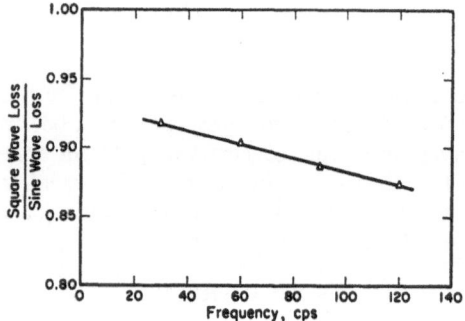

FIG. 1. Loss ratio vs frequency at 5 kgauss for Hipersil.

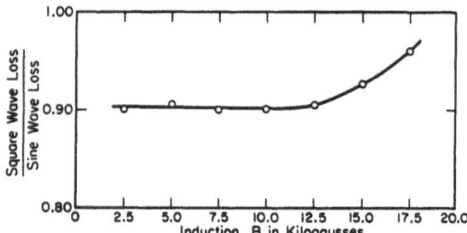

FIG. 2. Loss ratio vs induction at 60 cps for Hipersil.

To obtain the same peak flux density with a sine wave of $d\varphi/dt$, the amplitude must be $2\pi fAB_m$ by Eq. (2). The eddy current loss using a sinusoidal $d\varphi/dt$ is from Eq. (1), $P_{SI} = 2K\pi^2 f^2 A^2 B_m^2$, and similarly for a square wave of $d\varphi/dt$, $P_{SQ} = 16Kf^2 A^2 B_m^2$. Then the ratio $\xi$ of square-wave to sine-wave eddy current loss for the same peak flux density is

$$\xi = \frac{P_{SQ}}{P_{SI}} = \frac{8}{\pi^2} = 0.81. \tag{6}$$

## EXPERIMENTAL RESULTS

The analysis accounts for classical eddy current losses only. To study the behavior of total iron losses the ratio $\xi$ is measured for a 12-mil Hipersil specimen at various values of induction and frequency. To perform these tests the standard 25-cm Epstein frame is used in conjunction with a sensitive reflecting wattmeter and a negative feedback scheme[1] to preserve the desired waveforms of $d\varphi/dt$ at all values of flux density.

For an induction of 5 kgauss, a plot of $\xi$ vs frequency is given in Fig. 1. As frequency is increased from 30 to 120 cps, $\xi$ decreases from 0.92 to 0.87. This indicates that the portion of total core losses due to classical eddy currents increases as the frequency is raised. Fig. 2 is a curve of $\xi$ as a function of peak flux density for a frequency of 60 cps. At inductions below 12.5 kgauss $\xi$ is about 0.90. As the induction is increased from that value to 17.5 kgauss, $\xi$ increases to 0.96. The material behaves as though classical eddy current losses become less important as the peak flux density is increased. This observation agrees with previous studies of losses.[2] At low inductions, many domain walls are free to move at any time during the cycle but at high inductions only a few domain walls move during the high flux density portion of the cycle. If there were infinitely many walls moving, a $\xi$ of $8/\pi^2$ might exist, but as the number of walls available for motion decreases, $\xi$ increases. To what value $\xi$ might eventually rise as the peak flux density is increased toward saturation has not been determined in this investigation, because at inductions past 18 kgauss waveform distortion became intolerable.

[1] J. McFarlane and M. J. Harris, J. Inst. Elec. Engrs. (London) 105A, 395 (1958).
[2] W. J. Carr, J. Appl. Phys. 30, 90S (1959).

JOURNAL OF APPLIED PHYSICS     SUPPLEMENT TO VOL. 33, NO. 3     MARCH, 1962

# Low-Frequency Losses and Domain Boundary Movements in Silicon Iron

DANIEL A. WYCKLENDT

*Research Division, Allis-Chalmers Manufacturing Company, Milwaukee, Wisconsin*

Earlier observations of very low-frequency losses in (110) [001] textured silicon iron have been extended to include losses produced by excitation with a sinusoidal field intensity. Excitation frequencies of 0.005 to 10 cps and flux density amplitudes of 10, 15, and 17 kgauss were used in these measurements. The downward concavity of the loss-frequency curve near zero frequency, previously reported for sinusoidal induction, is greatly exaggerated by imposition of a sinusoidal field intensity.

To further investigate these low frequency loss anomalies, low-amplitude, high-frequency fields were impressed on the material during cyclic magnetization at low frequencies. The superposed field, most often of 320- or 1000-cps frequency, had an amplitude considerably smaller than the coercive force of the material. Waveforms of the gross, low-frequency magnetization and the incremental, high-frequency magnetization were accurately controlled by high-gain, direct-coupled, feedback amplifiers. The ratio of the $\dot{B}$ amplitude to the $H$ amplitude for the high-frequency components varied greatly within a cycle of magnetization. $|\dot{B}_h|/|H_h|$ was also found to vary with the peak value and waveform of the gross induction. Most significantly, the ratio increased greatly with frequency of the gross magnetization at very low frequencies. These changes can be attributed to an increase in the number of domain walls in simultaneous motion when the speed of gross magnetization is increased.

## INTRODUCTION

EARLIER measurements[1] of the losses of (110) [001] textured Si-Fe revealed a distinctive downward bend in the loss-frequency curves at near-zero frequencies. The purpose of the present work was to further study the low-frequency loss phenomena to secure experimental information pertinent to loss anomalies in (110) [001] textured Si-Fe.

Eddy current loss mechanisms associated with movements of domain walls have been advanced recently.[2,3] However, the change of domain structure during a magnetization cycle has been largely conjectured. It was hoped that this work would contribute information pertinent to dynamic configurational changes of domains.

## LOSSES AT LOW FREQUENCIES

Recently, techniques for the measurement of losses incident to magnetization with sinusoidal induction of high amplitude and very low frequencies have been developed. These techniques have been extended to include controlled field intensity, permitting measurement of losses associated with sinusoidal field intensities. The improved apparatus is similar to the feedback amplifier and bridge combination used by Wycklendt and Kay,[1] with provisions for obtaining controlling feedback proportional to the field intensity.

Four strips of commercial grade, (110) [001] textured Si-Fe, each 3×28 cm and nominally 0.014 in. thick comprised the test specimen. A mutual inductance bridge was used for measurements in the range 0.25 to 10 cps. Below 0.25 cps, losses were calculated from $B$-$H$ loops.

Observed losses as functions of frequency are shown in Fig. 1. The downward curvature of the loss-frequency curve, apparent enough for sinusoidal induction, is considerably exaggerated by imposition of sinusoidal field intensity.

## RESPONSE TO SUPERPOSED AUDIO FREQUENCY MAGNETIZATION

To learn something of the domain changes associated with the low frequency anomalies, small audio-frequency fields were impressed on the specimen during magnetization at low frequencies. This experiment is an extension of a technique used by Becker[4] in his study of switching reversals. The audio frequency response $\dot{B}_h$ is dependent on the active domain wall area.

The speed of an isolated wall separating domains of opposite magnetization[5] is given by $v \propto (H-H_c)$. If the speed of flux change in the direction of domain magnetization is measured, $\dot{B} \propto (H-H_c)$, or if a high frequency

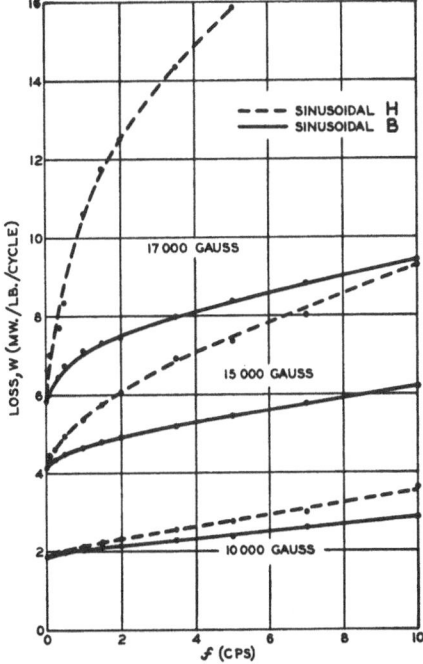

FIG. 1. Losses in oriented 3.2% Si-Fe at low frequencies.

[1] D. A. Wycklendt and R. M. Kay, J. Appl. Phys. **32**, 368S (1961).
[2] R. H. Pry and C. P. Bean, J. Appl. Phys. **29**, 532 (1958).
[3] W. J. Carr, J. Appl. Phys. **30**, 90S (1959).
[4] J. J. Becker, J. Appl. Phys. **30**, 401 (1959).
[5] H. J. Williams, W. Shockley, and C. Kittel, Phys. Rev. **80**, 1090 (1950).

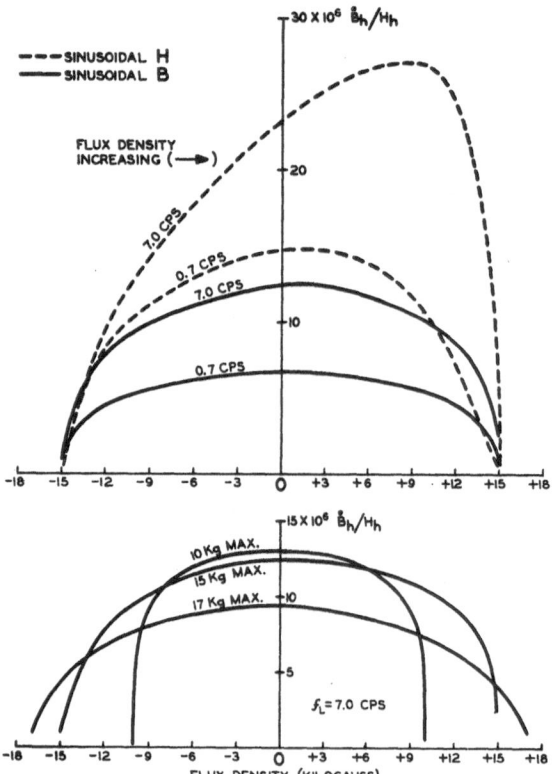

FIG. 2. Response to superposed audio-frequency
(320 cps) increment.

increment $H_h$ is superposed, $\dot{B}_h = kH_h$. The factor $k$ is proportional to the domain mobility and area. Assuming that the mobility is constant, $|\dot{B}_h|/|H_h|$ is a measure of the wall area participating in the magnetization.

In the simple domain model for (110) [001] textured Si-Fe strip, magnetization in the rolling direction [001] is attributed to motions of groups of walls which are roughly perpendicular to both the sheet surface and the rolling direction. The above relation of $|\dot{B}_h|/|H_h|$ to the total area of participating domain walls is thus applicable, provided the spacing is not too small. If the spacing is small, of the order of the sheet thickness, $|\dot{B}_h|/|H_h|$ is a more complicated function of the space parameter.

The amplifier systems used for loss measurements were adapted to the superposed audio-frequency tests by the expedient of obtaining the reference input from a summing amplifier in which the audio-frequency component was added to the gross magnetization signal. When the feedback was arranged to produce sinusoidal induction, the superposed audio frequency was constrained to sinusoidal induced voltage of constant amplitude. Alternatively, if the feedback was arranged to control the gross field intensity, the superposed audio-frequency magnetization was constrained to sinusoidal field intensity of constant amplitude. The super-

posed field was small compared to the coercive force. Occasionally, amplitudes of about one-fifth of $H_c$ occurred momentarily near peak gross magnetization. Audio frequencies of 320 or 1000 cps were most commonly used.

The audio-frequency response was separated from the lower frequency components associated with the gross magnetization by sharply tuned filters, and displayed on an oscilloscope as a function of the gross induction. Provisions were made to blank the oscilloscope during half the cycle.

The quantity $|\dot{B}_h|/|H_h|$ for the audio component was calculated from observed voltages and currents. $|\dot{B}_h|/|H_h|$ was found insensitive to the amplitude of the audio-frequency increment of field intensity. This quantity was negligibly sensitive to the particular audio frequency over most of the gross magnetization cycle except near the ends. Wherever $|\dot{B}_h|/|H_h|$ was substantially independent of the audio frequency, $\dot{B}_h$ and $H_h$ were in phase and the preceding interpretation of $|\dot{B}_h|/|H_h|$ as a measure of wall area was applicable.

Typical curves of $|\dot{B}_h|/|H_h|$ as a function of gross induction are shown in Fig. 2. The feature of greatest interest is the dependence of $|\dot{B}_h|/|H_h|$ on the speed or frequency of the gross, low-frequency induction. It can be inferred that when the speed of magnetization is increased, the area (number) of walls which participate in the magnetization also increases. The sinusoidal field process, characterized by rapid changes of induction, produces a notably exaggerated response. Some dependence on the gross induction amplitude was also found.

## DISCUSSION

The rough magnitude of the ratio of domain spacing $\lambda$ to sheet thickness $2d$ is

$$(\lambda/2d) = K|H_h|/|\dot{B}_h|; \quad (\lambda/2d) > 1.$$

For the specimen studied here, $K$ is $25 \times 10^6$ when emu are used. From Fig. 2, for example, this ratio for active domain walls (calculated for $B = 0$) was approximately 3.6 for 0.7 cps, sinusoidal $B$ of 15 000 gauss amplitude, and about 2 for 7 cps.

Applied to the behavior of the loss-frequency curves, the smaller number of active walls at lower frequencies implies that the walls must attain proportionally higher speeds to produce the imposed rate of magnetization. Because the eddy current loss varies with the square of the wall speed and approximately with the first power of the number of walls, the loss per cycle should rise most rapidly with frequency where the number of active walls is smallest. Therefore, the $|\dot{B}_h|/|H_h|$ measurements and the inferred numerical density of active domain walls provide an explanation for the shape of the loss-frequency curves.

# Iron Losses in Elliptically Rotating Fields*

R. D. STRATTAN† AND F. J. YOUNG

*Carnegie Institute of Technology, Pittsburgh, Pennsylvania*

Power frequency losses in silicon iron alloys due to an elliptically rotating magnetic field have been measured by a calorimetric technique and predicted approximately using a simple model. Previously reported experiments for measuring these losses fail to approach saturation and have doubtful field uniformities. A thin disk-shaped specimen is placed in the elliptically rotating field of a set of two phase air-cored rectangular Helmholtz-type coils. The 60-cps rotational loss is calculated from the initial rate of specimen temperature rise sensed by a copper-Constantan thermocouple attached to the disk. The thermocouple output is amplified and recorded with an accuracy of $10^{-3}$°C. The field coil system produces fields uniform to 2% in a 1.0-in.-diam region and of strengths up to 250 oe, which is sufficient to approach saturation in 0.005-in. thick by 0.875-in.-diam disks. The losses have been measured for grain oriented and single-crystal materials with all materials showing a decrease in loss for an increase in flux density for large, nearly circular fields. The losses for elliptical magnetization may be approximated with a simple model composed of a mixture of lossless domain rotation, eddy current loss, and alternating hysteresis loss proportional to an alternating flux. This model is intended only for the calculation of rotational hysteresis losses and not to explain the origin of these losses. The calculated and experimental losses agree for all amplitudes and eccentricities of magnetization to within 40% for rotation in the (100) plane with the worst discrepancy for circularly rotating magnetization.

## INTRODUCTION

SOME form of rotating magnetization occurs in many magnetic devices of engineering interest but, except for circular flux, only a small amount of work has gone into investigating the losses encountered. In general, the magnitude of the magnetization varies as it rotates so that elliptically rotating magnetization is an approximation of this complex rotation. The uniformity of the flux density in the previous experiments for measuring elliptical field losses[1,2] is questionable, and the losses for flux densities near saturation were not measured. The losses for flux densities approaching saturation are of interest because the losses decrease as circular saturation is approached.

The losses for elliptically rotating fields are measured using a calorimetric technique similar to that of Young and Schenk.[1] A disk-shaped specimen is placed in the elliptically rotating field of perpendicular Helmholtz-type coils. A simple model useful for calculating elliptical hysteresis loss is also proposed and its predicted losses are compared to the measured losses.

## EXPERIMENTAL APPARATUS AND RESULTS

The losses are calculated from the initial rate of temperature rise of the specimen. The temperature rise of the specimen is sensed with a copper-Constantan thermocouple attached to the surface of the specimen and recorded with a system using a dc microvolt amplifier and a recording potentiometer capable of measuring temperature differences of $10^{-3}$ °C.

The nearly elliptical flux is excited by placing the disk in an elliptical applied field. The applied magnetic field is produced by two perpendicular pairs of rectangular Helmholtz-type air-cored coils. The coils are capable of producing field strengths up to 250 oe and are powered from the 60-cps line. The fields are uniform to 2% in a 1.0-in.-diam region.

The specimens are in the form of 0.875-in.-diam disks cut from 0.004 to 0.015-in. thick sheet material. The flux density within the specimen is calculated from the dimensions of the disk and the voltage induced in a pickup coil wound around the disk's midsection.

Single crystal and grain oriented silicon iron alloys have been tested. The constant loss contours on the $B_x - B_y$ plane for (110)[001] textured grain oriented material are shown in Fig. 1. The losses as a function of peak flux density for constant peak flux density cross fields for a single crystal specimen with a (100) plane of magnetic rotation are shown in Fig. 2. The measured losses are the total iron losses including hysteresis and eddy current losses.

The losses for the grain-oriented material have the same general field dependence as the single crystal material except that the losses are less in the rolling direction than in the direction perpendicular to it, as

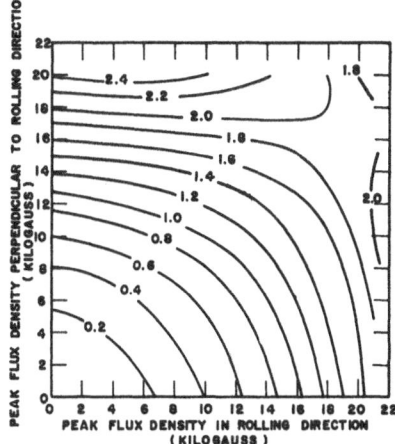

FIG. 1. Constant loss contours of measured elliptical field losses for 0.005-in. thick grain oriented material. 60-cps loss values are in w/lb.

---

* This is an abstract of a portion of the doctoral thesis submitted by R. D. Strattan to the Electrical Engineering Department of Carnegie Institute of Technology in partial fulfillment of the requirements for the Doctor of Philosophy degree.

† Present address: Boeing Company, Wichita, Kansas.

[1] F. J. Young and H. L. Schenk, J. Appl. Phys. **31**, 194S (1960).
[2] A. Kaplan, J. Appl. Phys. **32**, 370S (1961).

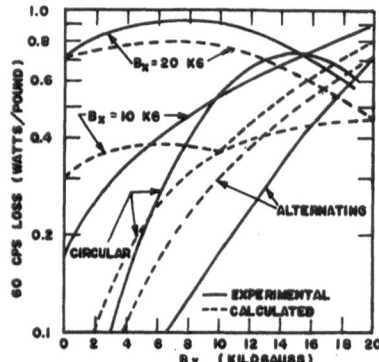

FIG. 2. Comparison of experimental data and losses calculated using the proposed model for (100) plane single-crystal 0.0075-in. thick material. The parameters used for the calculated losses are $P_1 = P_2 = 0.48$ w/lb, $P_3 = 0.22$ w/lb, and $B_s = 20$ kgauss.

would be expected. The most interesting characteristic of all the materials tested is the decrease of losses for increasing flux densities for nearly circular fields near saturation.

## MODEL OF ROTATIONAL HYSTERESIS

There is a lack of a method of calculating hysteresis loss for a rotating field. An analysis similar to that of Stoner and Wohlfarth is not valid for multidomain bulk material. A model that would accurately describe the domain wall motion and domain magnetization realignment that actually occurs would be ideal but the construction of such a model would be very complex and nearly unworkable. The model proposed here does not attempt to describe the basic magnetic phenomena responsible, but merely simulates the elliptical hysteresis loss with a simple model. The model has some physical bases but it is primarily based on the observed elliptical field losses. It is intended only to serve as a means of making approximate calculations of the losses for a specific application where rotating magnetization occurs once the alternating hysteresis loss of the material is known.

The proposed model assumes that the flux density

change accounted for by domain rotation is $B_2^2/B_s$ and the remainder is by 180° domain wall motion. $B_1$ is the major axis elliptical flux density, $B_2$ is the minor axis flux density, and $B_s$ is the saturation flux density. The peak alternating flux density in the major axis direction is then $B_1 - B_2^2/B_s$ and the peak alternating flux density in the minor axis direction is $B_2 - B_2^2/B_s$. The domain wall hysteresis loss is assumed to be proportional to the peak flux density and the rotational process is assumed to be lossless. If $P_1$ is the hysteresis loss for a peak alternating flux density of $B_s$ in the major axis direction, and $P_2$ is the corresponding loss in the minor axis direction, the total predicted elliptical hysteresis loss is

$$P_{\text{hys}} = P_1\left[\frac{B_1}{B_s} - \left(\frac{B_2}{B_s}\right)^2\right] + P_2\left[\frac{B_2}{B_s} - \left(\frac{B_2}{B_s}\right)^2\right].$$

The eddy current loss

$$P_{\text{e.c.}} = P_3\left[\left(\frac{B_1}{B_s}\right)^2 + \left(\frac{B_2}{B_s}\right)^2\right]$$

must be added to the hysteresis loss. The calculated losses are compared to the measured losses for the (100) plane single crystal material in Fig. 2. Although there are sizable differences in the calculated and measured losses, especially for circular fields, the general field dependence is similar.

## SUMMARY

The 60-cps power losses for disks of silicon iron alloy material subjected to elliptically rotating magnetic fields are measured. The initial rate of temperature rise of the disk-shaped specimen is used to calculate the elliptical field loss. Both single crystal and grain oriented materials have been tested. The most interesting feature of the losses is that the losses decrease as the magnetization approaches circular saturation. A simple model is proposed which provides a method of roughly calculating elliptical field losses for specific applications.

JOURNAL OF APPLIED PHYSICS     SUPPLEMENT TO VOL. 33, NO. 3     MARCH, 1962

# Antiferromagnetism and Resonance

## M. T. WEISS, *Chairman*

### Ferromagnetic Resonance Magnon Distribution in Yttrium Iron Garnet

T. J. MATCOVICH, H. S. BELSON, N. GOLDBERG, AND C. W. HAAS

*Remington Rand Univac, Division of Sperry Rand Corporation, Blue Bell, Pennsylvania*

The presence of excited magnons results in a shift of the ferromagnetic resonance frequency. Calculations of this frequency shift have been made by retaining terms in the Hamiltonian up to fourth order in the spin wave variables. The shift can be expressed in terms of the number of excited $k=0$ magnons and $k \neq 0$ magnons. Determination of this shift can then be used to study the magnon distribution during resonance. This frequency shift has been measured in four single-crystal yttrium iron garnet spheres at 9 Gc. The shifts are observed to be directly proportional to the rf power absorbed by the sample, positive when the dc magnetic field is in [100] direction, and negative when the dc magnetic field is in the [111] direction. These results require a distribution of $k \neq 0$ magnons dominated by magnons with $48° < \Theta_k < 60°$. If it is assumed that the $k \neq 0$ magnons are degenerate with the $k=0$ magnons, the total number of $k \neq 0$ magnons is approximately equal to the number of $k=0$ magnons, and magnons in the region centered about $k = 1-2 \times 10^5$ cm$^{-1}$ dominate. This result is in agreement with a prediction of the surface pit scattering model; namely, the relaxation of the $k=0$ magnons results in the production of $k \neq 0$ magnons with wavelength approximately equal to the pit size. The results are in disagreement with any model requiring a uniform distribution of degenerate modes.

## INTRODUCTION

IN a typical low power ferromagnetic resonance experiment the rf field excites a steady supply of $k=0$ magnons. These magnons can relax to the lattice either by relaxing first to other magnons and then to the lattice, or by relaxing directly to the lattice. In equilibrium, therefore, there is a magnon population composed of $k=0$ and $k \neq 0$ magnons in excess of the thermal number. Information about this magnon population would be very useful in determining the relative importance of various relaxation mechanisms. This paper describes an extension of a method previously reported[1] of studying the magnon population at resonance, and reports the results of measurements on yttrium iron garnet spheres by this method.

## THEORETICAL

Spin waves are the normal modes of excitation of a magnetic sample when terms in the Hamiltonian which are higher than second order in the spin wave annihilation and creation operators are ignored. The usual expression for the $k=0$ spin wave resonance frequency is appropriate to this approximation. The higher order terms describe interactions among spin waves; if these are retained, the resonance frequency will depend upon the degree of excitation of all of the spin waves.

The frequency shift has been calculated retaining terms in the Hamiltonian up to fourth order in the annihilation and creation operators. The shift in the resonance frequency from its low power value is given by

$$\Delta f = a^i n_0 + \sum_{k \neq 0} b_k^i n_k, \tag{1}$$

where $n_0$ is the number of $k=0$ magnons per unit volume, and $n_k$ is the number of magnons per unit volume with wave vector $\mathbf{k}$. The coefficients $a^i$ and $b_k^i$ depend upon the crystallographic direction $i$ along which the equilibrium magnetization lies.

In particular we find that for spherical samples with cubic symmetry

$$a^i = -9\alpha^i \frac{K}{M} \frac{\gamma^2 \hbar}{2\pi M} \tag{2}$$

$$b_k^i = \frac{\gamma^2 \hbar}{2\pi M} \frac{\gamma}{\omega_k} \left\{ \left[ H_0 + 2\alpha^i \frac{K}{M} \right. \right.$$
$$\left. + H_e a_0^2 k^2 - \frac{2\pi M}{3}(2-3\sin^2\theta) \right]$$
$$\times \left[ 2\pi M(2-3\sin^2\theta) - 2\alpha^i M \frac{d(K/M)}{dM} \right]$$
$$\left. - (2\pi M \sin^2\theta)^2 \right\}, \tag{3}$$

where

$$\alpha^i = \begin{cases} 1, & \text{if } i \text{ signifies [100] direction} \\ -\tfrac{2}{3}, & \text{if } i \text{ signifies [111] direction.} \end{cases}$$

In the above equation, $K$ is the first-order anisotropy constant (the effect of higher order anisotropy terms is negligible), $\gamma$ is the gyromagnetic ratio, $\hbar$ is Planck's constant, $M$ is the saturation magnetization at the ambient temperature, $\omega_k$ is the frequency of the spin wave with wave vector $\mathbf{k}$, $\theta_k$ is the angle the wave vector $\mathbf{k}$ makes with the dc magnetic field, $H_{ex}$ is the exchange field, and $a_0$ is the lattice constant.

[1] T. Matcovich, H. S. Belson, and N. Goldberg, J. Appl. Phys. **32**, 163S (1961).

TABLE I. Analysis of frequency shift data.

| Sample number | $\frac{\Delta f}{n_0} \times 10^{13}$ cps | | $\frac{n'}{n_0} \sum_{k \neq 0} b_k \rho_k \times 10^{13}$ cps | | $\theta_{eff}$ | $k_{eff} \times 10^{-5}(\text{cm}^{-1})$ | $n'/n_0$ | $\Delta H_{av}$(oe) | $D$(mm) |
|---|---|---|---|---|---|---|---|---|---|
| | [100] | [111] | [100] | [111] | | | | | |
| 1 | 2.81 | −2.94 | 1.50 | −2.07 | 54.6° | 1.2 | 1.6 | 1.26 | 0.51 |
| 2 | 1.89 | −1.76 | 0.58 | −0.89 | 54.7° | 1.2 | 0.7 | 1.03 | 0.66 |
| 3 | 1.82 | −1.47 | 0.51 | −0.60 | 54.3° | 1.2 | 0.5 | 0.96 | 0.76 |
| 4 | 1.80 | −0.89 | 0.49 | −0.02 | 48.5° | 1.9 | 0.3 | 0.50 | 0.51 |

## EXPERIMENTAL

The frequency shift measurements were made by sweeping the microwave frequency through resonance first at low power, then a few milliseconds later at high power. The positions of the peaks of the resonance absorption curves relative to fixed frequency markers were compared to determine the shift. This procedure minimized the effects of drifts in the ambient temperature, magnetic field, and frequency standard. In addition, one could vary the rate at which the frequency was swept so that it was possible to vary the sample heating due to absorption of rf power. This heating caused a small spurious frequency shift which decreased with increased sweep speed. The effect was eliminated by extrapolating our results to infinite sweep speed. All of the data were taken at power levels below the onset of nonlinear effects.

The value of $n_0$ was obtained in the following way: The rate at which $k=0$ magnons are destroyed is $\gamma \Delta H n_0$, where $\Delta H$ is the resonance linewidth. The rate at which $k=0$ magnons are created is $P_a/V\hbar\omega_0$, where $P_a$ is the power absorbed by the sample and $V$ is the sample volume. At equilibrium these rates are equal. From this equality we find

$$n_0 = P_a/\gamma\hbar\omega_0\Delta HV. \qquad (4)$$

Measurements were made on four spherical yttrium iron garnet samples at 9 Gc with the dc magnetic field along the [100] and [111] directions. The power absorbed by the samples was of the order of 10 mw. At this power level the typical frequency shift was 100 kc. The shifts were found to be directly proportional to rf power for all samples. The results, which were normalized by dividing by $n_0$, are tabulated in Table I.

## DISCUSSION AND CONCLUSIONS

Since two-magnon relaxation processes (e.g., surface pit scattering) dominate the relaxation of the $k=0$ magnons,[2] magnons degenerate with the $k=0$ magnons, ($S$ magnons) will be excited. These in turn relax to other magnons and the lattice. The $S$ magnons will therefore make up an important part of the $n'$ magnons and in fact we shall assume that only $S$ magnons are excited. The values of $b_k$ were calculated from Eq. (3) for $S$

magnons only, i.e., $\omega_k = \omega_0$. It is convenient to consider these values as a function of $\theta_k$. It was found that the $b_k^{[100]}$ were positive for all values of $\theta_k$, but that $b_k^{[111]}$ is positive for $\theta_k < 48°$, but negative for $\theta_k > 48°$. This is quite important in the extraction of information from the observations.

It is convenient to rewrite Eq. (1) as

$$\frac{n'^i}{n_0^i} \sum_{k \neq 0} b_k^i \rho_k^i = \frac{\Delta f^i}{n_0^i} - a^i, \qquad (6)$$

where $n'$ is the total number of $k \neq 0$ magnons per unit volume and $\rho_k$ is the fraction of $n'$ with wave vector $k$. The measurement of the frequency shift and the determination of $n_0$ by using Eq. (4) yield the value of the sum on the left. We find that this quantity is negative in the [111] direction for all of the samples. Since $b_k$ for the [111] direction is negative only in the region $\theta > 48°$, the predominant contribution to the sum must come from this region. This conclusion does not require the assumption that only magnons degenerate with the $k=0$ magnon are excited since the form of $b_k$ as a function of $\theta_k$ is only weakly dependent on the energy of the spin waves as can be seen by examining Eq. (3). In fact, it can be concluded from our measurements that the magnon population is dominated by magnons in the region $48° < \theta < 60°$ for any physically possible energy distribution.

It is reasonable to assume that $n'/n_0$ as well as $\rho_k$ are the same for a [100] and [111] resonance because of the approximate equality of the half-width in these directions. If we now imagine that all of the $S$ magnons have the same $k$, we can determine both this effective $k$ and the ratio of $n'/n_0$ from our data. These results are summarized in Table I. The half-width and the diameter of the samples are also included in Table I.

Sparks, Loudon, and Kittel[3] have calculated the surface pit scattering line width contribution and find good agreement with experimental results. As would be expected the uniform mode relaxes most readily to those spin wave states of wave length $\lambda = 2\pi/k \simeq R$ where the pit size $R$ is approximately the same as the size of the polish grit. The values of $k_{eff}$ in Table I are in agreement with this pit size relationship. As an additional consequence of their model, Sparks, Loudon, and Kittel propose that pit scattering brings about rapid equilibra-

[2] R. C. LeCraw and E. G. Spencer, J. Appl. Phys. 30, 185S (1959).

[3] M. Sparks, R. Louden, and C. Kittel, Phys. Rev. 122, 91 (1961).

tion of the $S$ modes so that all these degenerate modes are populated equally. Since we find that the distribution of magnons is dominated by magnons with $\theta_k > 48°$, our results do not agree with this prediction.

## ACKNOWLEDGMENTS

The authors wish to thank H. Callen for many helpful discussions and suggestions and W. Luciw for his assistance in the experimental work.

---

JOURNAL OF APPLIED PHYSICS    SUPPLEMENT TO VOL. 33, NO. 3    MARCH, 1962

# Pulsed Critical Field Measurements in Magnetic Systems

SIMON FONER AND SHOU-LING HOU*

*M.I.T. National Magnet Laboratory,† Cambridge 39, Massachusetts*

Pulsed field magnetic moment measurements in several magnetic systems are briefly summarized. Particularly high sensitivity is attained for systems which exhibit a nonlinear variation of magnetization versus applied field, e.g., "spin-flop." Changes of magnetization as small as 0.1 gauss could be observed with applied fields of over 100 kgauss under favorable circumstances. Critical fields for $Cr_2O_3$ and $(Cr_2O_3)_{0.9} \cdot (Al_2O_3)_{0.1}$ from 4.2°K to about 0.95 $T_N$ and the corresponding calculated values of $(2\lambda K)^{\frac{1}{2}}$ are presented, where $\lambda$ is the exchange constant and $K$ is the anisotropy energy. The values of $(2\lambda K)^{\frac{1}{2}}$ agree with earlier antiferromagnetic resonance measurements in similar crystals. Results of measurements in antiferromagnetic $MnF_2$, $CoF_2$, and $FeTiO_3$, and in metamagnetic $FeCl_2$ and $CoCl_2$ are briefly discussed. Application of the method to paramagnets is also indicated.

A T high magnetic fields the magnetic moment becomes a nonlinear function of applied field for many magnetic systems. These nonlinear effects can readily be observed even when they are superimposed on much larger linear effects when pulsed field inductive techniques are employed. The experimental method and some experimental observations of such nonlinear effects are briefly summarized in this paper.

The inductive magnetic moment measurement in a pulsed magnetic field is made by inserting a sample in a coil of effective area-turns $NA_s$. In order to eliminate the induced voltage due to the pulsed field, a series opposing coil of effective area-turns $NA_b$ is usually introduced in the active field volume. The ideal case of $NA_s = NA_b$ is rarely achieved. The ratio of voltage $E_s$ induced during a field pulse by the sample, to the background voltage, $E_b$, is given by

$$\frac{E_s}{E_b} = \frac{4\pi G_s A_s (dM/dH)}{A_b (1-R)}, \quad (1)$$

where $G_s$ is determined by the sample dimensions, $R$ is the ratio of $A_s/A_b$, $M$ is the magnetization, and $H$ is the applied magnetic field. In addition to the enhancement of $E_s/E_b$ by a nonlinear dependence of $M$ versus $H$, further improvement can be attained by filtering out the fundamental, low-frequency field pulse while examining the higher frequency components of $dM/dt$. The quantity observed during an experiment is $(E_s - E_b)$ versus time.

A striking nonlinear effect in antiferromagnets is the "spin-flop" phenomenon. If $H$ is applied parallel to the preferred direction of a uniaxial antiparallel spin system with negative anisotropy energy $K$, a sudden change in the magnetization parallel to $H$ is observed when $H$ equals a critical field $H_c$ given by

$$H_c^2 = 2\lambda K/(1-\alpha), \quad (2)$$

where $\lambda$ is the exchange constant and the ratio of susceptibilities for the spins parallel or perpendicular to $H$ is given by $\alpha = \chi_{\parallel}/\chi_{\perp}$. Equation (2) corresponds to the zero frequency antiferromagnetic resonance (AFMR). The change in magnetization, $\Delta M$, at $H = H_c$ is given by

$$\Delta M = [2K(1-\alpha)/\lambda]^{\frac{1}{2}}. \quad (3)$$

The magnitude of $dM/dt$ or $dM/dH$ depends on the

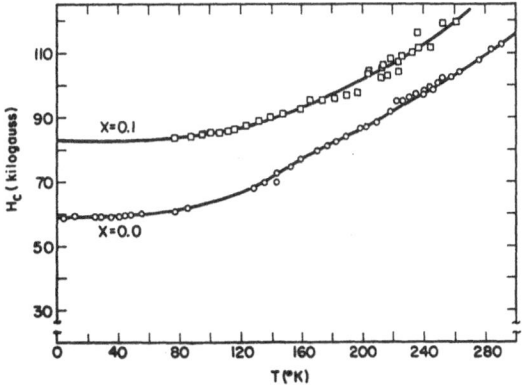

FIG. 1. Critical field for spin-flop $H_c$ versus temperature for $(Cr_2O_3)_{1-x} \cdot (Al_2O_3)_x$.

* Present address: Harvard University, Division of Engineering and Applied Physics, Cambridge, Massachusetts.

† Supported by the Air Force through the Air Force Office of Scientific Research.

accuracy with which $H$ is aligned parallel to the easy axis. For perfect alignment, the "spin-flop" would occur in an interval $\Delta t$ (or $\Delta H$) limited by the spin-relaxation time. When the angle $\theta$ between $H$ and the easy axis is not zero, the width of the spin-flop transition is a complex function of $\theta$. Numerical calculation shows that $\Delta H$ increases rapidly with increasing $\theta$, and that the maximum value of $dM/dH$ occurs at $H<H_c$ for $\theta>0$. Accurate alignment is therefore essential for these spin-flop experiments. When measurements are made with misaligned crystals, first-order corrections can be made if we assume $\Delta H$ arises from misalignment and then we compare the data with the calculated dependence of $\Delta H$ versus $\theta$. For a typical experiment (no filtering) $NA_s=15$ cm$^2$-turns, and $R=1.0002$, a $\Delta M=0.1$ gauss could be detected for $\Delta H=10^3$ gauss at $H=10^5$ gauss.

Spin-flop has been observed in CuCl$_2 \cdot$2H$_2$O with small dc fields and in MnF$_2$ at 4.2°K by Jacobs[1] recently with pulsed fields. Critical fields versus temperature for $(Cr_2O_3)_{1-x} \cdot (Al_2O_3)_x$, for which $\Delta M$ is much smaller, are plotted in Fig. 1. Values of $(2\lambda K)^{\frac{1}{2}}$ versus temperature, shown in Fig. 2, were calculated from equation (2) with independent susceptibility data obtained with similar materials.[2] The solid curves in Fig. 2 are the AFMR results[3] *plotted with no normalizing factors*. The agreement is within the experimental error (the slight deviation for $X=0.1$ at high temperatures is due to slightly different values of $X$ for the materials). For both samples $\Delta H$ was less than $10^3$ gauss ($\theta \leq 2$ degrees) and $H_c$ could be followed up to $T/T_N=0.95$. Similar experiments in MnF$_2$ agree with earlier AFMR data and Jacobs results at 4.2°K. These measurements show that $\chi$ is not dependent on $H$ and that the method can be used as an alternative for AFMR, particularly if the linewidth of the AFMR is too large for resonance observations. No spin-flop has yet been observed in FeTiO$_3$ for which AFMR is more complex. Experiments showed no spin-flop in CoF$_2$, but a nonlinear effect was observed at 4.2°K for fields up to 60 kgauss.

[1] I. S. Jacobs, J. Appl. Phys. **32**, 61S (1961). Additional references are given in this paper.

[2] S. Foner, J. Appl. Phys. **32**, 63S (1961); J. phys. radium **20**, 336 (1961).

[3] The $(Cr_2O_3)_{1-x} \cdot (Al_2O_3)_x$ single crystals for AFMR were grown at the Rutgers Department of Ceramics, New Brunswick, New Jersey. The spin-flop experiments were made with similar crystals grown at the Linde Corporation, Speedway Laboratories, Speedway, Indiana. Values of $\alpha$ for the former crystals were used here. Slightly smaller values of $\alpha$, measured for some of the latter crystals, would increase $(2\lambda K)^{\frac{1}{2}}$ by less than 3% at low temperatures.

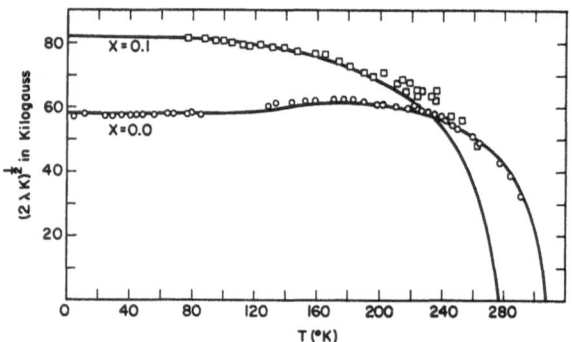

FIG. 2. Values of $(2\lambda K)^{\frac{1}{2}}$ versus temperature calculated from data in Fig. 1. The solid curves are the results of resonance experiments for comparable materials[2,3] plotted with no adjustable parameters.

Metamagnets also show sudden, nonlinear, large changes in M. A transition at about 13 kgauss was observed in FeCl$_2$ for $H$ parallel to the $c$ axis in agreement with earlier static measurements. The transition field was independent of temperature for 4.2°K $\leq T < T_N$. More complicated nonlinear effects were observed in CoCl$_2$ for $H$ perpendicular to the $c$ axis. A transition below 3 kgauss was seen in agreement with earlier neutron diffraction results.[4] A broad high-field transition (indicating a negative susceptibility for a 550-$\mu$sec half-period pulsed field) was also observed. AFMR and further experiments with dc magnetic fields are in progress. The details will be presented at a later date.

The above technique may also be applied to measurements of the zero field splittings in paramagnets which have converging energy levels and suitable relaxation times. Nonlinear effects at the energy level cross-over region can then be detected by induction measurements.

Small nonlinear variations of $M$ versus $H$ can readily be observed with pulsed field techniques. Some of these effects suggest new areas of interest and some confirm earlier results. The various experimental results can at times augment, compare favorably with, or surpass those of magnetic resonance experiments.

We are grateful to Mr. E. P. Warekois of Lincoln Laboratory for valuable assistance with sample preparation and x-ray orientation data, and to Mr. W. G. Fisher for assistance with the experiments. We are also indebted to Professors J. W. Stout, and C. Frondel, and Dr. M. K. Wilkinson for various single-crystal samples.

[4] W. C. Koehler, M. K. Wilkinson, J. W. Cable, and E. O. Wollan, Phys. Rev. **113**, 497 (1959).

JOURNAL OF APPLIED PHYSICS    SUPPLEMENT TO VOL. 33, NO. 3    MARCH, 1962

# Magnetic Susceptibility and Magnetostriction of CoO, MnO, and NiO

T. R. McGuire and W. A. Crapo

*IBM Research Center, Yorktown Heights, New York*

The experimentally observed increase in the magnetic susceptibility and the accompanying change in length of antiferromagnetic MnO and NiO single crystals were used to evaluate critical fields and anisotropy of these materials. For CoO the susceptibility varied over too small a range to be of significance in this work.

ANTIFERROMAGNETIC monoxides have been the subject of many studies but there still remain field dependent properties which have not been fully investigated. We have attempted to learn something about the domain motion in CoO, MnO, and NiO crystals by field dependent susceptibility and magnetostriction measurements on specimens from the same boule. These three monoxides are of the rocksalt type and have magnetic structures in which spins in (111) planes are ferromagnetic but couple antiferromagnetically with adjacent (111) planes. A crystallographic distortion accompanies the antiferromagnetic ordering. The distortion, a contraction of a [111] for MnO and NiO and a contraction of [100] for CoO, gives rise to twin walls which can be considered, in part, as marking off antiferromagnetic domains. Roth[1] and Slack[2,3] in a detailed study of NiO distinguish two types of domains; the $T$ wall which is associated with twin formation and the $S$ wall marking the way spins rotate in (111) planes. The field dependence of $\chi$ is the result of the rotation of spins into a direction more nearly perpendicular to the applied field, while the magnetostriction is the result of twin wall motion and the accompanying reorientation of the deformation axis.

The samples were made by the flame fusion method, NiO and CoO by Scott[4] and MnO by Nakazumi.[5] The NiO and CoO were annealed at 1400°C and slow cooled. No stress was applied during cooling. Both x-ray and cleaved surfaces were used to determine crystallographic directions.

The susceptibility was measured by a force method. For MnO and NiO the field was along a [111]; for CoO a [100]. Additional measurements taken with $H$ perpendicular to the original direction varied no more than $\pm 3\%$, indicating the absence of a preferred axis for the spin orientation. As shown in Fig. 1, an increase in $\chi$ with $H$ is exhibited by both MnO and NiO at 77°K. The small change in CoO is considered almost within the experimental error. The values of $\chi$ for MnO are smaller in magnitude than those of a crystal from another source.[6] In all three samples there is a

hysteresis effect; the susceptibility is slightly higher with decreasing fields than for increasing fields. This effect did not persist until zero field. The hysteresis may be due to weak spontaneous moments in domain walls caused by the lack of regular arrangement of spin as suggested by Néel.[7]

Magnetostriction measurements were made at 77°K and room temperature using the strain gauge technique. The maximum value of the applied field was about 14 000 gauss. The room temperature value of the gauge factor specified by the manufacturer was assumed to be valid at 77°K. Figure 2 shows the magnetostriction ($\lambda$) at 77°K plotted against the square of the applied field for both $H$ and the gauge along the same [110] direction. In each case $\lambda$ is proportional to $H^2$. No measurable hysteresis in $\lambda$ could be observed. A result obtained from this study was the marked increase in $\lambda$ found in annealed samples over that obtained in unannealed ones. Room temperature data on the NiO crystal indicated no threshold effects similar to that reported by Belov and Levitin[8] in their

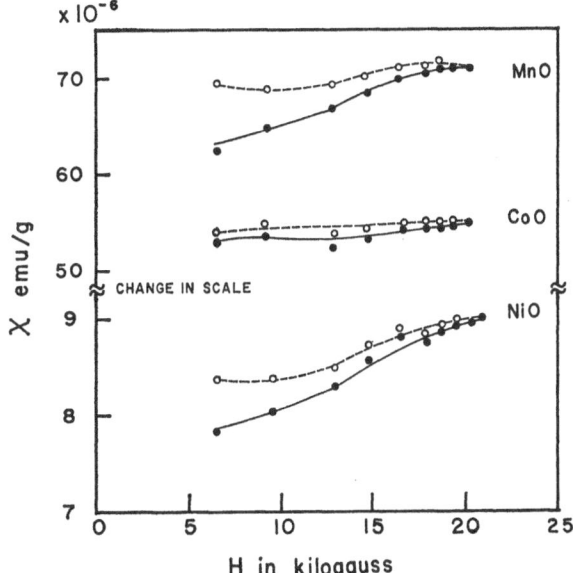

FIG. 1. Gram magnetic susceptibility as a function of field at 77°K along a [111] for MnO and NiO and along a [100] for CoO.

[1] W. L. Roth, J. Appl. Phys. **31**, 2000 (1960).

[2] W. L. Roth and G. A. Slack, J. Appl. Phys. **31**, 352S (1960).

[3] G. A. Slack, J. Appl. Phys. **31**, 1571 (1960).

[4] E. J. Scott, U. S. Naval Ordnance Laboratory, Silver Spring, Maryland.

[5] Y. Nakazumi, Tachigi Chemical Company, Osaka, Japan.

[6] T. R. McGuire and R. J. Happel, J. phys. radium **20**, 424 (1959).

[7] L. Néel, *Proceedings of the International Conference on Theoretical Physics, Tokyo and Kyoto, 1954* (Science Council of Japan, Tokyo, 1954).

[8] K. P. Belov and R. Z. Levitin, Soviet Phys.—JETP **10**, 400 (1960), translation from Russian.

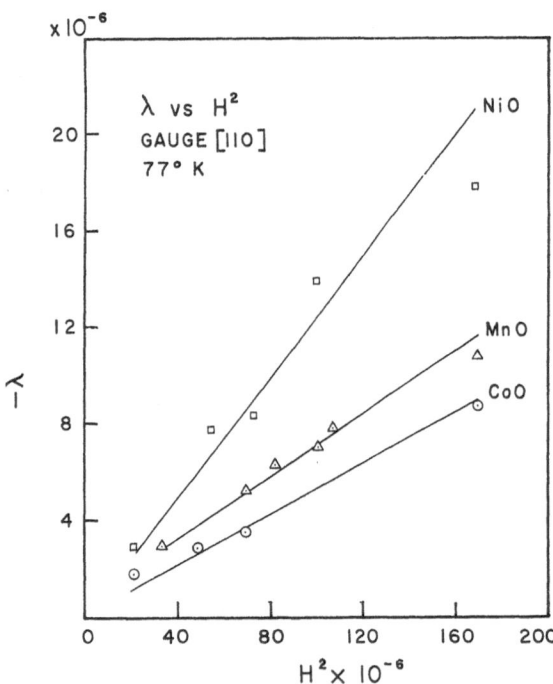

FIG. 2. The magnetostriction as a function of $H^2$ at 77°K along a [110] axis for each crystal.

TABLE I. Summary of susceptibility and magnetostriction results.

| | Distortion contractions[a] | Magneto-striction $\Delta\lambda$[b] | % of sample volume changed[a] | Suscepti-bility $\Delta\chi$[b] | % of sample volume changed[a] |
|---|---|---|---|---|---|
| MnO | 1 part in $10^3$ | $-7.8\times10^{-6}$ | 1% | $5\times10^{-6}$ | 12% |
| CoO | 1 in $10^2$ | $5.4\times10^{-6}$ | 0.1% | $0.5\times10^{-6}$ | 1.4% |
| NiO | 1 in $10^3$ | $-14\times10^{-6}$ | 1.5% | $0.5\times10^{-6}$ | 8.% |

[a] See text for explanation of these quantities.
[b] From 7000 to 14 000 gauss.

not distinguish between $T$-wall or $S$-wall domain motion and any increase in $\chi$ is due to both effects. Table I gives the volume percentage of the sample in which domain changes must occur (from 7000 to 14 000 gauss) in order that $\lambda$ and $\chi$ increase the experimentally observed amount.

The change in $\chi$ of CoO is too small to be of significance in this discussion. The fact that $\lambda$ is appreciable in CoO is probably due to the large distortion found in this compound.

From Table I the magnitude of the twin domain growth in NiO and MnO can be used to determine what part of the increase in $\chi$ is due to spin rotation in (111). With this information and using the susceptibility equation to Keffer and O'Sullivan[11] for $H > H_c$, where $H_c$ is the critical field for spin flopping, we obtain values of $H_c \approx 16\,800$ gauss, $K_2 \approx 10^4$ ergs/cc for MnO and $H_c \approx 18\,000$ gauss, $K \approx 3\times10^3$ ergs/cc for NiO. Further experiments are now in progress with samples which have been heat treated in controlled atmospheres to give the best stoichiometry. This could have an effect on the field dependent properties so that the above values of $H_c$ and $K_2$ are to be treated with some reservation.

Néel[7] has shown that the $T$-wall motion is governed by an $H^2$ term. This would account for $H^2$ dependence found for $\lambda$.

*Note added in proof.* The sign of the magnetostriction shown in Fig. 2 is incorrect. We now interpret $\lambda$ in the [110] as positive for both NiO and MnO and negative for CoO. The signs for $\Delta\lambda$ in Table I should also be reversed.

study of polycrystalline NiO. Our results for CoO are in agreement with values reported by Nakamichi and Yamamoto.[9]

In discussing the magnetostriction and susceptibility data for MnO and NiO we assume the magnetic structure model proposed by Kaplan[10] and further developed by Keffer and O'Sullivan[11] where spins are confined to (111) by a much larger anisotropy ($K_1$) than that governing their rotation within ($K_2$) a (111). We further assume that the magnetostriction is caused only by those domain movements which change a twin wall boundary. This is reasonable since the other alternative—$S$-wall motion—can take place without changing the [111] contracting axis. The change in length of the specimen when a field is applied is then proportional to the per cent volume of the $T$-wall domains. On the other hand, the susceptibility does

### ACKNOWLEDGMENT

We wish to thank H. R. Lilienthal for his help in obtaining some of these measurements.

[9] T. Nakamichi and M. Yamamoto, J. Phys. Soc. Japan **16**, 126 (1961).
[10] J. I. Kaplan, J. Chem. Phys. **22**, 1709 (1954).
[11] F. Keffer and W. O'Sullivan, Phys. Rev. **108**, 637 (1957).

# Impurity Ion Effects in the Ferrimagnetic Resonance of Ordered Lithium Ferrite

A. D. Schnitzler, V. J. Folen, and G. T. Rado

*U. S. Naval Research Laboratory, Washington 25, D. C.*

Experimental results are reported on the ferrimagnetic resonance linewidth, g factor, and magnetocrystalline anisotropy field of ionically ordered lithium ferrite monocrystals containing impurity ions. The measurements were made at frequencies ranging from 16.0 to 30.9 kMc and at temperatures ranging from 4.2° to 300°K. It is shown that: (a) the temperature dependence of the g factor is proportional to $1/T$ at "high" temperatures ($50°K \lesssim T \leq 300°K$) but less pronounced at "low" temperatures ($T \lesssim 50°K$); (b) the linewidth vs temperature curves exhibit a maximum in the transition region ($T \approx 50°K$) for all crystallographic directions which are not in the vicinity of [110]; (c) at 300°K the linewidth in the [111] direction increases linearly with increasing frequency; (d) all these experimental results can be interpreted on the basis of the theory of rapidly relaxing impurity ions proposed by Kittel and co-workers. In regard to the anisotropy field, it is shown experimentally that at 4.2°K anomalous peaks exist in the [110] and [112] directions of the (110) plane, and that the amplitude of the former is twice that of the latter. An analysis based on Kittel's theory of anomalous anisotropy fields is supplemented by specific assumptions to explain both of these observations as well as the other details of the angular dependence of the anisotropy field.

## INTRODUCTION

PURE crystals of ordered lithium ferrite and yttrium iron garnet (YIG) differ from other ferrimagnetic oxides in two important aspects: All the magnetic ions are identical and form an ionically ordered structure. The absence of irregular fluctuations of the crystalline and exchange fields suggests that the impurity ion effects in the ferrimagnetic resonance of these two materials are similar. This similarity is demonstrated below. Effects of the ionic order on the ferrimagnetic resonance[1] and magnetocrystalline anisotropy[2] of lithium ferrite were reported earlier.

## SAMPLES AND METHODS

The lithium ferrite monocrystals were grown in a PbO flux, using reagent grade chemicals. Spectrographic analysis indicated that the monocrystals contain about 0.1% Mn, 0.1% Ni, and 0.01% Co. Measurements were made on ionically ordered spherical samples highly polished with "Linde A" powder. The samples were mounted in tunable transmission cavities. Most data were taken near 24.2 kMc between 4.2° and 300°K.

## *g* FACTOR AND LINEWIDTH

Figure 1 shows our values of the experimental g factor and the linewidth ($\Delta H$) in the [111] direction as a function of temperature ($T$) at 24.2 kMc. While $g_{\text{exper}}$ is linear in $1/T$ at "high" temperatures ($50°K \lesssim T \leq 300°K$), it varies less strongly at "low" temperatures ($T \lesssim 50°K$). Extrapolation yields $g_{\text{exper}} = 2.0025 \pm 0.0005$ at $T = \infty$. The maximum of $\Delta H$ in the "transition region of the g factor" ($T \approx 50°K$) is characteristic of all crystallographic directions not in the vicinity of [110]. Measurements at 300°K in the [111] direction showed that $\Delta H$ decreases linearly from 16.3 oe to 10.7 oe as the frequency decreases from 30.9 to 16.0 kMc.[3] Ferrous ions are

probably not responsible for any of these results because oxidation of the samples for 20 hr at 700°K did not alter the data. The role of ferrous ions in the anomalous $\Delta H$ peak observed in a [110] direction of ordered samples was investigated earlier.[1]

We interpret our experimental results on both the g factor and $\Delta H$ on the basis of the theory[4] of rapidly relaxing impurity ions. Our assumption is that certain ions of the Mn, Ni, or Co impurities relax sufficiently rapidly in the crystalline and exchange fields of ordered (but probably not in disordered) lithium ferrite. If this is valid, then: (1) the $\Delta H$ vs $T$ curve exhibits a peak at

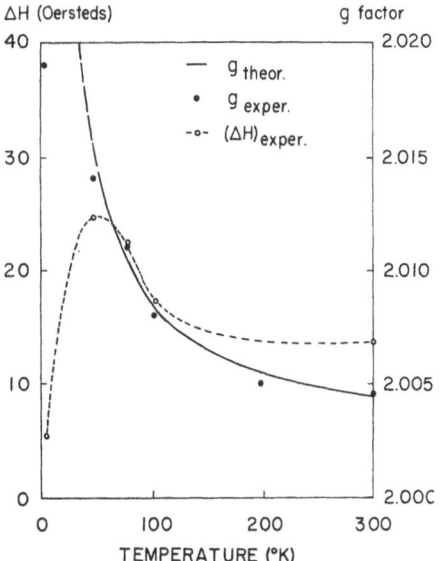

FIG. 1. Temperature dependence of the theoretical g factor $g_{\text{theor}}$ (which is linear in $1/T$), the experimental g factor $g_{\text{exper}}$, and the experimental linewidth $\Delta H$.

oe. Preliminary data obtained by F. W. Patten of this Laboratory by means of the parallel pump method at 300°K yield $\Delta H \approx 6$ oe for zero wave number spin waves propagating normal to the [111] direction at 4.7 kMc.

[4] C. Kittel, Phys. Rev. **115**, 1587 (1959); P. -de Gennes, C. Kittel, and A. M. Portis, Phys. Rev. **116**, 323 (1959); R. L. White, Phys. Rev. Letters **2**, 465 (1959).

[1] A. D. Schnitzler, V. J. Folen, and G. T. Rado, J. Appl. Phys **31**, 348S (1960).
[2] V. J. Folen, J. Appl. Phys. **31**, 166S (1960).
[3] Linear extrapolation of these data to 4.7 kMc yields $\Delta H = 6.2$

some temperature, and (2) the linewidth increases linearly with frequency, at least at sufficiently high frequencies. These expectations are fulfilled by the experimental results above. The location of the peak and the general form of our $\Delta H$ vs $T$ curve are comparable to those observed[5] in rare earth doped YIG.

Assuming that the magnetization of the rapidly relaxing (paramagnetic) impurity ions obeys the Curie law, we may expect that: (3) the $g$ factor varies linearly with $1/T$ at "high" temperatures, (4) the $g$ factor varies less strongly than $1/T$ at "low" temperatures, (5) the "transition region of the $g$ factor" occurs in the vicinity of the maximum of $\Delta H$, and (6) the value of $g_{exper}$ agrees with the $g$ value of the ferric lattice as given by the $g_{exper}$ of pure YIG. Expectations (3) through (6) are also fulfilled by our results.

The observation that $g_{exper}$ [see (3)] increases rather than decreases with $1/T$ may be interpreted by assuming the rapidly relaxing impurity ions are located on octahedrally coordinated sites so that their magnetization is parallel to that of the ferric lattice.

## ANOMALOUS ANISOTROPY

Figure 2 shows the variation of $H^{anis}$ at 4.2°K with direction in a (110) plane. $H^{anis}$ denotes that part of the

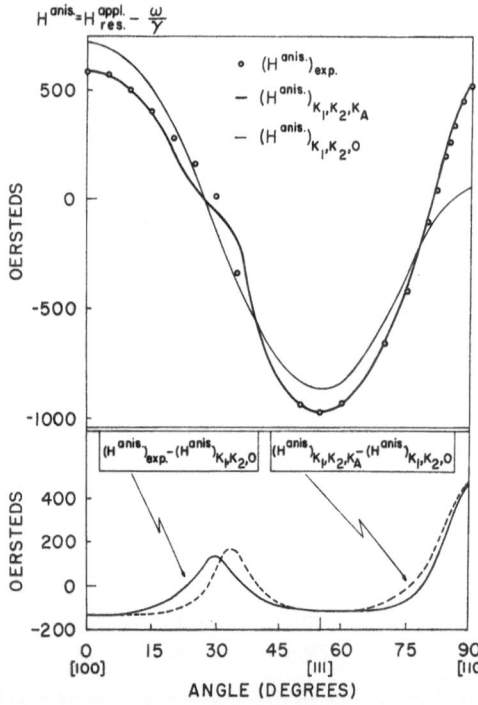

FIG. 2. Angular dependence, in a (110) plane, of various anisotropy fields and differences of anisotropy fields. The experimental data (circles) and the various curves are explained in the text. The peaks of the dashed curve in the lower part of the figure occur in the [110] and [112] directions of the (110) plane. The [112] direction is located at an abscissa of about 35°.

applied magnetic field for resonance, $H_{res}^{appl}$, which arises from magnetocrystalline anisotropy. The position of the zero on the ordinate depends on the value of $g$ and thus on the results of the calculations mentioned below. In the upper part of Fig. 2 is that theoretical curve (light line) which neglects the anomalous part (see below) of $H^{anis}$. This curve is calculated on the basis of: (1) the first and second order magnetocrystalline anisotropy constants $K_1$ and $K_2$ (which are adjustable parameters) taken from the calculation mentioned below, and (2) the correction[6] for incomplete alignment of the magnetization. The curve values are exceeded by the experimental $H^{anis}$ values mostly in the vicinity of specific directions ([110] and [112]), as shown by the solid curve in the lower part of Fig. 2, indicating that $H^{anis}$ in ordered[7] lithium ferrite contains a contribution from an anomalous anisotropy field also. Neither this contribution nor the remainder of the experimental $H^{anis}$ was influenced by the oxidation treatment mentioned previously.

We attribute the results mentioned above to the effect of impurity ions. To calculate the angular dependence of $H^{anis}$, we supplement the theory[8] of anomalous anisotropy fields with a specific model which assumes: (a) The effective impurity ions (of concentration $N$) are uniformly distributed over those crystallographic sites at which the uniaxial crystalline electric potential is associated with a [111] (i.e., trigonal) axis. (b) The ground state of these ions has a nonaccidental degeneracy when the angle $\alpha$ between the exchange field (i. e., the magnetization) and the trigonal axis is $\pi/2$. (c) The orientation-dependent part of the energy of the lowest states is given by $\pm C \cos\alpha$ for each ion; the parameter $C$ (which depends on the type of ion and the relative magnitudes of the perturbations acting on it) may represent an exchange or a spin-orbit splitting.

Calculated values[9] of the total $H^{anis}$ and of the anomalous part of $H^{anis}$ are shown, respectively, by the heavy line in the upper part of Fig. 2 and the dashed line in the lower part of Fig. 2. The agreement between the experimental and calculated values of $H^{anis}$ is reasonably good. The values (at 4.2°K) of the parameters used in the calculations are $K_1/M_s = -361$ oe, $K_2/M_s = -864$ oe, $4\pi M_s = 4160$ gauss,[10] $g = 2.019$, $N = 0.04\%$, and $C = 30$

[5] J. F. Dillon, Jr., and J. W. Nielsen, Phys. Rev. Letters **3**, 30 (1959).

[6] J. O. Artman, Proc. Inst. Radio Engrs. **44**, 1284 (1956).

[7] No anomalous anisotropy was observed in disordered lithium ferrite; see A. D. Schnitzler, thesis, University of Maryland, 1961. The $g$ factor and the temperature and frequency dependence of the linewidth of disordered lithium ferrite is reported by A. D. Schnitzler, Bull. Am. Phys. Soc. **7**, 54 (1962).

[8] C. Kittel, Phys. Rev. Letters **3**, 169 (1959); Phys. Rev. **117**, 681 (1960).

[9] Alternatively, we have assumed that the degeneracy mentioned in (b) is removed by a small perturbation $\delta$ so that the crossover of the lowest levels is a near (rather than an actual) crossover and the energy depends on $\delta$ as well as on the expression given in (c). This assumption appears to be promising, and a detailed calculation of the angular dependence of $H^{anis}$ is underway in conjunction with experiments involving doping with controlled amounts of cobalt impurities.

[10] G. T. Rado and V. J. Folen, J. Appl. Phys. **31**, 62 (1960).

cm$^{-1}$. Slight changes in the choice of $K_1$ and $K_2$ would raise or lower the two curves in the lower part of Fig. 2 without affecting the positions or relative heights of their peaks. Since the anomalous[11] anisotropy field

[11] The anomalous anisotropy field peaks reported in the present paper occur at different angles and are smaller in magnitude than those first observed by J. F. Dillon (see reference 5 and earlier papers quoted therein) in rare-earth doped YIG.

should exhibit peaks in directions for which $\alpha = \pi/2$ and since each [110] direction is perpendicular to two trigonal directions whereas each [112] direction is perpendicular to only one, not only the angular positions but also the relative heights of the observed anomalous anisotropy field peaks are in good agreement with the predictions of our model.

---

JOURNAL OF APPLIED PHYSICS    SUPPLEMENT TO VOL. 33, NO. 3    MARCH, 1962

# Microwave Resonance Linewidth in Single Crystals of Cobalt-Substituted Manganese Ferrite

R. W. TEALE

*Mullard Research Laboratories, Redhill, Surrey, England*

Measurements of the microwave resonance linewidth $\Delta H$ are reported for small spherical single crystal specimens of nominal composition $Co_xMn_{1-x}Fe_2O_4$ where, $x=0$, 0.02, 0.04, 0.08, 0.10. Accurate chemical analyses have been published. $\Delta H$ varies rapidly and linearly with $x$. At 294°K and 16.9 Gcps the gradients of $\Delta H$ against $x$ observed are $4.0 \times 10^3$ and $3.0 \times 10^3$ oe per cobalt ion for the [100] and [111] directions, respectively. At 9.6 Gcps the corresponding gradients are $2.7 \times 10^3$ and $2.1 \times 10^3$. Measurements on disk samples for $x=0.08$ gave results similar to those for a sphere. At 16.9 Gcps $\Delta H$ is plotted against temperature between 150° and 500°K for $x=0$, 0.04, and 0.08. Above 250°K $\Delta H$ varies only slowly with temperature. Below 150°K $\Delta H$ increases and becomes anisotropic rapidly; this is associated with anomalies in the field for resonance. The results suggest that magnon-magnon scattering by the cobalt does not make a significant contribution to $\Delta H$. The behavior observed bears a qualitative resemblance to that of yttrium iron garnet doped with certain rare-earth ions, but the fast relaxing ion model, successful for some of these, does not fit the present measurements.

## INTRODUCTION

THE mechanisms responsible for relaxation of the motion of the magnetization in ferrimagnetic insulators have recently received increased study, and understanding of this phenomenon is now developing. It is clear that two different mechanisms are important. In single crystals of pure YIG the uniform motion of the magnetization is damped through scattering processes which transfer energy to spin waves of short wavelength,[1] whereas in ferrimagnetic garnet crystals containing rare earth ions with a short relaxation time, these ions act to transfer energy directly from the uniform motion to lattice vibrations.[2]

White[3] has suggested that certain ions of the iron transition series may play a similar part in relaxation processes in spinel ferrites to that played by the rare earth ions mentioned above. The divalent cobalt ion when located on an octahedral lattice site is one of these ions.

Hass and Callen,[4] on the other hand, have predicted that the cobaltous ion may contribute significantly to relaxation by scattering energy into spin waves of short wavelength.

This paper reports experimental observations of the effect upon the microwave resonance linewidth of the substitution of cobalt into single crystals of manganese ferrite and compares the results with the above mentioned suggestions. In single crystal specimens the linewidth is expected to arise directly from relaxation processes.

## EXPERIMENTAL PROCEDURE

The single crystals employed were grown by the flame fusion process by Mr. J. Page. Extensive measurements of the magnetocrystalline anisotropy and $g$ factors of these crystals have already been reported.[5,6] The nominal chemical composition is $Co_xMn_{1-x}Fe_2O_4$, where $x$ ranges from 0 to 0.10. Chemical analyses have been published.[5]

Spherical specimens of approximately 0.4-mm diameter were ground, then polished using $\frac{1}{4}\mu$ diamond powder. Several disks of the crystal with $x \approx 0.08$ were cut and ground with a thickness of 0.1 mm and diameter 1.3 mm; one of these disks was polished.

## EXPERIMENTAL RESULTS

Figure 1(a) shows the values of $\Delta H$ measured at room temperature with the static magnetizing field oriented

[1] T. Kasuya and R. C. LeCraw, Phys. Rev. Letters 6, 223 (1961).
[2] P. G. de Gennes, C. Kittel, and A. M. Portis, Phys. Rev. 116, 323 (1959).
[3] R. L. White, Phys. Rev. Letters 2, 465 (1959).
[4] C. W. Hass and H. B. Callen, Phys. Rev. 122, 59 (1961).

[5] R. F. Pearson, Proc. Phys. Soc. (London) 74, 505 (1959).
[6] R. W. Teale and M. J. Hight, J. Appl. Phys. 32, 140S (1961).

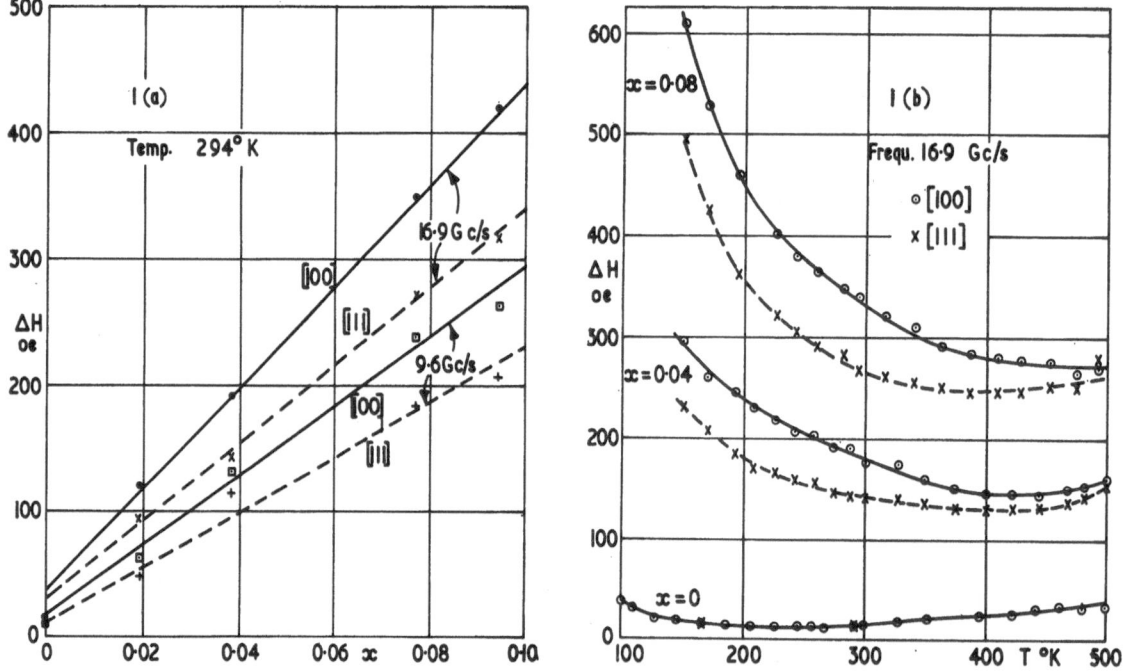

FIG. 1. Microwave resonance linewidth $\Delta H$ for $Co_x Mn_{1-x} Fe_2 O_4$. (a) $\Delta H$ vs $x$. (b) $\Delta H$ vs temperature.

along the [100] and [111] directions. For other crystallographic directions within a (110) plane $\Delta H$ lies between these extreme values. In Fig. 1(b) $\Delta H$ is plotted against temperature. The Curie temperatures of the crystals vary linearly with $x$ from 550°K for $x=0$ to 575°K for $x=0.10$.

Below 150°K the effect of the cobalt substitution upon $\Delta H$ becomes increasingly anisotropic. In a (110) plane sharp peaks develop in $\Delta H$ at 20° from [100]. At 84°K and 16.9 Gcps for $x=0.02$ $\Delta H$ is 310 oe in this direction compared with a minimum of 160 oe in [100] directions.

At 4.2°K and 16.9 Gcps a small peak in $\Delta H$ ($\sim$15 oe, high) was observed in a similar orientation for a specimen with $x=0.001$.

Measurements were carried out on three different disk samples with $x=0.08$ at 293°K and 16.9 Gcps yielding similar results. Polishing produced no change. With a disk magnetized perpendicular to its plane two absorption lines were clearly resolved, and there was a long tail to the low field side. The resonant field of the strongest line fitted accurately the resonant conditions for the uniform mode. This line was symmetrical down to the half height and 440 oe distant from the next resolved line. The linewidth was 320 oe. With the magnetizing field in the plane of the disk the linewidth was 340 oe. The orientation of the disks with respect to crystallographic axes was not determined.

## DISCUSSION

It seems clear that the main effects demonstrated are due to the cobalt substitution and that spurious effects such as the surface polish of the samples are not signifi-

cant. This is a consequence of the very large effect which small substitutions of cobalt have upon $\Delta H$.

The peak in $\Delta H$ at 20° from [100] observed below 150°K is clearly connected with the similarly located peak in the resonant field previously reported[6] for cobalt substituted manganese ferrite at 4.2°K. However, the effect upon $\Delta H$ is evident at much higher temperatures.

The maximum contribution to $\Delta H$ from cobalt calculated by Haas and Callen[4] is $1.6 \times 10^3 x$ oe. The observed contribution shown in Fig. 1(a) ranges from $4.0 \times 10^3 x$ oe to $2.1 \times 10^2 x$ oe. At 500°K the contribution is $3.26 \times 10^3 x$ oe at 16.9 Gcps. The observed contribution somewhat exceeds the maximum predicted contribution. Furthermore, comparison of the observed magnetocrystalline anisotropy at low temperatures with theoretical calculations suggests that the ground-state orbital degeneracy of the cobaltous ion is to some extent removed by the crystalline fields present on the octahedral sites in manganese ferrite.[6] In this case the calculated contribution to $\Delta H$ is several times smaller than the maximum given above.

Disks with the dimensions used in the experiments reported here when magnetized perpendicular to the plane should give $\Delta H$ roughly one-fifteenth of $\Delta H$ for a sphere if scattering to short wavelength spin waves is dominant.[4] No reduction in $\Delta H$ was observed for the disk specimens.

A slight narrowing of $\Delta H(\sim 7\%)$ was observed for a spherical specimen with $x=0.08$ when the uniform mode was lifted from degeneracy with long wavelength spin waves by the increase of $M_s$ beyond 400 emu.[7] Hence it

[7] C. R. Buffler, J. Appl. Phys. 30, 172S (1959).

seems that scattering to long but not to short wavelength spin waves affects $\Delta H$ slightly.

The rapid rise in $\Delta H$ below 200°K shown in Fig. 1(b) and the anisotropic character of $\Delta H$ and the field for resonance at low temperatures indicate a qualitative resemblance between the effect of the cobalt ion in manganese ferrite and the effect of the rare-earth ions with short relaxation times in the iron garnets.

Kittel[8] has suggested that the behavior of some of the rare earth iron garnets above roughly 300°K may be explained by assuming that the relaxation frequency of the ion concerned is much higher than any of the other frequencies involved, including the exchange frequency $\gamma H_e$, where $H_e$ is the effective exchange field acting on the rare-earth ion. This assumption leads[2,8] to the expressions $\Delta H/H_r = \gamma H_e \tau$ and $g_e = g_r M_s/(M_s - M_f)$, where $H_r$ is the static field for resonance, $\tau$ the relaxation time and $M_f$ the moment of the rare-earth ion, $g_e$

the $g$ factor measured at microwave frequencies, and $g_r$ the $g$ factor of the ions other than the rare-earth ions. $\tau$ is expected to decrease as $T$ increases. Measurements on holmium, erbium, and dysprosium iron garnets have shown agreement with these relationships.[9-11]

The observed ratios $g_r/g_e$ for $x = 0.10$ are 0.9814[6] and 0.9847 at 260° and 500°K, respectively. To fit these ratios to the above relationship it is necessary to assume unreasonably small values for $M_f$ for the cobalt ion. $\Delta H$ for cobalt rises more slowly between 9.6 and 16.9 Gc/s and between 500° and 250°K than the above relationship suggests.

Hence though direct relaxation to the lattice seems plausible, the fast relaxing ion assumptions are not justified for cobalt in manganese ferrite.

[8] C. Kittel, Phys. Rev. 115, 1587 (1959).

[9] G. P. Rodrigue, H. Meyer, and R. V. Jones, J. Appl. Phys. 31, 376S (1960).
[10] A. Vassiliev, J. Nicolas, and M. Hilderbrandt, Compt. rend. 252, 2681 (1961).
[11] A. Vassiliev, J. Nicolas, and M. Hilderbrandt, Compt. rend. 253, 242 (1961).

JOURNAL OF APPLIED PHYSICS      SUPPLEMENT TO VOL. 33, NO. 3      MARCH, 1962

# On the Possibility of Obtaining Large Amplitude Resonance in Very Thin Ferrimagnetic Disks

F. R. Morgenthaler*†
*Department of Electrical Engineering, Massachusetts Institute of Technology, Cambridge, Massachusetts*

A theoretical prediction is made that the uniform resonance and the spinwaves in a very thin ferrimagnetic disk can be made nondegenerate. If this condition can be achieved experimentally, the usual second order spinwave instabilities will be forbidden and cannot limit the component of the magnetization occurring at the resonance frequency ($\omega_0$). The precession cone angle ($\theta$) should then increase, in response to the usual rf driving field, until limited by third order instabilities involving spinwaves with frequencies equal to $\frac{3}{2}\omega_0$.

The state of nondegeneracy is to be achieved in the following manner: A thin disk, magnetized in a direction perpendicular to the plane, is excited by two transverse positive-circularly-polarized pulsed rf magnetic fields of specified amplitude. One of these fields has a frequency equal to $\omega_0$, the other a frequency detuned from $\omega_0$ by an amount comparable to the width of resonance. As $\theta$ increases from zero during the ensuing transient, there will occur a point at which the familar second order coupled spinwaves become unstable (the state of nondegeneracy has not yet been reached). Since there is a time lag before these modes can effect the uniform precession, the cone angle will continue to increase.

When some larger value of $\theta$ has been attained the degeneracy between the spin waves and the resonance is removed; the previously unstable spin waves are then decoupled from the uniform resonance.

The degeneracy is removed because, as the author has already shown theoretically, the resonance frequency of the thin disk should lie below the main spinwave band (volume band) provided $\theta > (2N_t)^{\frac{1}{2}}$, where $N_t$ (the transverse demagnetizing factor) is by assumption very small. This level of excitation would create the desired state of nondegeneracy, if it were not for the fact that a range of degenerate spinwaves (separate from the main band) still exists. These degenerate modes may be destroyed, however, by introducing the additional transverse rf field.

Large amplitude resonance would clearly be desirable for a variety of reasons. Optical modulation experiments, to give only one example, would be of great interest. The necessity for using thin disks or films is, in this case, a fortuitous circumstance. It also follows that any nonlinear device involving harmonic generation, frequency mixing, etc. would become more efficient.

## INTRODUCTION

IT is well known by now that the precession cone angle associated with the uniform component of the magnetization undergoing ferromagnetic resonance in a ferromagnet is limited to small values by the growth of unstable spin waves.[1] These instabilities usually involve

spinwaves varying at half the resonance frequency or (if these are forbidden) those which are degenerate with the resonance. Since the former type of instability may always be forbidden by choosing a large enough value of the resonance frequency, it is the degenerate spinwaves which are the real stumbling block to obtaining large amplitude resonance. Several methods have been proposed for stabilizing spin wave instabilities by weakening their parametric coupling to the uniform mode through modulation; one of these, due to Suhl,[2]

* Ford Postdoctoral Fellow; consultant to AFCRL Office of Aerospace Research USAF.
† This research was sponsored jointly by the Ford Foundation and the USAF.

[1] H. Suhl, Proc. Inst. Radio Engrs. 45, 1271 (1956).

[2] H. Suhl, Phys. Rev. Letters 6, 174 (1961).

suggests modulation of the dc magnetic field. The present author[3] has pointed out, however, that such a procedure frequency modulates the uniform precession and causes a dissolution of the transverse magnetization into many frequency (sideband) components. No *single frequency* component of the magnetization is made larger than that value which can be obtained in the absence of the modulation field. An alternate method was proposed[3] which should allow modulation of the spinwaves *without* at the same time causing modulation of the uniform precession. This should allow an increase in the resonant precession cone angle ($\theta$) without causing a spectral decomposition of the uniform precession. Unfortunately this method is what might be described as a "brute force" approach since the increase in $\theta$ is a function of the strength of the modulation. Power considerations place an upper bound on the degree of stabilization possible. The method proposed in this paper is superior in this respect.

### THE POSITION OF THE UNIFORM RESONANCE WITH RESPECT TO THE SPIN WAVE MANIFOLD

The differential equations governing the amplitudes and phases of the uniform precession and a single small amplitude (though arbitrary) spinwave have been obtained for the general case of an ellipsoid, magnetized in an arbitrary direction, when acted upon by an arbitrary number of spatially uniform rf fields.[4] The following equations are specialized for a spheroid magnetized along the axis of revolution in the absence of any rf fields.

$$\dot{\varphi}_k = \omega_H - N_z \omega_M \cos\theta + \lambda k^2 \omega_M \cos\theta$$
$$+ \sin^2\!\psi\,\omega_M \cos\theta \cos^2(\varphi_k - \xi)$$
$$- (\sin 2\psi)/2\,\omega_M \sin\theta[\cos(2\varphi_k - \varphi_0 - \xi) + \cos(\varphi_0 - \xi)]$$
$$+ \tfrac{1}{2}(\lambda k^2 + \cos^2\!\psi - N_t)$$
$$\times \omega_M \sin\theta \tan\theta[1 + \cos 2(\varphi_k - \varphi_0)], \quad (1)$$

$$\dot{\varphi}_0 = \omega_0 = \omega_H - N_z \omega_M \cos\theta + N_t \omega_M \cos\theta, \quad (2)$$

where $\varphi_0$ and $\varphi_k$ are the phases of the uniform precession and spinwave, respectively, $M$ is the saturation magnetization, $\omega_M = -\gamma\mu_0 M$, $\gamma$ is the gyromagnetic ratio (negative) including $g$ factor, $\xi$ is a phase constant, $\psi$ is the angle between the spin wave propagation vector **k** and the dc field (**H**), $\lambda$ is an exchange parameter, $\omega_H = -\gamma\mu_0 H$, and $N_z$ and $N_t$ are the longitudinal and transverse demagnetizing factors ($N_z + 2N_t = 1$).

A detailed study of these equations has been made in reference 4 and will not be repeated in its entirety here. It is sufficient to notice that for small values of $\theta$ the minimum spin wave frequency occurs when $\psi = 0$ and

$k = 0$. If $\langle\dot{\varphi}_k\rangle_{\mathrm{av}} \neq \langle\dot{\varphi}_0\rangle_{\mathrm{av}}(\omega_k \neq \omega_0)$ the average value of $\cos 2(\varphi_k - \varphi_0)$ will be zero or at least very small compared to unity,[5] and

$$(\omega_k)_{\min} = \omega_H - N_z \omega_M \cos\theta + (1 - N_t)/2\,\omega_M \sin\theta \tan\theta. \quad (3)$$

Notice that for vanishingly small $\theta$, $\omega_0$ always exceeds $(\omega_k)_{\min}$ by a frequency equal to $N_t \omega_M$.

The extremely interesting feature of Eqs. (2) and (3) is that $\omega_0$ will drop below $(\omega_k)_{\min}$ provided that $\tan^2\theta > 2N_t/(1 - N_t)$. For an extremely thin disk, $N_t$ will be very small and $\theta$ need not be large. This is reasonable since for a thin disk $\omega_0$ is nearly at the bottom of the manifold when $\theta = 0$.

In view of the assumptions used to obtain Eq. (3), $(\omega_k)_{\min}$ is the minimum *nondegenerate* spin wave frequency. To see if degenerate spin waves still exist even when the inequality is valid we assume $\omega_k = \omega_0$. This leads to

$$\theta^2 = 2N_t/(1 + \cos 2\alpha) > 2N_t, \quad (4)$$

where $\alpha$ is the phase angle between the uniform precession and $\psi = 0$, $k = 0$ spin wave. As $\theta$ is increased $\cos 2\alpha$ merely becomes more negative to satisfy Eq. (4). Figure 1 indicates the situation a little more clearly. An increase in $\theta$ causes an increase in $\omega_0$ but an even greater increase in $\omega_k$. When $\theta > (2N_t)^{\frac{1}{2}}$ the main spin wave band ($\omega_k \neq \omega_0$) has shifted *above* $\omega_0$, but because of Eq. (4) a degenerate range of "parasitic" spinwaves still exist. These degenerate modes (unlike the usual modes) require specific phase relationships between $\varphi_k$ and $\varphi_0$. Their instability thresholds are somewhat higher than normal[4] but presumably if excited during a transient experiment, they would act to reduce $\theta$ and bring about a return of normal degeneracy.

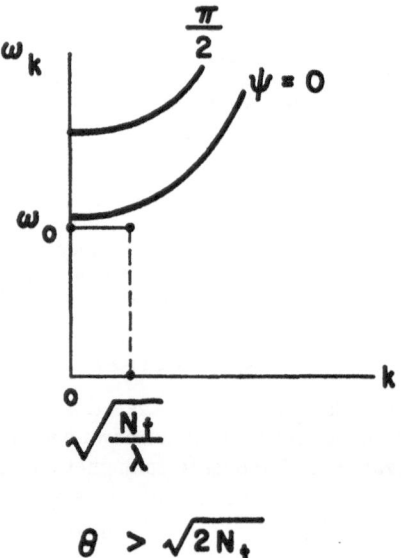

FIG. 1. Spinwave spectrum modified by finite amplitude uniform precession.

[3] F. R. Morgenthaler, F. A. Olson, and G. E. Bennett, "Suppression of spin wave instabilities associated with ferromagnetic resonance," Proceedings of the 1961 International Conference on Magnetism and Crystallography, Kyoto, Japan (to be published).

[4] F. R. Morgenthaler, Doctoral thesis, Department of Electrical Engineering, Massachusetts Institute of Technology, Cambridge, Massachusetts (May, 1960).

[5] This is true unless $|(\omega_k)_{\min} - \omega_0|$ is very small compared to $\theta^2 \omega_M$.

This degenerate band of phase sensitive spin waves can be destroyed by frequency modulating $\varphi_0$ without modulating $\varphi_k$.[6] The modulation can be obtained by exciting ferromagnetic resonance with two transverse rf magnetic fields of specified amplitude. One of these fields ($h_0$) would have a frequency equal to $\omega_0$, the other ($h'$) a frequency detuned from $\omega_0$ by an amount equal to $\omega_m$. It follows that for small values the modulating index $\delta$ approximately equals $-\gamma\mu_0 h'/(\theta\omega_m)$ and $\varphi_0 \simeq \omega_0 t + \delta \sin\omega_m t$. If $\omega_k = \omega_0$, the average value of $\cos 2(\varphi_k - \varphi_0)$ is now $J_0(2\delta)\cos 2\alpha$, where $J_0$ is the Bessel function of order zero. It is apparent that Eq. (4) fails to be satisfied when

$$\theta^2 > 2N_t/(1 - |J_0(2\delta)|). \qquad (5)$$

If $\theta$ is made larger than this critical value, the degenerate band of spin waves in Fig. 1 must disappear and one is left with the uniform precession below the entire spin wave band.[7] One may inquire why the uniform precession is apparently privileged in this respect. The answer lies in the assumption that only the uniform precession has a significant amplitude. It can raise the spin wave frequencies without in turn being effected by them.

## NONDEGENERATE SPIN WAVE INSTABILITIES

Assuming the state of nondegeneracy has been reached, it is instructive to ascertain which spin waves

---

[6] Since it will turn out that the amount of modulation required is relatively small there is no serious spectral decomposition of the uniform precession and it is not necessary to employ the more complicated procedure of modulating $\varphi_k$ without modulating $\varphi_0$.[3]

[7] This is true for volume spin waves extrapolated to zero $k$ value although the actual modes in the low $k$ region are magnetostatic. There are no magnetostatic modes below the extrapolated spin wave spectrum when $\theta = 0$ and it is reasonable to expect none when $\theta > 0$. This has not been definitely established as yet.

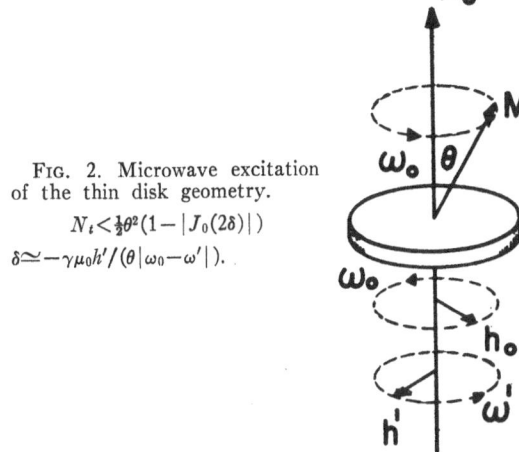

Fig. 2. Microwave excitation of the thin disk geometry.
$$N_t < \tfrac{1}{2}\theta^2(1 - |J_0(2\delta)|)$$
$$\delta \simeq -\gamma\mu_0 h'/(\theta|\omega_0 - \omega'|).$$

will finally limit $\theta$. Restrictions on space do not permit a discussion of this problem to be given here but one finds the following results. The modulating field needed to destroy the "parasitic" degenerate spin waves creates potentially unstable spin waves whose frequencies are $\omega_k = \omega_0 + (n/2)\omega_m$ $n = 1, 2, \cdots$. Independent of the modulating field, there exists third-order coupled spin waves whose frequencies are given by $\omega_k = \tfrac{3}{2}\omega_0$.

For a stable level of the large amplitude resonance to be maintained the critical value of $\theta$ given by Eq. (5) must be reached *before* either of the above mentioned instabilities occurs—this is possible if $N_t$ is made small enough. For YIG, the aspect ratio of the disk would have to be at least several hundred to one. The resonance cone angle could then increase nearly an order of magnitude before saturation effects took place.

JOURNAL OF APPLIED PHYSICS    SUPPLEMENT TO VOL. 33, NO. 3    MARCH, 1962

# Ferromagnetic Resonance Loss in Lithium Ferrite as a Function of Temperature

R. T. DENTON AND E. G. SPENCER

*Bell Telephone Laboratories, Incorporated, Murray Hill, New Jersey*

The ferromagnetic resonance linewidth as measured in ferrites is often dominated by loss due to ferrous ions. However, theoretical investigations of the scattering mechanisms in disordered magnetic materials such as ferrites have shown that volumetric disorder scattering should also play a significant role in determining the total linewidth of these materials. Lithium ferrite, $Li_{0.5}Fe_{2.5}O_4$, is one of the ferrites which in stoichiometric proportions would have all of the iron ions in the ferric state and which can have either an ordered or a disordered cystal structure at room temperature. It should thus be possible to reduce the contribution of ferrous ions so that the effects of disorder scattering can be observed by comparing the linewidths of ordered and disordered crystals. The data of this paper were taken on samples of lithium ferrite in which the linewidths are of the order of oersteds so that parallel pumping techniques can be used. The advantage of these measurements is that the role of surface scattering mechanisms is minimized and the resulting linewidth is the intrinsic linewidth for long-wavelength spin waves. Data are presented of intrinsic linewidth at microwave frequencies as a function of temperature from 290°K through a low-temperature maximum at 25°K. The value of the linewidth maximum in the disordered crystal proves to be quite anisotropic. Ordering the crystal increases the low-temperature maximum and alters the linewidth anisotropy. The effects of disorder scattering can be obtained from the linewidths above the low-temperature maximum and are shown to agree qualitatively with the predictions of theory.

T HEORETICAL investigations[1,2] of the scattering mechanisms in disordered magnetic materials such as ferrites have shown that disorder scattering should play a significant role in determining the total linewidth of these materials. Lithium ferrite, $Li_{0.5}Fe_{2.5}O_4$, is a magnetic material in which the effects of disorder scattering can be separated from other scattering mechanisms since it can exist in either an ordered or a disordered state.[3] The normal crystal structure of lithium ferrite at room temperature has an ordered arrangement of three ferric ions and one lithium ion along rows of octahedral sites extending in the [110] direction. However, the lithium ferrite structure becomes disordered above 750°C and will remain in the disordered state at room temperature if the crystal is quenched from above the transition temperature. In earlier measurements of the linewidth of ordered and disordered lithium ferrite[4] the contribution of disorder scattering was obscured by surface scattering due to imperfections and strain. Results are presented in this paper of measurements of the linewidth in ordered and disordered single crystals of lithium ferrite which show clearly the effects of disorder scattering. These measurements were made using the parallel pumping technique.[5] The linewidths obtained in this way are characteristic of low wave number spin waves or, alternatively, high order magnetostatic modes and are designated by the symbol $\Delta H_{k\rightarrow0}$.[6] These linewidths are not sensitive to surface scattering mechanisms but are affected by volumetric scattering.

Measurements were made of the linewidth of a polished single-crystal sphere of lithium ferrite[7] with the dc field along each of the principal crystallographic directions (1) as obtained from the melt, (2) after ordering, and (3) after disordering. Examination of an x-ray powder photograph of material from the same melt as the single crystal shows that only a small amount of long-range order exists before any ordering treatment. Ordering of the sample was obtained by heating in air to 700°C and cooling slowly to room temperature. The strong long-range ordering showed up clearly on a powder photograph. The crystal was disordered by quenching in a jet of cold air from 900°C. Figure 1 shows the variation of linewidth with temperature for the partially ordered sphere at a frequency of 10 880 Mc. The data are plotted for the dc field parallel to each of the principal crystallographic directions and it may be noted that the linewidth of 0.274 oe at 4.2°K with the dc field along a [100] axis is the lowest reported for

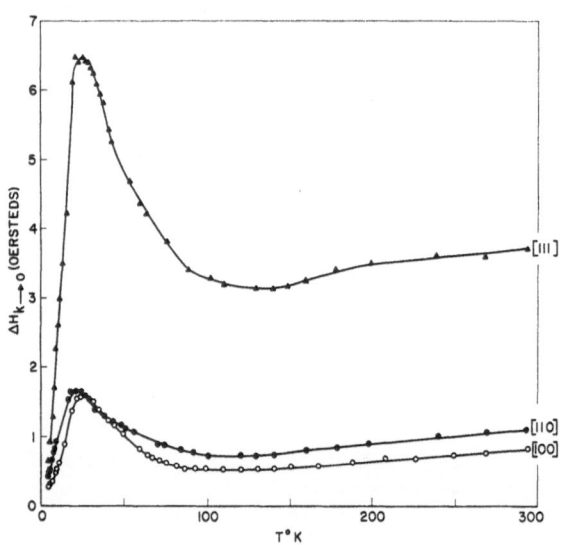

FIG. 1. Linewidth vs temperature for partially ordered sample of lithium ferrite.

[1] A. M. Clogston, H. Suhl, L. R. Walker, and P. W. Anderson, J. Phys. Chem. Solids **1**, 129 (1956).
[2] H. B. Callen and E. Pittelli, Phys. Rev. **119**, 1523 (1960).
[3] P. B. Braun, Nature **170**, 1123 (1952).

[4] A. D. Schnitzler, V. J. Folen, and G. T. Rado, J. Appl. Phys. **31**, 348S (1960).
[5] E. Schlomann, J. J. Green, and U. Milano, J. Appl. Phys. **31**, 386S (1960); also, F. R. Morgenthaler, J. Appl. Phys. 31, 95S (1960).
[6] E. G. Spencer, R. C. LeCraw, and R. C. Linares, Jr., Phys. Rev. **123**, 1937 (1961).
[7] The polished sphere of single-crystal lithium ferrite was obtained from the Airtron Division of Litton Industries.

ferrites. Figure 2 shows the variation of linewidth with temperature at 10 880 Mc for the same sphere after ordering. Figure 3 shows the linewidth as a function of temperature, again at 10 880 Mc, for the disordered crystal.

The data show a strong dependence of linewidth on ordering and exhibit several prominent characteristics:

(1) A low-temperature relaxation peak occurring at a temperature which is very nearly independent of crystal orientation. The amplitude of this peak is dependent on crystal orientation and is dramatically enchanced by ordering.

(2) A linewidth in the temperature region above the peak which in the disordered crystal is highly anisotropic and has only a small temperature-dependent component.

The low-temperature linewidth maximum for the disordered sample has an anisotropy which is similar to the anisotropy of the low-temperature maxima observed in nickel-ferrite[8] and in rare-earth doped yttrium iron garnet.[9] It is likely that in the data given here the low-temperature peak is caused by the presence of ferrous ions. Considerations which led us to this conclusion are the fact that magnetite is soluble in lithium ferrite in all proportions[10] and also that the resistivity of the sample used is 60 to 180 ohm cm. This would indicate the possibility of a sizeable ferrous ion content. A detailed interpretation of the linewidth anisotropy and the strong effects of ordering—in the region of the low-temperature maximum—is beyond the scope of this paper.

The characteristics of the linewidth at temperatures above the low-temperature maximum agree qualitatively with the detailed theory of disorder scattering developed by Callen and Pittelli. They calculate a line-

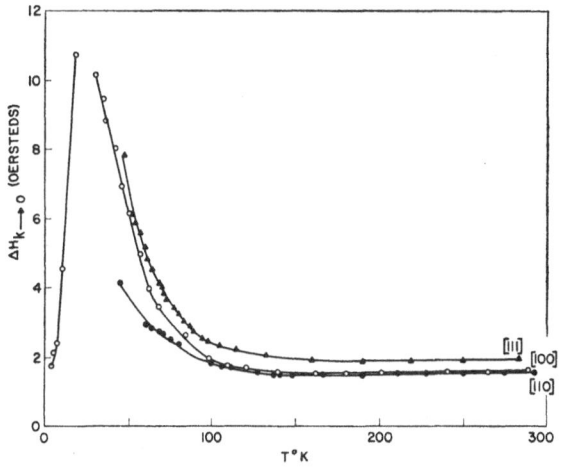

FIG. 2. Linewidth vs temperature for ordered sample of lithium ferrite.

[8] W. A. Yager, J. K. Galt, and F. R. Merritt, Phys. Rev. 26, 1203 (1955).
[9] J. F. Dillon, Preprints from International Conference on Magnetism and Crystallography 2, Part 1 (147-1), 1961.
[10] Structure Reports for 1950 13, 245 (1950).

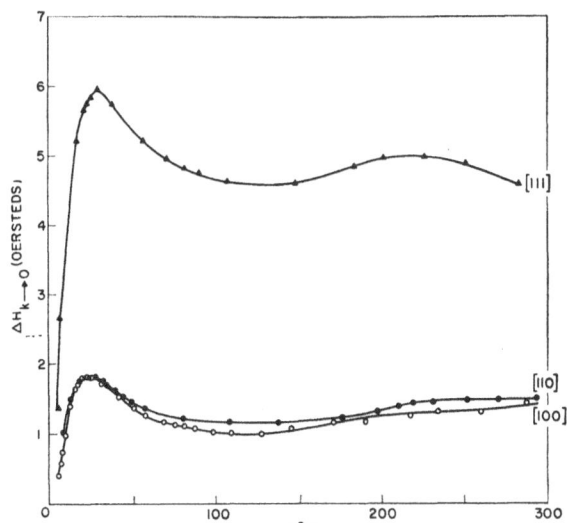

FIG. 3. Linewidth vs temperature for disordered sample of lithium ferrite.

width due to disorder scattering which is a maximum with the dc field in the [111] direction, zero with the dc field in the [100] direction, and which should be essentially independent of temperature. It may be observed from Figs. 1 and 3 that the linewidth of disordered lithium ferrite has the predicted anisotropy variation and only a weak temperature dependence. Figure 2 shows that after ordering this anisotropic contribution is greatly reduced and the linewidth at room temperature is determined mainly by a residual from the low-temperature maximum. The theoretical contribution of disorder scattering to the linewidth of lithium ferrite, as calculated by Callen and Pittelli, is five times larger than the measured value. However, Kittel[11] has suggested that the amount of disorder scattering may have been overestimated by the use of local fields appropriate to the $Fe^{2+}$ ion instead of the $Fe^{3+}$ ion in the derivation. Indeed, recent measurements[12] indicate that the use of the correct crystalline fields would result in a theoretical estimate perhaps several orders of magnitude lower than the measured value. Thus, there would appear to be quantitative disagreement between theory and experiment.

We are indebted to S. Geller for discussions on and interpretations of the ordering and its possible consequences. We are also indebted to L. R. Walker and J. K. Galt for discussions on the interpretation of the linewidth data. The competent and enthusiastic assistance of W. B. Snow in the microwave measurements is gratefully acknowledged. We would like to thank J. W. Nielsen of Airtron for providing the powder samples, information on the resistivity of the lithium ferrite, and a number of helpful discussions on the characteristics of this material.

[11] C. Kittel, Preprints from International Conference on Magnetism and Crystallography 2, Part 1, 161-3 (1961).
[12] V. J. Folen, J. Appl. Phys. 33, 1084 (1962), this issue.

JOURNAL OF APPLIED PHYSICS    SUPPLEMENT TO VOL. 33, NO. 3    MARCH, 1962

# Antiferromagnetic Resonance in MnTiO₃

J. J. STICKLER AND G. S. HELLER

*Lincoln Laboratory,\* Massachusetts Institute of Technology, Lexington 73, Massachusetts*

The effective internal field $(2H_E H_A)^{\frac{1}{2}}$ as well as its temperature dependence has been found from antiferromagnetic resonance experiments for two different single crystals of MnTiO₃, and from absorption edge data for a pure powder sample. The data are consistent with a uniaxial collinear two sublattice model as predicted from dc susceptibility and neutron diffraction measurements. The data for one crystal yielded results in good agreement with the powder material with a $g$ value of 2.1 and effective internal field of 52 kgauss at 4.2°K. All samples followed a Brillouin function for spin $\frac{5}{2}$ fairly well but with different Néel temperatures higher than that previously reported for this material.

FROM dc susceptibility measurements,[1] MnTiO₃, one of the class of materials with the ilmenite structure, has been shown to be antiferromagnetic below 41°K. Neutron diffraction studies[2] further showed this crystal to be uniaxial with spin directions along the $c$ axis. Based on a simple two sublattice model with the applied magnetic field $H$ along the $c$ axis, antiferromagnetic resonance should occur for frequencies $\omega$ such that[3]

$$\omega^{\pm}/\gamma = \left[ 2H_E H_A + \frac{H^2 \alpha^2}{4} \right]^{\frac{1}{2}} \pm H(1 - \alpha/2), \qquad (1)$$

where $\gamma$ is the gyromagnetic ratio, $H_A$ the effective uniaxial anisotropy field, $H_E$ the Weiss exchange field, and $\alpha$ the ratio of the parallel to perpendicular susceptibilities. As usual it has also been assumed that $H_A \ll H_E$. For temperatures low compared to the Néel temperature, $\alpha \ll 1$, and Eq. (1) assumes the simpler form

$$\omega^{\pm}/\gamma = (2H_E H_A)^{\frac{1}{2}} \pm H. \qquad (2)$$

Measurements of resonant field as a function of both frequency and temperature have been made utilizing both single crystal and powdered samples of MnTiO₃. In the case of the powdered sample, an absorption edge

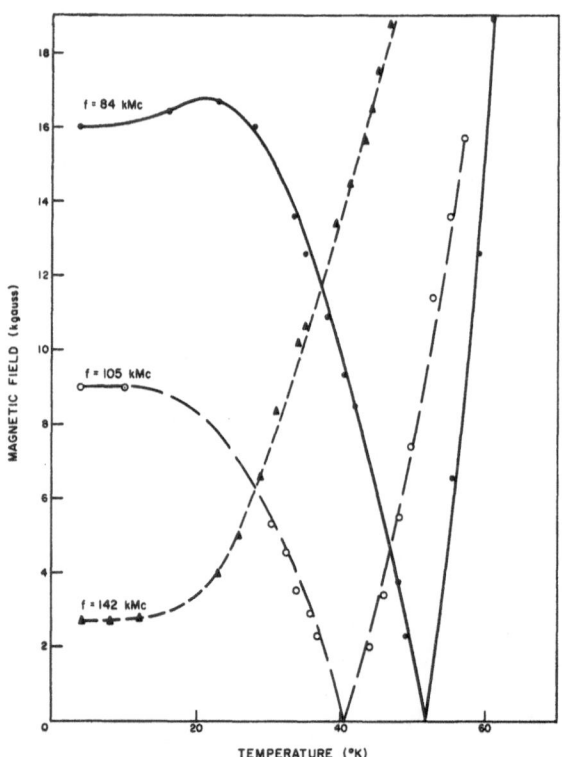

FIG. 2. Variation of resonant magnetic field with temperature for the single crystal $C_1$ of MnTiO₃ for several different frequencies.

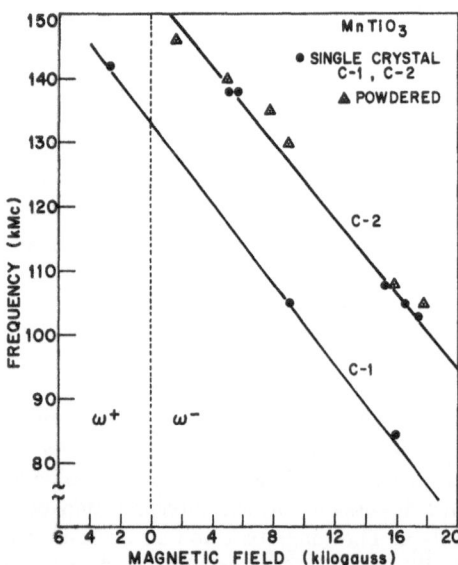

FIG. 1. Frequency versus resonant magnetic field for single crystals, $C_1$ and $C_2$, and powdered MnTiO₃ at 4.2°K.

is observed which can be related to the resonant field for a single crystal.[4] Two different single crystals, $C_1$ and $C_2$, were used for these measurements. $C_1$ was grown from the melt while $C_2$ was grown by the Bridgeman technique. Resonant field vs frequency at 4.2°K for these crystals as well as the powder data appear in Fig. 1.

\* Operated with support from the U. S. Army, Navy, and Air Force.

[1] Y. Ishikawa, J. Phys. Soc. Japan **13**, 1298 (1958).

[2] G. Shirane and S. J. Pickart, J. Phys. Soc. Japan **14**, 1352 (1959).

[3] See, e.g., C. Kittel, Phys. Rev. **82**, 565 (1951).

[4] G. S. Heller and J. J. Stickler, Bull. Am. Phys. Soc. **7**, 54 (1962).

TABLE I. Table of $g$ and $(2H_EH_A)^{\frac{1}{2}}$ values measured for single crystal and powdered MnTiO₃ at 4.2°K.

| Sample | | $g$ | $(2H_EH_A)^{\frac{1}{2}}$ (kgauss) |
|---|---|---|---|
| Single crystal | C-1 | 2.3 | 41 |
| | C-2 | 2.1 | 52 |
| Powder | | 2.1 | 52 |

A summary of the $g$ values and values of the effective internal field $(2H_EH_A)^{\frac{1}{2}}$ computed using Eq. (2) is given in Table I.

The single crystals were rather impure as well as inhomogeneous, while the powder sample was relatively pure. It is interesting to note the rather good agreement between the pure powder and the single crystal sample $C_2$. The line widths were of the order of 5 kgauss at 4.2°K.

Resonant field vs temperature data for the single crystal $C_1$ are shown in Fig. 2. The temperature behavior is similar to that observed in other uniaxial antiferro-

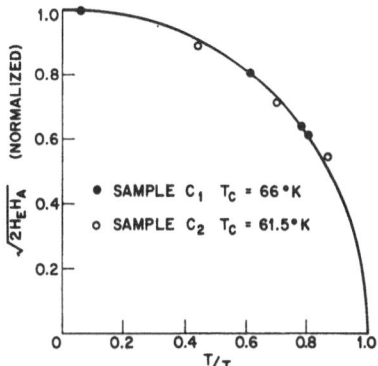

FIG. 4. Normalized values of $(2H_EH_A)^{\frac{1}{2}}$ as a function of normalized temperature $(T/T_c)$ for crystals $C_1$ and $C_2$. The solid line represents the theoretical Brillouin function for $S=\frac{5}{2}$.

magnetics such as Cr₂O₃.[5] For frequencies less than the critical frequency $\omega/\gamma = (2H_EH_A)^{\frac{1}{2}}$ the resonant field passes through zero as the temperature is increased. This is associated with the decrease in $(2H_EH_A)^{\frac{1}{2}}$ with temperature and the corresponding passage from the $\omega^-$ to the $\omega^+$ branch of Eq. (1). For frequencies higher than the critical frequency, only the $\omega^+$ branch can be observed and no zero field resonance occurs.

Resonant field vs temperature data for both the powdered sample and the single crystal sample are shown in Fig. 3. Within experimental error, the agreement between the powder and crystal samples is rather good. When resonance occurs at zero field then $\omega/\gamma = (2H_EH_A)^{\frac{1}{2}}$ as seen from Eq. (1). Values of $(2H_EH_A)^{\frac{1}{2}}$ as a function temperature normalized to the value at 4.2°K are compared with the Brillouin function for $S=\frac{5}{2}$ in Fig. 4. Values of Néel temperature chosen so as to give the best fit were 66°K for $C_1$ and 61.5°K for $C_2$, both Néel temperatures being higher than the previously reported 41°K from dc susceptibility measurements.[1]

Crystal $C_1$ was obtained from Dr. Arthur Linz, crystal $C_2$ was grown by Thomas Reed and Robert Fahey. The powder sample was prepared by Dr. Aaron Wold. The authors gratefully acknowledge the assistance of Mr. Richard Koch.

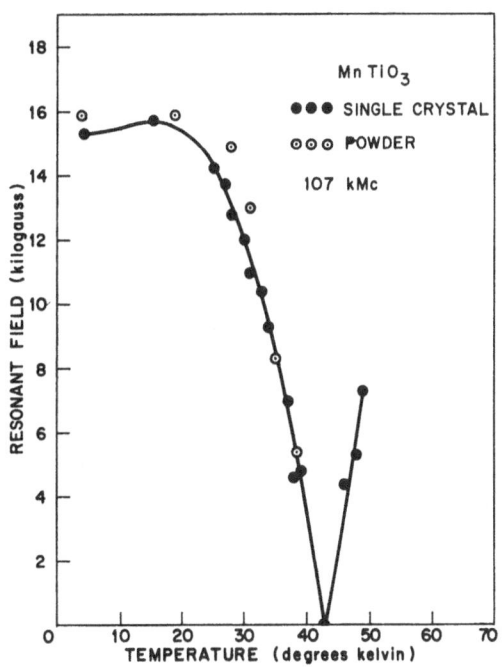

FIG. 3. Resonant magnetic field as a function of temperature for single crystal $C_2$ and powdered MnTiO₃ at 107 kMc.

[5] G. S. Heller, J. J. Stickler, and J. B. Thaxter, J. Appl. Phys. 32, 307S–312S (1961).

# Antiferromagnetic and Paramagnetic Resonance in $CuF_2 \cdot 2H_2O$

M. Peter and T. Moriya

*Bell Telephone Laboratories, Inc., Murray Hill, New Jersey*

Antiferromagnetic and paramagnetic resonance absorption has been measured in $CuF_2 \cdot 2H_2O$. At 1.4°K, the antiferromagnetic resonances in zero field occur at 95.84±0.05 kMc and 95.91±0.05 kMc. The easy direction of magnetization occurs at an angle of 3.5° with the $c$ axis. This orientation is in agreement with neutron diffraction and single-crystal susceptibility measurements. Due to strong exchange narrowing, narrow paramagnetic resonance lines were observed. The $g$ values are $g_\xi=2.07$, $g_\eta=2.08$, and $g_\zeta=2.42$. The principal axes are nearly coinciding with the ligand directions. From the $g$ values, the energy levels and wave functions of the $Cu^{++}$ ions are determined. A discussion of the anisotropy is given. The dipolar part is calculated numerically, and the anisotropic exchange of pseudodipolar type is determined, using the fact that $\omega_1 \sim \omega_2$ and the known easy magnetization direction. From the total anisotropy determined in this way, and the perpendicular susceptibility measured by Williams, the antiferromagnetic resonance frequency is predicted to occur at 130 kMc. The agreement with experiment is not unreasonable in view of the reduction of this frequency by the zero-point oscillations of the spin waves. A brief discussion of the superexchange interaction in this crystal is given.

THE crystallographic and magnetic properties of $CuF_2 \cdot 2H_2O$ have recently been investigated by several workers. Geller and Bond[1] have determined the crystal structure. The monoclinic crystal (space group $C_{2h}^3 - I2/m$) is essentially built up from layers parallel to the (101) plane. Figure 1 gives a projection along the $b$ axis. The unit cell contains two formula units with two $Cu^{2+}$ ions, $Cu_I$ and $Cu_{II}$, at the corner and body center position. Antiferromagnetism was indicated by the susceptibility measurements of Bozorth and Nielsen.[2] NMR studies[3] as well as specific heat measurements[4] placed the Néel temperature at 11°K. The magnetic unit cell determined by neutron diffraction[5] coincides with the chemical unit cell.

We made paramagnetic and antiferromagnetic resonance measurements on single crystals originally grown

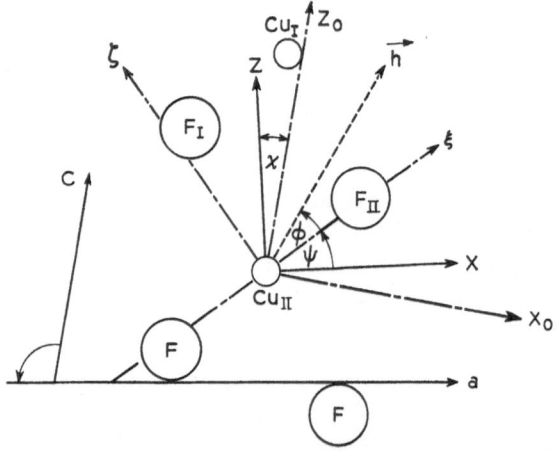

FIG. 1. Surrounding of Cu in (010) plane where $a=6.37$ A, $g=7.42$ A, $c=3.24$ A, $\beta=101°$ 15', $\psi=32°$, $\chi=14°$ 45'; $z_0$ is the easy direction of magnetization; $\xi$, $\eta$, $\zeta$ are the main axis of the $g$ tensor; and $x$, $y$, $z$ is an auxiliary system with $x\|a$. The angle $(F_I Cu_{II} F_{II})$ is 84°. $\mathbf{h}$ is the fictious field (see text).

by Nielsen. The $g$ values for $T=22°K$ were $g_\zeta=2.42$, $g_\xi=2.07$, and $g_\eta=2.08$. The directions of the main axes, $\xi$, $\eta$, $\zeta$, are indicated in Fig. 1. These axes coincide rather closely with the six ligands octahedrally surrounding the $Cu^{++}$ ion. From this and the fact that $g_\zeta > g_\xi \sim g_\eta$, we conclude that the ground state is the $e_g$ orbital of symmetry $\xi^2-\eta^2$.

The splitting of the excited $t_{2g}$ state by the axial perturbation can be estimated to be[6]:

$$E_1 - E_0 = -8\lambda/g_\zeta - 2 = 16\ 000 \text{ cm}^{-1}$$

and

$$E_2 - E_0 = E_3 - E_0 = -2\lambda/g_\xi - 2 = 21\ 000 \text{ cm}^{-1}.$$

Here, $E_1$ is the energy at the orbital $\xi\eta$, $E_2$ belongs to $\eta\zeta$, and $E_3$ to $\xi\zeta$. The antiferromagnetic resonance measurements were taken at 4.2° and 1.4°K. The resonance lines were about 30 gauss wide and showed multiple peaks similar to the ones observed by Johnson and Nethercot[7] in $MnF_2$. Since the crystal symmetry is monoclinic, two zero-field resonances $\omega_1$ and $\omega_2$ could be expected. These resonances were observed by interpolation from the resonances for finite field and by direct observation at zero field. The two resonances were found to coincide almost: $\omega_1=95.84\pm0.05$ kMc and $\omega_2=95.91\pm0.05$ kMc. These frequencies will be close to the frequencies at $T=0°$ since $\omega(T)$ is proportional to the sublattice magnetization at low temperature,[8] and the magnetization in turn depends very little on $T$ for $T < \hbar\omega/k$. At 4.2° the zero field resonances were found at $\omega_1 \sim \omega_2 = 93.5\pm0.2$ kMc.

The dependence of the resonance on the direction of the magnetic field could be closely fitted to the expressions given by Ubbink.[9] The anisotropy of the paramagnetic $g$ values was taken into account by transforming the observed field $\mathbf{H}$ into a fictitious field $\mathbf{h}=(g\mathbf{H})$.

[1] S. Geller and W. L. Bond, J. Chem. Phys. **29**, 925 (1958).
[2] R. M. Bozorth and J. W. Nielsen, Phys. Rev. **110**, 879 (1958).
[3] R. G. Shulman and B. J. Wyluda (to be published).
[4] F. Morin (private communications to R. G. Shulman).
[5] S. Abrahams (to be published).

[6] B. Bleaney and K. W. H. Stevens, Repts. Progr. Phys. **16**, 108, The Physical Society, London (1953).
[7] F. M. Johnson and A. H. Nethercot, Jr., Phys. Rev. **114**, 705 (1959).
[8] T. Oguchi and A. Honma, J. Phys. Soc. Japan **16**, 79 (1961).
[9] J. Ubbink, Physica **19**, 919 (1953).

The direction of minimum **h** gives the easy direction of magnetization. This direction is displaced by 3.5° from the $c$ axis. Both the low-temperature neutron diffraction study of Abrahams[5] and single-crystal susceptibility measurements by Sherwood and Williams[10] have indicated the same alignment. The latter measurements gave, for $T=1.5°$, $\chi_{||}\sim 0$ and $\chi_\perp=5.2\times10^{-3}$ per mole. The values of $\chi_{||}$ and $\chi_\perp$ are in agreement with the powder value given by Bozorth. The anisotropy constant is given by

$$K=[(\hbar\omega^2\chi/Ng^2\mu B^2]=0.042 \text{ cm}^{-1}.$$

Since the Cu⁺⁺ ion has a spin $\frac{1}{2}$, the anisotropy energy in CuF₂·2H₂O is composed of only two parts, namely the magnetic dipole-dipole interaction and the anisotropic exchange interaction. We neglect the antisymmetric part[11] of the latter since it has no influence on the antiferromagnetic spectrum of CuF₂·2H₂O on account of the inversion symmetry. The total energy is expressed as

$$E_{anis}=K_1\alpha^2+K_3\gamma^2+2K'\alpha\gamma,$$

where $\alpha$, $\beta$, and $\gamma$ are the directions of the antiferromagnetic ordering with respect to the $x$, $y$, $z$ axes. The contribution due to the magnetic dipole-dipole interaction was obtained using the experimental $g$ values and a computer calculation carried out and supplied to us by R. G. Shulman. Purely theoretical calculation of the various superexchange coupling constants is very difficult. We would rather calculate the anisotropic exchange coupling from the experimental results on the total anisotropy energy and the theoretical dipolar anisotropy. For this purpose, the following information is sufficient: (1) the easy direction of the antiferromagnetic ordering, denoted by $z_0$, is in the $xz$ plane, 14° 45′ from the $z$ axis, and (2) the anisotropy in the antiferromagnetic state has axial symmetry ($\omega_1\sim\omega_2$). In the $x_0$, $y$, $z_0$ coordinate system, the coefficient $K'$ vanishes. Since there is axial symmetry, in this system $K_1$ must also vanish. From these two conditions, we find the coefficient of the anisotropic exchange interaction for its main axes $\xi$ and $\zeta$: $K_{p\xi}=0.0085$ cm⁻¹ and $K_{p\zeta}=0.0278$ cm⁻¹. This determines in turn the total anisotropy energy: $E_{tot}=N/2(-0.096\gamma_0^2)$ cm⁻¹, and the antiferromagnetic resonance frequency is predicted to occur at 130 kMc. The agreement with the experimental value, $\omega=96$ kMc, is fair if we take into account that the antiferromagnetic resonance frequency is reduced by the zero-point motion of the sublattices. Since $\omega$ may be proportional to the sublattice magnetization, a 20% reduction of the latter from the classical value is needed to explain the observed discrepancy.

Finally, let us briefly discuss the superexchange interaction in this crystal. As the ground orbital wave function spreads in the (101) plane, the superexchange coupling between two neighboring Cu²⁺ ions in this plane is considered to be most important and of antiferromagnetic sign. Neutron diffraction experiments[5] have shown that the spins in the (101) plane form an antiferromagnetic network in agreement with our expectation. The exchange interaction between two spins neighboring in the (001) direction seems to be ferromagnetic since they are shown to be parallel below the Néel temperature.[5] This may be understood from a symmetry argument on the two Cu²⁺ ions and the intervening F⁻ ions. As is seen in Fig. 1, the angle (Cu F Cu) is nearly 90° so that the ground-state wave function of Cu$_{II}$ is nearly orthogonal to all the $s$ and $p$ wave functions of F$_I$. This means the transfer integral between the localized states around Cu$_I$ and Cu$_{II}$ is small, and therefore the antiferromagnetic superexchange contribution is small.[12] It is likely that the ferromagnetic direct exchange contribution dominates the total exchange interaction between these two Cu²⁺ ions. As for the magnitudes of these couplings, we expect the antiferromagnetic coupling in the (101) plane to be larger than the ferromagnetic coupling in the (001) direction.[13] These properties of the exchange interactions may provide a possible explanation for the maximum of the paramagnetic susceptibility at 26°K, which is more than twice as high as the Néel temperature. The Néel temperature may be determined by the interlayer coupling, while the depression of the susceptibility due to the short range order in the (101) plane may take place at much higher temperatures than the Néel temperature.

## ACKNOWLEDGMENTS

The authors wish to thank Miss B. Cetlin for computational help, C. F. Hempstead for loan of a 3-mm BWO, J. B. Mock for experimental assistance, R. C. Sherwood and H. J. Williams for communicating their susceptibility measurements, and R. G. Shulman for performing the dipole sum calculation.

[10] R. C. Sherwood and H. J. Williams (private communication).
[11] T. Moriya, Phys. Rev. **120**, 91 (1960).
[12] T. Moriya and K. Yosida, Progr. Theoret. Phys. **9**, 663 (1953); T. Nagamiya, K. Yosida, and R. Kubo, *Advances in Physics* (Taylor and Francis, Ltd., London, 1955), Vol. 4, p. 1.
[13] P. W. Anderson, Phys. Rev. **115**, 2 (1959).

JOURNAL OF APPLIED PHYSICS     SUPPLEMENT TO VOL. 33, NO. 3     MARCH, 1962

# dc Effects in Ferromagnetic Resonance in Thin Ferrite Films*

W. HEINZ AND L. SILBER

*Polytechnic Institute of Brooklyn, Brooklyn 1, New York*

The dc voltage which accompanies ferromagnetic resonance has been observed in thin films of magnetite. The dependence of this voltage on microwave power and static magnetic field is in qualitative agreement with theory, but larger by a factor of 3.

THIS paper reports some preliminary results on the observation of the dc voltage which accompanies ferromagnetic resonance in a ferrite. Under the conditions commonly employed to study ferromagnetic resonance in magnetic materials, a dc voltage is generated in the sample. This voltage is generated by the interaction of the precessing magnetization with the current induced in the sample by the microwave electric field, through the Hall effect and magnetoresistance anisotropy. This dc voltage has been observed in thin metallic films.[1] We report here its observation in films of a ferrite, $Fe_3O_4$. The conduction current in a magnetic material is related to the applied field by the equation[2]

$$\mathbf{J}=\sigma\mathbf{E}-(\Delta\rho/\rho M^2)(\mathbf{J}\cdot\mathbf{M})\mathbf{M}+R\sigma\mathbf{J}\times\mathbf{M}, \quad (1)$$

where $\Delta\rho$, the magnetoresistance anisotropy $\rho_{||}-\rho_\perp$ is the difference between the resistivities measured parallel to and perpendicular to the magnetization, and $R$ is the extraordinary Hall coefficient. Since $\mathbf{J}$ is a microwave quantity, and $\mathbf{M}$ has microwave components, the products indicated in Eq. (1) give rise to dc components, as well as those of higher frequency.

The simplest geometry to analyze is that of a thin film magnetized in its plane, with a plane electromagnetic wave normally incident on it, polarized with the electric field along the direction of the static magnetizing field. In order to define more closely the fields in the sample, Juretschke[3] has considered the geometry described above, in which the film is backed by a perfect short, and has calculated the dc voltage developed. For a metallic film of width $2a$, thickness $\Delta$, situated a distance $D$ above the short, he finds

$$V=\frac{ac}{4\pi\kappa_1\delta^2}H_1^2\frac{\sin2\kappa_1D[1+\tan^2\phi_4]^{\frac{1}{2}}}{\left[(\omega_0^2-\omega^2)^2+\frac{4\omega^2}{\tau^2}\right]^{\frac{1}{2}}}$$
$$\times\left[-\gamma^2 B\Delta\rho\sin(\phi_2+\phi_4)-\gamma MR\left(\omega^2+\frac{1}{\tau^2}\right)^{\frac{1}{2}}\right.$$
$$\left.\times\sin(\phi_2+\phi_4-\phi_1)\right], \quad (2)$$

where

$$\tan\phi_1=-\omega\tau, \quad \tan\phi_2=-\frac{2\omega/\tau}{\omega_0^2-\omega^2}, \quad \tan\phi_4=\frac{\Delta}{\kappa_1\delta^2}\tan\kappa_1D,$$

$\tau$ is the ferromagnetic relaxation time, $\kappa_1$ is the wave number in the medium between film and short, $H_1$ the magnetic field amplitude in this region, $\delta$ the skin depth in the film. The above result applies to a metal, for which the displacement current is negligible compared to the conduction current. In ferrites this approximation is valid only in magnetite and other ferrites containing appreciable amounts of ferrous ion. Making the necessary modification of Eq. (2), the voltage may be expressed in the following way, which makes more apparent the dependence on material properties:

$$V=\frac{a\sigma_2 H_1^2\sin2\kappa_1D}{2(\epsilon_1)^{\frac{1}{2}}}$$
$$\times\left[\frac{-\Delta\rho}{M}(\phi x'+\psi x'')+R(\phi\kappa''-\psi\kappa')\right], \quad (3)$$

where $x$ and $\kappa$ are the diagonal and off-diagonal elements of the magnetic susceptibility tensor,

$$\psi=1+\frac{\omega\epsilon_2\Delta}{2c(\epsilon_1)^{\frac{1}{2}}}\tan\kappa_1D; \quad \phi=\frac{2\pi\sigma_2}{c(\epsilon_1)^{\frac{1}{2}}}\tan\kappa_1D.$$

$\epsilon_2$ and $\sigma_2$ refer to the ferrite, $\epsilon_1$ and $\kappa_1$ to the region between ferrite and short.

For most values of $D$, $\phi\ll\psi$, so that the dc voltage is directly proportional to the conductivity of the film. Furthermore, under this assumption the magnetoresistance term will have the form of a resonant absorption curve as a function of static field, while the Hall effect term has the form of a dispersion curve. The maximum values of the permeability elements vary as $4\pi M_s/\Delta H$,

TABLE I. Properties of magnetite.

|  | Film | Bulk |
|---|---|---|
| Resistivity (ohm-cm) | $7.0\times10^{-1}$ | $3.96\times10^{-3}$ |
| Magnetoresistance anisotropy (ohm-cm) | $-3.5\times10^{-3}$ | ... |
| Extraordinary Hall coeff. $\left(\frac{\text{volt-cm}}{\text{amp-gauss}}\right)$ | $-1.5\times10^{-9}$ | $-3.28\times10^{-8}$ |
| Saturation magnetization (gauss) | 430 | 472 |

* The work reported in this paper was sponsored by the Air Force Office of Scientific Research, Office of Naval Research, U. S. Army Signal Research and Development Laboratory.
[1] W. Egan and H. J. Juretschke, Bull. Am. Phys. Soc. **3**, 194 (1958); P. E. Tannenwald and M. H. Seavey, Jr., J. phys. radium **20**, 323 (1959).
[2] J. P. Jan in *Solid State Physics* edited by F. Seitz and D. Turnbull (Academic Press Inc., New York, 1957), Vol. 5, Secs. II and V.
[3] H. J. Juretschke, J. Appl. Phys. **31**, 1401 (1960).

where $\Delta H$ is the ferromagnetic resonance linewidth, so narrow linewidth favors a large dc voltage. The only ferrites having appreciable conductivity, however, are those containing both di- and tri-valent *iron* ions, and the electrical conduction process provides a magnetic loss mechanism. Thus, one would expect the dc voltage to be much smaller in ferrites than in ferromagnetic metals.

The dc voltage discussed above was observed in two films of magnetite. These films were prepared by evaporating iron onto a ceramic substrate, oxidizing the metal film in air at 1000°C, and quenching in oil to preserve the $Fe_3O_4$ phase. Table I lists the properties of the films, and of bulk magnetite. Leads were attached either by using indium contacts, or air-drying silver paint. Direct-current voltage measurements indicated that the contacts were ohmic.

Because the dc voltage in ferrite films is so much smaller than in metals, the ferrite films were mounted in a cavity, rather than in a waveguide. The cavity used was a rectangular $TE_{104}$ cavity. The static magnetic field was perpendicular to the broad face of the cavity. With a ferrite film mounted near the point of maximum magnetic field, the unloaded $Q$ was 1000. The rf magnetic field was calculated to be 14.9 oe for an incident power of 1.0 kw. The measurements were performed at a frequency of 9318 Mc. The maximum power was limited to 3 kw by breakdown in the cavity. Fig. 1 shows the dc voltage developed in a magnetitte film of 2300 A thickness, as a function of static magnetic field. The incident power was 3.0 kw. $x''$, the imaginary part of the diagonal term of the susceptibility tensor, is shown for comparison. The maximum values are normalized to the same value to make the comparison of line shapes more apparent. The maximum dc voltage was 4.6 mv, and the maximum susceptibility was 1.7. It is apparent from the shape of the dc voltage curve that contributions from both magnetoresistive anisotropy (proportional to $x''$) and Hall effect (proportional to $\kappa'$) are present. It was not possible to measure $\kappa'$, and the shape of the $x''$ curve is so far from the theoretical shape that values of $\kappa'$ calculated from measured $x''$ values would not be very meaningful. For this reason, it was not possible to calculate the line shape of the dc voltage vs static field curve. A numerical comparison can be made at the field corresponding to ferromagnetic resonance because at this field $\kappa'$ should be zero. If one uses the dc values of Hall constant and magnetoresist-

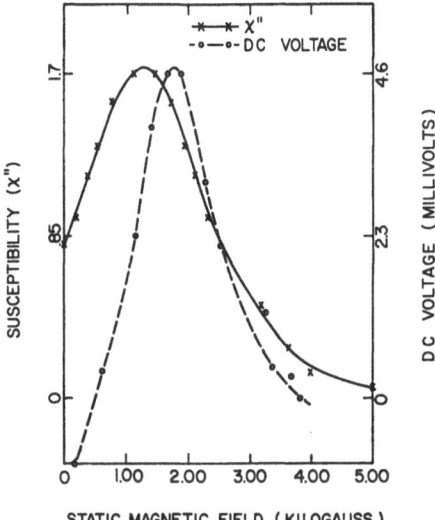

Fig. 1. dc voltage and susceptibility as function of static magnetic field.

ance anisotropy measured on the magnetite films, the Hall term should be smaller than the magnetoresistance term by a factor of several hundred. The curves of Fig. 1 indicate that the Hall term is not negligible, and is of the same order of magnitude as the magnetoresistance term.

The form of the dc signal vs static field curve is the same at all power levels, and the amplitude proportional to power over the range measured (1.0 to 3.0 kw peak). The signal reverses with reversal of the static field, as predicted by theory.

The amplitude of the dc signal at a field corresponding to ferromagnetic resonance is calculated to be 1.0 mv at a power level of 3.0 kw. The measured value was 2.8 mv. In view of the uncertainties in several of the factors in the theoretical expression, the agreement is good and perhaps fortuitous. In the calculation, the dc values of resistivity, Hall constant, and magnetoresistance are used. The microwave values may differ appreciably from these. We hope to extend the measurements of dc voltage by making measurements with films of a wide range of thickness, at various positions with respect to the end wall of the cavity. From these results it will be possible to determine microwave values of the resistivity, Hall coefficient, and magnetoresistance anisotropy.

# Permanent Magnets and Micromagnetics

## F. E. Luborsky, *Chairman*

### Failure of the Local-Field Concept for Hysteresis Calculations*

William Fuller Brown, Jr.
*Department of Electrical Engineering, University of Minnesota, Minneapolis 14, Minnesota*

The "local field" is the electric or magnetic field of external sources and of all molecules or particles except the one under examination, at the position of that one. The concept is useful in the molecular theory of dielectrics, where the basic relations are linear. It has also been applied, however, to magnetically interacting single-domain particles in a ferromagnetic powder; here the relations not only are nonlinear but involve discontinuous irreversible jumps, for which instability conditions are decisive. The local-field concept is then unreliable. In a two-particle system, for example, the local-field method examines the stability of one moment with respect to rotation in a fixed field due to the other moment and to external sources; this procedure imposes a constraint, for in the actual system the two moments may rotate simulatneously and independently. The incorrectness of the local-field method is demonstrated here by calculations for two- and three-particle chains of spheres and for two- and three-part composite cylinders. The concept of an interaction field that merely displaces the hysteresis loops of the particles without changing their widths, as in a common interpretation of the Preisach model, is thus erroneous.

THE "local-field" concept is useful in the molecular theory of dielectrics,[1] where the moment of one molecule is linearly related to the combined electric field intensity $\mathbf{E}^*$ of external sources and of the other molecules. The concept has also been applied to magnetically interacting single-domain particles in a ferromagnetic powder. Here the response of the moments to the fields is not linear or reversible, and the local-field concept can lead to false conclusions.

An illustration is a system of two identical single-domain spheres, with the applied field $H_0$ along their line of centers, and with no anisotropy other than that due to the magnetic interaction.[2] A large $H_0$ in direction $\rightarrow$ insures the alignment $\rightarrow\rightarrow$. As $H_0$ is decreased and reversed ($\leftarrow$), it ultimately reaches a critical value $H_c$ at which the initial state becomes unstable. A correct calculation tests the stability with respect to arbitrary independent rotations of the two moments ($\nearrow\nearrow$, $\nearrow\searrow$, $\nearrow\rightarrow$, etc.). If, on the other hand, the first particle is treated as if in a fixed field due to the external sources and the second particle, a test of its stability is a test of the stability of the system with respect to the restricted class of rotations $\nearrow\rightarrow$; a constraint has been imposed. This artificially stabilizes the system down to an $H_0$ algebraically smaller than the value at which instability of the type $\nearrow\searrow$ actually occurs. For if a rotation $\nearrow$ of particle 1 is accompanied by a rotation $\searrow$ of particle 2, the rotation $\searrow$ changes the field of 2 at position 1 in such a way as to help the rotation $\nearrow$.

Another calculable case is a circular cylinder of diameter $d$ and length $p^*d$, made up of $n$ identical single-domain subparticles, butted end to end, each of diameter $d$ and length $pd$, with $p=p^*/n$. Let subparticle $i$ be of volume $V_0$ and have magnetization $\mathbf{M}_i = M_{xi}\mathbf{i} + M_{yi}\mathbf{j} + M_{zi}\mathbf{k} = M_s(\alpha_i\mathbf{i} + \beta_i\mathbf{j} + \gamma_i\mathbf{k})$; $M_s$ is the spontaneous magnetization, $(\alpha_i, \beta_i, \gamma_i)$ are its direction cosines, and the cylinder axis is $0z$. The energy of the system is

$$W = \frac{1}{2}V_0 \sum_{i=1}^{n}\sum_{j=1}^{n} \left[ L_{ij}(M_{xi}M_{xj} + M_{yi}M_{yj}) + N_{ij}M_{zi}M_{zj} \right] - V_0 H_0 \sum_{i=1}^{n} M_{zi}; \quad (1)$$

$L_{ij}$ and $N_{ij}$ are constants. To test the stability of the state $\alpha_i = \beta_i = 0$, $\gamma_i = 1$, we express $W$ to the second order in $\alpha_i$ and $\beta_i$, with $\gamma_i = 1 - \frac{1}{2}(\alpha_i^2 + \beta_i^2)$:

$$w = W/(V_0 M_s^2) = \text{const} + \frac{1}{2}\sum_i \sum_j c_{ij}(\alpha_i\alpha_j + \beta_i\beta_j) + \frac{1}{2}h \sum_i (\alpha_i^2 + \beta_i^2), \quad (2)$$

with

$$c_{ij} = L_{ij} - \delta_{ij}\sum_k N_{ik}, \quad h = H_0/M. \quad (3)$$

Since there are no $\alpha\beta$ products, the rotations $\alpha$ and $\beta$ are independent; we consider the former ($\beta_i = 0$).

Since $w$ contains no terms linear in $\alpha_i$, the state $\alpha_i = 0$ is one of equilibrium. It is stable if $w > 0$ for arbitrary $\alpha_i$'s, i.e., if the quadratic form with matrix $(c_{ij} + h\delta_{ij})$ is positive definite. This condition requires that $h$ exceed the largest root of the determinantal equation

$$|c_{ij} + h\delta_{ij}| = 0. \quad (4)$$

Instability with respect to any specified set of deviations $(\alpha_1, \alpha_2, \cdots, \alpha_n)$ will occur at an $h$ that can be found by

---

* This work was assisted by a grant from the National Science Foundation.

[1] W. F. Brown, Jr., *Encyclopedia of Physics* (Springer-Verlag, Berlin, Germany, 1956), Vol. 17, pp. 47 ff.
[2] I. S. Jacobs and C. P. Bean, Phys. Rev. **100**, 1060–1067 (1955).

TABLE I. Magnitudes of reduced instability fields, $|h_c|$.

| | Chain of spheres | | Composite cylinder | |
| | $n=2$ | $n=3$ | $n=2$ | $n=3$ |
| --- | --- | --- | --- | --- |
| Rotation in unison | 1.571 | 2.225 | 5.02 | 5.02 |
| Local-field method | 1.047 | 1.178 | 4.63 | 3.97 |
| Rigorous method | 0.524 | 0.814 | 4.25 | 3.59 |
| Uncoupled particles | 0 | 0 | 3.87 | 2.86 |

setting $w=0$. The actual $h$ at first instability and the corresponding ratios $\alpha_1 : \alpha_2 : \cdots : \alpha_n$ can be found by maximizing this $h$ with respect to the $\alpha_i$'s under a normalization condition $\sum_i \alpha_i^2 = \text{const}$. The result is that the $\alpha_i$'s satisfy

$$\sum_i (c_{ij} + h_c \delta_{ij}) \alpha_j = 0 \quad (i = 1, 2, \cdots, n), \quad (5)$$

with $h_c$ the largest root of Eq. (4).

The $L_{ij}$'s and $N_{ij}$'s for the composite cylinder can be found from tables of the demagnetizing factors $L(p)$, $N(p)$ of a uniformly magnetized cylinder.[3] Clearly $L_{i,i+m-1} = L_{1m}$; and by comparison of different formulas for the energy of a uniformly magnetized composite cylinder,

$$L_{11} = L(p), \quad L_{12} = L(2p) - L(p), \quad (6)$$

$$L_{1m} = \tfrac{1}{2}\{mL(mp) - 2(m-1)L([m-1]p)$$
$$+ (m-2)L([m-2]p)\} \quad (m \geq 3); \quad (7)$$

similarly for the $N$'s.

Figure 1 shows $|h_c| = -h_c$ for the three-part cylinder as a function of $p^*(= 3p)$. Also shown are the values for rotation in unison over length $p^*d$ and for independent uncoupled cylinders of lengths $pd$. The fourth curve is calculated by the local-field method, with an end moment rotating in the fixed field of the others. The local-field model, $\nearrow\!\!\rightarrow\!\!\rightarrow$, is a good approximation not to the actual mode of deviation $\nearrow\searrow\nearrow$ but to the mode $\nearrow\!\!\rightarrow\searrow$ that satisfies Eqs. (5) with a larger $|h_c|$; at $p^*=6$, it gives 3.97, against 3.94 for $\nearrow\!\!\rightarrow\searrow$.

[3] W. F. Brown, Jr., Am. J. Phys. **28**, 542–551 (1960), Table I.

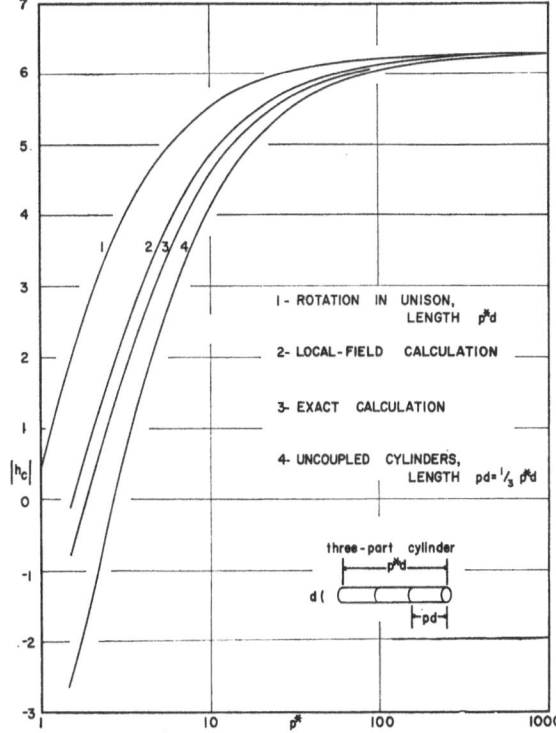

FIG. 1. Magnitude of reduced instability field, $|h_c| = -h_c = -H_c/M_s$, for three-part composite cylindrical particle of length-to-diameter ratio $p^* = 3p$.

Table I gives values for chains of spheres and for composite cylinders with $p^* = 6$.

The well-known Preisach model assumes magnetic units with rectangular hysteresis loops of widths $2a$ centered about field intensities $b$; $a$ and $b$ are random variables with constant distributions. The Preisach $b$ has sometimes been interpreted as the contribution of other particles to the field acting on a given particle.[4] We now see that the concept of a local field that merely shifts the hysteresis loop of a particle, without changing its width, is erroneous. When instability conditions are decisive, the local-field concept is unreliable.

[4] For references see D. F. Eldridge, J. Appl. Phys. **32**, 247S–249S (1961).

JOURNAL OF APPLIED PHYSICS    SUPPLEMENT TO VOL. 33, NO. 3    MARCH, 1962

# The Formation of Monocrystalline Alnico Magnets by Secondary Recrystalization Methods

E. STEINORT

*Centro-Magneti Permanenti, S. P. A., Milan, Italy*

AND

E. R. CRONK, S. J. GARVIN, AND H. TIDERMAN

*Thomas & Skinner Inc., Indianapolis, Indiana*

A brief review of the progress of the "Alnico Art" is given, tracing early attempts at increasing energy yield by changes in composition to present day techniques of promoting polycrystalline growth parallel to the magnetic axis. Energy product values have steadily risen to the present commercially available level of 7.5 (MGO). Since research has apparently reached a temporary impasse in further improvement in polycrystalline Alnico magnets, except by an uneconomic fabrication from a much larger initial casting or expensive foundry mold techniques an, attempt to take advantage of the 12.0 mega gauss-oe energy product available from single crystals of Alnico is an obvious solution. Two known methods of producing single crystals sufficiently large to be of practical value are described, and the obvious economic and production problems discussed. New work on a third method is described in which normal foundry techniques are used, but the formation of a single monocrystal of the entire magnet is promoted by secondary recrystallization. It is shown that if a sufficiently large grain edge strain can be induced in a polycrystalline aggregate, resulting in secondary recrystallization, formation of a single large crystal is possible. It is also shown that if a sufficiently large internal stress can be introduced by both mechanical and thermal means, the required grain-edge shift can be accomplished. The additions of normally prohibitive amounts of either carbon, nitrogen, manganese, or other "gamma-phase" promoting elements will serve to provide the necessary mechanical strain, and methods of controlling their effect on the magnetic results are shown. The thermal stresses are accomplished by a simple maintainance of 60–80°C gradient across the magnet poles. This process has yielded single crystals approaching $\frac{1}{4}$ lb in size whose energy product is 11.0 megagauss-oe. The practical nature of this process is discussed, and possible mass production techniques outlined.

THIS paper is based on the original work of Mr. Eberhard Steinort of Centro-Magneti Permanenti, Milan, Italy, and attempts to outline the basic concepts of his invention.

The materials presently known under the generic title of Alnico had their origin in the period 1931–1933 through studies by T. Mishima, K. Honda, and H. Matsumoto of the iron-nickel-aluminum, and iron-nickel-cobalt-titanium systems. Progress in developing magnets with progressively higher energy products was initially confined to changes in composition until heat-treating techniques to develop magnetic anisotropy were introduced in 1938. Even though very little attention was given to crystalline texture, energy product values of 5.0 MGO were common during World War 2.

Since the energy product values in directions perpendicular to the preferred magnetic axis fell as low as 0.6 MGO, it was felt for some time that the ultimate had been attained. However, R. E. Bernius and D. Ebling independently demonstrated that the properties could be further improved by obtaining preferred crystal orientation parallel to the desired magnetic axis. Thus, properties close to 8.0 MGO have been reported for castings with nearly 100% preferred crystal orientation.

It is clear now that a further step can be taken to improve the magnetic properties of an Alnico casting. This involves replacing the polycrystalline structure of the cast mass with a single crystal whose (100) axis corresponds to the preferred axis of magnetization. Measurements on individual large crystals cut from a larger polycrystalline mass indicate that energy products over 9.0 MGO are attainable.

Our studies have been concerned with a practical means of making monocrystalline Alnico magnets. The literature describes two methods of growing monocrystals of considerable size from a melt. The Bridgman method uses a crucible of a shape designed to provide a nucleus of crystallization with a chosen crystallographic orientation for the slow growth of a single crystal. This method can be applied to Alnico but can hardly be considered a practical manufacturing technique. The Czochralski method is based on dipping a previously prepared, properly oriented seed crystal into the melt and then lifting it slowly from the molten bath. This technique also can be used with Alnico type alloys but has, obviously, severe limitations for production purposes.

The literature also abounds with examples of recrystallization and growth induced in malleable metals such as pure iron. Here the polycrystalline metal is subjected to mechanical strains which induce critical internal directional stresses. By suitable heat treatment of the strained metal, the internal stresses are relieved through movements of the grain edges resulting in the phenomenon commonly described as recrystallization and growth. The factors that appear to influence the growth of large grains are:

(1) Internal mechanical strains causing large energy level differentials in the grain structure.

(2) A critical heat treating temperature.

(3) The presence of a marked thermal gradient during the early stages of heat treatment.

While mechanical straining of the Alnico alloys is impractical, it seemed to us that other means might be available for establishing the conditions necessary to promote the process of secondary recrystallization. The attainment of the critical heat treatment temperature presents no great difficulty and critical stresses can be induced metallurgically by the close proximity of two distinct crystalline textures with different lattice constants as well as by the effects of the differing thermal coefficients of expansion of the two phases. If a suitable thermal gradient can be induced as part of the annealing process, such a polycrystalline mass could be converted to a single crystal whose preferred (100) axis is parallel to the desired magnetic axis.

Thus, prompted by the random observation of the secondary recrystallization phenomenon in an Alnico casting, we have sought to establish the most favorable conditions for the formation of a monocrystal from a polycrystalline casting. The study sought to examine:

(a) The phases of the Alnico alloy system and their effect on grain growth.

(b) The most suitable composition for secondary recrystallization.

(c) The most suitable temperature range, time, and thermal gradient for recrystallization.

### A. THE PHASES OF THE ALNICO ALLOY SYSTEM

In the diagram of Fig. 1, the vertical dotted line represents an alloy of the Alnico 5 family. This alloy has a pure alpha-structure both above 1200°C as well as in the range 900–930°C. Since it is known that for good magnetic properties, gamma-phase precipitation must be avoided, heat treatment between 930° and 1180°C cannot be used unless this region is traversed very quickly. The deleterious effects of the presence of gamma phase on the magnetic properties are so great that a 7% content of gamma phase will lower energy values by about 25%. Preferred magnetic orientation is the result of the directional precipitation below 900°C of submicroscopic particles of alpha prime, a second alpha phase. Since the presence of even minor quantities of gamma-phase precipitates obstructs the proper alpha-prime formation and hence obstructs magnetic orientation, it is reasonable to conclude that these gamma precipitates can also cause major lattice distortions with

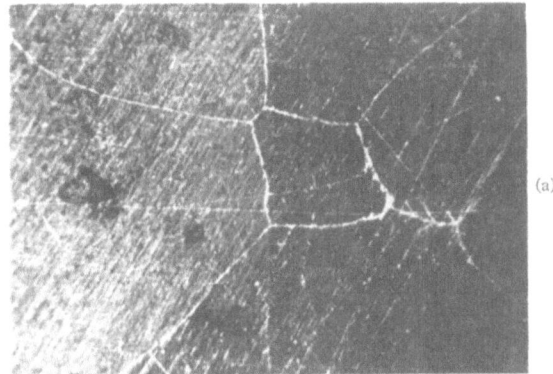

FIG. 2.

associated large internal mechanical strains, sufficient to induce large-crystal growth under suitable conditions.

It might be mentioned here that it may appear that we are proposing something of an absurdity since we first state that only magnets of pure alpha + alpha-prime phase content will give high magnetic properties, and then we propose extensive gamma-phase precipitation. However, it is part of the process to use the presence of the gamma phase to induce mechanical strains, and then when recrystallization into a monocrystal has occurred, to suppress the gamma phase by proper heat treatment.

According to the phase diagram in Fig. 1, the gamma phase should dissolve above 1200°C and the structure should revert to pure alpha phase. However, a certain time interval is required for this solution process, and it

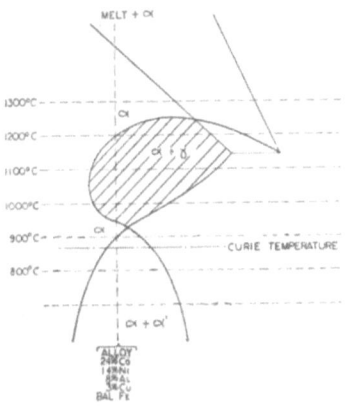

FIG. 1.

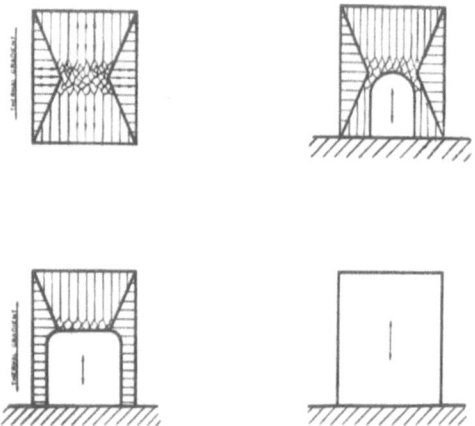

FIG. 3.

is therefore possible to raise the magnet to the crystallization temperature range of 1260° to 1310°C before the dissolution of the gamma phase has gone to completion. Thus, the two requirements for crystalline growth, i.e.,

(a) heat treatment at the critical recrystallization temperature and

(b) the presence of internal strains in the crystal lattice can be fulfilled.

In order to prevent gamma-phase dissolution before the recrystallization temperature is reached, it is necessary to either raise the temperature at an extremely fast rate, or provide increased stability of the gamma phase above 1200°C. It was decided that a combination of both methods would provide the best possible compromise.

### B. MOST SUITABLE COMPOSITION FOR SECONDARY RECRYSTALLIZATION

Certain elements, which may be referred to as gamma-phase precipitants, are known to widen the gamma-

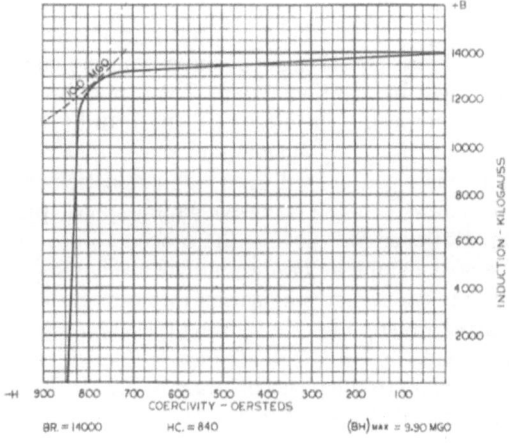

BR = 14000    HC = 840    (BH)max = 9.90 MGO

FIG. 4.

phase loop in the iron-phase diagram. These include carbon, nitrogen, manganese, ruthenium, rhodium, palladium, rhenium, osmium, iridium, platinum, and gold.

Such elements have heretofore been avoided in Alnico alloys, but we have found that small percentages added to the alloy will accelerate the gamma-phase precipitation at 930° to 1150°C, and increase the stability of this phase above 1200°C during the transition to the recrystallization temperature. We have also determined that once recrystallization to a monocrystal has occurred, it is possible to bring all of this gamma-phase precipitate into solution, and by proper cooling, traverse the precipitation temperature region at such a rate as to prevent further gamma-phase development.

### C. THE MOST SUITABLE TEMPERATURE RANGE, TIME, AND TEMPERATURE GRADIENT CONDITIONS FOR RECRYSTALLIZATION

As described, the cast alloy must first be heat treated at about 1000°C for periods of 30–60 min to promote the gamma-phase precipitation needed for straining of

FIG. 5.

the crystal lattices. Figs. 2(a), 2(b), and 2(c) show the progressive development of this precipitate. Once this precipitate has been formed, the temperature must then be rapidly raised to the critical recrystallization temperature range of 1250° to 1310°C. The gamma phase proceeds to dissolve at temperatures above 1200°C but the presence of gamma-phase stabilizing elements and the establishment of a thermal gradient of approximately 20°C/cm length from the base to the top of the casting, combined with a sufficiently rapid temperature rise, serves to promote monocrystal formation. The structure is free from all gamma phase in the 1250–1310°C temperature range, but the alpha-gamma equilibrium zone must then be traversed by rapid blast cooling to avoid gamma-phase precipitation during the cooling cycle.

In order to arrive at the best possible magnetic properties in a monocrystal, it is essential that the (100) axis of the crystal structure be parallel to the desired magnetic axis. This may be done by providing a properly oriented crystal nucleus at a preferred surface of the magnet body, and causing crystal growth to proceed from this nucleus. Since even unchilled normal sand castings have crystals lying close to the surface that are all oriented perpendicular to the surface, they can be used as the crystal nucleus. Figure 3 shows four steps in the growth of a monocrystal from a polycrystalline aggregate.

It has been found by experiment that if a temperature gradient in the order of 20°C/cm of magnet length is maintained in the direction of the preferred magnetic axis, the monocrystal will grow from the surface nucleation bed through the entire magnet. Such monocrystalline growth absorbs and converts to the monocrystalline structure not only the properly oriented crystals, but the transversely and randomly oriented crystals as well, until the entire casting becomes substantially a single crystal.

## CONCLUSIONS

We have described a process which is capable of producing monocrystalline Alnico magnets wtih superior magnetic properties.

Experiments with an alloy containing 24% cobalt, 14% nickel, 8% aluminum, 3% copper, the balance being iron, contaminated with 0.08% carbon or 0.35% manganese, following the procedure outlined, and then proceeding by conventional heat-treating methods to develop a preferred magnetic orientation parallel to the (100) axis of the monocrystalline casting, resulted in monocrystals of the proper orientation. Figure 4 shows graphically magnetic properties that have been obtained on such magnet castings and Fig. 5 shows photographically the internal appearance of a monocrystal.

While such properties have also been obtained by other laboratory methods, we feel that this procedure of exploiting the phenomenon of monocrystal formation from conventional polycrystalline castings offers the greatest scope for practical application.

---

JOURNAL OF APPLIED PHYSICS    SUPPLEMENT TO VOL. 33, NO. 3    MARCH, 1962

# The Preisach Diagram and Interaction Fields for Assemblies of $\gamma$-Fe$_2$O$_3$ Particles

G. BATE

*Development Laboratories, Data Systems Division, International Business Machines Corporation, Poughkeepsie, New York*

The Preisach diagram calculated for an assembly of $\gamma$-Fe$_2$O$_3$ particles at a packing density of 20% by volume has been used to obtain a distribution function of the particle remanence-coercivities in the absence of interaction fields. This function, which has a peak at 275 oe, is then compared with the function obtained by differentiating the remanence hysteresis loop, i.e., with the distribution function in the presence of interaction fields. It is found that the latter function is broader than the former; this can be qualitatively explained in terms of a two-particle model for the interacting particles. The magnitude of the interaction fields can be estimated from the Preisach diagram and is found to have a maximum value of roughly 300 oe.

## INTRODUCTION

THE Preisach diagram for assemblies of particles of $\gamma$-Fe$_2$O$_3$ in magnetic recording tape has been discussed recently by Daniel and Levine[1] and by Woodward and Della Torre.[2] In a previous paper,[3] the author has described an experiment which establishes the statistical stability of the diagram and shows that it can be used to predict accurately the results of other remanence experiments. In this paper, the Preisach diagram is used to find the distribution of remanence-coercivities $_pH_r$ of the particles in the absence of interaction fields. This distribution is then compared with that measured in the presence of interaction fields and, finally, the magnitude of these fields is discussed. The samples were of tape containing oxide particles (length$\sim$0.5–1$\mu$, axial ratio$\sim$7:1) at a packing density of 20% by volume; previous work[4,5] indicates that these particles should be single domains. The method of calculating the diagram has been described previously[3] and as we are concerned here with changes of remanence we need consider only that part of the diagram for which $a>o>b$, where $a$, $b$, are, respectively, the positive and negative remanence-coercivities of a particle.

## RESULTS

Figure 1 shows part of the Preisach diagram calculated from experimental results as described in reference 3 for a nonoriented sample; 86% of the particles whose representative points lie in this quadrant have both positive and negative switching fields less than 585 oe. By summing the numbers over either the vertical or the horizontal strips or by differentiating the descending

[1] E. D. Daniel and I. Levine, J. Acoust. Soc. Am. **32**, 1 (1960).
[2] J. G. Woodward and E. Della Torre, J. Appl. Phys. **31**, 56 (1960).
[3] G. Bate, J. Appl. Phys. (to be published).
[4] A. H. Morrish and S. P. Yu, J. Appl. Phys. **26**, 1049 (1955).
[5] A. H. Morrish and L. A. K. Watt, Phys. Rev. **105**, 1476 (1957).

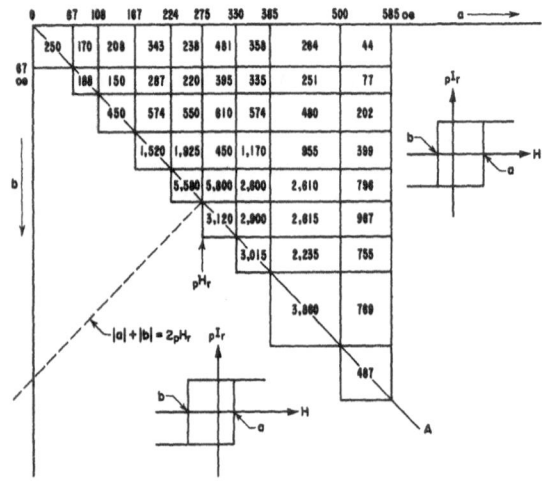

FIG. 1. Part of the Preisach diagram for nonoriented sample.

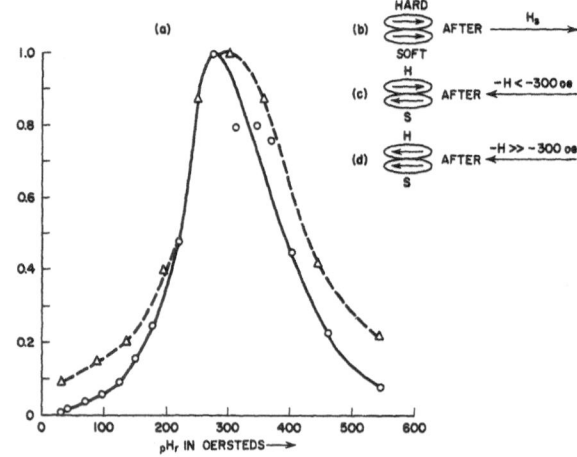

FIG. 2. Distribution functions of $_pH_r$. (i) full curve, without interaction fields; (ii) broken curve, with interaction fields.

major remanence curve, we can plot a distribution curve for $_pH_r$ in the presence of interaction fields. If we can assume that the effect of the interaction field is to move the remanence loop of a particle along the field axis without distorting the loop (the usefulness of the Preisach diagram is severely limited if this is not so), then we know that $|a|+|b| = \text{constant} = 2\,_pH_r$, where $_pH_r$ is the remanence coercivity of the isolated particle. Thus, a representative point $(a,b)$ is constrained to move only along the line $a+b = 2\,_pH_r$, i.e., at 45° to both $a$ and $b$ axes (Fig. 1). Hence, all particles of the same isolated $_pH_r$ will be found along the same diagonal regardless of their interaction fields. Then, by dividing the diagram into diagonal strips and summing the numbers in the strips, we can find distribution of $_pH_r$ [Fig. 2(a)] which the particles would have if there were no interactions. Along the other diagonal ($OA$ in Fig. 1) we have $a=b=\,_pH_r$, and, thus, we can find $_pH_r$ from the value of $a$ (or $b$) where the diagonal strip crosses the diagonal $OA$.

## DISCUSSION

From Fig. 2(a) we see that the effect of interaction fields is to broaden the apparent distribution without changing the position of the peak greatly. Néel[6] and Shtrikman and Treves[7] have considered a simple two-particle model in treating particle interactions and, in view of the appearance of dispersions of $\gamma$-Fe$_2$O$_3$ particles in electron micrographs, this seems to be an appropriate first approximation. In using this model, we consider the common case of a pair consisting of one hard and one soft particle originally parallel (Fig. 2(b)]. In order to switch the softer particle [Fig. 2(c)] we need a smaller field than its isolated $_pH_r$, since it experiences a field from the harder particle helping it to switch. Hence the remanence-coercivities of the softer particles will be decreased by interaction, as we see from Fig. 2(a). Similarly, the harder particle will require a field larger than its absolute $_pH_r$, since the field of the softer particle now opposes the switching [Fig. 2(d)], again agreeing with experiment. Figure 2(a) gives no idea of the actual magnitude of the interaction fields and to find these we have to examine the Preisach diagram itself. For example, for a particle with $a=600$ oe, $b\sim0$, and $_pH_r\sim300$ (near the peak), we see the interaction field must be 300 oe. This must be nearly the maximum interaction field for a particle in this quadrant, but by extrapolating the diagram into the first and third quadrants, we can find slightly larger values. Clearly, these interaction fields are much too small to explain the difference between the measured $H_r$ value for the assembly and that predicted from the shape of the particles. Perhaps the rotational mode is an incoherent one leading to reduced switching fields and, of course, one can never rule out the possible existence of some multidomain particles in the assembly.

### ACKNOWLEDGMENT

It is a pleasure to acknowledge the help of John R. Morrison who made the measurements.

[6] L. Néel, Compt. rend. **246**, 2313 (1958).
[7] S. Shtrikman and D. Treves, J. Appl. Phys. **31**, 58S (1960).

JOURNAL OF APPLIED PHYSICS    SUPPLEMENT TO VOL. 33, NO. 3    MARCH, 1962

# Saturation Magnetization of Swaged Mn-Al*

M. A. BOHLMANN

*Indiana Steel Products, Division of Indiana General Corporation, Valparaiso, Indiana*

Two years ago remarkable properties were reported for a new alloy based on manganese and aluminum. The value for saturation magnetization in Mn-Al had been extrapolated to infinite magnetizing field and was given as 6200 gauss. Our measurements indicate that the extrapolated figure is too low. Magnetization values of over 7100 gauss have been measured at 13 000 oe. Extrapolations of these values to infinite fields are not made because of anomalies in the magnetization curves at 7500 oe and 12 000 oe. The material having the high magnetization had an energy product of $4.0 \times 10^6$ gauss oe.

To reach the high values, it was necessary to adhere to processes and procedures having narrow limits. Saturation magnetization depended directly on composition, homogenizing, quenching rate, annealing, and deformation. The latter was most critical because a preferred axis of orientation is developed by this process. Mn-Al is very brittle, but the brittleness problem was circumvented by swaging the material in suitable jackets.

Rockwell tests indicate that hardness is directly related to $4\pi M$.

RECENTLY, Kono,[1] Koch et al.,[2] and Koester and Wachtel[3] have established the narrow composition range in which Mn-Al alloys can be made ferromagnetic. The magnetic phase is developed in alloys with 68–74% Mn; the best magnetic properties are obtained in alloys with about 55 at.% Mn.

To reach high values for magnetization, we found it necessary to adhere closely to certain procedures. We cast 72/28 Mn/Al in vacuum into iron or porcelain molds. The castings were first homogenized at 1150°C and then cooled to 600° in 20–40 sec. These quenched bars were very brittle. In order to deform them, we mounted the bars inside stainless steel jackets for swaging. Best magnetic properties were obtained by swaging to reduce the areas of the bars about 86%. An anneal of 2 hr at 400°C, when applied to adequately deformed specimens, raised $4\pi M$ to the highest values. The energy product also was increased by the annealing and reached a value of $4.0 \times 10^6$ gauss oe.

The magnetic alloy of Mn-Al is not saturated in technical magnetizing fields. Kono and Koch indicate that the magnetization continues to increase with increasing magnetizing force and they extrapolate to a $4\pi M_s$ value of 6200 gauss at infinite field. Our undeformed material can be extrapolated to the same value (see Fig. 1). Figure 1 also shows that our values measured for $4\pi M$ in swaged magnets exceed the extrapolated values. Our value for $4\pi M_s$ would be well over 7100 gauss, but no extrapolations are made for the swaged material because the left ends of the curves in Fig. 1 begin to rise steeply in high fields. The curves take a sharp turn upwards at about 12 000 oe.

The swaged magnets could not be removed from their jackets without disturbing the magnetic alignment of the material and, consequently, our results are given for magnets in jackets. To be sure that the anomalous

behavior was not contributed by the stainless steel jackets, we removed the swaged Mn-Al and measured the magnetic properties of the jackets alone. The jacket's contribution to $4\pi M$ was 1% or less of the total value and its $H_{ci}$ was less than 30 oe. Measurements on some Mn-Al swaged in brass tubes confirmed the observation that the curves in Fig. 1 apply to Mn-Al.

Our magnetic measurements were made in a modified Sanford-Bennett permeameter. Compensated coils were used to determine magnetization. The compensation was adjusted for each magnetizing force. Errors in compensation are less than $\frac{1}{2}$%. Since the areas of the magnets also enter directly into our $4\pi M$ values, we measured and checked areas carefully. We optically enlarged both ends of the magnets 20× and computed areas from the enlarged photographs. To be certain that the ends represented average cross sections we x rayed the magnets. The area values were further checked by density measurements. The magnet material was removed from the jacket and weighed. Swaged density was assumed to equal as-cast density, and the calculated areas confirmed the areas obtained by the enlarging method.

In order to get reproducible measurements of $B_r$, $H_{ci}$, and other points in a demagnetization curve, it is necessary to cycle a magnet along the major hysteresis

* Work was sponsored by the Aeronautical Systems Division of the Air Research and Development Command of the U. S. Air Force.

[1] H. Kono, J. Phys. Soc. Japan 13, 1444 (1958).
[2] A. J. J. Koch, P. Hokkeling, M. G. v. d. Steeg, and K. J. de Vos, J. Appl. Phys. 31, 75S (1960).
[3] W. Koester and E. Wachtel, Z. Metallk. 51, 271 (1960).

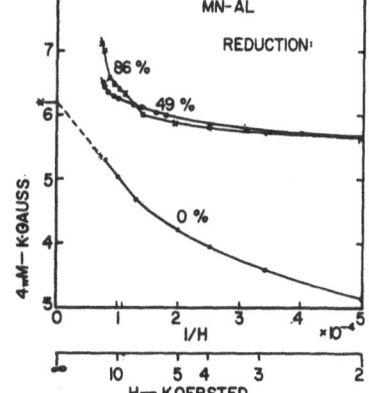

FIG. 1. Magnetization in Mn-Al magnets at room temperature.

loop. Since Mn-Al is not saturated in practical fields to be on a major loop, it was necessary to establish a minimum magnetizing field for magnetic measurements. We found that the demagnetization curve is reproduced as if it were on a major loop when a field exceeding 7500 oe is used for cycling the magnet. An anomaly in the magnetization curves appears at about 7500 oe. Magnetization $4\pi M$ after the fifth loop cycle is about 500 gauss higher than after the first cycle. At the second anomaly at 12 000 oe, the fifth loop is about 800 guass higher.

Cracks appeared in all the material when swaged because it was brittle. According to Koester and Wachtel[3] the material Mn-Al is least hard when the ferromagnetic phase is present. We measured the surface hardness of Mn-Al magnets and observed the same phenomena. The hardness was measured on the Rockwell D-scale, between the B- and C-scale of the tester. Hardness values ranged from about D25 to D55. The values varied inversely with $4\pi M$. Generally, hardness numbers had these values in material prepared as follows: as-cast = D55, homogenized and cooled = D44, annealed = D34. Hardness measurements on swaged material were unreliable because of the numerous cracks.

The author acknowledges the assistance of T. R. Schroeder and R. E. Quear in sample preparation and measurements. The suggestions by Dr. R. K. Tenzer and Dr. K. J. Kronenberg on this work are especially appreciated.

---

JOURNAL OF APPLIED PHYSICS       SUPPLEMENT TO VOL. 33, NO. 3       MARCH, 1962

# Studies of High Coercivity Cobalt-Phosphorous Electrodeposits

J. S. SALLO AND J. M. CARR
*Minneapolis–Honeywell Research Center, Hopkins, Minnesota*

Electrodeposited Co-P films have been prepared having coercivities of 1000 to 1300 oe in the plane of the film. The material has been studied by means of x-ray diffraction, electron microscopy, and the Craik technique. It is concluded that the material has a rod-like structure and that shape anisotropy is at least partially responsible for the observed coercivity. The magnetic properties of this material are compared with those of previously reported platelet materials. An attempt is made to explain the properties of both rod and platelet materials on the basis of single domain particle theory.

PREVIOUS studies of high coercivity electrodeposits have been done in the Co-Ni-P[1] and Co-P[2] systems. In these cases coercivities of 400 to 800 oe were found and the deposits were shown to have a single-domain platelet-like structure. X-ray studies showed the crystallographic uniaxial easy direction of magnetization to be in the plane of the film. Domain-type studies utilizing the Craik technique[3] confirmed the single-domain platelet model and showed that no flux leaves the plane of the film when the sample is in its remanence state.

Further studies in the Co-P system have been undertaken and coercivities in the range of 1000 to 1300 oe have been obtained in the plane of the film. In these cases, x-ray diffraction shows the crystallographic easy direction of magnetization to be normal to the plane of the film. Electron micrographs of the surface of these samples (Fig. 1, top) show a fairly large grained (two microns) deposit. No subdivision of these grains is found when conventional replication techniques are employed.

Bitter patterns of the surface of these deposits were obtained utilizing the Craik technique. The demagnetized surface produces a large circular or elliptical pattern which may represent flux leaving the film at grain boundaries (Fig. 1, bottom). The magnetized (remanance state) surface, in contrast to the previous work, also produces a Bitter pattern (Fig. 2, top). The pattern observed consists of circles having diameters of about 2000 to 3000 A. The pattern suggests that the sample is composed of rods of cobalt growing nearly normal to the substrate, the circles representing the tips of these rods. An extension of the Craik technique[1] shows the surface topography and the Bitter pattern simultaneously (Fig. 2, bottom), and indicates that the large grains are really bundles of these small rods. Bitter pattern studies of the cross section of the sample have also been done. Here the results are not entirely clear; however, the rod-like nature of the material is further indicated.

Most Co-Ni-P and Co-P electrodeposits having coercivities of 400 to 800 oe consist of platelets. The Stoner–Wohlfarth model of single-domain particles[4] predicts no shape anisotropy for rotation in the plane of the platelets. For hexagonal crystal systems such as are involved here, this model predicts a crystal anisotropy for platelet particles which is identical in treatment with the general shape anisotropy model. This theory predicts $I_r/I_s$ of 0.5 for a random compact of single-domain

[1] J. S. Sallo and K. H. Olsen, J. Appl. Phys. **32**, 203S (1961).
[2] J. S. Sallo and J. M. Carr, Detroit Meeting, Electrochemical Society, 1961.
[3] D. J. Craik and P. M. Griffiths, Brit. J. Appl. Phys. **9**, 279 (1958).
[4] E. C. Stoner and E. P. Wohlfarth, Trans. Roy. Soc. (London) **A240**, 599 (1948).

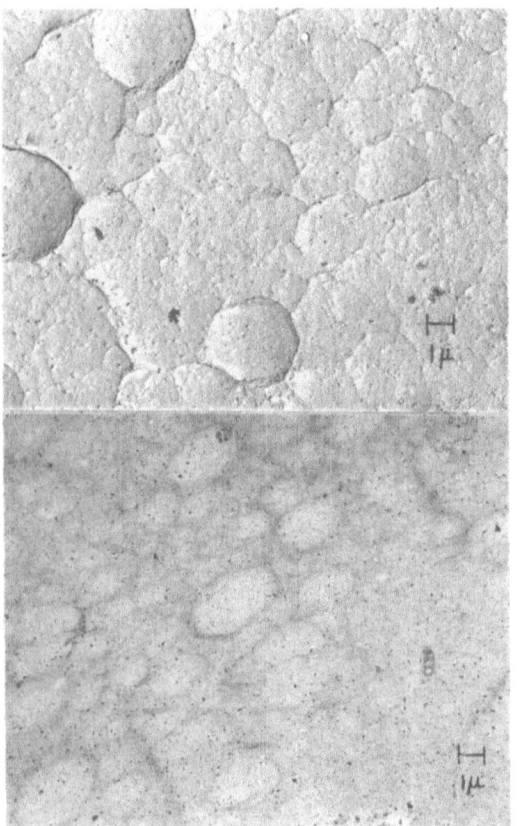

FIG. 1. (Top) electron micrograph of surface of Co-P electrodeposit. (Bottom) magnetite pattern of demagnetized surface of Co-P electrodeposit.

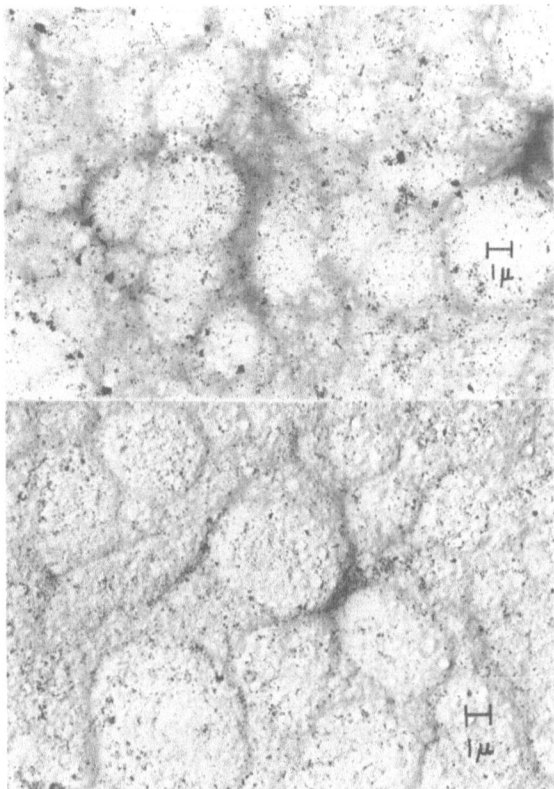

FIG. 2. (Top) magnetite pattern of magnetized surface of Co-P electrodeposit. (Bottom) carbon shadowed magnetite pattern of magnetized surface of Co-P electrodeposit.

platelets. Experimental results show $I_r/I_s$ values of 0.48 to 0.54 for both Co-Ni-P and Co-P materials in excellent agreement with the theory.

The higher coercivity Co-P material has a rod-like structure. In this case, shape anisotropy may contribute to the higher coercivities observed. Although the crystallographic easy direction of magnetization is normal to the plane of the film, the cross section studies indicate that the rods are not normal to the plane of the film. Both the Stoner–Wohlfarth model and the fanning model[5] of switching make identical predictions for $I_r/I_s$ ratios for applied field at large angles to the easy direction (major axis) of such rods. Experimental samples having $H_c$ of 1200 to 1300 oe have $I_r/I_s$ of 0.3 to 0.4. Samples at 500 oe have $I_r/I_s$ of 0.2. These values imply an easy direction at 70° or 80° from the plane of the film. If this value is correct, the coherent rotation (Stoner–Wohlfarth) model predicts an easy direction coercivity of about 6500 oe. This value, assuming the rods to be essentially pure cobalt, indicates that the rods

[5] I. S. Jacobs and C. P. Bean, Phys. Rev. **100**, 1060 (1955).

must be three to four times longer than their diameter, or about one micron long. Thus, the rods do not extend through the entire thickness of the deposit. The fanning mode of switching predicts an easy direction coercivity of 1600 oe and a maximum coercive force of about 1900 oe at 50° from the easy direction; however, the length of rod estimate is not altered.

The high coercivity Co-P electrodeposit differs from previously reported Co-Ni-P and Co-P deposits in that it has its easy direction of magnetization normal to the plane of the film. Shape anisotropy seems to be responsible for the high coercivity due to a rod-like single-domain structure. Conjecture as to the length of rods, maximum coercivity attainable, and mode of switching must be substantiated by further work.

### ACKNOWLEDGMENTS

The authors wish to acknowledge the assistance of Miss Judi Lund for the electron microscopy, Mr. T. J. Cebulla for the dc magnetic measurements, and Mr. R. A. Neenan for the x-ray diffraction reported in this work.

JOURNAL OF APPLIED PHYSICS     SUPPLEMENT TO VOL. 33, NO. 3     MARCH, 1962

# Remanent Torque Studies in Polycrystalline BaFe$_{12}$O$_{19}$*

P. J. FLANDERS AND S. SHTRIKMAN†

*The Franklin Institute Laboratories, Philadelphia, Pennsylvania*

Measurements of the anisotropy distribution in BaFe$_{12}$O$_{19}$ were carried out using the remanent torque technique. Two randomly oriented powder samples with coercive forces of 1500 and 4000 oe were tested. The remanent torque curves for both samples were essentially identical indicating an anisotropy distribution with an average of $2.1 \pm 0.2 \times 10^6$ ergs/cm$^3$ compared with a value of $3.3 \times 10^6$ ergs/cm$^3$ for single crystal BaFe$_{12}$O$_{19}$. The origin of this disagreement is unclear but it seems to be associated with splitting into domains.

RECENTLY[1,2] torque curves measured on randomly oriented $\gamma$-Fe$_2$O$_3$ powders with the field applied at a small angle to a previously saturating field were reported. It was shown that analysis of these remanent torque curves yields information on the magnetic anisotropies of the particles involved, which is more reliable than that obtained by other methods.[3,4] Also by comparing these torque curves[5] with remanence curves one can assess the relative importance of coherent and incoherent magnetization processes in the material studied.

A material which is especially interesting in connection with the remanence torque technique is BaFe$_{12}$O$_{19}$. Because of its high uniaxial crystal anisotropy one has $K = 3.3 \times 10^6$ ergs/cm$^3 \gg 1.6 \times 10^5 = I_s^2$ and therefore one is justified in applying a picture in which interparticle interactions have been neglected. Because magnetostatic effects can be neglected, one expects also that in this case the remanent torque curves will be an intrinsic property of the material; i.e., samples with different hysteretic properties should display the same remanent torque curves.

To test this, measurements were carried out on two BaFe$_{12}$O$_{19}$ randomly oriented powder samples (I and II) whose hysteresis loops as shown in Fig. 1 differed considerably. As expected, however, the torque curves shown in Fig. 2 are very similar.

One pecularity should, however, be noted. The torque traced when the field is decreased from a certain value $H$ to zero is different from that found when the field is subsequently increased from zero back to $H$[6] (looping). This is in variance with previous observations on $\gamma$-Fe$_2$O$_3$[1] as well as with the theoretical picture. The origin of this looping is not very clear at present and therefore it will be disregarded in the following analysis by considering only the ascending branches of the torque curves. It should be noted, however, that the looping seems to be tied up with domain wall effects as it can be simply associated with hysteresis loops observed by Kooy and Enz[7] on BaFe$_{12}$O$_{19}$ single crystals.

The results of the analysis of the torque curves together with the coercive force of the sample studied are given in Table I. Instead of reproducing the anisotropy distribution curve derived from the torque curves as previously described,[8] only the average anisotropy $K$ and the half-width at half-value[9] $\Delta K$ are given. It is interesting to compare the experimental coercive force with theoretical prediction for an assembly of non-interacting single domain particles (Stoner-Wohlfarth model), with which Table I has been supplemented. The experimental coercive force is much lower than the theoretical one indicating the importance of non-

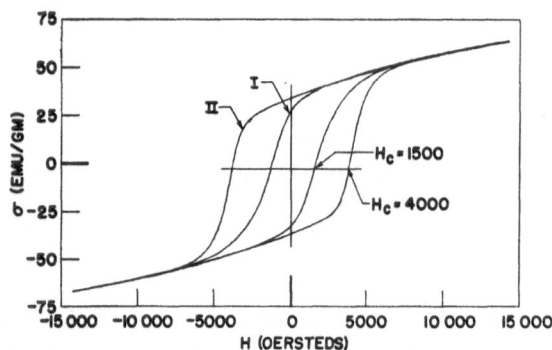

FIG. 1. Hysteresis loops for BaFe$_{12}$O$_{19}$ samples I and II.

* This research was supported by the United States Air Force under contract.

† On leave of absence from the Weizmann Institute of Science, Rehovoth, Israel.

[1] P. J. Flanders and S. Shtrikman, J. Appl. Phys. **33**, 216 (1962).

[2] P. J. Flanders and S. Shtrikman, Presented before the International Conference on Magnetism, Kyoto, Japan, September, 1961.

[3] A. E. Berkowitz and P. J. Flanders, Franklin Inst. Rept. No. F-2482, 1957. See also, E. P. Wohlfarth, Advances in Phys. **8**, 87 (1959).

[4] C. E. Johnson and W. F. Brown, Jr., J. Appl. Phys. **29**, 313 (1958); **29**, 1699 (1958); **30**, 136S (1959).

[5] A. E. Berkowitz and P. J. Flanders, Acta Meta. **8**, 823 (1960).

TABLE I. Coercive force, average anisotropy, and half-width of anisotropy distribution for 2 BaFe$_{12}$O$_{19}$ samples and theoretical S-W model.

| | $H_c$ Coercive force | $K$ Average anisotropy | $\Delta K$ Half-width at half-value |
|---|---|---|---|
| Sample I | 1500 | $2.1 \pm 0.2 \times 10^6$ ergs/cm$^3$ | $0.5 \pm 0.2 \times 10^6$ ergs/cm$^3$ |
| Sample II | 4000 | $2.1 \pm 0.2 \times 10^6$ ergs/cm$^3$ | $0.5 \pm 0.2 \times 10^6$ ergs/cm$^3$ |
| Stoner-Wohlfarth | 8500 | $3.3 \times 10^6$ ergs/cm$^3$ | |

[6] This torque loop is retraced when the field is cycled between $H$ and $O$ and is independent of history provided no fields larger than $H$ were applied.

[7] C. Kooy and U. Enz, Philips Research Repts. **15**, 7 (1960).

[8] In particular it should be emphasized that the curves were normalized in the same way as described in reference 1.

[9] Half the width of the anisotropy distribution measured at half of the peak anisotropy.

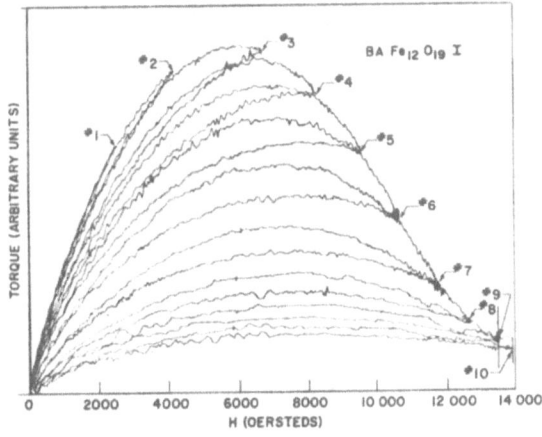

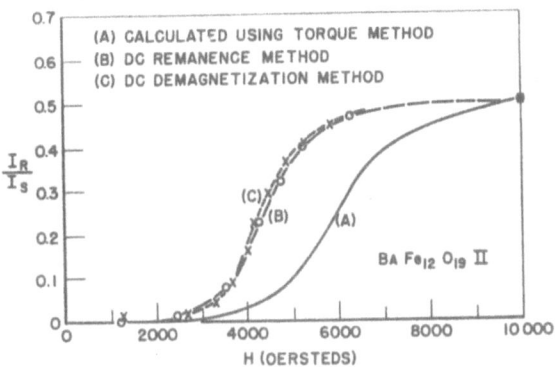

FIG. 3. Reduced remanence versus field curves
for BaFe$_{12}$O$_{19}$ sample II.

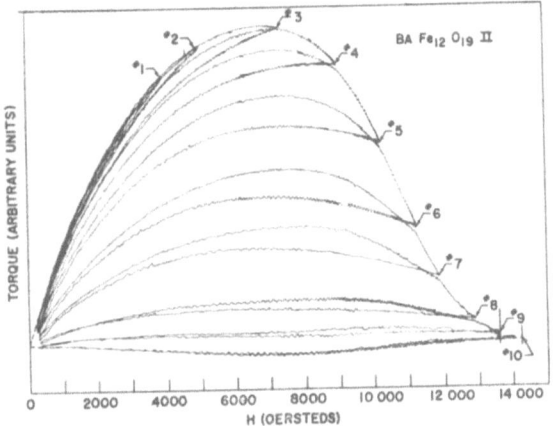

FIG. 2. Remanent torque vs field curves for BaFe$_{12}$O$_{19}$ samples
I and II whose hysteresis loops are shown in Fig. 1.

coherent magnetization processes in the determination of the coercive force, these being more important in sample I than in II. The agreement between theory and experiment is much better when the anisotropy is considered, indicating that the remanent torque curves can be described by the Stoner-Wohlfarth[10] model to a reasonable approximation. This method should, therefore, be a useful one for the determination of the uniaxial

anisotropy of materials available only in polycrystalline form, such as manganese-aluminum.[11]

The large width of the anisotropy distribution wants commenting. Magnetostatic effects which could cause it are expected, as already mentioned above, to be rather small in BaFe$_{12}$O$_{19}$. This is confirmed by the applicability of Wohlfarth[12] relations to the remanence curves in this material, as shown in Fig. 3, curves B and C. (A is the remanence curve calculated from the anisotropy distribution assuming coherent rotation. The wide discrepancy between the calculated remanence A and the experimental values B and C is due to the importance of incoherent rotation when the remanence is acquired.)

Splitting into domains might also result in broadening. If this takes place, the magnetization of some domains will be inclined to the remanence by an angle $\alpha > \pi/2$. Now the field $H_m$ above which the torque becomes reversible is equal to the critical jump field tabulated by Stoner and Wohlfarth, which varies with $\alpha$. A broad anisotropy distribution then means a distribution in $\alpha$ in agreement with what is expected, as $\alpha$ should be structure sensitive. Because $H_m$ has its highest value for $\alpha \doteq \pi/2$, this effect might also account for the discrepancy between the theoretical and observed anisotropy.

[10] E. C. Stoner and E. P. Wohlfarth, Trans. Roy. Soc. (London) A240, 599 (1948).

[11] A. J. J. Koch, P. Hokkeling, M. G. v. d. Steeg, and K. J. deVoss, J. Appl. Phys. 31, 75S (1960).
[12] E. P. Wohlfarth, J. Appl. Phys. 29, 595 (1958).

JOURNAL OF APPLIED PHYSICS     SUPPLEMENT TO VOL. 33, NO. 3     MARCH, 1962

# Permanent Magnetic Properties of Iron-Cobalt-Phosphides

K. J. de Vos, W. A. J. J. Velge, M. G. van der Steeg, and H. Zijlstra

*Metallurgical Laboratory, N. V. Philips' Gloeilampenfabrieken, Eindhoven, The Netherlands*

An investigation of the phase relations of the section $Fe_2P$-$Co_2P$ in the system Fe-Co-P showed that at high temperature all (Fe, Co)$_2$P alloys, including the final components $Fe_2P$ and $Co_2P$, probably form a continuous range of solid solutions having the hexagonal $Fe_2P$ structure. At low temperatures $Fe_2P$ and $Co_2P$ form an almost complete range of solid solutions with the orthorhombic $Co_2P$ lattice. The Curie temperature of the hexagonal phase increases from about room temperature for $Fe_2P$ to a value of nearly 190°C for (Fe, Co)$_2$P with an Fe:Co ratio of 70:30. The anisotropy field $H_A$ appeared to be a function of the Fe:Co ratio and possesses a maximum value of about 19 000 oe near an Fe:Co ratio of 90:10. Particles obtained by lixiviating an Fe-Co-Cu-P melt can possess a coercivity $_IH_C$ up to 2000 oe. This value can be substantially increased by annealing. The temperature coefficient of $_IH_C$ appears to be relatively large and seems to be linked to the change of $K$ with temperature. The coercivity decreases appreciably after mechanical loading, e.g. pressing, but can be restored by annealing; the anisotropy field $H_A$ as well as the saturation magnetization $\sigma_\infty$ changes in the same sense. Anisotropic permanent magnets have been made with a $BH_{max}$ value of about $2 \times 10^6$ gauss oe.

## PHASE RELATIONS IN THE SECTION $Fe_2P$-$Co_2P$

IN connection with the search for ferromagnetic materials with a high crystal anisotropy, intermetallic compounds of the type (Fe, Co)$_2$P were investigated. Berak[1] found that the section $Fe_2P$-$Co_2P$ in the system Fe-Co-P contains two regions of solid solution, one having the hexagonal $Fe_2P$ structure,[2] the other having the orthorhombic $Co_2P$ structure.[3] Preliminary investigations of (Fe, Co)$_2$P alloys by the authors gave rise to the supposition that the solubility of cobalt in the hexagonal $Fe_2P$ structure, at least at temperatures above 900°C, must be much larger than pointed out by Berak. Therefore a number of (Fe, Co)$_2$P alloys were prepared by melting, using iron-phosphor and cobalt-phosphor pre-alloys. All samples contained, besides (Fe, Co)$_2$P, small amounts of (Fe, Co)$_3$P or some excess of Co. The alloys were investigated by means of differential thermal

analysis, and the heat effects observed during heating and cooling are plotted in Fig. 1. At the line a–b the (Fe, Co)$_3$P constituents of the alloys are melting; the line b–c represents the formation of the Co-Co$_2$P eutectic. The heat effects along the line d–e are due to the transformation hexagonal ⇆ orthorhombic. This is checked by x-ray patterns from specimens quenched from several temperatures. Alloys with an Fe:Co ratio[4] 70:30, for instance, can be obtained completely in the hexagonal state after quenching from a temperature above 1130°C at a rate of more than 100°C/sec. When the alloys are quenched from temperatures well above e–f, the x-ray patterns always reveal the coexistence of the hexagonal and the orthorhombic phase. In $Co_2P$ itself, a transformation was found at about 1140°C. The $Co_2P$ high-temperature modification, not mentioned in the literature, cannot be supercooled to room temperature. As the heat effects taking place along line e–f have the same character as those along line d–e, it is believed that all (Fe, Co)$_2$P alloys, including the final components $Fe_2P$ and $Co_2P$, form a continuous range of solid solutions at high temperature. After annealing at 700°C during 24 hr all alloys with Fe:Co ratios ranging from 95:5 to 0:100 ($Co_2P$) have the orthorhombic structure, which means that at low temperatures $Fe_2P$ and $Co_2P$ form an almost complete range of solid solutions.

## MAGNETIC PROPERTIES OF THE (Fe, Co)$_2$P ALLOYS

For a number of (Fe, Co)$_2$P alloys,[5] the Curie temperatures are determined as the point of intersection of the temperature axis and the tangent to the point of inflection of the $\sigma$–$T$ curve; in all cases the measuring field is 5000 oe. It appeared that the Curie temperature $T_c$ of the samples in the hexagonal state is higher than in the orthorhombic state. For the hexagonal state the $T_c$ increases from 5°C for the $Fe_2P$ composition to 190°C for the alloys having the Fe:Co ratio 70:30.

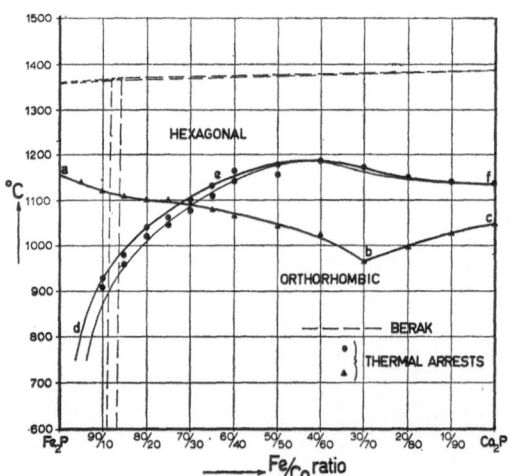

FIG. 1. The section $Fe_2P$-$Co_2P$ of the system Fe-Co-P; the small thermal arrests along the line a–b–c indicate that some excess of Fe and Co is present.

[1] J. Berak, Arch. Eisenhüttenw. **22**, 131 (1951).
[2] G. Hägg, Z. Krist. **68**, 470 (1928).
[3] S. Rundqvist, Acta Chem. Scand. **14**, 1961 (1960).
[4] All Fe:Co ratios are by weight.
[5] See also, Shū Chiba, J. Phys. Soc. Japan **15**, 581 (1960); M. C. Cadeville and A. Meyer, Compt. rend. **252**, 1124 (1961).

Moreover it was found that (Fe, Co)$_2$P alloys, prepared by lixiviating an Fe-Co-Cu-P melt in analogy with Maronneau's method,[6] have a somewhat lower $T_c$ in both modifications. This is probably attributable to Cu dissolved in the lattice.

Preparing the (Fe, Co)$_2$P alloys by this lixiviating method has the advantage that directly after leaching, fine particles are obtained. The coercive force $H_c$ of such powders is measured as a function of the Fe:Co ratio. It is found that $H_c$ is a function of the cooling rate of the Fe-Co-Cu-P melt, which is demonstrated by curves a and b of Fig. 2. Microscopic investigations indicate a smaller particle size with increasing cooling rate. Besides it is found that $H_c$ can be substantially increased by an annealing treatment in the temperature range between 650° and 300°C (Fig. 2, curve c). The highest coercive force is found for alloys with an Fe:Co ratio between 90:10 and 80:20. This is probably related to the maximum in the value of the anisotropy field $H_A$, derived from the approach to saturation measured for a number of alloys with a varying Fe:Co ratio (Fig. 2).

The temperature coefficient of the coercive force appears to be relatively large; e.g., $H_c$ of a 85:15 powder decreases from a value of 2000 oe at room temperature to 760 oe at 70°C and increases to 5580 oe at liquid nitrogen temperature. In this connection it is of interest to know the variation with temperature of the anisotropy constant $K$. From torque measurements at room temperature on a 85:15 powder aligned in a magnetic field, a value for $K$ of $3.1 \times 10^6$ ergs/cm³ can be derived. At liquid nitrogen temperature the same sample yields

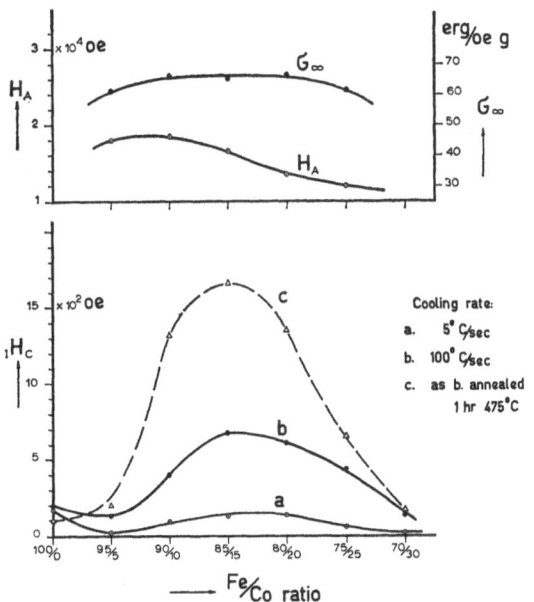

FIG. 2. Lower part: Intrinsic coercivity $_IH_C$ of (Fe,Co)$_2$P powders as a function of the Fe:Co ratio, after different treatments. Upper part: Anisotropy field $H_A$ and saturation magnetization $\sigma_\infty$ measured on the samples of curve b.

[6] G. Maronneau, Compt. rend. **130**, 657 (1900).

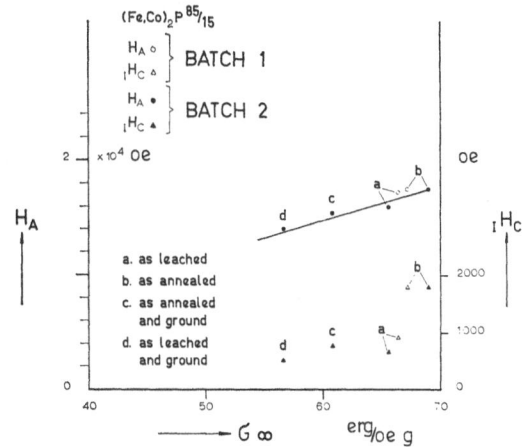

FIG. 3. Anisotropy field $H_A$, intrinsic coercivity $_IH_C$, and saturation magnetization $\sigma_\infty$ of an (Fe,Co)$_2$P powder with an Fe:Co ratio of 85:15, after different treatments.

a $K$ of $7.1 \times 10^6$ ergs/cm³. These values of the anisotropy constant $K$ must be seen as lower limits because the degree of alignment of the particles is uncertain, though hysteresis loops measured parallel and perpendicular to the aligning field have shown that the particles can be oriented to a high degree. All the same, it can be concluded that the value of $K$ changes with temperature in the same sense as the coercivity.

The anisotropy field $H_A$ and $\sigma_\infty$ as derived from the approach to saturation of a 85:15 powder, as well as the coercivity $H_c$, is given in Fig. 3 for two different batches in the as-leached and the annealed state. These quantities have also been measured for one of these batches after grinding in a vibrating mill. There seems to be a simple relation between $H_A$ and $\sigma_\infty$, but not between $H_c$ and $\sigma_\infty$. It should be noted that $H_c$ decreases appreciably after mechanical loading and increases by annealing.

## DISCUSSION OF THE RESULTS

The high coercivity of the (Fe, Co)$_2$P powders seems to be connected with the hexagonal structure. The increase in coercive force by an annealing treatment is not connected with visible changes in the x-ray patterns, but after prolonged annealing the appearance of the first traces of the orthorhombic modification is always accompanied by a decrease in $H_c$. Parallel with the improvement of the $H_c$ by the annealing treatment, the anisotropy field $H_A$ as well as the saturation magnetization $\sigma_\infty$, increase. A decrease of structural imperfections may play an important role in these phenomena. This is supported by the sensitivity of the $H_c$ to mechanical deformation. At present, however, there is no conclusive evidence for the real character of the imperfections.

## CONCLUSIONS

In regard to the application of iron-cobalt phosphides as permanent magnets, it must be considered that the constituent materials iron, cobalt, and phosphor are fairly inexpensive. In preparing fine particles by the method of lixiviating an iron-cobalt-copper-phosphor melt, the economy of the process is strongly dependent on the recovery of copper. In making anisotropic permanent magnets by pressing the particles in a magnetic field, the maximum density that can be attained seems to be 60–70%.

Magnetic properties that should be noticed are the moderate saturation magnetization of 69 erg/oe gauss

$(4\pi I_s = 6000$ gauss), the relatively low Curie point, the high-temperature coefficient of the coercivity, and finally the sensitivity to mechanical loading. Little is known about the corrosion resistance, but at room temperature it appeared to be excellent.

Anisotropic permanent magnets with the following properties in the preferred direction have been made: $B_r = 4000$ gauss, $_IH_C = 2200$ oe, $_BH_C = 1600$ oe, $BH_{max} \sim 2.0 \times 10^6$ gauss oe.

---

JOURNAL OF APPLIED PHYSICS    SUPPLEMENT TO VOL. 33, NO. 3    MARCH, 1962

# The Effect of Angular Variations in Field on Fine-Particle Remanence

Eric D. Daniel*
Memorex Corporation, Santa Clara, California

AND

R. Noble
M. S. S. Recording Company, Slough, Buckinghamshire, England

The Stoner–Wohlfarth model is used to calculate the remanence of a randomly-oriented assembly of single-domain particles when the applied field changes in direction. The cases considered are (1) fields applied successively in directions differing by multiples of $\frac{1}{2}\pi$ and (2) fields rotated through multiples of $\frac{1}{2}\pi$. Some consideration is also given to the anhysteretic remanence when the unidirectional and alternating fields are orthogonal. Experimental results obtained on disks of $\gamma Fe_2O_3$ recording tape are found to be in reasonably good qualitative agreement with the calculations.

## INTRODUCTION

THE remanence of a recorded element of magnetic tape is associated with a field which varies in direction as well as magnitude. The effect of the recording field rotation has usually been ignored in analyses of the recording process, not because it is unimportant, but because of a lack of information at the basic level. The work reported here was undertaken independently by the authors as a first step towards providing this information. The Stoner–Wohlfarth model has been used as being the simplest and probably the one most suited to the magnetic materials used in recording. From a more general point of view, however, calculations of the type described can be used as yet another means of comparing various models of coherent and incoherent magnetization processes.

## STATIC REMANENCE

The remanence of an assembly of noninteracting, randomly oriented, single-domain particles having uniaxial anisotropy has been investigated by Wohlfarth. Starting from the demagnetized state, a reduced field $h = H/N(m)I_0$ produces a reduced remanence

$$j_r(h) = I_r/I_0 = \tfrac{1}{2}\cos 2\theta, \qquad (1)$$

where $N(m)$ is the coefficient of shape anisotropy, $I_0$ is the saturation magnetization, and $\theta(h)$ is the orientation angle corresponding to the critical field strength[2] $h_0 = h$. Calculations of the components of remanence $j_x$ and $j_y$ produced by applying successive or rotating fields having directions in the $(x,y)$ plane can be made by summing the effect of reversals in those particles whose orientations with respect to $h$ lie between $\frac{1}{4}\pi - \theta(h)$ and $\frac{1}{4}\pi + \theta(h)$ at any instant during the process. Some of the simpler cases are listed below:

(a) Successive orthogonal fields $h_x$, $h_y$ of equal magnitude give

$$j_x = \tfrac{1}{2}\cos 2\theta - (1/\pi)[\cos^2\theta(\tfrac{1}{2}\pi - \sin^{-1}\tan\theta) \\ - \sin\theta(\cos 2\theta)^{\frac{1}{2}}] \quad (2)$$

$$j_y = \tfrac{1}{2}\cos 2\theta. \qquad (3)$$

(Solutions corresponding to $h_x \neq h_y$ have been worked out but the results are omitted for brevity.)

(b) Successive antiparallel fields[3] $h_x$, $-h'_x$ give $j_y = 0$ and

$$j_x = \tfrac{1}{2}\cos 2\theta - \cos 2\theta', h' \lesssim h, \qquad (4)$$

$$j_x = -\tfrac{1}{2}\cos 2\theta', h' \gtrsim h. \qquad (5)$$

(c) A field of constant strength $h$ rotated through $\frac{1}{2}\pi$,

---

* With Ampex Corporation, Redwood City, California, when this work was initiated.
[1] E. P. Wohlfarth, Research 8, S42 (1955).

[2] E. C. Stoner and E. P. Wohlfarth, Phil. Trans. Roy. Soc. London A240, 599 (1948).
[3] E. P. Wohlfarth, J. Appl. Phys. 29, 595 (1958).

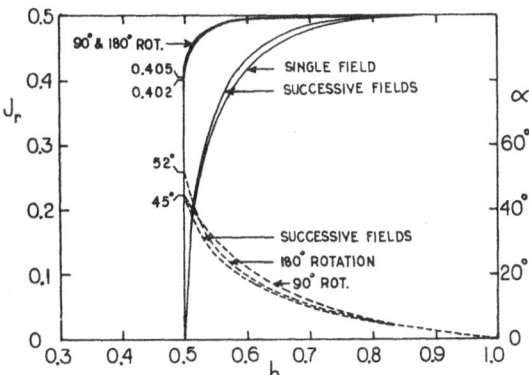

FIG. 1. Theoretical curves.——Magnitude of resultant remanence. - - - - Angle between remanence and final field direction.

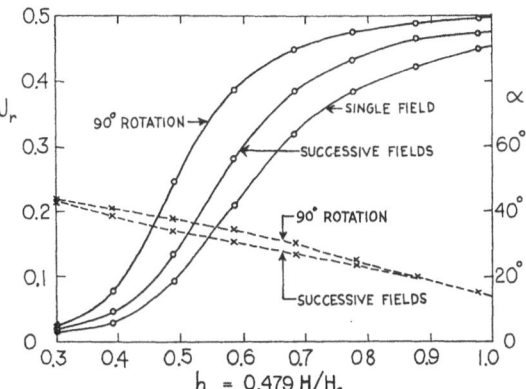

FIG. 2. Experimental curves.——Magnitude of resultant remanence. - - - - Angle between remanence and final field direction.

from $0x$ to $0y$ gives

$$j_x = (\tfrac{1}{2}\pi)[\sin2\theta + \sin\theta(\cos2\theta)^{\frac{1}{2}}$$
$$+ \cos^2\theta \sin^{-1}\tan\theta] \quad (6)$$
$$j_y = \tfrac{1}{2}\cos^2\theta + (\tfrac{1}{2}\pi)[\sin2\theta - \sin\theta(\cos2\theta)^{\frac{1}{2}}$$
$$- \cos^2\theta\sin^{-1}\tan\theta] \quad (7)$$

(d) A field of constant strength $h$ rotated through $\pi$ from $0x$ to $-0x$ gives[4]

$$j_x = -\tfrac{1}{2}\cos^2\theta, \quad j_y = (1/\pi)\sin2\theta. \quad (8)$$

Corresponding magnetization curves are shown in Fig. 1. Compared with the normal curve, the most obvious effect of changes in field direction is to hasten the rise to saturation. In fact, the resultant magnetization corresponding to 90° rotation rises discontinuously to the value $\sqrt{2}(\tfrac{1}{2}\pi + \tfrac{1}{8}) \approx 0.4018$ as soon as $h$ reaches the value 0.5, the minimum critical field strength, and lags exactly 45° behind the final field direction. The curve for 180° field rotation is similar, but the discontinuous rise is to the slightly higher value $(\tfrac{1}{4} + 1/\pi^2)^{\frac{1}{2}} \approx 0.4047$, and the initial angle of lag is 51.9°. It can be shown that the latter curve applies to any field rotation $\phi \gtrsim \pi$ since, in this range, the resultant magnetization becomes of constant magnitude and lags a fixed amount behind the field.

Some experimental results are shown in Fig. 2. The sample was a 0.75-in.-diam. disk of recording tape containing randomly oriented, acicular $\gamma\mathrm{Fe_2O_3}$ particles

about $1\,\mu$ long, of dimensional ratio about 6. Fields were applied in the plane of the disk and the remanence was measured by rotating the disk about its axis beneath an iron-cored pickup device. The field scale in Fig. 2 has been normalized by putting $h = 0.479\ H/H_c$.

## ANHYSTERETIC REMANENCE

An analysis of anhysteresis using the Stoner–Wohlfarth model leads[5] to an anhysteretic remanence:

$$j_a = j_r(h_b + h), \quad (9)$$

where $h_b$ is the initial amplitude of the slowly decreased alternating field and $h$ is the constant value of the unidirectional field. When $h \ll h_b$, (9) reduces to:

$$j_a \approx j_r(h) = \tfrac{1}{2}\cos\theta_b. \quad (10)$$

This result is unsatisfactory insofar as it implies that the initial anhysteretic susceptibility is infinite—a consequence of ignoring interactions between particles. Nevertheless, the result is of some value in predicting the way in which, for a given $h$, $j_a$ varies with $h_b$.

Similar calculations have been made for the case when $h_b$ and $h$ are orthogonal. The approximation corresponding to (10) is

$$j_a \approx \tfrac{1}{2} - 2\theta_b/\pi. \quad (11)$$

The remanence is still in the direction of $h$, but the $j_a$ vs $h_b$ curve begins to rise to the limiting value more gradually than when $h_b$ and $h$ are parallel. These conclusions are in keeping with experimental findings.

---

[4] S. Shtrikman and E. P. Wohlfarth, J. Appl. Phys. 32, 241S (1961), have made use of this result and calculated the corresponding solution for completely incoherent reversals.

[5] E. P. Wohlfarth, Phil. Mag. 2, 719 (1957).

JOURNAL OF APPLIED PHYSICS     SUPPLEMENT TO VOL. 33, NO. 3     MARCH, 1962

# Possibility of Domain Wall Nucleation by Thermal Agitation

Amikam Aharoni

*Department of Electronics, The Weizmann Institute of Science, Rehovoth, Israel*

Very fine particles reverse their magnetization spontaneously, passing the energy barrier by means of thermal activation. In the same way one could argue that small portions of larger particles could be spontaneously reversed, thereby serving as nucleii for reversed domains. These can then grow, because it is known that the energy favors subdivision into domains. In treating such a possibility, however, one should take into account the additional barrier for forming a wall around the "nucleus," which is usually left out in standard domain theory calculations when the wall energy is used. This energy is rather small but the barrier for formation of the wall is large.

The model studied is a small sphere, centrally located inside a spherical particle, the magnetization in which starts to rotate gradually from the $z$ direction in which the rest of the particle is magnetized. The angle between magnetization and the $z$ axis is assumed to vary linearly in the radial coordinate, from zero on the surface of the inner sphere, to its maximum value in the center. This value eventually reaches $\pi$, thus completing a wall around a reversed domain of radius zero, which should then start to grow by further supply of energy.

Studying the energy barrier for such a process, it is seen that the main contribution is from the exchange energy, which practically forbids the process. The probability for a domain wall to nucleate at room temperature is thus shown to be negligibly small, which justifies the neglection of temperature affects in Brown's equations.

B ROWN[1] has shown that once an ellipsoidal ferromagnetic particle is magnetized to saturation, no change in this state is possible, in particular no domains can nucleate, until a rather large negative field is applied, which is in disagreement with common observations. On the other hand it is known that for very small particles, thermal fluctuations can accumulate enough energy to overcome the barrier and the particles can then reverse their magnetization spontaneously.[2] It therefore seems rather plausible that small parts of a larger particles can reverse spontaneously, thereby forming a "nucleus" for a reversed domain. This possibility was excluded by Dijkstra[3] who estimated the energy barrier to be too high for thermal agitation to overcome, but his estimation of the energy presumes the existence of domain surrounded by a wall and avoids the main problem of how this is nucleated at all. (He has also devised a rather sophisticated mechanism for domain nucleation as a rotation process, but again without considering how the wall is formed to begin with. Such a calculation can be related to measurements in which the nuclei actually exist,[4] but not to our problem.) If the barrier consisted only of the magnetocrystalline anisotropy, as is the case for small particles, cubical regions whose linear dimensions are of the order of $10^{-6}$ cm would be reversed[2] at room temperature, which approaches the estimation for a domain wall dimension. One could therefore probably expect somewhat larger regions to be *gradually* reversed, forming a kind of a wall, which forms say near the surface, then move in, leaving a fully reversed domain behind it. However, such an argument does not take into account the work that should be done against the exchange energy, which is much more important over short distances since unlike small particles, small regions in large particles always have neighboring regions of the saturated material. This will be illustrated for a rather simple example which suggests that temperature effects can actually be neglected for ordinary room temperature experiments, except possibly when the field approaches the nucleation field, i.e., when the barrier becomes small and nucleation might therefore start at somewhat lower field than the static theory yields.

The effect of thermal activation has been studied by Néel,[5] by Stacey,[6] and by Brown[7] using different approaches. Their results are practically the same for the sake of following treatment, but I prefer to use the formula of Stacey,[6] since he uses essentially Bose-Einstein statistics rather than Boltzmann statistics and this seems better related to the ferromagnetic case. According to Stacey, the probability per unit time of jumping over an energy barrier $E$ at temperature $T$ is given by

$$\frac{\pi^2}{6\sqrt{3}} \cdot \frac{kT}{h} e^{-E/kT},$$

where $h$ is Planck's constant and $k$ is Boltzmann's constant. At room temperature, if $10^2$ sec is taken as a reasonable time for performing the experiment,[2] this expression implies a possibility of overcoming a barrier of about $10^{-12}$ ergs.

The main problem now remains, of estimating the energy barrier for domain nucleation. To do this, one should actually know the exact mechanism of this process and this is not known. The curling mode, which yields the lowest nucleation field in large enough ellipsoids of revolution,[8] is of no use in this respect, since it is essential to follow the energies to the barrier maximum. A certain mode is therefore guessed here, which in principle does not prove much, because even when there is a way to pass a barrier, there are always many ways

[1] W. F. Brown, Jr., Revs. Modern Phys. **17**, 15 (1945).

[2] C. P. Bean and J. D. Livingston, J. Appl. Phys. **30**, 120S (1959).

[3] L. J. Dijkstra, "A Nucleation Problem in Ferromagnetism" *Thermodynamics in Physical Metallurgy* (American Society for Metals, Cleveland, Ohio, 1950), pp. 271–281.

[4] S. Methfessel, S. Middelhoek, and H. Thomas, J. Appl. Phys. **32**, 294S (1961). See also L. F. Bates and D. H. Martin, Proc. Phys. Soc. (London) **B69**, 145 (1956) for an interesting study of "domain nucleation" where the samples were evidently not previously saturated completely.

[5] L. Néel, Ann. Géophys. **5**, 99 (1949).

[6] F. D. Stacey, Proc. Phys. Soc. (London) **73**, 136 (1959).

[7] W. F. Brown, Jr., J. Appl. Phys. **30**, 130S (1959).

[8] A. Aharoni, J. Appl. Phys. **30**, 70S (1959).

*not* to pass it. However, this mode contains the essence of the possible mechanism suggested above and its nucleation field will be seen to be not much higher than the curling mode, so that it seems to me that it is a reasonably fair approach to the problem.

The model consists of a sphere of a ferromagnetic material, magnetised to saturation in the $z$ direction, in which a smaller concentric sphere of radius $R$ is defined. The magnetization of the inner sphere is assumed to rotate in the $xz$ plane, in the positive direction of the coordinate $\theta$. The rotation is assumed to be gradual, actually linear, with a spherical symmetry, namely the angle between the magnetization and the $z$ axis is a linear function of the spherical coordinate $r$ and is zero on $r=R$ (and for $r>R$). This is continuous function with a discontinuous derivative. The justification for using such functions will be published elsewhere. It actually means that in spherical coordinate system the direction cosines of the magnetization are

$$\alpha_r = \sin\theta \ \sin\omega(R-r) \ \cos\varphi + \cos\theta \ \cos\omega(R-r), \quad (1a)$$

$$\alpha_\theta = \cos\theta \ \sin\omega(R-r) \ \cos\varphi - \sin\theta \ \cos\omega(R-r), \quad (1b)$$

$$\alpha_\varphi = -\sin\varphi \ \sin\omega(R-r), \quad (1c)$$

for $0 \leqslant r \leqslant R$, and

$$\alpha_r = \cos\theta, \quad \alpha_\theta = -\sin\theta, \quad \alpha_\varphi = 0 \quad (2)$$

for $R \leqslant r \leqslant \rho$. Here $\omega$ is a parameter giving the amplitude of rotation.

The divergence of the magnetization vector is not zero, and implies a certain potential which opposes the rotation, and which can be calculated analytically from the appropriate Poisson equation. On the other hand, the potential of the surface charges on the outer sphere, $r=\rho$, *favors* the rotation. The potentials can be calculated analytically and it turns out that the contributions of these two potentials to the energy are equal and therefore cancel out. There is thus no change in the self-magnetostatic energy by the rotation.

There is a loss in the magnetocrystalline anisotropy energy for the part of the material that rotates out of the $z$ axis, assumed an easy axis. Carrying out the integrations, one obtains for the case of unidirectional anisotropy the energy change

$$E_k = 2\pi K \left[ \frac{R^3}{3} - \frac{R}{2\omega^2} + \frac{\sin 2\omega R}{4\omega^3} \right]. \quad (3)$$

The exchange energy is given by squares of gradients, i.e., $r$ derivatives in this case, which implies

$$E_{\text{ex}} = (4\pi/3)R^3 A\omega^2. \quad (4)$$

The energy of rotation in an external field $H$ applied in the $z$ direction is

$$E_H = 4\pi H I_s \left[ \frac{R^3}{3} - \frac{2}{\omega^2}\left(R - \frac{\sin\omega R}{\omega}\right) \right]. \quad (5)$$

For small values of $\omega$, the sines in (3) and (5) can be expanded in a Taylor series and one obtains then that the first term that does not vanish contains $\omega^2$, the total energy being

$$E \approx (\pi/3)\omega^2 R^3 \{4A + (R^2/5)(2K + HI_s)\}.$$

This means that if nucleation is sought due to $H$ only, the nucleation field is

$$H_n + 2K/I_s = 20AR^{-2}I_s^{-1} = 20I_s(R/R_0)^{-2}. \quad (6)$$

The nucleation is thus more difficult than for the curling mode, but the difference is not especially significant for hard materials, if $R$ is not too small.

Consider now the probability of spontaneous nucleation in zero external field, against the barrier given by (3) and (4). It is readily seen that $\partial E/\partial R > 0$ and $\partial E/\partial \omega > 0$ for the region of interest $0 \leqslant \omega R \leqslant \pi$. It is therefore seen that there is no intermediate angle for which the barrier reaches its maximum, and one has to continue and supply energy until a whole "wall," with an angle $\pi$ in the center, is reached. Further energy will therefore have to be supplied for this wall to move and leave a central reversed domain, but this can be left out because the energy barrier up to the completion of the wall turns out to be already too large, since for $\omega R = \pi$ one obtains from (3) and (4) for the total energy

$$E = 2KR^3\{\pi/3 - (1/2\pi)\} + (4\pi^3/3)AR. \quad (7)$$

For $A = 10^{-6}$ erg/cm, the second term alone reaches the aforesaid limit of $10^{-12}$ erg for radius $R$ less than one lattice constant in all ferromagnetic materials, so that the model of gradual change over at least a few spins is definitely of question.

For the sake of comparison it is interesting to note also the barrier for coherent rotation of the inner sphere, of radius $R$, at an angle $\omega$. For estimation of the exchange energy, one can take the number of spins on the interface as $4\pi(R/a)^{-2}$, where $a$ is the lattice constant. These and only these have neighbors inclined to them at an angle $\omega$, so that the exchange energy is roughly,[9]

$$E_{\text{ex}} = -(4\pi A/a)R^2 \cos\omega. \quad (8)$$

The rest of the calculation (of magnetostatic and anisotropy energies) is straightforward. The energy barrier for spontaneous rotation in zero applied field is found to increase monotonously with $R$. As a function of $\omega$, it has a maximum if $R$ is larger than some hundred angstroms for hard materials[10] like MnBi or BaFe$_{12}$O$_{19}$. For smaller radii than these, the maximum is for $\omega = \pi$ which yields a barrier energy of $8\pi AR^2/a$ which again cannot be overcome by thermal agitation, even if $R$ is of the order of $a$.

It seems therefore that thermal agitation can be safely neglected in the study of the Brown paradox.

[9] C. Kittel, Revs. Modern Phys. **21**, 541 (1949).
[10] The exchange energy was calculated for a cubic crystal, while these materials are hexagonal. For sake of this rough estimation, the lattice constant $a$ can be taken as the average of the three unit cell dimensions, yielding about 5 A for MnBi, and 12 A for BaFe$_{12}$O$_{19}$.

# Study of Particle Arrangements and Magnetic Domains on the Surface of Permanent Magnets*

K. J. KRONENBERG

*Indiana Steel Products Division of Indiana General Corporation, Valparaiso, Indiana*

Some types of permanent magnets are compacts of very small particles. Their properties depend largely on mutual magnetic influence between particles or groups of them. To determine these, more should be known about packing arrangements, orientation, separating layers, and the homogeneity of the compact. A method is described which makes surface details visible and simultaneously reveals magnetic phenomena of such compacts. The size of the compacted particles mostly requires the resolving power of an electron microscope. Special replicating techniques combine shadow-graphing and colloid application even on rough and porous surfaces.

Examples are presented of such surface studies on barium ferrite and iron-cobalt particle magnets. Surfaces of barium ferrite magnets with crystal sizes of 0.2–1 $\mu$ show indications of magnetic domains smaller than observed before, sometimes with less than 0.1 $\mu$-width. On compacts of small elongated iron-cobalt particles, structures have become visible which indicate particle groupings varying with the methods of compacting. Colloid outlines surface pole arrangements in correlation to particle groups.

ON surfaces of permanent magnets which have multiphase structures of a size too small to allow the existence of domain walls, iron oxide colloid arranges itself into some types of domain patterns. Observations of this sort have been published repeatedly.[1–4] The seeming discrepancy with the common theory has been mentioned without an attempt of explanation. Later, the details of the colloid patterns on the surfaces of Alnico V had become one of the keys for the understanding of the mechanism of magnetization reversal.[5]

Craik and Isaac[6] have observed similar moving colloid patterns during the magnetization reversal on magnets of elongated single domain particles. Observations of these "magnetic interaction domains" indicate that the magnetic grouping of particles into such domains is of foremost importance to the magnetization process of the material.

However, the size of the phenomena is too small to yield details by light microscopy. Electron microscopy has to be applied in order to study the relation between material structure and magnetic behavior.

The two methods published to date[7,5] for the study of colloid patterns in the electron microscope are not successful on the porous surfaces of compacts of small particles. The colloid is swallowed up before it arranges itself to reveal domain configurations.

We developed a method which overcomes this difficulty. In a vacuum evaporator the surface of the sample is covered with a thin uniform layer of Victawet (by Victor Chemical Company). Such a layer is commonly evaporated between a sample and a carbon replica as a wetting agent to further the removal of the replica. A short heating process is necessary to transform the Victawet into the active wetting agent, sodium metaphosphate.[8] Before this transformation, the evaporated material closes the pores of the surface and allows a drop of aqueous iron oxide colloid solution to be spread and dried on the sample. A short heating of the sample surface transforms the Victawet-layer underneath the dry colloid into an active wetting agent. After platinum shadowing, the sample receives a conventional carbon layer. A water droplet lifts the carbon layer which contains the platinum shadow and the colloid like a blister off the sample surface. An electron microscope object carrier is slipped into this blister and the carbon replica lifted straight up.

We applied this replicating method first on two types of magnets: small-crystal barium ferrite magnets ($H_c$ about 4000 oe) and compacts of mercury-cathode deposited particles of Fe-Co alloy with various binders ($H_c$ about 1000 oe). Both materials had shown peculiar phenomena under the light microscope.

Barium ferrite with coercive forces of 4000 oe or more is composed of crystals the dimensions of which are 0.2 to $1 \times 10^{-4}$ cm. When colloid is spread over such a magnet, structures are found in the light microscope at the limit of its resolving power which resembled the domain patterns on larger crystals of barium ferrite. When the magnetization of the magnet is reversed the colloid shifts in sudden jumps, no matter how slowly the increase of the reversing field takes place. On oriented barium ferrite, sudden changes of the colloid arrangements occur over areas which contain hundreds of crystals.[2] Such magnetic cooperation of a multitude of crystals is as decisive for the magnetic properties of barium ferrite as the magnetic interaction domains are for the compacts of elongated single domain particles.

Figure 1(a) is made without colloid, it shows the

* Work was sponsored by the Aeronautical Systems Division of the Air Research and Development Command of the U. S. Air Force.

[1] E. A. Nesbitt and H. J. Williams, Phys. Rev. **80**, 112 (1950).

[2] K. J. Kronenberg, motion picture presented at the Conference for Magnetism and Magnetic Materials, Pittsburgh, 1955.

[3] W. Andrae, Ann. Phys. **19**, 10 (1956).

[4] A. Kussmann and J. H. Wollenberger, Z. angew. Phys. **8**, 213 (1956).

[5] K. J. Kronenberg and R. K. Tenzer, J. Appl. Phys. **29**, 299 (1958).

[6] D. J. Craik and E. D. Isaac, Proc. Phys. Soc. (London) **76**, 1160 (1960).

[7] D. J. Craik, Proc. Phys. Soc. (London) **B69**, 647 (1956).

[8] J. O. Stiegler and T. S. Noggle, Rev. Sci. Instr. **32**, 406 (1961).

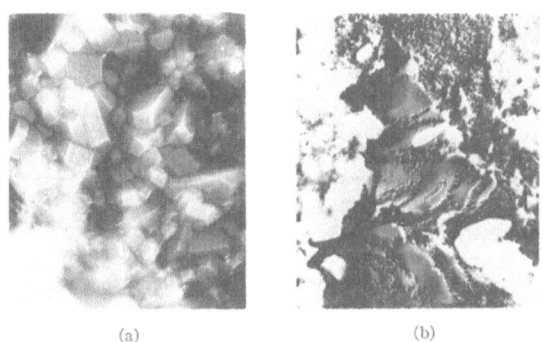

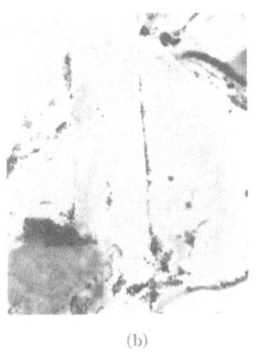

FIG. 1. Surface of small-crystal barium ferrite 1:20 000. (a) Pt-shadowed carbon replica. (b) Pt-shadowed carbon replica with colloid patterns.

FIG. 2. Surface of compact of elongated single domain particles. (a) Pt-shadowed carbon replica 1:10 000. (b) Pt-shadowed carbon replica with colloid alignment 1:20 000.

smooth clean surfaces of barium ferrite crystals. No surface roughness is found which could influence the arrangement of colloid. Figure 1(b) shows the colloid arrangements on the crystals of the same magnet. The colloid has arranged itself in a highly diversified way. On some crystal planes it has aligned intersecting the crystal or forming spikes. These shapes are very similar to the ones found on larger barium ferrite crystals on the side planes parallel to their hexagonal axis. On other crystal planes the colloid sits in dense arrangements of evenly spaced clusters or waves. This corresponds to the wave patterns found on larger barium ferrite crystals on their hexagonal top faces.[9] The main difference of the colloid arrangements on the small crystals is the distance between the arranged lines. On small crystals we find distances of 800 to 5000 A. This is in agreement with findings of Kooy and Enz[10] extending them to smaller sizes.

[9] K. J. Sixtus, K. J. Kronenberg, and R. K. Tenzer, J. Appl. Phys. 27, 1051 (1956).
[10] C. Kooy and U. Enz, Philips Research Repts. 15, 7 (1960).

Surfaces of compacts of mercury-deposited iron-cobalt particles without colloid show smooth flat areas of an often surprisingly regular, straight outline [Fig. 2(a)]. Within the flat areas, traces of an oriented alignment of close packed particles are visible. Between such smooth areas one finds rougher areas, the appearance of which depends largely on the type of binder used. Figure 2(b) is one example of the colloid alignment on a smooth surface of such an area. Such colloid alignments were found intersecting the smooth portions of the surfaces. It appears from this that compacts of elongated single domain particles do not have a uniform structure of separated particles evenly distributed within the binding matrix. The elongated particles seem more or less to be packed into bundles of high density and alignment. Within such bundles, magnetization changes are governed by magnetic interaction. Between the bundles seem to be areas of less regular shape filled with the binder material. More investigations should be conducted in order to establish the influence these arrangements in bundles may have for the properties of the compacts.

JOURNAL OF APPLIED PHYSICS    SUPPLEMENT TO VOL. 33, NO. 3    MARCH, 1962

# Exchange Anisotropy—A Review

W. H. MEIKLEJOHN

*General Electric Research Laboratory, Schenectady, New York*

Exchange anisotropy describes a magnetic interaction across the interface between two magnetic materials. A shifted hysteresis loop, $\sin\theta$ torque curve, and rotational hysteresis in magnetic fields greater than $2K/M_s$ may result from this interaction if one of the materials is antiferromagnetic. This interaction has been found to exist between ferro-antiferromagnetic materials, ferri-antiferromagnetic materials, and ferri-ferromagnetic materials. The work of various people is discussed in terms of the expected behavior in these exchange coupled systems. Some interesting results of the exchange interaction are reviewed. These include improved properties of fine particle magnets, a memory effect in a mixed ferrimagnetic spinel, a rotatable anisotropy in thin films, and an explanation for the reverse magnetization of a deposit in the earth's crust. The interfacial conditions necessary to obtain the interaction between the two magnetic systems are discussed, and it is shown that they are met in several cases. Models are presented which yield rotational hysteresis in magnetic fields greater than $2K/M_s$ as has been found in all of the exchange coupled systems.

## INTRODUCTION

THE term exchange anisotropy was coined to describe a magnetic interaction between an antiferromagnetic material and a ferromagnetic material. The interaction was first discovered in the Co-CoO system[1] and has been extended to other materials by various workers. The interaction has now been found between antiferromagnetic and ferrimagnetic phases and also between ferrimagnetic and ferromagnetic phases.

The origin of this exchange anisotropy is considered to be the coupling of a ferromagnetic spin system to an antiferromagnetic spin system. To obtain a preferred direction of the coupling it is usually necessary that the Curie temperature of the ferromagnet be greater than the Néel temperature of the antiferromagnet. If the material is exposed to a large magnetic field at a temperature above $T_N$, the spins of the ferromagnet will align with the field, and as the material is cooled through $T_N$, the spin planes of the antiferromagnet next to the ferromagnet will align ferromagnetically. The next spin plane in the antiferromagnet orders so as to be oppositely directed to the first spin plane, and when the antiferromagnet is fully ordered and has a high anisotropy, it holds the ferromagnet in the field cooling direction. This simple model qualitatively explains the general features of the interaction-namely a displaced hysteresis loop, $\sin\theta$ torque curve, and a high-field rotational hysteresis.

An almost ideal displaced hysteresis loop was found by Kouvel[2] in a $(Ni,Fe)_3Mn$ alloy when cooled in a magnetic field as shown by the solid curve in Fig. 1. The equal magnetization values in the highest measured magnetic fields indicate that this is not just a minor hysteresis loop.

Another feature of the ferro-antiferromagnetic coupled system is a $\sin\theta$ torque function. The shifted hysteresis loop can be derived from a model exhibiting this type of torque curve, and therefore they represent the same phenomenon. The $\sin\theta$ torque function indicates that exchange anisotropy is a unidirectional anisotropy and not a uniaxial anisotropy as represented by a $\sin 2\theta$ torque function. The $\sin\theta$ torque curve is displaced from the zero torque axis due to a rotational hysteresis loss. This loss at high magnetic fields is the second distinct feature of exchange anisotropy and is probably a more general characteristic of the effect than is a shifted hysteresis loop.

Rotational hysteresis exists for magnetic fields greater than $2K/M_s$, where $K$ and $M_s$ are the crystalline anisotropy constant and saturation magnetization of the ferromagnetic component. Rotational hysteresis loss occurs in a pure ferromagnetic particle with uniaxial anisotropy for magnetic fields greater than $K/M_s$ but less than $2K/M_s$. This type of loss is represented in

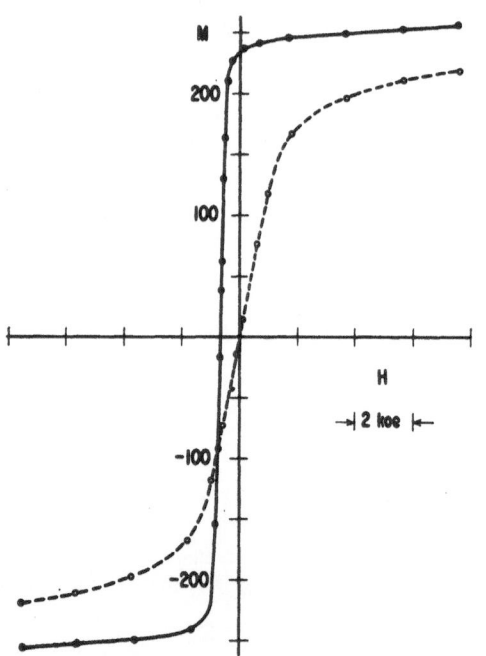

FIG. 1. Hysteresis loops measured at 4.2°K for 18.9 atomic % Fe in $(Ni,Fe)_3Mn$. Specimen cooled in +5-koe field (solid curves) or in zero field (dashed curve) (after Kouvel).

[1] W. H. Meiklejohn and C. P. Bean, Phys. Rev. **102**, 1413 (1956); **105**, 904 (1957).

[2] J. S. Kouvel, J. Phys. Chem. Solids **16**, 152 (1960).

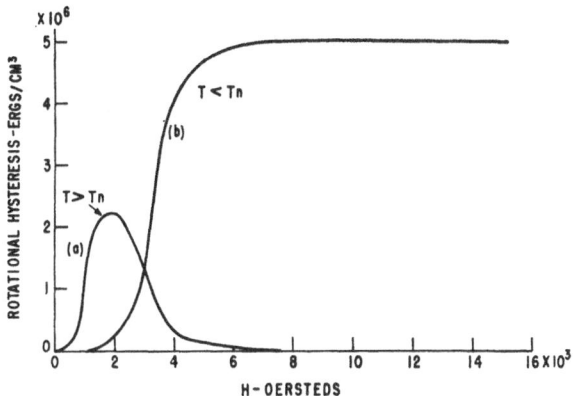

FIG. 2. Rotational hysteresis of 200-A cobalt particles that have a cobaltous oxide shell. Data for curves (a) and (b) taken at 300° and 77°K, respectively.

Co-CoO by curve "a" of Fig. 2, where the temperature of the antiferromagnet is greater than $T_N$, and therefore the exchange anisotropy is zero. Curve "b" of Fig. 2 shows the rotational hysteresis in Co-CoO when $T < T_N$, and the antiferromagnet is ordered and coupled to the ferromagnet. The rotational hysteresis is finite for applied magnetic fields up to 15 000 oe. The existence of a nonvanishing rotational hysteresis in magnetic fields greater than $2K/M_s$ is an indication of an exchange interaction. The origin of this rotational hysteresis is definitely not the consequence of a pure unidirectional anisotropy which accounts for the shifted loop and $\sin\theta$ torque function. The origin of this loss is discussed in the section on rotational hysteresis.

## ANTIFERROMAGNETIC-FERROMAGNETIC SYSTEMS

The majority of the literature on exchange anisotropy deals with the interaction between an antiferromagnet and a ferromagnet. In cases where the antiferromagnet is rigidly held to the lattice (very large magnetocrystalline anisotropy constant), the hysteresis loop of the material should be displaced, and the torque curve should be a $\sin\theta$ function. The rotational hysteresis loss may be nonzero even when measured in magnetic fields greater than $2K/M_s$, where $K$ and $M_s$ are the magnetocrystalline anisotropy constant and saturation magnetization, respectively, of the ferromagnet. In those cases where the antiferromagnet is loosely held to the lattice, the hysteresis loop of the material should be symmetrical, and the torque curve should be a $\sin 2\theta$ or higher multiple-angle function. The material should, however, exhibit nonvanishing rotational hysteresis in magnetic fields greater than $2J_K/M_s$, if the exchange anisotropy constant ($J_K$) of the ferro-antiferromagnetic coupling is greater than the magnetocrystalline anisotropy constant ($K_{AF}$) of the antiferromagnetic-lattice coupling. It will be shown later that this rotational hysteresis may be very small and appear to reduce to zero at high magnetic fields if the ratio $J_K/K_{AF}$ is large.

Very striking examples of ferro-antiferromagnetic interactions are given in the work of Kouvel et al.[2-8] in metallic systems that contain manganese. An important feature of these alloys is that the material is atomically disordered. The interacting regions are thought to be due to statistical fluctuations in the local Mn concentration. The magnetic character of these regions is due to the coexistence of antiferromagnetic and ferromagnetic interactions between Mn atoms of different separation. In some regions the separations between the Mn atoms within the region are such as to produce an antiferromagnetic group while an adjacent region may be a ferromagnetic group. The exchange interaction between these groups results in a shifted hysteresis loop when the material is cooled in a magnetic field. The conceptual separation of these regions into antiferromagnetic and ferromagnetic regions may not be strictly valid when these are intermingled ferro-antiferromagnetic interactions. Kouvel[7] has considered an Ising model of a lattice with antiferromagnetic interactions between nearest neighbors and ferromagnetic interactions between next-nearest neighbor Mn atoms which qualitatively yields the magnetic behavior of these Mn alloy systems.

An example of an almost ideal shifted hysteresis loop is shown by the solid curve in Fig. 1. The hysteresis loop shown for the material when cooled in a zero field (dashed curve) is also ideal in that magnetization has been locked in a random fashion by the exchange anisotropy, and therefore the remanence is zero.

These Mn alloy systems also have a very interesting

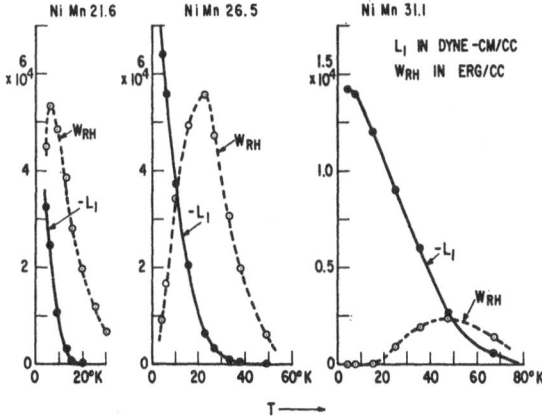

FIG. 3. Amplitude of $\sin\theta$ component of torque, $L_1$, and rotational hysteresis loss, $W_{RH}$, of NiMn alloys measured at 2000 oe. Specimens cooled to 4.2°K in 5000 oe. Numbers denote atomic percent manganese (after Kouvel and Graham).

[3] J. S. Kouvel, C. D. Graham, Jr., and I. S. Jacobs, J. phys. radium **20**, 198 (1959).

[4] J. S. Kouvel and C. D. Graham, Jr., J. Phys. Chem. Solids **11**, 220 (1959).

[5] J. S. Kouvel and C. D. Graham, Jr., J. Appl. Phys. **30**, 312S (1959).

[6] J. S. Kouvel, J. Phys. Chem. Solids **16**, 107 (1960).

[7] J. S. Kouvel, J. Appl. Phys. **31**, 142S, (1960).

[8] J. S. Kouvel, J. Phys. Chem. Solids **21**, 57 (1961).

rotational hysteresis behavior as shown for the Ni-Mn system in Fig. 3. The rotational hysteresis is very small for the 26.5 and 31.1% manganese alloys at very low temperatures when the unidirectional anisotropy is very large as shown by the curves $-L_1$. This result is in agreement with the ideal case where the rotational hysteresis should be identically zero for all fields. As the temperature increases, as pointed out by Kouvel, the crystalline anisotropy ($K_{AF}$) of the antiferromagnetic clusters decreases and becomes comparable with the exchange anisotropy constant $J_K$. At this temperature the antiferromagnet is rotated in an irreversible manner due to its exchange coupling with the ferromagnetic component. This produces a rotational hysteresis loss which is accompanied by the disappearance of the unidirectional anisotropy as represented by $-L_1$. The nature of the dependence of the rotational hysteresis on the ratio $J_K/K_{AF}$ (exchange anisotropy to crystalline anisotropy of the antiferromagnet) is discussed in the section on rotational hysteresis.

The work of Roth[9] on FeO containing interstitial cations is another example of a shifted loop due to an exchange interaction brought about by a statistical distribution.

In some unpublished work Roth and Brun[10] have obtained shifted loops, $\sin\theta$ torque curves, and nonvanishing rotational hysteresis at large fields in non-stoichiometric $LaMnO_3$. The $Mn^{+4}$ concentration was varied from 0 to ~35%.

Exchange anisotropy effects have been found in Fe-FeS by Greiner et al.,[11] in Fe-Al by Kouvel,[12] in oxidized Fe-Co by Darnell,[13] in $Mn_{1-x}Cr_xSb$ by Pry et al.,[14] in cold-worked stainless steel by Meiklejohn,[15] in Fe-FeO by Meiklejohn,[16] in $[Co_x, Ni_{1-x}]$-$[(CoO)_x, (NiO)_{1-x}]$ by Schmid,[17] and in $CrO_2$-$Cr_2O_3$ by Darnell and Cloud.[18]

Hysteresis loops that are shifted from the origin have been obtained in $UMn_2$ by Lin and Kaufmann,[19] in $\alpha Fe_2O_3$ by Iwata et al.,[20] and in $LaFeO_3$ by Watanabe.[21] There is a possibility that exchange anisotropy is operative in these systems although the authors have not considered such an explanation. Exchange anisotropy can cause a hysteresis loop to be shifted along the

$B$ axis and not along the $H$ axis. If the specimen contains extremely small ferromagnetic clusters in which all the spins are rigidly coupled to the antiferromagnet which in turn is also rigidly coupled to the lattice, then there will be a polarization which cannot be reversed in laboratory fields. The material $\alpha Fe_2O_3$ may represent a slight modification of this system. Suppose the $\alpha Fe_2O_3$ contains cation vacancies which cause small ferrimagnetic clusters. It appears possible that these small clusters (several atoms) may be aligned by cooling in a magnetic field and be coupled to the antiferromagnetic so strongly that they are not reversed in accessible applied magnetic fields. Now if the antiferromagnet has a parasitic ferromagnetism due to canted spins as proposed by Dzyaloshinsky,[22] then the hysteresis loop, which is a result of the contributions of the magnetization of these two components, would be shifted along the $B$ axis but not along the $H$ axis.

Magnetic systems containing NiO are the best examples of low-crystalline anisotropy antiferromagnetic materials that should show a nonvanishing rotational hysteresis at high fields but no shifted loop. The work of Roth and Slack[23,24] has shown that the spin system in antiferromagnetic NiO can be rotated in the (111) plane by fields greater than 2400 oe. This work explains why neither Schmid[17] nor I[25] obtained a shifted hysteresis loop in oxidized nickel particles. The very interesting work of Schmid shows a rotational hysteresis at 8000 oe that goes to zero at the Néel temperature of NiO. The $\sin\theta$ torque curves obtained at 8000 oe are unexpected in view of the work of Roth and Slack. According to Roth,[26] the field required to rotate the spins in the (111) is quite strain-sensitive, and this may account for the $\sin\theta$ curve at 8000 oe.

The work of Prosen et al.[27] showing a rotatable anisotropy in oxidized permalloy films and the work of Lommel and Graham[28] on nickel films are probably other examples of ferro-antiferromagentic interaction as pointed out by the authors.

We can make a simple calculation of the field in excess of the coercive field of the pure material needed to rotate the NiO spin system if we assume that the NiO is rotated only because of its coupling to the nickel. In this model it is assumed that the applied magnetic field causes a torque on the ferromagnetic nickel which in turn causes a torque on the NiO by an exchange anisotropy. The NiO will turn around at the same field at which rotational hysteresis occurs. It is shown in the section on rotational hysteresis that the threshold field is given by $HM/J_K=1$. On rewriting this equation we

[9] W. Roth, J. Appl. Phys. 30, 303S (1959).

[10] W. L. Roth and B. L. Brun (unpublished work, General Electric Research Lab. Rept. 58-RL-2109).

[11] J. H. Greiner, I. M. Croll, and M. Sulich, J. Appl. Phys. 32, 188S (1961).

[12] J. S. Kouvel, J. Appl. Phys. 30, 313S (1959).

[13] F. J. Darnell, J. Appl. Phys. 32, 186S (1961).

[14] R. H. Pry, J. S. Kouvel, and E. S. Miksch, J. Appl. Phys. 31, 162S (1960).

[15] W. H. Meiklejohn, J. Appl. Phys. 32, 274S (1961).

[16] W. H. Meiklejohn, J. Appl. Phys. 29, 454 (1958).

[17] H. Schmid, Cobalt 6, 8 (1960), published by Centre D'Information Du Cobalt, Brussels 35, rue des Colonies, Brussels 1, Belgium.

[18] F. J. Darnell and W. H. Cloud (unpublished work).

[19] S. T. Lin and A. R. Kaufmann, Phys. Rev. 108, 1171 (1957).

[20] T. Iwata, M. Iwata, and M. Yamumoto, J. Phys. Soc. Japan 14, 855 (1959).

[21] H. Watanabe, J. Phys. Soc. Japan 14, 511 (1959).

[22] I. Dzyaloshinsky, J. Phys. Chem. Solids 4, 241 (1958).

[23] W. L. Roth and G. A. Slack, J. Appl. Phys. 31, 352S (1960).

[24] W. L. Roth, J. Appl. Phys. 31, 2000 (1960).

[25] W. H. Meiklejohn (unpublished work).

[26] W. L. Roth (private communication).

[27] R. J. Prosen, J. O. Holmen, and B. E. Gran, J. Appl. Phys. 32, 91S (1961).

[28] J. M. Lommel and C. D. Graham, Jr., J. Appl. Phys. 33, 1160 (1962), this issue.

obtain $H = (J_K/K_{AF})(K_{AF}/M)$. From the work of Roth and Slack[28] on NiO we obtain $K_{AF}/M \sim 1$, where $M$ is the magnetization of Ni. As long as $J_K/K_{AF} > 1$, then the NiO will rotate. If there is an appreciable rotational hysteresis in the system, then $J_K/K_{AF} < 4$ as shown in the section on rotational hysteresis. Therefore, the increased field above the coercive field should be about 1 to 4 oe, which appears to be in reasonable agreement with the work of Prosen et al. If a large field is required to reverse the magnetization of the pure ferromagnetic material, then the volume torque exerted directly on the NiO by this large field becomes important. This torque attempts to align the spins perpendicular to the field and may greatly increase the threshold field.

The $\sin 2\theta$ torque curves for fields less than the threshold field is probably due to the nucleation of domain walls in the thick ferromagnetic film. The ferromagnetic material between this Block wall and the NiO is then the only ferromagnetic material exerting an appreciable torque on the NiO. There should be a small $\sin\theta$ component and an extremely small shift in the hysteresis loop for these low fields.

### ANTIFERROMAGNETIC-FERRIMAGNETIC SYSTEMS

The exchange coupled ferri-antiferromagnetic systems have essentially the same properties as the ferro-antiferromagnetic systems. There is the intriguing possibility of the ferrimagnetic material having a compensation point which would result in the hysteresis loop being shifted down or to the right.

Examples of ferri-antiferromagnetic interactions in materials having a relatively fixed antiferromagnet are contained in the work of Jacobs and Kouvel,[29] Roth and Cooper,[30] and Meiklejohn and Carter.[31] Mixed manganites with the formula $(D_x Mn_{1-x})$ $Mn_2O_4$, where $D_x$ represents diamagnetic ions $Zn^{2+}$ and $Mg^{2+}$, were investigated by Jacobs and Kouvel and showed displaced hysteresis loops with almost equal magnetizations at $\pm 10$ koe. Pulsed fields of $-140$ koe would not make the loop symmetrical. The authors believe that random distribution of the $Mn^{2+}$, $Zn^{2+}$, and $Mg^{2+}$ ions on the tetrahedral sites results in antiferromagnetic regions and ferrimagnetic regions. This model is similar to that proposed for disordered alloys. They also find an "extraordinary" viscosity as demonstrated by the measured pulsed loop being not completely outside the quasi-static loop. This may indicate a breakdown of the unidirectional anisotropy.

Roth and Cooper,[30] in some unpublished work, have found a shifted loop in hexagonal mixed barium-potassium ferrites, $[x BaFe_{12}O_{19}(1-x)KFe_{11}O_{17}]$. $\sin\theta$ torque curves and rotational hysteresis were obtained in

magnetic fields up to the highest measured field of 15 000 oe. The displaced loop is believed to be due to ferro-antiferromagnetic exchange interactions between spinel-like blocks which are separated by layers containing either barium or potassium. The displaced hysteresis loop at composition $x > 0.5$ appears to be associated with a new $A$ phase.

Meiklejohn and Carter[31] found a hysteresis loop displaced to the right of the $B$ axis in a solid solution $x FeTiO_3(1-x)Fe_2O_3$. This material is a component of the Haruna deposit in Japan which is magnetized in a direction opposite to the earth's field. The opposite displacement from that found in the Co-CoO system was not due to a compensation point in the ferrimagnetic phase. The proposed explanation for this "reverse" shifted loop was that the antiferromagnetic phase (rich in $Fe_2O_3$) ordered with the spins nearly parallel to the field direction due to its parasitic ferromagnetism. The ferrimagnetic phase ordered at a lower temperature in such a manner that the net magnetization preferred to be opposite to the field cooling direction. No detailed model for this interaction was proposed. The existence of ferrimagnetic and antiferromagnetic phases was based upon the work of Uyeda.[32] Due to the presence of a transition,[33] it was possible to cool the material from 500°C to room temperature in a magnetic field and obtain a loop shifted to the right, and then cool the material in the same field down to 77°K and obtain a loop shifted to the left. This lower temperature shift is believed due to the $Fe_2O_3$ spins ordering at $-20$°C so as to be perpendicular to their direction above $-20$°C. Since the ferrimagnet is ordered at $-20$°C, the $Fe_2O_3$ spins order so as to hold the ferrimagnet, and a "normally" shifted hysteresis loop occurs.

Menyuk et al.[34] have reported a hysteresis loop shifted below the $H$ axis in a cobalt vanadate. The material has a compensation point at 70°K, and they explain the displaced loop as due to the cobalt spins being held to the lattice by a strong crystalline anisotropy.

An example of a ferrimagnetic system coupled to a low-crystalline anisotropy antiferromagnet is represented by the work of Jacobs and Lawrence.[35] They found that they could magnetize a ferrimagnetic spinel $(\sim NiFe_{0.15}Cr_{1.85}O_4)$ at a temperature below its Curie point, heat it above the Curie point in a zero field, and then have it regain 25% of its low-temperature remanence when cooled below the Curie point in a zero field. The explanation of this memory effect involves an exchange coupling to NiO impurity. The NiO is oriented by the large magnetic field at the low temperature and then acts as the nucleating site when the material is cooled from above the Curie temperature of the spinel.

[29] I. S. Jacobs and J. S. Kouvel, Phys. Rev. **122**, 412 (1961).
[30] W. L. Roth and A. S. Cooper (unpublished work, General Electric Research Lab. Rept. 60-RL-2461M).
[31] W. H. Meiklejohn and R. E. Carter, J. Appl. Phys. **30**, 2020 (1959); **31**, 164S (1960).

[32] S. Uyeda, J. Geomagnet. Geoelec. **9**, 61 (1957).
[33] F. J. Morin, Phys. Rev. **78**, 819 (1950).
[34] N. Menyuk, K. Dwight, and D. G. Wickham, Phys. Rev. Letters **4**, 119 (1960).
[35] I. S. Jacobs and P. E. Lawrence, J. Appl. Phys. **31**, 1388 (1960).

## FERRIMAGNETIC-FERROMAGNETIC SYSTEMS

Exchange coupled ferri-ferromagnetic systems should not exhibit a shifted hysteresis loop because the anisotropy fields of both systems are usually less than available laboratory magnetic fields. On the same basis one would expect the rotational hysteresis to be zero for applied magnetic fields greater than these individual anisotropy fields. However, experiments on oxidized iron ($Fe_3O_4$) particles show a rotational hysteresis in an applied magnetic field equal to about four times the maximum anisotropy field.

Iron particles of about 200-A-diam were oxidized until about 50% by volume was $Fe_3O_4$. The particles then consisted of a core of iron and a thick shell of $Fe_3O_4$. The particles were cooled in a magnetic field from 500°C to liquid nitrogen temperature (77°K). A torque curve measured on this sample at 15 000 oe is shown in curve "b" of Fig. 4. The material has a large sin $2\theta$ torque and a rotational hysteresis represented by the displacement of the torque curve from the axis. The sin $2\theta$ torque is probably due to a low-temperature magnetic anneal. When natural crystals of $Fe_3O_4$ are cooled in a magnetic field below −110°C, it has been shown[36] that a uniaxial anisotropy develops due to a cubic to orthorhombic transformation. Apparently the presence of excess iron causes the $Fe_3O_4$ to behave more like natural $Fe_3O_4$ than like synthetic $Fe_3O_4$. The synthetic material develops a very small sin $2\theta$ torque as shown in curve "a" of Fig. 4, which is in agreement with

previous work.[36] The important feature of curve "a" is the zero displacement of the torque curve which indicates a zero rotational hysteresis as expected. The nonzero rotational hysteresis of curve "b" is unexpected.

Curve "c" of Fig. 4 shows that the nonvanishing rotational hysteresis in this system occurs above room temperature. The rapid rise at the lowest temperatures may be associated with the cubic to orthohombic transition as proposed by Pearson and Cooper.[37] Although the rotational hysteresis appears to be heading to zero at the $T_N$ of $Fe_3O_4$, higher temperature measurements are necessary in order to resolve this point. The origin of this rotational hysteresis in magnetic fields greater than $2K/M_s$ appears to be due to an exchange anisotropy acting in such a manner as to make the net magnetization of the ferrimagnetic $Fe_3O_4$ opposite to that of the iron. Calculation of the rotational hysteresis for such a model has been presented[38] and is given in more detail in a later section. A similar result might be obtained between two coupled ferrimagnetic systems.

If the ferrimagnetic system is coupled to the ferromagnetic system such that the net magnetization of the ferrimagnet wants to be in the same direction as the ferromagnetic magnetization, then the rotational hysteresis should be zero in fields exceeding the larger of the two values of $2K/M_s$. Thus, if the ferrimagnetic material has a larger crystalline anisotropy than the ferromagnetic material, then the net anisotropy of the system should be greater than the crystalline anisotropy of the ferromagnet. This should manifest itself as an increased coercive force. The work of Falk and Hooper[39] on oxidized iron-cobalt fine particles appears to be an example of this type of interaction. The oxidization of these particles produced a ferrimagnetic coating which increased the coercive force by 500 oe.

## INTERFACIAL CRYSTALLOGRAPHIC CONDITIONS

Experiments on partially oxidized thin films of cobalt by Bean[40] have been very important in showing that exchange anisotropy is an interfacial property. Figure 5 shows the sin $\theta$ torque and rotational hysteresis torque of a 300-A cobalt film as a function of the oxide layer thickness. The independence of the torque on oxide thickness over the range of 40 to 200 A shows that the effect is indeed a surface property.

The proposed explanation for exchange anisotropy in the Co-CoO system requires that the spin planes of the oxide be coherent with the metal. It would seem unnecessary to have the spin direction in the metal parallel to the spin direction in the oxide if the exchange anisotropy field is large compared to the crystalline anisotropy field in the metal. This appears to be true in fine

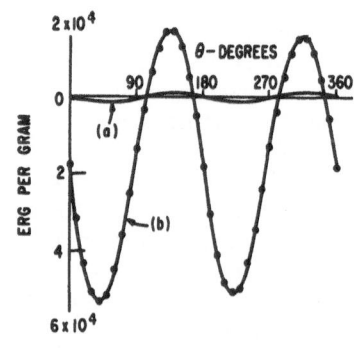

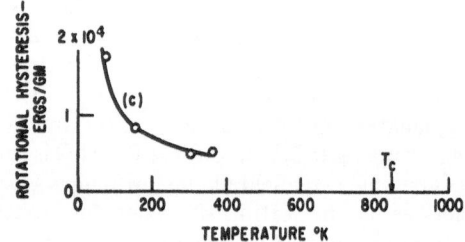

Fig. 4. Torque curves taken at 77°K on pure $Fe_3O_4$ powder; (a) and oxidized 200 A-iron particles, (b) after cooling from 500°C in 15000 oe, measurements taken at 15000 oe, and (c) rotational hysteresis of the oxidized iron particles (50% $Fe_3O_4$ shell).

[36] H. J. Williams and R. M. Bozorth, Revs. Modern Phys. **25**, 79 (1953), and references therein.

[37] R. F. Pearson and R. Cooper, Proc. Phys. Soc. (London) **78**, 17 (1961).
[38] W. H. Meiklejohn, Bull. Am. Phys. Soc. II **6**, 54 (1961).
[39] R. B. Falk and G. D. Hooper, J. Appl. Phys. **32**, 190S (1961).
[40] C. P. Bean, *Structure and Properties of Thin Films* (John Wiley & Sons, Inc., New York, 1959), p. 331.

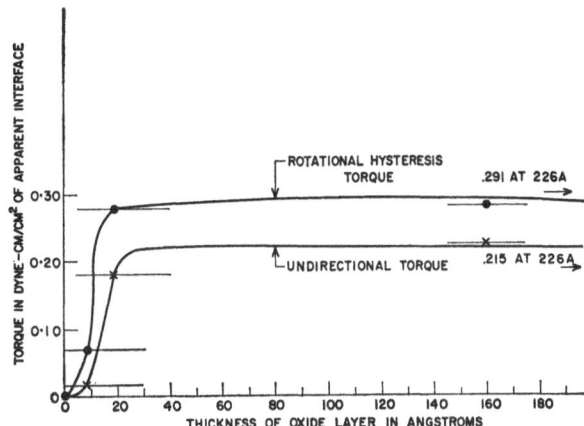

FIG. 5. Development of rotational hysteresis and unidirectional torque on a 300-A cobalt film initially reduced in hydrogen and subsequently heated (oxidized) in air (after Bean).

particles of cobalt where the measured coercive force is about 1000 oe, while the hysteresis loop can be shifted by 5000 oe. If we assume as a first approximation that the metal is magnetically isotropic, then we can concern ourselves only with the problem of the coherency of the plane in the oxide containing parallel spins. The ferromagnetic planes of NiO, CoO, and FeO have been shown by the neutron diffraction work of Shull et al.[41] and Roth[42] to be the (111).

Newkirk and Martin[43] have investigated the crystallographic orientation relationship between Ni and NiO and between Co and CoO. Although $Co_3O_4$ was formed instead of CoO, the authors believe that the results are directly applicable to CoO. They found that the (111) of CoO was parallel to the (00.1) of the Co. The (111) plane of NiO was found to be parallel to the (111) of Ni. These are the desired coherency planes according to the exchange anisotropy model.

Some very good work on these orientation relationships has also been done by Schmid.[17] The technique used in this investigation was to form the oxide on very thin metal sheets and then observe the relative metal and oxide orientations by electron-transmission diffraction without removal of the oxide. The results obtained in this work are in agreement with the work of Newkirk and Martin. Schmid also formed $\alpha$-$Fe_2O_3$ on iron and deduced that the crystallographic orientation of FeO could be determined from the $\alpha$-$Fe_2O_3$ orientation. He found that the (111) of FeO is parallel to the (111) plane of the metal. This again is the desired coherency plane for exchange anisotropy in this system. Schmid also found on one sample of Co-CoO that the (110) of CoO was parallel to the (00.1) of Co and suggests that this result explains the occurrence of a symmetrical loop when the material is cooled in zero field. Such an orien-

tation is not necessary to explain the symmetrical loop when the material is cooled in zero field. Such a symmetrical loop can be explained as due the ferromagnetic magnetization of each particle being oriented parallel to the nearest (111) in the randomly oriented CoO.

Bean and Doyle[44] have performed some interesting experiments on thin cobalt films that bear directly on this problem. They evaporated cobalt on the (111) face of a heated MgO single crystal and formed fcc cobalt. When this was partially oxidized to form the (111) of CoO parallel to the (111) plane of the fcc cobalt, they obtained a $\sin\theta$ torque curve as expected. However, when they evaporated Co on the (100) face of the MgO single crystal to form a (100) plane of cobalt and then oxidized the surface of the cobalt, they did not obtain a $\sin\theta$ torque. Presumably this is due to the fact that there are alternating spins on the coherent (100) plane of the CoO.

## ROTATIONAL HYSTERESIS

The rotational hysteresis in a single domain ferromagnetic particle has been shown by Bean and Meiklejohn[45] to have a finite value only when the magnetic field is greater than $K/M_s$ and less than $2K/M_s$. This behavior is shown for oxidized cobalt particles as curve "a" of Fig. 2. The temperature of measurement is greater than $T_N$ for the oxide, and there is no exchange interaction, so the behavior is that of pure cobalt particles. However, at temperatures less than $T_N$ the CoO is ordered, and there is an interaction that causes the rotational hysteresis to have a finite value at magnetic fields greater than $2K/M_s$ as shown by curve "b" of Fig. 2. This nonvanishing rotational hysteresis at high fields is a more general feature of exchange anisot-

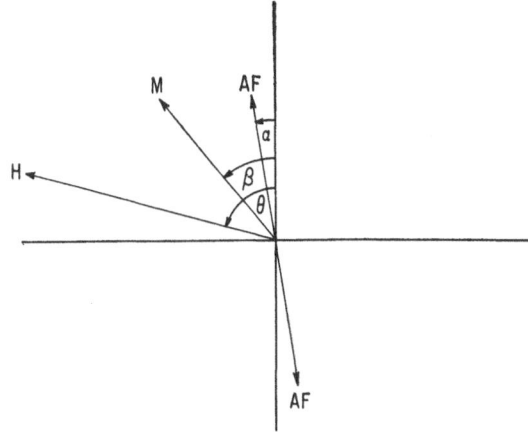

$$E = -HM \cos(\theta-\beta) + K_{AF} \sin^2\alpha - J_K \cos(\beta-\alpha)$$

FIG. 6. Vector diagram and energy function used to calculate the rotational hysteresis loss shown in Fig. 7.

[41] C. G. Shull, W. A. Strauser, and E. D. Wollan, Phys. Rev. 83, 333 (1951).
[42] W. L. Roth, Phys. Rev. 110, 1333 (1958); 111, 772 (1958).
[43] J. B. Newkirk and W. G. Martin, Trans. Met. Soc. AIME 212, 398 (1958).

[44] C. P. Bean and M. V. Doyle (private communication).
[45] C. P. Bean and W. H. Meiklejohn, Bull. Am. Phys. Soc. 1, 148 (1956).

ropy than is a shifted hysteresis loop. A shifted hysteresis loop will not occur in an exchange-coupled ferri-ferromagnetic system. Nonvanishing rotational hysteresis at high magnetic fields is not proof of an exchange anisotropy because it must be possible to have losses in certain materials by other means.

On the other hand, a pure unidirectional anisotropy which accounts for a shifted hysteresis loop and a $\sin\theta$ torque function yields no rotational hysteresis at any magnetic field.

In the work on the Co-CoO system by Meiklejohn and Bean[1] the nonvanishing rotational hysteresis at high fields was attributed to the rotation of the antiferromagnetic phase in some particles where the oxide was very thin and did not have the high crystalline anisotropy of bulk material. A similar explanation was invoked by Kouvel[2] in the alloy system. We will now examine such a model as shown by a vector diagram in Fig. 6. The volumes of ferromagnetic material and antiferromagnetic material are assumed to be equal and separated by a plane interface. No crystalline anisotropy has been introduced for the ferromagnet in this model as shown by the energy expression in Fig. 6. Therefore, any rotational hysteresis is due to the assumed exchange interaction between the single domain ferromagnet ($M$) and the single domain antiferromagnet ($AF$). The antiferromagnet is coupled only to the ferromagnet ($M$), and therefore $\alpha$ is completely determined by $\beta$. As the magnetic field $H$ is rotated, $M$ is pulled around and in turn pulls around $AF$. If $H > J_K/M$ and $J_K > K_{AF}$, then at some critical angle $\theta_c$ the value of $\alpha$ and $\beta$ will change discontinuously and irreversibly to larger values. The difference in energy between the two equilibrium positions is the rotational hysteresis loss per half cycle. The energy and torque equations are given in Appendix I.

The reduced loss $W_r/J_K$ as a function of the reduced field $HM/J_K$ is shown in Fig. 7. Curve "a" shows the loss when the exchange anisotropy ($J_K$) is equal to twice the crystalline anisotropy of the antiferromagnet ($K_{AF}$). Although the loss decreases very rapidly with

magnetic field, it appears to be finite at all fields below that which breaks up the antiferromagnet. Curve "b" shows that if $J_K$ is nearly equal to $K_{AF}$, the loss does not fall off as rapidly with field and even appears to be relatively constant, in agreement with some measured curves.[1] However, as shown in curve "c" of Fig. 7, there is a restricted range of values of $J_K/K_{AF}$ where the rotational hysteresis has an appreciable value. It appears that the work of Kouvel shown in Fig. 3 brings out this fact quite clearly. At low temperatures $J_K/K_{AF}$ is probably less than 1, and there is no rotational hysteresis. At temperatures where the shift has almost disappeared, $J_K/K_{AF}$ is probably slightly greater than 1, and the rotational hysteresis has its maximum value.

On the other hand, this is probably not the explanation for oxidized cobalt. The work of Bean on oxidized thin cobalt films (Fig. 5) shows that the CoO still has a large anisotropy even when only 30-A thick. Moreover, the restricted range of $J_K/K_{AF}$, where the losses have an appreciable value, makes this explanation unlikely. The origin of the loss in Co-CoO may be due to screwing in domain walls in the antiferromagnet or ferromagnet. Preliminary calculations on a model which forbids domain walls greater than 180° to be screwed into the ferromagnet does yield rotational hysteresis at high fields.

The vector diagram of a model[36] which may explain the rotational hysteresis found in the Fe-Fe$_3$O$_4$ system (Fig. 4) is shown in Fig. 8. The ferromagnetic and ferrimagnetic systems are assumed to be so coupled that the net magnetization of the ferrimagnet wants to be opposite to that of the ferromagnet. This coupling is introduced as a $-\gamma M_1$ field acting on $M_2$, where $M_1$ and $M_2$ represent the ferromagnetic and ferrimagnetic magnetizations respectively. This coupling is considered to be weak enough so that the magnetizations can be less than 180° apart in laboratory fields. If we follow the magnetizations shown in Fig. 8 as $H$ is rotated, a critical angle $\theta_c$ will occur at which $M_1$ will jump ahead of $H$, and $M_2$ will jump behind $H$ (i.e., smaller $\beta$). This will occur in a discontinuous irreversible manner. Again, the rotational hysteresis per half cycle is the difference in energy of these two configurations. This loss will occur if the magnetic field is greater than the smallest of the four "fields" $\gamma M_1$, $\gamma M_2$, $K_1/M_1$, $K_2/M_2$, and less than the greatest of the four "fields" $\gamma M_1$, $\gamma M_2$, $2K_1/M_1$, and $2K_2/M_2$. The rotational hysteresis loss for $\gamma M_2/\gamma M_1 = 0.2$, $(2K_1/M_1)/\gamma M_1 = 0.8$, and $(2K_2/M_2)/\gamma M_1 = 0.8$ is shown in Fig. 8. The energy and torque equations are given in Appendix II. It is apparent that this model can yield nonvanishing rotational hysteresis for magnetic fields greater than the maximum $2K/M_s$ of the system.

## ACKNOWLEDGMENTS

I wish to express my appreciation to C. P. Bean, I. S. Jacobs, J. S. Kouvel, and W. L. Roth for many helpful discussions. I am also indebted to C. P. Bean, M. V.

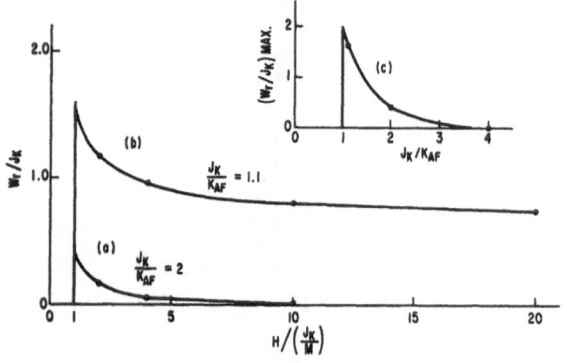

FIG. 7. Reduced rotational hysteresis loss calculated from the vector model in Fig. 6 for: (a) $J_K/K_{AF}=2$, and (b) $J_K/K_{AF}=1.1$ as a function of the reduced field. Curve (c) shows the maximum reduced loss (taken for $HM/J_K=1$) as a function of the ratio $J_K/K_{AF}$.

Doyle, and to W. L. Roth for permission to quote some of their unpublished work.

## APPENDIX I

The simple model used to calculate the rotational hysteresis loss consists of a single domain antiferromagnet in juxtaposition to a single domain ferromagnet with a plane interface. A similar problem has been discussed,[1] and we now extend the calculations to obtain data as a function of magnetic field. The coupling at the interface is considered to be weaker than in either the ferromagnet or the antiferromagnet. The vector diagram for this model is given in Fig. 6. The energy is given by

$$E = -HM \cos(\theta-\beta) + K_{AF} \sin^2\alpha - J_K \cos(\beta-\alpha),$$

where $E$ = energy per unit volume of ferromagnet, $H$ = applied magnetic field, $M$ = magnetic moment per unit volume of ferromagnet, $K_{AF}$ = crystalline anisotropy energy per unit volume of *ferromagnet*, and $J_K$ = exchange anisotropy energy per unit volume of *ferromagnet*.

From the partial derivatives of the energy with respect to $\alpha$ and $\beta$, one obtains the following minimum energy equations:

$$(J_K/K_{AF}) \sin(\beta-\alpha) = \sin 2\alpha$$
$$(HM/J_K) \sin(\theta-\beta) = \sin(\beta-\alpha).$$

These equations were solved numerically for certain values of $J_K/K_{AF}$ and $HM/J_K$. Curves of $\alpha$ and $\beta$ as a function of $\theta$ were plotted from the numerical solutions. These curves were "S" curves, and, therefore, three solutions were obtained for certain ranges of $\theta$. An inspection of these solutions showed that the lowest branch of the "S" curves represented the stable states with a jump to the higher branch occurring at the first "knee" of the curve. The energy difference between these two states is the rotational hysteresis loss per half cycle. This reduced loss is plotted as a function of the reduced field in Fig. 7.

## APPENDIX II

The simple model used to calculate the rotational hysteresis loss consists of a single domain ferrimagnet $(M_2)$ in juxtaposition to a single domain ferromagnet $(M_1)$ with a plane interface. The exchange coupling across the interface is such that the ferrimagnet experiences a torque tending to align the net magnetization

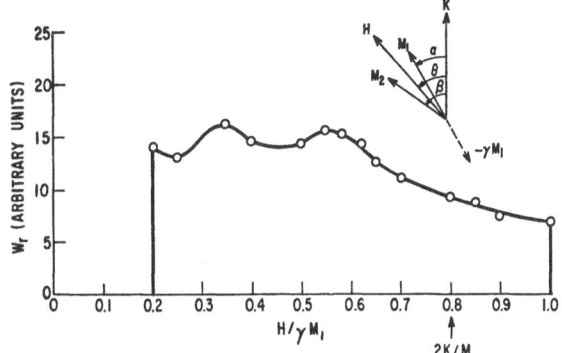

FIG. 8. Vector diagram of the model used to calculate the rotational hysteresis loss shown in the figure. The points shown on the curve are obtained from a numerical calculation. The energy expression for the vector model is given in Appendix II.

antiparallel to the magnetization of the ferromagnet. This torque is introduced in the model by considering that the ferrimagnet $(M_2)$ experiences a field $-\gamma M_1$, as shown in the vector model of Fig. 8. The energy of the system is given by:

$$E = -M_1 \cdot H - M_2 \cdot H - M_2 \cdot (-\gamma M_1) + K_1 \sin^2\alpha + K_2 \sin^2\beta,$$

where $E$ = total energy of the system per unit volume of ferromagnet, $M_1$ = magnetic moment of the ferromagnet per unit volume of the ferromagnet, $M_2$ = magnetic moment of the ferrimagnet per unit volume of the *ferromagnet*, $H$ = applied field, $\gamma$ = exchange coupling constant, $K_1$ = uniaxial crystalline anisotropy energy of the ferromagnet, and $K_2$ = uniaxial crystalline anisotropy energy of the ferrimagnet per unit volume of the *ferromagnet*.

The minimum energy conditions are given by the following equations:

$$h \sin(\theta-\alpha) - M \sin(\beta-\alpha) - k_1 \sin 2\alpha = 0$$
$$h \sin(\beta-\theta) - \sin(\beta-\alpha) + k_2 \sin 2\beta = 0,$$

where

$$h = H/\gamma M_1 \quad M = M_2/M_1$$
$$k_1 = K_1/\gamma M_1^2 \quad k_2 = K_2/\gamma M_1 M_2.$$

Numerical solutions of these equations were obtained for various values of the reduced parameters $h$, $M$, $k_1$, and $k_2$. Curves of $\alpha$ and $\beta$ as a function of $\theta$ were obtained. The curves contained discontinuities within restricted values of the parameters. These discontinuities represented a rotational hysteresis loss. The loss as a function of the reduced field is shown in Fig. 8 where the points represent a numerical calculation.

JOURNAL OF APPLIED PHYSICS     SUPPLEMENT TO VOL. 33, NO. 3     MARCH, 1962

# Alloys and Compounds

## J. A. OSBORN, *Chairman*

## Alloys of the First Transition Series with Pd and Pt*

S. J. PICKART

*U. S. Naval Ordnance Laboratory, Silver Spring, Maryland and Brookhaven National Laboratory, Upton, New York*

AND

R. NATHANS

*Brookhaven National Laboratory, Upton, New York*

Neutron diffraction measurements on the ordered fcc alloys $FePd_3$, $MnPt_3$, and $Cr_{0.3}Pt_{0.7}$ support the existence of localized magnetic moments on the Pd and Pt atoms. The individual moments were determined by combining the difference in moment, based on reasonable assumptions about the magnetic form factors, with the total moment as determined from magnetization measurements. The results for $FePd_3$ are consistent with suggestions based on magnetization measurements; at 300°K Fe has a moment of $2.73 \pm 0.13$ $\mu_B$ and Pd of $0.51 \pm 0.05$ $\mu_B$. In $MnPt_3$ at 77°K a small moment on the Pt is observed ($0.17 \pm 0.04$ $\mu_B$), with most of the moment residing on the Mn ($3.60 \pm 0.09$ $\mu_B$). At room temperature the magnetic ordering in $Cr_{0.3}Pt_{0.7}$ is ferrimagnetic, with a Cr moment (if assumed equal on both Cr and Pt sites) of $2.56 \pm 0.10$ $\mu_B$ and a Pt moment of $0.47 \pm 0.05$ oppositely aligned.

## I. INTRODUCTION

THE variation of atomic magnetic moment with composition in dilute alloys of Fe in Pd led Crangle[1] to suggest that Pd has a localized moment of about 0.4 $\mu_B$ in this alloy system. These measurements were an extension of earlier work by Fallot,[2] who reported that the ordered $FePd_3$ phase was ferromagnetic with a moment of about 4.0 $\mu_B$ per formula unit. It is possible to verify the Pd moment in this phase by measuring the magnetic component of the neutron superstructure scattering, which is proportional to the difference between the Fe and Pd moments.

The ordered composition $FePt_3$, which is antiferromagnetic, was examined by neutron diffraction by Bacon,[3] who concluded that the Pt had no localized moment in this phase. The ordered alloys $MnPt_3$ and $CrPt_3$, reported to be ferromagnetic by Auwärter and Kussmann[4] and Friederich and Kussmann,[5] also suggest themselves as interesting topics for a neutron study aimed at determining the atomic moments.

## II. EXPERIMENTAL

The alloys were prepared by arc melting and rolling, followed by homogenizing heat treatments. The lattice parameters determined by x-ray measurements and the annealing temperatures are listed in Table I. The alloys were of the stoichiometric composition except for $CrPt_3$, which had the composition $Cr_{0.3}Pt_{0.7}$. All the samples as prepared were single phase, but showed preferred orientation effects in the neutron diagram. The data to be presented here, however, were based on field on-off measurements and so may be scaled to the nuclear intensities without the need for absolute standardization.

All the samples showed a high degree of $Cu_3Au$-type ordering, in which the iron group element occupies the corner of the fcc cell and the Pd or Pt the three face-centered positions. The degree of ordering was determined from a measurement of the nuclear scattering of the superlattice peaks relative to the fundamental, the comparison being confined to successive orders of a given reflection to make it independent of the preferred orientation. We found that the ordering was almost complete for all the samples considered.

## III. NEUTRON DIFFRACTION RESULTS

The determination of the magnetic moments on the constituent atoms in these binary-ordered alloys may be made from measurements of the magnetic scattering in the superlattice reflections. If we make the assumption that the moment values of the atoms do not depend strongly on their immediate environment, we may write for the complete structure factor of such a reflection

$$F^2{}_{\text{superlattice}} = S^2[(b_\mathrm{I} - b_\mathrm{II})^2 + 2/3(p_\mathrm{I} - p_\mathrm{II})^2],$$

where $S$ is the long-range order parameter, $b_\mathrm{I}$, $b_\mathrm{II}$, and $p_\mathrm{I}$, $p_\mathrm{II}$ the nuclear and magnetic scattering amplitudes for the two kinds of atoms. The magnetic scattering amplitudes are proportional to the magnetic moments on the individual atoms. Measurements made with and without a magnetic field sufficient to orient the spins along the scattering vector (8–9000 oe) allow determination of $(p_\mathrm{I} - p_\mathrm{II})^2$ with respect to $(b_\mathrm{I} - b_\mathrm{II})^2$, inde-

* Work performed under the auspices of the U. S. Atomic Energy Commission.
[1] J. Crangle, Phil. Mag. **5**, 335 (1960).
[2] M. Fallot, Ann. Phys. **10**, 291 (1938).
[3] G. E. Bacon, Bull. Am. Phys. Soc. **5**, 458 (1960).
[4] M. Auwärter and A. Kussmann, Ann. Physik **7**, 169 (1950).
[5] E. Friederich and A. Kussmann, Physik. Z. **36**, 185 (1935).

TABLE I. Properties of iron group alloys with Pd and Pt.

| Alloy | Lattice parameter | Annealing temperature | Temperature of measurement | Total moment per molecule | Moment difference | Individual moments |
|---|---|---|---|---|---|---|
| FePd$_3$ | 3.855 A | 810°C | 300°K | 4.26±0.08 $\mu_B$ | 2.42±0.15 $\mu_B$ | $\mu_{Fe}$=2.73±0.13 $\mu_B$<br>$\mu_{Pd}$=0.51±0.04 $\mu_B$ |
| Cr$_{1.2}$Pt$_{2.8}$ | 3.866 A | 1200°C | 300°K | 1.76±0.04 $\mu_B$ | 2.38±0.09 $\mu_B$ | $\mu_{Cr}$[a]=2.56±0.10 $\mu_B$<br>$\mu_{Pt}$=−0.47±0.05 $\mu_B$ |
| MnPt$_3$ | 3.890 A | 1100°C | 77°K | 4.12±0.08 $\mu_B$ | 3.60±0.09 $\mu_B$ | $\mu_{Mn}$=3.60±0.09 $\mu_B$<br>$\mu_{Pt}$=0.17±0.04 $\mu_B$ |

[a] Calculated assuming the excess 0.2 Cr atoms per molecule occupy the Pt position at random, with the same moment as the remaining Cr.

pendent of the ordering parameter $S$. The proper choice of sign for $(p_I - p_{II})$ is made by a subsidiary measurement using polarized neutrons; the difference in moment is combined with the value for the total moment of the same samples as measured by McGuire.[6]

Since the magnetic moment $\mu$ is proportional to $p/f$, where $f$ is the appropriately averaged form factor, some knowledge of the angular dependence of the form factor is required for determination of the difference moment. In the absence of any definite information, we assumed in calculating $\mu_I - \mu_{II}$ that the spin density of Fe in FePd$_3$ is similar to pure Fe,[7] and that of Pd is similar to Mo$^{3+}$, the only reported measurement[8] of a 4d transition element. Since the $f_{4d}$ is roughly one-half the $f_{3d}$ at the first reflection, it seems reasonable to assume further that $f_{5d}$ is very nearly zero at this point. For Cr and Mn we used $f_{Mn^{2+}}$, which was found to be approximately correct for metallic Cr[9] and alloys of Mn.[10]

The individual atomic moments determined by this procedure are given in Table I. The probable errors quoted do not include the uncertainty due to the lack of knowledge of the form factor, but are based on statistical and reproducibility factors. Although the exact value for the moments is influenced by the form factor used, the over-all conclusions are not expected to change, since the angle of scattering of the first superlattice reflection is small relative to $f_{3d}$ and large relative to $f_{4d}$ and $f_{5d}$. Experiments using polarized neutrons are in progress to refine the form factor values further, and preliminary analysis indicates that the values for the moment difference in Table I are not off by more than about 10% for this reason.

## IV. DISCUSSION

The observation of a substantial moment on the Pd atoms in FePd$_3$ is consistent with the suggestion of

Crangle[1] and Bozorth et al.[11] that a moment is induced on the Pd near-neighbor atoms in the dilute concentrations of Fe group atoms in palladium. This result is to be contrasted with the case of FePt$_3$, for which the neutron study showed an antiferromagnetic alignment with essentially no moment on the platinum atoms, although it has the same type of ordering and a nearly equal lattice parameter. This would seem to indicate that the predominant interaction in the FePd$_3$ alloy is a ferromagnetic interaction between near-neighbor Fe and Pd atoms, which overcomes the negative interaction between the next-nearest neighbor Fe atoms responsible for the ordering in FePt$_3$.

Perhaps the most unexpected result is the indicated ferrimagnetic alignment in CrPt$_3$, which would imply a negative coupling between the Cr and Pt atoms. Sato[12] has attempted to ascertain the sign and relative strength of the various interactions present in the alloys of Pt with Fe group metals by an analysis of the relevant magnetization data. He concludes that the interaction between the Pt and Fe series element is positive and increases as the atomic number decreases from Co to Cr, which is inconsistent with the above result. However, it should be noted that this result depends on assigning the same moment to the Cr on both the Cr and Pt sites. The magnetization measurements[4,5] show that as more Cr or Mn is added, producing Cr-Cr and Mn-Mn near-neighbor pairs, the Curie point of the alloy rises, while the over-all moment decreases. This would follow if the Cr-Cr and Mn-Mn near-neighbor interactions are negative, as suggested by the magnetic alignment in pure Cr and $\gamma$-Mn. Adopting this model lowers the Cr moment to 2.22 $\mu_B$, with the negative Pt site moment having its origin in the reversed Cr spins. A positive Cr-Cr next-nearest neighbor interaction must be presumed to be responsible in this case for the ferromagnetic alignment in the stoichiometric CrPt$_3$.

It is of interest to attempt to correlate the available neutron data with the experimental observations of ferromagnetism in dilute solutions of the Fe group metals in Pd and Pt. The finding of a larger induced moment of the Pd than on the Pt atoms is consistent

[6] T. R. McGuire (private communication).

[7] R. Nathans, C. G. Shull, G. Shirane, and A. Andresen, J. Phys. Chem. Solids 10, 138 (1959).

[8] M. K. Wilkinson, E. O. Wollan, H. R. Child, and J. W. Cable, Phys. Rev. 121, 74 (1961).

[9] L. M. Corliss, J. M. Hastings, and R. J. Weiss, Phys. Rev. Letters 3, 211 (1959).

[10] S. J. Pickart and R. Nathans, J. Appl. Phys. 30, 280S (1959); J. S. Kasper and J. S. Kouvel, J. Phys. Chem. Solids 11, 231 (1959).

[11] R. M. Bozorth, P. A. Wolff, D. D. Davis, V. B. Compton, and J. H. Wernick, Phys. Rev. 122, 1157 (1961).

[12] H. Sato, J. Appl. Phys. 31, 327S (1960).

with magnetization measurements of Bozorth and his co-workers.[11,13] Their measurements also showed that the substitution of Mn for Fe in a 1% Fe in Pt alloy produces a lowering of the Curie point and a slight increase in the net moment, which is not inconsistent with the Mn moment and the positive Mn-Pt interaction found here. No data are available for comparison with CrPt$_3$, but the absence of a Pt moment in antiferromagnetic FePt$_3$ would be hard to reconcile with the magnetization results. The possibility of course exists that the mechanism of polarization of the Pt or Pd atom in the dilute solutions is different from that in the ordered compositions.

[13] R. M. Bozorth, D. D. Davis, and J. H. Wernick, International Conference on Magnetism and Crystallography, Kyoto, Japan, September, 1961 (to be published in J. Phys. Soc. Japan).

The polarized neutron measurements of the polycrystalline form factors, now in progress, should give some insight into the radial spin density of the atoms in these alloys. Further measurements on single crystals, if such become available, will allow more detailed features of the spin density of the induced Pd and Pt moments to be investigated.

### ACKNOWLEDGMENTS

We are greatly indebted to Dr. D. W. Ernst of the U. S. Naval Ordnance Laboratory and Dr. H. J. Albert of Engelhard Industries, Inc., for furnishing the alloys, and to Dr. T. R. McGuire for communicating the results of his magnetization measurements. Dr. Henri Boutin assisted in the early phases of the neutron diffraction measurements.

---

JOURNAL OF APPLIED PHYSICS     SUPPLEMENT TO VOL. 33, NO. 3     MARCH, 1962

# Structural and Magnetic Properties of Copper-Substituted Manganese-Aluminum Alloys

MAKOTO SUGIHARA AND ICHIRO TSUBOYA

*Electrical Communication Laboratory, Nippon Telegraph and Telephone Public Corporation, Musashino-shi, Tokyo, Japan*

Magnetic properties of ternary Cu(0-12.5 at. %)−Mn(55-42.5 at. %)−Al(45%) alloys have been studied in relation to the change of the tetragonal structure from the binary Mn-Al alloy to the ternary alloys. The substitution of Mn by Cu stabilizes the ferromagnetic tetragonal phase of the Mn-Al alloys. An increase of the Cu content brings about a decrease of the Curie temperature, the intrinsic coercive force, and the ratio of remanence to saturation induction. Correspondingly, the introduction of Cu into Mn-Al alloys first reduces the axial ratio $c/a$ of the tetragonal phase from 1.30 in the 55 at. % Mn-Al alloy to 1.20 at 10 at. % Cu, and then converts the tetragonal structure into body-centered cubic at more than 10 at. % Cu. In the latter case, the ternary alloys are ferromagnetic disregarding aging conditions.

IN the Mn-Al system, there has been observed a ferromagnetic phase by Kôno,[1] Koch *et al.*[2] and Köster and Wachtel.[3] This ferromagnetic phase has a body-centered tetragonal structure at the composition of 45 at. % Al. The magnetic phase is metastable and special heat treatments such as controlled cooling[1,2] or aging after quenching[3] are necessary to obtain this phase. Recently, a new magnetic phase $\kappa$ has been observed in ternary systems of Mn-Al with Cu,[4] Ni,[5] Fe,[6] or Co.[7] This $\kappa$ phase has the stable bcc structure with 40–50 at. % Al.

This paper reports the change of the structural and magnetic properties of the tetragonal phase of Mn-Al

system by substituting Cu for Mn and the connection between this phase and the $\kappa$ phase in the Mn-Al-Cu system.[8]

The nominal composition of the specimens was 45 at. % Al, $(55-x)$ at. % Mn, and $x$ at. % Cu, where $0 \leq x \leq 12.5$. In the alloy preparation, electrolytic Mn and Cu and high purity Al were melted in an alumina crucible by a Tammann furnace in air and cast into rod form. The ingots were homogenized at 900°C for three hours in vacuum and quenched into water. Different specimens of each alloy were aged at several temperatures, 300°, 400°, 500°, 600°, 700°, or 800°C for several hours. After these heat treatments the magnetic properties were measured, and the crystal structure analyzed by x-ray powder technique.

According to the magnetic testing, specimens with $0 \leq x < 10$ are ferromagnetic after aging at temperatures from 400° to 700°C and nonmagnetic after aging at the other temperatures. In the case of $x \geq 10$, specimens are ferromagnetic disregarding aging conditions. The crystal structure of ferromagnetic specimens with $0 \leq x \leq 10$

[1] H. Kôno, J. Phys. Soc. Japan 13, 1444 (1958).
[2] A. J. J. Koch, P. Hokkeling, M. G. v. d. Steeg, and K. J. de Vos, J. Appl. Phys. 31, 75S (1960).
[3] W. Köster and E. Wachtel, Z. Metallkde. 51, 271 (1960).
[4] I. Tsuboya and M. Sugihara, J. Phys. Soc. Japan 16, 571 (1961).
[5] I. Tsuboya and M. Sugihara, J. Phys. Soc. Japan 16, 1257 (1961).
[6] I. Tsuboya and M. Sugihara, J. Phys. Soc. Japan 15, 1534 (1960).
[7] I. Tsuboya and M. Sugihara (to be published).

[8] I. Tsuboya, J. Phys. Soc. Japan 16, 1875 (1961).

remains to be body-centered tetragonal as the 55 at.% Mn-Al alloy. The body-centered cubic $\kappa$ phase is observed for $x > 10$. The coexistence of cubic and tetragonal phases is observed in the range $10 \leq x \leq 12$. The tetragonal phase in the Mn-Al system is metastable as it decomposes into $\beta$-Mn and Al rich $\eta$ phase after extensive aging above 400° C. The stability of the tetragonal phase, however, increases for $x > 0$. The lattice constants of the tetragonal phase are $c = 3.58$ A and $a = 2.78$ A for $x = 0$, but $c$ decreases to 3.20 A and $a$ increases to 2.85 A at $x = 12.5$. For $x > 12.5$, only a cubic phase with the lattice constant being 2.98 A is observed. $(c/a - 1)$ versus composition measured after aging for 1 hour at 500°C is shown in Fig. 1. $(c/a - 1) = 0.29$ for $x = 0$, decreases with an increase of $x$ to 0.20 at $x = 10$, and becomes zero for $x > 10$. This tendency is not affected by the time of aging. These results suggest that the tetragonality of the ferromagnetic phase of the Mn-Al system is decreased by substituting Cu for Mn.

The Curie temperature $T_c$ is 350°C for $x = 0$ and decreases with an increase of $x$ as shown in Fig. 1. $T_c$ of the tetragonal phase is higher than that of the $\kappa$ phase and two Curie temperatures are observed at the same time for $x = 7.5$ to 10. These temperatures correspond to the two phases that may coexist in this composition range. The intrinsic coercive force $_iH_C$ depends on the

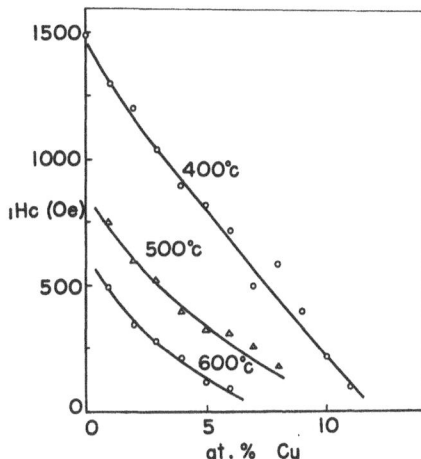

FIG. 2. Relation between $_iH_C$ and content ($x$ at. %) of Cu substituting for Mn.

aging temperature, the aging time, and the composition of the alloys. When aged at 500° to 700°C, $_iH_C$ decreases with an increase of aging time, while $_iH_C$ increases when the aging temperature is 400°C. In both cases, it becomes almost constant after aging for several hours. The composition dependence of $_iH_C$ measured after aging for 16 hr is shown in Fig. 2 for several aging temperatures. $_iH_C$ monotonously decreases with an increase of $x$ from 1500 oe at $x = 0$ to 100 oe at $x = 10$ for specimens aged at 400°C. The ratio of remanence to saturation induction, $B_r/B_s$ monotonously decreases with an increase of Cu content, for example, from 0.5 for $x = 0$ to 0.4 for $x = 10$, under the aging condition of 500°C for 30 min.

The tetragonal phase showing high coercive force was found in Cu substituted Mn-Al alloys after aging at 400° to 700°C. The tetragonal phase transforms abruptly to the cubic $\kappa$ phase at about 10 at. % Cu. Thus the change of magnetic properties by substituting Cu for Mn is accompanied by a decrease of the tetragonality and the tetragonal→cubic phase transformation.

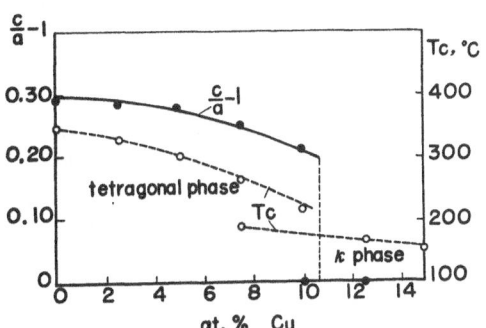

FIG. 1. Relations of $(c/a - 1)$ and Curie temperature $T_c$ to content ($x$ at. %) of Cu substituting Mn after aging at 500°C for 1 hour.

JOURNAL OF APPLIED PHYSICS    SUPPLEMENT TO VOL. 33, NO. 3    MARCH, 1962

# Neutron Diffraction Investigations of Ferromagnetic Palladium and Iron Group Alloys

J. W. Cable, E. O. Wollan, W. C. Koehler, and M. K. Wilkinson

*Oak Ridge National Laboratory, Oak Ridge, Tennessee*

In order to account for the magnetic properties of alloys it becomes important to determine the individual magnetic moments of the constituent atoms. This determination can be accomplished by the combination of neutron diffraction and magnetic induction measurements. Such measurements were made on the following ferromagnetic alloys: $Pd_3Fe$, $PdFe$, $Pd_3Co$, $PdCo$, $Ni_3Co$, and $NiCo$. The average moment values were obtained from magnetic induction measurements while the differences in the atomic moments were determined from either the ferromagnetic diffuse scattering of the disordered alloys or the superlattice reflections of the ordered alloys.

THE magnetic moments of the constituent atoms in alloys are directly related to the electronic structures of those atoms and hence are an important consideration in any theory of metals. The determination of these individual atomic moments requires neutron diffraction measurements and, for ferromagnetic alloys, can best be made by a combination of neutron diffraction and magnetic induction measurements. This report describes such measurements for the ferromagnetic alloys PdCo, $Pd_3Co$, NiCo, $Ni_3Co$, PdFe, and $Pd_3Fe$.

The alloys were prepared by arc melting of stoichiometric mixtures of the 99.9% pure metals. Neutron and x-ray diffraction analyses showed a single face-centered cubic phase for each of the alloys. In an attempt to obtain ordered samples the alloys were heated in a vacuum to 900°C and slowly cooled to 300°C over a 10-day period. The furnace was then turned off and the samples were allowed to cool to room temperature. After this treatment the neutron diffraction patterns showed appreciable short-range order in the Pd-Co alloys, about 90% long-range order in the Pd-Fe alloys and complete disorder in the Ni-Co alloys. Mechanical filing of these materials produced disorder in the Pd-Fe alloys although short-range order effects were still apparent. It also destroyed some of the short-range order in $Pd_3Co$, but had little or no effect on the other alloys.

The magnetic induction measurements were made by movement of the magnetized sample through a coil connected to a ballistic galvanometer. A nickel sample was used for calibration, and saturation moments at infinite field were obtained by matching slopes of the sample magnetization curves to the nickel curve. Measurements were made at room temperature and at 78°K, and corrected to 0°K with a Brillouin function. The results are given in column two of Table I, and represent the average moment per atom at 0°K and infinite field.

The neutron diffraction data were taken at room temperature using a neutron wavelength of 1.08 A. Three types of magnetic scattering are obtained: (1) Normal lattice reflections for which the intensities are proportional to the square of the average moment. These yield the same information as a magnetic induction measurement, but with less precision; (2) superlattice reflections with intensities proportional to $\zeta^2(\mu_1 f_1 - \mu_2 f_2)^2$ in which $\zeta$ is the long-range order parameter, while $\mu$ and $f$ are the moment and form factor of the constituents 1 and 2; (3) ferromagnetic disorder scattering which is proportional to $(\mu_1 f_1 - \mu_2 f_2)^2$. In this case, the measurements extend to small scattering angles and, in the absence of short-range order, can be readily extrapolated to the forward direction to yield the difference in the moment values. If short-range order is present in the sample, the departure of the magnetic diffuse scattering from a smooth form factor behavior introduces some uncertainty in the extrapolation procedure. For an ordered alloy, the long-range order parameter is obtained from the nuclear superlattice intensities. The error in this measurement is primarily due to uncertainties in the form factors applicable to the specific alloy.

The results for the seven alloys in the present investigation are given in Table I. The quoted moment values should be regarded as preliminary results with errors for the iron and cobalt moments of about 0.1 $\mu_B$, and about 0.05 $\mu_B$ for palladium and nickel. Further refinement is being carried out along with additional studies of the more dilute iron in palladium alloys. Details will be published elsewhere.

TABLE I. Magnetic moment values of some palladium and iron group alloys.

|  | $\bar{\mu}$ | $\mu_{Co, Fe}$ | $\mu_{Pd, Ni}$ |
|---|---|---|---|
| PdCo | 1.16 | 2.0 | 0.35 |
| $Pd_3Co$ | 0.85 | 2.0 | 0.45 |
| NiCo | 1.20 | 1.8 | 0.60 |
| $Ni_3Co$ | 0.90 | 1.9 | 0.60 |
| PdFe | 1.60 | 2.9 | 0.30 |
| $Pd_3Fe$ | 0.95 | 3.0 | 0.30 |
| $Pd_3Fe$ (ordered) | 1.09 | 3.0 | 0.45 |

JOURNAL OF APPLIED PHYSICS    SUPPLEMENT TO VOL. 33, NO. 3    MARCH, 1962

# Magnetic Properties of $Cr_5S_6$ in Chromium Sulfides

K. Dwight, R. W. Germann, N. Menyuk, and A. Wold

*Lincoln Laboratory,\* Massachusetts Institute of Technology, Lexington 73, Massachusetts*

Several samples of $CrS_x$ with $1.145 < x < 1.200$ were carefully prepared and chemically analyzed. The expected crystal structures were verified by x-ray analysis, which failed to indicate the presence of any impurities. Conductivity measurements indicated the existence of metallic conductivity in these samples between 77° and 300°K. Magnetic measurements supported the suggestion that the ferrimagnetism of $CrS_x$ materials arises from the presence of the $Cr_5S_6$ phase.

The magnetization curves for $x = 1.194$ demonstrated the following salient properties of the $Cr_5S_6$ phase: a ferrimagnetic Curie point of 305°K, a peak magnetization of 6.9 emu/gm in an external field of 8500 oe, a ferrimagnetic-antiferromagnetic transition over a 5° temperature interval, a transition temperature in the neighborhood of 158°K which is subject both to thermal hysteresis and to shifting by an external magnetic field, and an antiferromagnetic susceptibility below this transition. The transition temperature, as well as the peak magnetization, decreases with increasing chromium content. A simple model is tentatively proposed, and appears to be consistent with the observed behavior.

W E have investigated the magnetic properties of several chromium sulfides $CrS_x$ with $1.145 < x < 1.200$. Over this range of compositions, the spontaneous magnetization increases sharply with increasing $x$, vanishing for $x = 1.145$ and attaining its maximum for $x = 1.200$. This result strongly supports Jellinek's[1] suggestion that all the spontaneous magnetization observed in chromium sulfides arises from the presence of the $Cr_5S_6$ phase (NiAs with ordered vacancies). Our measurements indicate that the vanishing of the magnetic moment[2,3] below 158°K is due to a ferrimagnetic-antiferromagnetic transition which is subject to appreciable thermal hysteresis, thereby casting doubt on any explanation in terms of canted spins.[3] Furthermore, we do not observe any anomalous peak in the magnetic susceptibility near 100°K, as previously reported[3] for an apparently unanalyzed sample.

Our samples were prepared from spectroscopically pure metallic chromium powder and from sublimed sulfur powder pressed into pellets. After being weighed and mixed, some samples (quenched) were sealed in evacuated quartz tubes, heated at 50°/hr to 1000°C, held there for 3 days, and then quenched in ice water. After regrinding, this process was repeated except that the powders were held at 1000°C for a week, and were pressed into bars after cooling. Other samples (slow cooled) were sealed, heated to 400°C at 17°/hr, held there for 24 hr, heated to 1000°C at 25°/hr, held there for 24 hr, and then cooled to room temperature at 30°/hr. After regrinding, these samples were pressed into bars, resealed, heated at 50°/hr to 1000°C, held there for 30 hr, and finally cooled at 30°/hr to room temperature. After their preparation, all the above samples were carefully analyzed for total chromium content by standard oxidation-reduction techniques, the sulfur content being represented by the difference. The results of the chemical analyses of our four samples are shown in Table I.

The ordering of vacancies in the $Cr_5S_6$ phase results in a superlattice structure discernible by x-ray diffraction.[1] Using Cr $K\alpha$ radiation, we examined our samples both by diffractometer and powder photograph techniques, which showed only the NiAs and superlattice structures to be present. The relative strengths of the principal superlattice line, given in Table I, indicate the appearance of the $Cr_5S_6$ phase for values of $x$ only slightly greater than 1.145 ($Cr_7S_8$), in agreement with Jellinek's findings.[1] A comparison of our results for samples C and D suggests that the slow cooling process yields more of the ordered phase than quenching for values of $x$ close to 1.145.

The resistivities of our bar-shaped samples were measured by the four-terminal method using a Keithley milliohmmeter, and the values at 77° and 300°K are included in Table I. In agreement with previous findings,[4] the room-temperature values indicate metallic conductivity, and the resistivities decreased gradually with decreasing temperature. The lack of variation with composition is worth noting; evidently the conduction mechanism is not greatly affected by the ordering of vacancies. It should be pointed out that the existence of metallic conductivity implies that extreme caution must be exercised in the use of ionic models to explain the magnetic behavior.

We used a vibrating-coil magnetometer to measure the magnetizations of our samples continuously from

TABLE I. Characterization of $CrS_x$ samples.

| Sample | Preparation | $x$ | x-ray superlattice | Resistivity (ohm-cm) 300°K | 77°K |
|---|---|---|---|---|---|
| A | slow cooled | 1.194 | strong | $1.06 \times 10^{-3}$ | $0.50 \times 10^{-3}$ |
| B | quenched | 1.178 | strong | $1.00 \times 10^{-3}$ | $0.40 \times 10^{-3}$ |
| C | slow cooled | 1.155 | very faint[a] | $0.94 \times 10^{-3}$ | $0.58 \times 10^{-3}$ |
| D | quenched | 1.155 | none[a] | $0.94 \times 10^{-3}$ | $0.58 \times 10^{-3}$ |

[a] From powder photographs.

\* Operated with support from the U. S. Army, Navy, and Air Force.

[1] F. Jellinek, Acta Cryst. **10**, 620 (1957).

[2] H. Haraldsen and A. Neuber, Z. anorg. u. allgem. Chem. **234**, 337 and 372 (1937).

[3] M. Yuzuri, T. Hirone, H. Watanabe, S. Nagasaki, and S. Maeda, J. Phys. Soc. Japan **12**, 385 (1957).

[4] T. Kamigaichi, K. Masumoto, and T. Hihara, J. Phys. Soc. Japan **15**, 1355 (1960).

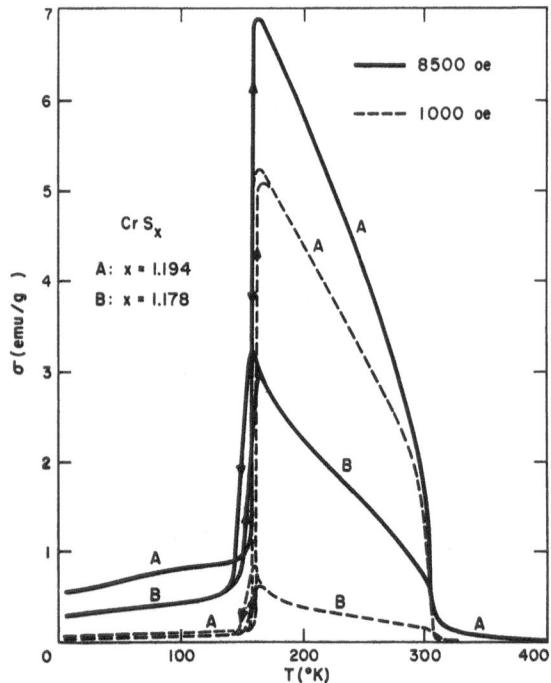

FIG. 1. The temperature variation of the magnetization of samples of $CrS_x$ with $x=1.194$ (A) and $x=1.178$ (B) in external fields of 1000 and 8500 oe.

4.2° to 400°K in external fields of 1000 and 8500 oe. Our magnetization curves for sample A are given in Fig. 1. This sample has a ferrimagnetic Curie point at 305°K. Upon cooling in 1000 oe, the magnetization increases to 5.2 emu/g, drops sharply over a 5° temperature interval centered at 158°K, and then decreases gradually with decreasing temperature. Upon warming, the magnetization increases over a 5° interval centered at 162°K to 5.1 emu/g, and then rejoins the cooling curve. The cooling and warming rates were slow enough to eliminate any experimental temperature lag, so that the 4° thermal hysteresis represents a property of the material.

Upon warming in 8500 oe, the transition occurred at a temperature $T_t(w)=158°K$, which is 4°K lower than $T_t(w)$ in 1000 oe. This shift implies a delicate energy balance in the transition region. Comparison of the 1000 and 8500 oe curves demonstrates the existence of a spontaneous magnetization above $T_t$ and of a pure susceptibility below $T_t$. The fact that this susceptibility decreases with decreasing temperature indicates that the low temperature magnetic phase is antiferromagnetic.

Analogous curves for sample B are included in Fig. 1 for comparison. For this composition, $T_t(c)=151°K$ and $T_t(w)=157°K$ in 1000 oe, showing a decrease in the transition temperature and an increase in the thermal hysteresis. In addition, the transition interval is broadened to about 8°, and $T_t(c)=147°K$ and $T_t(w)=153°K$ in 8500 oe. Although the ferrimagnetism in sample B has the same Curie temperature (305°K) as in sample A, the striking difference between its 1000

and 8500 oe curves indicates either a drastic reduction in the spontaneous magnetization, or a drastic increase in anisotropic effects, or both. It is not clear whether this aspect is due entirely to the compositional difference, or whether the different preparative technique is involved. Nevertheless, $\sigma_{max}$ and $T_t(c)$ for this sample agree closely with the values given in reference 3 for a slow-cooled material.

The magnetization of sample C also showed a definite transition, broadened to about 15°, with $T_t(c)\cong68°K$ in 1000 oe and 58°K in 8500 oe. Here the shift with applied field amounted to about 10°K and the peak magnetization in 8500 oe was only 0.25 emu/g.

From the above data, it appears that the $Cr_5S_6$ phase is responsible for the observed magnetic properties. Increasing the chromium content decreases the total magnetization, lowers and broadens the transition, increases the hysteresis, and increases the shift in applied fields. Hence, our measurements on sample A should closely represent the actual properties of interest, and our peak value of $\sigma=6.9$ emu/g in 8500 oe leads to $0.56\ \mu_B/Cr_5S_6$, which will be an underestimate if appreciable anisotropy exists.[1]

The ordered structure[1] of $Cr_5S_6$ suggests a possible model for its magnetic behavior. In NiAs structures, the dominant exchange interaction is an antiferromagnetic one along the $c$ axis and leads to antiferromagnetic chains along this axis. In $Cr_5S_6$, there are two types of such chains in which every fourth member is missing.[1] Furthermore, these chains will possess net moments if the interactions across the gaps are ferromagnetic. Given these ferrimagnetic chains, a ferromagnetic coupling between them will produce a ferrimagnetic material, whereas an antiferromagnetic coupling will yield antiferromagnetism through a cancellation of the net moments. Thus a shift from ferromagnetic to antiferromagnetic interactions between the two types of chains could give rise to the observed transition.

These chains are connected by both direct cation-cation and indirect superexchange interactions, which can be reasonably expected to compete and give rise to the sensitive balance indicated by the observed shift of $T_t$ with applied field. The antiferromagnetic direct interaction is more sensitive to lattice spacing than the indirect one, and these materials have an appreciable ($\cong2\times10^{-5}$) coefficient of thermal expansion.[3] Hence the lattice contraction could reasonably cause a low-temperature shift from a net ferromagnetic coupling to a net antiferromagnetic one. Furthermore, the room temperature lattice spacings increase with decreasing $x$,[1] thereby increasing the amount of contraction required and lowering the transition temperature, as observed. Consequently our model appears attractive, although a more detailed evaluation will require further investigation.

### ACKNOWLEDGMENTS

We wish to thank E. R. Whipple for performing some of the necessary chemical analyses.

JOURNAL OF APPLIED PHYSICS    SUPPLEMENT TO VOL. 33, NO. 3    MARCH, 1962

# Anomalous Magnetic Moments and Transformations in the Ordered Alloy FeRh

J. S. KOUVEL AND C. C. HARTELIUS

*General Electric Research Laboratory, Schenectady, New York*

Magnetization and electrical resistivity measurements on an iron-rhodium alloy of approximate composition FeRh, having CsCl-type structure, confirm recent x-ray and neutron diffraction evidence for a first-order antiferromagnetic-ferromagnetic transition at about 350°K. From 350°K down to 77°K, an essentially constant, field-independent susceptiblity of about $1 \times 10^{-4}$ emu/g is observed. Above 350°K, the alloy behaves like a normal ferromagnet with a Curie point of 675°K and a saturation magnetization (extrapolated to 0°K) of 130 emu/g. This $\sigma_0$ value is equivalent to an average atomic moment of 1.85 $\mu_B$, or to 3.85 $\mu_B$ per iron atom if the rhodium moment is assumed to be zero. Alternatively, if the iron moment is taken to be 2.2 $\mu_B$, the rhodium moment has to be about 1.5 $\mu_B$. Qualitative results are given for the effect of high magnetic fields and pressures on the first-order transition temperature of this alloy.

EARLY measurements by Fallot and Hocart[1] on iron-rhodium alloys of about 50 at.% Rh revealed an abrupt increase in magnetization as the temperature increased through a critical value, the critical temperature rising with increasing rhodium content. They also observed a temperature hysteresis associated with this magnetization change, which reinforced the suggestion of a first-order transition. Moreover, from x-ray measurements they deduced that above the transition temperature each of their Fe-Rh specimens had an ordered CsCl-type structure and that below it had partially transformed to a fcc structure. However, recent x-ray diffraction experiments by de Bergevin and Muldawer[2] have demonstrated that the CsCl-type phase is retained below the transition temperature and that for increasing temperature the transition consists of a rapid yet uniform expansion of this ordered cubic structure; the estimated volume change was about 1%.

According to subsequent neutron diffraction work reported by de Bergevin and Muldawer[3], the first-order transition in ordered FeRh corresponds to an antiferromagnetic-ferromagnetic transformation. For the ferromagnetic state above the transition temperature, the atomic moment of iron was claimed to be roughly the same as in pure iron but the rhodium moment was unspecified. Unfortunately, the qualitative results presented by Fallot and Hocart[1] for the magnetization of this alloy offer no opportunity for comparison of ferromagnetic moments. The peculiar character of their magnetization versus temperature curves, probably arising from the use of low magnetizing fields, further emphasized the need for a quantitative and more conclusive magnetic study of this unusual material.

We therefore prepared (by induction-melting under argon) an iron-rhodium ingot, for which chemical analysis gave 52.0 at.% Rh. A cylindrical specimen for magnetic measurements was machined to size, $\frac{1}{2}$ in. long by $\frac{1}{4}$ in. in diameter; one of the turnings of fairly uniform cross section was selected for electrical resistivity measurements. Both these specimens plus filings taken from the same ingot were annealed for one day at 950°C and cooled slowly to room temperature. X-ray diffraction from the annealed filings indicated a highly ordered CsCl-type structure ($a_0 = 2.99$ A) with faint traces of a fcc phase ($a_0 = 3.74$ A).

Magnetization measurements at various fields were made by a standard coil-insertion method as the specimen was taken slowly around a complete temperature cycle between 77° and 770°K. The magnetization per gram, $\sigma$, measured in the maximum applied field of 5 koe, is shown plotted versus temperature in Fig. 1(a), and it is evident that a remarkably sharp transition occurs at about 350°K with appreciable temperature hysteresis. At all temperatures below this transition, the magnetization rises slowly and linearly with field, the gram susceptibility being about $1 \times 10^{-4}$ emu. Above the transition, the magnetization rises rapidly and attains

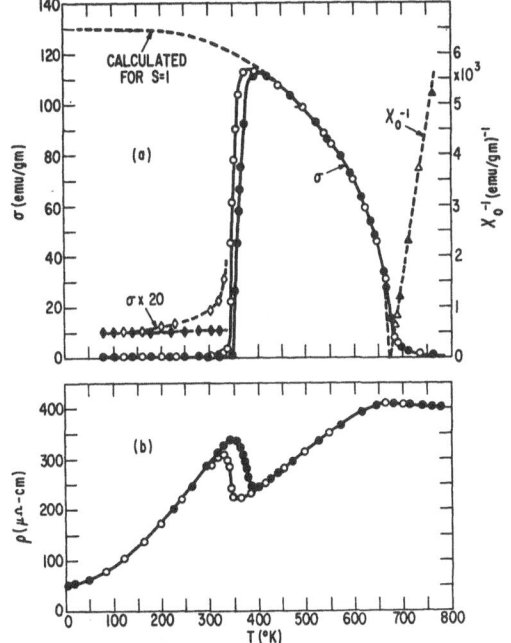

FIG. 1. (a) Magnetization (at 5 koe) and inverse initial susceptibility and (b) electrical resistivity of ordered FeRh for increasing and decreasing temperature (closed and open symbols, respectively).

[1] M. Fallot and R. Hocart, Rev. Sci. **77**, 498 (1939).
[2] F. de Bergevin and L. Muldawer, Compt. rend. **252**, 1347 (1961).
[3] F. de Bergevin and L. Muldawer, Bull. Am. Phys. Soc. **6**, 159 (1961).

a near-saturation value at 5 koe, which decreases with increasing temperature in a manner characteristic of a normal ferromagnet with a Curie point of about 675°K. Electrical resistivity measurements were made by conventional potentiometric techniques over a similar temperature cycle, extending in this case down to 4.2°K. Although the absolute resistivity values are fairly inaccurate due to the necessarily rough estimate of the specimen dimensions, the qualitative features of the resistivity versus temperature curves in Fig. 1(b) suffice to show a clear correlation with the results of our magnetic measurements. The abrupt changes in the resistivity as well as in the magnetization at about 350°K are consistent with the first-order antiferromagnetic-ferromagnetic transition reported earlier, whereas the more gradual variations of both these properties at about 675°K are indicative of the usual second-order transition associated with a ferromagnetic Curie temperature. This behavior closely resembles that found[4] in Cr-modified $Mn_2Sb$, whose abrupt antiferromagnetic-ferrimagnetic transition has been attributed to an exchange-inversion mechanism.[5]

As shown in Fig. 1(a), our magnetization data between 350° and 675°K is fitted very closely by the magnetization-temperature curve computed from the Brillouin function for $S=1$ and normalized to a value at 0°K of 130 emu/g. This $\sigma_0$ value gives an average atomic moment of 1.85 $\mu_B$. Thus, if rhodium is taken to have no moment, the average moment per iron atom is 3.85 $\mu_B$, an abnormally high value. It seems more reasonable to assume that the iron moment in this alloy is about 2.2 $\mu_B$, as in pure iron; each rhodium atom must then contribute about 1.5 $\mu_B$ to the ferromagnetically aligned moment. Furthermore, from the linear temperature dependence of the inverse susceptibility above the Curie point, indicated in Fig. 1(a), we compute an average effective paramagnetic moment, $g\mu_B\{S(S+1)\}^{\frac{1}{2}}$, of 3.2 $\mu_B$, which is consonant with the large ferromagnetic moment of this alloy. Similarly large ferromagnetic moments can be deduced from Fallot's data[6] on disordered bcc Fe-Rh alloys of lower Rh concentration, suggesting an appreciable average moment for rhodium in these alloys as well.

In collaboration with P. E. Lawrence, we have observed metamagnetic transitions to a ferromagnetic state induced in ordered FeRh by large pulsed fields (up to 140 koe) at temperatures below 350°K, the threshold field increasing with decreasing temperature. This is equivalent to a decrease of the first-order transition temperature with increasing field. We have also found, in association with W. R. Tully, that the application of pressures (up to 60 kilobar) increases this transition temperature. Both these preliminary results are qualitatively consistent with the magnetization and volume changes associated with a temperature excursion through this first-order transition at constant field and pressure. The variations with temperature of the lattice parameter of this ordered structure are being studied in detail by L. M. Osika. Also in progress is a magnetic and structural investigation of this material after different heat treatments.

We are grateful to J. E. Hilliard and I. S. Jacobs for the availability of their high pressure and high-pulsed field equipments, respectively, and for useful advice on these experiments. We also acknowledge helpful discussions with C. P. Bean, J. S. Kasper, and D. S. Rodbell.

[4] T. J. Swoboda, W. H. Cloud, T. A. Bither, M. S. Sadler, and H. S. Jarrett, Phys. Rev. Letters 4, 509 (1960).
[5] C. Kittel, Phys. Rev. 120, 335 (1960).
[6] M. Fallot, Ann. phys. 10, 291 (1938).

JOURNAL OF APPLIED PHYSICS    SUPPLEMENT TO VOL. 33, NO. 3    MARCH, 1962

# Age-Hardened Gold Permalloy and Gold Perminvar

E. A. NESBITT AND E. M. GYORGY

*Bell Telephone Laboratories, Inc., Murray Hill, New Jersey*

This work is an extension of previous work which showed that the addition of gold to Permalloy caused the precipitation of a gold-rich phase and thereby permitted the control of the coercive force by heat treatment. This addition of a second phase to Permalloy greatly improved the switching speed of the alloy. Some compositions of gold Permalloy had switching speeds four times faster than that of the standard composition. In the present work, some of the details of the precipitation process were studied. It was found that for some specimens annealed at 650°C, the precipitation of the gold-rich phase depended strongly upon cold working. When some alloys containing gold were quenched from the solution temperature and annealed at 650°C for 2 hr, there was little increase in coercivity and the electron micrographs showed practi-cally no precipitation of particles of 1000-A diameter or larger. However, when the material was quenched from the solution temperature, drastically cold rolled and annealed at 650°C, the coercivity was increased substantially and the electron micro-graphs revealed the presence of many particles having a diameter of 1000 A. Seven percent gold was also added to the Perminvar composition of 43% Ni-34% Fe-23% Co. The alloy behaved in a manner somewhat similar to the gold-Permalloy compositions. Age-hardening appeared in the vicinity of 600°C and square hysteresis loops were obtained. However, the threshold for rapid rotational flux reversal was substantially higher than the coercive force. Consequently, these alloys appeared less desirable for switching applications than Permalloy.

R ECENTLY[1], it was shown that the addition of gold to 78.5 Permalloy caused the precipitation of a gold-rich phase and thereby permitted the control of the coercive force by heat treatment. This is important because these alloys are of interest for magnetic switch-ing devices which sometimes require coercive forces higher than that normally obtained after annealing. This addition of a second phase to Permalloy greatly improved the switching speed of the alloy. Some of the gold Permalloy compositions had switching speeds four

650°C for 2 hr, there was little increase in coercivity and the electron micrograph revealed practically no pre-cipitation of particles of 1000-A diameter or larger. However, when the material was quenched from 900°C, drastically cold rolled and annealed at 650°C, the coer-civity was increased substantially and the electron micrograph revealed the presence of many particles of precipitate approximately 1000 A in diameter. The 14%

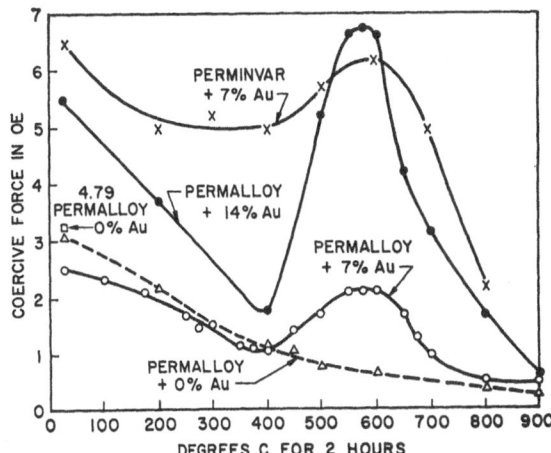

FIG. 1. Coercive force $H_c$ vs temperature of anneal for alloys of Permalloy and Perminvar (tape 0.125×0.000125 in.) containing gold. Note absence of age-hardening peak at 575°C for the 0% Au specimens.

times faster than that of the standard composition. In the present work, some of the details of the precipitation process were investigated. It was found that for some of the specimens annealed at 650°C, the precipitation of the gold-rich phase depended strongly upon cold work-ing. For example, when the 7% Au alloy was quenched from the solution temperature (900°C) and annealed at

[1] E. A. Nesbitt and E. M. Gyorgy, J. Appl. Phys. **32**, 1305 (1961).

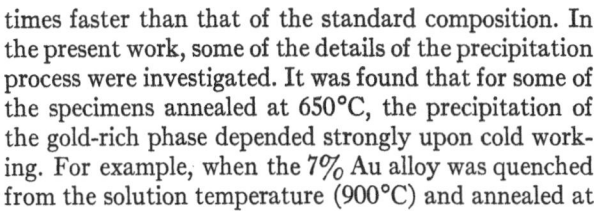

FIG. 2. Reciprocal of the switching time vs applied field for the gold Perminvar alloy after annealing at 600°, 700°, and 800°C. For the latter heat treatment, the specimen was annealed in a magnetic field.

gold Permalloy had a more normal behavior. After it was solution heat treated and aged, substantial precipitation was observed without cold working the alloy.

If the size of the precipitate is appreciably smaller than 1000 A, it will be ineffective in impeding the motion of the domain wall. In general, the procedure for obtaining the precipitated structure was to anneal the tape at the solution temperature of 900°C and then cool rapidly. This was followed by cold rolling the tape from 0.014-in. thickness to 0.00013 in. which is the thickness for switching applications. The gold-rich phase was then precipitated by annealing from 600–650°C for 2 hr. Under these conditions, alloys containing 7 and 14 percent gold yield square hysteresis loops having coercive forces of at least 1.7 and 4.3 oe, respectively.

Seven percent gold was also added to the Perminvar composition of 43% Ni-23% Co-34% Fe. A comparison of its coercive force with that of the gold Permalloy alloys for various annealing temperatures is shown in Fig. 1. The Perminvar curve has approximately the same shape as that for the 7 percent gold Permalloy but its coercive force values are much higher. This is probably the result of a higher strain anisotropy since the magnetostriction of Perminvar is much higher than that of Permalloy. The coercive force is 6.5 oe after cold

rolling and it drops to a value of 5.0 oe after annealing in the region of 300° to 400°C. Upon annealing at 600°C, it rises again probably the result of precipitation hardening. At still higher temperatures the coercive force drops rapidly. The switching characteristics of this alloy were investigated and the results obtained after annealing at 600°, 700°, and 800°C are shown in Fig. 2.

The curves of $1/\tau$ as a function of $H$ are linear at sufficiently large fields. The reciprocal of the slope, i.e., the switching coefficient is approximately 0.30 oe $\mu$sec which is in good agreement with the value predicted by the nonuniform rotational model.[1] The high value of the threshold for the 800°C sample can be attributed to a magnetic annealing treatment. The 800°C sample was cooled in a circumferential field. This magnetic heat treatment is expected to increase the value of the uniaxial anisotropy[2] and consequently the threshold for rapid rotational flux reversal.[3] The extrapolation of the linear region to the field axis, i.e., the threshold for rapid flux reversal is substantially higher than the coercive force. Consequently, these alloys appear less desirable for switching applications than Permalloy.

[2] G. A. Kelsall, Physics **5**, 169 (1934).
[3] E. M. Gyorgy and D. Treves, J. Appl. Phys. **33**, 1222 (1962), this issue.

---

JOURNAL OF APPLIED PHYSICS      SUPPLEMENT TO VOL. 33, NO. 3      MARCH, 1962

# New Modified Mn₂Sb Compositions Showing Exchange Inversion

T. A. BITHER, P. H. L. WALTER, W. H. CLOUD, T. J. SWOBODA, AND P. E. BIERSTEDT

*Central Research Department, Experimental Station, E. I. du Pont de Nemours and Company, Wilmington, Delaware*

Transformations from ferrimagnetic to antiferromagnetic ordering with decreasing temperature have now been observed in manganese antimonides modified with V, Co, Cu, Zn, Ge, and As as well as with Cr. The magnetic structure change is accompanied by a first-order change in the volume of the unit cell as well as a change in electrical resistivity. Crystallographic data for these materials in the exchange inversion region and at room temperature are presented as a function of the concentration of modifying element.

TRANSFORMATIONS from ferrimagnetic to antiferromagnetic ordering with decreasing temperature, previously reported for the compounds $Mn_{2-x}Cr_xSb$,[1,2] have now been observed in manganese antimonides modified with the elements V, Co, Cu, Zn, Ge, and As. In these compounds, as with chromium modified $Mn_2Sb$, the exchange inversion temperature $T_s$, at which this transition takes place, depends on the concentration of the modifying element as shown in Fig. 1. The saturation magnetization curves are similar to those of $Mn_{2-x}Cr_xSb$, falling somewhat below that of unmodified $Mn_2Sb$. Weak residual magnetism usually observed in the antiferromagnetic state is due to the presence of some MnSb precipitate.

These materials have the same tetragonal crystal structure as $Mn_2Sb$ ($P4/nmm$) in both the ferrimagnetic and antiferromagnetic states. The magnetic structure change is accompanied however by a first-order change in the volume of the unit cell. These modifying elements act similarly to chromium; they shrink the structure in such a way that the interlayer distance along the tetragonal axis is decreased so that exchange inversion can take place. This interlayer distance is a function of the modifying element, and in these compositions it also depends on concentration. In contrast, $Mn_{2-x}Cr_xSb$ compounds of varying chromium content were previously observed to transform at the same critical lattice constant.[1] In Table I, the $c$-axis length at the temperature at which the ferrimagnetic state disappears upon cooling, is listed for several compositions.

Figure 2 gives the $c$-axis length at room temperature as a function of concentration for various compositions. Discontinuities in the curves for Cr, Ge, and As modified antimonides indicate the change from the ferrimagnetic

[1] T. J. Swoboda, W. H. Cloud, T. A. Bither, M. S. Sadler, and H. S. Jarrett, Phys. Rev. Letters **4**, 509 (1960).
[2] W. H. Cloud, T. A. Bither, and T. J. Swoboda, J. Appl. Phys. **32**, 55S (1961).

TABLE I. *c*-axis length in ferrimagnetic state at exchange inversion temperature.

| | $Mn_{2-x}A_xSb$ | | | | | | $Mn_2Z_xSb_{1-x}$ | |
| | Cr | | Co | | Cu | | Ge | As |
| $x$ | 0.05 | 0.10 | 0.10 | 0.20 | 0.10 | 0.15 | 0.05 | 0.20 |
| $T$, °C | −73 | 36 | −210 | −121 | −82 | −34 | −88 | −45 |
| $c$, Å | 6.530 | 6.531 | 6.465 | 6.427 | 6.520 | 6.515 | 6.530 | 6.487 |

to antiferromagnetic state. The binary compounds Cu₂Sb and Mn₂As have the same tetragonal crystal structure as Mn₂Sb and are antiferromagnetic.[3-5]

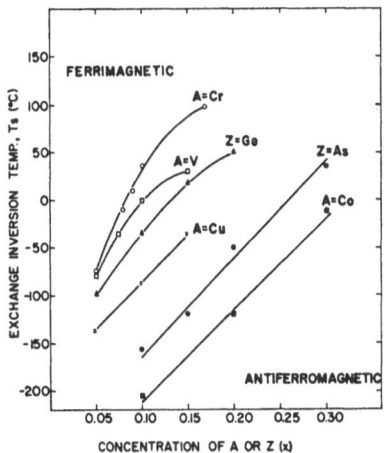

FIG. 1. Ferrimagnetic-antiferromagnetic transition temperatures of $Mn_{2-x}A_xSb$ and $Mn_2Z_xSb_{1-x}$ compounds (*x* between 0.05–0.30).

The anomalous increase in lattice constant in the copper modified system just above the composition Cu₀.₁₅Mn₁.₈₅Sb reflects the simultaneous formation of

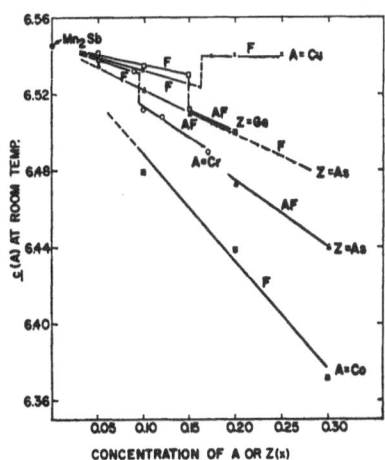

FIG. 2. Room temperature *c*-axis data for $Mn_{2-x}A_xSb$ and $Mn_2Z_xSb_{1-x}$ compounds (*x* between 0.05–0.30). F=ferrimagnetic, AF=antiferromagnetic state.

³ M. Yuzuri, J. Phys. Soc. Japan **15**, 2007 (1960).
⁴ M. Yuzuri and M. Yamada, J. Phys. Soc. Japan **15**, 1845 (1960).
⁵ A. E. Austin, E. Adelson, and W. H. Cloud, J. Appl. Phys. **33**, 1356 (1962), this issue.

the intermediate cubic compound CuMnSb[6] and the copper content of the tetragonal phase in this region drops to approximately Cu₀.₀₅Mn₁.₉₅Sb.

Magnetic transitions in manganese antimonides containing arsenic are much broader (about 100° wide) than those in the other modified systems. Crystallographic measurements on the compound Mn₂As₀.₂₀Sb₀.₈₀ verified this observation as shown in Fig. 3. From *c*-axis length data, both the antiferromagnetic and ferrimagnetic states are seen to coexist over a 90° temperature range. In contrast, extremely sharp transitions less than one degree in width have been obtained in cobalt-modified Mn₂Sb compounds.

As shown previously for $Mn_{2-x}Cr_xSb$ compounds, exchange inversion also causes significant changes in electrical resistivity which can be accounted for by the

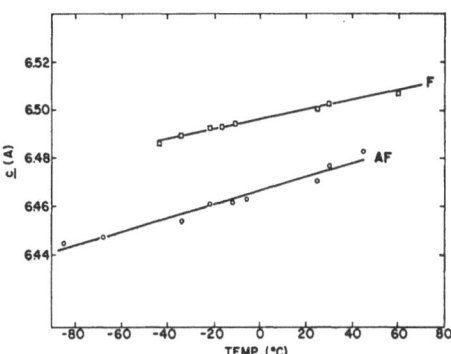

FIG. 3. *c*-axis of Mn₂As₀.₂₀Sb₀.₈₀ vs temperature. F=ferrimagnetic, AF=antiferromagnetic state.

scattering of conduction electrons by magnons.[7] During the transition from ferrimagnetic to antiferromagnetic ordering, an increase in resistivity is also observed in these materials. Thus, in Mn₁.₈₀Co₀.₂₀Sb, the resistivity increased sharply from 180 to 310 μohm-cm and in Mn₁.₉₀Cu₀.₁₀Sb from 160 to 260 μohm-cm.

It was previously demonstrated, in the case of chromium-modified Mn₂Sb, that this element substitutes for manganese in the Mn(I) position of the Mn₂Sb structure.[2] In contrast, it has been found for germanium-modified Mn₂Sb that the germanium substitutes for antimony in this structure.

Mr. Keith Babcock performed the single crystal x-ray measurements reported in Table I and Fig. 3.

⁶ H. Nowotny and B. Glatzl, Monatsch. Chem. **83**, 237 (1952).
⁷ H. S. Jarrett, P. E. Bierstedt, F. J. Darnell, and M. Sparks, J. Appl. Phys. **32**, 57S (1961).

# Magnetic Characteristics of Hydrogenated Holmium*

Y. KUBOTA AND W. E. WALLACE

*Department of Chemistry, University of Pittsburgh, Pittsburgh 13, Pnensylvania*

Various physical and chemical properties of the lanthanide hydrides suggest that they are saline in nature, the hydrogen existing as a negatively charged ion. These properties further suggest that the anionic charge is obtained from the conduction band of the lanthanide and that in the fully hydrogenated metal the conduction band is completely depopulated. Since coupling of the lanthanides occurs via the conduction electrons, this should lead to very considerable suppression of the interactions in the hydrides compared to the element, but an unchanged effective atomic moment. To determine whether this is the case susceptibility measurements have been made on a series of nine Ho-H alloys. These show for the dihydride, which still contains some conduction electrons, a Néel point at around 8°K (instead of 135°K for the element), whereas for the trihydride, which lacks conduction electrons, no Néel point is observed down to 4°K. Susceptibilities of the hydrides above about 15°K obey the Curie-Weiss law with a much smaller Weiss constant than that for elemental holmium. The atomic moment is found, however, to be virtually unaffected by hydrogenation.

## INTRODUCTION

THE lanthanide metals have long been known to be capable of absorbing copious quantities of hydrogen, forming what are usually termed "hydrides." Studies by Stalinski,[1] Peltz and Wallace,[2] and Warf and Hardcastle[3] have shown that in the La-, Ce-, Pr-, Dy-, Ho-, and Yb-H systems metallic conduction disappears when the metal is hydrogenated to saturation. This behavior together with the observed stoichiometries of the lanthanide hydrides at saturation indicates, as has been pointed out elsewhere,[4] that these materials are saline in nature, the hydrogen present existing as an anion. Studies of the magnetic properties of the Ce hydrides by Stalinski[1] and $GdH_2$ by Trombe[5] showed that the atomic moment is unchanged upon hydrogenation, indicating that the hydrogen does not acquire its electrons from the metallic ion core. These findings together with the crystallographic results of Pebler and Wallace[4] indicate that the hydride anion is formed by consumption of electrons in the conduction band and they furthermore strongly suggest that in the fully hydrogenated material the conduction band is completely depopulated.

It appears that exchange in the lanthanide metals occurs via the conduction electrons.[6,7] If so, progressive alteration in magnetic behavior is anticipated with increasing hydrogen content. For example, the tendency toward magnetic alignment is expected to be greatly suppressed in the hydride as compared to the element. To ascertain whether this is the case and to determine whether the atomic moment is unaffected by hydrogenation for metals other than Ce and Gd, a series of investigations involving lanthanide hydrides is being carried out in this Laboratory. The present study, involving holmium hydrides, is the first of these to reach completion.

## EXPERIMENTAL

The holmium employed was h.p. grade obtained from the Nuclear Corporation of America and was stated to be 99.9% pure. The hydrogen employed was purified by diffusing it through a palladium barrier. Samples were prepared using a procedure which has been described elsewhere.[4]

Susceptibilities were measured from 4°K to room temperature using equipment and techniques that have been described.[8,9] Measurements were made in each case at 2.84 and 6.08 koe to test for possible field dependence of the susceptibility. Results were the same within experimental error.

## RESULTS AND DISCUSSION

Included in the study were elemental holmium and nine hydrides designated as $HoH_x$, in which $x=1.51$, 1.60, 1.82, 2.00, 2.31, 2.48, 2.62, 2.93, and 3.06. X-ray diffraction studies of the holmium hydrides showed[4] the following: When holmium (which is cph) is hydrogenated to approximately the dihydride the system adopts a structure in which the holmium ions are in a fcc arrangement. This cubic phase persists from about $x=1.9$ to 2.2. For $x>2.6$ the structure reverts to cph, whereas for $x$ between 2.2 and 2.6 the system is a mixture of the cubic and hexagonal material. Thus in the trihydride the holmium arrangement is like that in the element except that $c/a=1.802$ in the former. For $x$ between 0 and 1.9 the system is a mixture of elemental holmium and (hydrogen-poor) dihydride. Hence on the basis of the diffraction studies the alloys $HoH_{1.51}$, $HoH_{1.60}$, and probably $HoH_{1.82}$ were mixtures containing elemental holmium.

Ho and $HoH_{1.51}$ clearly showed a Néel point at 135°K. In contrast with this $HoH_{1.62}$ and $HoH_{1.82}$ did not show

*This work was assisted by a contract with the U. S. Atomic Energy Commission.

[1] B. Stalinski, Bull. acad. Pol. Sci. **5**, 1001 (1957); **7**, 269 (1959).

[2] Unpublished measurements.

[3] J. C. Warf and K. Hardcastle, Final Report on Office of Naval Research, August, 1961.

[4] A. Pebler and W. E. Wallace, J. Phys. Chem. (to be published).

[5] F. Trombe, Compt. rend. **219**, 182 (1944).

[6] G. W. Pratt, Phys. Rev. **106**, 53 (1957); **108**, 1233 (1957).

[7] A. Arrott (to be published).

[8] K. Nassau, L. V. Cherry, and W. E. Wallace, J. Phys. Chem. Solids **16**, 131 (1960).

[9] R. A. Butera, R. S. Craig, and L. V. Cherry, Rev. Sci. Instr. **32**, 708 (1961).

TABLE I. Results of magnetic studies of holmium and holmium hydrides.

| Sample | $T_N{}^a$ (°K) | $T_d{}^b$ (°K) | $\chi^c \times 10^4$ | | | $\Theta^d$ (°K) | $\mu_{eff}{}^e$ (Bohr magnetons) |
| | | | 4.2°K | 78°K | 300°K | | |
|---|---|---|---|---|---|---|---|
| Ho | 135 | 135 | | 18.3 | 3.93 | 86 | 10.4 |
| HoH$_{1.60}$ | 8 | 10 | 35.8 | 7.81 | 2.35 | −15 | 10.0 |
| HoH$_{1.82}$ | 8 | 12 | 39.4 | 8.54 | 2.34 | −5 | 9.9 |
| HoH$_{2.00}$ | 8 | 12 | 41.9 | 8.33 | 2.38 | −5 | 9.9 |
| HoH$_{2.31}$ | 8 | 12 | 41.6 | 8.47 | 2.27 | −9 | 9.6 |
| HoH$_{2.48}$ | ? | 12 | 39.6 | 9.09 | ⋯ | −9 | 10.4 |
| HoH$_{2.62}$ | N.O. | 12 | 38.3 | 8.46 | 2.39 | −9 | 9.8 |
| HoH$_{2.93}$ | N.O. | 12 | 40.1 | 8.62 | 2.45 | −8 | 10.1 |
| HoH$_{3.06}$ | N.O. | 12 | 40.7 | 8.81 | 2.53 | −8 | 10.1 |

[a] $T_N$ =Néel temperature (N.O. means not observed).
[b] $T_d$ =temperature below which departures from Curie-Weiss behavior become noticeable.
[c] $\chi$ =susceptibility per gram.
[d] $\Theta$ =Weiss constant.
[e] $\mu_{eff}$ =effective atomic moment.

a Néel point in this range[10] nor did any of the other six samples which contained larger amounts of hydrogen. The hydrides with $x=1.6$ to 2.31 showed, however, a rather sharp maximum in the susceptibility, suggesting a Néel point, at 8–9°K. The maximum is absent for alloys with $x>2.62$ and its presence is doubtful in HoH$_{2.48}$. These results indicate that antiferromagnetism is absent in fully hydrogenated holmium, which is to be expected if the coupling of the moments requires the existence of conduction electrons.

Above a certain temperature, which is called $T_d$, the temperature dependence of the susceptibilities follows the Curie-Weiss Law. Values for $T_d$, the Weiss constant $\Theta$, the effective moment per holmium atom, and susceptibilities at several temperatures are given[11] in

[10] The absence of a Néel point at about 135°K for HoH$_{1.60}$ and HoH$_{1.82}$ is not surprising since these samples contain only 6 and 1% of elemental holmium, respectively.
[11] The $\chi$ values at 4.2° and 78°K were obtained under "static" conditions, that is with temperature constant, whereas that for 300°K and $\Theta$ and $\mu_{eff}$ were derived from dynamic experiments with temperature rising 1–2 deg/min. Fig. 1 shows a result of such dynamic experiments. Due to less than perfect thermal bonding between sample and thermometer, in the dynamic experiments the thermometer lagged behind the sample by an almost constant amount. This, of course, does not affect the computed magneton number but it does lead to a small overestimate of the Weiss constant and a slight underestimate of $\chi$.

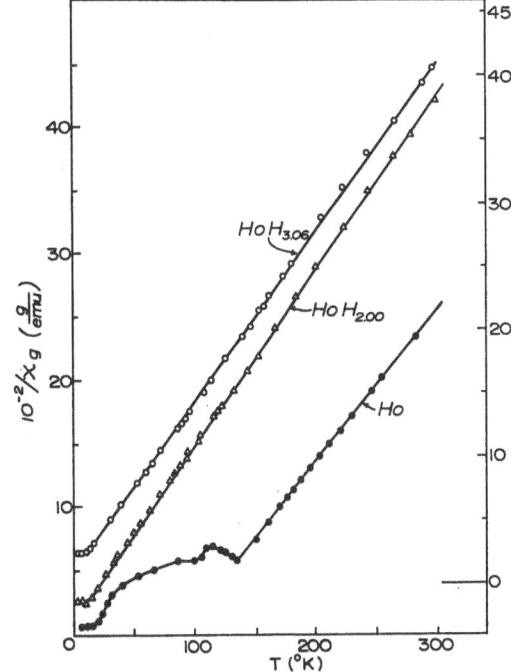

FIG. 1. Reciprocal of susceptibility versus temperature for holmium and holmium hydrides. The scale to the right applies to HoH$_{3.06}$, that to the left to the other samples.

Table I. Plots of $1/\chi_g$ for Ho, HoH$_{2.00}$, and HoH$_{3.06}$ are shown in Fig. 1. These results show that while the hydrogenation reduces $\chi$ substantially, there is little, if any, accompanying change in $\mu_{eff}$. This is thus in keeping with the earlier observations on Ce and Gd hydrides. This is also consistent with the notion that the holmium core electrons are unaffected by hydrogenation. If either the Weiss constant or the Néel temperature is taken as a measure of the interactions, one must conclude that the magnetic interactions are substantially weaker in the hydrides than in the pure metal, which again is anticipated if the interactions occur via the conduction electrons.

JOURNAL OF APPLIED PHYSICS    SUPPLEMENT TO VOL. 33, NO. 3    MARCH, 1962

# Magnetic Susceptibility and Internal Friction of Tetragonal Manganese-Copper Alloys Containing 70 Percent Manganese

A. E. Schwaneke and J. W. Jensen

*Rolla Metallurgy Research Center, Bureau of Mines, U. S. Department of the Interior, Rolla, Missouri*

Relation of structure to magnetic susceptibility and internal friction is described for a 70% manganese-30% copper alloy. The aged alloy exhibits a reversible fct to fcc transition at temperatures from 20° to 140°C, depending on the amount of aging. Magnetic susceptibility apparently obeys the Curie-Weiss law but a change is observed in the temperature-susceptibility curve at the structure transition temperature. The low-stress internal friction of this alloy shows a resonance peak at the temperature of the structure change.

Aging experiments with the 70 and 80% manganese alloys give evidence that quenching of alloys containing more than 80% manganese from solid-solution temperatures does not prevent changes typical of alloys aged to initiate precipitation of alpha manganese. Thus the fct⇌fcc transformation that others have associated with the disappearance of antiferromagnetic ordering also is associated with an early stage of the alpha-manganese precipitation.

THE magnetic susceptibility and internal friction of manganese-copper alloys containing from 60 to about 80% manganese are being investigated by the Bureau of Mines to correlate these properties with crystallographic structure. It is reported that similar properties are observed in alloys containing more than 80% manganese, but there are significant differences.[1] Internal-friction and magnetic measurements, as related to crystallographic structure, have been made on a manganese-copper alloy containing 70 wt% manganese. This alloy is typical of those having face-centered tetragonal crystal structures at room temperature and above and containing from 60 to 80 wt % manganese. The tetragonal structure results after a partial precipitation of alpha-manganese produced by aging the solution-treated alloy. Comprehensive studies of these alloys have shown that the aging process causes similar changes in structure and properties to occur for any composition within the range. Notable changes are: A cubic to tetragonal transformation; a decrease in electrical resistivity; the appearance of extremely high vibration damping; increases in hardness, tensile strength, and yield strength; and a decrease in Young's modulus.

Determinations of the structure, internal friction, and magnetic susceptibility were made on 12 specimens of a high-purity 70% manganese-30% copper alloy. Details of the melting, casting, and forming procedures have been described previously.[2] The specimens used for the experimental tests were in the form of $\frac{1}{8}$-in.-diam rods about $2\frac{1}{2}$ in. long. All specimens were subjected to a solution heat treatment for 1 hr at 810°C in a helium atmosphere. A subsequent heat treatment at 450°C produced the partial precipitation of alpha-manganese, accompanied by a change of the cubic solid-solution matrix to tetragonal. Experience showed that an aging temperature of 450°C permitted good control of the precipitation. The original solid-solution

structure could be reproduced with little grain growth by reheating at the solid-solution temperature.

The tetragonal or cubic structure of the rods was determined at various temperatures from 20° to 200°C with an x-ray diffractometer. Magnetic susceptibility over the same temperature range was determined by the Guoy method using fields of 10 to 13 kgauss. Internal friction at temperatures listed was measured at a frequency of 33 kc with a quartz composite oscillator such as that described by Marx.[3]

The x-ray diffractometer showed that the alloy rods in the solution-treated condition had a cubic structure at room temperature and above. When subjected to an aging treatment as short as 10 min they exhibited a tetragonal structure at room temperature. However, when the specimen was heated a few degrees above room temperature the structure changed back to cubic. Upon cooling to room temperature again the alloy specimen exhibited the tetragonal structure. An increase in the aging time raised the temperature at which the tetragonal-cubic change took place. For an aging time of 30 min the transition occurred at about 65°C; for 1 hr, at 95°C; for 2 hr, at about 130°C; and for 3 hr, back down at 95°C. The transformation was not abrupt but was a gradual change in the tetragonality of the lattice toward the cubic form. On cooling, the specimen always reverted to the tetragonal structure at temperatures below the transition point. The maximum tetragonality at room temperature and the highest transition temperature occurred for specimens subjected to an aging of 2 hr at 450°C. More or less aging time resulted in lower transition temperatures. This tetragonal-cubic transition corresponds to the one observed by Basinski and Christian[4] in solution-treated manganese-copper alloys.

The magnetic susceptibility at temperatures from 20° to 200°C for the alloy in the solution-treated condition and after being subjected to different periods of heat-

[1] R. Street, J. Appl. Phys. **31**, 310S (1960).
[2] J. A. Rowland, C. E. Armantrout, and D. F. Walsh, Bur. Mines Rept. of Investigations 5127, 20 pp. (1955).
[3] J. Marx, Rev. Sci. Instr. **22**, 503 (1951).
[4] Z. S. Basinski and J. W. Christian, J. Inst. Metals **80**, 659 (1951–52).

treatment at 450°C is shown in graph (a), Fig. 1. The top curve shows the susceptibility of the alloy aged for 2 hr at 450°C. The alloy was paramagnetic, obeying the Curie-Weiss law. However, near the temperature of the structure transition a departure from the C-W curve occurred. The dotted line shows the location of the C-W curve. This deviation is not as prominent for the specimens aged for less than 2 hr.

The internal friction versus temperature for the same conditions is shown in graph (b). Here the peaks in internal friction occurred at about the same temperatures at which the structural changes occurred. Note that although the x-ray data indicated the maximum tetragonality was produced by a 2-hr aging time, the maximum decrement at the peak results from an aging period of only 40 min.

In order to complete the picture of the internal friction properties of the alloy, the variation in Young's modulus is shown in graph (c). The modulus was determined from the resonant frequency of the specimen being tested in the composite oscillator. The delta-M effect described by Nowick is evident.[5] The temperature at which the modulus was lowest is again the approximate temperature of the tetragonal-cubic transformation.

In two respects these data are in contrast to results reported by other investigators. First, the tetragonal structure and the tetragonal-cubic transformation, observed by Basinski and Christian[4] in alloys of 70 to 95% manganese that were quenched from solid-solution temperatures, occur at temperatures of 20° to 140°C in alloys with less than 80% manganese only after aging. Extensive data on the alloys with 65 to 80% manganese has shown that the 5 changes in properties listed in the first paragraph, which were observed on aging the 70% manganese alloy, are observed in quenched alloys as composition is increased from 75 to 80% manganese. As a consequence of this behavior and because of this evidence, it is postulated that alloys above 82% in manganese content cannot be quenched from solid-

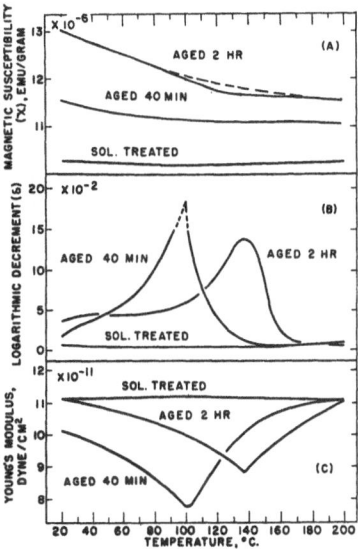

FIG. 1. The effect of aging at 450°C on (a) magnetic susceptibility, (b) internal friction, and (c) Young's modulus for a 70% manganese-30% copper alloy. The dotted line in (a) shows the location of the Curie-Weiss curve for the alloy aged 2 hr.

---

[5] A. S. Nowick in *Progress in Metal Physics* (Pergamon Press, New York, 1953), Vol. 4, pp. 1–70.

solution without the occurrence of changes characteristic of nucleation and precipitation of alpha-manganese.

The second contrasting result is that the alloy with 70% manganese, even though tetragonal, was paramagnetic. It had been anticipated that it would be antiferromagnetic. The assumption had been made that the high damping capacity of the aged, tetragonal alloys could be associated with antiferromagnetic ordering in view of the effect of stress on the anisotropy of magnetic susceptibility of an 82% alloy reported by Bacon, and others.[1] It appears that there is a fundamental difference in magnetic properties of the tetragonal aged 70% alloy and the tetragonal quenched 85% alloy.

The experimental evidence shows that the magnetic susceptibility, internal friction, and Young's modulus respond together with noticeable changes at about the temperature where the tetragonal-cubic transformation is completed. It should be noted that torsion-pendulum data at high stresses, which could not be included here, show a gradual decrease in damping capacity as the temperature of the 70% alloy is increased, and the tetragonal lattice gradually reverts to cubic.

JOURNAL OF APPLIED PHYSICS    SUPPLEMENT TO VOL. 33, NO. 3    MARCH, 1962

# Magnetic Susceptibilities and Exchange Effects in Four Organic Free Radicals*

J. H. BURGESS, R. S. RHODES, M. MANDEL, AND A. S. EDELSTEIN

*Department of Physics, Stanford University, Stanford, California*

Studies of EPR intensity and linewidth in four free radicals at temperatures between 300° and 1.5°K have shown the presence of antiferromagnetic exchange coupling. In two cases a low-temperature paramagnetism occurs following a Curie-Weiss law with positive Weiss constant. A model involving isolated spins in a magnetically dilute system is proposed to explain these results.

THE influence of electron exchange interaction on the magnetic properties of organic free radicals is well known, particularly the "exchange narrowing" of the EPR absorption. Various susceptibility measurements[1] have yielded negative Weiss constants at higher temperatures indicative of antiferromagnetic coupling but in no case has a sharp cooperative transition been observed. To gain more information about the magnetic ordering process we have studied the EPR absorption intensity and linewidth in four free radicals at frequencies ranging from 25 Mc to 10 kMc and temperatures from 300° to 1.5°K. The materials examined were 1,1 diphenyl-2-picryl hydrazyl (DPPH), 1,3 bisdiphenylene-2-phenyl allyl (BDPA), picryl-$n$-amino carbazyl (PAC), and Wursters blue perchlorate (WB). The results, some of which have been reported elsewhere[2-4] are summarized in Table I.

The mass susceptibilities are inferred from the intensity of the EPR absorption lines at 25 Mc and are normalized to the room temperature values of static susceptibility made on the same samples by Duffy.[1] At high temperatures all four materials exhibit Curie-Weiss law behavior, $\chi = C_H/(T-\theta_H)$. The values of the Weiss constants are in agreement with the results of static susceptibility measurements with the exception of PAC for which previous data are lacking. In WB, PAC, and BDPA the susceptibility passes smoothly through a maximum at a temperature $T_m$ below which it decreases by an order of magnitude. At still lower temperatures the susceptibility of WB and PAC reaches a minimum and again follows a Curie-Weiss law

$\chi = C_L/(T-\theta_L)$, this time with positive values of the Weiss constant $\theta_L$ indicating weak ferromagnetic coupling. The ratio $C_L/C_H$ is about 1/80 for WB and 1/140 for PAC. Measurements below 1.5°K are needed to determine if BDPA exhibits a similar low temperature behavior. The situation in DPPH is different in that no susceptibility maximum separates the two Curie-Weiss law regions and the Weiss constant is negative in both. In agreement with Singer and Spencer[5] we find a transition region between 77° and 4.2°K in which the Curie constant decreases by about a factor of 2. The susceptibility does appear to reach a maximum near 2°K.

The EPR linewidths at 25 Mc increase as the temperature is lowered approaching at 1.5°K values several times those at room temperature. More significant however is the appearance in WB, PAC, and BDPA of a sharp linewidth peak at a temperature $T_{\Delta H}$ which in each case falls below the temperature of maximum susceptibility $T_m$.

Models involving pairs or triplets of spins coupled antiferromagnetically by exchange have been used before[2,3] to explain susceptibilities which decrease at low temperatures. These schemes fail to give a minimum in the susceptibility and so are not applicable to the free radicals. Large clusters of odd numbers of spins can give a susceptibility minimum. However, the ratio of the high and low temperature Curie constants indicate an average cluster of about 50 spins which seems hard to justify without a definite structural reason. Crystal structure studies should help clarify this point.

A model having the qualitative properties of the observed data has been given by Sato *et al.*[6] Using an Ising model for spin $\frac{1}{2}$ and with random distribution of the magnetic and nonmagnetic members of the sample, they obtain a formula for the magnetic susceptibility which gives a broad maximum and at lower temperatures passes through a minimum and rises again. This low temperature paramagnetism is attributed to isolated spins which do not have magnetic neighbors. The nature of the defects or spin vacancies which might isolate the spin is not known. For some of these radicals the parent nonparamagnetic compounds have similar structure and could substitute for a paramagnetic molecule. In the

TABLE I. Summary of susceptibility data.[a]

| Free radicals | $\theta_H$ | $\theta_L$ (°K) | $T_m$ (°K) | $T_{\Delta H}$ (°K) | $c$ |
|---|---|---|---|---|---|
| WB | −39±15 | 0.6±0.1 | 180 | 78 | 0.95 |
| PAC | −130±10 | 2.1±0.1 | 75 | 12 | |
| BDPA | −2.2±1.0 | | 6 | 1.8 | 0.96 |
| DPPH | −15±10 | −0.4±0.3 | 2 | | 0.92 |

[a] $\theta_H$ and $\theta_L$ are the high and low temperature Weiss constants, $T_m$ is the temperature of the susceptibility maximum, $T_{\Delta H}$ is the temperature of the EPR linewidth peak, and $c$ is the magnetic concentration.

* Supported by the Air Force Office of Scientific Research.

[1] W. Duffy, J. Chem. Phys. (to be published).

[2] R. S. Rhodes, J. H. Burgess, and A. S. Edelstein, Phys Rev. Letters 6, 422 (1961).

[3] A. S. Edelstein and M. Mandel, J. Chem. Phys. 35, 1130 (1961).

[4] M. E. Anderson, R. S. Rhodes, and G. E. Pake. J. Chem. Phys. 35, 1527 (1961).

[5] L. S. Singer and E. G. Spencer, J. Chem. Phys. 21, 939 (1953).

[6] H. Sato, A. Arrott, and R. Kikuchi, J. Phys. Chem. Solids 10, 19 (1959).

model, only nearest neighbor interactions are included so that these spins are truly isolated. In practice some interaction, although small, must persist and contribute to the observed value of $\theta_L$. This model requires a high concentration of spins to produce susceptibility behavior like that observed. This is consistent with the concentrations observed by Duffy in static susceptibility measurements (Table I).

In addition the model predicts Néel temperatures greater than zero if $c(z-1)>1$, where $c$ is the magnetic concentration and $z$ is the number of nearest neighbors. As they point out, the existence of a maximum in the susceptibility does not necessarily imply antiferromagnetism. In fact, unless the Néel temperature is sufficiently high, a true cooperative transition may produce little change in the susceptibility curve.[6]

Because of this difficulty it is important to examine the specific heat. The model predicts a Schottky type anomaly associated with the broad susceptibility maximum and a sharp cooperative peak at the Néel point.

Recent measurements[7] on BDPA have shown a Schottky anomaly at 3°K and preliminary results[8] indicate a peak near 1.8°K. Specific heat measurements have not yet been made in PAC and WB. This peak correlates well with the anomaly at 1.8°K in the EPR linewidth which is expected at the Néel point.[9]

It appears then that this "lonesome" spin model can account for many of the properties of the free radicals WB, PAC, and BDPA. Further experiments are in progress to determine if DPPH behaves similarly at still lower temperatures.

We wish to express our gratitude to Professor W. Duffy, Jr., and Mr. W. O. Hamilton for permission to quote their results prior to publication and to Professor G. E. Pake for many helpful discussions.

[7] J. P. Goldsborough, M. Mandel, and G. E. Pake, Proc. 7th Intern. Conf. on Low Temp. Physics, Toronto, 1960.
[8] W. Hamilton, Stanford University (private communication).
[9] F. M. Johnson and A. H. Nethercot, Phys. Rev. 119, 705 (1959).

---

JOURNAL OF APPLIED PHYSICS    SUPPLEMENT TO VOL. 33, NO. 3    MARCH, 1962

# Paramagnetic Behavior of Iron-Rich Iron-Vanadium Alloys

Sigurds Arajs, R. V. Colvin, Henry Chessin, and J. M. Peck
*Edgar C. Bain Laboratory for Fundamental Research, United States Steel Corporation Research Center, Monroeville, Pennsylvania*

Magnetic susceptibilities of 14 iron-vanadium alloys (to 26 at.% V) have been studied between their Curie temperatures and 1500°K. These alloys exhibit unusual paramagnetic behavior. Plots of the magnetic susceptibility vs vanadium concentration have a maximum which decreases with increasing temperatures. The curve of the paramagnetic Curie temperature vs at.% V also exhibits a maximum at about 10 at.% V. The paramagnetic moment per iron atom decreases gradually with the amount of vanadium in iron.

THERE are no completely satisfactory theories of paramagnetism of ferromagnetic materials of the first transition metal series. It is questionable whether a satisfactory theory will precede acquisition of substantially complete experimental data. Therefore, we have recently begun extensive investigations of various ferromagnetic iron-base binary solid solutions above their Curie temperatures to gain a better knowledge about their magnetic behavior. Our studies on iron-silicon alloys have already been reported.[1] This paper presents the magnetic susceptibility measurements on iron-vanadium alloys from their Curie temperatures to 1500°K.

The iron-vanadium alloys were prepared by a levitation melting technique using iron (99.9+%, Cat. No 849) obtained from Johnson–Matthey and Company and vanadium (99.6%) from Electrometallurgical Company. These alloys were homogenized at 1420°K for 12 hr. The apparatus and experimental techniques used in this investigation have been described elsewhere.[1]

The inverse paramagnetic susceptibility $(1/\chi)$ as a function of temperature $(T)$ of some iron-vanadium alloys and of pure iron (dotted curve) is shown in Fig. 1. For purposes of clarity the data on 4.5, 5.4, 8.2, 8.6, 9.7, 10.8, and 12.5 at.% V alloys have been omitted from this figure. Two facts should be noticed from these measure-

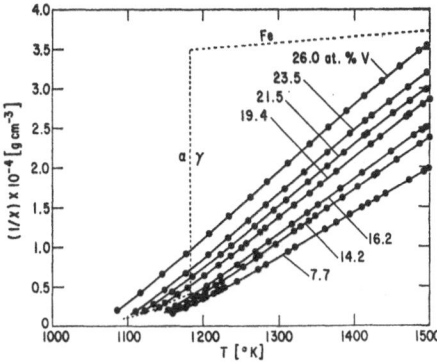

FIG. 1. Inverse paramagnetic susceptibility of some iron-vanadium alloys as a function of temperature.

[1] S. Arajs and D. S. Miller, J. Appl. Phys. 31, 986 (1960).

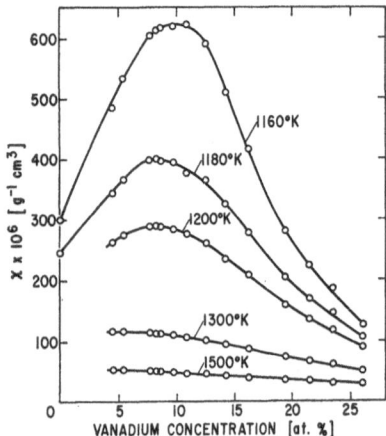

FIG. 2. Paramagnetic susceptibility of iron-vanadium alloys at various temperatures.

ments. First, the alloys of lower vanadium content obey the Curie–Weiss relationship while those containing 19.4, 21.5, 23.5, and 26.0 at.% V show deviations from this law. A possible reason for these deviations may be the $s$-electron magnetization and the contribution from the orbital moment of $d$ electrons to the total magnetic susceptibility.[2] Second, the addition of small amounts of vanadium to bcc iron increases its paramagnetic susceptibility. This unusual effect is shown more clearly in Fig. 2. The maximum in the $\chi$ vs at.% V curves decreases with increasing temperatures. At 1160°K the maximum occurs at about 10 at.% V. Higher temperatures shift this susceptibility maximum to lower vanadium concentrations. The usual influence of vanadium (whose own paramagnetism is believed to be of the Pauli type) on the magnetic behavior of iron has practically disappeared at 1500°K. These observations were found to be independent of heating and cooling rates used during the experiments. Some quenching and annealing studies on samples having vanadium concen-

trations exhibiting the maximum in $\chi$ vs at.% V curve gave negative results. It appears that the observed effect is an intrinsic property of the iron-vanadium system. It may be remarked that vanadium in the above-mentioned alloys was found to be in solid solution. Also, x-ray studies gave no indications of the presence of $\sigma$ phase in the alloy containing 26.0 at.% V.

The paramagnetic Curie temperatures ($\theta_p$) and the Curie constants ($C$) of iron-vanadium alloys and of pure iron[3] are shown in Fig. 3. The curve of $\theta_p$ vs at.% V also shows a maximum at 10 at.% V. The Curie constant gradually decreases with vanadium content. If it is assumed that the magnetic moment results from iron atoms only, then the calculations of the paramagnetic moment per iron atom ($\mu_{Fe}$) gives $\mu_{Fe}$ as a gradually decreasing function of vanadium amount in iron. Simple dilution effect from the viewpoint of the paramagnetic moment, for example, found in iron-silicon alloys[1] up to 15 at.% Si does not occur in iron-vanadium alloys. Figure 3 also shows the ferromagnetic Curie temperatures ($\theta_f$) of iron-vanadium alloys.[4] The expected similarity between $\theta_p$ and $\theta_f$ curves, since these are related to the interatomic exchange interactions, can be observed. It may be remarked that a simple theory[5] for binary alloys with one magnetic component based on the classical Heisenberg model predicts $\theta_p$ to be proportional to the atomic fraction of magnetic atoms. The expected $\theta_p$ on this model is shown in Fig. 3. Obviously iron-vanadium alloys are not describable by such an approximation. Moreover, even simpler systems such as iron-silicon,[1,6] iron-aluminum,[7] and iron-tin[6] do not follow this prediction.

At the present time it appears to be difficult to explain the above-described observations quantitatively (qualitatively one can make some arguments using the Bethe–Slater curve with various competing interactions and the clustering of vanadium atoms) using the existing theories based either on the Heisenberg model or simple energy band arguments. Further work with the purpose of obtaining additional information on the magnetic behavior and other physical properties of iron-vanadium system and other binary iron alloys is in progress.

### ACKNOWLEDGMENTS

The authors are grateful to A. Arrott of Ford Motor Company, P. Beck of the University of Illinois, D. R. Behrendt of Lewis Research Center, NASA, and C. W. Chen of Westinghouse Research Laboratories for stimulating discussions.

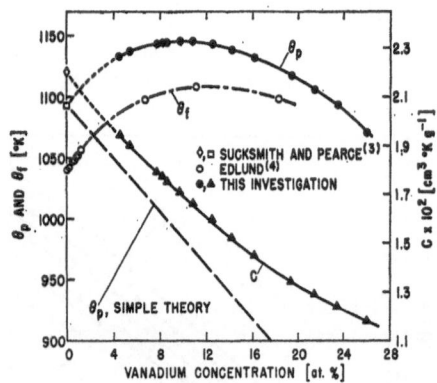

FIG. 3. Curie temperatures and Curie constants of iron-vanadium alloys.

[3] M. Shimizu, J. Phys. Soc. Japan **16**, 1114 (1961).

[3] W. Sucksmith and R. R. Pearce, Proc. Roy. Soc. (London) **A167**, 189 (1938).
[4] D. L. Edlund, Ph.D. thesis, Massachusetts Institute of Technology (1934).
[5] J. S. Smart, J. Phys. Chem. Solids **16**, 169 (1960).
[6] L. Néel, Ann. Phys. **17**, 5 (1932).
[7] M. Fallot, J. phys. radium **5**, 153 (1944).

JOURNAL OF APPLIED PHYSICS    SUPPLEMENT TO VOL. 33, NO. 3    MARCH, 1962

# Magnetic Moment of Co-Cu Solid Solutions with 40 to 85% Cu

E. KNELLER

*IBM Research Center, Yorktown Heights, New York*

Metastable Co-Cu solid solutions in the concentration range between 12% and 95% Cu were obtained by simultaneous evaporation of Co and Cu on a cold substrate. The spin moment of the as-deposited alloys is equal to that of Fe-Ni, Co-Ni, Ni-Cu solid solutions with the same average number of electrons per atom. During annealing at 700°C for a suitable time the spin moment increases to the value calculated for the two-phase equilibrium state of the alloys. For an alloy with 65% Cu the Curie temperature is 580°K. All alloys have fcc structure before and after annealing.

IN the binary system Co-Cu no thermodynamically stable solid solutions exist at any temperature in the concentration range between 12 and 95 at.% Cu.[1] This paper reports the preparation of metastable solid solutions in this concentration range by simultaneous evaporation of Cu and Co on a substrate which was kept at a low enough temperature to prevent decomposition.

Polycrystalline films with thicknesses of ~1000 Å were evaporated on a quartz substrate at room temperature from independently controllable sources of 99.6 Co and 99.99 Cu in a vacuum of $10^{-5}$ mm Hg. The compositions of the films were determined spectrophotometrically with an accuracy of ±2%. The film thickness was measured by the Tolansky interferometric method. The thermodynamic equilibrium state (two phases) could be obtained by annealing the films for one hour at temperatures above 600°C in a vacuum of $10^{-7}$ mm Hg. From electron micrographs an average grain diameter of about 100 A was determined for the as-deposited films. Electron diffraction analyses showed that all alloys were fcc before and after annealing, the lattice constant being close to that of Cu. As-deposited pure Co films also had fcc structure.

The saturation magnetization $I_s$ was determined as a function of temperature between 77° and 300°K from vertical torque $L(\psi)$ curves at small angles $\psi$ between the field direction and the plane of the film using the equation[2]

$$(\psi/L)H = 1/I_s V + H/2K_\perp V, \qquad (1)$$

where $V$ is the volume of the film, and $K_\perp$ the perpendicular anisotropy constant. $K_\perp$ is equal to $2\pi I_s^2$ for a homogeneous film with shape anisotropy only.

From $I_s$ values extrapolated to 0°K, the mean spin moment of the as-deposited films was calculated using the g factors of Co-Ni alloys with the same number of electrons per atom as the Co-Cu alloys, or $g_{Ni}$ for alloys with more than 28 electrons per atom, as given by Meyer and Asch.[3] During annealing at 600° to 700°C the alloys decompose into almost pure Cu and 90 Co 10 Cu.[1] Hence for the annealed films the g factor of the isoelectronic alloy 80 Co 20 Ni was used. In Fig. 1 the results are plotted versus the mean electron number. For com-

parison the unbroken line represents the experimental data for Fe-Ni, Co-Ni, or Ni-Cu alloys. The broken line was calculated for the two-phase equilibrium state.

Figure 1 shows that the spin moment of the as-deposited Co-Cu alloys is, within the accuracy of the measurement, equal to the spin moment of fcc Fe-Ni, Co-Ni, or Ni-Cu solid solutions with corresponding numbers of electrons per atom.[3] The spin moment, which corresponds to the average number of holes in the 3d-shell, decreases linearly with increasing mean electron number, i.e., Cu concentration and disappears near 28.5 electrons per atom (~76% Cu). In terms of the band model, each Cu atom adds two electrons to the alloy, which fill up the d band of Co. Since this occurs only in solid solutions, it can be concluded that the as-deposited Co-Cu alloys are solid solutions.

When the alloys decompose during annealing at 600° to 700°C, their average spin moment increases according to Fig. 1 to its value for the equilibrium mixture of Cu and 90 Co 10 Cu.

The Curie temperature in the as-deposited state was measured for an alloy with 64.5 at.% Cu to be approximately 580°K, which is higher than for the Ni-Cu alloy

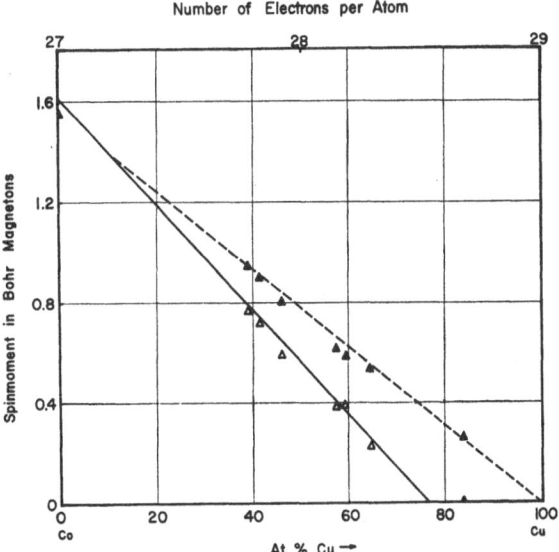

FIG. 1. Spin moment of Co-Cu alloys versus electron number, respectively. Cu concentration; (△) as deposited, (▲) annealed, (——) (————) theoretical.

[1] M. Hansen and K. Anderko, *Constitution of Binary Alloys* (McGraw–Hill Book Company, Inc., New York, 1958).

[2] S. Chikazumi, J. Appl. Phys. **32**, 81S (1961).

[3] A. J. P. Meyer and G. Asch, J. Appl. Phys. **32**, 330S (1961).

with the same number of electrons per atom. At lower Cu concentrations the Curie temperature becomes too high to be measured without causing decomposition.

From the difference in color of the two surfaces of films, which were annealed for one hour at 700°C, it appears that the equilibrium phases are not finely dispersed, but rather form a Cu layer and a 90-Co 10-Cu layer, i.e., a sandwich film with the Cu-colored surface always being on the air side. An explanation for this situation can be seen in a preferred nucleation of the Co-rich phase on the quartz side, which may occur when (for technical reasons or because of different sticking coefficients for Co and Cu) a thin Co-richer layer is formed initially. When, namely, a 20- to 30-Å Cu layer

is deposited on the quartz before evaporating the alloy, then the Cu color appears on the quartz side.

A more detailed analysis of the structure of the annealed films is possible by an analysis of the perpendicular anisotropy constant $K_\perp$ as determined from Eq. (1). This, together with a study of the kinetics of the decomposition process, will be subject of a later paper.

Concluding, it may be said that the above results seem to encourage a simple way for the study of alloys in metastable states, which cannot be realized by classical methods.

The author wishes to thank Miss E. I. Alessandrini and Mr. W. C. Kateley for experimental help and Dr. A. E. Berkowitz for many stimulating discussions.

---

JOURNAL OF APPLIED PHYSICS      SUPPLEMENT TO VOL. 33, NO. 3      MARCH, 1962

# Magnetic Structures of Mn₂As and Mn₂Sb₀.₇As₀.₃

A. E. Austin and E. Adelson
*Battelle Memorial Institute, Columbus, Ohio*

and

W. H. Cloud
*Central Research Department, E. I. du Pont de Nemours and Company, Wilmington, Delaware*

Neutron diffraction has been performed on powder specimens of Mn₂As and Mn₂Sb₀.₇As₀.₃. Mn₂As has the antiferromagnetic structure found in Mn₁.₉Cr₀.₁Sb below the exchange inversion temperature. The atomic moments are 3.7 and 3.5 Bohr magnetons for Mn(I) and Mn(II), respectively. The moments lie perpendicular to the tetragonal axis. The magnetic structure of Mn₂Sb₀.₇As₀.₃, which undergoes exchange inversion over the 35° to 115°C temperature range, is the same as that of Mn₂As at room temperature and is the same as that of ferrimagnetic Mn₂Sb at 150°C. The atomic moments at room temperature (antiferromagnetic) are 2.3 and 2.8 Bohr magnetons, respectively, for Mn(I) and Mn(II). The moments are perpendicular to the tetragonal axis at room temperature and parallel to it at 150°C.

THE work reported in the paper by Bither *et al.*[1] has shown that exchange inversion in substituted Mn₂Sb compounds can be produced by a variety of substituents, both for Mn and Sb. The compounds for which As has been substituted for Sb are of particular interest because Mn₂As and Mn₂Sb have the same crystal structure but are antiferromagnetic and ferrimagnetic, respectively. Neutron diffraction studies of Mn₂Sb have been made by Wilkinson *et al.*[2] The ferrimagnetic to antiferromagnetic transition in Mn₁.₉Cr₀.₁Sb has been studied by neutron diffraction.[3,4] Magnetic susceptibility and specific heat measurements of Yuzuri and Yamada[5] indicate that Mn₂As had a Néel temperature of 573°K, but its magnetic structure has not been

determined. We have therefore undertaken neutron diffraction studies of Mn₂As and also Mn₂Sb₀.₇As₀.₃ which undergoes exchange inversion over the 35–115°C range in order to determine if the antiferromagnetic and ferrimagnetic structures of Mn₂Sb₀.₇As₀.₃ are the same as those of Mn₂As and Mn₂Sb, respectively.

Since suitable single crystals of both Mn₂As and Mn₂Sb₀.₇As₀.₃ have proved very difficult to prepare, the neutron diffraction studies were made on powder specimens. Quantitative determinations of scattering amplitudes were made by comparing integrated intensities with those observed from a powdered specimen of Mn₁.₉Cr₀.₁Sb for which the scattering amplitudes were known from single crystal studies.[3,4] The absorption and temperature factors were assumed to be the same for the three powders. The experimental constant[6] was determined from the diffraction data of a measured amount of Mn₁.₉Cr₀.₁Sb.

[1] T. A. Bither, P. H. L. Walter, W. H. Cloud, T. J. Swoboda, and P. E. Bierstedt, J. Appl. Phys. **33**, 1346 (1962).

[2] M. K. Wilkinson, N. S. Gingrich, and C. G. Shull, J. Phys. Chem. Solids **2**, 289 (1957).

[3] W. H. Cloud, H. S. Jarrett, A. E. Austin, and E. Adelson, Phys. Rev. **120**, 1969 (1960).

[4] W. H. Cloud, T. A. Bither, and T. J. Swoboda, J. Appl. Phys. **32**, 55S (1961).

[5] M. Yuzuri and M. Yamada, J. Phys. Soc. Japan **15**, 1845 (1960).

[6] G. E. Bacon, *Neutron Diffraction* (Clarendon Press, Oxford, England, 1955), p. 93.

TABLE I. Neutron diffraction intensities at room temperature.

| Compound | Amount of powder | Integrated intensities (arbitrary scale) | | Structure factor $|F|^2$ (barns) | |
|---|---|---|---|---|---|
| | | $00\frac{3}{2}$ | $10\frac{1}{2}$ | $00\frac{3}{2}$ | $10\frac{1}{2}$ |
| $Mn_{1.9}Cr_{0.1}Sb$ | 19.00 g | 3.16 | 1.96 | 13.22 | 3.16 |
| $Mn_2As$ | 14.66 g | 7.57 | 3.02 | 40.89 | 6.73 |
| $Mn_2Sb_{0.7}As_{0.3}$ | 15.09 g | 4.00 | 1.95 | 20.09 | 3.90 |

## Mn₂As

Neutron diffraction studies at room temperature show that the antiferromagnetic structure of $Mn_2As$ is the same as that of $Mn_{1.9}Cr_{0.1}Sb$ below the exchange inversion temperature [Fig. 1(B)]. The magnetic $c$ axis is twice that of the x-ray cell, which has the dimensions: $a=3.78$ A, $c=6.28$ A. The magnetic $00\frac{3}{2}$ and $10\frac{1}{2}$ reflections (index based on the x-ray cell) are clearly resolved. Their integrated intensities are shown in Table I. The $00\frac{3}{2}$ intensity, which is roughly proportional to the sum of the Mn(I) and Mn(II) moments, is clearly larger than that of $Mn_{1.9}Cr_{0.1}Sb$.

In the $Mn_2As$ structure, the Mn(I) atoms are at positions 0, 0, 0 and $\frac{1}{2}$, $\frac{1}{2}$, 0. The Mn(II) and As atoms are at positions 0, $\frac{1}{2}$, $z$ and $\frac{1}{2}$, 0, $\bar{z}$. Nowotny and Halla[7] have shown that the positions of the Mn(II) and As atoms in $Mn_2As$ should be very close to those in $Fe_2As$ which had been found to be $z=0.330$ and $z=-0.265$ for Fe(II) and As, respectively. Using x-ray powder intensities measured on a diffractometer we find that these values of the $z$ parameters in $Mn_2As$ give a discrepancy of less than 20% between observed and calculated structure factors of ten reflections for which $l\neq0$.

Using $z=0.33$ for Mn(II), the moments were determined to be 3.7 and 3.5 Bohr magnetons for Mn(I) and Mn(II), respectively (see Table II). The values of magnetic form factors used in Table II were the same as those used by Wilkinson et al. for $Mn_2Sb$.[2] The moments in $Mn_2As$ are perpendicular to the tetragonal axis.

TABLE II. Atomic moments at room temperature.

| Compound | $|F_0|^{2a}=|F|^2/f^2$ (barns) | | Mn(I) | Mn(II) |
|---|---|---|---|---|
| | $00\frac{3}{2}$ | $10\frac{1}{2}$ | | |
| $Mn_{1.9}Cr_{0.1}Sb^b$ | 18.72 | 4.58 | 1.7 | 2.5 |
| $Mn_2Sb_{0.7}As_{0.3}$ | 27.81 | 5.88 | 2.3 | 2.8 |
| $Mn_2As$ | 57.95 | 10.65 | 3.7 | 3.5 |
| $Mn_2Sb$ (reference 2) | | | 1.8 | 2.9 |

ᵃ $f$=magnetic form factor.
ᵇ Recent neutron diffraction studies have shown that the atomic moments given in reference 4 are slightly in error due to extinction effects. The values given here were obtained from neutron diffraction of a thin crystal for which extinction was negligible.

[7] H. Nowotny and F. Halla, Z. physik. Chem. **36B**, 322 (1937).

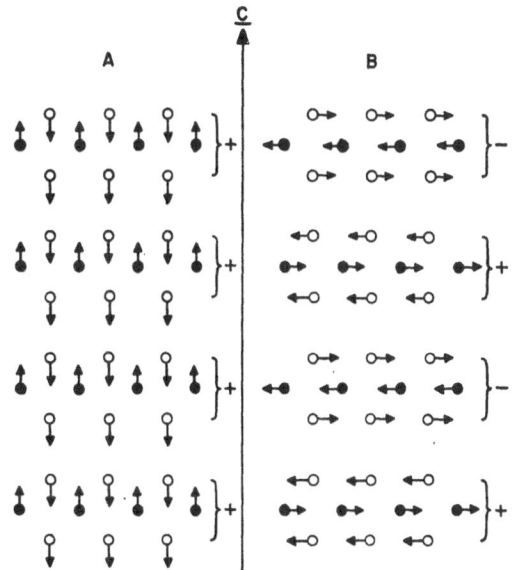

FIG. 1. Magnetic structures: (A), ferrimagnetic state of $Mn_2Sb$, $Mn_{1.9}Cr_{0.1}Sb$ at 50°C, and $Mn_2Sb_{0.7}As_{0.3}$ at 150°C; (B), antiferromagnetic state at room temperature of $Mn_2As$, $Mn_{1.9}Cr_{0.1}Sb$, and $Mn_2Sb_{0.7}As_{0.3}$. Closed circles denote Mn(I) layers; open circles denote Mn(II) layers. Braces indicate three-layer sets. Arrow $c$ gives direction of tetragonal axis.

## Mn₂Sb₀.₇As₀.₃

Neutron diffraction measurements on a powder specimen were made at room temperature and at 150°C. At room temperature the magnetic structure is the antiferromagnetic structure of Fig. 1(B) and at 150°C it is the ferrimagnetic structure of Fig. 1(A) in exact analogy to the $Mn_{1.9}Cr_{0.1}Sb$ system. The intensities of the $00\frac{3}{2}$ and $10\frac{1}{2}$ reflections are shown in Table I. X-ray powder data give $a=4.02$ A, $c=6.44$ A for $Mn_2Sb_{0.7}As_{0.3}$ at room temperature. Assuming that the $z$ parameter for Mn(II) is 0.295 as it is in $Mn_2Sb$[8] the Mn(I) and Mn(II) moments were found to be 2.3 and 2.8 Bohr magnetons respectively in the room-temperature antiferromagnetic state (see Table II).

The ferrimagnetic and antiferromagnetic structures of the $Mn_2Sb$ (or $Mn_2As$) crystal structure involve ferromagnetic or antiferromagnetic ordering of ferrimagnetic layers as shown in Fig. 1. These are the ferri and anti 2 structures suggested by Yuzuri and Yamada.[5] Within a layer the Mn(I), moments are always antiparallel to the Mn(II) moments. Partial substitution of As into $Mn_2Sb$ increases the Mn(I) moment but causes little change in the Mn(II) moment; conversely substitution of Cr for Mn decreases the Mn(II) moment.

### ACKNOWLEDGMENTS

The samples of $Mn_2As$ and $Mn_2Sb_{0.7}As_{0.3}$ were prepared by T. A. Bither and P. H. L. Walter.

[8] L. Heaton and N. S. Gingrich, Acta. Cryst. **8**, 207 (1955).

# Oxides–2 and Crystals

R. L. White, *Chairman*

## Investigations of Spin-Wave Interactions by the Parallel Pumping Technique*

J. J. Green and E. Schlömann†

*Research Division, Raytheon Company, Waltham, Massachusetts*

The imaginary part of the parallel pump susceptibility has been measured over a wide range of power levels and applied dc fields. Some of the structure present in the susceptibility curves is explained by a confluence process between two parametrically excited spin waves. The square of the frequency of the relaxation oscillations, which are frequently present in parallel pump excitation, has been found to increase linearly with the square of the magnetic field.

THE technique of parallel pumping[1,2] has been used successfully to study spin-wave relaxation in yttrium iron garnet.[3,4] The great value of the technique lies in the fact that one can selectively excite spin waves of a given wave number through the proper choice of dc field ($H$). For applied dc fields smaller than a certain characteristic value, $H_c = 4\pi M (N_z - \frac{1}{2}) + [(2\pi M)^2 + (\omega/2\gamma)^2]^{\frac{1}{2}}$, the excited spin waves have a wave number $k \neq 0$ and the angle $\theta_k$ between the wave vector $\mathbf{k}$ and the dc field equals $\pi/2$. Here $4\pi M$ is the saturation magnetization, $N_z$ is the longitudinal demagnetizing factor, $\omega$ is the pump frequency, and $\gamma$ is the gyromagnetic ratio. From measurements of the critical field $h_{\text{crit}}$ for $H < H_c$ the dependence of the spin-wave linewidth $\Delta H_k$ on $k$ has been intensively investigated in single-crystal yttrium iron garnet and has been well correlated to theoretical calculations[3-6] of damping due to scattering processes involving magnons and phonons. One of the important scattering processes, a confluence process, involves two magnons which coalesce to form a third magnon. If the three magnons have wave vectors $\mathbf{k}_1$, $\mathbf{k}_2$, and $\mathbf{k}_3$, then in such a process it is necessary to have $\mathbf{k}_1 + \mathbf{k}_2 = \mathbf{k}_3$ and $\omega_{k_1} + \omega_{k_2} = \omega_{k_3}$.

It has been pointed out by Gottlieb and Suhl[7] that a special situation arises when the two coalescing spin waves are both parametrically excited by the pump field. In this case $k_1 = k_2$ and $\omega_{k_1} = \omega_{k_2} = \omega/2$. This is only possible for $H$ less than a certain value of dc field which we will denote by $H_{3m}$, $3m$ standing for three magnons. It may be shown[8] that the difference between $H_c$ and

$H_{3m}$ is given by

$$\frac{H_c - H_{3m}}{4\pi M} = \frac{1}{6}\left\{\left[1 + 4\left(\frac{\omega}{\gamma 4\pi M}\right)^2\right] - \left[1 + \left(\frac{\omega}{\gamma 4\pi M}\right)^2\right]\right\}. \quad (1)$$

For $H = H_{3m}$ the conservation laws are satisfied by $\mathbf{k}_1 = \mathbf{k}_2$, while for $H < H_{3m}$, $\mathbf{k}_1$ and $\mathbf{k}_2$ can be equal in magnitude, but not have the same direction. Thus we see that for $H < H_{3m}$ it is possible to have confluence processes in which two parametrically excited spin waves coalesce to form a third spin wave. It has been shown theoretically[7,8] that these confluence processes tend to decrease the imaginary part of the parallel susceptibility at high power levels.

Our experiments were performed using slightly undercoupled reflection cavities in the $TE_{101}$ mode. The samples were all spherical and mounted several sphere diameters from the end wall. The threshold fields were determined by noting the power levels at which the reflection coefficient of the cavity changed. The susceptibility was determined from the change in the reflection coefficient.

The susceptibility measurements (Fig. 1) were made at constant input power and except for curves C and B′ the power level was approximately 9 db above the threshold at $H = H_c$. In the case of sample C the threshold was so high (35 oe) that it was not possible to go 9 db above the threshold. Sample A, whose $\chi_{11}''$ shows the most structure, was pure yttrium iron garnet, while samples B and C contained approximately 0.5% Ho and 0.3% Tb, respectively. In the vicinity of $H_c$, $H_{3m}$, and $H_s (H_s = N_z 4\pi M)$ the susceptibility curves have peaks. Here $H_s$ is the applied dc field at which the internal dc field becomes practically zero. For $H > H_c$, $\chi_{11}''$ drops off rapidly since the threshold increases very rapidly with increasing $H$. This is not the case for $H < H_c$ since the threshold increases only for sample A and then only very slowly. A possible explanation of the decrease in $\chi_{11}''$ below $H_c$ could be the spin wave phonon interaction.[9,10] At this frequency the parametrically excited

* Supported in part by the U. S. Army Signal Corp Research and Development Laboratory under contract.

† Presently on leave at the W. W. Hansen Laboratories of Physics, Stanford University, Stanford, California.

[1] E. Schlömann, J. J. Green, and U. Milano, J. Appl. Phys. **31**, 386S (1960).
[2] F. R. Morgenthaler, J. Appl. Phys. **31**, 95S (1960).
[3] T. Kasuya and R. C. LeCraw, Phys Rev. Letters **6**, 223 (1961).
[4] J. J. Green and E. Schlömann, J. Appl. Phys. **32**, 168S (1961).
[5] E. Schlömann, Phys. Rev. **121**, 1312 (1961).
[6] M. Sparks, R. Loudon, and C. Kittel, Phys. Rev. **122**, 791 (1961).
[7] P. E. Gottlieb and H. Suhl, J. Appl. Phys. (to be published).
[8] E. Schlömann, J. Appl. Phys. (to be published).
[9] E. H. Turner, Phys. Rev. Letters **5**, 100 (1960).
[10] E. Schlömann, J. Appl. Phys. **31**, 1647 (1960).

spin waves would be degenerate with phonons of the same frequency and wave number when $H$ is about 40 oe less than $H_c$. The decline of $\chi_{11}''$ below $H_{3m}$ is presumably due to the confluence of two parametrically excited magnons. In curve B' this decline is not present, presumably because the parametrically excited spin waves do not yet have sufficient population to produce any noticeable increase in their own relaxation rates. The value of $H_{3m}$ as determined by Eq. (1) is a function of frequency. To make a check on the confluence hypothesis we have also measured $\chi_{11}''$ for a sample A at 5.5 and 18 kMc. The value of $H_{3m}$ at these frequencies was in excellent agreement with Eq. (1). For $H < H_s$ the samples will no longer be magnetically saturated and this presumably is the cause for the rapid decrease in $\chi_{11}''$ for $H < H_s$. It is interesting to note that the magnitudes of $\chi_{11}''$ are the same for the three samples even though the relaxation rates are quite different. We have measured $\chi_{11}''$ for sample A at even higher power levels and have found even more structure, but at the moment have no explanation for this structure.

The threshold field $h_{crit}$ for samples B and C was found to be nearly independent of dc field for $H < H_c$ and for $H > H_c$ was very closely proportional to $\sin^{-2}\theta_k$. This implies that the contribution of the rare-earth ions to the relaxation rate is the same for all spin waves regardless of wave number or direction.

Measurements on sample A have shown that the reflected power is amplitude modulated at a frequency of the order of one megacycle as soon as the power level exceeds the threshold by several db. In samples B and C this effect is barely noticeable probably due to the shorter spin-wave relaxation time. Hartwick et al.[11] have referred to these oscillations as relaxation oscillations. We have found that the frequency of these oscillations increases smoothly with power level.

The physical origin of these oscillations is not completely settled at this moment. We believe that the

[11] T. S. Hartwick, E. R. Peressini, and M. T. Weiss, J. Appl. Phys. **32**, 223S (1961).

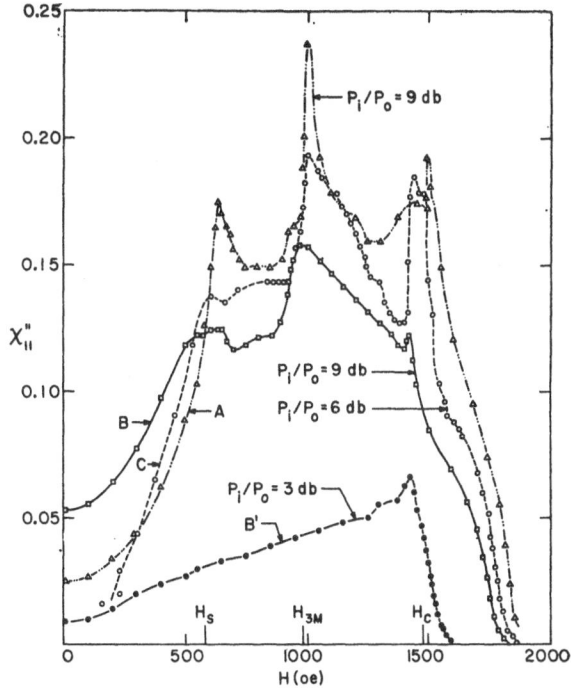

FIG. 1. Parallel pump susceptibility versus applied dc field at room temperature and a frequency of 9.1 kMc. △ sample A (YIG), $h_{crit} = 1.3$ oe; ●□ sample B (YIG+0.5% Ho), $h_{crit} = 22.5$ oe; ○ sample C (YIG+0.3% Tb), $h_{crit} = 35$ oe; $P_i =$ incident power, $P_o =$ threshold at $H = H_c$.

oscillations are at best in part caused by the change in the dc magnetic moment that accompanies the spin-wave excitation. Because of this effect a spin wave which begins to grow under the influence of the rf field will gradually become detuned and will therefore decay back. A theoretical calculation based on this picture and believed to be applicable when the applied dc field is considerably larger than $H_c$ indicates that the square of the frequency of the relaxation oscillations should be linearly dependent on the square of the rf magnetic field. The experimental data are in reasonably good agreement with the theoretical predictions.

JOURNAL OF APPLIED PHYSICS     SUPPLEMENT TO VOL. 33, NO. 3     MARCH, 1962

# Resonance Properties of Single-Crystal Hexagonal Ferrites

C. R. Buffler

*Bomac Laboratories, Varian Associates, Beverly, Massachusetts*

Ferromagnetic resonance measurements have been performed on single-crystal samples of the hexagonal ferrites $BaFe_{12}O_{19}$, $Zn_2Y$, and $Co_2Y$. Values of linewidth, $g$ value, and anisotropy field are given as a function of temperature from 77°K to 600°K at various frequencies between 8 and 75 kMc. $Zn_2Y$ is singularly interesting because it exhibits an extremely narrow room temperature linewidth at $X$ band. Values as low as 16 oe, with anisotropy fields of 9000 oe are reported. The theory of the resonance behavoir is developed and compared to the experimental results.

## INTRODUCTION

THE monocrystalline forms of the hexagonal ferrites have become more and more interesting not only because of their own intrinsic properties but because of possible applications to microwave devices. Their extremely high effective anisotropy fields along with low loss make them suitable for experiments with pulsed microwave generators,[1] millimeter wave frequency multipliers, or passive limiters.[2] In order to explore these possibilities, the resonance properties must be known and understood. Measurements of the parameters; linewidth, $\Delta H$; magnetogyric ratio, $\gamma$; field for resonance, $H_0$; and anisotropy fields $H_a = 2K_1/M$, and $H_a' = 36K_3/M$ have been made as a function of temperature and frequency on single-crystal samples of materials designated as $Zn_2Y$ ($Ba_2Zn_2Fe_{12}O_{22}$), $Co_2Y$ ($Ba_2Co_2Fe_{12}O_{22}$), and M ($BaFe_{12}O_{19}$).

## THEORY

In order to calculate several of the above resonance parameters, the theoretical behavior of resonance frequency as a function of field must be known. This behavior can easily be derived from the expression for the energy surface of the magnetization in a manner first introduced by Smit and Beljers.[3] The energy surface for the calculation of the resonance behavior of the hexagonal ferrites can be written in three terms as a function of the standard polar angles:

$$F(\theta,\phi) = -\mathbf{M} \cdot \mathbf{H}_0 + \tfrac{1}{2}\mathbf{M} \cdot \mathbf{N} \cdot \mathbf{M}$$
$$+ K_1 \sin^2\theta + K_3 \sin6\phi \sin^6\theta. \quad (1)$$

Here $F(\theta,\phi)$ is the free energy, $\mathbf{M}$ the magnetization vector, $\mathbf{H}_0$ the external applied field, $\mathbf{N}$ the demagnetizing tensor, and $K_1$ and $K_3$ the first- and third-order anisotropy constants. The second-order anisotropy term $K_2 \sin^4\theta$ and the third-order polar anisotropy term $K_3' \sin^6\theta$ are neglected throughout. The resonance frequency in terms of these variables is shown to be

$$\omega/\gamma = (F_{\theta\theta}F_{\phi\phi} - F_{\theta\phi}^2)^{\frac{1}{2}}/M \sin\theta, \quad (2)$$

where $F_{\theta\theta}$ etc. are second partial derivatives with respect to the indicated variables. Evaluating (2) for four special cases neglecting the second- and third-order anisotropies

gives:

(a) Uniaxial crystal with applied field in hard plane.

$$\left[\frac{\omega}{\gamma}\right]^2 = \left\{1 - \frac{(N_y - N_x)M}{H_a - (N_z - N_y)M}\right\}$$
$$\times \{[H_a - (N_z - N_y)M]^2 - H_0^2\},$$

for $H_0 \leq H_a - (N_z - N_y)M$ and $H_a > 4\pi M$; (3a)

$$\left[\frac{\omega}{\gamma}\right]^2 = \{H_0 - (N_y - N_x)M\}$$
$$\times \{H_0 - [H_a - (N_z - N_y)M]\},$$

for $H_0 \geq H_a - (N_z - N_y)M$ and $H_a > 4\pi M$. (3b)

In this case the hexagonal easy or $c$ axis is taken as the $z$ direction and the applied field $H_0$ is taken along the $y$ axis.

(b) Uniaxial crystal with applied field along easy axis.

$$\left[\frac{\omega}{\gamma}\right]^2 = [(H_0 + H_a) - (N_z - N_x)M]$$
$$\times [(H_0 + H_a) - (N_z - N_y)M]. \quad (4)$$

In this case the $c$ axis and the applied field are both taken along the $z$ axis. Note the similarity to the well-known Kittel equation.

(c) Planar crystals with applied field in easy plane.

$$\left[\frac{\omega}{\gamma}\right]^2 = [H_0 - (N_y - N_x)M][H_0 + H_a + (N_z - N_y)M]. \quad (5)$$

Here the $c$ axis is taken along $z$ and the applied field along $y$.

(d) Planar crystal with applied field in hard direction.

$$\left[\frac{\omega}{\gamma}\right]^2 = [(H_0 - H_a) - (N_z - N_x)M]$$
$$[(H_0 - H_a) - (N_z - N_y)M], \quad (6)$$

for $H_0 > H_a + N_z M$.

In his case, both the $c$ axis and the applied field are taken along $z$.

[1] M. W. Muller, Proc. Inst. Radio Engrs. **49**, 957 (1961).
[2] I. Bady and E. Schlömann, J. Appl. Phys. **33**, 1377 (1962).
[3] J. Smit and H. Beljers, Philips Research Repts. **10**, 113 (1955).

## EXPERIMENT

Single crystals of the materials, $Zn_2Y$, $Co_2Y$, and M were kindly supplied by A. Tauber of the U. S. Army Signal Research and Development Laboratory at Ft. Monmouth, New Jersey. Since the natural growth habit of these materials is in the form of thin platelets,[4] it was decided to fabricate small disk samples for the measurements. These disks were ground to a diameter of 1.025 in. and a thickness of 0.004 in. In all cases, measurements were made by the standard cavity perturbation technique using a $TE_{n10}$ rectangular cavity. The x direction mode number n was adjusted to give an appropriate filling factor.

Room temperature resonance measurements of the planar ferrite $Zn_2Y$ as a function of frequency were made with the applied field in the easy plane in order to determine the values of $\gamma$ and $H_a$. The experimental data fitted to Eq. (5) from 8 to 70 kMc for a value of $\gamma = 2.62$ Mc/gauss and $H_a = 9900$ oe. For this measurement $4\pi M$ at room temperature was taken as 2850 gauss.[5] Measurements in the basal plane indicate that any third order anisotropy field $H_a'$ must be less than 5 oe. The measurements of $H_a$ and $H_a'$ agree with the static measurements of Smit and Wijn.[5] The line widths measured ranged from 60 to 16 oe depending upon the sample chosen and the surface polish. In a group of samples measured, the narrowest line width was 20 oe. This value was reduced to 16 oe by successive polishes with compounds with $10\mu$ to $0.1\mu$ particle sizes (c.f., reference 6). As a function of frequency, the linewidth increased from 20 oe at X band to about 200 oe at 70 kMc. Since the sample size remained the same, it is presumed that a large factor in the increase was due to the inhomogeneity of the driving field at the higher frequencies.

Linewidth measurements as a function of temperature were made at three frequencies. At 8.26 kMc, $\Delta H$ increased from its room temperature value to 300 oe at 77°K. At 11.69 kMc, the increase at 77°K was only to 80 oe, while at 23.9 kMc, no increase at all was noted, $\Delta H$ remaining constant as a function of temperature. This behavior is presently being analyzed in the light of the recent work of Bady and Schlömann.[2] Field for resonance measurements as a function of temperature indicate that the value for $\gamma$ changes only slightly, possibly decreasing from its value of 2.62 Mc/gauss at 300°K to 2.55 Mc/gauss at 77°K.

Similar room temperature resonance measurements were made on disk samples of the planar ferrite $Co_2Y$. Using the value $4\pi M = 2300$ gauss[5] and fitting Eq. (5) to the experimental data from 8 to 70 kMc, the values $H_a = 28\,000$ oe and $\gamma = 3.75$ Mc/gauss were obtained. Measurements at 23.6 kMc and 300°K in the basal

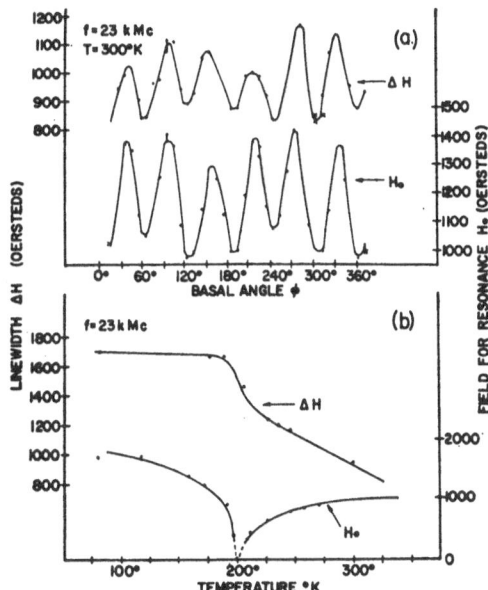

FIG. 1. Field for resonance and linewidth of $Co_2Y$ (a) as a function of basal angle and (b) as a function of temperature.

plane give a value of $H_a' = 200$ oe. The value of $H_a$ is in good agreement with static measurements.[5] The measured value of $H_a'$ is somewhat higher than the static value of 155 oe measured by Smit and Wijn.[5]

Linewidths in the easy plane depended markedly on the anisotropy, following exactly the same hexagonal angular dependence as the field for resonance, that is $\Delta H$ is a maximum or minimum at the same angle $\phi$ at which $H_0$ is a maximum or a minimum. This is in contrast to most measurements of $\Delta H$ versus $\phi$ for other materials.[7] An approximate expression for the linewidth can be written $\Delta H = (950 \text{ oe}) + (150 \text{ oe}) \cos 6\phi$. The measured field for resonance and $\Delta H$ as a function of basal angle $\phi$ are shown in Fig. 1(a).

As a function of frequency, the linewidth of $Co_2Y$ behaves similarly to $Zn_2Y$, increasing from an average value of 900 oe at 10 kMc to an average value of 1300 oe at 70 kMc. Again the reason is probably the large sample size compared to the wave length involved.

The temperature behavior of $Co_2Y$ is extremely interesting in that the material is planar (i.e., has an easy plane of magnetization at $\theta = \pi/2$ above a critical temperature equal to approximately 220°K. As the temperature is decreased below 220°K, the easy plane at $\theta = \pi/2$ becomes an easy cone at an angle $\theta$ which decreases as the temperature is lowered, reaching a value $\theta = 70°$ at approximately 77°K.[5] The resonance behavior as a function of temperature also exhibits a singularity at the critical temperature 220°K. The field for resonance for example decreases to a sharp minimum at this point, and then increases again as the temperature is lowered. The linewidth increases linearly as the temperature is decreased to 220°K, and then jumps sharply

[4] A. Tauber, R. O. Savage, R. J. Gambino, and C. J. Wainfrey, J. Appl. Phys. **33**, 1381 (1962).

[5] J. Smit and H. Wijn, *Ferrites* (John Wiley & Sons, Inc., New York, 1959).

[6] R. LeCraw, E. Spencer, and C. Porter, Phys. Rev. **110**, 1311 (1958).

[7] C. Buffler, J. Appl. Phys. **31**, 222S (1960).

as the magnetization goes out of the basal plane. Below 220°, $\Delta H$ remains fairly constant down to 77°K. The above behavior is illustrated in Fig. 1(b). The sharp dip in $H_0$ at 220°K cannot be explained using the static values of $4\pi M$ and $H_a$ versus temperature from Smit and Wijn[5] alone. Further knowledge of the effect of $H_a' = 36K_3/M$ on $H_0$ and the behavior of this parameter versus temperature must be known before a satisfactory answer can be given.

Because of the extremely high anisotropy fields in $M$ ($BaFe_{12}O_{19}$), it is diffiuclt to obtain information from resonance measurements below 40 kMc. A fit of Eq. (4) to the experimental data for both single and polycrystalline samples gives $H_a = 17\,000$ oe and $\gamma = 2.87$ Mc/gauss.

$H_a$ is in agreement with static measurements[5] and previous resonance measurements of Smit et al.[3] It is in slight disagreement, however, with the recent measurements of Wang et al.[8] It is possible that the above quoted values may be slightly more accurate since the Eq. (4) was fitted from 38 to 70 kMc while the Wang curve was fitted only in the region of 68 kMc.

### ACKNOWLEDGMENTS

I would like to express my appreciation to T. Litchfield for his technical assistance and to T. Pappalardo and R. Pelletier for the use of their facilities.

[8] F. Wang, K. Ishii, and J. Tsui, J. Appl. Phys. **32**, 1621 (1961).

---

JOURNAL OF APPLIED PHYSICS      SUPPLEMENT TO VOL. 33, NO. 3      MARCH, 1962

# Growth of Yttrium Iron Garnet on a Seed from a Molten Salt Solution

R. A. LAUDISE, R. C. LINARES, AND E. F. DEARBORN
*Bell Telephone Laboratories, Inc., Murray Hill, New Jersey*

Yttrium iron garnet has been crystallized on a seed crystal from molten $BaO\text{-}xB_2O_3$, where $x$ is 0.61. Two methods were used: (1) slow cooling of a melt in which a rapidly rotating seed was suspended, and (2) growth in a temperature gradient where excess yttrium iron garnet was maintained in a hotter part of the system, and a rapidly rotating seed was suspended in a cooler region. The geometry of the furnace, crucible, baffle, stirrer, and circulator system required to produce controlled growth in each of these systems is described, and the nature of the rate limiting step in each of the systems is discussed. The dependence of growth rate in the [110] direction on stirring rate, cooling rate, and solvent weight is presented. Good quality growth with rates as high as 50–75 mil/day was achieved in favorable cases.

### INTRODUCTION

THE preferred method for the growth of $Y_3Fe_5O_{12}$ is growth by the slow cooling of $PbO$[1] or $PbO$-$PbF_2$[2] fluxes. To improve the size and perfection of the yttrium iron garnet (YIG) crystals, to more easily effect doping, and to shorten or eliminate the chemical leaching step, it would be advantageous to grow garnet on a seed crystal under controlled conditions. One system which suggests itself is the growth of garnet in a temperature gradient where excess garnet is maintained in the hotter lower portion of a vessel, and a garnet seed is suspended in the top. In such a system convection would aid in circulation. A requisite for ease of operation of such a system is that garnet sink in the solvent. Since garnet floats in molten $PbO$, it is not an attractive solvent for such an experiment.

Therefore a search for another solvent for YIG was instituted, and one of the authors discovered that the system $BaO\text{-}xB_2O_3$ was nearly ideal.[3] Solubility and phase-equilibrium information for this system will be reported elsewhere. It was determined that garnet was congruently saturating[4] between 1000° and 1190° when $x$ was 0.61, and solubilities were determined over the range of interest to this work. In this solvent system garnet was congruently saturating and denser than the solvent, the vapor pressure was low, and the solvent was nontoxic.

Two classes of experiments were decided upon: growth of garnet by *slow cooling* of the saturated solvent in contact with a seed and growth of a seed in a *temperature gradient*. Miller has used the slow cooling method to grow potassium niobate on a seed.[5]

In the temperature gradient method excess garnet was placed in the bottom hot portion of a crucible containing the solvent, and a stirred seed was maintained in the upper cooler region. This method has been used by Timofeeva and Zalesskii[6] and by Reynolds and Guggenheim.[7]

[1] J. W. Nielsen and E. F. Dearborn, J. Phys. Chem. Solids 4, 202 (1958).
[2] J. W. Nielsen, J. Appl. Phys. **31**, 51S (1960).
[3] R. C. Linares, J. Am. Ceram. Soc. (to be published).

[4] "Congruently saturating" is used in this work in the sense defined by Findlay. See, for example, *The Phase Rule* by A. Findlay, A. N. Campbell, and N. O. Smith (Dover Publications, New York, 1951), p. 346.
[5] C. E. Miller, J. Appl. Phys. **29**, 233 (1958).
[6] V. A. Timofeeva and A. V. Zalesskii, *Rost Kristallov*, edited by A. V. Shubnikov and N. N. Sheftal (U. S. S. R. Academy of Sciences Press, Moscow, 1959), Vol. 2, p. 88; *Crystal Growth*, (translated by Consultants Bureau, New York, 1959), Vol. 2, p. 69.
[7] G. F. Reynolds and H. Guggenheim, J. Phys. Chem. **65**, 1655 (1961).

## EXPERIMENTAL

In all experiments reagent grade starting materials were used. Crystalline garnet nutrient, which was placed in the lower hotter region of the crucible for the gradient experiment, was prepared by slow cooling $PbO$-$PbF_2$ melts by Nielsen's method.[2] A cylindrical platinum crucible was placed in a platinum-wound, ceramic core furnace which could be adjusted in order to make the crucible isothermal or to cause it to experience a gradient. In the gradient experiments the gradient was partly localized by means of a platinum baffle of the sort used in hydrothermal crystal growth[8] and first applied to molten salt systems by Reynolds and Guggenheim.[7] A typical baffle had 30% open area. The baffle was fastened to a motor-driven stirring rod. To this rod was attached stirring paddles above and below the baffle and the seed crystal above the baffle. All parts in contact with the flux were platinum. In the slow cooling experiments no nutrient, baffle, or paddles were present, but the seed was mounted on the stirring rod alone. The stirrer was designed to rotate alternately 30 sec clockwise and 30 sec counterclockwise. Its speed was adjustable from 0–400 rpm. Temperature control was effected by conventional contactor and saturable core-type controllers.

## RESULTS AND DISCUSSION

In a series of slow cooling experiments in a 3-in.-diam×5-in. long crucible with a 135-g melt on seed plates whose principal faces were (110),[9] it was shown that the growth rate depended on the stirring rate until a critical stirring rate was exceeded. The greater the cooling rate, the greater the critical stirring rate. Rates as high as 3 mil/hr at 40 rpm and 5°/hr were achieved. The growth rate was found to be linear with cooling rate up to cooling rates of 10°/hr, provided the critical stirring rate was exceeded. The rate was also found to depend linearly on the mass of solution up to weights of solution near 1400 g, provided the critical stirring rate was exceeded. It was not surprising that the rate was found to depend on the stirring speed, since, if a diffusion process were involved in the rate-limiting step, stirring would shorten the diffusion path. Thus, it would be expected that successive increments of stirring would be

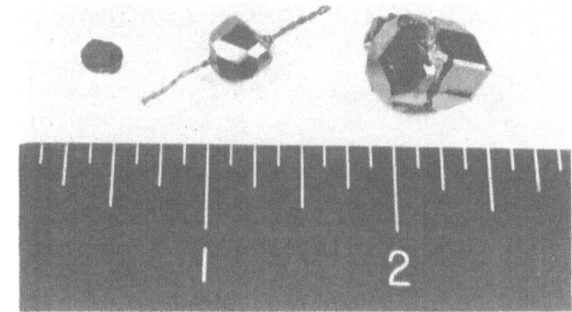

FIG. 1. YIG crystals together with typical seed grown by temperature gradient method (scale in inches).

less effective until a critical stirring speed where the effect of diffusion on the rate is negligible would be reached. Under conditions of high supersaturation or for large melt volumes, the diffusion step is apparently more severely rate limiting, and more rapid stirring is required to exceed the critical stirring speed. Spontaneous nucleation is negligible under the experimental conditions so the seed is the principal sink for YIG.

YIG was grown in the temperature gradient experiments under the following conditions: dissolving temperature, 1130°C; temperature gradient, 40°; growth temperature, 1170°C; stirring speed, 200 rpm; crucible diameter 3 in.; rate in ⟨110⟩, 0.5–1.7 mils/hr.

When the crucible diameter was less than 3 in. or the stirring speed was lower, the rate tended to decrease with time since, as the YIG nutrient sintered and its surface area decreased, the dissolving step tended to become rate limiting. Under such conditions when several grams of finely powdered high-surface area YIG were added to the nutrient zone, the rate recovered. However, a crucible diameter of 3 in. and 200 rpm stirring speed served to provide rapid enough dissolving that the rate was nearly independent of the time of the experiment. Figure 1 shows typical crystals grown in this study.

Further studies of the kinetics of YIG growth, attempts to grow large crystals, and the application of this technique to other materials are under way. Details of this work will be reported elsewhere.[10]

### ACKNOWLEDGMENTS

The authors would like to thank G. F. Reynolds and H. Guggenheim for several discussions concerning the role of baffles in molten salt systems. We would also like to thank R. L. Barns for many useful discussions and M. Tanenbaum for several suggestions concerning the treatment of results.

[8] G. T. Kohman, U. S. Patent No. 2895812. See also R. A. Laudise and J. W. Nielsen in *Solid State Physics*, edited by D. Turnbull and F. A. Seitz (Academic Press Inc., New York, Vol. XIII to be published).

[9] In all experiments the seed plate on which growth took place had two large surfaces parallel to (110) and consequently grew in both the [110] and [1̄10] directions. The rates reported here are half the increase in thickness of the seed plate divided by the time of the experiment and consequently are the rates in the [110] direction. It should be noted that the area of the seed did not change substantially during any of the experiments.

[10] R. A. Laudise in *The Art and Science of Crystal Growth* (to be published).

JOURNAL OF APPLIED PHYSICS    SUPPLEMENT TO VOL. 33, NO. 3    MARCH, 1962

# Cobalt Ferrite Crystal Growth from the Ternary Flux System $Na_2O$-$CoO$-$Fe_2O_3$

W. Kunnmann, A. Wold, and E. Banks*

*Lincoln Laboratory†, Massachusetts Institute of Technology, Lexington 73, Massachusetts*

It is shown that the simple ferrite spinels ($M^{2+}Fe_2O_4$) crystallize readily from the ternary flux system $MO$-$Na_2O$-$Fe_2O_3$. The liquidus-solidus phase diagrams for the pseudobinary of $Na_2Fe_2O_4$-$CoFe_2O_4$ and the ternary $Na_2O$-$CoO$-$Fe_2O_3$ are presented. A general discussion of flux growth of ferrite crystals of both the simple spinel type and the more complex layered structures ($BaFe_{12}O_{19}$) is given. Crystals of $CoFe_2O_4$ were grown from a mixture of 1.5 moles of $CoFe_2O_4$ and 1 mole of $Na_2Fe_2O_4$ by utilizing a platinum wire cold finger.

## INTRODUCTION

RECENT increased interest in the magnetic and electrical properties of pure ferrites has created a need for single crystals of these materials. The growth of stoichiometric ferrite crystals is difficult because the melting points of these compounds are higher than the dissociation temperature of ferric oxide in oxygen at atmospheric pressure. This problem can be overcome if a suitable flux can be found which will form a system with the ferrite, possessing a eutectic with a temperature lower than the dissociation point of ferric oxide.

Ferrites have been successfully grown from molten borax,[1] lead oxide,[2] sodium carbonate,[3-5] and ferrite melts under high oxygen pressures.[6] Little information is available concerning the pertinent phase diagrams, although Nielsen[7] has reported some information concerning the primary crystallization products in the system $PbO$-$Fe_2O_3$-$Y_2O_3$. Gambino[5] has published the liquidus curve for the $BaFe_{12}O_{19}$-$Na_2O$ system. Summergrad and Banks,[3] and Mones and Banks,[4] found that molten sodium carbonate was an effective flux for crystal growth of $BaFe_{12}O_{19}$ and its analogues $La_{0.5}M_{0.5}Fe_{12}O_{19}$, where M is an alkali metal. In these investigations homogeneous single crystals of $BaFe_{12-x}Al_xO_{19}$ were grown from a sodium carbonate flux. Gambino and Leonhard[5] have reported that crystals of barium ferrite grown from a sodium carbonate flux are spectrochemically free from sodium contamination. It is important to note that simple ferrites of the type $MFe_2O_4$, where M is either an alkaline earth or transition metal ion, cannot be grown from sodium carbonate. In most cases the sodium carbonate reacts with the ferrite to form sodium ferrite and MO. We have found that the ferrites $MFe_2O_4$ crystallize readily in the presence of excess ferric oxide, which reacts with the sodium carbonate to form the flux sodium ferrite.

## EXPERIMENTAL

*Materials.* Cobalt carbonate, anhydrous reagent grade, was dried at 100°C for 24 hr and analyzed by converting to cobaltous oxide at 1000°C in a nitrogen atmosphere. Ferric oxide was 99.9% anhydrous reagent grade. Sodium carbonate was Baker's anhydrous reagent grade.

*Methods.* A modified system which is capable of performing both D. T. A. and thermal analysis was utilized to detect the various phase changes in the $Na_2O$-$CoO$-$Fe_2O_3$ system. The voltage from a thermocouple imbedded in a standard was plotted on one axis of an X–Y recorder and the voltage of the sample thermocouple on the other axis. By this technique both a thermal analysis trace (on an arbitrary time base) and a D. T. A. trace on the diagonal were simultaneously obtained. The sensing thermocouples were Pt-90% Pt, 10% Rh standardized against the melting point of pure gold. The thermocouple outputs were amplified and fed to an X–Y type recorder which had a sensitivity of 1 mv/in. A hollow Globar element, which gave a 24-in. heating zone, was utilized in the construction of the furnace. Platinum crucibles were used to contain the melts. Several determinations of each liquidus point were made by reheating the melt to 1400°C. The melt was maintained at this temperature for several hours and then cooled slowly at a rate of 3° per minute. Measurements were reproduced to within ±20 $\mu v$ (approximately 2°C). However, the absolute temperatures reported probably have a considerably greater error because of the inherent difficulty of accurate temperature measurement in this range. The various phases were identified by quenching particular compositions from temperatures below which the thermal trace indicated the existence of a solid phase. The samples were then examined by x-ray techniques, utilizing chromium radiation.

## GROWTH OF $CoFe_2O_4$ CRYSTALS

It was determined experimentally that large pure cobalt ferrite crystals could be grown from melts of cobalt ferrite-sodium ferrite mixtures containing 1.5 moles of $CoFe_2O_4$ and 1.0 mole of $Na_2Fe_2O_4$. The mixtures are heated to 1350°C and slow-cooled at the rate

* Present address: Department of Chemistry, Polytechnic Institute of Brooklyn, Brooklyn 1, New York.
† Operated with support from the U. S. Army, Navy, and Air Force.
[1] J. K. Galt, B. T. Matthias, and J. P. Remeika, Phys. Rev. **79**, 391 (1950).
[2] J. P. Remeika, J. Am. Chem. Soc. **78**, 4259 (1956); J. W. Nielsen, J. Appl. Phys. **29**, 390 (1958).
[3] R. Summergrad and E. Banks, J. Phys. Chem. Solids **2**, 312 (1957).
[4] A. Mones and E. Banks, J. Phys. Chem. Solids **4**, 217 (1958).
[5] R. J. Gambino and F. J. Leonhard, J. Am. Ceram. Soc. **44**, 221 (1961).
[6] A. Ferretti, R. J. Arnott, E. Delaney, and A. Wold, J. Appl. Phys. **32**, 905 (1961).
[7] J. W. Nielsen and E. F. Dearborn, J. Phys. Chem. Solids **5**, 202 (1958).

of 2°C/hr. A platinum wire cold finger is suspended into the melt. The position of the crucible in the heat zone is such that the sides and bottom of the crucible are at constant temperature, and cooling takes place at the cold finger as a result of radiation losses at the top of the crucible. The cold finger is maintained by this manner at only several degrees below the temperature of the melt. Crystal nucleation forms at the cold finger resulting in several large crystals having a conical shape. The crystals grown were approximately 1 in. in diameter and $\frac{3}{4}$ in. thick. Chemical analysis indicated the absence of $Fe^{++}$, and there appears to be no sodium present when crystals are grown by this technique.

## RESULTS AND DISCUSSION

### The $Na_2Fe_2O_4$-$CoFe_2O_4$ Pseudobinary System

The experimental results of the liquidus system $Na_2O$-$CoO$-$Fe_2O_3$ are summarized in Fig. 1. A portion of this ternary diagram indicates that a simple pseudobinary system is formed, with a eutectic point at the composition $0.6Na_2Fe_2O_4$-$0.4CoFe_2O_4$ and temperature of 1150°C. The eutectic halt was not observed at low $CoFe_2O_4$ compositions. X-ray diffraction studies of crystals grown from $Na_2Fe_2O_4$ melts showed no detectable shift in the lattice constant of $CoFe_2O_4$. This would indicate that solid solution does not occur to any appreciable extent (less than 1%), and is in accord with the spectrochemical analysis of $BaFe_{12}O_{19}$ crystals grown by Gambino and Leonhard,[5] which showed the absence of sodium. In addition, x-ray patterns of quenched samples show the presence of only two phases, $CoFe_2O_4$ and $Na_2Fe_2O_4$, over the compositional range indicated. It is of interest to note that the freezing point depression resulting from the addition of $CoFe_2O_4$ to $Na_2Fe_2O_4$ is only slightly less than when $Fe_2O_3$ is added to $Na_2Fe_2O_4$. The results obtained for the system $Fe_2O_3$-$Na_2Fe_2O_4$

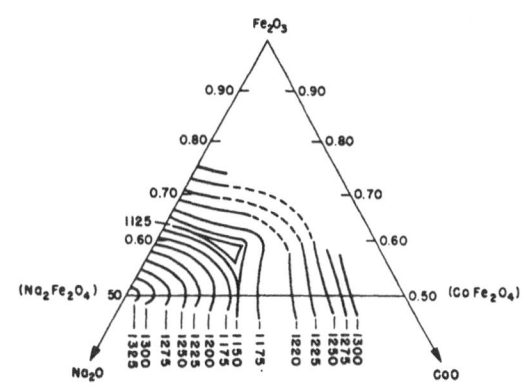

FIG. 1. Liquidus-solidus isotherms for $Na_2O$-$CoO$-$Fe_2O_3$ ternary system.

confirm those reported by Knick and Kohlmeyer.[8] However, the freezing point depression of $CoFe_2O_4$ by $Na_2Fe_2O_4$ is substantially greater than that of $Fe_2O_3$ by $Na_2Fe_2O_4$. The melting points of pure $CoFe_2O_4$ and $Fe_2O_3$ are approximately of the same order of magnitude. There was no evidence for the formation of either sodium cobaltite or ferrous ion over the range of compositions and temperatures studied.

### The $Na_2O$-$CoO$-$Fe_2O_3$ Ternary System

Experimental results for a portion of the liquidus system $Na_2O$-$CoO$-$Fe_2O_3$ are summarized in Fig. 1. The liquidus data indicated that a simple ternary system is formed for the members $Na_2Fe_2O_4$-$CoFe_2O_4$-$Fe_2O_3$. Only in the regions of high concentration of $CoFe_2O_4$ and $Fe_2O_3$, where decomposition and other effects notably change the system, are there complications. In the region immediately below 50% $Na_2O$-50% $CoO$ (see Fig. 1), no serious deviations occur. However, beyond approximately 40% $Fe_2O_3$ the data were not reproducible.

[8] R. Knick and E. J. Kohlmeyer, Z. anorg. allgem. Chem. 244, 77 (1940).

# Hexagonal Ferrites for Use at *X*- to *V*-Band Frequencies*

G. P. RODRIGUE, J. E. PIPPIN, AND M. E. WALLACE

*Sperry Microwave Electronics Company, Clearwater, Florida*

A series of uniaxial materials has been prepared and studied in which the effective anisotropy field can be controlled and varied over the range of 0 to 12 700 oe. The materials are solid solutions of the NiW and CoW compounds and have the chemical formula $BaO \cdot 2[Ni_{1-x}Co_xO] \cdot 7.8Fe_2O_3$ with $x$ varying from zero, where the anisotropy field is 12 700 oe, to 0.5 where the anisotropy field is vanishingly small. The temperature dependence of the net resultant anisotropy field, linewidth, and $g$ factor of these materials is discussed. By using oriented polycrystalline samples of these materials, the effective internal anisotropy field can be used to augment external applied fields. Different materials of this series have been used to construct resonance isolators at *V*- and *K*-band frequencies with bandwidths of the order of 20% and peak isolation ratios of greater than 30 to 1.

## INTRODUCTION

ALL ferromagnetic crystals exhibit a preferred direction of magnetization with respect to the crystallographic axes. Such a preference is described phenomenologically by the magnetocrystalline anisotropy energy. The presence of this anisotropy results in a torque on the magnetization and is therefore often represented by an effective anisotropy field, defined to have the magnitude and direction that would be required of an external field to produce the same torque on the magnetization. In the uniaxial hexagonal ferrite crystal, the preferred direction of magnetization is along the *C* axis, while the planar hexagonal ferrite crystals have a preferred plane of magnetization (the basal plane). The effective anisotropy fields of all crystallites in a polycrystalline sample can be made to cooperate by aligning the *C* axes of the individual crystallites along the same direction in space. Such aligned uniaxial hexagonal ferrites, which generally have large anisotropy fields, can be used to make nonreciprocal microwave devices requiring little or no external magnetic field.

The anisotropy fields of the barium and strontium ferrite systems have been varied over a wide range by the substitution of aluminum for iron[1] and the simultaneous substitution of titanium and cobalt for iron.[2,3] Substituted barium ferrite materials have been used in fabricating millimeter wave devices requiring minimum externally applied fields.[4]

Materials reported here are based on uniaxial hexagonal materials of the W[5] structure and consist of solid solutions of the nickel W and cobalt W compositions. Nickel W is uniaxial with an anisotropy field of 12 700 oe. Cobalt W is planar. By preparing solid solutions of cobalt W in nickel W, it is possible to obtain uniaxial materials having anisotropy fields controllable between 12 700 oe and zero.

## MATERIALS

Solid solutions of NiW and CoW were synthesized with the chemical formula $BaO \cdot 2Ni_{1-x}Co_xO \cdot 7.8Fe_2O_3$ with $x=0$, 0.05, 0.1, 0.2, 0.25, 0.3, 0.4, and 0.5. The materials were prepared by conventional ceramic techniques and allowance was made for iron pickup in the preparation process. The *C* axes of individual crystallites were aligned by pressing presintered material in the presence of an orienting dc magnetic field. The orienting field used was approximately equal to 0.7 $H_{an}$, where $H_{an}$ is the anticipated value of the anisotropy field for the material in preparation. Sludge press techniques were employed, and it was found that the higher the degree of dispersion of the presintered material in the sludge, the better the alignment of the final product.

The degree of orientation achieved can be qualitatively judged from the asymmetry in shrinkage that occurs on firing, and from metallographic inspection. A numerical index of alignment on a scale of 0 to 1.0 (1.0 being perfect alignment) can be obtained from x-ray powder diffraction data. This index is the same quality factor suggested by Lotgering.[6] Since the orienting torque exerted on individual crystallites varies directly with anisotropy field, one would expect (for a given suspension) the degree of alignment to fall off with anisotropy field. This is indeed found to be the case. Most nearly complete alignment (index of 0.95) was obtained on materials having the largest anisotropy field and the best suspension in the sludge. For the higher cobalt content samples with low anisotropies, a substantial improvement in alignment was obtained by using the Lotgering chemical alignment technique.[6]

Values of anisotropy field were obtained from ferromagnetic resonance data taken with a conventional ferromagnetic spectrometer[7] using a reflection cavity resonant at 37.5 kMc. The imaginary component of the susceptibility tensor was measured as a function of

* This work was supported in part by the Rome Air Development Command, Rome, New York.

[1] D. J. DeBitetto, F. K. DuPré, and F. G. Brockman, *Millimeter Waves* (Polytechnic Press, Brooklyn, 1959), p. 95.
[2] J. Smit and H. P. J. Wijn, *Ferrites* (John Wiley & Sons, Inc., New York, 1959), p. 177.
[3] D. J. DeBitetto, F. K. DuPré, and D. W. Krautkopf, Bull. Am. Phys. Soc. 7, 54 (1962).
[4] H. G. Beljers, Philips Tech. Rev. 22, 11 (1960).
[5] G. H. Jonker, H. P. J. Wijn, and P. B. Braun, *Conference on Magnetism and Magnetic Materials* (American Institute of Electrical Engineers, New York, 1957), p. 478.
[6] F. K. Lotgering, J. Inorg. & Nuclear Chem. 9, 113 (1959).
[7] J. O. Artman and P. E. Tannenwald, J. Appl. Phys. 26, 1124 (1955).

applied dc field with $H_{dc}$ both perpendicular and parallel to the *C* axis of the spherical sample. From these data, the linewidth in both directions, *g* factor, and anisotropy field were determined. The variation of anisotropy field with cobalt content is shown in Fig. 1 to be almost linear. The effective *g* factor remained essentially constant over this range of cobalt substitution, at least within the experimental accuracy of approximately 5%, at $1.90 \pm 0.10$. Resonance linewidths observed for easy and hard directions of magnetization ($H_{dc}$ parallel and perpendicular to *C* axis, respectively) were approximately equal, and were also roughly independent of composition. Linewidths were found to be highly dependent on degree of alignment and hence were lower for the more anisotropic samples. This is, however, felt to be a dependence on alignment rather than on cobalt content. The most highly aligned samples exhibited linewidths of 1500 to 1800 oe, moderate alignments resulted in linewidths 2000—3000 oe, and poor alignment produced resonance lines of arbitrarily large breadth (>7000 oe).

Resonance measurements were conducted as a function of temperature over the most frequently encountered range of room temperature to 125°C. The observed dependence of anisotropy field on temperature for three different compositions is shown in Fig. 2. The anisotropy field of these compounds is found to increase almost linearly with temperature, and the positive slope of the lines is larger for the higher cobalt content materials. The positive slope and the increase in slope with increasing cobalt content both indicate that the anisotropy constant of cobalt W is a stronger function of temperature than is that of nickel W. This behavior is

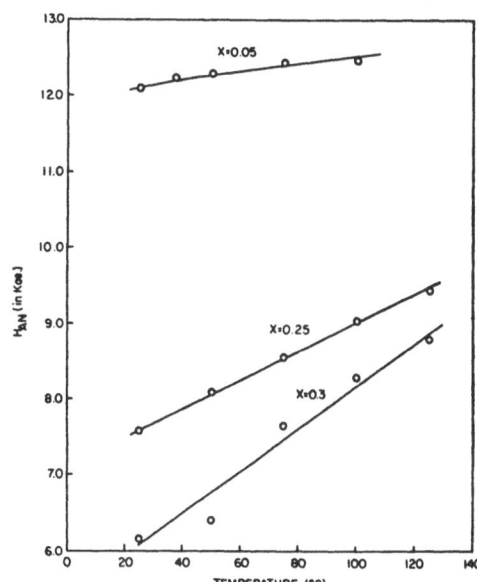

FIG. 2. Variation of anisotropy field with temperature in three compositions of nickel cobalt W. The parameter is *x* in BaO·2 [$Ni_{1-x}Co_xO$]·7.8[$Fe_2O_3$].

similar to that of the spinel ferrites where the anisotropy of cobalt ferrite[8] falls off more rapidly with temperature than that of nickel ferrite.[9] The linewidth of the materials studied is found to be essentially constant over the temperature range covered; the effective *g* factor increases with temperature and approaches 2.0 near 125°C.

## DEVICES

The 10% cobalt substituted nickel W compound with an anisotropy field of approximately 11 000 oe was selected for use in a *V*-band resonance isolator. Several different geometries were used. A rectangular rod of material mounted against a similar rod of Stycast produced a maximum isolation ratio of 37 db to 1 db at 36 kMc with 3200 oe applied. Isolation loss greater than 10 db with insertion loss less than 1 db was maintained over a 1.6 kMc bandwidth. A slab of material magnetized in its plane and mounted on a slab of dielectric ($\epsilon=9$) produced maximum isolation ratios of 26 db to 1 db with approximately 300 oe applied. The 10 to 1 bandwidth was 3 kMc with a center frequency of 37 kMc. The wide variation in applied field required for these geometries is due to the differing demagnetizing factors present, and these resonance fields may be compared to the 12 000 to 15 000 oe required for ordinary ferrites. A well-aligned sample of the 5% cobalt substituted NiW material gave peak isolation ratios of 20 to 1 with no external applied field whatsoever. The 10 to 1 bandwidth was 2.5 kMc centered at 37 kMc.

At *K*-band frequencies the 25% cobalt substituted

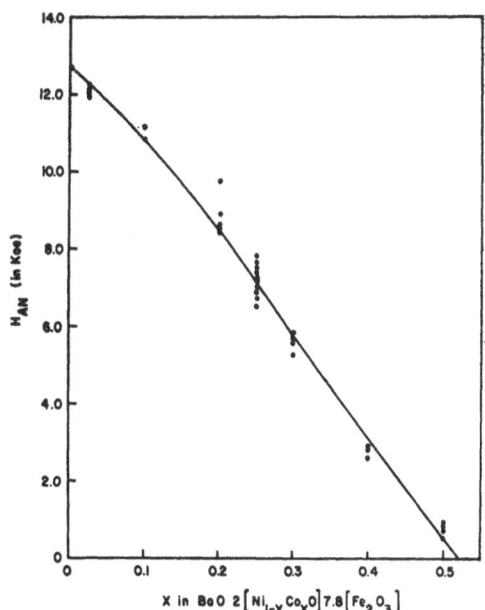

FIG. 1. Measured uniaxial anisotropy fields as a function of cobalt content in nickel-cobalt W compounds.

[8] H. Shenker, Navord Report 3858, U. S. Naval Ordnance Laboratory (1955).

[9] D. W. Healy, Jr., Phys. Rev. **86**, 1009 (1952).

NiW compound with an anisotropy field of 7000 oe is appropriate for resonance isolator fabrication. A slab of material magnetized in its plane and mounted on a dielectric slab produced a maximum isolation ratio of 35/1 at a frequency of 23 kMc with an applied field of less than 1000 oe supplied by small ceramic magnets. The 10 to 1 bandwidth was in this case greater than 6 kMc, or about a 25% bandwidth.

While the anisotropy field is shown in Fig. 2 to be a function of temperature, the broad linewidth of these materials should render them relatively insensitive to temperature variations. Moreover, in cases where the internal anisotropy field is used in conjunction with an externally applied field supplied by a permanent magnet, the increasing anisotropy field should be compensated to some extent by the decreasing external field at elevated temperatures.

The 40% cobalt substituted NiW material ($H_{an} = 2800$) was evaluated at $X$ band, but isolation ratios of no better than 8 to 1 were obtained. This material had a linewidth of 2700 oe and it is expected that improved orientation will so reduce the linewidth as to make considerably higher isolation ratios feasible.

### ACKNOWLEDGMENTS

The authors are indebted to D. Taft who designed the $K$-band isolator, to Dr. A. L. Stanford for his suggestions in the early phases of this work, and to D. Dudley and H. Burkhardt for technical assistance.

JOURNAL OF APPLIED PHYSICS          SUPPLEMENT TO VOL. 33, NO. 3          MARCH, 1962

# High-Power Characteristics of Single-Crystal Ferrites with Planar Anisotropy

Samuel Dixon, Jr.

*U. S. Army Signal Research and Development Laboratory, Fort Monmouth, New Jersey*

Single crystal $(ZnY)(Ba_2Zn_2Fe_{12}O_{22})$ was used in an investigation of the variation of the magnetic susceptibility with microwave magnetic field. The large anisotropy field in these materials makes them attractive for applications in microwave devices. Measurements were made in the frequency range of 8.6 to 37.7 kMc. The microwave magnetic field in the cavities extended up to 35 oe. Measurements taken at a constant frequency indicate that there is an appreciable shift in the field required for resonance as a function of power. This shift in the resonance field is superimposed on the usual decline of the main resonance. Measurements indicate that the shift effect of the resonance field decreases as the linewidth and frequency are increased. In the frequency range of 8.6 to 16.6 kMc, the critical field increases with frequency. Above this frequency, the critical field tends to decrease. At $X$-band frequencies, the critical field compares favorably with that of YIG material.

## INTRODUCTION

SINGLE-crystal ferroxplanar materials, as described by Tauber *et al.*,[1] were used in an investigation of the magnetic susceptibility with the microwave magnetic field. The saturation magnetization of this material was 2000 gauss. The effective anisotropy was approximately 8000 oe. This large anisotropy field, which tends to keep the magnetization in the easy plane, reduces the field required for ferromagnetic resonance. A series of experiments was performed over the frequency range 8.6 to 37.7 kMc. The rf magnetic field intensities extended up to 35 oe. It is the purpose of this paper to describe some of the charactersitics at high rf magnetic field intensities.

## EXPERIMENTAL DETAIL

Susceptibility measurements were made on an unpolished ferroxplanar sphere with a diameter of 0.020 in. The same sample was used over the frequency range of 8.6 to 37.7 kMc. The susceptibility was measured using the well-known cavity technique.[2] Magnetrons were used as the rf power source. A low-duty cycle of $3 \times 10^{-5}$ was used to avoid heating

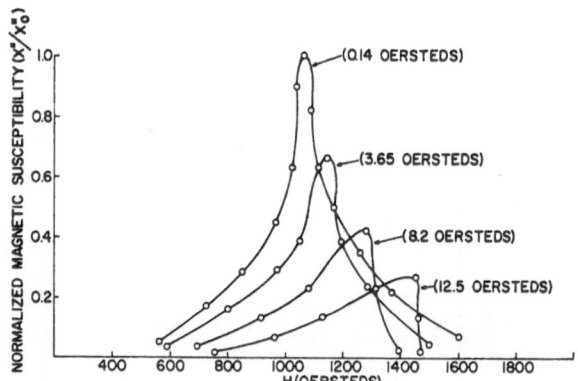

FIG. 1. ZnY single-crystal ferroxplanar. Variation of the normalized susceptibility and the field required for resonance with the incident rf power as a parameter: frequency=8957 Mc, $\Delta H = 85$ oe.

[1] A. Tauber, R. O. Savage, R. J. Gambino, and C. G. Whinfrey, J. Appl. Phys. **33**, 1381 (1962), this issue.

[2] J. L. Carter, S. Dixon, Jr., and I. Reingold, IRE Trans. on Microwave Theory Tech. **MMT-9**, No. 2, 195 (1961).

effects. The cavities used were of the transmission type operating in the $TE_{10n}$ mode. The length of the cavities varied from 1 to 15 wavelengths. The multiple wavelength cavities made it possible to make measurements at more than one frequency, using the same cavity. The sample was placed in the center of the cavity at a position of maximum rf magnetic field. The orientation of the sample was such that the rf magnetic field and the applied magnetic field were perpendicular to each other, and in the easy plane.

### EXPERIMENTAL RESULTS

Normalized magnetic susceptibility, as a function of applied dc magnetic field, is shown in Fig. 1. It should be noted that the susceptibility, the field required for resonance, and the shape of the resonance curve vary as a function of rf magnetic field intensity. The shift of the resonance field as a function of rf magnetic field, with frequency as a parameter, is shown in Fig. 2. The

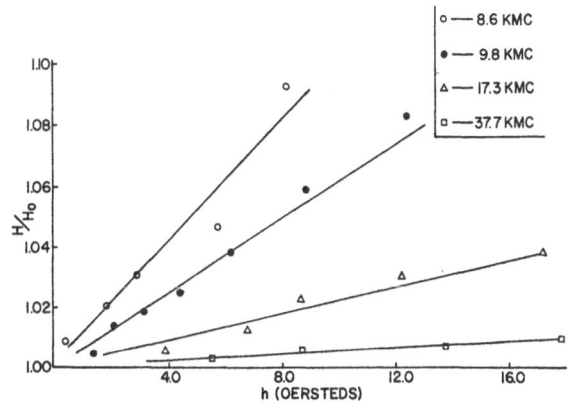

FIG. 2. ZnY ($Ba_2Zn_2Fe_{12}O_{22}$) single-crystal ferroxplanar. Shift in the field required for resonance vs rf magnetic field—with frequency as a parameter.

TABLE I. Results of measurements made in the frequency range 8.6 to 37.7 kMc.

| Frequency (kMc) | $H_0$ computed (oe) | $H_0$ measured (oe) | $\Delta H$ (oe) | $h_c$ (oe) | $Q$ |
|---|---|---|---|---|---|
| 8.6 | 1050 | 1190 | 60 | 0.16 | 19.7 |
| 8.9 | 1120 | 1225 | 70 | 0.30 | 16.4 |
| 9.8 | 1330 | 1455 | 70 | 1.0 | 20.8 |
| 16.6 | 3160 | 3250 | 75 | 1.8 | 43.4 |
| 17.3 | 3360 | 3460 | 75 | 1.5 | 46.2 |
| 37.7 | 10050 | 9680 | 100 | 1.4 | 96.8 |

observed results indicate that the shift of the resonance field decreases as the frequency is increased. The shift of the resonance field also decreases as the linewidth of the sample increases. The results of the measurements are presented in Table I. The anisotropy field, computed from the resonance field, is 8000 oe. When this value of anisotropy field is used in an equation derived previously by Bady,[3]

$$(\omega/\gamma)^2 = H_0(H_0 + H_a)$$

the computed and measured value of the applied field ($H_0$) is in good agreement. The measured linewidth varies from 60 oe at 8.6 kMc to 100 oe at 37.7 kMc. The "$Q$" of the spherical sample varies appreciably over the frequency range. The critical field was determined by using the method described by Green and Schlömann.[4] Using an equation derived previously by

Bady and Schlömann,[5] computations indicate that the first-order nonlinear process is dominant in the frequency range 7.5 to 8.9 kMc. Coincidence conditions also exist in this frequency range. At higher frequencies, the second-order process is responsible for the saturation effect. The measurements at 8.6 and 8.9 kMc would fall in the region of the first-order process. At the other frequencies, the saturation effect is caused by the second-order process. It should be noted that in the frequency range of 8.6 to 16.6 kMc, the critical field increases with frequency. Above this range, the critical field tends to decrease.

### DISCUSSION

Weiss,[6] and others, have pointed out that under certain conditions a shift of the resonance field occurs at sufficiently high-power levels in cubic crystals. This effect was attributed to crystalline anisotropy. The magnitude of the shift observed in this material is much larger than that measured in cubic crystals. The higher anisotropy field may account in part for the observed results.

The reduction in the applied magnetic field and the increase in "$Q$" as a function of frequency indicate that the planar ferrites should have important applications in nonlinear microwave devices operating at millimeter wavelengths.

Ferroxplanar materials with linewidths less than 20 oe have been reported. At $X$-band frequencies the critical fields in this material compare favorable with those of highly polished single crystal YIG. This should result in improved nonlinear microwave devices.

[3] I. Bady, I.R.E. Transactions on Microwave Theory Tech. **MTT-9**, 52 (January, 1961).

[4] J. J. Green and E. Schlömann, Raytheon Technical Memoranda T-168, June, 1959.

[5] I. Bady and E. Schlöman, J. Appl. Phys. **33**, 1377 (1962), this issue.

[6] M. T. Weiss, J. Appl. Phys. **31**, 103S (1960).

JOURNAL OF APPLIED PHYSICS     SUPPLEMENT TO VOL. 33, NO. 3     MARCH, 1962

# Nickel Aluminum Gallium Ferrites for Use at High Signal Levels

J. W. Nielsen and J. E. Zneimer

*Airtron Inc., Division of Litton Industries, Morris Plains, New Jersey*

A new nickel ferrite has been developed which contains simultaneously aluminum and gallium. Compositions have been prepared according to the formula $NiAl_yGa_xFe_{2-(x+y)}O_4$. The total substitutions of gallium and aluminum have been varied from 0 to 1.2. Several ratios of aluminum to gallium have been made. The unique feature of this system is that since the gallium substitutes in the tetrahedral sites and aluminum in the octahedral sites, the compensation point normally associated with nickel aluminum spinels can be shifted to much higher nonmagnetic ion substitutions. There exists a composition for which there is no compensation point. Measurements of saturation magnetization, Curie temperature, linewidth, and $g_{eff}$ factor have been made. A C-band circulator-switch using one of these materials has been constructed and successfully tested at 75-kw peak power with a 0.001 duty cycle. This device was also successfully tested at a power level of 475 w cw in an ambient temperature of 85°C.

## INTRODUCTION

IN the past ten years many spinels and garnets have been developed for use at microwave frequencies. With the recent additional requirement for devices to operate at high signal levels and at high ambient temperatures particularly at C band and below, new materials are required. There were no materials available with the four requirements: saturation magnetization between 500 and 1300 gauss, effective $g_{eff}$ factors less than 3, a Curie temperature over 200°C, and high power handling capabilities. However, the early work of Suhl,[1] the more recent work of Schlömann, Saunders, and Sirvetz,[2] and that of Reingold, Dixon, and Carter,[3] plus the device experience of Wantuch,[4] all suggest that nickel ferrite shows a high resistance to high power effects. We will discuss a nickel ferrite system which satisfies the four requirements given above.

### NICKEL ALUMINUM GALLIUM FERRITE

This program is in part an extension of the work of Maxwell and Pickart.[5] They found that gallium substitutes in tetrahedral sites in nickel ferrite while aluminum substitutes in the octrahedral sites. Thus a simultaneous substitution of gallium and aluminum will shift the compensation point to very high concentrations of substituents. In the system $NiAl_yGa_xFe_{2-(x+y)}O_4$ there are series of compositions that will have no compensation point at all.

Figure 1 is a plot of saturation magnetization versus gallium substitution and aluminum substitution in nickel ferrite, and combinations of gallium and aluminum substitutions. Gallium substitutions up to 0.75 moles in nickel ferrite will increase the $4\pi M_s$. Additional substitutions will then decrease the saturation magnetization. The weighted average of these two curves will be of some intermediate value. The actual saturation magnetization will depend on:

(1) The ratio of gallium to aluminum.
(2) The tenacity with which gallium and aluminum stay in the tetrahedral and octrahedral sites respectively.
(3) The general weakening of magnetic interaction as the nonmagnetic ion concentration increases, especially the $A$-$B$ interaction.

Two ratios of gallium ($x$) to aluminum ($y$) have been plotted; $x=y$, and $2x=y$. Extrapolating these curves to zero magnetization indicates that the compensation point has been shifted to $x+y$ substitutions equal to 1.8 and 1.4 for the above ratios of gallium to aluminum. This compares to compensation point composition of approximately 0.67 in nickel aluminum ferrite.

The effect of gallium on the Curie temperature is the same as that of aluminum. This is very important since it is in contrast to zinc which lowers the Curie temperature more severely. Figure 2 is a plot of Curie temperature versus composition in nickel aluminum, nickel gallium, and nickel aluminum gallium spinels. The Curie

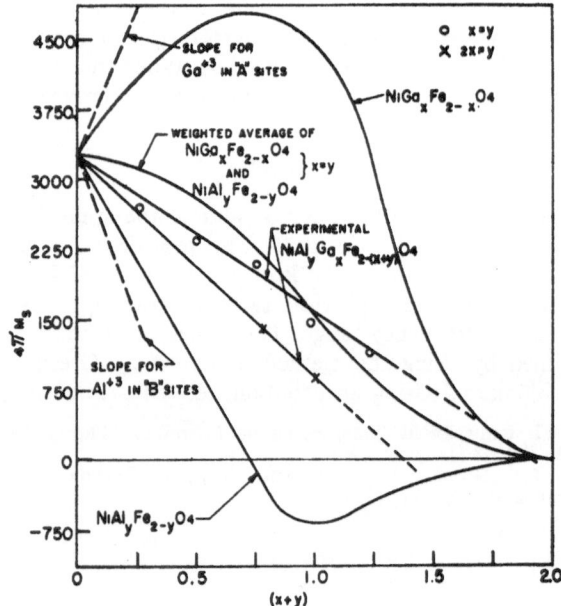

Fig. 1. Variation of saturation magnetization with composition.

[1] H. Suhl, Proc. Inst. Radio Engrs. 44, 1270 (1956).
[2] E. Schlömann, J. H. Saunders, and M. H. Sirvetz, I.R.E. Trans. on Microwave Theory Tech. MTT-8, 96 (1960).
[3] I. Reingold, S. Dixon, Jr., and J. L. Carter, USASBLD Tech. Rept. #2143, January, 1960.
[4] E. Wantuch (private communications).
[5] L. R. Maxwell and S. J. Pickart, Phys. Rev. 92, 1120 (1953).

temperatures of the nickel aluminum gallium spinels are within quantitative agreement with the single substituted spinels. The latter data are those of Maxwell and Pickart.[5]

In Fig. 2 is also plotted the variation of linewidth and $g_{eff}$ (at 9.2 kMc) with composition. The linewidth gradually decreases for both ratios of aluminum to gallium. The $g_{eff}$ increases slightly with increasing substitution. As the ratio of aluminum to gallium is increased, and the compensation point is moved to lower total substitutions, the $g_{eff}$ has a further increase.

## HIGH POWER TEST

In order to evaluate the high power characteristics of this material, a $Y$ circulator-switch was constructed. This device operated at $C$ band with a 10% bandwidth. The insertion loss of the device was 0.5 db or less and had greater than 20-db isolation. The material had the following characteristics; $4\pi M_s$ less than 1200 gauss, $T_c$ greater than 300°C, tan δ at 20 Mc less than 0.002, dielectric constant at 20 Mc, approximately 7, linewidth ~325 oe, and a $g_{eff}$ factor of 2.5. The device was tested under the following high power conditions with no deterioration of properties. The peak power applied was 75 kw with a duty cycle of 0.001 for an average power of 75 w. This peak power was estimated to be equivalent to 25 rf oe. Four hundred and seventy-five watts cw was passed through the circulator with the device in an oven at 85°C. The drive current remained constant at 1.2 amp from 25° to 85°C. This device has an rf switching time of 100 μsec. Additional experiments are in progress to determine the power level for the onset of nonlinearity.

## CONCLUSIONS

This new class of spinels satisfies the original requirements of allowing one to vary the saturation magnetization of a material by changing the total amount of

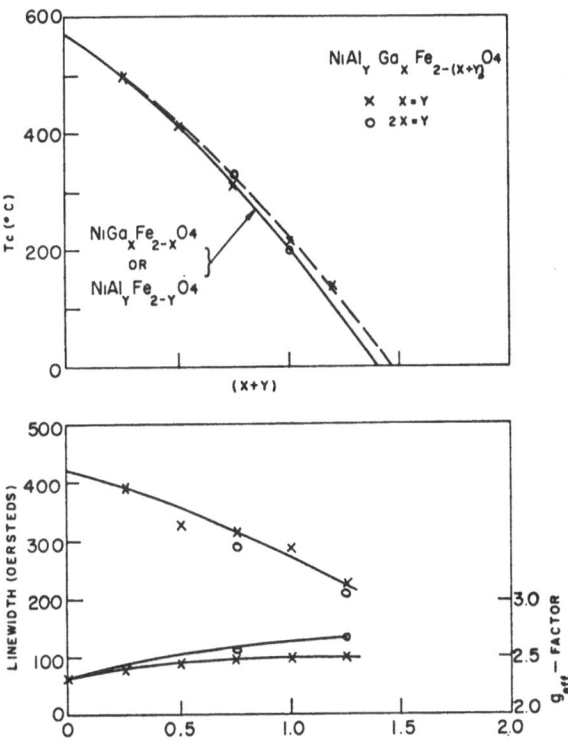

FIG. 2. Variation of Curie temperature, linewidth, and $g_{eff}$ factor at 9.2 kMc with composition.

substitution or by varying the ratio of the substituted nonmagnetic ions. It is possible to obtain low loss materials with a magnetization of 650 gauss or greater, Curie temperatures in excess of 200°C, $g_{eff}$ factors of less than 3, and reasonably low linewidths. With the wide range of properties that can be obtained in the system, and the ability of devices utilizing the system to operate at high rf power levels, nickel gallium aluminum ferrites hold out the promise of extending significantly the life of ferrite devices in the high power field.

JOURNAL OF APPLIED PHYSICS    SUPPLEMENT TO VOL. 33, NO. 3    MARCH, 1962

# Subsidiary Absorption Effects in Ferrimagnetics*

J. H. SAUNDERS AND J. J. GREEN

*Research Division, Raytheon Company, Waltham, Massachusetts*

Subsidiary absorption measurements on polycrystalline yttrium aluminum iron garnet and nickel aluminum ferrite yield values of $\Delta H_k$ which are the order of several oersteds. These spin wave linewidths appear to be independent of aluminum content. At power levels several db in excess of the power level for the onset of subsidiary absorption, a power dependent zero field loss has been found.

RECENTLY there have been many measurements of spin wave relaxation rates in polycrystalline materials by means of saturation of the main resonance. The usefulness of this type of experiment is limited in two respects: (1) only one set of spin waves go unstable; (2) for high power applications the region of interest occurs at dc fields below resonance rather than at resonance because of considerations of average power dissipation. Furthermore, when the linewidth of the uniform mode is not Lorentzian as assumed by Suhl[1] or when the resonance susceptibility, $\chi_{res}''$, does not go into a clear $h^{-1}$ dependence at high rf field strengths thereby permitting extrapolation to determine the critical rf field strength,[2] the relaxation rates obtained from saturation of the main resonance are questionable. In parallel pumping[3,4] these disadvantages do not exist because the uniform mode is not involved. Although the uniform mode is involved in subsidiary absorption by perpendicular pumping,[1] there are no complications due to inhomogeneity broadening provided one is biased to the dc fields far enough below resonance. For both parallel and perpendicular pumping, spin waves can be selectively excited through variation of the dc field, and

the onset of spin wave instability is sharp so that there is no difficulty in deciding what the critical rf field is.

Our measurements were made on spherical samples at 9.2 kMc and at room temperature. The sample absorption was determined by measuring the change in reflection coefficient of a slightly undercoupled $TE_{101}$ cavity.

We have studied the subsidiary absorption in polycrystalline materials which are formed from nickel ferrite and yttrium iron garnet by the partial substitution of aluminum for iron. In both types of material the aluminum causes a reduction of the saturation magnetization. Since at the subsidiary absorption the threshold field is a function of the applied dc field,[1,3] we have listed in Table I the linewidth for the spin waves which have the lowest threshold. The linewidths are all a few oersteds and appear to be independent of aluminum content with NiF being the single exception. For certain samples, values of $\Delta H_{k\ res}$ are not listed because it was impossible to saturate the main resonance. For NiF the perpendicular pump subsidiary absorption was too close to the main resonance to measure. The values of $\Delta H_{k\perp}(\omega/2)$ and $\Delta H_{k\parallel}(\omega/2)$ are approximately equal, which is not surprising since the two spin waves involved are equal in frequency and wave number and differ only in the angle their wave vector makes with the dc field. The presence of rare impurities in the yttrium aluminum samples is suggested by the independence of rf threshold field upon dc field at low values of dc field which we previously have found to be characteristic of rare-earth doped samples. This would help account for values of $\Delta H_k$ which are an order of magnitude larger in these materials than in comparable single-crystal samples. However, the frequency dependence [compare $\Delta H_{k\parallel}(\omega/2)$ and $\Delta H_{k\ res}(\omega)$] does not agree with the theory of rare-earth relaxation by deGennes et al.[5] The threshold for the YAlIG (12% Al) sample is high and this we believe is due to the presence of some rare-earth ions. There are two values of $\Delta H_{k\ res}(\omega)$ listed for the NiAlF. These arise from two different extrapolations to determine $h_{crit}$ in the $\chi_{res}''$ versus $h^{-1}$ plot. In these plots $\chi_{res}''$ appears to be going into a $h^{-1}$ dependence at the point of maximum available power. At power levels slightly below the maximum available power level, $\chi_{res}''$ shows a linear relation of the type $\chi_0''(h_0h^{-1}+1)(h_0h_c^{-1}+1)^{-1}$. Here $\chi_0''$ is the low power resonance

TABLE I. Spin wave linewidth for aluminum substituted YIG and NiF.

| Sample | %Al | $4\pi M$ | $\Delta H_{k\parallel}(\omega/2)$[a] | $\Delta H_{k\perp}(\omega/2)$[b] | $\Delta H_{k\ res}(\omega)$[cd] |
|--------|-----|----------|------------------|------------------|------------------|
| YIG | 0 | 1740 | 2.30 | 2.67 | 4.73 |
| YAlIG | 4 | 1420 | 3.21 | 3.60 | 3.67 |
| YAlIG | 8 | 1180 | 3.98 | 5.24 | 2.97 |
| YAlIG | 12 | 880 | 13.05 | 14.9 | |
| YAlIG | 16 | 704 | 7.00 | 4.05 | |
| NiF | 0 | 3027 | 18.8 | | 7.35 |
| NiAlF | 3.5 | 2650 | 8.75 | 7.11 | 8.90, 16.8[e] |
| NiAlF | 6.5 | 2330 | 7.95 | 6.06 | 6.40, 16.4[e] |
| NiAlF | 9.5 | 1955 | 7.10 | 5.72 | 4.05, 12.15[e] |
| NiAlF | 12.5 | 1675 | 6.68 | 6.76 | 7.15, 18.1[e] |
| NiAlF | 15.5 | 1425 | 6.67 | 7.05 | |
| NiAlF | 19 | 1050 | 7.00 | 5.84 | |

[a] $\Delta H_{k\parallel}(\omega/2)$ = linewidth from parallel pump subsidiary absorption.
[b] $\Delta H_{k\perp}(\omega/2)$ = linewidth from perpendicular pump subsidiary absorption.
[c] $\Delta H_{k\ res}(\omega/2)$ = linewidth from saturation of the main resonance.
[d] $\omega$ = the pump frequency and the frequency of the excited spin waves is indicated in parentheses following the linewidths.
[e] See text for explanation of double values.

* Supported in part by the U. S. Army Signal Corps Research and Development Laboratory under contract.

[1] H. Suhl, J. Phys. Chem. Solid **1**, 209 (1957).
[2] J. J. Green and E. Schlömann, IRE Trans. on Microwave Theory Tech. **MTT-8**, 100 (1960).
[3] E. Schlömann, J. J. Green, and U. Milano, J. Appl. Phys. **31**, 386S (1960).
[4] F. R. Morgenthaler, J. Appl. Phys. **31**, 95S (1960).
[5] P. G. deGennes, C. Kittel, and A. M. Portis, Phys. Rev. **116**, 323 (1959).

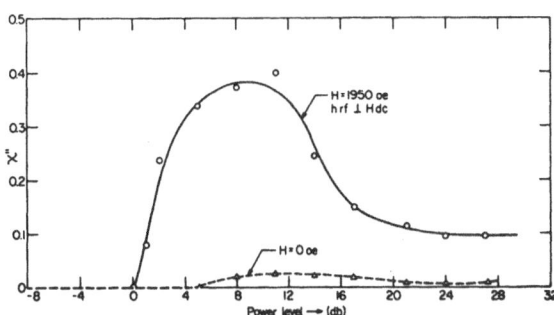

FIG. 1. Subsidiary absorption and zero field loss in single-crystal YIG. (Spherical sample, $T = 20°C$, $\nu = 9.2$ kMc).

susceptibility, $h_0$ is a positive constant field, and $h_c$ is the critical field to which one would extrapolate. If this $h_c$ is taken for the critical field then one gets the smaller values of $\Delta H_{k\ res}(\omega)$ listed in Table I. If one believes that $\chi_{res}''$ at the highest power level is going into a $h^{-1}$ dependence and makes the appropriate extrapolation then one gets the higher values listed in Table I. This is an example of the difficulties encountered in the extrapolation method when there is not a clear $h^{-1}$ dependence in the saturation curve.

During the course of these subsidiary absorption measurements a power dependent zero field loss was observed. This loss is not to be confused with the low power zero field loss which arises from resonance in the anisotropy field and demagnetizing fields of the various domains.[6] This latter loss did not occur in our materials because of their low magnetization. The power dependent zero field loss has been observed in both polycrystalline and single-crystal materials and usually appears at power levels several db higher than that for the onset of subsidiary absorption. In Fig. 1 a comparison is made between the loss at a dc field of 1950 oe (subsidiary absorption) and 0 oe (power dependent zero field loss). The single-crystal sample represented in Fig. 1 had a low threshold ($\simeq 1.5$ oe) so that it was possible to make the comparison over a large range of power levels ($\simeq 27$ db). All our observations indicate that this zero field loss is intimately related to the subsidiary absorption and probably involves the excitation of some half-frequency spin waves.

The authors are indebted to E. Schlömann for many valuable discussions.

[6] D. Polder and J. Smit, Revs. Modern Phys. 25, 89 (1953).

---

JOURNAL OF APPLIED PHYSICS    SUPPLEMENT TO VOL. 33, NO. 3    MARCH, 1962

# Magnetic Properties of Yttrium-Gadolinium-Aluminum-Iron Garnets*

E. A. MAGUIRE AND J. J. GREEN

*Research Division, Raytheon Company, Waltham, Massachusetts*

The magnetic and microwave properties of polycrystalline garnets of the $Gd_{3-x}Y_xFe_{5-y}Al_yO_{12}$ type were investigated. Compositions of low magnetization ($\sim 600$ gauss) and narrow linewidth ($\sim 125$ oe) were found which are suitable for operation in devices where temperature compensation is required.

IN the garnet structure, with the composition $Z_3Fe_5O_{12}$, the magnetic moment of the rare-earth ions ($Z^{3+}$) on the $c$ sites is coupled antiferromagnetically[1] ($H_{ex} \cong 10^5$ oe) to the resultant moment of the $a-d$ sublattice ($Fe^{3+}$). The $a$ and $d$ sublattices are strongly antiferromagnetically coupled ($H_{ex} \cong 10^6$ oe) and have a net moment due to the excess of $d$ sites per unit cell (16 $a$ sites, 24 $c$ sites, and 24 $d$ sites per unit cell). If the magnetic moment per unit cell of the rare-earth ions ($Z^{3+}$) is larger than the $a-d$ moment per unit cell, then at low temperatures, where there is complete alignment of the $c$ sublattice, the net magnetic moment will be dominated by the moment of the $c$ sublattice. At high temperatures, where the alignment on th $c$ sublattice is small, the $a-d$ moment will dominate. At some intermediate temperature ($T_{cp}$) there will be a compensation point, a temperature at which the net moment is zero.

At some temperature between $T_{cp}$ and the Curie temperature ($T_c$) the total magnetization ($M$) will be temperature compensated (i.e., $dM/dT = 0$). If the rare-earth ion is gadolinium (Gd) then $T_{cp}$ occurs at room temperature (290°K) and $dM/dT = 0$ at 470°K.

It is possible to shift the position of $T_{cp}$ by altering the moments on the $a-d$ and $c$ sublattices. If diamagnetic yttrium ($Y^{3+}$) is substituted for $Gd^{3+}$, then the moment of the $c$ sublattice will be proportional to the percentage of $Gd^{3+}$ present and this will shift $T_{cp}$ to lower temperatures. If nonmagnetic aluminum ($Al^{3+}$) is substituted for $Fe^{3+}$, the $Al^{3+}$ will go primarily on the $d$ sites[2] thus lowering the $a-d$ moment. The lower $a-d$ moment will result in a smaller exchange field at the $c$ sites and hence a smaller $c$ moment.

Polycrystalline garnets were made in which both yttrium and aluminum substitutions occurred. The general composition of these materials can be written as $Gd_{3-x}Y_xFe_{5-y}Al_yO_{12}$. Compositions with more than 40

* Supported in part by the U. S. Army Signal Corps Research and Development Laboratory under contract.

[1] R. Pauthenet, Ann. phys. 13, 424 (1958).

[2] S. Geller, J. Appl. Phys. 31, 30S (1960).

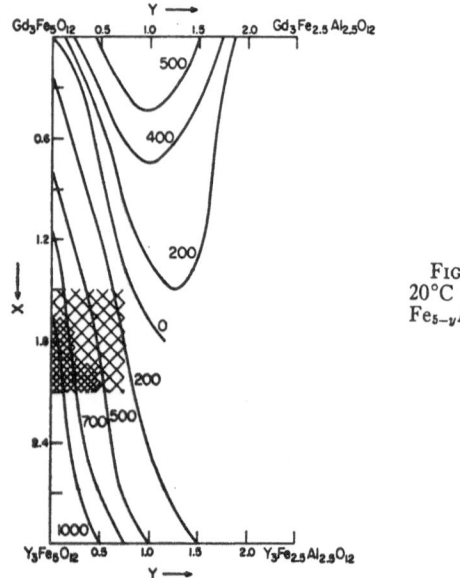

FIG. 1. $4\pi M_s$ at 20°C for $Gd_{3-x}Y_x$ $Fe_{5-y}Al_yO_{12}$ garnets.

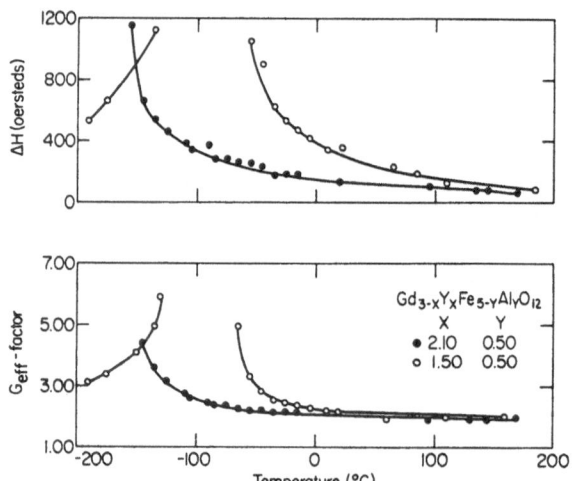

FIG. 3. Linewidths and effective $g$ factors for several samples of the $Gd_{3-x}Y_xFe_{4.5}Al_{0.5}O_{12}$ type.

percent aluminum were not investigated because for these compositions the Curie temperature would be less than room temperature.

Figure 1 shows the magnetization at room temperature for materials in this system. As can be seen in this diagram, the substitution of either yttrium or aluminum increases the room temperature magnetization of gadolinium iron garnet. To the left of the line representing zero gauss, the $a$–$d$ moment dominates. To the right of this line, the dominant moment is that of the $c$ sublattice. Therefore temperature compensation cannot be expected from compositions to the right of the zero gauss line. The shaded region of this diagram was found to contain the most promising compositions for temper-

ature compensation. Narrow linewidth values (125 oe or less) further restrict the interesting materials to the area of heavier shading. The boundaries of this region are not exact as illustrated and require more investigation in order to be definitive.

Magnetization versus temperature is shown in Fig. 2 for some compositions which are temperature compensated between room temperature and the Curie temperature. In these materials, the amount of aluminum was maintained constant and the gadolinum content progressively decreased by the substitution of yttrium. Note the lack of any indications of compensation until the yttrium content is large enough to place the composition in the area where the $a$–$d$ moment is dominant. Temperature compensation occurs in a useful temperature range with compositions containing between 50 and 70 percent yttrium.

In devices with nonspherical geometries, it is desirable that the effective $g$ and saturation magnetization and therefore the resonance frequency be temperature independent. The temperature dependence of the linewidths and the effective $g$ factors of some of these materials is shown in Fig. 3. In the regions where $dM/dT=0$, the effective $g$ factors can be seen to vary slowly with temperature.

The potential value of these materials is for use in devices which must work over an extended temperature range. The low magnetizations suggest that they will probably be most valuable for low frequency applications while the narrow linewidths mean that with these materials it should be possible to design devices which will have reasonable performance characteristics.

The authors are indebted to E. Schlömann for many valuable discussions.

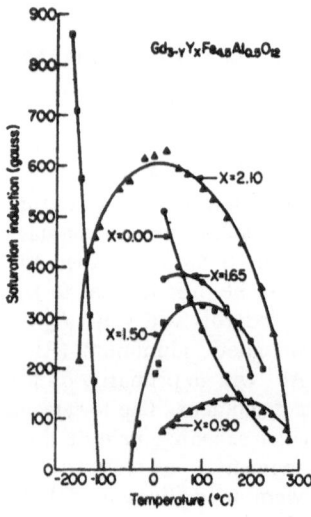

FIG. 2. Magnetization versus temperature for several samples of the $Gd_{3-x}$ $Y_xFe_{4.5}Al_{0.5}O_{12}$ type.

JOURNAL OF APPLIED PHYSICS    SUPPLEMENT TO VOL. 33, NO. 3    MARCH, 1962

# Temperature Stable Microwave Hybrid Garnets*

GORDON R. HARRISON AND L. R. HODGES, JR.

*Applied Physics Section, Sperry Microwave Electronics Company, Clearwater, Florida*

Temperature stability of the magnetization of microwave garnet materials is very desirable provided in obtaining this stability the linewidth, g factor, and Curie temperature of the material are still suitable for microwave applications. Most rare-earth garnets possess compensation points in their magnetization vs temperature characteristics. These compensation points when suitably controlled by ionic substitutions can be used to produce microwave garnet materials whose magnetizations are quite temperature independent. The microwave properties of the following compositions have been investigated:

$$3[(1-X-Z)Y_2O_3 \cdot XGd_2O_3 \cdot ZDy_2O_3] \cdot 5[(1-W)Fe_2O_3 \cdot WAl_2O_3],$$

for $X=0$ to 1.00, $Z=0$ to 0.1, and $W=0$ to 0.1. The magnetizations of these compositions have been studied from $-195°C$ to the Curie temperatures. The compensation points of the magnetization are controlled by varying the rare-earth and aluminum content of the material. The result is a series of materials possessing magnetizations in the region of 1200 to 300 gauss which vary no more than $\pm 10\%$ over the temperature range of $-25°$ to $+125°C$. The other associated properties of these compositions are very suitable for microwave applications. The thresholds for nonlinear effects in these materials are relatively high particularly for those compositions containing dysprosium. Many temperature stable microwave components have been developed for high-power applications using these materials.

## INTRODUCTION

THIS study is concerned with the microwave properties of the series of compositions, $3[(1.00\text{-}X\text{-}Z)Y_2O_3 \cdot XGd_2O_3 \cdot ZDy_2O_3] \cdot 5[(1.00\text{-}W)Fe_2O_3 \cdot WAl_2O_3]$, for $X=0$ to 1, $Z=0$ to 0.1, and $W=0$ to 0.1. Gadolinium iron garnet[1] possesses a high magnetization at low temperatures and a compensation point near room temperature ($\sim 15°C$); dysprosium iron garnet has similar properties with a lower compensation point; yttrium iron garnet has no compensation point. Solid solutions of these compositions[2-5] in conjunction with aluminum substitutions[5-9] allow for the selective positioning of the compensation point to produce the best stability of the magnetization in the temperature region of $-25°C$ to $+125°C$.

## EXPERIMENTAL RESULTS

The method of preparation and the property measuring techniques of these polycrystalline samples were the same as those described previously.[5]

Polycrystalline garnets of the form $3[(1.00\text{-}X)Y_2O_3 \cdot XGd_2O_3] \cdot 5Fe_2O_3$ were prepared for values of $X$ ranging from 0 to 1. The magnetization data from $-195°C$ to the Curie temperature of the compositions are presented in Fig. 1. Definite compensation points were observed for each composition. The data of Anderson *et al.*,[3] showed only "dips" in the curves. Data collected on other samples by the authors reveal similar characteristics which were attributed to nonuniformity and inhomogenity in the sample.[10] Extrapolated data on the positions of the compensation points reveal a value of $X=0.12$ for a compensation point at $0°K$. This is in somewhat disagreement with a calculated value of $X=0.24$.

Particularly, note the data for $X=0.6$. This composition produces a relatively stable magnetization over a large temperature range centered around $50°C$. Other microwave properties of this composition are as follows: linewidth $\Delta H$ ($25°C$, $X$ band) $=200$ oe; $g_{eff}$ factor ($25°C$,

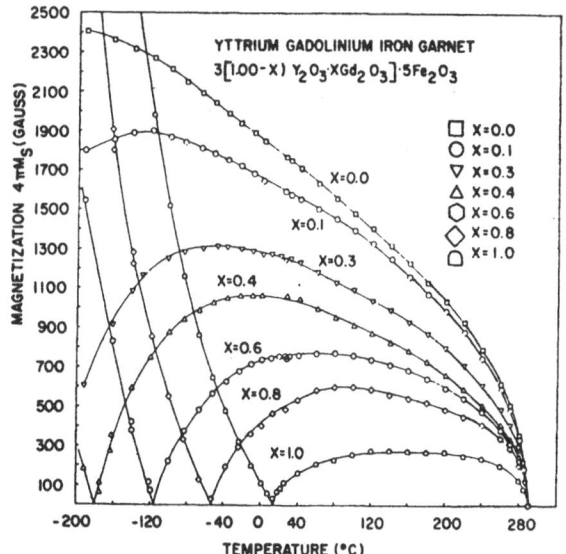

FIG. 1. The variation of the saturation magnetization with temperature for yttrium gadolinium iron garnet.

* This work was supported in part by the Air Force Systems Command and Rome Air Development Center.

[1] R. Pauthenet, Comp. rend. **243**, 1499 (1956); J. Appl. Phys. **29**, 253 (1958).

[2] M. H. Sirvetz and J. E. Zneimer, J. Appl. Phys. **29**, 431 (1958).

[3] Elmer E. Anderson, J. Richard Cunningham, Jr., and G. E. McDuffie, Phys. Rev. **116**, 624 (1959).

[4] G. P. Rodrigue, "Ferromagnetic resonance in yttrium and rare earth iron garnets," Thesis, Harvard University (1959).

[5] Gordon R. Harrison and L. R. Hodges, Jr., J. Am. Ceram. Soc. **44**, 214 (1961); Microwave J. **4**, No. 6, 53 (1961).

[6] M. A. Gilleo and S. Geller, Phys. Rev. **110**, 73, (1958).

[7] H. S. Yoder and M. L. Keith, Am. Mineralogist **36**, 519 (1951).

[8] G. Villers, R. Pauthenet, and J. Loriers, J. Phys. radium **20**, 382 (1959).

[9] Gordon R. Harrison and L. R. Hodges, Jr., Proceedings of the 1961 Sperry Research Symposium (to be published).

[10] S. Geller, H. J. Williams, and R. C. Sherwood, Phys. Rev. **123**, 1692 (1961).

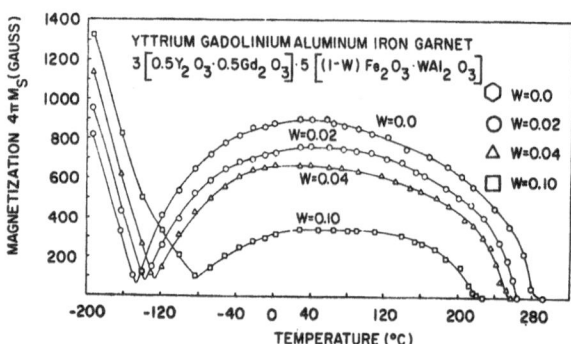

FIG. 2. The variation of the saturation magnetization with temperature for yttrium gadolinium aluminum iron garnet.

$X$ band)$=2.08$; dielectric loss tangent (1 Gc)$=0.004$; dielectric constant (1 Gc)$=13$.

To obtain temperature stabilized magnetizations at various values the following series of compositions was investigated:

$$3[0.5Y_2O_3 \cdot 0.5Gd_2O_3] \cdot 5[(1-W)Fe_2O_3 \cdot WAl_2O_3],$$
$$\text{for } W=0 \text{ to } 0.1.$$

The magnetization versus temperature characteristics are shown in Fig. 2. For small substitutions aluminum locates predominately on the $24d$ site in the garnet structure. This reduces the resultant magnetization of the iron sublattices and thus the magnetization at 25°C, reduces the Curie temperature,[6] and positions the compensation point to higher temperatures as $W$ increases. The result is a series of materials with temperature stable magnetizations from 900 gauss downward. Other microwave properties of these compositions are as follows: linewidth $\Delta H(25°C, X$ band)$=165$ to 300 oe; $g_{\text{eff}}$ factor (25°C, $X$ band)$=2.04$ to 2.10; dielectric loss tangent (1Gc)$=0.004$. True compensation points were not observed for these compositions due presumably to a slight nonuniformity in composition from grain to grain. The compensation points did increase with increasing aluminum content as expected.

To obtain a series of materials with greater high power thresholds,[11] solid solutions of the above compositions with dysprosium iron garnet were studied. The

[11] E. Schlömann, J. J. Green, and U. Milano, J. Appl. Phys. **31**, 386S (1960).

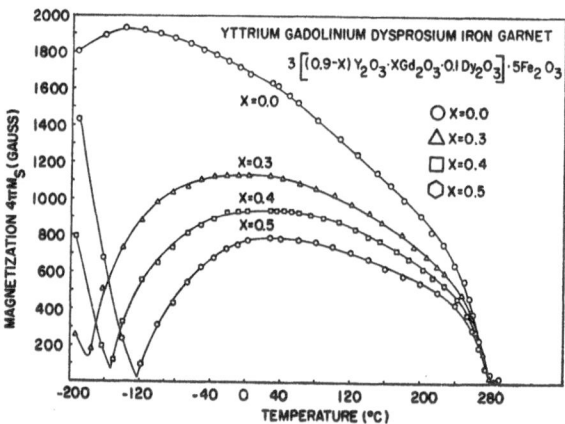

FIG. 3. The variation of the saturation magnetization with temperature for yttrium gadolinium dysprosium iron garnet.

magnetic properties of the compositions

$$3[(0.9-X)Y_2O_3 \cdot XGd_2O_3 \cdot 0.1Dy_2O_3] \cdot 5Fe_2O_3$$
$$\text{for } X=0 \text{ to } 0.5$$

are shown in Fig. 3.

Temperature stability is again observed in compositions for $X=0.3$ to 0.5; these materials also exhibit superior microwave high power handling capability. Other microwave properties of these compositions are: linewidth $\Delta H(25°C, X$ band)$=450$ to 800 oe; $g_{\text{eff}}$ factor (25°C, $X$ band)$=2.00$ to 1.80; dielectric loss tangent (1 Gc)$=0.004$. Aluminum may also be added to these materials for further control of the magnetization.

The data presented here illustrate the technique of positioning compensation points by "molecular engineering" to temperature stabilize the magnetization of microwave hybrid garnets. These materials have been utilized in developing many superior high power, temperature stabilized microwave devices.

### ACKNOWLEDGMENTS

The authors gratefully acknowledge the assistance of Dr. R. T. Arnold in the magnetometer measurements and H. W. Burghardt in the collection of data. Many stimulating discussions with Dr. G. P. Rodrigue concerning the data were extremely valuable. The assistance rendered by the personnel of the Materials Laboratory is also gratefully acknowledged.

JOURNAL OF APPLIED PHYSICS    SUPPLEMENT TO VOL. 33, NO. 3    MARCH, 1962

# Spin Wave Excitation in Planar Ferrites

Isidore Bady

*United States Army Signal Research and Development Laboratory, Fort Monmouth, New Jersey*

AND

Ernst Schlömann*

*Research Division, Raytheon Company, Waltham, Massachusetts*

The large magnetic anisotropy field that is encountered in many planar ferrites markedly affects the rf magnetic field at which spin wave instability sets in. When the rf magnetic field is applied perpendicular to the dc field, spin wave instability can occur through the first- or second-order process. At ferromagnetic resonance, the first-order process dominates if the frequency is smaller than a characteristic frequency $\omega_c$. This characteristic frequency has been calculated for ellipsoidal samples. For most sample shapes, the anisotropy field tends to increase $\omega_c$, thus favoring the occurrence of the first-order process. In addition, the anisotropy tends to decrease the critical field at which the spin wave instability sets in. Assuming the spin wave relaxation rates to be equal in the two situations compared, we find that the second-order threshold is reduced (approximately) in the ratio of $(8\pi M/H_a)^{\frac{1}{2}}$ for relatively high frequencies; and in the ratio of $(\omega/\gamma H_a)^{\frac{1}{2}}$ for relatively low frequencies. In parallel pumping, the threshold field is reduced in the ratio of $4\pi M/(4\pi M + H_a)$ at low biasing fields.

## I. INTRODUCTION

PLANAR ferrites are ferrites with a preferential plane of magnetization. They were first described by Jonker et al.,[1] who also presented some of their radio-frequency properties. Microwave properties were discussed by Bady.[2] The large magnetic anisotropy field $H_a$ that is encountered in many of these materials markedly affects the rf magnetic field at which spin wave instability sets in. It is the purpose of this paper to discuss some of these effects.

The preferred plane will be taken as lying in the XZ plane, and the biasing field will be applied in the Z direction. $\theta$ is the angle that the propagation vector makes with the Z axis, and $\varphi$ is the angle its XY projection makes with the X axis.

## II. DISPERSION RELATION

The formula for the resonant frequency, $\omega_k$, of the spin waves, is given below:

$$\left(\frac{\omega_k}{\gamma}\right)^2 = (H + H_a + Dk^2 + 4\pi M \sin^2\varphi \sin^2\theta)$$
$$\times (H + Dk^2 + 4\pi M \cos^2\varphi \sin^2\theta)$$
$$- (4\pi M)^2 \sin^4\theta \sin^2\varphi \cos^2\varphi. \quad (1)$$

Here $H$ is the internal dc magnetic field, $\gamma$ is the gyromagnetic ratio, and $k$ is the wave number of the spin wave. The term $Dk^2$ arises from the exchange interaction and has been slightly oversimplified in this treatment. Actually, $Dk^2$ should be replaced by $D(k_x^2 + k_z^2) + D'k_y^2$ since the Y direction is not equi-

valent to the X and Z directions. We have also assumed by way of approximation that the gyromagnetic ratio is isotropic. The principal effect of the anisotropy field is that $\omega_k$ is now dependent on $\varphi$, whereas it was independent of $\varphi$ in the case of isotropic ferrites. $\omega_k$ is larger at $\varphi = 0$ than at $\varphi = \pi/2$.

## III. PERPENDICULAR PUMPING, FIRST-ORDER PROCESS

Spin wave instability can occur through the first-order or the second-order nonlinear process. In the former, the spin wave frequency is half the pump frequency; in the latter, the spin wave frequency is equal to the pump frequency.

The threshold rf magnetic field, required for spin wave instability, is inherently lower for the first-order process than for the second-order process, provided both are allowed to occur at ferromagnetic resonance. However, there is an upper limit $\omega_c$ to the frequency at which the first-order process can occur at resonance. This is given in Eq. (2):

$$\left(\frac{\omega_c}{\gamma 4\pi M}\right)^2 = \frac{2}{9}\{3(N_x - N_y)\zeta + (N_x + N_y)^2$$
$$+ 6N_xN_y + (N_x + N_y)g\}, \quad (2)$$

where

$$g = [9\zeta^2 + 6\zeta(N_x - N_y) + (N_x + N_y)^2 + 12N_xN_y]^{\frac{1}{2}}$$

and

$$\zeta = H_a/4\pi M.$$

In the case of a very thin disk cut parallel to the preferential plane ($N_x = 0$, $N_y = 1$), the above equation reduces to

$$\omega_c/\gamma 4\pi M = \tfrac{2}{3}(1 - 3\zeta)^{\frac{1}{2}}, \quad \text{for} \quad \zeta < \tfrac{1}{3}$$
$$= 0, \quad \text{for} \quad \zeta > \tfrac{1}{3}. \quad (3)$$

Thus we see that for a value of $\zeta$ greater than $\tfrac{1}{3}$, the first-order process will not be possible at any frequency.

* Presently on leave at W. W. Hansen Laboratories of Physics, Stanford University, Stanford, California. E. Schlömann's contribution was partially supported by the United States Army Signal Research and Development Laboratory.
[1] G. H. Jonker, H. P. J. Wijn, and P. B. Braun, Philips Tech. Rev. 18, 145 (1956).
[2] I. Bady, I.R.E. Trans. on Microwave Theory Tech MTT-9, 52 (1961).

In other geometrical configurations, however, the anisotropy field greatly increases the maximum frequency at which the first-order process is possible at resonance. For instance, for the case of a very thin disk cut perpendicular to the preferential plane, and magnetized in its plane, the characteristic frequency $\omega_c$ is given by

$$\omega_c/\gamma 4\pi M = \tfrac{2}{3}(1+3\zeta)^{\frac{1}{2}}. \qquad (4)$$

For $Co_2Y(Ba_2Co_2Fe_{12}O_{22})$, with $4\pi M = 2300$ gauss and $H_a = 28\,000$ oe, the characteristic frequency is 26 400 Mc. In order for an isotropic ferrite to have the same characteristic frequency, it would need a saturation magnetization of 14 000 gauss.

We note parenthetically that there exists a lower limit to the frequencies at which ferromagnetic resonance can be observed. In the case of a very thin disk cut perpendicular to the preferential plane, and magnetized in its plane, this low frequency limit is given by

$$\omega_l = \gamma(4\pi M H_a)^{\frac{1}{2}}, \qquad (5)$$

i.e., approximately 22 500 Mc for $Co_2Y$.

It can be shown that the ratio $\omega_c/\omega_l$ will approach $(\tfrac{4}{9})^{\frac{1}{2}}$ for very large values of $\zeta$ for all geometrical configurations for which the first-order process is possible at resonance. The ratio increases as $\zeta$ becomes smaller.

The threshold field required for spin wave instability over the frequency range where the first-order process can occur at resonance is shown in Eq. (6). It is assumed that $\zeta \gg 1$.

$$h_x = A\Delta H \frac{\Delta H_k}{4\pi M} \cdot \frac{\omega}{\gamma H_a}, \qquad (6)$$

where $h_x$ is a linearly polarized field in the $X$ direction. The value of $A$ depends on geometry and the ratio $\omega/\omega_l$. For a ratio of one, $A$ is approximately equal to 0.9 for a sphere and to 0.6 for a thin slab cut perpendicular to the preferred plane. For a ratio of 1.1, the corresponding values of $A$ are 1.5 and 0.9. In the above cases, the spin waves easiest to excite will propagate at an angle $\varphi = 0$.

## IV. SECOND-ORDER PROCESS

In the following discussion, it will be assumed that $\zeta \gg 1$ and that the ferrite is at ferromagnetic resonance. At very high frequencies the threshold field is given by the following approximate equation:

$$h_x = \Delta H(2\Delta H_k/H_a)^{\frac{1}{2}} = h_{x_i}(8\pi M/H_a)^{\frac{1}{2}}, \qquad (7)$$

where $h_{x_i}$ is the field required for the onset of spin waves in an isotropic ferrite with the same properties as the planar ferrites except that $H_a = 0$. At very low

frequencies, the threshold field is given by the following approximate equation

$$h_x = h_{x_i}(\omega/\gamma H_a)^{\frac{1}{2}}. \qquad (8)$$

At frequencies where equation 8 is valid, the spin waves easiest to excite propagate parallel to the biasing field. At higher frequencies, the direction of propagation of the spin waves easiest to excite varies with geometry, $\zeta$, and frequency.

## V. PARALLEL PUMPING

The frequency of the spin waves generated in a sufficiently large rf magnetic field applied parallel to the dc field is equal to half the pump frequency. The instability threshold is lowest for waves whose propagation vectors lie in the plane that contains the dc field and the hard axis of the crystal. We find that the ratio of the threshold rf field $h_z$ and the spin wave linewidth $\Delta H_k$ is given by

$$h_z/\Delta H_k = \omega/\gamma(4\pi M + H_a), \quad \text{if} \quad H < H_c$$

$$= \frac{H\omega/\gamma}{(\omega/2\gamma)^2 - H^2}, \quad \text{if} \quad H_c < H < H_d. \qquad (9)$$

$$H_c = -\tfrac{1}{2}(4\pi M + H_a)$$
$$\qquad + [(\omega/2\gamma)^2 + \tfrac{1}{4}(4\pi M + H_a)^2]^{\frac{1}{2}} \qquad (10)$$

is the internal dc field below which the unstable spin waves have a finite wave number. Similarly,

$$H_d = -\tfrac{1}{2}H_a + [(\omega/2\gamma)^2 + \tfrac{1}{4}H_a^2]^{\frac{1}{2}} \qquad (11)$$

is the internal dc field above which all spin wave frequencies are higher than half the pump frequency. At $H = H_d$ the critical rf field increases abruptly from $\Delta H_k \omega/\gamma H_a$ to a much larger value (infinity in the present approximation). This behavior is quite different from that observed in the absence of anisotropy where the critical field gradually increases to infinity with increasing dc field. It is a direct consequence of the fact that due to crystalline anisotropy the spin precession is elliptical even for those spin waves that have the lowest frequency ($Z$-directed spin waves).

## ACKNOWLEDGMENTS

Mr. Bady wishes to thank Dr. R. Denton of Bell Laboratories for helpful discussions. The work done by Mr. Bady, though part of a Signal Corps project, will also be used in connection with graduate studies at Rutgers University. E. Schlömann is greatly indebted to Mr. R. I. Joseph of the Raytheon Research Division for his help in the calculations.

JOURNAL OF APPLIED PHYSICS    SUPPLEMENT TO VOL. 33, NO. 3    MARCH, 1962

# Effect of Mechanical, Thermal, and Chemical Treatment of the Ferrimagnetic Resonance Linewidth on Lithium Ferrite Crystals

J. W. Nielsen, D. A. Lepore, J. Zneimer, and G. B. Townsend

*Airtron Division of Litton Industries, Morris Plains, New Jersey*

Ferrimagnetic resonance linewidths as low as 2.6 oe can be obtained in lithium ferrite spheres at room temperature and 5 kMc by a series of treatments including polishing, heat treatment at 800°C, addition of lithium, and ordering. There exist optimum times of processing for each crystal. Freshly grown crystals show unusual variations in linewidth with processing, and this behavior is tentatively explained on the basis of strains in the crystal and nonuniformity of lithium content within the spherical sample.

## INTRODUCTION

FERRIMAGNETIC resonance linewidths of 3.0 oe or less can be obtained in lithium ferrite crystals at room temperature and 5 kMc. The "large and unexpected effects" of heat treatment and polishing on the linewidth which were mentioned by Schnitzler, Folen, and Rado,[1] and attributed to plastic deformation in a thin layer below the surface, were observed. A detailed explanation of our observations is not made, but some of the effects are qualitatively understood.

## LINEWIDTH MEASUREMENTS

Linewidth measurements were made at 5 kMc and room temperature using a technique described by LeCraw and Spencer[2] in which the sphere orients itself in the magnetic field. To eliminate possible measurement errors, the linewidths of "standard" crystals which had been measured before the work began were determined periodically. The $\Delta H$ of a standard YIG crystal was measured by R. C. LeCraw of the Bell Telephone Laboratories and agreed with ours to within 0.04 of an oersted. Repeated measurements of the YIG crystal and the lithium ferrite crystals were reproducible within 5%.

## 1. Crystal Preparation

Lithium ferrite crystals were grown from molten lead oxide solutions which were cooled slowly from 1250°C. Chemical analyses of the crystals showed them to have the formula $Li_{0.51}Fe_{2.5}O_4$. Room temperature $4\pi M_s$ was 3550±40. The dc resistivities were ~100 ohm cm.

## 2. The Change in $\Delta H$ with Polishing

Rough ground spheres had linewidths over 40 oe. Upon polishing with successively finer grits, the linewidth decreased. After four or five hours of polishing, 0.25 $\mu$ diamond grit was used. This decreased the linewidth to about 7.5 oe. With prolonged polishing $\Delta H$ rose and fell in a periodic manner as shown in Fig. 1. The finest $Al_2O_3$ grit appeared to increase $\Delta H$, though the

periodic change was still observed. The surface continued to improve as observed under a microscope.

The crystals at the first "polishing minimum" were pitted in an irregular manner. YIG in the same state of polish exhibits a $\Delta H$ of several oersteds.

Following Schnitzler *et al.*,[1] we attributed this to strain and accidents in polishing and subjected the spheres to a heat treatment.

## 3. The Change in $\Delta H$ with Heat Treatment

All crystals, even rough ground spheres which were heated at 800°C and quenched in air showed decreased linewidths, and after about five hours at 800°C, a minimum in linewidth was reached.

The curves in Fig. 2 are for spheres which had been polished to or beyond the "polishing minimum". Note that $\Delta H$ decreases to almost the same value regardless of the initial value. This averaged about 2 oe less than the minimum $\Delta H$ obtained by polishing. The decrease in $\Delta H$ does not appear to depend seriously on the oxidation of divalent iron at the surface since the decrease in $\Delta H$ is not consistently faster in oxygen.

With prolonged treatment at 800°C the linewidth rises. $\Delta H$ of samples heated in oxygen did not increase as rapidly as those heated in air. This suggests that the rise in $\Delta H$ under prolonged heating at 800°C may be caused by oxygen loss and the formation of $Fe^{2+}$ at the

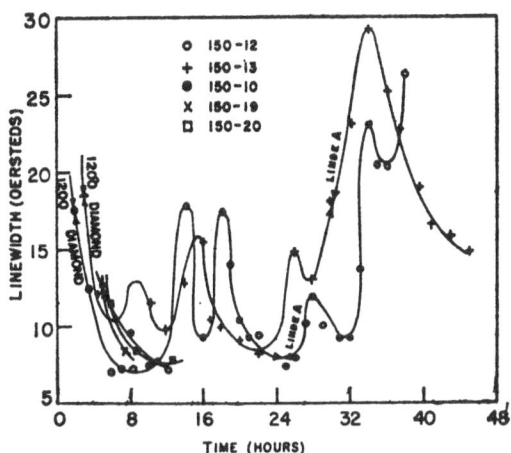

FIG. 1. Variation of $\Delta H$ with polishing.

[1] A. D. Schnitzler, V. J. Folen, and G. T. Rado, J. Appl. Phys. 31, 348S (1960).

[2] R. C. LeCraw and E. G. Spencer, J. Appl. Phys. 30, 185S (1959); see also R. C. LeCraw, E. G. Spencer, and C. S. Porter, Phys. Rev. 110, 1311 (1958).

surface of the sphere. However, there is the possibility of lithium loss at 800°C.

## 4. Change in $\Delta H$ with Lithium Treatment

To see if the increase in $\Delta H$ with prolonged heating was caused by lithium loss, three spheres having $\Delta H$'s of 6.35 oe, 7.1 oe, and 11 oe, were coated with lithium carbonate by evaporating a saturated solution from their surfaces. They were heated for three hours at 700°C and polished to give approximately constant $\Delta H$, which was 3 oe. Thus lithium treatment plus further polishing reduced the linewidth, suggesting that the increase in $\Delta H$ with heating was at least partially due to loss of $Li^+$.

## 5. Change in $\Delta H$ with Ordering

Braun[3] first reported ordering of lithium ions in lithium ferrite. In agreement with Schnitzler et al.,[1] we could observe no effect of ordering on $\Delta H$ before heat treatment. After Denton and Geller's[4] observation of ordering in one of our samples, we ordered spheres at 700°C which had been heated previously at 800°C and quenched. The effect of ordering was apparent. The $\Delta H$ decreased between 0.4 and 1.6 oe. A decrease had also been reported by Schnitzler et al.[1] for a (111) oriented sample. One sample which had also been given a previous lithium treatment had $\Delta H$ of 2.6 oe, the lowest we observed.

### DISCUSSION AND SUMMARY

The above experiments suggest that distribution of lithium in a freshly grown lithium ferrite crystal accounts for a large portion of the ferrimagnetic resonance linewidth. Surface strain contributes another portion. The initial decrease in $\Delta H$ with polishing results from the removal of large pits and strains in the crystal surface. After the surface is partly smooth, the observed erratic changes in linewidth with polishing are probably caused by a combination of periodic compositional changes in the crystal and the accidental introduction and removal of strain during the polishing operation. We suggest that the lithium content varies in the crystals in a periodic way as a result of temperature fluctuations during growth. The grinding removes layers of

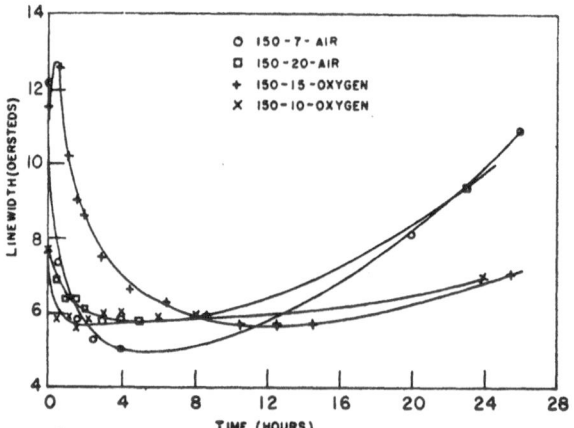

Fig. 2. $\Delta H$ versus time at 800°C.

material, thus the magnons "see" a different surface after each grinding period. This, plus the strained surface caused by polishing (lithium ferrite is much softer than YIG), can account for the data in Fig. 1. However, the effect of large pits is not nearly so pronounced as it is in YIG.

Treatment at 800°C permits the mobile $Li^+$ ions to "smooth out" the periodic composition and simultaneously relieves strain. Further polishing provides evidence for this in that $\Delta H$ varies only slightly with polishing after treatment at 800°C. We attribute the slight variations to strain.

The increase in $\Delta H$ with prolonged heating at 800°C appears to be the result of evaporation of lithium coupled with oxygen loss. The decrease in $\Delta H$ with lithium treatment and a slower increase in $\Delta H$ in $O_2$ at 800°C support this.

The effect of ordering has not been observed until after heat treatment at 800°C in agreement with Schnitzler et al.[1] Then treatment at 700°C will order the crystal as Braun found, and $\Delta H$ decreases.

### ACKNOWLEDGMENTS

One of the authors (J. W. N.) wishes to thank R. T. Denton for communicating much of his work before publication, especially the x-ray data obtained in collaboration with S. Geller. Discussions with R. T. Denton and E. G. Spencer were both helpful and stimulating. We wish also to thank R. C. LeCraw for his measurement of the "standard" YIG crystal.

[3] P. B. Braun, Nature 170, 1123 (1952).
[4] R. T. Denton (private communication).

JOURNAL OF APPLIED PHYSICS    SUPPLEMENT TO VOL. 33, NO. 3    MARCH, 1962

# Growth of Single-Crystal Hexagonal Ferrites Containing Zn

Arthur Tauber, R. O. Savage, R. J. Gambino,* and C. G. Whinfrey†

*U. S. Army Signal Research and Development Laboratory, Fort Monmouth, New Jersey*

Single crystals of $Ba_2Zn_2Fe_{12}O_{22}(ZnY)$, $BaZn_2Fe_{16}O_{27}(ZnW)$, $Ba_3Zn_2Fe_{24}O_{41}(ZnZ)$, $Ba_4Zn_2Fe_{36}O_{60}$ $(ZnM_2Y)$, and $Ba_2Zn_2Fe_{28}O_{46}(ZnM_2S)$, as well as $ZnFe_2O_4$, have been grown using a $Na_2FeO_4$ flux by recrystallization from the melt in platinum crucibles. Crystallization of these compounds is considered within the quaternary subsystem $Na_2FeO_4$-$Fe_2O_3$-$BaO$-$ZnO$. The compositions investigated all lie on four different planes. Hexagonal crystals obtained from this system are characteristically black with highly reflecting natural faces and easy, almost perfect basal cleavage. Crystals up to $\frac{1}{2}$ in. long and $\frac{1}{8}$ in. thick have been grown. Ferrimagnetic resonance linewidth measurements made on single-crystal disks of $(ZnY)$ at 9 kMc have given values as low as 18 oe.

## INTRODUCTION

FERRIMAGNETIC compounds with hexagonal and rhombohedral structures related to magneto-plumbite were discovered by Wijn in 1952.[1] Jonker et al.[2] and Gorter[3] described compounds containing Zn in 1956. No consistent growth of large, good quality crystals of these compounds has been reported. Wijn[1,4] has obtained crystals by induction melting techniques; these usually consisted, however, of parallel inter-growths of several phases. Growth of these crystals from their own melt would be desirable, but suppression of $Fe^{2+}$ at the temperature of melting is extremely difficult. Flux systems such as $PbO$-$PbF_2$ have the attractive property of lowering the temperature to which the charge must be heated and at which the crystals grow.

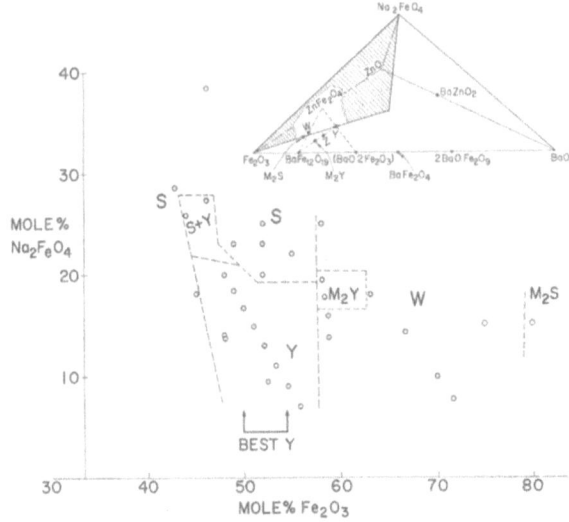

FIG. 1. Single-crystal fields in the BaO:ZnO 1:1 plane. Compositions are starting compositions. $s$=spinel ($ZnFe_2O_4$), $Y=Ba_2Zn_2Fe_{12}O_{22}$, $W=BaZn_2Fe_{16}O_{27}$, $M_2Y=Ba_4Zn_2Fe_{36}O_{60}$ and $M_2S=Ba_2Zn_2Fe_{28}O_{46}$. *Insert*—Composition tetrahedron for the subsystem $Na_2FeO_4$-$Fe_2O_3$-$BaO$-$ZnO$. The part of the BaO:ZnO 1:1 plane which was investigated is hatched.

Unfortunately, $Pb^{2+}$ easily substitutes for $Ba^{2+}$, and this leads to heavily lead-substituted crystals. Mones and Banks[5] have shown that single crystals of $BaFe_{12}O_{19}$ and substituted barium ferrites (substitution for $Fe^{3+}$) could be grown from melts containing $Na_2CO_3$ as the flux. This work was extended by Gambino and Leonhard.[6] Based on these results, it seemed feasible to grow the hexagonal crystals from the quaternary system $Na_2O$-$Fe_2O_3$-$BaO$-$ZnO$. The results of crystallization experiments are described herein.

Early in this investigation it was recognized that $Na_2FeO_4$ was the stable fourth component up to the melting point. As a matter of fact, the stable sodium-iron oxide in the system $Na_2O$-$Fe_2O_3$ in air at temperatures much above the melting points observed in the quaternary system here, is sodium ferrate. We have therefore adopted the subsystem $Na_2FeO_4$-$Fe_2O_3$-$BaO$-$ZnO$ for purposes of identifying compositions and fields of crystalline compounds.

## EXPERIMENTAL PROCEDURE

The principle experiments consisted of exploratory crystallization runs on starting compositions (not necessarily equilibrium compositions) in quaternary planes which are defined by constant BaO:ZnO ratios. The insert in Fig. 1 shows in perspective the tetrahedral plane corresponding to the BaO:ZnO ratio of 1:1, on which lay the initial compositions of $Ba_2Zn_2Fe_{12}O_{22}$ $(ZnY)$ and $Ba_2Zn_2Fe_{28}O_{46}(ZnM_2S)$. Other planes investigated, but not shown, are 2:1 for $Ba_4Zn_2Fe_{36}O_{60}$ $(ZnM_2Y)$, 3:2 for $Ba_3Zn_2Fe_{24}O_{41}(ZnZ)$, and 1:2 for $BaZn_2Fe_{16}O_{27}(ZnW)$, from which the indicated crystals were grown.

Mixtures of starting reagents (Merck $Na_2CO_3$, Fischer C. P. ZnO and $BaCO_3$, and C. K. Williams EPR-50 $Fe_2O_3$) were blended and ground together with a liquid dispersing agent (ethyl alcohol or ethyl acetate), packed into 15–100 cc platinum crucibles (charges up to 70 g), and heated to 1250–1375°C in a vertical Kanthal or horizontal globar furnace. The crucibles were positioned so that the top was at least 20° to 30°C

* Present address: IBM, Yorktown Heights, New York.
† Deceased.
[1] H. P. J. Wijn, Nature 170, 707 (1952).
[2] G. H. Jonker, H. P. J. Wijn, and P. B. Braun, Philips Tech. Rev. 18, 145 (1956).
[3] E. W. Gorter, Proc. Inst. Elec. Engrs. S104, 9 (1956).
[4] P. B. Braun, Philips Research Repts. 12, 491 (1957).

[5] A. H. Mones and E. Banks, J. Phys. Chem. Solids 4, 217 (1958).
[6] R. J. Gambino and F. Leonhard, J. Am. Ceram. Soc. 44, 221 (1961).

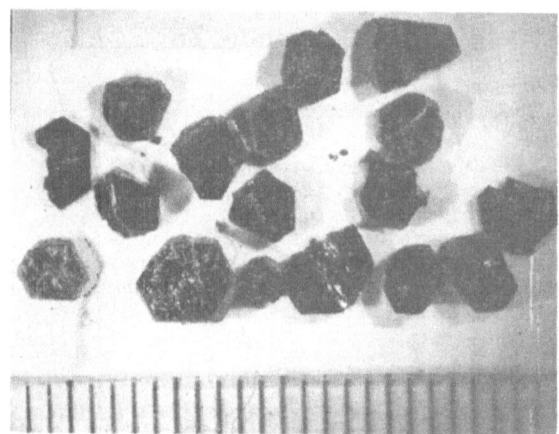

FIG. 2. $(Zn)Y(Ba_2Zn_2Fe_{12}O_{22})$ crystals; scale in millimeters.

hotter than the bottom. The melts were cooled at a rate of 0.75–4°C/hr, to between 1150° and 1050°C, and then cooled more rapidly in the furnace to room temperature.

Crystals were leached from the crucibles with dilute (10–20% by volume) nitric acid. In order to increase the size and yield of crystals for a given run, a recharging of crucible technique was used.[7,8] $Na_2FeO_4$ was prepared directly from $Na_2CO_3$ and $Fe_2O_3$ and used as the flux in several experiments. Spectrochemical analysis indicates a small amount of Na in crystals; a maximum value of 0.1% by weight has been detected. Except for Al (up to 0.1%), other impurities present were each less than 0.025%. The Na in these crystals is probably not present within the structure but as discrete inclusions.

The crystals were identified by Weissenberg, Laue, and Debye-Scherrer powder x-ray diffraction methods employing $FeK$ radiation.

## RESULTS AND DISCUSSION

The experimental runs for BaO:ZnO 1:1 are plotted on the appropriate composition plane of the quaternary tetrahedron in Fig. 1. The values given for $Na_2FeO_4$ and $Fe_2O_3$ are initial and not necessarily equilibrium compositions. The dashed lines, which are somewhat arbitrary, demark the composition fields in which single crystals of the indicated compounds were obtained. To date, all of the crystals grown during this investigation, excepting (ZnZ), have at least been detected in the 1:1 plane. The composition fields for single crystals appear to be extremely narrow along the $Fe_2O_3$ composition axis. The best (ZnY) crystals were obtained in the indicated 5 mole % $Fe_2O_3$ range (Fig. 1). In this range, the charges melted at 1100° to 1200°C as the flux concentration was decreased.

For almost all compositions studied in the 1:1 plane, spinel $ZnFe_2O_4$ (probably with a small amount of $Na^{1+}$ substituting for $Zn^{2+}$)[9] was obtained at least as a polycrystalline or small, 0.1 mm, single-crystal product. For low concentrations of $Fe_2O_3$ and/or high concentrations of $Na_2FeO_4$, single crystals of spinel up to $\frac{3}{8}$ in. on an edge were obtained. The spinel field appears to be quite large. For all investigations employing more than 30 mole % $Na_2FeO_4$, regardless of the BaO:ZnO ratio, spinel was observed as the primary phase.

All the crystals grown are characteristically black with highly reflecting faces. Except for spinel, which occurs in a truncated octahedral habit, all crystals exhibit an easy, almost perfect basal cleavage. The most detailed information has been obtained on (ZnY) crystals. These are generally tabular, with well-developed basal pinacoids (Fig. 2); a few specimens varied from hexagonal prismatic to pyramidal. Crystals up to $\frac{1}{2}$ in. long and $\frac{1}{8}$ in. thick have been obtained. The average size of large crystals is about $\frac{3}{8}$ in. long and $\frac{1}{8}$ in. thick. Pycnometric measurements on 20 (ZnY) crystals yielded an average value of 5.44±0.04 g/cm³ (x-ray density 5.46 g/cm³).

Ferrimagnetic resonance linewidths at 9 kMc, obtained on disks ground from (ZnY) crystals, have been found to be as low as 18 oe by other workers.[10,11] High power experiments employing spheres ground from crystals having linewidths of 60–300 oe show a large shift in the applied dc field for resonance.[11] When 2000-w peak power was applied to a cavity containing a (ZnY) sphere, a shift of 600 oe was observed. Thin sections (transparent, red) of crystals exhibiting this behavior have been found to contain octahedral or truncated octahedral negative crystals or inclusions. The [111] axes of the octahedral are oriented along the [0001] axis of the host crystal. If these are inclusions, they are probably spinel, $ZnFe_2O_4$; there are however, other compounds in the system which could crystallize in an octahedral or pseudo-octahedral habit. Although crystals with lower linewidths showed a larger shift effect, no correlation between these measurements and the presence or concentration of "inclusions" has been made.

## ACKNOWLEDGMENTS

The authors gratefully acknowledge the assistance of Dr. J. A. Kohn for crystallographic determinations and critical review of the manuscript. We are indebted to J. W. Mellichamp and R. K. Buder for spectrographic analysis of many samples.

[7] J. W. Nielsen, J. Appl. Phys. 31, 51S (1960).
[8] D. Barry and R. W. Roberts, J. Appl. Phys. 32, 1405 (1961).

[9] A. H. Mones and E. Banks, J. Phys. Chem. Solids 6, 267 (1958).
[10] C. R. Buffler (private communication).
[11] S. Dixon (private communication).

JOURNAL OF APPLIED PHYSICS    SUPPLEMENT TO VOL. 33, NO. 3    MARCH, 1962

# Author Index to the Seventh Symposium on Magnetism and Magnetic Materials, November 13–16, 1961

JOURNAL OF APPLIED PHYSICS     SUPPLEMENT TO VOL. 33, NO. 3     MARCH, 1962

# Subject Index to the Seventh Symposium on Magnetism and Magnetic Materials, November 13–16, 1961

## Crystal Fields

## Crystals and Crystal Growth

## Devices